XanEdu Custom Solutions
Simple Solutions for Tomorrow's Learning

Variational Principles in Classical Mechanics

Massachusetts College of Liberal Arts

Fall 2022

Variational Principles in Classical Mechanics, Fall 2022

Massachusetts College of Liberal Arts

VARIATIONAL PRINCIPLES

in

CLASSICAL MECHANICS

SECOND EDITION

Douglas Cline

VARIATIONAL PRINCIPLES
IN
CLASSICAL MECHANICS
SECOND EDITION

Douglas Cline
University of Rochester

24 November 2018

ii

©2018, 2017 by Douglas Cline

ISBN: 978-0-9988372-6-0 e-book (Adobe PDF)
ISBN: 978-0-9988372-7-7 print (Paperback)

Variational Principles in Classical Mechanics, 2nd edition

Contributors
    Author: Douglas Cline
    Illustrator: Meghan Sarkis

Published by University of Rochester River Campus Libraries
    University of Rochester
    Rochester, NY 14627

Version 2.0

Contents

CONTENTS

Examples

Preface

The goal of this book is to introduce the reader to the intellectual beauty, and philosophical implications, of the fact that nature obeys variational principles plus Hamilton's Action Principle which underlie the Lagrangian and Hamiltonian analytical formulations of classical mechanics. These variational methods, which were developed for classical mechanics during the $18^{th} - 19^{th}$ century, have become the preeminent formalisms for classical dynamics, as well as for many other branches of modern science and engineering. The ambitious goal of this book is to lead the reader from the intuitive Newtonian vectorial formulation, to introduction of the more abstract variational principles that underlie Hamilton's Principle and the related Lagrangian and Hamiltonian analytical formulations. This culminates in discussion of the contributions of variational principles to classical mechanics and the development of relativistic and quantum mechanics. The broad scope of this book attempts to unify the undergraduate physics curriculum by bridging the chasm that divides the Newtonian vector-differential formulation, and the integral variational formulation of classical mechanics, as well as the corresponding philosophical approaches adopted in classical and quantum mechanics. This book introduces the powerful variational techniques in mathematics, and their application to physics. Application of the concepts of the variational approach to classical mechanics is ideal for illustrating the power and beauty of applying variational principles.

The development of this textbook was influenced by three textbooks: *The Variational Principles of Mechanics* by Cornelius Lanczos (1949) [La49], *Classical Mechanics* (1950) by Herbert Goldstein[Go50], and *Classical Dynamics of Particles and Systems* (1965) by Jerry B. Marion[Ma65]. Marion's excellent textbook was unusual in partially bridging the chasm between the outstanding graduate texts by Goldstein and Lanczos, and a bevy of introductory texts based on Newtonian mechanics that were available at that time. The present textbook was developed to provide a more modern presentation of the techniques and philosophical implications of the variational approaches to classical mechanics, with a breadth and depth close to that provided by Goldstein and Lanczos, but in a format that better matches the needs of the undergraduate student. An additional goal is to bridge the gap between classical and modern physics in the undergraduate curriculum. The underlying philosophical approach adopted by this book was espoused by Galileo Galilei "You cannot teach a man anything; you can only help him find it within himself."

This book was written in support of the physics junior/senior undergraduate course P235W entitled "Variational Principles in Classical Mechanics" that the author taught at the University of Rochester between $1993 - 2015$. Initially the lecture notes were distributed to students to allow pre-lecture study, facilitate accurate transmission of the complicated formulae, and minimize note taking during lectures. These lecture notes evolved into the present textbook. The target audience of this course typically comprised $\approx 70\%$ junior/senior undergraduates, $\approx 25\%$ sophomores, $\leq 5\%$ graduate students, and the occasional well-prepared freshman. The target audience was physics and astrophysics majors, but the course attracted a significant fraction of majors from other disciplines such as mathematics, chemistry, optics, engineering, music, and the humanities. As a consequence, the book includes appreciable introductory level physics, plus mathematical review material, to accommodate the diverse range of prior preparation of the students. This textbook includes material that extends beyond what reasonably can be covered during a one-term course. This supplemental material is presented to show the importance and broad applicability of variational concepts to classical mechanics. The book includes 164 worked examples to illustrate the concepts presented. Advanced group-theoretic concepts are minimized to better accommodate the mathematical skills of the typical undergraduate physics major. To conform with modern literature in this field, this book follows the widely-adopted nomenclature used in "Classical Mechanics" by Goldstein[Go50], with recent additions by Johns[Jo05].

The second edition of this book has revised the presentation and includes recent developments in the field. The book is broken into four major sections, the first of which presents a brief historical introduction (chapter 1), followed by a review of the Newtonian formulation of mechanics plus gravitation (chapter 2), linear oscillators and wave motion (chapter 3), and an introduction to non-linear dynamics and chaos (chapter 4). The second section introduces the variational principles of analytical mechanics that underlie this book. It includes an introduction to the calculus of variations (chapter 5), the Lagrangian formulation of mechanics with applications to holonomic and non-holonomic systems (chapter 6), a discussion of symmetries, invariance, plus Noether's theorem (chapter 7). This book presents an introduction to the Hamiltonian, the Hamiltonian formulation of mechanics, the Routhian reduction technique, and a discussion of the subtleties involved in applying variational principles to variable-mass problems.(Chapter 8). The second edition of this book presents a unified introduction to Hamiltons Principle, introduces a new approach for applying Hamilton's Principle to systems subject to initial boundary conditions, and discusses how best to exploit the hierarchy of related formulations based on action, Lagrangian/Hamiltonian, and equations of motion, when solving problems subject to symmetries (chapter 9). A consolidated introduction to the application of the variational approach to nonconservative systems is presented (chapter 10). The third section of the book, applies Lagrangian and Hamiltonian formulations of classical dynamics to central force problems (chapter 11), motion in non-inertial frames (chapter 12), rigid-body rotation (chapter 13), and coupled linear oscillators (chapter 14). The fourth section of the book introduces advanced applications of Hamilton's Action Principle, Lagrangian mechanics and Hamiltonian mechanics. These include Poisson brackets, Liouville's theorem, canonical transformations, Hamilton-Jacobi theory, the action-angle technique (chapter 15), and classical mechanics in the continua (chapter 16). This is followed by a brief review of the revolution in classical mechanics introduced by Einstein's theory of relativistic mechanics. The extended theory of Lagrangian and Hamiltonian mechanics is used to apply variational techniques to the Special Theory of Relativity, followed by a discussion of the use of variational principles in the development of the General Theory of Relativity (chapter 17). The book finishes with a brief review of the role of variational principles in bridging the gap between classical mechanics and quantum mechanics, (chapter 18). These advanced topics extend beyond the typical syllabus for an undergraduate classical mechanics course. They are included to stimulate student interest in physics by giving them a glimpse of the physics at the summit that they have already struggled to climb. This glimpse illustrates the breadth of classical mechanics, and the pivotal role that variational principles have played in the development of classical, relativistic, quantal, and statistical mechanics.

The front cover picture of this book shows a sailplane soaring high above the Italian Alps. This picture epitomizes the unlimited horizon of opportunities provided when the full dynamic range of variational principles are applied to classical mechanics. The adjacent pictures of the galaxy, and the skier, represent the wide dynamic range of applicable topics that span from the origin of the universe, to everyday life. These cover pictures reflect the beauty and unity of the foundation provided by variational principles to the development of classical mechanics.

Information regarding the associated P235 undergraduate course at the University of Rochester is available on the web site at http://www.pas.rochester.edu/~cline/P235/index.shtml. Information about the author is available at the Cline home web site: http://www.pas.rochester.edu/~cline/index.html.

The author thanks Meghan Sarkis who prepared many of the illustrations, Joe Easterly who designed the book cover plus the webpage, and Moriana Garcia who organized publication. Andrew Sifain developed the diagnostic workshop questions. The author appreciates the permission, granted by Professor Struckmeier, to quote his published article on the extended Hamilton-Lagrangian formalism. The author acknowledges the feedback and suggestions made by many students who have taken this course, as well as helpful suggestions by his colleagues; Andrew Abrams, Adam Hayes, Connie Jones, Andrew Melchionna, David Munson, Alice Quillen, Richard Sarkis, James Schneeloch, Steven Torrisi, Dan Watson, and Frank Wolfs. These lecture notes were typed in LATEX using Scientific WorkPlace (MacKichan Software, Inc.), while Adobe Illustrator, Photoshop, Origin, Mathematica, and MUPAD, were used to prepare the illustrations.

Douglas Cline,
University of Rochester, 2018

Prologue

Two dramatically different philosophical approaches to science were developed in the field of classical mechanics during the 17th - 18th centuries. This time period coincided with the Age of Enlightenment in Europe during which remarkable intellectual and philosophical developments occurred. This was a time when both philosophical and causal arguments were equally acceptable in science, in contrast with current convention where there appears to be tacit agreement to discourage use of philosophical arguments in science.

Snell's Law: The genesis of two contrasting philosophical approaches to science relates back to early studies of the reflection and refraction of light. The velocity of light in a medium of refractive index n equals $v = \frac{c}{n}$. Thus a light beam incident at an angle θ_1, to the normal of a plane interface between medium 1 and medium 2, is refracted at an angle θ_2 in medium 2, where the angles are related by Snell's Law.

$$\frac{\sin \theta_1}{\sin \theta_2} = \frac{v_1}{v_2} = \frac{n_2}{n_1} \qquad \text{(Snell's Law)}$$

Ibn Sahl of Bagdad (984) first described the refraction of light, while Snell (1621) derived his law mathematically. Both of these scientists used the "vectorial approach" where the light velocity v is considered to be a vector pointing in the direction of propagation.

Fermat's Principle: Fermat's principle of least time (1657), which is based on the work of Hero of Alexandria (~ 60) and Ibn al-Haytham (1021), states that *"light travels between two given points along the path of shortest time"*. The transit time τ of a light beam between two locations A and B, in a medium with position-dependent refractive index $n(s)$, is given by

$$\tau = \int_{t_A}^{t_B} dt = \frac{1}{c} \int_A^B n(s)ds \qquad \text{(Fermat's Principle)}$$

Fermat's Principle leads to the derivation of Snell's Law.

Philosophically the physics underlying the contrasting vectorial and Fermat's Principle derivations of Snell's Law are dramatically different. The vectorial approach is based on differential relations between the velocity vectors in the two media, whereas Fermat's variational approach is based on the fact that the light preferentially selects a path for which the integral of the transit time between the initial location A and the final location B is minimized. That is, the first approach is based on "vectorial mechanics" whereas Fermat's approach is based on variational principles in that the path between the initial and final locations is varied to find the path that minimizes the transit time. Fermat's enunciation of variational principles in physics played a key role in the historical development, and subsequent exploitation, of the principle of least action in analytical formulations of classical mechanics as discussed below.

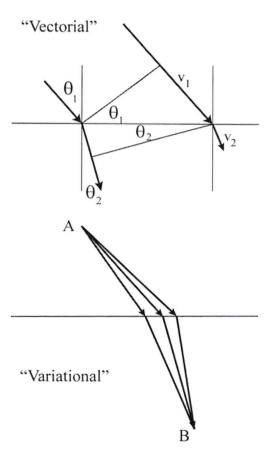

Figure 1: Vectorial and variational representations of Snell's Law for refraction of light.

Newtonian mechanics: Momentum and force are vectors that underlie the Newtonian formulation of classical mechanics. Newton's monumental treatise, entitled *"Philosophiae Naturalis Principia Mathematica"*, published in 1687, established his three universal laws of motion, the universal theory of gravitation, the derivation of Kepler's three laws of planetary motion, and the development of calculus. Newton's three universal laws of motion provide the most intuitive approach to classical mechanics in that they are based on vector quantities like momentum, and the rate of change of momentum, which are related to force. Newton's equation of motion

$$\mathbf{F} = \frac{d\mathbf{p}}{dt} \qquad \text{(Newton's equation of motion)}$$

is a vector differential relation between the instantaneous forces and rate of change of momentum, or equivalent instantaneous acceleration, all of which are vector quantities. Momentum and force are easy to visualize, and both cause and effect are embedded in Newtonian mechanics. Thus, if all of the forces, including the constraint forces, acting on the system are known, then the motion is solvable for two body systems. The mathematics for handling Newton's "vectorial mechanics" approach to classical mechanics is well established.

Analytical mechanics: Variational principles apply to many aspects of our daily life. Typical examples include; selecting the optimum compromise in quality and cost when shopping, selecting the fastest route to travel from home to work, or selecting the optimum compromise to satisfy the disparate desires of the individuals comprising a family. Variational principles underlie the analytical formulation of mechanics. It is astonishing that the laws of nature are consistent with variational principles involving the principle of least action. Minimizing the action integral led to the development of the mathematical field of variational calculus, plus the analytical variational approaches to classical mechanics, by Euler, Lagrange, Hamilton, and Jacobi.

Leibniz, who was a contemporary of Newton, introduced methods based on a quantity called *"vis viva"*, which is Latin for *"living force"* and equals twice the kinetic energy. Leibniz believed in the philosophy that God created a perfect world where nature would be thrifty in all its manifestations. In 1707, Leibniz proposed that the optimum path is based on minimizing the time integral of the *vis viva*, which is equivalent to the action integral of Lagrangian/Hamiltonian mechanics. In 1744 Euler derived the Leibniz result using variational concepts while Maupertuis restated the Leibniz result based on teleological arguments. The development of Lagrangian mechanics culminated in the 1788 publication of Lagrange's monumental treatise entitled *"Mécanique Analytique"*. Lagrange used d'Alembert's Principle to derive Lagrangian mechanics providing a powerful analytical approach to determine the magnitude and direction of the optimum trajectories, plus the associated forces.

The culmination of the development of analytical mechanics occurred in 1834 when Hamilton proposed his Principle of Least Action, as well as developing Hamiltonian mechanics which is the premier variational approach in science. Hamilton's concept of least action is defined to be the time integral of the Lagrangian. Hamilton's Action Principle (1834) minimizes the action integral S defined by

$$S = \int_A^B L(\mathbf{q}, \dot{\mathbf{q}}, t) dt \qquad \text{(Hamilton's Principle)}$$

In the simplest form, the Lagrangian $L(\mathbf{q}, \dot{\mathbf{q}}, t)$ equals the difference between the kinetic energy T and the potential energy U. Hamilton's Least Action Principle underlies Lagrangian mechanics. This Lagrangian is a function of n generalized coordinates q_i plus their corresponding velocities $\dot{q}_i$. Hamilton also developed the premier variational approach, called Hamiltonian mechanics, that is based on the Hamiltonian $H(\mathbf{q}, \mathbf{p}, t)$ which is a function of the n fundamental position q_i plus the conjugate momentum p_i variables. In 1843 Jacobi provided the mathematical framework required to fully exploit the power of Hamiltonian mechanics. Note that the Lagrangian, Hamiltonian, and the action integral, all are scalar quantities which simplifies derivation of the equations of motion compared with the vector calculus used by Newtonian mechanics.

Figure 2 presents a philosophical roadmap illustrating the hierarchy of philosophical approaches based on Hamilton's Action Principle, that are available for deriving the equations of motion of a system. The primary **Stage1** uses Hamilton's Action functional, $S = \int_{t_i}^{t_f} L(\mathbf{q}, \dot{\mathbf{q}}, t) dt$ to derive the Lagrangian, and Hamiltonian functionals which provide the most fundamental and sophisticated level of understanding. **Stage1** involves specifying all the active degrees of freedom, as well as the interactions involved. **Stage2** uses the Lagrangian or Hamiltonian functionals, derived at **Stage1**, in order to derive the equations of motion for the system of

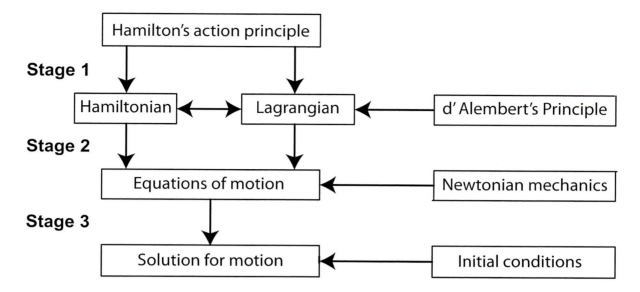

Figure 2: Philosophical road map of the hierarchy of stages involved in analytical mechanics. Hamilton's Action Principle is the foundation of analytical mechanics. Stage 1 uses Hamilton's Principle to derive the Lagranian and Hamiltonian. Stage 2 uses either the Lagrangian or Hamiltonian to derive the equations of motion for the system. Stage 3 uses these equations of motion to solve for the actual motion using the assumed initial conditions. The Lagrangian approach can be derived directly based on d'Alembert's Principle. Newtonian mechanics can be derived directly based on Newton's Laws of Motion. The advantages and power of Hamilton's Action Principle are unavailable if the Laws of Motion are derived using either d'Alembert's Principle or Newton's Laws of Motion.

interest. *Stage*3 then uses these derived equations of motion to solve for the motion of the system subject to a given set of initial boundary conditions. Note that Lagrange first derived Lagrangian mechanics based on d' Alembert's Principle, while Newton's Laws of Motion specify the equations of motion used in Newtonian mechanics.

The analytical approach to classical mechanics appeared contradictory to Newton's intuitive vectorial treatment of force and momentum. There is a dramatic difference in philosophy between the vector-differential equations of motion derived by Newtonian mechanics, which relate the instantaneous force to the corresponding instantaneous acceleration, and analytical mechanics, where minimizing the scalar action integral involves integrals over space and time between specified initial and final states. Analytical mechanics uses variational principles to determine the optimum trajectory, from a continuum of tentative possibilities, by requiring that the optimum trajectory minimizes the action integral between specified initial and final conditions.

Initially there was considerable prejudice and philosophical opposition to use of the variational principles approach which is based on the assumption that nature follows the principles of economy. The variational approach is not intuitive, and thus it was considered to be speculative and "metaphysical", but it was tolerated as an efficient tool for exploiting classical mechanics. This opposition to the variational principles underlying analytical mechanics, delayed full appreciation of the variational approach until the start of the 20th century. As a consequence, the intuitive Newtonian formulation reigned supreme in classical mechanics for over two centuries, even though the remarkable problem-solving capabilities of analytical mechanics were recognized and exploited following the development of analytical mechanics by Lagrange.

The full significance and superiority of the analytical variational formulations of classical mechanics became well recognised and accepted following the development of the Special Theory of Relativity in 1905. The Theory of Relativity requires that the laws of nature be invariant to the reference frame. This is not satisfied by the Newtonian formulation of mechanics which assumes one absolute frame of reference and a separation of space and time. In contrast, the Lagrangian and Hamiltonian formulations of the principle of least action remain valid in the Theory of Relativity, if the Lagrangian is written in a relativistically-invariant form in space-time. The complete invariance of the variational approach to coordinate frames is precisely the formalism necessary for handling relativistic mechanics.

Hamiltonian mechanics, which is expressed in terms of the conjugate variables $(\mathbf{q}, \mathbf{p})$, relates classical mechanics directly to the underlying physics of quantum mechanics and quantum field theory. As a consequence, the philosophical opposition to exploiting variational principles no longer exists, and Hamiltonian mechanics has become the preeminent formulation of modern physics. The reader is free to draw their own conclusions regarding the philosophical question "is the principle of economy a fundamental law of classical mechanics, or is it a fortuitous consequence of the fundamental laws of nature?"

From the late seventeenth century, until the dawn of modern physics at the start of the twentieth century, classical mechanics remained a primary driving force in the development of physics. Classical mechanics embraces an unusually broad range of topics spanning motion of macroscopic astronomical bodies to microscopic particles in nuclear and particle physics, at velocities ranging from zero to near the velocity of light, from one-body to statistical many-body systems, as well as having extensions to quantum mechanics. Introduction of the Special Theory of Relativity in 1905, and the General Theory of Relativity in 1916, necessitated modifications to classical mechanics for relativistic velocities, and can be considered to be an extended theory of classical mechanics. Since the 1920's, quantal physics has superseded classical mechanics in the microscopic domain. Although quantum physics has played the leading role in the development of physics during much of the past century, classical mechanics still is a vibrant field of physics that recently has led to exciting developments associated with non-linear systems and chaos theory. This has spawned new branches of physics and mathematics as well as changing our notion of causality.

Goals: The primary goal of this book is to introduce the reader to the powerful variational-principles approaches that play such a pivotal role in classical mechanics and many other branches of modern science and engineering. This book emphasizes the intellectual beauty of these remarkable developments, as well as stressing the philosophical implications that have had a tremendous impact on modern science. A secondary goal is to apply variational principles to solve advanced applications in classical mechanics in order to introduce many sophisticated and powerful mathematical techniques that underlie much of modern physics.

This book starts with a review of Newtonian mechanics plus the solutions of the corresponding equations of motion. This is followed by an introduction to Lagrangian mechanics, based on d'Alembert's Principle, in order to develop familiarity in applying variational principles to classical mechanics. This leads to introduction of the more fundamental Hamilton's Action Principle, plus Hamiltonian mechanics, to illustrate the power provided by exploiting the full hierarchy of stages available for applying variational principles to classical mechanics. Finally the book illustrates how variational principles in classical mechanics were exploited during the development of both relativisitic mechanics and quantum physics. The connections and applications of classical mechanics to modern physics, are emphasized throughout the book in an effort to span the chasm that divides the Newtonian vector-differential formulation, and the integral variational formulation, of classical mechanics. This chasm is especially applicable to quantum mechanics which is based completely on variational principles. Note that variational principles, developed in the field of classical mechanics, now are used in a diverse and wide range of fields outside of physics, including economics, meteorology, engineering, and computing.

This study of classical mechanics involves climbing a vast mountain of knowledge, and the pathway to the top leads to elegant and beautiful theories that underlie much of modern physics. This book exploits variational principles applied to four major topics in classical mechanics to illustrate the power and importance of variational principles in physics. Being so close to the summit provides the opportunity to take a few extra steps beyond the normal introductory classical mechanics syllabus to glimpse the exciting physics found at the summit. This new physics includes topics such as quantum, relativistic, and statistical mechanics.

Chapter 1

A brief history of classical mechanics

1.1 Introduction

This chapter reviews the historical evolution of classical mechanics since considerable insight can be gained from study of the history of science. There are two dramatically different approaches used in classical mechanics. The first is the vectorial approach of Newton which is based on vector quantities like momentum, force, and acceleration. The second is the analytical approach of Lagrange, Euler, Hamilton, and Jacobi, that is based on the concept of least action and variational calculus. The more intuitive Newtonian picture reigned supreme in classical mechanics until the start of the twentieth century. Variational principles, which were developed during the nineteenth century, never aroused much enthusiasm in scientific circles due to philosophical objections to the underlying concepts; this approach was merely tolerated as an efficient tool for exploiting classical mechanics. A dramatic advance in the philosophy of science occurred at the start of the 20th century leading to widespread acceptance of the superiority of using variational principles.

1.2 Greek antiquity

The great philosophers in ancient Greece played a key role by using the astronomical work of the Babylonians to develop scientific theories of mechanics. **Thales of Miletus (624 - 547BC)**, the first of the seven great greek philosophers, developed geometry, and is hailed as the first true mathematician. **Pythagorus (570 - 495BC)** developed mathematics, and postulated that the earth is spherical. **Democritus (460 - 370BC)** has been called the father of modern science, while **Socrates (469 - 399BC)** is renowned for his contributions to ethics. **Plato (427-347 B.C.)** who was a mathematician and student of Socrates, wrote important philosophical dialogues. He founded the Academy in Athens which was the first institution of higher learning in the Western world that helped lay the foundations of Western philosophy and science. **Aristotle (384-322 B.C.)** is an important founder of Western philosophy encompassing ethics, logic, science, and politics. His views on the physical sciences profoundly influenced medieval scholarship that extended well into the Renaissance. He presented the first implied formulation of the principle of virtual work in statics, and his statement that "what is lost in velocity is gained in force" is a veiled reference to kinetic and potential energy. He adopted an Earth centered model of the universe. **Aristarchus (310 - 240 B.C.)** argued that the Earth orbited the Sun and used measurements to imply the relative distances of the Moon and the Sun. The greek philosophers were relatively advanced in logic and mathematics and developed concepts that enabled them to calculate areas and perimeters. Unfortunately their philosophical approach neglected collecting quantitative and systematic data that is an essential ingredient to the advancement of science.

 Archimedes (287-212 B.C.) represented the culmination of science in ancient Greece. As an engineer he designed machines of war, while as a scientist he made significant contributions to hydrostatics and the principle of the lever. As a mathematician, he applied infinitessimals in a way that is reminiscent of modern integral calculus, which he used to derive a value for π. Unfortunately much of the work of the brilliant Archimedes subsequently fell into oblivion. **Hero of Alexandria (10 - 70 A.D.)** described the principle of reflection that light takes the shortest path. This is an early illustration of variational principle

1

of least time. **Ptolemy (83 - 161 A.D.)** wrote several scientific treatises that greatly influenced subsequent philosophers. Unfortunately he adopted the incorrect geocentric solar system in contrast to the heliocentric model of Aristarchus and others.

1.3   Middle Ages

The decline and fall of the Roman Empire in ∼410 A.D. marks the end of Classical Antiquity, and the beginning of the Dark Ages in Western Europe (Christendom), while the Muslim scholars in Eastern Europe continued to make progress in astronomy and mathematics. For example, in Egypt, **Alhazen (965 - 1040 A.D.)** expanded the principle of least time to reflection and refraction. The Dark Ages involved a long scientific decline in Western Europe that languished for about 900 years. Science was dominated by religious dogma, all western scholars were monks, and the important scientific achievements of Greek antiquity were forgotten. The works of Aristotle were reintroduced to Western Europe by Arabs in the early 13th century leading to the concepts of forces in static systems which were developed during the fourteenth century. This included concepts of the work done by a force, and the virtual work involved in virtual displacements. **Leonardo da Vinci (1452-1519)** was a leader in mechanics at that time. He made seminal contributions to science, in addition to his well known contributions to architecture, engineering, sculpture, and art.

Nicolaus Copernicus (1473-1543) rejected the geocentric theory of Ptolomy and formulated a scientifically-based heliocentric cosmology that displaced the Earth from the center of the universe. The Ptolomic view was that heaven represented the perfect unchanging divine while the earth represented change plus chaos, and the celestial bodies moved relative to the fixed heavens. The book, *De revolutionibus orbium coelestium* (On the Revolutions of the Celestial Spheres), published by Copernicus in 1543, is regarded as the starting point of modern astronomy and the defining epiphany that began the Scientific Revolution. The book *De Magnete* written in 1600 by the English physician **William Gilbert (1540-1603)** presented the results of well-planned studies of magnetism and strongly influenced the intellectual-scientific evolution at that time.

Johannes Kepler (1571-1630), a German mathematician, astronomer and astrologer, was a key figure in the 17th century Scientific Revolution. He is best known for recognizing the connection between the motions in the sky and physics. His laws of planetary motion were developed by later astronomers based on his written work *Astronomia nova*, *Harmonices Mundi*, and *Epitome of Copernican Astrononomy*. Kepler was an assistant to **Tycho Brahe (1546-1601)** who for many years recorded accurate astronomical data that played a key role in the development of Kepler's theory of planetary motion. Kepler's work provided the foundation for Isaac Newton's theory of universal gravitation. Unfortunately Kepler did not recognize the true nature of the gravitational force.

Galileo Galilei (1564-1642) built on the Aristotle principle by recognizing the law of inertia, the persistence of motion if no forces act, and the proportionality between force and acceleration. This amounts to recognition of work as the product of force times displacement in the direction of the force. He applied virtual work to the equilibrium of a body on an inclined plane. He also showed that the same principle applies to hydrostatic pressure that had been established by Archimedes, but he did not apply his concepts in classical mechanics to the considerable knowledge base on planetary motion. Galileo is famous for the apocryphal story that he dropped two cannon balls of different masses from the Tower of Pisa to demonstrate that their speed of descent was independent of their mass.

1.4   Age of Enlightenment

The Age of Enlightenment is a term used to describe a phase in Western philosophy and cultural life in which reason was advocated as the primary source and legitimacy for authority. It developed simultaneously in Germany, France, Britain, the Netherlands, and Italy around the 1650's and lasted until the French Revolution in 1789. The intellectual and philosophical developments led to moral, social, and political reforms. The principles of individual rights, reason, common sense, and deism were a revolutionary departure from the existing theocracy, autocracy, oligarchy, aristocracy, and the divine right of kings. It led to political revolutions in France and the United States. It marks a dramatic departure from the Early Modern period which was noted for religious authority, absolute state power, guild-based economic systems, and censorship of ideas. It opened a new era of rational discourse, liberalism, freedom of expression, and scientific method. This new environment led to tremendous advances in both science and mathematics in addition to music, literature, philosophy, and art. Scientific development during the 17th century included the pivotal advances made by Newton and Leibniz at the beginning of the revolutionary Age of Enlightenment, culminating in the development of variational calculus and analytical mechanics by Euler and Lagrange. The scientific advances of this age include publication of two monumental books *Philosophiae Naturalis Principia Mathematica* by Newton in 1687 and *Mécanique analytique* by Lagrange in 1788. These are the definitive two books upon which classical mechanics is built.

René Descartes (1596-1650) attempted to formulate the laws of motion in 1644. He talked about conservation of motion (momentum) in a straight line but did not recognize the vector character of momentum. **Pierre de Fermat (1601-1665)** and René Descartes were two leading mathematicians in the first half of the 17th century. Independently they discovered the principles of analytic geometry and developed some initial concepts of calculus. Fermat and **Blaise Pascal (1623-1662)** were the founders of the theory of probability.

Isaac Newton (1642-1727) made pioneering contributions to physics and mathematics as well as being a theologian. At 18 he was admitted to Trinity College Cambridge where he read the writings of modern philosophers like Descartes, and astronomers like Copernicus, Galileo, and Kepler. By 1665 he had discovered the generalized binomial theorem, and began developing infinitessimal calculus. Due to a plague, the university closed for two years in 1665 during which Newton worked at home developing the theory of calculus that built upon the earlier work of Barrow and Descartes. He was elected Lucasian Professor of Mathematics in 1669 at the age of 26. From 1670 Newton focussed on optics leading to his *Hypothesis of Light* published in 1675 and his book *Opticks* in 1704. Newton described light as being made up of a flow of extremely subtle corpuscles that also had associated wavelike properties to explain diffraction and optical interference that he studied. Newton returned to mechanics in 1677 by studying planetary motion and gravitation that applied the calculus he had developed. In 1687 he published his monumental treatise entitled *Philosophiae Naturalis Principia Mathematica* which established his three universal laws of motion, the universal theory of gravitation, derivation of Kepler's three laws of planetary motion, and was his first publication of the development of calculus which he called "the science of fluxions". Newton's laws of motion are based on the concepts of force and momentum, that is, force equals the rate of change of momentum. Newton's postulate of an invisible force able to act over vast distances led him to be criticized for introducing "occult agencies" into science. In a remarkable achievement, Newton completely solved the laws of mechanics. His theory of classical mechanics and of gravitation reigned supreme until the development of the Theory of Relativity in 1905. The followers of Newton envisioned the Newtonian laws to be absolute and universal. This dogmatic reverence of Newtonian mechanics prevented physicists from an unprejudiced appreciation of the analytic variational approach to mechanics developed during the 17th through 19th centuries. Newton was the first scientist to be knighted and was appointed president of the Royal Society.

Gottfried Leibniz (1646-1716) was a brilliant German philosopher, a contemporary of Newton, who worked on both calculus and mechanics. Leibniz started development of calculus in 1675, ten years after Newton, but Leibniz published his work in 1684, which was three years before Newton's Principia. Leibniz made significant contributions to integral calculus and developed the notation currently used in calculus. He introduced the name calculus based on the Latin word for the small stone used for counting. Newton and Leibniz were involved in a protracted argument over who originated calculus. It appears that Leibniz saw drafts of Newton's work on calculus during a visit to England. Throughout their argument Newton was the ghost writer of most of the articles in support of himself and he had them published under nonde-plume of his friends. Leibniz made the tactical error of appealing to the Royal Society to intercede on his behalf. Newton, as president of the Royal Society, appointed his friends to an "impartial" committee to investigate this issue, then he wrote the committee's report that accused Leibniz of plagiarism of Newton's work on calculus, after which he had it published by the Royal Society. Still unsatisfied he then wrote an anonymous review of the report in the Royal Society's own periodical. This bitter dispute lasted until the death of Leibniz. When Leibniz died his work was largely discredited. The fact that he falsely claimed to be a nobleman and added the prefix "von" to his name, coupled with Newton's vitriolic attacks, did not help his credibility. Newton is reported to have declared that he took great satisfaction in "breaking Leibniz's heart." Studies during the 20th century have largely revived the reputation of Leibniz and he is recognized to have made major contributions to the development of calculus.

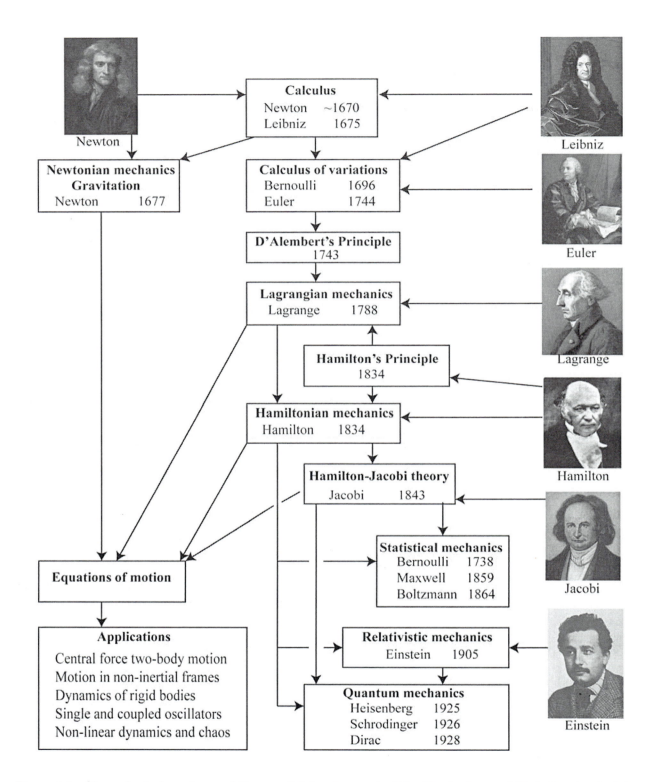

Figure 1.1: Chronological roadmap of the parallel development of the Newtonian and Variational-principles approaches to classical mechanics.

1.5 Variational methods in physics

Pierre de Fermat (1601-1665) revived the principle of least time, which states that *light travels between two given points along the path of shortest time* and was used to derive Snell's law in 1657. This enunciation of variational principles in physics played a key role in the historical development of the variational principle of least action that underlies the analytical formulations of classical mechanics.

Gottfried Leibniz (1646-1716) made significant contributions to the development of variational principles in classical mechanics. In contrast to Newton's laws of motion, which are based on the concept of momentum, Leibniz devised a new theory of dynamics based on kinetic and potential energy that anticipates the analytical variational approach of Lagrange and Hamilton. Leibniz argued for a quantity called the *"vis viva"*, which is Latin for *living force,* that equals twice the kinetic energy. Leibniz argued that the change in kinetic energy is equal to the work done. In 1687 Leibniz proposed that the optimum path is based on minimizing the time integral of the vis viva, which is equivalent to the action integral. Leibniz used both philosophical and causal arguments in his work which were acceptable during the Age of Enlightenment. Unfortunately for Leibniz, his analytical approach based on energies, which are scalars, appeared contradictory to Newton's intuitive vectorial treatment of force and momentum. There was considerable prejudice and philosophical opposition to the variational approach which assumes that nature is thrifty in all of its actions. The variational approach was considered to be speculative and "metaphysical" in contrast to the causal arguments supporting Newtonian mechanics. This opposition delayed full appreciation of the variational approach until the start of the 20th century.

Johann Bernoulli (1667-1748) was a Swiss mathematician who was a student of Leibniz's calculus, and sided with Leibniz in the Newton-Leibniz dispute over the credit for developing calculus. Also Bernoulli sided with the Descartes' vortex theory of gravitation which delayed acceptance of Newton's theory of gravitation in Europe. Bernoulli pioneered development of the calculus of variations by solving the problems of the catenary, the brachistochrone, and Fermat's principle. Johann Bernoulli's son Daniel played a significant role in the development of the well-known Bernoulli Principle in hydrodynamics.

Pierre Louis Maupertuis (1698-1759) was a student of Johann Bernoulli and conceived the universal hypothesis that in nature there is a certain quantity called action which is minimized. Although this bold assumption correctly anticipates the development of the variational approach to classical mechanics, he obtained his hypothesis by an entirely incorrect method. He was a dilettante whose mathematical prowess was behind the high standards of that time, and he could not establish satisfactorily the quantity to be minimized. His teleological[1] argument was influenced by Fermat's principle and the corpuscle theory of light that implied a close connection between optics and mechanics.

Leonhard Euler (1707-1783) was the preeminent Swiss mathematician of the 18th century and was a student of Johann Bernoulli. Euler developed, with full mathematical rigor, the calculus of variations following in the footsteps of Johann Bernoulli. Euler used variational calculus to solve minimum/maximum isoperimetric problems that had attracted and challenged the early developers of calculus, Newton, Leibniz, and Bernoulli. Euler also was the first to solve the rigid-body rotation problem using the three components of the angular velocity as kinematical variables. Euler became blind in both eyes by 1766 but that did not hinder his prolific output in mathematics due to his remarkable memory and mental capabilities. Euler's contributions to mathematics are remarkable in quality and quantity; for example during 1775 he published one mathematical paper per week in spite of being blind. Euler implicitly implied the principle of least action using vis visa which is not the exact form explicitly developed by Lagrange.

Jean le Rond d'Alembert (1717-1785) was a French mathematician and physicist who had the clever idea of extending use of the principle of virtual work from statics to dynamics. d'Alembert's Principle rewrites the principle of virtual work in the form

$$\sum_{i=1}^{N}(\mathbf{F}_i - \dot{\mathbf{p}}_i)\delta\mathbf{r}_i = 0$$

where the inertial reaction force $\dot{\mathbf{p}}$ is subtracted from the corresponding force $\mathbf{F}$. This extension of the principle of virtual work applies equally to both statics and dynamics leading to a single variational principle.

Joseph Louis Lagrange (1736-1813) was an Italian mathematician and a student of Leonhard Euler. In 1788 Lagrange published his monumental treatise on analytical mechanics entitled *Mécanique Analytique*

---

[1] Teleology is any philosophical account that holds that final causes exist in nature, meaning that — analogous to purposes found in human actions — nature inherently tends toward definite ends.

which introduces his Lagrangian mechanics analytical technique which is based on d'Alembert's Principle of Virtual Work. Lagrangian mechanics is a remarkably powerful technique that is equivalent to minimizing the action integral S defined as

$$S = \int_{t_1}^{t_2} L dt$$

The Lagrangian L frequently is defined to be the difference between the kinetic energy T and potential energy V. His theory only required the analytical form of these scalar quantities. In the preface of his book he refers modestly to his extraordinary achievements with the statement "The reader will find no figures in the work. The methods which I set forth do not require either constructions or geometrical or mechanical reasonings: but only algebraic operations, subject to a regular and uniform rule of procedure." Lagrange also introduced the concept of undetermined multipliers to handle auxiliary conditions which plays a vital part of theoretical mechanics. William Hamilton, an outstanding figure in the analytical formulation of classical mechanics, called Lagrange the "Shakespeare of mathematics," on account of the extraordinary beauty, elegance, and depth of the Lagrangian methods. Lagrange also pioneered numerous significant contributions to mathematics. For example, Euler, Lagrange, and d'Alembert developed much of the mathematics of partial differential equations. Lagrange survived the French Revolution, and, in spite of being a foreigner, Napoleon named Lagrange to the Legion of Honour and made him a Count of the Empire in 1808. Lagrange was honoured by being buried in the Pantheon.

Carl Friedrich Gauss (1777-1855) was a German child prodigy who made many significant contributions to mathematics, astronomy and physics. He did not work directly on the variational approach, but Gauss's law, the divergence theorem, and the Gaussian statistical distribution are important examples of concepts that he developed and which feature prominently in classical mechanics as well as other branches of physics, and mathematics.

Simeon Poisson (1781-1840), was a brilliant mathematician who was a student of Lagrange. He developed the Poisson statistical distribution as well as the Poisson equation that features prominently in electromagnetic and other field theories. His major contribution to classical mechanics is development, in 1809, of the Poisson bracket formalism which featured prominently in development of Hamiltonian mechanics and quantum mechanics.

The zenith in development of the variational approach to classical mechanics occurred during the 19th century primarily due to the work of Hamilton and Jacobi.

William Hamilton (1805-1865) was a brilliant Irish physicist, astronomer and mathematician who was appointed professor of astronomy at Dublin when he was barely 22 years old. He developed the Hamiltonian mechanics formalism of classical mechanics which now plays a pivotal role in modern classical and quantum mechanics. He opened an entirely new world beyond the developments of Lagrange. Whereas the Lagrange equations of motion are complicated second-order differential equations, Hamilton succeeded in transforming them into a set of first-order differential equations with twice as many variables that consider momenta and their conjugate positions as independent variables. The differential equations of Hamilton are linear, have separated derivatives, and represent the simplest and most desirable form possible for differential equations to be used in a variational approach. Hence the name "canonical variables" given by Jacobi. Hamilton exploited the d'Alembert principle to give the first exact formulation of the principle of least action which underlies the variational principles used in analytical mechanics. The form derived by Euler and Lagrange employed the principle in a way that applies only for conservative (scleronomic) cases. A significant discovery of Hamilton is his realization that classical mechanics and geometrical optics can be handled from one unified viewpoint. In both cases he uses a "characteristic" function that has the property that, by mere differentiation, the path of the body, or light ray, can be determined by the same partial differential equations. This solution is equivalent to the solution of the equations of motion.

Carl Gustave Jacob Jacobi (1804-1851), a Prussian mathematician and contemporary of Hamilton, made significant developments in Hamiltonian mechanics. He immediately recognized the extraordinary importance of the Hamiltonian formulation of mechanics. Jacobi developed canonical transformation theory and showed that the function, used by Hamilton, is only one special case of functions that generate suitable canonical transformations. He proved that any complete solution of the partial differential equation, without the specific boundary conditions applied by Hamilton, is sufficient for the complete integration of the equations of motion. This greatly extends the usefulness of Hamilton's partial differential equations. In 1843 Jacobi developed both the Poisson brackets, and the Hamilton-Jacobi, formulations of Hamiltonian mechanics. The latter gives a single, first-order partial differential equation for the action function in terms of the n generalized coordinates which greatly simplifies solution of the equations of motion. He also derived a principle of least action for time-independent cases that had been studied by Euler and Lagrange. Jacobi developed a superior approach to the variational integral that, by eliminating time from the integral, determined the path without saying anything about how the motion occurs in time.

James Clerk Maxwell (1831-1879) was a Scottish theoretical physicist and mathematician. His most prominent achievement was formulating a classical electromagnetic theory that united previously unrelated observations, plus equations of electricity, magnetism and optics, into one consistent theory. Maxwell's equations demonstrated that electricity, magnetism and light are all manifestations of the same phenomenon, namely the electromagnetic field. Consequently, all other classic laws and equations of electromagnetism were simplified cases of Maxwell's equations. Maxwell's achievements concerning electromagnetism have been called the "second great unification in physics". Maxwell demonstrated that electric and magnetic fields travel through space in the form of waves, and at a constant speed of light. In 1864 Maxwell wrote "A Dynamical Theory of the Electromagnetic Field" which proposed that light was in fact undulations in the same medium that is the cause of electric and magnetic phenomena. His work in producing a unified model of electromagnetism is one of the greatest advances in physics. Maxwell, in collaboration with **Ludwig Boltzmann (1844-1906)**, also helped develop the Maxwell–Boltzmann distribution, which is a statistical means of describing aspects of the kinetic theory of gases. These two discoveries helped usher in the era of modern physics, laying the foundation for such fields as special relativity and quantum mechanics. Boltzmann founded the field of statistical mechanics and was an early staunch advocate of the existence of atoms and molecules.

Henri Poincaré (1854-1912) was a French theoretical physicist and mathematician. He was the first to present the Lorentz transformations in their modern symmetric form and discovered the remaining relativistic velocity transformations. Although there is similarity to Einstein's Special Theory of Relativity, Poincaré and Lorentz still believed in the concept of the ether and did not fully comprehend the revolutionary philosophical change implied by Einstein. Poincaré worked on the solution of the three-body problem in planetary motion and was the first to discover a chaotic deterministic system which laid the foundations of modern chaos theory. It rejected the long-held deterministic view that if the position and velocities of all the particles are known at one time, then it is possible to predict the future for all time.

The last two decades of the 19th century saw the culmination of classical physics and several important discoveries that led to a revolution in science that toppled classical physics from its throne. The end of the 19th century was a time during which tremendous technological progress occurred; flight, the automobile, and turbine-powered ships were developed, Niagara Falls was harnessed for power, etc. During this period, **Heinrich Hertz (1857-1894)** produced electromagnetic waves confirming their derivation using Maxwell's equations. Simultaneously he discovered the photoelectric effect which was crucial evidence in support of quantum physics. Technical developments, such as photography, the induction spark coil, and the vacuum pump played a significant role in scientific discoveries made during the 1890's. At the end of the 19th century, scientists thought that the basic laws were understood and worried that future physics would be in the fifth decimal place; some scientists worried that little was left for them to discover. However, there remained a few, presumed minor, unexplained discrepancies plus new discoveries that led to the revolution in science that occurred at the beginning of the 20th century.

1.6 The 20th century revolution in physics

The two greatest achievements of modern physics occurred at the beginning of the 20th century. The first was Einstein's development of the Theory of Relativity; the Special Theory of Relativity in 1905 and the General Theory of Relativity in 1915. This was followed in 1925 by the development of quantum mechanics.

Albert Einstein (1879-1955) developed the Special Theory of Relativity in 1905 and the General Theory of Relativity in 1915; both of these revolutionary theories had a profound impact on classical mechanics and the underlying philosophy of physics. The Newtonian formulation of mechanics was shown to be an approximation that applies only at low velocities, while the General Theory of Relativity superseded Newton's Law of Gravitation and explained the Equivalence Principle. The Newtonian concepts of an absolute frame of reference, plus the assumption of the separation of time and space, were shown to be invalid at relativistic velocities. Einstein's postulate that the laws of physics are the same in all inertial frames requires a revolutionary change in the philosophy of time, space and reference frames which leads to a breakdown in the Newtonian formalism of classical mechanics. By contrast, the Lagrange and Hamiltonian variational formalisms of mechanics, plus the principle of least action, remain intact using a relativistically invariant Lagrangian. The independence of the variational approach to reference frames is precisely the formalism necessary for relativistic mechanics. The invariance to coordinate frames of the basic field equations also must remain invariant for the General Theory of Relativity which also can be derived in terms of a relativistic action principle. Thus the development of the Theory of Relativity unambiguously demonstrated the superiority of the variational formulation of classical mechanics over the vectorial Newtonian formulation, and thus the considerable effort made by Euler, Lagrange, Hamilton, Jacobi, and others in developing the analytical variational formalism of classical mechanics finally came to fruition at the start of the 20th century. Newton's two crowning achievements, the Laws of Motion and the Laws of Gravitation, that had reigned supreme since published in the Principia in 1687, were toppled from the throne by Einstein.

Emmy Noether (1882-1935) has been described as "the greatest ever woman mathematician". In 1915 she proposed a theorem that a conservation law is associated with any differentiable symmetry of a physical system. Noether's theorem evolves naturally from Lagrangian and Hamiltonian mechanics and she applied it to the four-dimensional world of general relativity. Noether's theorem has had an important impact in guiding the development of modern physics.

Other profound developments that had revolutionary impacts on classical mechanics were quantum physics and quantum field theory. The 1913 model of atomic structure by **Niels Bohr (1885-1962)** and the subsequent enhancements by **Arnold Sommerfeld (1868-1951),** were based completely on classical Hamiltonian mechanics. The proposal of wave-particle duality by **Louis de Broglie (1892-1987)**, made in his 1924 thesis, was the catalyst leading to the development of quantum mechanics. In 1925 **Werner Heisenberg (1901-1976)**, and **Max Born (1882-1970)** developed a matrix representation of quantum mechanics using non-commuting conjugate position and momenta variables.

Paul Dirac (1902-1984) showed in his Ph.D. thesis that Heisenberg's matrix representation of quantum physics is based on the Poisson Bracket generalization of Hamiltonian mechanics, which, in contrast to Hamilton's canonical equations, allows for non-commuting conjugate variables. In 1926 **Erwin Schrödinger (1887-1961)** independently introduced the operational viewpoint and reinterpreted the partial differential equation of Hamilton-Jacobi as a wave equation. His starting point was the optical-mechanical analogy of Hamilton that is a built-in feature of the Hamilton-Jacobi theory. Schrödinger then showed that the wave mechanics he developed, and the Heisenberg matrix mechanics, are equivalent representations of quantum mechanics. In 1928 Dirac developed his relativistic equation of motion for the electron and pioneered the field of quantum electrodynamics. Dirac also introduced the Lagrangian and the principle of least action to quantum mechanics, and these ideas were developed into the path-integral formulation of quantum mechanics and the theory of electrodynamics by **Richard Feynman(1918-1988).**

The concepts of wave-particle duality, and quantization of observables, both are beyond the classical notions of infinite subdivisions in classical physics. In spite of the radical departure of quantum mechanics from earlier classical concepts, the basic feature of the differential equations of quantal physics is their self-adjoint character which means that they are derivable from a variational principle. Thus both the Theory of Relativity, and quantum physics are consistent with the variational principle of mechanics, and inconsistent with Newtonian mechanics. As a consequence Newtonian mechanics has been dislodged from the throne it occupied since 1687, and the intellectually beautiful and powerful variational principles of analytical mechanics have been validated.

The 2015 observation of gravitational waves is a remarkable recent confirmation of Einstein's General Theory of Relativity and the validity of the underlying variational principles in physics. Another advance in physics is the understanding of the evolution of chaos in non-linear systems that have been made during the past four decades. This advance is due to the availability of computers which has reopened this interesting branch of classical mechanics, that was pioneered by Henri Poincaré about a century ago. Although classical mechanics is the oldest and most mature branch of physics, there still remain new research opportunities in this field of physics.

The focus of this book is to introduce the general principles of the mathematical variational principle approach, and its applications to classical mechanics. It will be shown that the variational principles, that were developed in classical mechanics, now play a crucial role in modern physics and mathematics, plus many other fields of science and technology.

References:
Excellent sources of information regarding the history of major players in the field of classical mechanics can be found on Wikipedia and the book "Variational Principle of Mechanics" by Lanczos.[La49]

Chapter 2

Review of Newtonian mechanics

2.1 Introduction

It is assumed that the reader has been introduced to Newtonian mechanics applied to one or two point objects. This chapter reviews Newtonian mechanics for motion of many-body systems as well as for macroscopic sized bodies. Newton's Law of Gravitation also is reviewed. The purpose of this review is to ensure that the reader has a solid foundation of elementary Newtonian mechanics upon which to build the powerful analytic Lagrangian and Hamiltonian approaches to classical dynamics.

Newtonian mechanics is based on application of Newton's Laws of motion which assume that the concepts of distance, time, and mass, are absolute, that is, motion is in an inertial frame. The Newtonian idea of the complete separation of space and time, and the concept of the absoluteness of time, are violated by the Theory of Relativity as discussed in chapter 17. However, for most practical applications, relativistic effects are negligible and Newtonian mechanics is an adequate description at low velocities. Therefore chapters $2 - 16$ will assume velocities for which Newton's laws of motion are applicable.

2.2 Newton's Laws of motion

Newton defined a vector quantity called *linear momentum* **p** which is the product of mass and velocity.

$$\mathbf{p} = m\dot{\mathbf{r}} \tag{2.1}$$

Since the mass m is a scalar quantity, then the velocity vector $\dot{\mathbf{r}}$ and the linear momentum vector **p** are colinear.

Newton's laws, expressed in terms of linear momentum, are:

1 *Law of inertia:* A body remains at rest or in uniform motion unless acted upon by a force.

2 *Equation of motion:* A body acted upon by a force moves in such a manner that the time rate of change of momentum equals the force.

$$\mathbf{F} = \frac{d\mathbf{p}}{dt} \tag{2.2}$$

3 *Action and reaction:* If two bodies exert forces on each other, these forces are equal in magnitude and opposite in direction.

Newton's second law contains the essential physics relating the force **F** and the rate of change of linear momentum **p**.

Newton's first law, the law of inertia, is a special case of Newton's second law in that if

$$\mathbf{F} = \frac{d\mathbf{p}}{dt} = 0 \tag{2.3}$$

then **p** is a constant of motion.

Newton's third law also can be interpreted as a statement of the conservation of momentum, that is, for a two particle system with no external forces acting,

$$\mathbf{F}_{12} = -\mathbf{F}_{21} \tag{2.4}$$

9

If the forces acting on two bodies are their mutual action and reaction, then equation 2.4 simplifies to

$$\mathbf{F}_{12} + \mathbf{F}_{21} = \frac{d\mathbf{p}_1}{dt} + \frac{d\mathbf{p}_2}{dt} = \frac{d}{dt}(\mathbf{p}_1 + \mathbf{p}_2) = 0 \qquad (2.5)$$

This implies that the total linear momentum ($\mathbf{P} = \mathbf{p}_1 + \mathbf{p}_2$) is a constant of motion.

Combining equations 2.1 and 2.2 leads to a second-order differential equation

$$\mathbf{F} = \frac{d\mathbf{p}}{dt} = m\frac{d^2\mathbf{r}}{dt^2} = m\ddot{\mathbf{r}} \qquad (2.6)$$

Note that the force on a body $\mathbf{F}$, and the resultant acceleration $\mathbf{a} = \ddot{\mathbf{r}}$ are colinear. Appendix $C2$ gives explicit expressions for the acceleration $\mathbf{a}$ in cartesian and curvilinear coordinate systems. The definition of force depends on the definition of the mass m. Newton's laws of motion are obeyed to a high precision for velocities much less than the velocity of light. For example, recent experiments have shown they are obeyed with an error in the acceleration of $\Delta a \leq 5 \times 10^{-14} m/s^2$.

2.3   Inertial frames of reference

An inertial frame of reference is one in which Newton's Laws of motion are valid. It is a non-accelerated frame of reference. An inertial frame must be homogeneous and isotropic. Physical experiments can be carried out in different inertial reference frames. The Galilean transformation provides a means of converting between two inertial frames of reference moving at a constant relative velocity. Consider two reference frames O and O' with O' moving with constant velocity $\mathbf{V}$ at time t. Figure 2.1 shows a Galilean transformation which can be expressed in vector form.

$$\begin{aligned} \mathbf{r}' &= \mathbf{r} - \mathbf{V}t \qquad (2.7) \\ t' &= t \end{aligned}$$

Equation 2.7 gives the boost, assuming Newton's hypothesis that the time is invariant to change of inertial frames of reference. The time differential of this transformation gives

$$\begin{aligned} \dot{\mathbf{r}}' &= \dot{\mathbf{r}} - \mathbf{V} \qquad (2.8) \\ \ddot{\mathbf{r}}' &= \ddot{\mathbf{r}} \end{aligned}$$

Note that the forces in the primed and unprimed inertial frames are related by

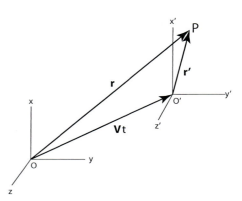

Figure 2.1: Frame O' moving with a constant velocity V with respect to frame O at the time t.

$$\mathbf{F} = \frac{d\mathbf{p}}{dt} = m\ddot{\mathbf{r}} = m\ddot{\mathbf{r}}' = \mathbf{F}' \qquad (2.9)$$

Thus Newton's Laws of motion are invariant under a Galilean transformation, that is, the inertial mass is unchanged under Galilean transformations. *If Newton's laws are valid in one inertial frame of reference, then they are valid in any frame of reference in uniform motion with respect to the first frame of reference.* This invariance is called **Galilean invariance.** There are an infinite number of possible inertial frames all connected by Galilean transformations.

Galilean invariance violates Einstein's Theory of Relativity. In order to satisfy Einstein's postulate that the laws of physics are the same in all inertial frames, as well as satisfy Maxwell's equations for electromagnetism, it is necessary to replace the Galilean transformation by the Lorentz transformation. As will be discussed in chapter 17, the Lorentz transformation leads to Lorentz contraction and time dilation both of which are related to the parameter $\gamma \equiv \frac{1}{\sqrt{1-\left(\frac{v}{c}\right)^2}}$ where c is the velocity of light in vacuum. Fortunately, most situations in life involve velocities where $v << c$; for example, for a body moving at $25,000$m.p.h. ($11,111$ m/s) which is the escape velocity for a body at the surface of the earth, the γ factor differs from unity by about $6.8x10^{-10}$ which is negligible. Relativistic effects are significant only in nuclear and particle physics as well as some exotic conditions in astrophysics. Thus, for the purpose of classical mechanics, usually it is reasonable to assume that the Galilean transformation is valid and is well obeyed under most practical conditions.

2.4 First-order integrals in Newtonian mechanics

A fundamental goal of mechanics is to determine the equations of motion for an $n-$body system, where the force $\mathbf{F}_i$ acts on the individual mass m_i where $1 \leq i \leq n$. Newton's second-order equation of motion, equation 2.6 must be solved to calculate the instantaneous spatial locations, velocities, and accelerations for each mass m_i of an n-body system. Both $\mathbf{F}_i$ and $\ddot{\mathbf{r}}_i$ are vectors, each having three orthogonal components. The solution of equation 2.6 involves integrating second-order equations of motion subject to a set of initial conditions. Although this task appears simple in principle, it can be exceedingly complicated for many-body systems. Fortunately, solution of the motion often can be simplified by exploiting three first-order integrals of Newton's equations of motion, that are related directly to conservation of either the linear momentum, angular momentum, or energy of the system. In addition, for the special case of these three first-order integrals, the internal motion of any many-body system can be factored out by a simple transformations into the center of mass of the system. As a consequence, the following three first-order integrals are exploited extensively in classical mechanics.

2.4.1 Linear Momentum

Newton's Laws can be written as the differential and integral forms of the *first-order time integral* which equals the change in linear momentum. That is

$$\mathbf{F}_i = \frac{d\mathbf{p}_i}{dt} \qquad \int_1^2 \mathbf{F}_i dt = \int_1^2 \frac{d\mathbf{p}_i}{dt} dt = (\mathbf{p}_2 - \mathbf{p}_1)_i \qquad (2.10)$$

This allows Newton's law of motion to be expressed directly in terms of the linear momentum $\mathbf{p}_i = m_i \dot{\mathbf{r}}_i$ of each of the $1 < i < n$ bodies in the system. This first-order time integral features prominently in classical mechanics since it connects to the important concept of linear momentum $\mathbf{p}$. This first-order time integral gives that the total linear momentum is a constant of motion when the sum of the external forces is zero.

2.4.2 Angular momentum

The angular momentum $\mathbf{L}_i$ of a particle i with linear momentum $\mathbf{p}_i$ with respect to an origin from which the position vector $\mathbf{r}_i$ is measured, is defined by

$$\mathbf{L}_i \equiv \mathbf{r}_i \times \mathbf{p}_i \qquad (2.11)$$

The torque, or moment of the force $\mathbf{N}_i$ with respect to the same origin is defined to be

$$\mathbf{N}_i \equiv \mathbf{r}_i \times \mathbf{F}_i \qquad (2.12)$$

where $\mathbf{r}_i$ is the position vector from the origin to the point where the force $\mathbf{F}_i$ is applied. Note that the torque $\mathbf{N}_i$ can be written as

$$\mathbf{N}_i = \mathbf{r}_i \times \frac{d\mathbf{p}_i}{dt} \qquad (2.13)$$

Consider the time differential of the angular momentum, $\frac{d\mathbf{L}_i}{dt}$

$$\frac{d\mathbf{L}_i}{dt} = \frac{d}{dt}(\mathbf{r}_i \times \mathbf{p}_i) = \frac{d\mathbf{r}_i}{dt} \times \mathbf{p}_i + \mathbf{r}_i \times \frac{d\mathbf{p}_i}{dt} \qquad (2.14)$$

However,

$$\frac{d\mathbf{r}_i}{dt} \times \mathbf{p}_i = m\frac{d\mathbf{r}_i}{dt} \times \frac{d\mathbf{r}_i}{dt} = 0 \qquad (2.15)$$

Equations $2.13 - 2.15$ can be used to write the first-order time integral for angular momentum in either differential or integral form as

$$\frac{d\mathbf{L}_i}{dt} = \mathbf{r}_i \times \frac{d\mathbf{p}_i}{dt} = \mathbf{N}_i \qquad \int_1^2 \mathbf{N}_i dt = \int_1^2 \frac{d\mathbf{L}_i}{dt} dt = (\mathbf{L}_2 - \mathbf{L}_1)_i \qquad (2.16)$$

Newton's Law relates torque and angular momentum about the same axis. When the torque about any axis is zero then angular momentum about that axis is a constant of motion. If the total torque is zero then the total angular momentum, as well as the components about three orthogonal axes, all are constants.

2.4.3   Kinetic energy

The third first-order integral, that can be used for solving the equations of motion, is the *first-order spatial integral* $\int_1^2 \mathbf{F}_i \cdot d\mathbf{r}_i$. Note that this spatial integral is a scalar in contrast to the first-order time integrals for linear and angular momenta which are vectors. The work done on a mass m_i by a force $\mathbf{F}_i$ in transforming from condition 1 to 2 is defined to be

$$[W_{12}]_i \equiv \int_1^2 \mathbf{F}_i \cdot d\mathbf{r}_i \tag{2.17}$$

If $\mathbf{F}_i$ is the net resultant force acting on a particle i, then the integrand can be written as

$$\mathbf{F}_i \cdot d\mathbf{r}_i = \frac{d\mathbf{p}_i}{dt} \cdot d\mathbf{r}_i = m_i \frac{d\mathbf{v}_i}{dt} \cdot \frac{d\mathbf{r}_i}{dt} dt = m_i \frac{d\mathbf{v}_i}{dt} \cdot \mathbf{v}_i dt = \frac{m_i}{2} \frac{d}{dt} \left(\mathbf{v}_i \cdot \mathbf{v}_i \right) dt = d \left(\frac{1}{2} m_i v_i^2 \right) = d\left[T \right]_i \tag{2.18}$$

where the kinetic energy of a particle i is defined as

$$[T]_i \equiv \frac{1}{2} m_i v_i^2 \tag{2.19}$$

Thus the work done on the particle i, that is, $[W_{12}]_i$ equals the change in kinetic energy of the particle if there is no change in other contributions to the total energy such as potential energy, heat dissipation, etc. That is

$$[W_{12}]_i = \left[\frac{1}{2} m v_2^2 - \frac{1}{2} m v_1^2 \right]_i = [T_2 - T_1]_i \tag{2.20}$$

Thus the differential, and corresponding first integral, forms of the kinetic energy can be written as

$$\mathbf{F}_i = \frac{dT_i}{d\mathbf{r}_i} \qquad\qquad \int_1^2 \mathbf{F}_i \cdot d\mathbf{r}_i = (T_2 - T_1)_i \tag{2.21}$$

If the work done on the particle is positive, then the final kinetic energy $T_2 > T_1$. Especially noteworthy is that the kinetic energy $[T]_i$ is a scalar quantity which makes it simple to use. This first-order spatial integral is the foundation of the analytic formulation of mechanics that underlies Lagrangian and Hamiltonian mechanics.

2.5   Conservation laws in classical mechanics

Elucidating the dynamics in classical mechanics is greatly simplified when conservation laws are applicable. In nature, isolated many-body systems frequently conserve one or more of the first-order integrals for linear momentum, angular momentum, and mass/energy. Note that mass and energy are coupled in the Theory of Relativity, but for non-relativistic mechanics the conservation of mass and energy are decoupled. Other observables such as lepton and baryon numbers are conserved, but these conservation laws usually can be subsumed under conservation of mass for most problems in non-relativistic classical mechanics. The power of conservation laws in calculating classical dynamics makes it useful to combine the conservation laws with the first integrals for linear momentum, angular momentum, and work-energy, when solving problems involving Newtonian mechanics. These three conservation laws will be derived assuming Newton's laws of motion, however, these conservation laws are fundamental laws of nature that apply well beyond the domain of applicability of Newtonian mechanics.

2.6   Motion of finite-sized and many-body systems

Elementary presentations in classical mechanics discuss motion and forces involving single point particles. However, in real life, single bodies have a finite size introducing new degrees of freedom such as rotation and vibration, and frequently many finite-sized bodies are involved. A finite-sized body can be thought of as a system of interacting particles such as the individual atoms of the body. The interactions between the parts of the body can be strong which leads to rigid body motion where the positions of the particles are held fixed with respect to each other, and the body can translate and rotate. When the interaction between the bodies is weaker, such as for a diatomic molecule, additional vibrational degrees of relative motion between the individual atoms are important. Newton's third law of motion becomes especially important for such many-body systems.

2.7 Center of mass of a many-body system

A finite sized body needs a reference point with respect to which the motion can be described. For example, there are 8 corners of a cube that could server as reference points, but the motion of each corner is complicated if the cube is both translating and rotating. The treatment of the behavior of finite-sized bodies, or many-body systems, is greatly simplified using the concept of **center of mass**. The center of mass is a particular fixed point in the body that has an especially valuable property; that is, the translational motion of a finite sized body can be treated like that of a point mass located at the center of mass. In addition the translational motion is separable from the rotational-vibrational motion of a many-body system when the motion is described with respect to the center of mass. Thus it is convenient at this juncture to introduce the concept of center of mass of a many-body system.

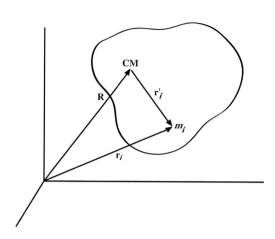

Figure 2.2: Position vector with respect to the center of mass.

For a many-body system, the position vector $\mathbf{r}_i$, defined relative to the laboratory system, is related to the position vector $\mathbf{r}'_i$ with respect to the center of mass, and the center-of-mass location $\mathbf{R}$ relative to the laboratory system. That is, as shown in figure 2.2

$$\mathbf{r}_i = \mathbf{R} + \mathbf{r}'_i \tag{2.22}$$

This vector relation defines the transformation between the laboratory and center of mass systems. For discrete and continuous systems respectively, the *location of the center of mass is uniquely defined as being where*

$$\sum_i^n m_i \mathbf{r}'_i = \int \mathbf{r}' \rho dV = 0. \qquad \text{(Center of mass definition)}$$

Define the total mass M as

$$M = \sum_i^n m_i = \int_{body} \rho dV \qquad \text{(Total mass)}$$

The average location of the system corresponds to the location of the center of mass since $\frac{1}{M}\sum_i m_i \mathbf{r}'_i = 0$, that is

$$\frac{1}{M}\sum_i m_i \mathbf{r}_i = \mathbf{R} + \frac{1}{M}\sum_i m_i \mathbf{r}'_i = \mathbf{R} \tag{2.23}$$

The vector $\mathbf{R}$, which describes the location of the center of mass, depends on the origin and coordinate system chosen. For a continuous mass distribution the location vector of the center of mass is given by

$$\mathbf{R} = \frac{1}{M}\sum_i m_i \mathbf{r}_i = \frac{1}{M}\int \mathbf{r}\rho dV \tag{2.24}$$

The center of mass can be evaluated by calculating the individual components along three orthogonal axes.

The *center-of-mass frame of reference* is defined as the frame for which the center of mass is stationary. This frame of reference is especially valuable for elucidating the underlying physics which involves only the relative motion of the many bodies. That is, the trivial translational motion of the center of mass frame, which has no influence on the relative motion of the bodies, is factored out and can be ignored. For example, a tennis ball ($0.06kg$) approaching the earth ($6 \times 10^{24}kg$) with velocity v could be treated in three frames, (a) assume the earth is stationary, (b) assume the tennis ball is stationary, or (c) the center-of-mass frame. The latter frame ignores the center of mass motion which has no influence on the relative motion of the tennis ball and the earth. The center of linear momentum and center of mass coordinate frames are identical in Newtonian mechanics but not in relativistic mechanics as described in chapter 17.4.3.

2.8   Total linear momentum of a many-body system

2.8.1   Center-of-mass decomposition

The total linear momentum $\mathbf{P}$ for a system of n particles is given by

$$\mathbf{P} = \sum_i^n \mathbf{p}_i = \frac{d}{dt} \sum_i^n m_i \mathbf{r}_i \tag{2.25}$$

It is convenient to describe a many-body system by a position vector $\mathbf{r}'_i$ with respect to the center of mass.

$$\mathbf{r}_i = \mathbf{R} + \mathbf{r}'_i \tag{2.26}$$

That is,

$$\mathbf{P} = \sum_i^n \mathbf{p}_i = \frac{d}{dt} \sum_i^n m_i \mathbf{r}_i = \frac{d}{dt} M\mathbf{R} + \frac{d}{dt} \sum_i^n m_i \mathbf{r}'_i = \frac{d}{dt} M\mathbf{R} + 0 = M\dot{\mathbf{R}} \tag{2.27}$$

since $\sum_i^n m_i \mathbf{r}'_i = 0$ as given by the definition of the center of mass. That is;

$$\mathbf{P} = M\dot{\mathbf{R}} \tag{2.28}$$

Thus the total linear momentum for a system is the same as the momentum of a single particle of mass $M = \sum_i^n m_i$ *located at the center of mass of the system.*

2.8.2   Equations of motion

The force acting on particle i, in an n-particle many-body system, can be separated into an external force $\mathbf{F}_i^{Ext}$ plus internal forces $\mathbf{f}_{ij}$ between the n particles of the system

$$\mathbf{F}_i = \mathbf{F}_i^E + \sum_{\substack{j \\ i \neq j}}^n \mathbf{f}_{ij} \tag{2.29}$$

The origin of the external force is from outside of the system while the internal force is due to the mutual interaction between the n particles in the system. Newton's Law tells us that

$$\dot{\mathbf{p}}_i = \mathbf{F}_i = \mathbf{F}_i^E + \sum_{\substack{j \\ i \neq j}}^n \mathbf{f}_{ij} \tag{2.30}$$

Thus the rate of change of total momentum is

$$\dot{\mathbf{P}} = \sum_i^n \dot{\mathbf{p}}_i = \sum_i^n \mathbf{F}_i^E + \sum_i^n \sum_{\substack{j \\ i \neq j}}^n \mathbf{f}_{ij} \tag{2.31}$$

Note that since the indices are dummy then

$$\sum_i \sum_{\substack{j \\ i \neq j}}^n \mathbf{f}_{ij} = \sum_j \sum_{\substack{i \\ i \neq j}}^n \mathbf{f}_{ji} \tag{2.32}$$

Substituting Newton's third law $\mathbf{f}_{ij} = -\mathbf{f}_{ji}$ into equation 2.32 implies that

$$\sum_i \sum_{\substack{j \\ i \neq j}}^n \mathbf{f}_{ij} = \sum_j \sum_{\substack{i \\ i \neq j}}^n \mathbf{f}_{ji} = -\sum_i \sum_{\substack{j \\ i \neq j}}^n \mathbf{f}_{ij} = 0 \tag{2.33}$$

which is satisfied only for the case where the summations equal zero. That is, for every internal force, there is an equal and opposite reaction force that cancels that internal force.

Therefore the first-order integral for linear momentum can be written in differential and integral forms as

$$\dot{\mathbf{P}} = \sum_i^n \mathbf{F}_i^E \qquad \int_1^2 \sum_i^n \mathbf{F}_i^E dt = \mathbf{P}_2 - \mathbf{P}_1 \qquad (2.34)$$

The reaction of a body to an external force is equivalent to a single particle of mass M located at the center of mass assuming that the internal forces cancel due to Newton's third law.

Note that the total linear momentum $\mathbf{P}$ is conserved if the net external force $\mathbf{F}^E$ is zero, that is

$$\mathbf{F}^E = \frac{d\mathbf{P}}{dt} = 0 \qquad (2.35)$$

Therefore the $\mathbf{P}$ of the center of mass is a constant. Moreover, if the component of the force along any direction $\widehat{\mathbf{e}}$ is zero, that is,

$$\mathbf{F}^E \cdot \widehat{\mathbf{e}} = \frac{d\mathbf{P} \cdot \widehat{\mathbf{e}}}{dt} = 0 \qquad (2.36)$$

then $\mathbf{P} \cdot \widehat{\mathbf{e}}$ is a constant. This fact is used frequently to solve problems involving motion in a constant force field. For example, in the earth's gravitational field, the momentum of an object moving in vacuum in the vertical direction is time dependent because of the gravitational force, whereas the horizontal component of momentum is constant if no forces act in the horizontal direction.

2.1 *Example: Exploding cannon shell*

Consider a cannon shell of mass M moves along a parabolic trajectory in the earths gravitational field. An internal explosion, generating an amount E of mechanical energy, blows the shell into two parts. One part of mass kM, where $k < 1$, continues moving along the same trajectory with velocity v' while the other part is reduced to rest. Find the velocity of the mass kM immediately after the explosion.

It is important to remember that the energy release E is given in the center of mass. If the velocity of the shell immediately before the explosion is v and v' is the velocity of the kM part immediately after the explosion, then energy conservation gives that $\frac{1}{2}Mv^2 + E = \frac{1}{2}kMv'^2 T$. The conservation of linear momentum gives $Mv = kMv'$. Eliminating v from these equations gives

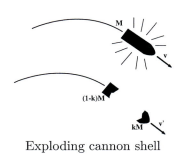

$$v' = \sqrt{\frac{2E}{[k(1-k)M]}}$$

Exploding cannon shell

2.2 *Example: Billiard-ball collisions*

A billiard ball with mass m and incident velocity v collides with an identical stationary ball. Assume that the balls bounce off each other elastically in such a way that the incident ball is deflected at a scattering angle θ to the incident direction. Calculate the final velocities v_f and V_f of the two balls and the scattering angle ϕ of the target ball. The conservation of linear momentum in the incident direction x, and the perpendicular direction give

$$mv = mv_f \cos\theta + mV_f \cos\phi \qquad\qquad 0 = mv_f \sin\theta - mV_f \sin\phi$$

Energy conservation gives .

$$\frac{m}{2}v^2 = \frac{m}{2}v_f^2 + \frac{m}{2}V_f^2$$

Solving these three equations gives $\phi = 90^0 - \theta$, that is, the balls bounce off perpendicular to each other in the laboratory frame. The final velocities are

$$v_f = v\cos\theta \qquad\qquad V_f = v\sin\theta$$

2.9   Angular momentum of a many-body system

2.9.1   Center-of-mass decomposition

As was the case for linear momentum, for a many-body system it is possible to separate the angular momentum into two components. One component is the angular momentum about the center of mass and the other component is the angular motion of the center of mass about the origin of the coordinate system. This separation is done by describing the angular momentum of a many-body system using a position vector $\mathbf{r}_i'$ *with respect to the center of mass plus the vector location* $\mathbf{R}$ *of the center of mass.*

$$\mathbf{r}_i = \mathbf{R} + \mathbf{r}_i' \tag{2.37}$$

The total angular momentum

$$
\begin{aligned}
\mathbf{L} &= \sum_i^n \mathbf{L}_i = \sum_i^n \mathbf{r}_i \times \mathbf{p}_i \\
&= \sum_i^n (\mathbf{R} + \mathbf{r}_i') \times m_i \left(\dot{\mathbf{R}} + \dot{\mathbf{r}}_i' \right) \\
&= \sum_i^n m_i \left[\mathbf{r}_i' \times \dot{\mathbf{r}}_i' + \mathbf{r}_i' \times \dot{\mathbf{R}} + \mathbf{R} \times \dot{\mathbf{r}}_i' + \mathbf{R} \times \dot{\mathbf{R}} \right]
\end{aligned}
\tag{2.38}
$$

Note that if the position vectors are with respect to the center of mass, then $\sum_i^n m_i \mathbf{r}_i' = 0$ resulting in the middle two terms in the bracket being zero, that is;

$$\mathbf{L} = \sum_i^n \mathbf{r}_i' \times \mathbf{p}_i' + \mathbf{R} \times \mathbf{P} \tag{2.39}$$

The total angular momentum separates into two terms, the angular momentum about the center of mass, plus the angular momentum of the center of mass about the origin of the axis system. This factoring of the angular momentum only applies for the center of mass. This is called Samuel König's first theorem.

2.9.2   Equations of motion

The time derivative of the angular momentum

$$\dot{\mathbf{L}}_i = \frac{d}{dt} \mathbf{r}_i \times \mathbf{p}_i = \dot{\mathbf{r}}_i \times \mathbf{p}_i + \mathbf{r}_i \times \dot{\mathbf{p}}_i \tag{2.40}$$

But

$$\dot{\mathbf{r}}_i \times \mathbf{p}_i = m_i \dot{\mathbf{r}}_i \times \dot{\mathbf{r}}_i = 0 \tag{2.41}$$

Thus the torque N_i acting on mass i is given by

$$\mathbf{N}_i = \dot{\mathbf{L}}_i = \mathbf{r}_i \times \dot{\mathbf{p}}_i = \mathbf{r}_i \times \mathbf{F}_i \tag{2.42}$$

Consider that the resultant force acting on particle i in this n-particle system can be separated into an external force $\mathbf{F}_i^{Ext}$ plus internal forces between the n particles of the system

$$\mathbf{F}_i = \mathbf{F}_i^E + \sum_{\substack{j \\ i \neq j}}^n \mathbf{f}_{ij} \tag{2.43}$$

The origin of the external force is from outside of the system while the internal force is due to the interaction with the other $n-1$ particles in the system. Newton's Law tells us that

$$\dot{\mathbf{p}}_i = \mathbf{F}_i = \mathbf{F}_i^E + \sum_{\substack{j \\ i \neq j}}^n \mathbf{f}_{ij} \tag{2.44}$$

The rate of change of total angular momentum is

$$\dot{\mathbf{L}} = \sum_i \dot{\mathbf{L}}_i = \sum_i \mathbf{r}_i \times \dot{\mathbf{p}}_i = \sum_i \mathbf{r}_i \times \mathbf{F}_i^E + \sum_i \sum_{\substack{j \\ i \neq j}} \mathbf{r}_i \times \mathbf{f}_{ij} \tag{2.45}$$

Since $\mathbf{f}_{ij} = -\mathbf{f}_{ji}$ the last expression can be written as

$$\sum_i \sum_{\substack{j \\ i \neq j}} \mathbf{r}_i \times \mathbf{f}_{ij} = \sum_i \sum_{\substack{j \\ i < j}} (\mathbf{r}_i - \mathbf{r}_j) \times \mathbf{f}_{ij} \tag{2.46}$$

Note that $(\mathbf{r}_i - \mathbf{r}_j)$ is the vector $\mathbf{r}_{ij}$ connecting j to i. *For central forces* the force vector $\mathbf{f}_{ij} = f_{ij}\widehat{\mathbf{r}_{ij}}$ thus

$$\sum_i \sum_{\substack{j \\ i < j}} (\mathbf{r}_i - \mathbf{r}_j) \times \mathbf{f}_{ij} = \sum_i \sum_{\substack{j \\ i < j}} \mathbf{r}_{ij} \times f_{ij}\widehat{\mathbf{r}_{ij}} = 0 \tag{2.47}$$

That is, for central internal forces the total internal torque on a system of particles is zero, and the rate of change of total angular momentum for central internal forces becomes

$$\dot{\mathbf{L}} = \sum_i \mathbf{r}_i \times \mathbf{F}_i^E = \sum_i \mathbf{N}_i^E = \mathbf{N}^E \tag{2.48}$$

where $\mathbf{N}^E$ is the net external torque acting on the system. Equation 2.48 leads to the differential and integral forms of the first integral relating the total angular momentum to total external torque.

$$\dot{\mathbf{L}} = \mathbf{N}^E \qquad\qquad \int_1^2 \mathbf{N}^E dt = \mathbf{L}_2 - \mathbf{L}_1 \tag{2.49}$$

Angular momentum conservation occurs in many problems involving zero external torques $\mathbf{N}^E = 0$, plus two-body central forces $\mathbf{F} = f(r)\hat{\mathbf{r}}$ since the torque on the particle about the center of the force is zero

$$\mathbf{N} = \mathbf{r} \times \mathbf{F} = f(r)[\mathbf{r} \times \hat{\mathbf{r}}] = 0 \tag{2.50}$$

Examples are, the central gravitational force for stellar or planetary systems in astrophysics, and the central electrostatic force manifest for motion of electrons in the atom. In addition, the component of angular momentum about any axis $\mathbf{L}.\hat{\mathbf{e}}$ is conserved if the net external torque about that axis $\mathbf{N}.\hat{\mathbf{e}} = 0$.

2.3 Example: Bolas thrown by gaucho

*Consider the bolas thrown by a gaucho to catch cattle. This is a system with conserved linear and angular momentum about certain axes. When the bolas leaves the gaucho's hand the center of mass has a linear velocity **V** plus an angular momentum about the center of mass of **L**. If no external torques act, then the center of mass of the bolas will follow a typical ballistic trajectory in the earth's gravitational field while the angular momentum vector **L** is conserved, that is, both in magnitude and direction. The tension in the ropes connecting the three balls does not impact the motion of the system as long as the ropes do not snap due to centrifugal forces.*

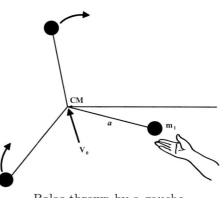

Bolas thrown by a gaucho

2.10   Work and kinetic energy for a many-body system

2.10.1   Center-of-mass kinetic energy

For a many-body system the position vector $\mathbf{r}'_i$ with respect to the center of mass is given by.

$$\mathbf{r}_i = \mathbf{R} + \mathbf{r}'_i \tag{2.51}$$

The location of the center of mass is uniquely defined as being at the location where $\int \rho \mathbf{r}'_i dV = 0$. The velocity of the ith particle can be expressed in terms of the velocity of the center of mass $\dot{\mathbf{R}}$ plus the velocity of the particle with respect to the center of mass $\dot{\mathbf{r}}'_i$. That is,

$$\dot{\mathbf{r}}_i = \dot{\mathbf{R}} + \dot{\mathbf{r}}'_i \tag{2.52}$$

The total kinetic energy T is

$$T = \sum_i^n \frac{1}{2} m_i v_i^2 = \sum_i^n \frac{1}{2} m_i \dot{\mathbf{r}}_i \cdot \dot{\mathbf{r}}_i = \sum_i^n \frac{1}{2} m_i \dot{\mathbf{r}}'_i \cdot \dot{\mathbf{r}}'_i + \left(\frac{d}{dt} \sum_i m_i \mathbf{r}'_i \right) \cdot \dot{\mathbf{R}} + \sum_i \frac{1}{2} m_i \dot{\mathbf{R}} \cdot \dot{\mathbf{R}} \tag{2.53}$$

For the special case of the center of mass, the middle term is zero since, by definition of the center of mass, $\sum_i m_i \dot{\mathbf{r}}'_i = 0$. Therefore

$$T = \sum_i^n \frac{1}{2} m_i v_i'^2 + \frac{1}{2} M V^2 \tag{2.54}$$

Thus the total kinetic energy of the system is equal to the sum of the kinetic energy of a mass M moving with the center of mass velocity plus the kinetic energy of motion of the individual particles relative to the center of mass. This is called Samuel König's second theorem.

Note that for a fixed center-of-mass energy, the total kinetic energy T has a minimum value of $\sum_i^n \frac{1}{2} m_i v_i'^2$ when the velocity of the center of mass $V = 0$. For a given internal excitation energy, the minimum energy required to accelerate colliding bodies occurs when the colliding bodies have identical, but opposite, linear momenta. That is, when the center-of-mass velocity $V = 0$.

2.10.2   Conservative forces and potential energy

In general, the line integral of a force field $\mathbf{F}$, that is, $\int_1^2 \mathbf{F} \cdot d\mathbf{r}$, is both path and time dependent. However, an important class of forces, called **conservative forces,** exist for which the following two facts are obeyed.

1) **Time independence:**
The force depends only on the particle position $\mathbf{r}$, that is, it does not depend on velocity or time.
2) **Path independence:**
For any two points 1 and 2, the work done by $\mathbf{F}$ is independent of the path taken between 1 and 2.

If forces are path independent, then it is possible to define a scalar field, called potential energy, denoted by $U(\mathbf{r})$, that is only a function of position. The path independence can be expressed by noting that the integral around a closed loop is zero. That is

$$\oint \mathbf{F} \cdot d\mathbf{r} = 0 \tag{2.55}$$

Applying Stokes theorem for a path-independent force leads to the alternate statement that the curl is zero. See appendix $G.3.3$.

$$\boldsymbol{\nabla} \times \mathbf{F} = 0. \tag{2.56}$$

Note that the vector product of two del operators ∇ acting on a scalar field U equals

$$\boldsymbol{\nabla} \times \boldsymbol{\nabla} U = 0 \tag{2.57}$$

Thus it is possible to express a path-independent force field as the gradient of a scalar field, U, that is

$$\mathbf{F} = -\boldsymbol{\nabla} U \tag{2.58}$$

Then the spatial integral

$$\int_1^2 \mathbf{F} \cdot d\mathbf{r} = -\int_1^2 (\boldsymbol{\nabla} U) \cdot d\mathbf{r} = U_1 - U_2 \tag{2.59}$$

Thus for a path-independent force, the work done on the particle is given by the change in potential energy if there is no change in kinetic energy. For example, if an object is lifted against the gravitational field, then work is done on the particle and the final potential energy U_2 exceeds the initial potential energy, U_1.

2.10.3   Total mechanical energy

The **total mechanical energy** E of a particle is defined as the sum of the kinetic and potential energies.

$$E = T + U \tag{2.60}$$

Note that the potential energy is defined only to within an additive constant since the force $\mathbf{F} = -\boldsymbol{\nabla} U$ depends only on difference in potential energy. Similarly, the kinetic energy is not absolute since any inertial frame of reference can be used to describe the motion and the velocity of a particle depends on the relative velocities of inertial frames. Thus the total mechanical energy $E = T + U$ is not absolute.

If a single particle is subject to several path-independent forces, such as gravity, linear restoring forces, etc., then a potential energy U_i can be ascribed to each of the m forces where for each force $\mathbf{F}_i = -\boldsymbol{\nabla} U_i$. In contrast to the forces, which add vectorially, these scalar potential energies are additive, $U = \sum_i^m U_i$. Thus the total mechanical energy for m potential energies equals

$$E = T + U(\mathbf{r}) = T + \sum_i^m U_i(\mathbf{r}) \tag{2.61}$$

The time derivative of the total mechanical energy $E = T + U$, equals

$$\frac{dE}{dt} = \frac{dT}{dt} + \frac{dU}{dt} \tag{2.62}$$

Equation 2.18 gave that $dT = \mathbf{F} \cdot d\mathbf{r}$. Thus, the first term in equation 2.62 equals

$$\frac{dT}{dt} = \mathbf{F} \cdot \frac{d\mathbf{r}}{dt} \tag{2.63}$$

The potential energy can be a function of both position and time. Thus the time difference in potential energy due to change in both time and position is given as

$$\frac{dU}{dt} = \sum_i \frac{\partial U}{\partial x_i} \frac{dx_i}{dt} + \frac{\partial U}{\partial t} = (\boldsymbol{\nabla} U) \cdot \frac{d\mathbf{r}}{dt} + \frac{\partial U}{\partial t} \tag{2.64}$$

The time derivative of the total mechanical energy is given using equations 2.63, 2.64 in equation 2.62.

$$\frac{dE}{dt} = \frac{dT}{dt} + \frac{dU}{dt} = \mathbf{F} \cdot \frac{d\mathbf{r}}{dt} + (\boldsymbol{\nabla} U) \cdot \frac{d\mathbf{r}}{dt} + \frac{\partial U}{\partial t} = [\mathbf{F} + (\boldsymbol{\nabla} U)] \cdot \frac{d\mathbf{r}}{dt} + \frac{\partial U}{\partial t} \tag{2.65}$$

Note that if the field is path independent, that is $\boldsymbol{\nabla} \times \mathbf{F} = 0$, then the force and potential are related by

$$\mathbf{F} = -\boldsymbol{\nabla} U \tag{2.66}$$

Therefore, for *path independent forces,* the first term in the time derivative of the total energy in equation 2.65 is zero. That is,

$$\frac{dE}{dt} = \frac{\partial U}{\partial t} \tag{2.67}$$

In addition, when the potential energy U is not an explicit function of time, then $\frac{\partial U}{\partial t} = 0$ and thus the total energy is conserved. That is, for the combination of (a) path independence plus (b) time independence, then *the total energy of a conservative field is conserved.*

Note that there are cases where the concept of potential still is useful even when it is time dependent. That is, if path independence applies, i.e. $\mathbf{F} = -\nabla U$ at any instant. For example, a Coulomb field problem where charges are slowly changing due to leakage etc., or during a peripheral collision between two charged bodies such as nuclei.

2.4 Example: Central force

A particle of mass m moves along a trajectory given by $x = x_0 \cos \omega_1 t$ and $y = y_0 \sin \omega_2 t$.

a) Find the x and y components of the force and determine the condition for which the force is a central force.

Differentiating with respect to time gives

$$\dot{x} = -x_0 \omega_1 \sin(\omega_1 t) \qquad\qquad \ddot{x} = -x_0 \omega_1^2 \cos(\omega_1 t)$$
$$\dot{y} = -y_0 \omega_2 \cos(\omega_2 t) \qquad\qquad \ddot{y} = -y_0 \omega_2^2 \sin(\omega_2 t)$$

Newton's second law gives

$$\mathbf{F} = m(\ddot{x}\hat{\imath} + \ddot{y}\hat{\jmath}) = -m\left[x_0 \omega_1^2 \cos(\omega_1 t)\,\hat{\imath} + y_0 \omega_2^2 \sin(\omega_2 t)\,\hat{\jmath}\right] = -m\left[\omega_1^2 x\hat{\imath} + \omega_2^2 y\hat{\jmath}\right]$$

Note that if $\omega_1 = \omega_2 = \omega$ then

$$\mathbf{F} = = -m\omega^2 \left[x\hat{\imath} + y\hat{\jmath}\right] = -m\omega^2 \mathbf{r}$$

That is, it is a central force if $\omega_1 = \omega_2 = \omega$.

b) Find the potential energy as a function of x and y.

Since

$$\mathbf{F} = -\nabla U = -\left[\frac{\partial U}{\partial x}\hat{\imath} + \frac{\partial U}{\partial y}\hat{\jmath}\right]$$

then

$$U = \frac{1}{2}m\left(\omega_1^2 x^2 + \omega_2^2 y^2\right)$$

assuming that $U = 0$ at the origin.

c) Determine the kinetic energy of the particle and show that it is conserved.

The total energy

$$E = T + U = \frac{1}{2}m\left(\dot{x}^2 + \dot{y}_2\right) + \frac{1}{2}m\left(\omega_1^2 x^2 + \omega_2^2 y^2\right) = \frac{1}{2}m\left(x_0^2\omega_1^2 + y_0^2\omega_2^2\right)$$

since $\cos^2\theta + \sin^2\theta = 1$. Thus the total energy E is a constant and is conserved.

2.10.4   Total mechanical energy for conservative systems

Equation 2.20 showed that, using Newton's second law, $\mathbf{F} = \frac{d\mathbf{p}}{dt}$, the first-order spatial integral gives that the work done, W_{12}, is related to the change in the kinetic energy. That is,

$$W_{12} \equiv \int_1^2 \mathbf{F} \cdot d\mathbf{r} = \frac{1}{2}mv_2^2 - \frac{1}{2}mv_1^2 = T_2 - T_1 \tag{2.68}$$

The work done W_{12} also can be evaluated in terms of the known forces $\mathbf{F}_i$ in the spatial integral.

Consider that the resultant force acting on particle i in this n-particle system can be separated into an external force $\mathbf{F}_i^{Ext}$ plus internal forces between the n particles of the system

$$\mathbf{F}_i = \mathbf{F}_i^E + \sum_{\substack{j \\ i \neq j}}^n \mathbf{f}_{ij} \tag{2.69}$$

The origin of the external force is from outside of the system while the internal force is due to the interaction with the other $n - 1$ particles in the system. Newton's Law tells us that

$$\dot{\mathbf{p}}_i = \mathbf{F}_i = \mathbf{F}_i^E + \sum_{\substack{j \\ i \neq j}}^n \mathbf{f}_{ij} \tag{2.70}$$

The work done on the system by a force moving from configuration $1 \rightarrow 2$ is given by

$$W_{1\rightarrow 2} = \sum_i^n \int_1^2 \mathbf{F}_i^E \cdot d\mathbf{r}_i + \sum_i^n \sum_{\substack{j \\ i \neq j}}^n \int_1^2 \mathbf{f}_{ij} \cdot d\mathbf{r}_i \tag{2.71}$$

Since $\mathbf{f}_{ij} = -\mathbf{f}_{ji}$ then

$$W_{1\rightarrow 2} = \sum_i^n \int_1^2 \mathbf{F}_i^E \cdot d\mathbf{r}_i + \sum_i^n \sum_{\substack{j \\ i < j}}^n \int_1^2 \mathbf{f}_{ij} \cdot (d\mathbf{r}_i - d\mathbf{r}_j) \tag{2.72}$$

where $d\mathbf{r}_i - d\mathbf{r}_j = d\mathbf{r}_{ij}$ is the vector from j to i.

Assume that both the external and internal forces are conservative, and thus can be derived from time independent potentials, that is

$$\mathbf{F}_i^E = -\boldsymbol{\nabla}_i U_i^{Ext} \tag{2.73}$$

$$\mathbf{f}_{ij} = -\boldsymbol{\nabla}_i U_{ij}^{Int} \tag{2.74}$$

Then

$$
\begin{aligned}
W_{1\rightarrow 2} &= -\sum_i^n \int_1^2 \boldsymbol{\nabla}_i U_i^{Ext} \cdot d\mathbf{r}_i - \sum_i^n \sum_{\substack{j \\ i < j}}^n \int_1^2 \boldsymbol{\nabla}_i U_{ij}^{Int} \cdot d\mathbf{r}_{ij} \\
&= \sum_i^n U_i^{Ext}(1) - \sum_i^n U_i^{Ext}(2) + \sum_i^n U_i^{Int}(1) - \sum_i^n U_i^{Int}(2) \\
&= U^{Ext}(1) - U^{Ext}(2) + U^{Int}(1) - U^{Int}(2)
\end{aligned}
\tag{2.75}
$$

Define the total external potential energy,

$$U^{Ext} = \sum_i^n U_i^{Ext} \tag{2.76}$$

and the total internal energy

$$U^{Int} = \sum_i^n U_i^{Int} \tag{2.77}$$

Equating the two equivalent equations for $W_{1\rightarrow 2}$, that is 2.68 and 2.75.gives that

$$W_{1\rightarrow 2} = T_2 - T_1 = U^{Ext}(1) - U^{Ext}(2) + U^{Int}(1) - U^{Int}(2) \tag{2.78}$$

Regroup these terms in equation 2.78 gives

$$T_1 + U^{Ext}(1) + U^{Int}(1) = T_2 + U^{Ext}(2) + U^{Int}(2)$$

This shows that, for *conservative forces*, the total energy is conserved and is given by

$$E = T + U^{Ext} + U^{Int} \tag{2.79}$$

The three first-order integrals for linear momentum, angular momentum, and energy provide powerful approaches for solving the motion of Newtonian systems due to the applicability of conservation laws for the corresponding linear and angular momentum plus energy conservation for conservative forces. In addition, the important concept of center-of-mass motion naturally separates out for these three first-order integrals. Although these conservation laws were derived assuming Newton's Laws of motion, these conservation laws are more generally applicable, and *these conservation laws surpass the range of validity of Newton's Laws of motion.* For example, in 1930 Pauli and Fermi postulated the existence of the neutrino in order to account for non-conservation of energy and momentum in β-decay because they did not wish to relinquish the concepts of energy and momentum conservation. The neutrino was first detected in 1956 confirming the correctness of this hypothesis.

2.11   Virial Theorem

The Virial theorem is an important theorem for a system of moving particles both in classical physics and quantum physics. The Virial Theorem is useful when considering a collection of many particles and has a special importance to central-force motion. For a general system of mass points with position vectors $\mathbf{r}_i$ and applied forces $\mathbf{F}_i$, consider the scalar product G

$$G \equiv \sum_i \mathbf{p}_i \cdot \mathbf{r}_i \tag{2.80}$$

where i sums over all particles. The time derivative of G is

$$\frac{dG}{dt} = \sum_i \mathbf{p}_i \cdot \dot{\mathbf{r}}_i + \sum_i \dot{\mathbf{p}}_i \cdot \mathbf{r}_i \tag{2.81}$$

However,

$$\sum_i \mathbf{p}_i \cdot \dot{\mathbf{r}}_i = \sum_i m\dot{\mathbf{r}}_i \cdot \dot{\mathbf{r}}_i = \sum_i mv^2 = 2T \tag{2.82}$$

Also, since $\dot{\mathbf{p}}_i = \mathbf{F}_i$

$$\sum_i \dot{\mathbf{p}}_i \cdot \mathbf{r}_i = \sum_i \mathbf{F}_i \cdot \mathbf{r}_i \tag{2.83}$$

Thus

$$\frac{dG}{dt} = 2T + \sum_i \mathbf{F}_i \cdot \mathbf{r}_i \tag{2.84}$$

The time average over a period τ is

$$\frac{1}{\tau} \int_0^\tau \frac{dG}{dt} dt = \frac{G(\tau) - G(0)}{\tau} = \langle 2T \rangle + \left\langle \sum_i \mathbf{F}_i \cdot \mathbf{r}_i \right\rangle \tag{2.85}$$

where the $\langle \rangle$ brackets refer to the time average. Note that if the motion is periodic and the chosen time τ equals a multiple of the period, then $\frac{G(\tau)-G(0)}{\tau} = 0$. Even if the motion is not periodic, if the constraints and velocities of all the particles remain finite, then there is an upper bound to G. This implies that choosing $\tau \to \infty$ means that $\frac{G(\tau)-G(0)}{\tau} \to 0$. In both cases the left-hand side of the equation tends to zero giving the *Virial theorem*

$$\langle T \rangle = -\frac{1}{2} \left\langle \sum_i \mathbf{F}_i \cdot \mathbf{r}_i \right\rangle \tag{2.86}$$

The right-hand side of this equation is called the *Virial of the system*. For a single particle subject to a conservative central force $\mathbf{F} = -\boldsymbol{\nabla} U$ the Virial theorem equals

$$\langle T \rangle = \frac{1}{2} \langle \boldsymbol{\nabla} U \cdot \mathbf{r} \rangle = \frac{1}{2} \left\langle r \frac{\partial U}{\partial r} \right\rangle \tag{2.87}$$

If the potential is of the form $U = kr^{n+1}$ that is, $F = -k(n+1)r^n$, then $r\frac{\partial U}{\partial r} = (n+1)U$. Thus for a single particle in a central potential $U = kr^{n+1}$ the Virial theorem reduces to

$$\langle T \rangle = \frac{n+1}{2} \langle U \rangle \tag{2.88}$$

The following two special cases are of considerable importance in physics.

   Hooke's Law: Note that for a linear restoring force $n = 1$ then

$$\langle T \rangle = + \langle U \rangle \tag{$n = 1$}$$

You may be familiar with this fact for simple harmonic motion where the average kinetic and potential energies are the same and both equal half of the total energy.

Inverse-square law: The other interesting case is for the inverse square law $n = -2$ where

$$\langle T \rangle = -\frac{1}{2} \langle U \rangle \qquad\qquad (n = -2)$$

The Virial theorem is useful for solving problems in that knowing the exponent n of the field makes it possible to write down directly the average total energy in the field. For example, for $n = -2$

$$\langle E \rangle = \langle T \rangle + \langle U \rangle = -\frac{1}{2} \langle U \rangle + \langle U \rangle = \frac{1}{2} \langle U \rangle \qquad\qquad (2.89)$$

This occurs for the Bohr model of the hydrogen atom where the kinetic energy of the bound electron is half of the potential energy. The same result occurs for planetary motion in the solar system.

2.5 *Example: The ideal gas law*

The Virial theorem deals with average properties and has applications to statistical mechanics. Consider an ideal gas. According to the Equipartition theorem the average kinetic energy per atom in an ideal gas is $\frac{3}{2}kT$ where T is the absolute temperature and k is the Boltzmann constant. Thus the average total kinetic energy for N atoms is $\langle KE \rangle = \frac{3}{2}NkT$. The right-hand side of the Virial theorem contains the force F_i. For an ideal gas it is assumed that there are no interaction forces between atoms, that is the only force is the force of constraint of the walls of the pressure vessel. The pressure P is force per unit area and thus the instantaneous force on an area of wall dA is $d\mathbf{F}_i = -\hat{\mathbf{n}} P dA$ where $\hat{n}$ designates the unit vector normal to the surface. Thus the right-hand side of the Virial theorem is

$$-\frac{1}{2} \left\langle \sum_i \mathbf{F}_i \cdot \mathbf{r}_i \right\rangle = \frac{P}{2} \int \hat{\mathbf{n}} \cdot \mathbf{r}_i dA$$

Use of the divergence theorem thus gives that $\int \hat{\mathbf{n}} \cdot \mathbf{r}_i dA = \int \boldsymbol{\nabla} \cdot \mathbf{r} dV = 3 \int dV = 3V$. Thus the Virial theorem leads to the ideal gas law, that is

$$NkT = PV$$

2.6 *Example: The mass of galaxies*

The Virial theorem can be used to make a crude estimate of the mass of a cluster of galaxies. Assuming a spherically-symmetric cluster of N galaxies, each of mass m, then the total mass of the cluster is $M = Nm$. A crude estimate of the cluster potential energy is

$$\langle U \rangle \approx \frac{GM^2}{R} \qquad\qquad (\alpha)$$

where R is the radius of a cluster. The average kinetic energy per galaxy is $\frac{1}{2}m \langle v \rangle^2$ where $\langle v \rangle^2$ is the average square of the galaxy velocities with respect to the center of mass of the cluster. Thus the total kinetic energy of the cluster is

$$\langle KE \rangle \approx \frac{Nm \langle v \rangle^2}{2} = \frac{M \langle v \rangle^2}{2} \qquad\qquad (\beta)$$

The Virial theorem tells us that a central force having a radial dependence of the form $F \propto r^n$ gives $\langle KE \rangle = \frac{n+1}{2} \langle U \rangle$. For the inverse-square gravitational force then

$$\langle KE \rangle = -\frac{1}{2} \langle U \rangle \qquad\qquad (\gamma)$$

Thus equations α, β and γ give an estimate of the total mass of the cluster to be

$$M \approx \frac{R \langle v \rangle^2}{G}$$

This estimate is larger than the value estimated from the luminosity of the cluster implying a large amount of "dark matter" must exist in galaxies which remains an open question in physics.

2.12   Applications of Newton's equations of motion

Newton's equation of motion can be written in the form

$$\mathbf{F} = \frac{d\mathbf{p}}{dt} = m\frac{d\mathbf{v}}{dt} = m\frac{d^2\mathbf{r}}{dt^2} \tag{2.90}$$

A description of the motion of a particle requires a solution of this second-order differential equation of motion. This equation of motion may be integrated to find $\mathbf{r}(t)$ and $\mathbf{v}(t)$ if the initial conditions and the force field $\mathbf{F}(t)$ are known. Solution of the equation of motion can be complicated for many practical examples, but there are various approaches to simplify the solution. It is of value to learn efficient approaches to solving problems.

The following sequence is recommended
a) Make a vector diagram of the problem indicating forces, velocities, etc.
b) Write down the known quantities.
c) Before trying to solve the equation of motion directly, look to see if a basic conservation law applies. That is, check if any of the three first-order integrals, can be used to simplify the solution. The use of conservation of energy or conservation of momentum can greatly simplify solving problems.

The following examples show the solution of typical types of problem encountered using Newtonian mechanics.

2.12.1   Constant force problems

Problems having a constant force imply constant acceleration. The classic example is a block sliding on an inclined plane, where the block of mass m is acted upon by both gravity and friction. The net force $\mathbf{F}$ is given by the vector sum of the gravitational force $\mathbf{F}_g$, normal force $\mathbf{N}$ and frictional force $\mathbf{f}_f$.

$$\mathbf{F} = \mathbf{F}_g + \mathbf{N} + \mathbf{f}_f = m\mathbf{a} \tag{2.91}$$

Taking components perpendicular to the inclined plane in the y direction

$$-F_g \cos\theta + N = 0 \tag{2.92}$$

That is, since $F_g = mg$,

$$N = mg\cos\theta \tag{2.93}$$

Similarly, taking components along the inclined plane in the x direction

$$F_g \sin\theta - f_f = m\frac{d^2x}{dt^2} \tag{2.94}$$

Using the concept of coefficient of friction μ,

$$f_f = \mu N \tag{2.95}$$

Thus the equation of motion can be written as

$$mg\left(\sin\theta - \mu\cos\theta\right) = m\frac{d^2x}{dt^2} \tag{2.96}$$

The block accelerates if $\sin\theta > \mu\cos\theta$, that is, $\tan\theta > \mu$. The acceleration is constant if μ and θ are constant, that is

$$\frac{d^2x}{dt^2} = g\left(\sin\theta - \mu\cos\theta\right) \tag{2.97}$$

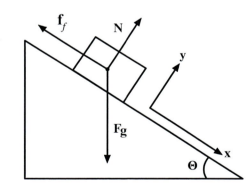

Figure 2.3: Block on an inclined plane

Remember that if the block is stationary, the friction coefficient balances such that $(\sin\theta - \mu\cos\theta) = 0$, that is, $\tan\theta = \mu$. However, there is a maximum static friction coefficient μ_S beyond which the block starts sliding. The kinetic coefficient of friction μ_K is applicable for sliding friction and usually $\mu_K < \mu_S$.

Another example of constant force and acceleration is motion of objects free falling in a uniform gravitational field when air drag is neglected. Then one obtains the simple relations such as $v = u + at$, etc.

2.12.2 Linear Restoring Force

An important class of problems involve a linear restoring force, that is, they obey Hooke's law. The equation of motion for this case is

$$F(x) = -kx = m\ddot{x} \tag{2.98}$$

It is usual to define

$$\omega_0^2 \equiv \frac{k}{m} \tag{2.99}$$

Then the equation of motion then can be written as

$$\ddot{x} + \omega_0^2 x = 0 \tag{2.100}$$

which is the equation of the harmonic oscillator. Examples are small oscillations of a mass on a spring, vibrations of a stretched piano string, etc.

The solution of this second order equation is

$$x(t) = A \sin(\omega_0 t - \delta) \tag{2.101}$$

This is the well known sinusoidal behavior of the displacement for the simple harmonic oscillator. The angular frequency ω_0 is

$$\omega_0 = \sqrt{\frac{k}{m}} \tag{2.102}$$

Note that this linear system has no dissipative forces, thus the total energy is a constant of motion as discussed previously. That is, it is a conservative system with a total energy E given by

$$\frac{1}{2}m\dot{x}^2 + \frac{1}{2}kx^2 = E \tag{2.103}$$

The first term is the kinetic energy and the second term is the potential energy. The Virial theorem gives that for the linear restoring force the average kinetic energy equals the average potential energy.

2.12.3 Position-dependent conservative forces

The linear restoring force is an example of a conservative field. The total energy E is conserved, and if the field is time independent, then the conservative forces are a function only of position. The easiest way to solve such problems is to use the concept of potential energy U illustrated in Figure 2.4.

$$U_2 - U_1 = -\int_1^2 \mathbf{F} \cdot \mathbf{dr} \tag{2.104}$$

Consider a conservative force in one dimension. Since it was shown that the total energy $E = T + U$ is conserved for a conservative field, then

$$E = T + U = \frac{1}{2}mv^2 + U(x) \tag{2.105}$$

Therefore:

$$v = \frac{dx}{dt} = \pm\sqrt{\frac{2}{m}[E - U(x)]} \tag{2.106}$$

Integration of this gives

$$t - t_0 = \int_{x_0}^{x} \frac{\pm dx}{\sqrt{\frac{2}{m}[E - U(x)]}} \tag{2.107}$$

where $x = x_0$ when $t = t_0$. Knowing $U(x)$ it is possible to solve this equation as a function of time.

It is possible to understand the general features of the solution just from inspection of the function $U(x)$. For example, as shown in figure 2.4 the motion for energy E_1 is periodic between the turning points x_a and x_b. Since the potential energy curve is approximately parabolic between these limits the motion will exhibit simple harmonic motion. For E_0 the turning point coalesce to x_0, that is there is no motion. For total energy E_2 the motion is periodic in two independent regimes, $x_c \leq x \leq x_d$, and $x_e \leq x \leq x_f$. Classically the particle cannot jump from one pocket to the other. The motion for the particle with total energy E_3 is that it moves freely from infinity, stops and rebounds at $x = x_g$ and then returns to infinity. That is the particle bounces off the potential at x_g. For energy E_4 the particle moves freely and is unbounded. For all these cases, the actual velocity is given by the above relation for $v(x)$. Thus the kinetic energy is largest where the potential is deepest. An example would be motion of a roller coaster car.

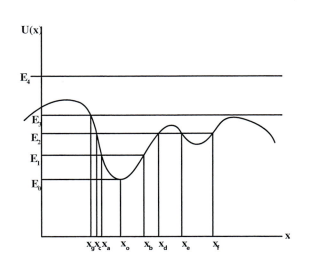

Figure 2.4: One-dimensional potential $U(x)$.

Position-dependent forces are encountered extensively in classical mechanics. Examples are the many manifestations of motion in gravitational fields, such as interplanetary probes, a roller coaster, and automobile suspension systems. The linear restoring force is an especially simple example of a position-dependent force while the most frequently encountered conservative potentials are in electrostatics and gravitation for which the potentials are;

$$U(r) = \frac{1}{4\pi\epsilon_0} \frac{q_1 q_2}{r_{12}^2} \qquad \text{(Electrostatic potential energy)}$$

$$U(r) = -G\frac{m_1 m_2}{r_{12}^2} \qquad \text{(Gravitational potential energy)}$$

Knowing $U(r)$ it is possible to solve the equation of motion as a function of time.

2.7 *Example: Diatomic molecule*

An example of a conservative field is a vibrating diatomic molecule which has a potential energy dependence with separation distance x that is described approximately by the Morse function

$$U(x) = U_0 \left[1 - e^{-\frac{(x-x_0)}{\delta}} \right]^2 - U_0$$

where $U_0, x_0,$ and δ are parameters chosen to best describe the particular pair of atoms. The restoring force is given by

$$F(x) = -\frac{dU(x)}{dx} = 2\frac{U_0}{\delta} \left[1 - e^{-\frac{(x-x_0)}{\delta}} \right] \left[e^{-\frac{(x-x_0)}{\delta}} \right]$$

This has a minimum value of $U(x_0) = U_0$ at $x = x_0$.

Note that for small amplitude oscillations, where

$$(x - x_0) << \delta$$

the exponential term in the potential function can be expanded to give

$$U(x) \approx U_0 \left[1 - (1 - -\frac{(x-x_0)}{\delta}) \right]^2 - U_0 \approx \frac{U_0}{\delta^2}(x-x_0)^2 - U_0$$

This gives a restoring force

$$F(x) = -\frac{dU(x)}{dx} = -2\frac{U_0}{\delta}(x - x_0)$$

That is, for small amplitudes the restoring force is linear.

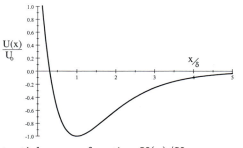

Potential energy function $U(x)/U_0$ versus x/δ for the diatomic molecule.

2.12.4 Constrained motion

A frequently encountered problem involving position dependent forces, is when the motion is constrained to follow a certain trajectory. Forces of constraint must exist to constrain the motion to a specific trajectory. Examples are, the roller coaster, a rolling ball on an undulating surface, or a downhill skier, where the motion is constrained to follow the surface or track contours. The potential energy can be evaluated at all positions along the constrained trajectory for conservative forces such as gravity. However, the additional forces of constraint that must exist to constrain the motion, can be complicated and depend on the motion. For example, the roller coaster must always balance the gravitational and centripetal forces. Fortunately forces of constraint $\mathbf{F}_C$ often are normal to the direction of motion and thus do not contribute to the total mechanical energy since then the work done $\mathbf{F}_C \cdot d\mathbf{l}$ is zero. Magnetic forces $\mathbf{F} = q\mathbf{v} \times \mathbf{B}$ exhibit this feature of having the force normal to the motion.

Solution of constrained problems is greatly simplified if the other forces are conservative and the forces of constraint are normal to the motion, since then energy conservation can be used.

2.8 *Example: Roller coaster*

Consider motion of a roller coaster shown in the adjacent figure. This system is conservative if the friction and air drag are neglected and then the forces of constraint are normal to the direction of motion.

The kinetic energy at any position is just given by energy conservation and the fact that

$$E = T + U$$

where U depends on the height of the track at any the given location. The kinetic energy is greatest when the potential energy is lowest. The forces of constraint can be deduced if the velocity of motion on the track is known. Assuming that the motion is confined to a vertical plane, then one has a centripetal force of constraint $\frac{mv^2}{\rho}$ normal to the track inwards towards the center of the radius of curvature ρ, plus the gravitation force downwards of mg.

The constraint force is $\frac{mv_T^2}{\rho} - mg$ upwards at the top of the loop, while it is $\frac{mv_B^2}{\rho} + mg$ downwards at the bottom of the loop. To ensure that the car and occupants do not leave the required trajectory, the force upwards at the top of the loop has to be positive, that is, $v_T^2 \geq \rho g$. The velocity at the bottom of the loop is given by $\frac{1}{2}mv_B^2 = \frac{1}{2}mv_T^2 + 2mg\rho$ assuming that the track has a constant radius of curvature ρ. That is; at a minimum $v_B^2 = \rho g + 4\rho g = 5\rho g$. Therefore the occupants now will feel an acceleration downwards of at least $\frac{v_B^2}{\rho} + g = 6g$ at the bottom of the loop. The

Roller coaster (CCO Public Domain)

first roller coaster was built with such a constant radius of curvature but an acceleration of 6g was too much for the average passenger. Therefore roller coasters are designed such that the radius of curvature is much larger at the bottom of the loop, as illustrated, in order to maintain sufficiently low g loads and also ensure that the required constraint forces exist.

Note that the minimum velocity at the top of the loop, v_T, implies that if the cart starts from rest it must start at a height $h \geq \frac{\rho}{2}$ above the top of the loop if friction is negligible. Note that the solution for the rolling ball on such a roller coaster differs from that for a sliding object since one must include the rotational energy of the ball as well as the linear velocity.

Looping the loop in a sailplane involves the same physics making it necessary to vary the elevator control to vary the radius of curvature throughout the loop to minimize the maximum g load.

2.12.5   Velocity Dependent Forces

Velocity dependent forces are encountered frequently in practical problems. For example, motion of an object in a fluid, such as air, where viscous forces retard the motion. In general the retarding force has a complicated dependence on velocity. A quadrative-velocity drag force in air often can be expressed in the form,

$$\mathbf{F}_D(v) = -\frac{1}{2}c_D\rho A v^2 \hat{\mathbf{v}} \tag{2.108}$$

where c_D is a dimensionless drag coefficient, ρ is the density of air, A is the cross sectional area perpendicular to the direction of motion, and v is the velocity. Modern automobiles have drag coefficients as low as 0.3. As described in chapter 16, the drag coefficient c_D depends on the Reynold's number which relates the inertial to viscous drag forces. Small sized objects at low velocity, such as light raindrops, have low Reynold's numbers for which c_D is roughly proportional to v^{-1} leading to a linear dependence of the drag force on velocity, i.e. $F_D(v) \propto v$. Larger objects moving at higher velocities, such as a car or sky-diver, have higher Reynold's numbers for which c_D is roughly independent of velocity leading to a drag force $F_D(v) \propto v^2$. This drag force always points in the opposite direction to the unit velocity vector. Approximately for air

$$\mathbf{F}_D(v) = -\left(c_1 v + c_2 v^2\right)\hat{\mathbf{v}} \tag{2.109}$$

where for spherical objects of diameter D, $c_1 \approx 1.55\times 10^{-4}D$ and $c_2 \approx 0.22D^2$ in MKS units. Fortunately, the equation of motion usually can be integrated when the retarding force has a simple power law dependence. As an example, consider free fall in the Earth's gravitational field.

2.9 *Example: Vertical fall in the earth's gravitational field.*

Linear regime $c_1 \gg c_2 v$
For small objects at low-velocity, i.e. low Reynold's number, the drag approximately has a linear dependence on velocity. Then the equation of motion is

$$-mg - c_1 v = m\frac{dv}{dt}$$

Separate the variables and integrate

$$t = \int_{v_0}^{v} \frac{m\,dv}{-mg - c_1 v} = -\frac{m}{c_1}\ln\left(\frac{mg + c_1 v}{mg + c_1 v_0}\right)$$

That is

$$v = -\frac{mg}{c_1} + \left(\frac{mg}{c_1} + v_0\right)e^{-\frac{c_1}{m}t}$$

Note that for $t \gg \frac{m}{c_1}$ the velocity approaches a terminal velocity of $v_\infty = -\frac{mg}{c_1}$. The characteristic time constant is $\tau = \frac{m}{c_1} = \frac{v_\infty}{g}$. Note that if $v_0 = 0$, then

$$v = v_\infty\left(1 - e^{-\frac{t}{\tau}}\right)$$

For the case of small raindrops with $D = 0.5mm$, then $v_\infty = 8m/s$ ($18mph$) and time constant $\tau = 0.8\,\sec$. Note that in the absence of air drag, these rain drops falling from $2000m$ would attain a velocity of over 400 m.p.h. It is fortunate that the drag reduces the speed of rain drops to non-damaging values. Note that the above relation would predict high velocities for hail. Fortunately, the drag increases quadratically at the higher velocities attained by large rain drops or hail, and this limits the terminal velocity to moderate values. For the United States these velocities still are sufficient to do considerable crop damage in the mid-west.

Quadratic regime $c_2 v \gg c_1$
For larger objects at higher velocities, i.e. high Reynold's number, the drag depends on the square of the velocity making it necessary to differentiate between objects rising and falling. The equation of motion is

$$-mg \pm c_2 v^2 = m\frac{dv}{dt}$$

where the positive sign is for falling objects and negative sign for rising objects. Integrating the equation of motion for falling gives

$$t = \int_{v_0}^{v} \frac{mdv}{-mg + c_2 v^2} = \tau \left(\tanh^{-1} \frac{v_0}{v_\infty} - \tanh^{-1} \frac{v}{v_\infty} \right)$$

where $\tau = \sqrt{\frac{m}{c_2 g}}$ and $v_\infty = \sqrt{\frac{mg}{c_2}}$. That is, $\tau = \frac{v_\infty}{g}$. For the case of a falling object with $v_0 = 0$, solving for velocity gives

$$v = v_\infty \tanh \frac{t}{\tau}$$

As an example, a $0.6kg$ basket ball with $D = 0.25m$ will have $v_\infty = 20m/s$ (43 m.p.h.) and $\tau = 2.1sec$.

Consider President George H.W. Bush skydiving. Assume his mass is 70kg and assume an equivalent spherical shape of the former President to have a diameter of $D = 1m$. This gives that $v_\infty = 56m/s$ (120mph) and $\tau = 5.6sec$. When Bush senior opens his 8m diameter parachute his terminal velocity is estimated to decrease to 7m/s (15 mph) which is close to the value for a typical (8m) diameter emergency parachute which has a measured terminal velocity of 11mph in spite of air leakage through the central vent needed to stabilize the parachute motion.

2.10 *Example: Projectile motion in air*

Consider a projectile initially at $x = y = 0$ at $t = 0$, that is fired at an initial velocity v_0 at an angle θ to the horizontal. In order to understand the general features of the solution, assume that the drag is proportional to velocity. This is incorrect for typical projectile velocities, but simplifies the mathematics. The equations of motion can be expressed as

$$m\ddot{x} = -km\dot{x}$$

$$m\ddot{y} = -km\dot{y} - mg$$

where k is the coefficient for air drag. Take the initial conditions at $t = 0$ to be $x = y = 0$, $\dot{x} = v_o \cos\theta$, $\dot{y} = v_o \sin\theta$.

Solving in the x coordinate,

$$\frac{d\dot{x}}{dt} = -k\dot{x}$$

Therefore

$$\dot{x} = v_o \cos\theta e^{-kt}$$

That is, the velocity decays to zero with a time constant $\tau = \frac{1}{k}$.

Integration of the velocity equation gives

$$x = \frac{v_o}{k} \left(1 - e^{-kt} \right)$$

Note that this implies that the body approaches a value of $x = \frac{v_o}{k}$ as $t \to \infty$.

The trajectory of an object is distorted from the parabolic shape, that occurs for $k = 0$, due to the rapid drop in range as the drag coefficient increases. For realistic cases it is necessary to use a computer to solve this numerically.

2.12.6 Systems with Variable Mass

Classic examples of systems with variable mass are the rocket, a falling chain, and nuclear fission. Consider the problem of vertical rocket motion in a gravitational field using Newtonian mechanics. When there is a vertical gravitational external field, the vertical momentum is not conserved due to both gravity and the ejection of rocket propellant. In a time dt the rocket ejects propellant dm_p vertically with exhaust velocity relative to the rocket of u. Thus the momentum imparted to this propellant is

$$dp_p = -udm_p \tag{2.110}$$

Therefore the rocket is given an equal and opposite increase in momentum dp_R

$$dp_R = +udm_p \tag{2.111}$$

In the time interval dt the net change in the linear momentum of the rocket plus fuel system is given by

$$dp = (m - dm_p)(v + dv) + dm_p(v - u) - mv = mdv - udm_p \tag{2.112}$$

The rate of change of the linear momentum thus equals

$$F_{ex} = \frac{dp}{dt} = m\frac{dv}{dt} - u\frac{dm_p}{dt} \tag{2.113}$$

Consider the problem for the special case of vertical ascent of the rocket against the external gravitational force $F_{ex} = -mg$. Then

$$-mg + u\frac{dm_p}{dt} = m\frac{dv}{dt} \tag{2.114}$$

This can be rewritten as

$$-mg + u\dot{m}_p = m\dot{v}$$

The second term comes from the variable rocket mass where the loss of mass of the rocket equals the mass of the ejected propellant. Assuming a constant fuel burn $\dot{m}_p = \alpha$ then

$$\dot{m} = -\dot{m}_p = -\alpha \tag{2.115}$$

where $\alpha > 0$. Then the equation becomes

$$dv = \left(-g + \frac{\alpha}{m}u\right) dt \tag{2.116}$$

Since

$$\frac{dm}{dt} = -\alpha \tag{2.117}$$

then

$$-\frac{dm}{\alpha} = dt \tag{2.118}$$

Inserting this in the above equation gives

$$dv = \left(\frac{g}{\alpha} - \frac{u}{m}\right) dm \tag{2.119}$$

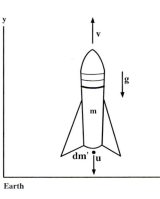

Figure 2.5: Vertical motion of a rocket in a gravitational field

Integration gives

$$v = -\frac{g}{\alpha}(m_0 - m) + u\ln\left(\frac{m_0}{m}\right) \tag{2.120}$$

But the change in mass is given by

$$\int_{m_0}^{m} dm = -\alpha \int_0^t dt \tag{2.121}$$

That is

$$m_0 - m = \alpha t \tag{2.122}$$

Thus

$$v = -gt + u\ln\left(\frac{m_0}{m}\right) \tag{2.123}$$

Note that once the propellant is exhausted the rocket will continue to fly upwards as it decelerates in the gravitational field. You can easily calculate the maximum height. Note that this formula assumes that the acceleration due to gravity is constant whereas for large heights above the Earth it is necessary to use the true gravitational force $-G\frac{Mm}{r^2}$ where r is the distance from the center of the earth. In real situations it is necessary to include air drag which requires a computer to numerically solve the equations of motion. The highest rocket velocity is attained by maximizing the exhaust velocity and the ratio of initial to final mass. Because the terminal velocity is limited by the mass ratio, engineers construct multistage rockets that jettison the spent fuel containers and rockets. The variational-principle approach applied to variable mass problems is discussed in chapter 8.7

2.12.7 Rigid-body rotation about a body-fixed rotation axis

The most general case of rigid-body rotation involves rotation about some body-fixed point with the orientation of the rotation axis undefined. For example, an object spinning in space will rotate about the center of mass with the rotation axis having any orientation. Another example is a child's spinning top which spins with arbitrary orientation of the axis of rotation about the pointed end which touches the ground about a static location. Such rotation about a body-fixed point is complicated and will be discussed in chapter 13. Rigid-body rotation is easier to handle if the orientation of the axis of rotation is fixed with respect to the rigid body. An example of such motion is a hinged door.

For a rigid body rotating with angular velocity ω, the total angular momentum $\mathbf{L}$ is given by

$$\mathbf{L} = \sum_{i}^{n} \mathbf{L}_i = \sum_{i}^{n} \mathbf{r}_i \times \mathbf{p}_i \tag{2.124}$$

For rotation equation appendix $D29$ gives

$$\mathbf{v}_i = \boldsymbol{\omega} \times \mathbf{r}_i \tag{2.94}$$

thus the angular momentum can be written as

$$\mathbf{L} = \sum_{i}^{n} \mathbf{r}_i \times \mathbf{p}_i = \sum_{i}^{n} m_i \mathbf{r}_i \times \boldsymbol{\omega} \times \mathbf{r}_i \tag{2.125}$$

The vector triple product can be simplified using the vector identity equation $B.24$ giving

$$\mathbf{L} = \sum_{i}^{n} \left[\left(m_i r_i^2 \right) \boldsymbol{\omega} - (\mathbf{r}_i \cdot \boldsymbol{\omega}) m_i \mathbf{r}_i \right] \tag{2.126}$$

Rigid-body rotation about a body-fixed symmetry axis

The simplest case for rigid-body rotation is when the body has a symmetry axis with the angular velocity $\boldsymbol{\omega}$ parallel to this body-fixed symmetry axis. For this case then $\mathbf{r}_i$ can be taken perpendicular to $\boldsymbol{\omega}$, for which the second term in equation 2.126, i.e. $(\mathbf{r}_i \cdot \boldsymbol{\omega}) = 0$, thus

$$\mathbf{L}_{sym} = \sum_{i}^{n} \left(m_i r_i^2 \right) \boldsymbol{\omega} \qquad (\mathbf{r}_i \text{ perpendicular to } \boldsymbol{\omega})$$

The **moment of inertia** about the symmetry axis is defined as

$$I_{sym} = \sum_{i}^{n} m_i r_i^2 \tag{2.127}$$

where r_i is the perpendicular distance from the axis of rotation to the body, m_i. For a continuous body the moment of inertia can be generalized to an integral over the mass density ρ of the body

$$I_{sym} = \int \rho r^2 dV \tag{2.128}$$

where r is perpendicular to the rotation axis. The definition of the moment of inertia allows rewriting the angular momentum about a symmetry axis $\mathbf{L}_{sym}$ in the form

$$\mathbf{L}_{sym} = I_{sym} \boldsymbol{\omega} \tag{2.129}$$

where the moment of inertia I_{sym} is taken about the symmetry axis and assuming that the angular velocity of rotation vector is parallel to the symmetry axis.

Rigid-body rotation about a non-symmetric body-fixed axis

In general the fixed axis of rotation is not aligned with a symmetry axis of the body, or the body does not have a symmetry axis, both of which complicate the problem.

For illustration consider that the rigid body comprises a system of n masses m_i located at positions $\mathbf{r}_i$, with the rigid body rotating about the z axis with angular velocity $\boldsymbol{\omega}$. That is,

$$\boldsymbol{\omega} = \omega_z \hat{\mathbf{z}} \tag{2.130}$$

In cartesian coordinates the fixed-frame vector for particle i is

$$\mathbf{r}_i = (x_i, y_i, z_i) \tag{2.131}$$

using these in the cross product (2.94) gives

$$\mathbf{v}_i = \boldsymbol{\omega} \times \mathbf{r}_i = \begin{pmatrix} -\omega_z y_i \\ \omega_z x_i \\ 0 \end{pmatrix} \tag{2.132}$$

which is written as a column vector for clarity. Inserting $\mathbf{v}_i$ in the cross-product $\mathbf{r}_i \times \mathbf{v}_i$ gives the components of the angular momentum to be

$$\mathbf{L} = \sum_i^n m_i \mathbf{r}_i \times \mathbf{v}_i = \sum_i^n m_i \omega_z \begin{pmatrix} -z_i x_i \\ -z_i y_i \\ x_i^2 + y_i^2 \end{pmatrix}$$

That is, the components of the angular momentum are

$$L_x = -\left(\sum_i^n m_i z_i x_i \right) \omega_z \equiv I_{xz} \omega_z \tag{2.133}$$

$$L_y = -\left(\sum_i^n m_i z_i y_i \right) \omega_z \equiv I_{yz} \omega_z$$

$$L_z = \left(\sum_i^n m_i \left[x_i^2 + y_i^2 \right] \right) \omega_z \equiv I_{zz} \omega_z$$

Note that the perpendicular distance from the z axis in cylindrical coordinates is $\rho = \sqrt{x_i^2 + y_i^2}$, thus the angular momentum L_z about the z axis can be written as

$$L_z = \left(\sum_i^n m_i \rho^2 \right) \omega_z = I_{zz} \omega_z \tag{2.134}$$

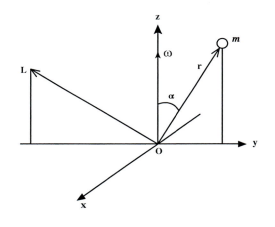

Figure 2.6: A rigid rotating body comprising a single mass m attached by a massless rod at a fixed angle α shown at the instant when m happens to lie in the yz plane. As the body rotates about the $z-$ axis the mass m has a velocity and momentum into the page (the negative x direction). Therefore the angular momentum $\mathbf{L} = \mathbf{r} \times \mathbf{p}$ is in the direction shown which is not parallel to the angular velocity ω.

where (2.134) gives the elementary formula for the moment of inertia $I_{zz} = I_{sym}$ about the z axis given earlier in (2.129).

The surprising result is that L_x and L_y are non-zero implying that *the total angular momentum vector* $\mathbf{L}$ *is in general not parallel with* $\boldsymbol{\omega}$. This can be understood by considering the single body m shown in figure 2.6. When the body is in the y, z plane then $x = 0$ and $L_x = 0$. Thus the angular momentum vector $\mathbf{L}$ has a component along the $-y$ direction as shown which is not parallel with $\boldsymbol{\omega}$ and, since the vectors $\boldsymbol{\omega}, \mathbf{L}, \mathbf{r}_i$ are coplanar, then $\mathbf{L}$ must sweep around the rotation axis $\boldsymbol{\omega}$ to remain coplanar with the body as it rotates about the z axis. Instantaneously the velocity of the body $\mathbf{v}_i$ is into the plane of the paper and, since

$\mathbf{L}_i = m_i \mathbf{r}_i \times \mathbf{v}_i$, then $\mathbf{L}_i$ is at an angle $(90° - \alpha)$ to the z axis. This implies that a torque must be applied to rotate the angular momentum vector. This explains why your automobile shakes if the rotation axis and symmetry axis are not parallel for one wheel.

The first two moments in (2.133) are called **products of inertia** of the body designated by the pair of axes involved. Therefore, to avoid confusion, it is necessary to define the diagonal moment, which is called the **moment of inertia**, by two subscripts as I_{zz}. Thus in general, a body can have three moments of inertia about the three axes plus three products of inertia. This group of moments comprise the **inertia tensor** which will be discussed further in chapter 13. If a body has an axis of symmetry along the z axis then the summations will give $I_{xz} = I_{yz} = 0$ while I_{zz} will be unchanged. That is, for rotation about a symmetry axis the angular momentum and rotation axes are parallel. For any axis along which the angular momentum and angular velocity coincide is called a **principal axis** of the body.

2.11 *Example: Moment of inertia of a thin door*

Consider that the door has width a and height b and assume the door thickness is negligible with areal density $\sigma kg/m2$. Assume that the door is hinged about the y axis. The mass of a surface element of dimension $dx.dy$ at a distance x from the rotation axis is $dm = \sigma dxdy$, thus the mass of the complete door is $M = \sigma ab$. The moment of inertia about the y axis is given by

$$I = \int_{x=0}^{a} \int_{y=0}^{b} \sigma x^2 dy dx = \frac{1}{3}\sigma ba^3 = \frac{1}{3}Ma^2$$

2.12 *Example: Merry-go-round*

A child of mass m jumps onto the outside edge of a circular merry-go-round of moment of inertia I, and radius R and initial angular velocity ω_0. What is the final angular velocity ω_f?

If the initial angular momentum is L_0 and, assuming the child jumps with zero angular velocity, then the conservation of angular momentum implies that

$$
\begin{aligned}
L_0 &= L_f \\
I\omega_0 &= I\omega + m v_f R \\
I\frac{v_0}{R} &= \frac{v_f}{R}(I + mR^2)
\end{aligned}
$$

That is

$$\frac{v_f}{v_0} = \frac{\omega_f}{\omega_0} = \frac{I}{I + mR^2}$$

Note that this is true independent of the details of the acceleration of the initially stationary child.

2.13 *Example: Cue pushes a billiard ball*

Consider a billiard ball of mass M and radius R is pushed by a cue in a direction that passes through the center of gravity such that the ball attains a velocity v_0. The friction coefficient between the table and the ball is μ. How far does the ball move before the initial slipping motion changes to pure rolling motion?

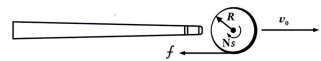

Cue pushing a billiard ball horizontally at the height of the centre of rotation of the ball.

Since the direction of the cue force passes through the center of mass of the ball, it contributes zero torque to the ball. Thus the initial angular momentum is zero at $t = 0$. The friction force f points opposite to the direction of motion and causes a torque N_s about the center of mass in the direction $\hat{s}$.

$$\mathbf{N}_s = \mathbf{f} \cdot \mathbf{R} = \mu M g R$$

Since the moment of inertia about the center of a uniform sphere is $I = \frac{2}{5}MR^2$ then the angular acceleration of the ball is

$$\dot{\omega} = \frac{\mu M g R}{I} = \frac{\mu M g R}{\frac{2}{5}MR^2} = \frac{5}{2}\frac{\mu g}{R} \qquad (\alpha)$$

Moreover the frictional force causes a deceleration a_s of the linear velocity of the center of mass of

$$a_s = -\frac{f}{M} = -\mu g \qquad (\beta)$$

Integrating α from time zero to t gives

$$\omega = \int_0^t \dot{\omega} dt = \frac{5}{2}\frac{\mu g}{R}t$$

The linear velocity of the center of mass at time t is given by integration of equation β

$$v_s = \int_0^t a_s dt = v_0 - \mu g t$$

The billiard ball stops sliding and only rolls when $v_s = \omega R$, that is, when

$$\frac{5}{2}\frac{\mu g}{R}tR = v_0 - \mu g t$$

That is, when

$$t_{roll} = \frac{2}{7}\frac{v_0}{\mu g}$$

Thus the ball slips for a distance

$$s = \int_0^{t_{roll}} v_s dt = v_0 t_{roll} - \frac{\mu g t_{roll}^2}{2} = \frac{12}{49}\frac{v_0^2}{\mu g}$$

Note that if the ball is pushed at a distance h above the center of mass, besides the linear velocity there is an initial angular momentum of

$$\omega = \frac{M v_0 h}{\frac{2}{5}MR^2} = \frac{5}{2}\frac{v_0 h}{R^2}$$

For the special case $h = \frac{2}{5}R$, the ball immediately assumes a pure non-slipping roll. For $h < \frac{2}{5}R$ one has $\omega < \frac{v_0}{R}$ while $h > \frac{2}{5}R$ corresponds to $\omega > \frac{v_0}{R}$. In the latter case the frictional force points forward.

2.12.8   Time dependent forces

Many problems involve action in the presence of a time dependent force. There are two extreme cases that are often encountered. One case is an impulsive force that acts for a very short time, for example, striking a ball with a bat, or the collision of two cars. The second case involves an oscillatory time dependent force. The response to impulsive forces is discussed below whereas the response to oscillatory time-dependent forces is discussed in chapter 3.

Translational impulsive forces

An impulsive force acts for a very short time relative to the response time of the mechanical system being discussed. In principle the equation of motion can be solved if the complicated time dependence of the force, $F(t)$, is known. However, often it is possible to use the much simpler approach employing the concept of an impulse and the principle of the conservation of linear momentum.

Define the linear impulse **P** to be the first-order time integral of the time-dependent force.

$$\mathbf{P} \equiv \int \mathbf{F}(t) dt \qquad (2.135)$$

Since $\mathbf{F}(t) = \frac{d\mathbf{p}}{dt}$ then equation 2.135 gives that

$$\mathbf{P} = \int_0^t \frac{d\mathbf{p}}{dt'}dt' = \int_0^t d\mathbf{p} = \mathbf{p}(t) - \mathbf{p}_0 = \Delta\mathbf{p} \tag{2.136}$$

Thus the impulse $\mathbf{P}$ is an unambiguous quantity that equals the change in linear momentum of the object that has been struck which is independent of the details of the time dependence of the impulsive force. Computation of the spatial motion still requires knowledge of $F(t)$ since the 2.136 can be written as

$$\mathbf{v}(t) = \frac{1}{m}\int_0^t \mathbf{F}(t')dt' + \mathbf{v}_0 \tag{2.137}$$

Integration gives

$$\mathbf{r}(t) - \mathbf{r}_0 = \mathbf{v}_0 t + \int_0^t \left[\frac{1}{m}\int_0^{t''} \mathbf{F}(t')dt'\right]dt" \tag{2.138}$$

In general this is complicated. However, for the case of a constant force $\mathbf{F}(t) = \mathbf{F}_0$, this simplifies to the constant acceleration equation

$$\mathbf{r}(t) - \mathbf{r}_0 = \mathbf{v}_0 t + \frac{1}{2}\frac{\mathbf{F}_0}{m}t^2 \tag{2.139}$$

where the constant acceleration $\mathbf{a} = \frac{\mathbf{F}_0}{m}$.

Angular impulsive torques

Note that the principle of impulse also applies to angular motion. Define an impulsive torque $\mathbf{T}$ as the first-order time integral of the time-dependent torque.

$$\mathbf{T} \equiv \int \mathbf{N}(t)dt \tag{2.140}$$

Since torque is related to the rate of change of angular momentum

$$\mathbf{N}(t) = \frac{d\mathbf{L}}{dt} \tag{2.141}$$

then

$$\mathbf{T} = \int_0^t \frac{d\mathbf{L}}{dt'}dt' = \int_0^t d\mathbf{L} = \mathbf{L}(t) - \mathbf{L}_0 = \Delta\mathbf{L} \tag{2.142}$$

Thus the impulsive torque $\mathbf{T}$ equals the change in angular momentum $\Delta\mathbf{L}$ of the struck body.

2.14 *Example: Center of percussion of a baseball bat*

When an impulsive force P strikes a bat of mass M at a distance s from the center of mass, then both the linear momentum of the center of mass, and angular momenta about the center of mass, of the bat are changed. Assume that the ball strikes the bat with an impulsive force $P = \Delta p^{ball}$ perpendicular to the symmetry axis of the bat at the strike point S which is a distance s from the center of mass of the bat. The translational impulse given to the bat equals the change in linear momentum of the ball as given by equation 2.136 coupled with the conservation of linear momentum

$$\mathbf{P} = \Delta\mathbf{p}_{cm}^{bat} = M\Delta\mathbf{v}_{cm}^{bat}$$

Similarly equation 2.142 gives that the angular impulse T equals the change in angular momentum about the center of mass to be

$$\mathbf{T} = \mathbf{s} \times \mathbf{P} = \Delta\mathbf{L} = I_{cm}\Delta\omega_{cm}$$

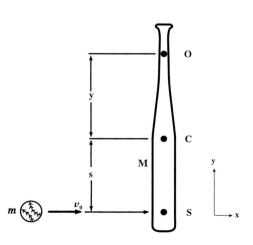

The above equations give that

$$\Delta \mathbf{v}_{cm}^{bat} = \frac{\mathbf{P}}{M}$$

$$\Delta \boldsymbol{\omega}_{cm}^{bat} = \frac{\mathbf{s} \times \mathbf{P}}{I_{cm}}$$

Assume that the bat was stationary prior to the strike, then after the strike the net translational velocity of a point O along the body-fixed symmetry axis of the bat at a distance y from the center of mass, is given by

$$\mathbf{v}\left(y\right) = \Delta \mathbf{v}_{cm} + \Delta \boldsymbol{\omega}_{cm} \times \mathbf{y} = \frac{\mathbf{P}}{M} + \frac{1}{I_{cm}}\left(\left(\mathbf{s} \times \mathbf{P}\right) \times \mathbf{y}\right) = \frac{\mathbf{P}}{M} + \frac{1}{I_{cm}}\left[\left(\mathbf{s} \cdot \mathbf{y}\right)\mathbf{P} - \left(\mathbf{s} \cdot \mathbf{P}\right)\mathbf{y}\right]$$

It is assumed that P and s are perpendicular and thus $(\mathbf{s} \cdot \mathbf{P}) = 0$ which simplifies the above equation to

$$\mathbf{v}\left(y\right) = \Delta \mathbf{v}_{cm} + \Delta \boldsymbol{\omega}_{cm} \times \mathbf{y} = \frac{\mathbf{P}}{M}\left(1 + \frac{M\left(\mathbf{s} \cdot \mathbf{y}\right)}{I_{cm}}\right)$$

Note that the translational velocity of the location O, along the bat symmetry axis at a distance y from the center of mass, is zero if the bracket equals zero, that is, if

$$\mathbf{s} \cdot \mathbf{y} = -\frac{I_{cm}}{M} = -k_{cm}^2$$

where k_{cm} is called the radius of gyration of the body about the center of mass. Note that when the scalar product $s \cdot y = -\frac{I_{cm}}{M} = -k_{cm}^2$ then there will be no translational motion at the point O. This point on the y axis lies on the opposite side of the center of mass from the strike point S, and is called the center of percussion corresponding to the impulse at the point S. The center of percussion often is referred to as the "sweet spot" for an object corresponding to the impulse at the point S. For a baseball bat the batter holds the bat at the center of percussion so that they do not feel an impulse in their hands when the ball is struck at the point S. This principle is used extensively to design bats for all sports involving striking a ball with a bat, such as, cricket, squash, tennis, etc. as well as weapons such of swords and axes used to decapitate opponents.

2.15 *Example: Energy transfer in charged-particle scattering*

Consider a particle of charge $+e_1$ moving with very high velocity v_0 along a straight line that passes a distance b from another charge $+e_2$ and mass m. Find the energy Q transferred to the mass m during the encounter assuming the force is given by Coulomb's law electrostatics. Since the charged particle e_1 moves at very high speed it is assumed that charge 2 does not change position during the encounter. Assume that charge 1 moves along the $-y$ axis through the origin while charge 2 is located on the x axis at $x = b$. Let us consider the impulse given to charge 2 during the encounter. By symmetry the y component must cancel while the x component is given by

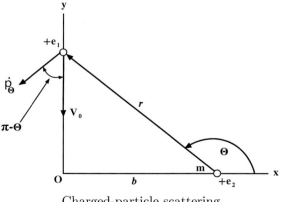

Charged-particle scattering

$$dp_x = F_x dt = -\frac{e_1 e_2}{4\pi\epsilon_0 r^2}\cos\theta \, dt = -\frac{e_1 e_2}{4\pi\epsilon_0 r^2}\cos\theta \frac{dt}{d\theta}d\theta$$

But

$$r\dot{\theta} = -v_0 \cos\theta$$

where

$$\frac{b}{r} = \cos(\pi - \theta) = -\cos\theta$$

Thus

$$dp_x = -\frac{e_1 e_2}{4\pi\epsilon_0 b v_0}\cos\theta d\theta$$

Integrate from $\frac{\pi}{2} < \theta < \frac{3\pi}{2}$ gives that the total momentum imparted to e_2 is

$$p_x = -\frac{e_1 e_2}{4\pi\epsilon_0 b v_0}\int_{\frac{\pi}{2}}^{\frac{3\pi}{2}}\cos\theta d\theta = \frac{e_1 e_2}{2\pi\epsilon_0 b v_0}$$

Thus the recoil energy of charge 2 is given by

$$E_2 = \frac{p_x^2}{2m} = \frac{1}{2m}\left(\frac{e_1 e_2}{2\pi\epsilon_0 b v_0}\right)^2$$

2.13 Solution of many-body equations of motion

The following are general methods used to solve Newton's many-body equations of motion for practical problems.

2.13.1 Analytic solution

In practical problems one has to solve a set of equations of motion since the forces depend on the location of every body involved. For example one may be dealing with a set of coupled oscillators such as the many components that comprise the suspension system of an automobile. Often the coupled equations of motion comprise a set of coupled second-order differential equations. The first approach to solve such a system is to try an analytic solution comprising a general solution of the inhomogeneous equation plus *one particular* solution of the inhomogeneous equation. Another approach is to employ numeric integration using a computer.

2.13.2 Successive approximation

When the system of coupled differential equations of motion is too complicated to solve analytically, one can use the method of successive approximation. The differential equations are transformed to integral equations. Then one starts with some initial conditions to make a first order estimate of the functions. The functions determined by this first order estimate then are used in a second iteration and this is repeated until the solution converges. An example of this approach is when making Hartree-Foch calculations of the electron distributions in an atom. The first order calculation uses the electron distributions predicted by the one-electron model of the atom. This result then is used to compute the influence of the electron charge distribution around the nucleus on the charge distribution of the atom for a second iteration etc.

2.13.3 Perturbation method

The perturbation technique can be applied if the force separates into two parts $F = F_1 + F_2$ where $F_1 >> F_2$ and the solution is known for the dominant F_1 part of the force. Then the correction to this solution due to addition of the perturbation F_2 usually is easier to evaluate. As an example, consider that one of the Space Shuttle thrusters fires. In principle one has all the gravitational forces acting plus the thrust force of the thruster. The perturbation approach is to assume that the trajectory of the Space Shuttle in the earth's gravitational field is known. Then the perturbation to this motion due to the very small thrust, produced by the thruster, is evaluated as a small correction to the motion in the Earth's gravitational field. This perturbation technique is used extensively in physics, especially in quantum physics. An example from my own research is scattering of a $1GeV$ ^{208}Pb ion in the Coulomb field of a ^{197}Au nucleus. The trajectory for elastic scattering is simple to calculate since neither nucleus is excited and the total energy and momenta are conserved. However, usually one of these nuclei will be internally excited by the electromagnetic interaction. This is called Coulomb excitation. The effect of the Coulomb excitation usually can be treated as a perturbation by assuming that the trajectory is given by the elastic scattering solution and then calculate the excitation probability assuming the Coulomb excitation of the nucleus is a small perturbation to the trajectory.

2.14   Newton's Law of Gravitation

Gravitation plays a fundamental role in classical mechanics as well as being an important example of a conservative central $\left(\frac{1}{r}\right)^2$ force. Although you may not be familiar with use of vector calculus for the gravitational field **g**, it is assumed that you have met the identical approach for studies of the electric field **E** in electrostatics. The primary difference is that mass m replaces charge e, and gravitational field **g** replaces the electric field **E**. This chapter reviews the concepts of vector calculus as used for study of conservative inverse-square law central fields.

In 1666 Newton formulated the Theory of Gravitation which he eventually published in the Principia in 1687. Newton's Law of Gravitation states that each mass particle attracts every other particle in the universe with a force that varies directly as the product of the mass and inversely as the square of the distance between them. That is, the force on a gravitational point mass m_G produced by a mass M_G

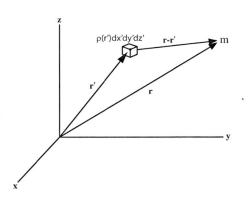

Figure 2.7: Gravitational force on mass m due to an infinitessimal volume element of the mass density distribution.

$$\mathbf{F}_m = -G\frac{m_G M_G}{r^2}\widehat{\mathbf{r}} \qquad (2.143)$$

where $\widehat{\mathbf{r}}$ is the unit vector pointing from the gravitational mass M_G to the gravitational mass m_G as shown in figure 2.7. Note that the force is attractive, that is, it points toward the other mass. This is in contrast to the repulsive electrostatic force between two similar charges. Newton's law was verified by Cavendish using a torsion balance. The experimental value of $G = (6.6726 \pm 0.0008) \times 10^{-11} N \cdot m^2/kg^2$.

The gravitational force between point particles can be extended to finite-sized bodies using the fact that the gravitational force field satisfies the superposition principle, that is, the net force is the vector sum of the individual forces between the component point particles. Thus the force summed over the mass distribution is

$$\mathbf{F}\left(\mathbf{r}\right)_m = -Gm_G \sum_{i=1}^{n} \frac{m_{Gi}}{r_i^2}\widehat{\mathbf{r}}_i \qquad (2.144)$$

where $\mathbf{r}_i$ is the vector from the gravitational mass m_{Gi} to the gravitational mass m_G at the position $\mathbf{r}$.

For a continuous gravitational mass distribution $\rho_G\left(\mathbf{r}'\right)$, the net force on the gravitational mass m_G at the location $\mathbf{r}$ can be written as

$$\mathbf{F}_m\left(\mathbf{r}\right) = -Gm_G \int_V \frac{\rho_G\left(\mathbf{r}'\right)\left(\widehat{\mathbf{r}} - \widehat{\mathbf{r}'}\right)}{\left(\overline{\mathbf{r}} - \overline{\mathbf{r}}'\right)^2}dv' \qquad (2.145)$$

where dv' is the volume element at the point $\mathbf{r}'$ as illustrated in figure 2.7.

2.14.1   Gravitational and inertial mass

Newton's Laws use the concept of *inertial mass* $m_I \equiv m$ in relating the force **F** to acceleration **a**

$$\mathbf{F} = m_I\mathbf{a} \qquad (2.146)$$

and momentum **p** to velocity **v**

$$\mathbf{p} = m_I\mathbf{v} \qquad (2.147)$$

That is, *inertial mass is the constant of proportionality relating the acceleration to the applied force.*

The concept of *gravitational mass* m_G *is the constant of proportionality between the gravitational force and the amount of matter.* That is, on the surface of the earth, the gravitational force is assumed to be

$$\mathbf{F}_G = m_G \left[-G \sum_{i=1}^{n} \frac{m_{Gi}}{r_i^2}\widehat{\mathbf{r}}_i\right] = m_G\mathbf{g} \qquad (2.148)$$

where $\mathbf{g}$ is the *gravitational field* which is a position-dependent force per unit gravitational mass pointing towards the center of the Earth. The gravitational mass is measured when an object is weighed.

Newton's Law of Gravitation leads to the relation for the gravitational field $\mathbf{g}(\mathbf{r})$ at the location $\mathbf{r}$ due to a gravitational mass distribution at the location $\mathbf{r}'$ as given by the integral over the gravitational mass density ρ_G

$$\mathbf{g}(\mathbf{r}) = -G \int_V \frac{\rho_G(\mathbf{r}')\left(\widehat{\mathbf{r}} - \widehat{\mathbf{r}'}\right)}{(\overline{\mathbf{r}} - \overline{\mathbf{r}}')^2} dv' \tag{2.149}$$

The acceleration of matter in a gravitational field relates the gravitational and inertial masses

$$\mathbf{F}_G = m_G \mathbf{g} = m_I \mathbf{a} \tag{2.150}$$

Thus

$$\mathbf{a} = \frac{m_G}{m_I} \mathbf{g} \tag{2.151}$$

That is, the acceleration of a body depends on the gravitational strength g and the ratio of the gravitational and inertial masses. It has been shown experimentally that all matter is subject to the same acceleration in vacuum at a given location in a gravitational field. That is, $\frac{m_G}{m_I}$ is a constant common to all materials. Galileo first showed this when he dropped objects from the Tower of Pisa. Modern experiments have shown that this is true to 5 parts in 10^{13}.

The exact equivalence of gravitational mass and inertial mass is called the **weak principle of equivalence** which underlies the General Theory of Relativity as discussed in chapter 17. It is convenient to use the same unit for the gravitational and inertial masses and thus they both can be written in terms of the common mass symbol m.

$$m_I = m_G = m \tag{2.152}$$

Therefore the subscripts G and I can be omitted in equations 2.150 and 2.152. Also the local acceleration due to gravity $\mathbf{a}$ can be written as

$$\mathbf{a} = \mathbf{g} \tag{2.153}$$

The gravitational field $\mathbf{g} \equiv \frac{\mathbf{F}}{m}$ has units of N/kg in the MKS system while the acceleration $\mathbf{a}$ has units m/s^2.

2.14.2 Gravitational potential energy U

Chapter 2.10.2 showed that a conservative field can be expressed in terms of the concept of a potential energy $U(\mathbf{r})$ which depends on position. The potential energy difference $\Delta U_{a \to b}$ between two points $\mathbf{r}_a$ and $\mathbf{r}_b$, is the work done moving from a to b against a force $\mathbf{F}$. That is:

$$\Delta U_{a \to b} = U(\mathbf{r}_b) - U(\mathbf{r}_a) = -\int_{r_a}^{r_b} \mathbf{F} \cdot d\mathbf{l} \tag{2.154}$$

In general, this line integral depends on the path taken.

Consider the gravitational field produced by a single point mass m_1. The work done moving a mass m_0 from r_a to r_b in this gravitational field can be calculated along an arbitrary path shown in figure 2.8 by assuming Newton's law of gravitation. Then the force on m_0 due to point mass m_1 is;

$$\mathbf{F} = -G \frac{m_1 m_0}{r^2} \widehat{\mathbf{r}} \tag{2.155}$$

Expressing $d\mathbf{l}$ in spherical coordinates $d\mathbf{l} = dr\widehat{\mathbf{r}} + rd\theta\widehat{\boldsymbol{\theta}} + r\sin\theta d\phi\widehat{\boldsymbol{\phi}}$ gives the path integral (2.154) from $(r_a \theta_a \phi_a)$ to $(r_b \theta_b \phi_b)$ is

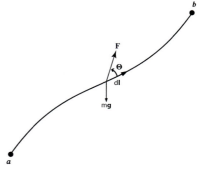

Figure 2.8: Work done against a force field moving from a to b.

$$\begin{aligned}\Delta U_{a \to b} &= -\int_a^b \mathbf{F} \cdot d\mathbf{l} = \int_a^b \left[G \frac{m_1 m_0}{r^2} (\widehat{\mathbf{r}} \cdot \widehat{\mathbf{r}} dr + \widehat{\mathbf{r}} \cdot \widehat{\boldsymbol{\theta}} d\theta + r\sin\theta \widehat{\mathbf{r}} \cdot \widehat{\boldsymbol{\phi}} d\phi) \right] = G \int_\alpha^b \frac{m_1 m_0}{r^2} \widehat{\mathbf{r}} \cdot \widehat{\mathbf{r}} dr \\ &= -G m_1 m_0 \left[\frac{1}{r_b} - \frac{1}{r_a} \right]\end{aligned} \tag{2.156}$$

since the scalar product of the unit vectors $\hat{\mathbf{r}} \cdot \hat{\mathbf{r}} = 1$. Note that the second two terms also cancel since $\hat{\mathbf{r}} \cdot \hat{\boldsymbol{\theta}} = \hat{\mathbf{r}} \cdot \hat{\boldsymbol{\phi}} = 0$ since the unit vectors are mutually orthogonal. *Thus the line integral just depends only on the starting and ending radii and is independent of the angular coordinates or the detailed path taken between* $(r_a \theta_a \phi_a)$ *and* $(r_b \theta_b \phi_b)$.

Consider the Principle of Superposition for a gravitational field produced by a set of n point masses. The line integral then can be written as:

$$\Delta U^{net}_{a \to b} = - \int_{r_a}^{r_b} \mathbf{F}_{net} \cdot d\mathbf{l} = - \sum_{i=1}^{n} \int_{r_a}^{r_b} \mathbf{F}_i \cdot d\mathbf{l} = \sum_{i=1}^{n} \Delta U^i_{a \to b} \tag{2.157}$$

Thus the net potential energy difference is the sum of the contributions from each point mass producing the gravitational force field. Since each component is conservative, then the total potential energy difference also must be conservative. For a *conservative force, this line integral is independent of the path taken*, it depends only on the starting and ending positions, $\mathbf{r}_a$ and $\mathbf{r}_b$. That is, the potential energy is a local function dependent only on position. The usefulness of gravitational potential energy is that, since the gravitational force is a conservative force, it is possible to solve many problems in classical mechanics using the fact that the sum of the kinetic energy and potential energy is a constant. Note that the gravitational field is conservative, since the potential energy difference $\Delta U^{net}_{a \to b}$ is *independent of the path taken*. It is conservative because the force is *radial* and time independent, it is not due to the $\frac{1}{r^2}$ dependence of the field.

2.14.3   Gravitational potential ϕ

Using $\mathbf{F} = m_0 \mathbf{g}$ gives that the change in potential energy due to moving a mass m_0 from a to b in a gravitational field $\mathbf{g}$ is:

$$\Delta U^{net}_{a \to b} = -m_0 \int_{r_a}^{r_b} \mathbf{g}_{net} \cdot d\mathbf{l} \tag{2.158}$$

Note that the probe mass m_0 factors out from the integral. It is convenient to define a new quantity called *gravitational potential* ϕ where

$$\Delta \phi^{net}_{a \to b} = \frac{\Delta U^{net}_{a \to b}}{m_0} = - \int_{r_a}^{r_b} \mathbf{g}_{net} \cdot d\mathbf{l} \tag{2.159}$$

That is; *gravitational potential difference is the work that must be done, per unit mass, to move from a to b with no change in kinetic energy.* Be careful not to confuse the gravitational potential *energy* difference $\Delta U_{a \to b}$ and gravitational potential difference $\Delta \phi_{a \to b}$, that is, ΔU has units of energy, *Joules*, while $\Delta \phi$ has units of *Joules/kg*.

The gravitational potential is a property of the gravitational force field; it is given as minus the line integral of the gravitational field from a to b. The change in gravitational potential energy for moving a mass m_0 from a to b is given in terms of gravitational potential by:

$$\Delta U^{net}_{a \to b} = m_0 \Delta \phi^{net}_{a \to b} \tag{2.160}$$

Superposition and potential

Previously it was shown that the gravitational force is conservative for the superposition of many masses. To recap, if the gravitational field

$$\mathbf{g}_{net} = \mathbf{g}_1 + \mathbf{g}_2 + \mathbf{g}_3 \tag{2.161}$$

then

$$\phi^{net}_{a \to b} = - \int_{r_a}^{r_b} \mathbf{g}_{net} \cdot d\mathbf{l} = - \int_{r_a}^{r_b} \mathbf{g}_1 \cdot d\mathbf{l} - \int_{r_a}^{r_b} \mathbf{g}_2 \cdot d\mathbf{l} - \int_{r_a}^{r_b} \mathbf{g}_3 \cdot d\mathbf{l} = \Sigma_i^n \phi^i_{a \to b} \tag{2.162}$$

Thus gravitational potential is a simple additive scalar field because the Principle of Superposition applies. The gravitational potential, between two points differing by h in height, is gh. Clearly, the greater g or h, the greater the energy released by the gravitational field when dropping a body through the height h. The unit of gravitational potential is the $\frac{Joule}{kg}$.

2.14.4 Potential theory

The gravitational force and electrostatic force both obey the inverse square law, for which the field and corresponding potential are related by:

$$\Delta\phi_{a\rightarrow b} = -\int_{r_a}^{r_b} \mathbf{g}\cdot d\mathbf{l} \tag{2.163}$$

For an arbitrary infinitessimal element distance $d\mathbf{l}$ the change in gravitational potential $d\phi$ is

$$d\phi = -\mathbf{g}\cdot d\mathbf{l} \tag{2.164}$$

Using cartesian coordinates both $\mathbf{g}$ and $d\mathbf{l}$ can be written as

$$\mathbf{g} = \widehat{\mathbf{i}}g_x + \widehat{\mathbf{j}}g_y + \widehat{\mathbf{k}}g_z \qquad\qquad d\mathbf{l} = \widehat{\mathbf{i}}dx + \widehat{\mathbf{j}}dy + \widehat{\mathbf{k}}dz \tag{2.165}$$

Taking the scalar product gives:

$$d\phi = -\mathbf{g}\cdot d\mathbf{l} = -g_x dx - g_y dy - g_z dz \tag{2.166}$$

Differential calculus expresses the change in potential $d\phi$ in terms of partial derivatives by:

$$d\phi = \frac{\partial\phi}{\partial x}dx + \frac{\partial\phi}{\partial y}dy + \frac{\partial\phi}{\partial z}dz \tag{2.167}$$

By association, 2.166 and 2.167 imply that

$$g_x = -\frac{\partial\phi}{\partial x} \qquad\qquad g_y = -\frac{\partial\phi}{\partial y} \qquad\qquad g_z = -\frac{\partial\phi}{\partial z} \tag{2.168}$$

Thus on each axis, the gravitational field can be written as minus the gradient of the gravitational potential. In three dimensions, the gravitational field is minus the total gradient of potential and the gradient of the scalar function ϕ can be written as:

$$\mathbf{g} = -\boldsymbol{\nabla}\phi \tag{2.169}$$

In cartesian coordinates this equals

$$\mathbf{g} = -\left[\widehat{\mathbf{i}}\frac{\partial\phi}{\partial x} + \widehat{\mathbf{j}}\frac{\partial\phi}{\partial y} + \widehat{\mathbf{k}}\frac{\partial\phi}{\partial z}\right] \tag{2.170}$$

Thus the gravitational field is just the gradient of the gravitational potential, which always is perpendicular to the equipotentials. Skiers are familiar with the concept of gravitational equipotentials and the fact that the line of steepest descent, and thus maximum acceleration, is perpendicular to gravitational equipotentials of constant height. The advantage of using potential theory for inverse-square law forces is that *scalar potentials* replace the more complicated *vector forces*, which greatly simplifies calculation. Potential theory plays a crucial role for handling both gravitational and electrostatic forces.

2.14.5 Curl of the gravitational field

It has been shown that the gravitational field is conservative, that is $\Delta U_{a\rightarrow b}$ is independent of the path taken between a and b. Therefore, equation 2.159 gives that the gravitational potential is independent of the path taken between two points a and b. Consider two possible paths between a and b as shown in figure 2.9. The line integral from a to b via route 1 is equal and opposite to the line integral back from b to a via route 2 if the gravitational field is conservative as shown earlier.

A better way of expressing this is that the line integral of the gravitational field is zero around any closed path. Thus the line integral between a and b, via path 1, and returning back to a, via path 2, are equal and opposite. That is, the net line integral for a closed loop is zero

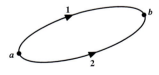

Figure 2.9: Circulation of the gravitational field.

$$\oint \mathbf{g}_{net} \cdot d\mathbf{l} = 0 \tag{2.171}$$

which is a measure of the *circulation* of the gravitational field. The fact that the circulation equals zero corresponds to the statement that the gravitational field is *radial* for a point mass.

Stokes Theorem, discussed in appendix $H3$, states that

$$\oint_C \mathbf{F} \cdot d\mathbf{l} = \int_{\substack{Area \\ bounded \\ by \\ C}} (\mathbf{\nabla} \times \mathbf{F}) \cdot d\mathbf{S} \tag{2.172}$$

Thus the zero circulation of the gravitational field can be rewritten as

$$\oint_C \mathbf{g} \cdot d\mathbf{l} = \int_{\substack{Area \\ bounded \\ by \\ C}} (\mathbf{\nabla} \times \mathbf{g}) \cdot d\mathbf{S} = 0 \tag{2.173}$$

Since this is independent of the shape of the perimeter C, therefore

$$\mathbf{\nabla} \times \mathbf{g} = 0 \tag{2.174}$$

That is, the gravitational field is a curl-free field.

A property of any curl-free field is that it can be expressed as the gradient of a scalar potential ϕ since

$$\mathbf{\nabla} \times \mathbf{\nabla}\phi = 0 \tag{2.175}$$

Therefore, the curl-free gravitational field can be related to a scalar potential ϕ as

$$\mathbf{g} = -\mathbf{\nabla}\phi \tag{2.176}$$

Thus ϕ is consistent with the above definition of gravitational potential ϕ in that the scalar product

$$\Delta\phi_{a\to b} = -\int_a^b \mathbf{g}_{net} \cdot d\mathbf{l} = \int_a^b (\mathbf{\nabla}\phi) \cdot d\mathbf{l} = \int_a^b \sum_i \frac{\partial\phi}{\partial x_i} dx_i = \int_a^b d\phi \tag{2.177}$$

An identical relation between the electric field and electric potential applies for the inverse-square law electrostatic field.

Reference potentials:

Note that only *differences* in potential energy, U, and gravitational potential, ϕ, are meaningful, the absolute values depend on some arbitrarily chosen reference. However, often it is useful to measure gravitational potential with respect to a particular arbitrarily chosen reference point ϕ_o such as to sea level. Aircraft pilots are required to set their altimeters to read with respect to sea level rather than their departure airport. This ensures that aircraft leaving from say both Rochester, $559' msl$, and Denver $5000' msl$, have their altimeters set to a common reference to ensure that they do not collide. The gravitational force is the gradient of the gravitational field which only depends on differences in potential, and thus is independent of any constant reference.

Gravitational potential due to continuous distributions of charge   Suppose mass is distributed over a volume v with a density ρ at any point within the volume. The gravitational potential at any field point p due to an element of mass $dm = \rho dv$ at the point p' is given by:

$$\Delta\phi_{\infty\to p} = -G \int_v \frac{\rho(p')dv'}{r_{p'p}} \tag{2.178}$$

This integral is over a scalar quantity. Since gravitational potential ϕ is a scalar quantity, it is easier to compute than is the vector gravitational field $\mathbf{g}$. If the scalar potential field is known, then the gravitational field is derived by taking the gradient of the gravitational potential.

2.14.6 Gauss's Law for Gravitation

The flux Φ of the gravitational field $\mathbf{g}$ through a surface S, as shown in figure 2.10, is defined as

$$\Phi \equiv \int_S \mathbf{g} \cdot d\mathbf{S} \qquad (2.179)$$

Note that there are two possible perpendicular directions that could be chosen for the surface vector $d\mathbf{S}$. Using Newton's law of gravitation for a point mass m the flux through the surface S is

$$\Phi = -Gm \int_S \frac{\widehat{\mathbf{r}} \cdot d\mathbf{S}}{r^2} \qquad (2.180)$$

Note that the solid angle subtended by the surface dS at an angle θ to the normal from the point mass is given by

$$d\Omega = \frac{\cos\theta dS}{r^2} = \frac{\widehat{\mathbf{r}} \cdot d\mathbf{S}}{r^2} \qquad (2.181)$$

Thus the net gravitational flux equals

$$\Phi = -Gm \int_S d\Omega \qquad (2.182)$$

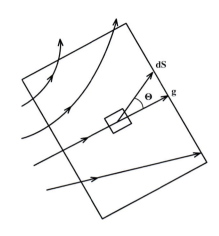

Figure 2.10: Flux of the gravitational field through an infinitessimal surface element dS.

Consider a *closed surface where the direction of the surface vector $d\mathbf{S}$ is defined as outwards*. The net flux *out* of this closed surface is given by

$$\Phi = -Gm \oint_S \frac{\widehat{\mathbf{r}} \cdot d\mathbf{S}}{r^2} = -Gm \oint_S d\Omega = -Gm4\pi \qquad (2.183)$$

This is independent of where the point mass lies within the closed surface or on the shape of the closed surface. Note that the solid angle subtended is zero if the point mass lies outside the closed surface. Thus the flux is as given by equation 2.183 if the mass is enclosed by the closed surface, while it is zero if the mass is outside of the closed surface.

Since the flux for a point mass is independent of the location of the mass within the volume enclosed by the closed surface, and using the principle of superposition for the gravitational field, then for n *enclosed* point masses the net flux is

$$\Phi \equiv \int_S \mathbf{g} \cdot d\mathbf{S} = -4\pi G \sum_i^n m_i \qquad (2.184)$$

This can be extended to continuous mass distributions, with local mass density ρ, giving that the net flux

$$\Phi \equiv \int_S \mathbf{g} \cdot d\mathbf{S} = -4\pi G \int_{\substack{enclosed \\ volume}} \rho dv \qquad (2.185)$$

Gauss's Divergence Theorem was given in appendix $H2$ as

$$\Phi = \oint_S \mathbf{F} \cdot d\mathbf{S} = \int_{\substack{Enclosed \\ volume}} \boldsymbol{\nabla} \cdot \mathbf{F} dv \qquad (2.186)$$

Applying the Divergence Theorem to Gauss's law gives that

$$\Phi = \oint_S \mathbf{g} \cdot d\mathbf{S} = \int_{\substack{Enclosed \\ volume}} \boldsymbol{\nabla} \cdot \mathbf{g} dv = -4\pi G \int_{\substack{enclosed \\ volume}} \rho dv$$

or

$$\int_{\substack{Enclosed \\ volume}} \left[\boldsymbol{\nabla} \cdot \mathbf{g} + 4\pi G\rho \right] dv = 0 \qquad (2.187)$$

This is true independent of the shape of the surface, thus the divergence of the gravitational field

$$\mathbf{\nabla} \cdot \mathbf{g} = -4\pi G\rho \tag{2.188}$$

This is a statement that the gravitational field of a point mass has a $\frac{1}{r^2}$ dependence.

Using the fact that the gravitational field is conservative, this can be expressed as the gradient of the gravitational potential ϕ,

$$\mathbf{g} = -\mathbf{\nabla}\phi \tag{2.189}$$

and Gauss's law, then becomes

$$\mathbf{\nabla} \cdot \mathbf{\nabla}\phi = 4\pi G\rho \tag{2.190}$$

which also can be written as Poisson's equation

$$\nabla^2\phi = 4\pi G\rho \tag{2.191}$$

Knowing the mass distribution ρ allows determination of the potential by solving Poisson's equation. A special case that often is encountered is when the mass distribution is zero in a given region. Then the potential for this region can be determined by solving Laplace's equation with known boundary conditions.

$$\nabla^2\phi = 0 \tag{2.192}$$

For example, Laplace's equation applies in the free space between the masses. It is used extensively in electrostatics to compute the electric potential between charged conductors which themselves are equipotentials.

2.14.7   Condensed forms of Newton's Law of Gravitation

The above discussion has resulted in several alternative expressions of Newton's Law of Gravitation that will be summarized here. The most direct statement of Newton's law is

$$\mathbf{g}\left(\mathbf{r}\right) = -G \int_V \frac{\rho\left(\mathbf{r}'\right)\left(\widehat{\mathbf{r}} - \widehat{\mathbf{r}'}\right)}{\left(\mathbf{r} - \mathbf{r}'\right)^2} dv' \tag{2.193}$$

An elegant way to express Newton's Law of Gravitation is in terms of the flux and circulation of the gravitational field. That is,

Flux:

$$\Phi \equiv \int_S \mathbf{g} \cdot d\mathbf{S} = -4\pi G \int_{\substack{enclosed \\ volume}} \rho \, dv \tag{2.194}$$

Circulation:

$$\oint \mathbf{g}_{net} \cdot d\mathbf{l} = 0 \tag{2.195}$$

The flux and circulation are better expressed in terms of the vector differential concepts of divergence and curl.

Divergence:

$$\mathbf{\nabla} \cdot \mathbf{g} = -4\pi G\rho \tag{2.196}$$

Curl:

$$\mathbf{\nabla} \times \mathbf{g} = 0 \tag{2.197}$$

Remember that the flux and divergence of the gravitational field are statements that the field between point masses has a $\frac{1}{r^2}$ dependence. The circulation and curl are statements that the field between point masses is radial.

Because the gravitational field is conservative it is possible to use the concept of the scalar potential field ϕ. This concept is especially useful for solving some problems since the gravitational potential can be evaluated using the scalar integral

$$\Delta\phi_{\infty \to p} = -G \int_v \frac{\rho(p')dv'}{r_{p'p}} \tag{2.198}$$

An alternate approach is to solve Poisson's equation if the boundary values and mass distributions are known where Poisson's equation is:

$$\nabla^2 \phi = 4\pi G \rho \tag{2.199}$$

These alternate expressions of Newton's law of gravitation can be exploited to solve problems. The method of solution is identical to that used in electrostatics.

2.16 *Example: Field of a uniform sphere*

Consider the simple case of the gravitational field due to a uniform sphere of matter of radius R and mass M. Then the volume mass density

$$\rho = \frac{3M}{4\pi R^3}$$

The gravitational field and potential for this uniform sphere of matter can be derived three ways;
   a) The field can be evaluated by directly integrating over the volume

$$\mathbf{g}(\mathbf{r}) = -G \int_S \frac{\rho(\mathbf{r}')\left(\hat{\mathbf{r}} - \hat{\mathbf{r}'}\right)}{(\mathbf{r} - \mathbf{r}')^2} dV'$$

   b) The potential can be evaluated directly by integration of

$$\Delta\phi_{\infty\to p} = -G \int_S \frac{\rho(p')dV'}{r_{p'p}}$$

and then

$$\mathbf{g} = -\boldsymbol{\nabla}\phi$$

   c) The obvious spherical symmetry can be used in conjunction with Gauss's law to easily solve this problem.

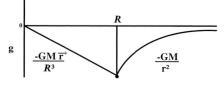

$$\int_S \mathbf{g} \cdot d\mathbf{S} = -4\pi G \int_{\substack{enclosed \\ volume}} \rho\, dv$$

$$4\pi r^2 g(r) = -4\pi GM \qquad \text{(r>R)}$$

That is: for $r > R$

$$\mathbf{g} = -G\frac{M}{r^2}\hat{\mathbf{r}} \qquad \text{(r>R)}$$

Similarly, for $r < R$

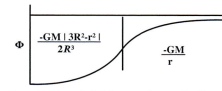

$$4\pi r^2 g(r) = \frac{4\pi}{3}r^3\rho \qquad \text{(r<R)}$$

That is:

Gravitational field **g** and gravitational potential Φ of a uniformly-dense spherical mass distribution of radius R.

$$\mathbf{g} = -G\frac{M}{R^3}\mathbf{r} \qquad \text{(r<R)}$$

The field inside the Earth is radial and is proportional to the distance from the center of the Earth. This is Hooke's Law, and thus ignoring air drag, any body dropped down a hole through the center of the Earth will undergo harmonic oscillations with an angular frequency of $\omega_0 = \sqrt{\frac{GM}{R^3}} = \sqrt{\frac{g}{R}}$. *This gives a period of oscillation of 1.4 hours, which is about the length of a P235 lecture in classical mechanics, which may seem like a long time.*

   Clearly method (c) is much simpler to solve for this case. In general, look for a symmetry that allows identification of a surface upon which the magnitude and direction of the field is constant. For such cases use Gauss's law. Otherwise use methods (a) or (b) whichever one is easiest to apply. Further examples will not be given here since they are essentially identical to those discussed extensively in electrostatics.

2.15 Summary

Newton's Laws of Motion:

A cursory review of Newtonian mechanics has been presented. The concept of inertial frames of reference was introduced since Newton's laws of motion apply only to inertial frames of reference.

Newton's Law of motion

$$\mathbf{F} = \frac{d\mathbf{p}}{dt} \tag{2.6}$$

leads to second-order equations of motion which can be difficult to handle for many-body systems.

Solution of Newton's second-order equations of motion can be simplified using the three first-order integrals coupled with corresponding conservation laws. The first-order time integral for linear momentum is

$$\int_1^2 \mathbf{F}_i dt = \int_1^2 \frac{d\mathbf{p}_i}{dt} dt = (\mathbf{p}_2 - \mathbf{p}_1)_i \tag{2.10}$$

The first-order time integral for angular momentum is

$$\frac{d\mathbf{L}_i}{dt} = \mathbf{r}_i \times \frac{d\mathbf{p}_i}{dt} = \mathbf{N}_i \qquad\qquad \int_1^2 \mathbf{N}_i dt = \int_1^2 \frac{d\mathbf{L}_i}{dt} dt = (\mathbf{L}_2 - \mathbf{L}_1)_i \tag{2.16}$$

The first-order spatial integral is related to kinetic energy and the concept of work. That is

$$\mathbf{F}_i = \frac{dT_i}{d\mathbf{r}_i} \qquad\qquad \int_1^2 \mathbf{F}_i \cdot d\mathbf{r}_i = (T_2 - T_1)_i \tag{2.21}$$

The conditions that lead to conservation of linear and angular momentum and total mechanical energy were discussed for many-body systems. The important class of conservative forces was shown to apply if the position-dependent force do not depend on time or velocity, and if the work done by a force $\int_1^2 \mathbf{F}_i \cdot d\mathbf{r}_i$ is independent of the path taken between the initial and final locations. The total mechanical energy is a constant of motion when the forces are conservative.

It was shown that the concept of center of mass of a many-body or finite sized body separates naturally for all three first-order integrals. The center of mass is that point about which

$$\sum_i^n m_i \mathbf{r}'_i = \int \mathbf{r}' \rho dV = 0. \qquad\qquad \text{(Centre of mass definition)}$$

where $\mathbf{r}'_i$ is the vector defining the location of mass m_i with respect to the center of mass. The concept of center of mass greatly simplifies the description of the motion of finite-sized bodies and many-body systems by separating out the important internal interactions and corresponding underlying physics, from the trivial overall translational motion of a many-body system..

The Virial theorem states that the time-averaged properties are related by

$$\langle T \rangle = -\frac{1}{2} \left\langle \sum_i \mathbf{F}_i \cdot \mathbf{r}_i \right\rangle \tag{2.86}$$

It was shown that the Virial theorem is useful for relating the time-averaged kinetic and potential energies, especially for cases involving either linear or inverse-square forces.

Typical examples were presented of application of Newton's equations of motion to solving systems involving constant, linear, position-dependent, velocity-dependent, and time-dependent forces, to constrained and unconstrained systems, as well as systems with variable mass. Rigid-body rotation about a body-fixed rotation axis also was discussed.

It is important to be cognizant of the following limitations that apply to Newton's laws of motion:

1) Newtonian mechanics assumes that all observables are measured to unlimited precision, that is t, E, $\mathbf{p}, \mathbf{r}$ are known exactly. Quantum physics introduces limits to measurement due to wave-particle duality.

2) The Newtonian view is that time and position are absolute concepts. The Theory of Relativity shows that this is not true. Fortunately for most problems $v \ll c$ and thus Newtonian mechanics is an excellent approximation.

3) Another limitation, to be discussed later, is that it is impractical to solve the equations of motion for many interacting bodies such as all the molecules in a gas. Then it is necessary to resort to using statistical averages, this approach is called statistical mechanics.

Newton's work constitutes a theory of motion in the universe that introduces the concept of **causality**. Causality is that there is a one-to-one correspondence between cause of effect. Each force causes a known effect that can be calculated. Thus the causal universe is pictured by philosophers to be a giant machine whose parts move like clockwork in a predictable and predetermined way according to the laws of nature. This is a deterministic view of nature. There are philosophical problems in that such a deterministic viewpoint appears to be contrary to free will. That is, taken to the extreme it implies that you were predestined to read this book because it is a natural consequence of this mechanical universe!

Newton's Laws of Gravitation

Newton's Laws of Gravitation and the Laws of Electrostatics are essentially identical since they both involve a central inverse square-law dependence of the forces. The important difference is that the gravitational force is attractive whereas the electrostatic force between identical charges is repulsive. That is, the gravitational constant G is replaced by $-\frac{1}{4\pi\epsilon_0}$, and the mass density ρ becomes the charge density for the case of electrostatics. As a consequence it is unnecessary to make a detailed study of Newton's law of gravitation since it is identical to what has already been studied in your accompanying electrostatic courses. Table 2.1 summarizes and compares the laws of gravitation and electrostatics. For both gravitation and electrostatics the field is central and conservative and depends as $\frac{1}{r^2}\hat{\mathbf{r}}$.

The laws of gravitation and electrostatics can be expressed in a more useful form in terms of the flux and circulation of the gravitational field as given either in the vector integral or vector differential forms. The radial independence of the flux, and corresponding divergence, is a statement that the fields are radial and have a $\frac{1}{r^2}\hat{\mathbf{r}}$ dependence. The statement that the circulation, and corresponding curl, are zero is a statement that the fields are radial and conservative.

Table 2.1; Comparison of Newton's law of gravitation and electrostatics.

	Gravitation	Electrostatics
Force field	$\mathbf{g} \equiv \frac{\mathbf{F}_G}{m}$	$\mathbf{E} \equiv \frac{\mathbf{F}_E}{q}$
Density	Mass density $\rho(\mathbf{r}')$	Charge density $\rho(\mathbf{r}')$
Conservative central field	$\mathbf{g}(\mathbf{r}) = -G \int_V \frac{\rho(\mathbf{r}')(\widehat{\hat{\mathbf{r}}-\mathbf{r}'})}{(\mathbf{r}-\mathbf{r}')^2} dv'$	$\mathbf{E}(\bar{\mathbf{r}}) = \frac{1}{4\pi\epsilon_0} \int_V \frac{\rho(\mathbf{r}')(\widehat{\hat{\mathbf{r}}-\mathbf{r}'})}{(\mathbf{r}-\mathbf{r}')^2} dv'$
Flux	$\Phi \equiv \int_S \mathbf{g} \cdot d\mathbf{S} = -4\pi G \int_{enclosed \atop volume} \rho dv$	$\Phi \equiv \int_S \mathbf{E} \cdot d\mathbf{S} = \frac{1}{\epsilon_0} \int_{enclosed \atop volume} \rho dv$
Circulation	$\oint \mathbf{g}_{net} \cdot d\mathbf{l} = 0$	$\oint \mathbf{E}_{net} \cdot d\mathbf{l} = 0$
Divergence	$\nabla \cdot \mathbf{g} = -4\pi G \rho$	$\nabla \cdot \mathbf{E} = \frac{1}{\epsilon_0}\rho$
Curl	$\nabla \times \mathbf{g} = 0$	$\nabla \times \mathbf{E} = 0$
Potential	$\Delta\phi_{\infty \to p} = -G \int_v \frac{\rho(p')dv'}{r_{p'p}}$	$\Delta\phi_{\infty \to p} = \frac{1}{4\pi\epsilon_0} \int_v \frac{\rho(p')dv'}{r_{p'p}}$
Poisson's equation	$\nabla^2 \phi = 4\pi G \rho$	$\nabla^2 \phi = -\frac{1}{\epsilon_0}\rho$

Both the gravitational and electrostatic central fields are conservative making it possible to use the concept of the scalar potential field ϕ. This concept is especially useful for solving some problems since the potential can be evaluated using a scalar integral. An alternate approach is to solve Poisson's equation if the boundary values and mass distributions are known. The methods of solution of Newton's law of gravitation are identical to those used in electrostatics and are readily accessible in the literature.

Workshop exercises

1. Spend a few minutes looking over the following problems, paying particular attention to the problems that you think you might have trouble with. All of the problems are taken from an introductory physics course on mechanics, so this should seem like review material. After you have had some time to look over the problems, you will take turns stepping up to the board to solve one. When it is your turn, you may pick ANY of the problems that have not already been solved. Depending on the number of students in the recitation, you may be asked to solve more than one problem. Good luck!

   (a) Justin fires a 12-gram bullet into a block of wood. The bullet travels at 190 m/s, penetrates the 2.0-kg block of wood, and emerges going 150 m/s. If the block is stationary on a frictionless surface when hit, how fast does it move after the bullet emerges?

   (b) A mass m at the end of a spring vibrates with a frequency of 0.88 Hz; when an additional 1.25 kg mass is added to m, the frequency is 0.48 Hz. What is the value of m?

   (c) Dan has a new chandelier in his living room. The chandelier is 27-kg and it hangs from the ceiling on a vertical 4.0-m-long wire. What horizontal force would Dan need to use to displace its position 0.10 m to one side? What will be the tension in the wire?

   (d) Dianne has a new spring with a spring constant of 900 N/m that she bought at Springs-R-Us. She places it vertically on a table and compresses it by 0.150 m. What upward speed can it give to a 0.300-kg ball when released?

   (e) A tiger leaps horizontally from a 6.5-m-high rock with a speed of 4.0 m/s. How far from the base of the rock will she land?

   (f) How much work must SuperRyan do to stop a 1300-kg car traveling at 100 km/hr?

   (g) Jason catches a baseball 3.1 s after throwing it vertically upward. With what speed did he throw it and what height did it reach?

   (h) Laura is practicing her figure skating and during her finale she can increase her rotation rate from an initial rate of 1.0 rev every 2.0 s to a final rate of 3.0 rev/s. If her initial moment of inertia was 4.6 kg·m^2, what is her final moment of inertia?

   (i) On an icy day in Rochester (imagine that!), you worry about parking your car in your driveway, which has an incline of 12°. Your neighbor Emily's driveway has an incline of 9°, and Brian's driveway across the street has one of 6°. The coefficient of static friction between tire rubber and ice is 0.15. Which driveway(s) will be safe to park a car?

2. Two particles are projected from the same point with velocities v_1 and v_2, at elevations α_1 and α_2, respectively $(\alpha_1 > \alpha_2)$. Show that if they are to collide in mid-air the interval between the firings must be

$$\frac{2v_1 v_2 \sin(\alpha_1 - \alpha_2)}{g(v_1 \cos\alpha_1 + v_2 \cos\alpha_2)}.$$

   (If you don't have time to solve this problem completely, then at least give an outline of how you would go about solving the problem.)

3. Read each of the following statements and, without consulting anyone else, mark them true or false. If you are unsure of any of them, make a guess. Once everyone has answered each of the statements individually, break into small groups and compare your answers. Try to come to an agreement as a group. The Teaching Assistant will then make sure everyone has the correct answer. Good luck!

   (a) The conservation of linear momentum is a consequence of translational symmetry, or the homogeneity of space.

   (b) For an isolated system with no external forces acting on it, the angular momentum will remain constant in both magnitude and direction.

   (c) A reference frame is called an inertial frame if Newton's laws are valid in that frame.

   (d) Newtonian mechanics and the laws of electromagnetism are invariant under Galilean transformations.

(e) The law of conservation of angular momentum is a consequence of rotational symmetry, or the isotropy of space.

(f) The center of mass of a system of particles moves like a single particle of mass M (total mass of the system) acted on by a single force F that is equal to the sum of all the external forces acting on the system.

(g) If Newton's laws are valid in one reference frame, then they are also valid in any reference frame accelerated with respect to the first system.

(h) The law of conservation of energy is a consequence of inversion symmetry, or the invertibility of space.

4. The teeter totter comprises two identical weights which hang on drooping arms attached to a peg as shown. The arrangement is unexpectedly stable and can be spun and rocked with little danger of toppling over.

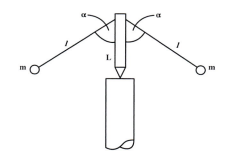

(a) Find an expression for the potential energy of the teeter toy as a function of θ when the teeter toy is cocked at an angle θ about the pivot point. For simplicity, consider only rocking motion in the vertical plane.

(b) Determine the equilibrium values(s) of θ.

(c) Determine whether the equilibrium is stable, unstable, or neutral for the value(s) of θ found in part (b).

(d) How could you determine the answers to parts (b) and (c) from a graph of the potential energy versus θ?

(e) Expand the expression for the potential energy about $\theta = 0$ and determine the frequency of small oscillations.

5. For each of the situations described below, determine which of the four functional forms of the force is most appropriate. Consider motion only along one dimension.

- Constant force: $F = constant$
- Time-dependent force: $F = F(t)$
- Velocity-dependent force: $F = F(v)$
- Distance-dependent force: $F = F(x)$

Go around the room and take turns answering a question. When it is your turn, pick a functional form and explain why you chose the one you did. If you are unsure, make a guess or ask a question to get help from the rest of the workshop. There may be more than one answer depending on your interpretation of the situation, so be sure to explore all of the possibilities.

(a) A mass resting on a frictionless table is attached to a spring, which in turn is attached to a wall. The mass is pulled to the side and executes simple harmonic motion in the horizontal direction.

(b) A freely-falling body subject to a constant gravitational field with no air resistance.

(c) An electron, initially at rest (treat it classically!), encounters an incoming electromagnetic wave of electric field intensity E given by $E = E_0 \sin(\omega t + \phi)$.

(d) A large mass is affected by the gravitational field of another mass a distance d away.

   (e) A freely-falling body subject to a constant gravitational field with air resistance.

   (f) A charged point particle is affected by the presence of another charged point particle a distance d away.

6. A particle of mass m is constrained to move on the frictionless inner surface of a cone of half-angle α.

   (a) Find the restrictions on the initial conditions such that the particle moves in a circular orbit about the vertical axis.

   (b) Determine whether this kind of orbit is stable. A particle of mass m is constrained to move on the frictionless inner surface of a cone of half-angle α, as shown in the figure.

7. Consider a thin rod of length L and mass M.

   (a) Draw gravitational field lines and equipotential lines for the rod. What can you say about the equipotential surfaces of the rod?

   (b) Calculate the gravitational potential at a point P that is a distance r from one end of the rod and in a direction perpendicular to the rod.

   (c) Calculate the gravitational field at P by direct integration.

   (d) Could you have used Gauss's law to find the gravitational field at P? Why or why not?

8. Consider a single particle of mass m.

   (a) Determine the position r and velocity v of a particle in spherical coordinates.

   (b) Determine the total mechanical energy of the particle in potential V.

   (c) Assume the force is conservative. Show that $F = -\nabla V$. Show that it agrees with Stoke's theorem.

   (d) Show that the angular momentum $L = r \times p$ of the particle is conserved. Hint: $\frac{d}{dt}(A \times B) = A \times \frac{dB}{dt} + \frac{dA}{dt} \times B$.

9. Consider a fluid with density ρ and velocity v in some volume V. The mass current $J = \rho v$ determines the amount of mass exiting the surface per unit time by the integral $\int_S J \cdot dA$.

   (a) Using the divergence theorem, prove the continuity equation, $\nabla \cdot J + \frac{\partial \rho}{\partial t} = 0$

10. A rocket of initial mass M burns fuel at constant rate k (kilograms per second), producing a constant force f. The total mass of available fuel is m_o. Assume the rocket starts from rest and moves in a fixed direction with no external forces acting on it.

   (a) Determine the equation of motion of the rocket.

   (b) Determine the final velocity of the rocket.

   (c) Determine the displacement of the rocket in time.

Problems

1. Consider a solid hemisphere of radius a. Compute the coordinates of the center of mass relative to the center of the spherical surface used to define the hemisphere.

2. A 2000kg Ford was travelling south on Mt. Hope Avenue when it collided with your 1000kg sports car travelling west on Elmwood Avenue. The two badly-damaged cars became entangled in the collision and leave a skid mark that is 20 meters long in a direction $14°$ to the west of the original direction of travel of the Excursion. The wealthy Excursion driver hires a high-powered lawyer who accuses you of speeding through the intersection. Use your P235 knowledge, plus the police officer's report of the recoil direction, the skid length, and knowledge that the coefficient of sliding friction between the tires and road is $\mu = 0.6$, to deduce the original velocities of both cars. Were either of the cars exceeding the 30mph speed limit?

3. A particle of mass m moving in one dimension has potential energy $U(x) = U_0[2(\frac{x}{a})^2 - (\frac{x}{a})^4]$, where U_0 and a are positive constants.

   a) Find the force $F(x)$ that acts on the particle.

   b) Sketch $U(x)$. Find the positions of stable and unstable equilibrium.

   c) What is the angular frequency ω of oscillations about the point of stable equilibrium?

   d) What is the minimum speed the particle must have at the origin to escape to infinity?

   e) At $t = 0$ the particle is at the origin and its velocity is positive and equal to the escape velocity. Find $x(t)$ and sketch the result.

4. a) Consider a single-stage rocket travelling in a straight line subject to an external force F^{ext} acting along the same line where v_{ex} is the exhaust velocity of the ejected fuel relative to the rocket. Show that the equation of motion is

   $$m\dot{v} = -\dot{m}v_{ex} + F^{ext}$$

   b) Specialize to the case of a rocket taking off vertically from rest in a uniform gravitational field g. Assume that the rocket ejects mass at a constant rate of $\dot{m} = -k$ where k is a positive constant. Solve the equation of motion to derive the dependence of velocity on time.

   c) The first couple of minutes of the launch of the Space Shuttle can be described roughly by; initial mass $= 2 \times 10^6$ kg, mass after 2 minutes $= 1 \times 10^6$ kg, exhaust speed $v_{ex} = 3000m/s$, and initial velocity is zero. Estimate the velocity of the Space Shuttle after two minutes of flight.

   d) Describe what would happen to a rocket where $\dot{m}v_{ex} < mg$.

5. A time independent field F is conservative if $\nabla \times F = 0$. Use this fact to test if the following fields are conservative, and derive the corresponding potential U.

   a) $F_x = ayz + bx + c, F_y = axz + bz, F_z = axy + by$

   b) $F_x = -ze^{-x}, F_y = \ln z, F_z = e^{-x} + \frac{y}{z}$

6. Consider a solid cylinder of mass m and radius r sliding without rolling down the smooth inclined face of a wedge of mass M that is free to slide without friction on a horizontal plane floor. Use the coordinates shown in the figure.

   a) How far has the wedge moved by the time the cylinder has descended from rest a vertical distance h ?

   b) Now suppose that the cylinder is free to roll down the wedge without slipping. How far does the wedge move in this case if the cylinder rolls down a vertical distance h ?

   c) In which case does the cylinder reach the bottom faster? How does this depend on the radius of the cylinder?

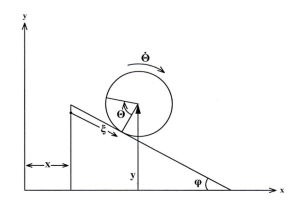

7. If the gravitational field vector is independent of the radial distance within a sphere, find the function describing the mass density $\rho(r)$ of the sphere.

Chapter 3

Linear oscillators

3.1 Introduction

Oscillations are a ubiquitous feature in nature. Examples are periodic motion of planets, the rise and fall of the tides, water waves, pendulum in a clock, musical instruments, sound waves, electromagnetic waves, and wave-particle duality in quantal physics. Oscillatory systems all have the same basic mathematical form although the names of the variables and parameters are different. The classical linear theory of oscillations will be assumed in this chapter since: (1) The linear approximation is well obeyed when the amplitudes of oscillation are small, that is, the restoring force obeys Hooke's Law. (2) The Principle of Superposition applies. (3) The linear theory allows most problems to be solved explicitly in closed form. This is in contrast to non-linear system where the motion can be complicated and even chaotic as discussed in chapter 4.

3.2 Linear restoring forces

An oscillatory system requires that there be a stable equilibrium about which the oscillations occur. Consider a conservative system with potential energy U for which the force is given by

$$\mathbf{F} = -\boldsymbol{\nabla} U \tag{3.1}$$

Figure 3.1 illustrates a conservative system that has three locations at which the restoring force is zero, that is, where the gradient of the potential is zero. Stable oscillations occur only around locations 1 and 3 whereas the system is unstable at the zero gradient location 2. Point 2 is called a *separatrix* in that an infinitessimal displacement of the particle from this separatrix will cause the particle to diverge towards either minimum 1 or 3 depending on which side of the separatrix the particle is displaced.

The requirements for stable oscillations about any point x_0 are that the potential energy must have the following properties.

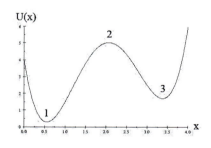

Figure 3.1: Stability for a one-dimensional potential U(x).

Stability requirements

1) The potential has a stable position for which the restoring force is zero, i.e. $\left(\frac{dU}{dx}\right)_{x=x_0} = 0$

2) The potential U must be positive and an even function of displacement $x - x_0$. That is. $\left(\frac{d^n U}{dx_n}\right)_{x_0} > 0$ where n is even.

The requirement for the restoring force to be linear is that the restoring force for perturbation about a stable equilibrium at x_0 is of the form

$$\mathbf{F} = -\alpha(x - x_0) = m\ddot{x} \tag{3.2}$$

The potential energy function for a linear oscillator has a pure parabolic shape about the minimum location, that is,

$$U = \frac{1}{2}k(x - x_0)^2 \tag{3.3}$$

53

where x_0 is the location of the minimum.

Fortunately, oscillatory systems involve small amplitude oscillations about a stable minimum. For weak non-linear systems, where the amplitude of oscillation Δx about the minimum is small, it is useful to make a Taylor expansion of the potential energy about the minimum. That is

$$U(\Delta x) = U(x_0) + \Delta x \frac{dU(x_0)}{dx} + \frac{\Delta x^2}{2!} \frac{d^2U(x_0)}{dx^2} + \frac{\Delta x^3}{3!} \frac{d^3U(x_0)}{dx^3} + \frac{\Delta x^4}{4!} \frac{d^4U(x_0)}{dx^4} + ... \qquad (3.4)$$

By definition, at the minimum $\frac{dU(x_0)}{dx} = 0$, and thus equation 3.3 can be written as

$$\Delta U = U(\Delta x) - U(x_0) = \frac{\Delta x^2}{2!} \frac{d^2U(x_0)}{dx^2} + \frac{\Delta x^3}{3!} \frac{d^3U(x_0)}{dx^3} + \frac{\Delta x^4}{4!} \frac{d^4U(x_0)}{dx^4} + ... \qquad (3.5)$$

For small amplitude oscillations, the system is linear if the second-order $\frac{\Delta x^2}{2!} \frac{d^2U(x_0)}{dx^2}$ term in equation 3.2 is dominant.

The linearity for small amplitude oscillations greatly simplifies description of the oscillatory motion and complicated chaotic motion is avoided. Most physical systems are approximately linear for small amplitude oscillations, and thus the motion close to equilibrium approximates a linear harmonic oscillator.

3.3   Linearity and superposition

An important aspect of linear systems is that the solutions obey the *Principle of Superposition*, that is, for the superposition of different oscillatory modes, the amplitudes add linearly. The linearly-damped linear oscillator is an example of a linear system in that it involves only linear operators, that is, it can be written in the operator form (appendix *F*.2)

$$\left(\frac{d^2}{dt^2} + \Gamma \frac{d}{dt} + \omega_o^2 \right) x(t) = A \cos \omega t \qquad (3.6)$$

The quantity in the brackets on the left hand side is a linear operator that can be designated by $\mathbb{L}$ where

$$\mathbb{L} x(t) = F(t) \qquad (3.7)$$

An important feature of linear operators is that they obey the *principle of superposition*. This property results from the fact that linear operators are distributive, that is

$$\mathbb{L}(x_1 + x_2) = \mathbb{L}(x_1) + \mathbb{L}(x_2) \qquad (3.8)$$

Therefore if there are two solutions $x_1(t)$ and $x_2(t)$ for two different forcing functions $F_1(t)$ and $F_2(t)$

$$\begin{aligned} \mathbb{L}x_1(t) &= F_1(t) \\ \mathbb{L}x_2(t) &= F_2(t) \end{aligned} \qquad (3.9)$$

then the addition of these two solutions, with arbitrary constants, also is a solution for linear operators.

$$\mathbb{L}(\alpha_1 x_1 + \alpha_2 x_2) = \alpha_1 F_1(t) + \alpha_2 F_2(t) \qquad (3.10)$$

In general then

$$\mathbb{L}\left(\sum_{n=1}^{N} \alpha_n x_n(t) \right) = \left(\sum_{n=1}^{N} \alpha_n F_n(t) \right) \qquad (3.11)$$

The left hand bracket can be identified as the linear combination of solutions

$$x(t) = \sum_{n=1}^{N} \alpha_n x_n(t) \qquad (3.12)$$

while the driving force is a linear superposition of harmonic forces

$$F(t) = \sum_{n=1}^{N} \alpha_n F_n(t) \qquad (3.13)$$

Thus these linear combinations also satisfy the general linear equation

$$\mathbb{L}x(t) = F(t) \tag{3.14}$$

Applicability of the Principle of Superposition to a system provides a tremendous advantage for handling and solving the equations of motion of oscillatory systems.

3.4 Geometrical representations of dynamical motion

The powerful pattern-recognition capabilities of the human brain, coupled with geometrical representations of the motion of dynamical systems, provide a sensitive probe of periodic motion. The geometry of the motion often can provide more insight into the dynamics than inspection of mathematical functions. A system with n degrees of freedom is characterized by locations q_i, velocities $\dot{q}_i$, and momenta p_i, in addition to the time t and instantaneous energy $H(t)$. Geometrical representations of the dynamical correlations are illustrated by the configuration space and phase space representations of these $2n + 2$ variables.

3.4.1 Configuration space (q_i, q_j, t)

A configuration space plot shows the correlated motion of two spatial coordinates q_i and q_j averaged over time. An example is the two-dimensional linear oscillator with two equations of motion and solutions

$$m\ddot{x} + k_x x = 0 \qquad\qquad m\ddot{y} + k_y y = 0 \tag{3.15}$$

$$x\left(t\right) = A\cos\left(\omega_x t\right) \qquad\qquad y\left(t\right) = B\cos\left(\omega_y t - \delta\right) \tag{3.16}$$

where $\omega = \sqrt{\frac{k}{m}}$. For unequal restoring force constants, $k_x \neq k_y$, the trajectory executes complicated Lissajous figures that depend on the angular frequencies ω_x, ω_y, and the phase factor δ. When the ratio of the angular frequencies along the two axes is rational, that is $\frac{\omega_x}{\omega_y}$ is a rational fraction, then the curve will repeat at regular intervals as shown in figure 3.2, and this shape depends on the phase difference. Otherwise the trajectory uniformly traverses the whole rectangle.

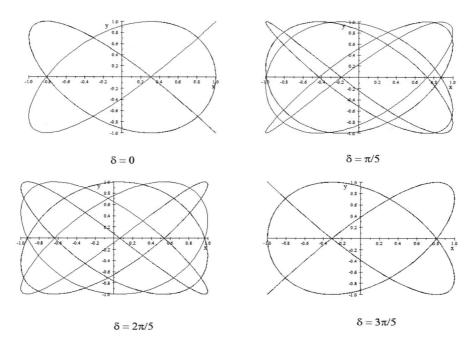

Figure 3.2: Configuration plots of (x, y) where $x = \cos(4t)$ and $y = \cos(5t - \delta)$ at four different phase values δ. The curves are called Lissajous figures

3.4.2   State space, $(q_i, \dot{q}_i, t)$

Visualization of a trajectory is enhanced by correlation of configuration q_i and it's corresponding velocity $\dot{q}_i$ which specifies the direction of the motion. The state space representation[1] is especially valuable when discussing Lagrangian mechanics which is based on the Lagrangian $L(\mathbf{q}, \dot{\mathbf{q}}, t)$.

The free undamped harmonic oscillator provides a simple illustration of state space. Consider a mass m attached to a spring with linear spring constant k for which the equation of motion is

$$-kx = m\ddot{x} = m\dot{x}\frac{d\dot{x}}{dx} \tag{3.17}$$

By integration this gives

$$\frac{1}{2}m\dot{x}^2 + \frac{1}{2}kx^2 = E \tag{3.18}$$

The first term in equation 3.18 is the kinetic energy, the second term is the potential energy, and E is the total energy which is conserved for this system. This equation can be expressed in terms of the state space coordinates as

$$\frac{\dot{x}^2}{\left(\frac{2E}{m}\right)} + \frac{x^2}{\left(\frac{2E}{k}\right)} = 1 \tag{3.19}$$

This corresponds to the equation of an ellipse for a state-space plot of $\dot{x}$ versus x as shown in figure 3.3*upper*. The elliptical paths shown correspond to contours of constant total energy which is partitioned between kinetic and potential energy. For the coordinate axis shown, the motion of a representative point will be in a clockwise direction as the total oscillator energy is redistributed between potential to kinetic energy. The area of the ellipse is proportional to the total energy E.

3.4.3   Phase space, (q_i, p_i, t)

Phase space, which was introduced by J.W. Gibbs for the field of statistical mechanics, provides a fundamental graphical representation in classical mechanics. The phase space coordinates $q_i p_i$ are the conjugate coordinates $(\mathbf{q}, \mathbf{p})$ and are fundamental to Hamiltonian mechanics which is based on the Hamiltonian $H(\mathbf{q}, \mathbf{p}, t)$. For a conservative system, only one phase-space curve passes through any point in phase space like the flow of an incompressible fluid. This makes phase space more useful than state space where many curves pass through any location. Lanczos [La49] defined an extended phase space using four-dimensional relativistic space-time as discussed in chapter 17.

Since $p_x = m\dot{x}$ for the non-relativistic, one-dimensional, linear oscillator, then equation 3.19 can be rewritten in the form

$$\frac{p_x^2}{2mE} + \frac{x^2}{\left(\frac{2E}{k}\right)} = 1 \tag{3.20}$$

This is the equation of an ellipse in the phase space diagram shown in Fig.3.3-*lower* which looks identical to Fig 3.3-*upper* where the ordinate variable $p_x = m\dot{x}$. That is, the only difference is the phase-space coordinates (x, p_x) replace the state-space coordinates $(x, \dot{x})$. State space plots are used extensively in this chapter to describe oscillatory motion. Although phase space is more fundamental, both state space and phase space plots provide useful representations for characterizing and elucidating a wide variety of motion in classical mechanics. The following discussion of the undamped simple pendulum illustrates the general features of state space.

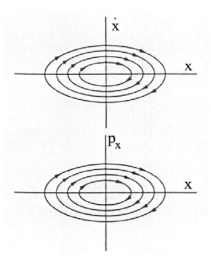

Figure 3.3: State space (upper), and phase space (lower) diagrams, for the linear harmonic oscillator.

---

[1]A universal name for the $(\mathbf{q}, \dot{\mathbf{q}})$ representation has not been adopted in the literature. Therefore this book has adopted the name "state space" in common with reference [Ta05]. Lanczos [La49] uses the term "state space" to refer to the extended phase space $(\mathbf{q}, \mathbf{p}, t)$ discussed in chapter 17.

3.4.4 Plane pendulum

Consider a simple plane pendulum of mass m attached to a string of length l in a uniform gravitational field g. There is only one generalized coordinate, θ. Since the moment of inertia of the simple plane-pendulum is $I = ml^2$, then the kinetic energy is

$$T = \frac{1}{2}ml^2\dot{\theta}^2 \tag{3.21}$$

and the potential energy relative to the bottom dead center is

$$U = mgl\left(1 - \cos\theta\right) \tag{3.22}$$

Thus the total energy equals

$$E = \frac{1}{2}ml^2\dot{\theta}^2 + mgl(1 - \cos\theta) = \frac{p_\theta^2}{2ml^2} + mgl\left(1 - \cos\theta\right) \tag{3.23}$$

where E is a constant of motion. Note that the angular momentum p_θ is not a constant of motion since the angular acceleration $\dot{p}_\theta$ explicitly depends on θ.

It is interesting to look at the solutions for the equation of motion for a plane pendulum on a $\left(\theta, \dot{\theta}\right)$ state space diagram shown in figure 3.4. The curves shown are equally-spaced contours of constant total energy. Note that the trajectories are ellipses only at very small angles where $1 - \cos\theta \approx \theta^2$, the contours are non-elliptical for higher amplitude oscillations. When the energy is in the range $0 < E < 2mgl$ the motion corresponds to oscillations of the pendulum about $\theta = 0$. The center of the ellipse is at $(0,0)$ which is a stable equilibrium point for the oscillation. However, when $|E| > 2mgl$ there is a phase change to rotational motion about the horizontal axis, that is, the pendulum swings around and over top dead center, i.e. it rotates continuously in one direction about the horizontal axis. The phase change occurs at $E = 2mgl$. and is designated by the separatrix trajectory.

Figure 3.4 shows two cycles for θ to better illustrate the cyclic nature of the phase diagram. The closed loops, shown as fine solid lines, correspond to pendulum oscillations about $\theta = 0$ or 2π for $E < 2mgl$. The dashed lines show rolling motion for cases where the total energy $E > 2mgl$. The broad solid line is the separatrix that separates the rolling and oscillatory motion. Note that at the separatrix, the kinetic energy and $\dot{\theta}$ are zero when the pendulum is at top dead center which occurs when $\theta = \pm\pi$. The point $(\pi, 0)$ is an unstable equilibrium characterized by phase lines that are hyperbolic to this unstable equilibrium point. Note that $\theta = +\pi$ and $-\pi$ correspond to the same physical point, that is, the phase diagram is better presented on a cylindrical phase space representation since θ is a cyclic variable that cycles around the cylinder whereas $\dot{\theta}$ oscillates equally about zero having both positive and negative values. The state-space diagram can be wrapped around a cylinder, then the unstable and stable equilibrium points will be at diametrically opposite locations on the surface of the cylinder at $\dot{\theta} = 0$. For small oscillations about equilibrium, also called librations, the correlation be-

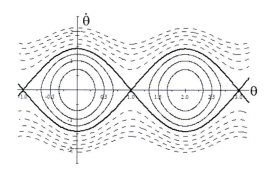

Figure 3.4: State space diagram for a plane pendulum. The θ axis is in units of π radians. Note that $\theta = +\pi$ and $-\pi$ correspond to the same physical point, that is the phase diagram should be rolled into a cylinder connected at $\theta = \pm\pi$.

tween $\dot{\theta}$ and θ is given by the clockwise closed loops wrapped on the cylindrical surface, whereas for energies $|E| > 2mgl$ the positive $\dot{\theta}$ corresponds to counterclockwise rotations while the negative $\dot{\theta}$ corresponds to clockwise rotations.

State-space diagrams will be used for describing oscillatory motion in chapters 3 and 4. Phase space is used in statistical mechanics in order to handle the equations of motion for ensembles of $\sim 10^{23}$ independent particles since momentum is more fundamental than velocity. Rather than try to account separately for the motion of each particle for an ensemble, it is best to specify the region of phase space containing the ensemble. If the number of particles is conserved, then every point in the initial phase space must transform to corresponding points in the final phase space. This will be discussed in chapters 8.3 and 15.2.7.

3.5   Linearly-damped free linear oscillator

3.5.1   General solution

All simple harmonic oscillations are damped to some degree due to energy dissipation via friction, viscous forces, or electrical resistance etc. The motion of damped systems is not conservative in that energy is dissipated as heat. As was discussed in chapter 2 the damping force can be expressed as

$$\mathbf{F}_D(v) = -f(v)\widehat{\mathbf{v}} \tag{3.24}$$

where the velocity dependent function $f(v)$ can be complicated. Fortunately there is a very large class of problems in electricity and magnetism, classical mechanics, molecular, atomic, and nuclear physics, where the damping force depends linearly on velocity which greatly simplifies solution of the equations of motion. This chapter discusses the special case of linear damping.

Consider the free simple harmonic oscillator, that is, assuming no oscillatory forcing function, with a linear damping term $\mathbf{F}_D(v) = -b\mathbf{v}$ where the parameter b is the damping factor. Then the equation of motion is

$$-kx - b\dot{x} = m\ddot{x} \tag{3.25}$$

This can be rewritten as

$$\ddot{x} + \Gamma\dot{x} + \omega_0^2 x = 0 \tag{3.26}$$

where the **damping parameter**

$$\Gamma = \frac{b}{m} \tag{3.27}$$

and the **characteristic angular frequency**

$$\omega_0 = \sqrt{\frac{k}{m}} \tag{3.28}$$

The general solution to the linearly-damped free oscillator is obtained by inserting the complex trial solution $z = z_0 e^{i\omega t}$. Then

$$(i\omega)^2 z_0 e^{i\omega t} + i\omega\Gamma z_0 e^{i\omega t} + \omega_0^2 z_0 e^{i\omega t} = 0 \tag{3.29}$$

This implies that

$$\omega^2 - i\omega\Gamma - \omega_0^2 = 0 \tag{3.30}$$

The solution is

$$\omega_\pm = i\frac{\Gamma}{2} \pm \sqrt{\omega_0^2 - \left(\frac{\Gamma}{2}\right)^2} \tag{3.31}$$

The two solutions $\omega_\pm$ are complex conjugates and thus the solutions of the damped free oscillator are

$$z = z_1 e^{i\left(i\frac{\Gamma}{2} + \sqrt{\omega_0^2 - \left(\frac{\Gamma}{2}\right)^2}\right)t} + z_2 e^{i\left(i\frac{\Gamma}{2} - \sqrt{\omega_0^2 - \left(\frac{\Gamma}{2}\right)^2}\right)t} \tag{3.32}$$

This can be written as

$$z = e^{-\left(\frac{\Gamma}{2}\right)t}\left[z_1 e^{i\omega_1 t} + z_2 e^{-i\omega_1 t}\right] \tag{3.33}$$

where

$$\omega_1 \equiv \sqrt{\omega_o^2 - \left(\frac{\Gamma}{2}\right)^2} \tag{3.34}$$

Underdamped motion $\omega_1^2 \equiv \omega_o^2 - \left(\frac{\Gamma}{2}\right)^2 > 0$

When $\omega_1^2 > 0$, then the square root is real so the solution can be written taking the real part of z which gives that equation 3.33 equals

$$x(t) = Ae^{-\left(\frac{\Gamma}{2}\right)t} \cos\left(\omega_1 t - \beta\right) \tag{3.35}$$

Where A and β are adjustable constants fit to the initial conditions. Therefore the velocity is given by

$$\dot{x}(t) = -Ae^{-\frac{\Gamma}{2}t}\left[\omega_1 \sin\left(\omega_1 t - \beta\right) + \frac{\Gamma}{2}\cos\left(\omega_1 t - \beta\right)\right] \tag{3.36}$$

This is the damped sinusoidal oscillation illustrated in figure 3.5*upper*. The solution has the following characteristics:

a) The oscillation amplitude decreases exponentially with a time constant $\tau_D = \frac{2}{\Gamma}$.

b) There is a small reduction in the frequency of the oscillation due to the damping leading to $\omega_1 = \sqrt{\omega_o^2 - \left(\frac{\Gamma}{2}\right)^2}$

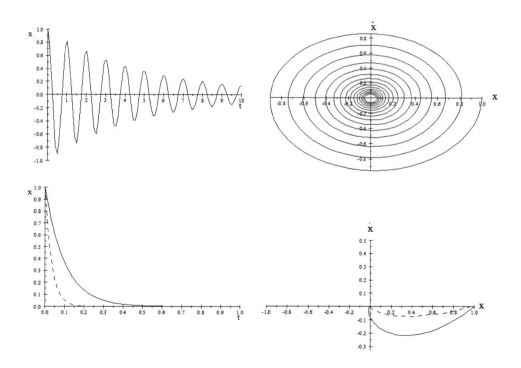

Figure 3.5: The amplitude-time dependence and state-space diagrams for the free linearly-damped harmonic oscillator. The upper row shows the underdamped system for the case with damping $\Gamma = \frac{\omega_0}{5\pi}$. The lower row shows the overdamped ($\frac{\Gamma}{2} > \omega_0$) [solid line] and critically damped ($\frac{\Gamma}{2} = \omega_0$) [dashed line] in both cases assuming that initially the system is at rest.

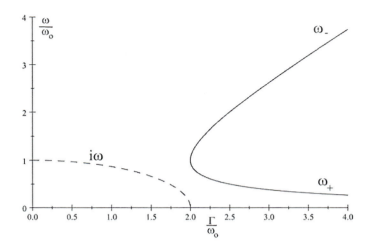

Figure 3.6: Real and imaginary solutions $\omega_\pm$ of the damped harmonic oscillator. A phase transition occurs at $\Gamma = 2\omega_0$. For $\Gamma < 2\omega_0$ (dashed) the two solutions are complex conjugates and imaginary. For $\Gamma > 2\omega_0$, (solid), there are two real solutions ω_+ and ω_- with widely different decay constants where ω_+ dominates the decay at long times.

Overdamped case $\omega_1^2 \equiv \omega_o^2 - \left(\frac{\Gamma}{2}\right)^2 < 0$

In this case the square root of ω_1^2 is imaginary and can be expressed as $\omega_1' = i\sqrt{\left(\frac{\Gamma}{2}\right)^2 - \omega_o^2}$. Therefore the solution is obtained more naturally by using a real trial solution $z = z_0 e^{\omega t}$ in equation 3.33 which leads to two roots

$$\omega_\pm = -\left[-\frac{\Gamma}{2} \pm \sqrt{\left(\frac{\Gamma}{2}\right)^2 - \omega_o^2}\right]$$

Thus the exponentially damped decay has two time constants ω_+ and ω_-.

$$x(t) = \left[A_1 e^{-\omega_+ t} + A_2 e^{-\omega_- t}\right] \tag{3.37}$$

The time constant $\frac{1}{\omega_-} < \frac{1}{\omega_+}$ thus the first term $A_1 e^{-\omega_+ t}$ in the bracket decays in a shorter time than the second term $A_2 e^{-\omega_- t}$. As illustrated in figure 3.6 the decay rate, which is imaginary when underdamped, i.e. $\frac{\Gamma}{2} < \omega_o$, bifurcates into two real values $\omega_\pm$ for overdamped, i.e. $\frac{\Gamma}{2} > \omega_o$. At large times the dominant term when overdamped is for ω_+ which has the smallest decay rate, that is, the longest decay constant $\tau_+ = \frac{1}{\omega_+}$. There is no oscillatory motion for the overdamped case, it slowly moves monotonically to zero as shown in fig 3.5*lower*. The amplitude decays away with a time constant that is longer than $\frac{2}{\Gamma}$.

Critically damped $\omega_1^2 \equiv \omega_o^2 - \left(\frac{\Gamma}{2}\right)^2 = 0$

This is the limiting case where $\frac{\Gamma}{2} = \omega_o$ For this case the solution is of the form

$$x(t) = (A + Bt)\, e^{-\left(\frac{\Gamma}{2}\right)t} \tag{3.38}$$

This motion also is non-sinusoidal and evolves monotonically to zero. As shown in figure 3.5 the critically-damped solution goes to zero with the shortest time constant, that is, largest ω. Thus analog electric meters are built almost critically damped so the needle moves to the new equilibrium value in the shortest time without oscillation.

It is useful to graphically represent the motion of the damped linear oscillator on either a state space $(\dot{x}, x)$ diagram or phase space (p_x, x) diagram as discussed in chapter 3.4. The state space plots for the undamped, overdamped, and critically-damped solutions of the damped harmonic oscillator are shown in figure 3.5. For underdamped motion the state space diagram spirals inwards to the origin in contrast to critical or overdamped motion where the state and phase space diagrams move monotonically to zero.

3.5.2 Energy dissipation

The instantaneous energy is the sum of the instantaneous kinetic and potential energies

$$E = \frac{1}{2}m\dot{x}^2 + \frac{1}{2}kx^2 \tag{3.39}$$

where x, and $\dot{x}$ are given by the solution of the equation of motion.

Consider the total energy of the underdamped system

$$E = \frac{1}{2}m\dot{x}^2 + \frac{1}{2}m\omega_0^2 x^2 \tag{3.40}$$

where $k = m\omega_0^2$. The average total energy is given by substitution for x and $\dot{x}$ and taking the average over one cycle. Since

$$x(t) = Ae^{-\left(\frac{\Gamma}{2}\right)t}\cos\left(\omega_1 t - \beta\right) \tag{3.41}$$

Then the velocity is given by

$$\dot{x}(t) = -Ae^{-\frac{\Gamma}{2}t}\left[\omega_1\sin\left(\omega_1 t - \beta\right) + \frac{\Gamma}{2}\cos\left(\omega_1 t - \beta\right)\right] \tag{3.42}$$

Inserting equations 3.41 and 3.42 into 3.40 gives a small amplitude oscillation about an exponential decay for the energy E. Averaging over one cycle and using the fact that $\langle\sin\theta\cos\theta\rangle = 0$, and $\left\langle[\sin\theta]^2\right\rangle = \left\langle[\cos\theta]^2\right\rangle = \frac{1}{2}$, gives the time-averaged total energy as

$$\langle E\rangle = e^{-\Gamma t}\left(\frac{1}{4}mA^2\omega_1^2 + \frac{1}{4}mA^2\left(\frac{\Gamma}{2}\right)^2 + \frac{1}{4}mA^2\omega_0^2\right) \tag{3.43}$$

which can be written as

$$\langle E\rangle = E_0 e^{-\Gamma t} \tag{3.44}$$

Note that the *energy* of the linearly damped free oscillator decays away with a time constant $\tau = \frac{1}{\Gamma}$. That is, the *intensity* has a time constant that is half the time constant for the decay of the *amplitude* of the transient response. Note that the average kinetic and potential energies are identical, as implied by the Virial theorem, and both decay away with the same time constant. This relation between the mean life τ for decay of the damped harmonic oscillator and the damping width term Γ occurs frequently in physics.

The damping of an oscillator usually is characterized by a single parameter Q called the **Quality Factor** where

$$Q \equiv \frac{\text{Energy stored in the oscillator}}{\text{Energy dissipated per radian}} \tag{3.45}$$

The energy loss per radian is given by

$$\Delta E = \frac{dE}{dt}\frac{1}{\omega_1} = \frac{E\Gamma}{\omega_1} = \frac{E\Gamma}{\sqrt{\omega_o^2 - \left(\frac{\Gamma}{2}\right)^2}} \tag{3.46}$$

where the numerator $\omega_1 = \sqrt{\omega_o^2 - \left(\frac{\Gamma}{2}\right)^2}$ is the frequency of the free damped linear oscillator.

Thus the Quality factor Q equals

$$Q = \frac{E}{\Delta E} = \frac{\omega_1}{\Gamma} \tag{3.47}$$

The larger the Q factor, the less damped is the system, and the greater is the number of cycles of the oscillation in the damped wave train. Chapter 3.11.3 shows that the longer the wave train, that is the higher is the Q factor, the narrower is the frequency distribution around the central value. The Mössbauer effect in nuclear physics provides a remarkably long wave train that can be used to make high precision measurements. The high-Q precision of the LIGO laser interferometer was used in the first successful observation of gravity waves in 2015.

Typical Q factors	
Earth, for earthquake wave	250-1400
Piano string	3000
Crystal in digital watch	10^4
Microwave cavity	10^4
Excited atom	10^7
Neutron star	10^{12}
LIGO laser	10^{13}
Mössbauer effect in nucleus	10^{14}

Table 3.1: Typical Q factors in nature.

3.6   Sinusoidally-drive, linearly-damped, linear oscillator

The linearly-damped linear oscillator, driven by a harmonic driving force, is of considerable importance to all branches of science and engineering. The equation of motion can be written as

$$\ddot{x} + \Gamma\dot{x} + \omega_0^2 x = \frac{F(t)}{m} \tag{3.48}$$

where $F(t)$ is the driving force. For mathematical simplicity the driving force is chosen to be a sinusoidal harmonic force. The solution of this second-order differential equation comprises two components, the complementary solution (*transient response*), and the particular solution (*steady-state response*).

3.6.1   Transient response of a driven oscillator

The transient response of a driven oscillator is given by the complementary solution of the above second-order differential equation

$$\ddot{x} + \Gamma\dot{x} + \omega_0^2 x = 0 \tag{3.49}$$

which is identical to the solution of the free linearly-damped harmonic oscillator. As discussed in section 3.5, the solution of the linearly-damped free oscillator is given by the real part of the complex variable z where

$$z = e^{-\frac{\Gamma}{2}t}\left[z_1 e^{i\omega_1 t} + z_2 e^{-i\omega_1 t}\right] \tag{3.50}$$

and

$$\omega_1 \equiv \sqrt{\omega_o^2 - \left(\frac{\Gamma}{2}\right)^2} \tag{3.51}$$

Underdamped motion $\omega_1^2 \equiv \omega_o^2 - \frac{\Gamma}{2}^2 > 0$:   When $\omega_1^2 > 0$, then the square root is real so the transient solution can be written taking the real part of z which gives

$$x(t)_T = \frac{F_0}{m} e^{-\frac{\Gamma}{2}t} \cos(\omega_1 t) \tag{3.52}$$

The solution has the following characteristics:

   a) The amplitude of the transient solution decreases exponentially with a time constant $\tau_D = \frac{2}{\Gamma}$ while the energy decreases with a time constant of $\frac{1}{\Gamma}$.

   b) There is a small downward frequency shift in that $\omega_1 = \sqrt{\omega_o^2 - \left(\frac{\Gamma}{2}\right)^2}$.

Overdamped case $\omega_1^2 \equiv \omega_o^2 - \left(\frac{\Gamma}{2}\right)^2 < 0$:   In this case the square root is imaginary, which can be expressed as $\omega_1' \equiv \sqrt{\left(\frac{\Gamma}{2}\right)^2 - \omega_o^2}$ which is real and the solution is just an exponentially damped one

$$x(t)_T = \frac{F_0}{m} e^{-\frac{\Gamma}{2}t} \left[e^{\omega_1' t} + e^{-\omega_1' t}\right] \tag{3.53}$$

There is no oscillatory motion for the overdamped case, it slowly moves monotonically to zero. The total energy decays away with two time constants greater than $\frac{1}{\Gamma}$.

Critically damped $\omega_1^2 \equiv \omega_o^2 - \left(\frac{\Gamma}{2}\right)^2 = 0$:   For this case, as mentioned for the damped free oscillator, the solution is of the form

$$x(t)_T = (A + Bt)e^{-\frac{\Gamma}{2}t} \tag{3.54}$$

The critically-damped system has the shortest time constant.

3.6.2 Steady state response of a driven oscillator

The particular solution of the differential equation gives the important steady state response, $x(t)_S$ to the forcing function. Consider that the forcing term is a single frequency sinusoidal oscillation.

$$F(t) = F_0 \cos(\omega t) \tag{3.55}$$

Thus the particular solution is the real part of the complex variable z which is a solution of

$$\ddot{z} + \Gamma \dot{z} + \omega_0^2 z = \frac{F_0}{m} e^{i\omega t} \tag{3.56}$$

A trial solution is

$$z = z_0 e^{i\omega t} \tag{3.57}$$

This leads to the relation

$$-\omega^2 z_0 + i\omega \Gamma z_0 + \omega_0^2 z_0 = \frac{F_0}{m} \tag{3.58}$$

Multiplying the numerator and denominator by the factor $(\omega_0^2 - \omega^2) - i\Gamma\omega$ gives

$$z_0 = \frac{\frac{F_0}{m}}{(\omega_0^2 - \omega^2) + i\Gamma\omega} = \frac{\frac{F_0}{m}}{(\omega_0^2 - \omega^2)^2 + (\Gamma\omega)^2} \left[(\omega_0^2 - \omega^2) - i\Gamma\omega \right] \tag{3.59}$$

The steady state solution $x(t)_S$ thus is given by the real part of z, that is

$$x(t)_S = \frac{\frac{F_0}{m}}{(\omega_0^2 - \omega^2)^2 + (\Gamma\omega)^2} \left[(\omega_0^2 - \omega^2) \cos \omega t + \Gamma\omega \sin \omega t \right] \tag{3.60}$$

This can be expressed in terms of a phase δ defined as

$$\tan \delta \equiv \left(\frac{\Gamma\omega}{\omega_0^2 - \omega^2} \right) \tag{3.61}$$

As shown in figure 3.7, the hypotenuse of the triangle equals $\sqrt{(\omega_0^2 - \omega^2)^2 + (\Gamma\omega)^2}$. Thus

$$\cos \delta = \frac{\omega_0^2 - \omega^2}{\sqrt{(\omega_0^2 - \omega^2)^2 + (\Gamma\omega)^2}} \tag{3.62}$$

and

$$\sin \delta = \frac{\Gamma\omega}{\sqrt{(\omega_0^2 - \omega^2)^2 + (\Gamma\omega)^2}} \tag{3.63}$$

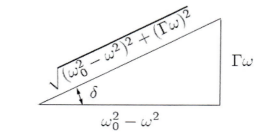

Figure 3.7: Phase between driving force and resultant motion.

The phase δ represents the phase difference between the driving force and the resultant motion. For a fixed ω_0 the phase $\delta = 0$ when $\omega = 0$, and increases to $\delta = \frac{\pi}{2}$ when $\omega = \omega_0$. For $\omega > \omega_0$ the phase $\delta \to \pi$ as $\omega \to \infty$.

The steady state solution can be re-expressed in terms of the phase shift δ as

$$
\begin{aligned}
x(t)_S &= \frac{\frac{F_0}{m}}{\sqrt{(\omega_0^2 - \omega^2)^2 + (\Gamma\omega)^2}} \left[\cos \delta \cos \omega t + \sin \delta \sin \omega t \right] \\
&= \frac{\frac{F_0}{m}}{\sqrt{(\omega_0^2 - \omega^2)^2 + (\Gamma\omega)^2}} \cos (\omega t - \delta)
\end{aligned}
\tag{3.64}
$$

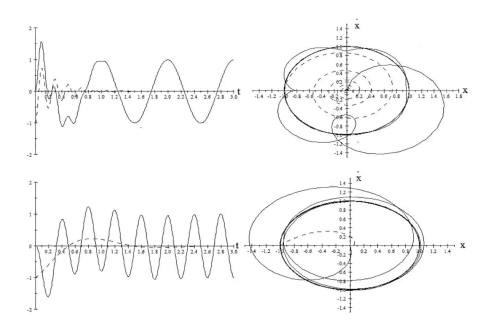

Figure 3.8: Amplitude versus time, and state space plots of the transient solution (dashed) and total solution (solid) for two cases. The upper row shows the case where the driving frequency $\omega = \frac{\omega_1}{5}$ while the lower row shows the same for the case where the driving frequency $\omega = 5\omega_1$.

3.6.3   Complete solution of the driven oscillator

To summarize, the total solution of the sinusoidally forced linearly-damped harmonic oscillator is the sum of the transient and steady-state solutions of the equations of motion.

$$x(t)_{Total} = x(t)_T + x(t)_S \tag{3.65}$$

For the underdamped case, the transient solution is the complementary solution

$$x(t)_T = \frac{F_0}{m} e^{-\frac{\Gamma}{2}t} \cos{(\omega_1 t - \beta)} \tag{3.66}$$

where $\omega_1 = \sqrt{\omega_o^2 - \left(\frac{\Gamma}{2}\right)^2}$.   The steady-state solution is given by the particular solution

$$x(t)_S = \frac{\frac{F_0}{m}}{\sqrt{\left(\omega_0^2 - \omega^2\right)^2 + \left(\Gamma\omega\right)^2}} \cos{(\omega t - \delta)} \tag{3.67}$$

Note that the frequency of the transient solution is ω_1 which in general differs from the driving frequency ω. The phase shift $\beta - \delta$ for the transient component is set by the initial conditions. The transient response leads to a more complicated motion immediately after the driving function is switched on.   Figure 3.8 illustrates the amplitude time dependence and state space diagram for the transient component, and the total response, when the driving frequency is either $\omega = \frac{\omega_1}{5}$ or $\omega = 5\omega_1$. Note that the modulation of the steady-state response by the transient response is unimportant once the transient response has damped out leading to a constant elliptical state space trajectory. For cases where the initial conditions are $x = \dot{x} = 0$ then the transient solution has a relative phase difference $\beta - \delta = \pi$ radians at $t = 0$ and relative amplitudes such that the transient and steady-state solutions cancel at $t = 0$.

The characteristic sounds of different types of musical instruments depend very much on the admixture of transient solutions plus the number and mixture of oscillatory active modes. Percussive instruments, such as the piano, have a large transient component. The mixture of transient and steady-state solutions for forced oscillations occurs frequently in studies of RLC networks in electrical circuit analysis.

3.6.4 Resonance

The discussion so far has discussed the role of the transient and steady-state solutions of the driven damped harmonic oscillator which occurs frequently is science, and engineering. Another important aspect is resonance that occurs when the driving frequency ω approaches the natural frequency ω_1 of the damped system. Consider the case where the time is sufficient for the transient solution to have decayed to zero.

Figure 3.9 shows the amplitude and phase for the *steady-state response* as ω goes through a resonance as the driving frequency is changed. The steady-states solution of the driven oscillator follows the driving force when $\omega << \omega_0$ in that the phase difference is zero and the amplitude is just $\frac{F_0}{k}$. The response of the system peaks at resonance, while for $\omega >> \omega_0$ the harmonic system is unable to follow the more rapidly oscillating driving force and thus the phase of the induced oscillation is out of phase with the driving force and the amplitude of the oscillation tends to zero.

Note that the resonance frequency for a driven damped oscillator, differs from that for the undriven damped oscillator, and differs from that for the undamped oscillator. The natural frequency for an **undamped harmonic oscillator** is given by

$$\omega_0^2 = \frac{k}{m} \tag{3.68}$$

The transient solution is the same as **damped free oscillations** of a damped oscillator and has a frequency of the system ω_1 given by

$$\omega_1^2 = \omega_0^2 - \left(\frac{\Gamma}{2}\right)^2 \tag{3.69}$$

That is, damping slightly reduces the frequency.

For the **driven oscillator** the maximum value of the steady-state amplitude response is obtained by taking the maximum of the function $x(t)_S$, that is when $\frac{dx_S}{d\omega} = 0$. This occurs at the resonance angular frequency ω_R where

$$\omega_R^2 = \omega_0^2 - 2\left(\frac{\Gamma}{2}\right)^2 \tag{3.70}$$

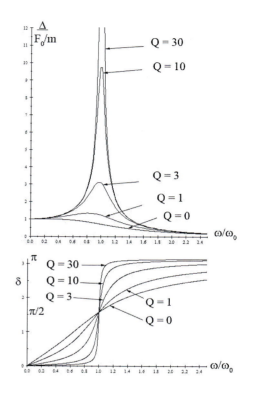

Figure 3.9: Resonance behavior for the linearly-damped, harmonically driven, linear oscillator.

No resonance occurs if $\omega_0^2 - 2\left(\frac{\Gamma}{2}\right)^2 < 0$ since then ω_R is imaginary and the amplitude decreases monotonically with increasing ω. Note that the above three frequencies are identical if $\Gamma = 0$ but they differ when $\Gamma > 0$ and $\omega_R < \omega_1 < \omega_0$.

For the driven oscillator it is customary to define the **quality factor Q** as

$$Q \equiv \frac{\omega_R}{\Gamma} \tag{3.71}$$

When $Q >> 1$ the system has a narrow high resonance peak. As the damping increases the quality factor decreases leading to a wider and lower peak. The resonance disappears when $Q < 1$.

3.6.5 Energy absorption

Discussion of energy stored in resonant systems is best described using the steady state solution which is dominant after the transient solution has decayed to zero. Then

$$x(t)_S = \frac{\frac{F_0}{m}}{\left(\omega_0^2 - \omega^2\right)^2 + \left(\Gamma\omega\right)^2} \left[\left(\omega_0^2 - \omega^2\right)\cos\omega t + \Gamma\omega\sin\omega t\right] \tag{3.72}$$

This can be rewritten as

$$x(t)_S = A_{el} \cos \omega t + A_{abs} \sin \omega t \tag{3.73}$$

where the **elastic amplitude**

$$A_{el} = \frac{\frac{F_0}{m}}{\left(\omega_0^2 - \omega^2\right)^2 + (\Gamma\omega)^2} \left(\omega_0^2 - \omega^2\right) \tag{3.74}$$

while the **absorptive amplitude**

$$A_{abs} = \frac{\frac{F_0}{m}}{\left(\omega_0^2 - \omega^2\right)^2 + (\Gamma\omega)^2}\Gamma\omega \tag{3.75}$$

Figure 3.10 shows the behavior of the absorptive and elastic amplitudes as a function of angular frequency ω. The absorptive amplitude is significant only near resonance whereas the elastic amplitude goes to zero at resonance. Note that the *full width at half maximum of the absorptive amplitude peak equals* Γ.

The work done by the force $F_0 \cos \omega t$ on the oscillator is

$$W = \int F dx = \int F \dot{x} dt \tag{3.76}$$

Thus the **absorbed power** $P(t)$ is given by

$$P(t) = \frac{dW}{dt} = F\dot{x} \tag{3.77}$$

The steady state response gives a velocity

$$\dot{x}(t)_S = -\omega A_{el} \sin \omega t + \omega A_{abs} \cos \omega t \tag{3.78}$$

Thus the steady-state instantaneous power input is

$$P(t) = F_0 \cos \omega t \left[-\omega A_{el} \sin \omega t + \omega A_{abs} \cos \omega t\right] \tag{3.79}$$

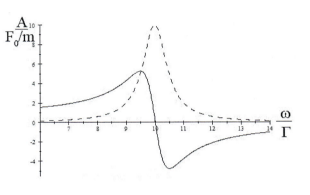

Figure 3.10:   Elastic (solid) and absorptive (dashed) amplitudes of the steady-state solution for $\Gamma = 0.10\omega_0$.

The absorptive term steadily absorbs energy while the elastic term oscillates as energy is alternately absorbed or emitted. The time average over one cycle is given by

$$\langle P \rangle = F_0 \left[-\omega A_{el} \langle \cos \omega t \sin \omega t \rangle + \omega A_{abs} \left\langle (\cos \omega t)^2 \right\rangle\right] \tag{3.80}$$

where $\langle \cos \omega t \sin \omega t \rangle$ and $\langle \cos \omega t^2 \rangle$ are the time average over one cycle. The time averages over one complete cycle for the first term in the bracket is

$$-\omega A_{el} \langle \cos \omega t \sin \omega t \rangle = 0 \tag{3.81}$$

while for the second term

$$\left\langle \cos \omega t^2 \right\rangle = \frac{1}{T} \int_{t_o}^{t_0+T} \cos \omega t^2 dt = \frac{1}{2} \tag{3.82}$$

Thus the time average power input is determined by only the absorptive term

$$\langle P \rangle = \frac{1}{2} F_0 \omega A_{abs} = \frac{F_0^2}{2m} \frac{\Gamma\omega^2}{\left(\omega_0^2 - \omega^2\right)^2 + (\Gamma\omega)^2} \tag{3.83}$$

This shape of the power curve is a classic Lorentzian shape. Note that the maximum of the average kinetic energy occurs at $\omega_{KE} = \omega_0$ which is different from the peak of the amplitude which occurs at $\omega_1^2 = \omega_0^2 - \left(\frac{\Gamma}{2}\right)^2$. The potential energy is proportional to the amplitude squared, i.e. x_S^2 which occurs at the same angular frequency as the amplitude, that is, $\omega_{PE}^2 = \omega_R^2 = \omega_0^2 - 2\left(\frac{\Gamma}{2}\right)^2$. The kinetic and potential energies resonate at different angular frequencies as a result of the fact that the driven damped oscillator is not conservative because energy is continually exchanged between the oscillator and the driving force system in addition to the energy dissipation due to the damping.

When $\omega \sim \omega_0 >> \Gamma$, then the power equation simplifies since

$$\left(\omega_0^2 - \omega^2\right) = \left(\omega_0 + \omega\right)\left(\omega_0 - \omega\right) \approx 2\omega_0\left(\omega_0 - \omega\right) \tag{3.84}$$

Therefore

$$\langle P \rangle \simeq \frac{F_0^2}{8m} \frac{\Gamma}{\left(\omega_0 - \omega\right)^2 + \left(\frac{\Gamma}{2}\right)^2} \tag{3.85}$$

This is called the Lorentzian or Breit-Wigner shape. The half power points are at a frequency difference from resonance of $\pm\Delta\omega$ where

$$\Delta\omega = |\omega_0 - \omega| = \pm\frac{\Gamma}{2} \tag{3.86}$$

Thus the *full width at half maximum of the Lorentzian curve equals* Γ. Note that the Lorentzian has a narrower peak but much wider tail relative to a Gaussian shape. At the peak of the absorbed power, the absorptive amplitude can be written as

$$A_{abs}(\omega = \omega_0) = \frac{F_0}{m}\frac{Q}{\omega_0^2} \tag{3.87}$$

That is, the peak amplitude increases with increase in Q. This explains the classic comedy scene where the soprano shatters the crystal glass because the highest quality crystal glass has a high Q which leads to a large amplitude oscillation when she sings on resonance.

The mean lifetime τ of the free linearly-damped harmonic oscillator, that is, the time for the energy of free oscillations to decay to $1/e$, was shown to be related to the damping coefficient Γ by

$$\tau = \frac{1}{\Gamma} \tag{3.88}$$

Therefore we have the **classical uncertainty principle for the linearly-damped harmonic oscillator** that the measured full-width at half maximum of the energy resonance curve for forced oscillation and the mean life for decay of the energy of a free linearly-damped oscillator are related by

$$\tau\Gamma = 1 \tag{3.89}$$

This relation is correct only for a linearly-damped harmonic system. Comparable relations between the lifetime and damping width exist for different forms of damping.

One can demonstrate the above line width and decay time relationship using an acoustically driven electric guitar string. Similarly, the width of the electromagnetic radiation is related to the lifetime for decay of atomic or nuclear electromagnetic decay. This classical uncertainty principle is exactly the same as the one encountered in quantum physics due to wave-particle duality. In nuclear physics it is difficult to measure the lifetime of states when $\tau < 10^{-13}s$. For shorter lifetimes the value of Γ can be determined from the shape of the resonance curve which can be measured directly when the damping is large.

3.1 *Example: Harmonically-driven series RLC circuit*

The harmonically-driven, resonant, series RLC circuit, is encountered frequently in AC circuits. Kirchhoff's Rules applied to the series RLC circuit lead to the differential equation

$$L\ddot{q} + R\dot{q} + \frac{q}{C} = V_0 \sin\omega t$$

where q is charge, L is the inductance, C is the capacitance, R is the resistance, and the applied voltage across the circuit is $V(\omega) = V_0 \sin\omega t$. The linearity of the network allows use of the phasor approach which assumes that the current $I = I_0 e^{i\omega t}$, the voltage $V = V_0 e^{i(\omega t + \delta)}$, and the impedance is a complex number

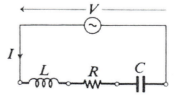

$Z = \frac{V_0}{I_0} e^{i\delta}$ *where δ is the phase difference between the voltage and the current. For this circuit the impedance is given by*

$$Z = R + i \left(\omega L - \frac{1}{\omega C} \right)$$

Because of the phases involved in this RLC circuit, at resonance the maximum voltage across the resistor occurs at a frequency of $\omega_R = \omega_0$, across the capacitor the maximum voltage occurs at a frequency $\omega_C^2 = \omega_0^2 - \frac{R^2}{2L^2}$, and across the inductor L the maximum voltage occurs at a frequency $\omega_L^2 = \frac{\omega_0^2}{1 - \frac{R^2}{2L^2}}$, where $\omega_0^2 = \frac{1}{LC}$ is the resonance angular frequency when $R = 0$. Thus these resonance frequencies differ when $R > 0$.

3.7   Wave equation

Wave motion is a ubiquitous feature in nature. Mechanical wave motion is manifest by transverse waves on fluid surfaces, longitudinal and transverse seismic waves travelling through the Earth, and vibrations of mechanical structures such as suspended cables. Acoustical wave motion occurs on the stretched strings of the violin, as well as the cavities of wind instruments. Wave motion occurs for deformable bodies where elastic forces acting between the nearest-neighbor atoms of the body exert time-dependent forces on one another. Electromagnetic wave motion includes wavelengths ranging from $10^5 m$ radiowaves, to $10^{-13} m$ γ-rays. Matter waves are a prominent feature of quantum physics. All these manifestations of waves exhibit the same general features of wave motion. Chapter 14 will introduce the collective modes of motion, called the normal modes, of coupled, many-body, linear oscillators which act as independent modes of motion. The basic elements of wavemotion are introduced at this juncture because the equations of wave motion are simple, and wave motion features prominently in several chapters throughout this book.

Consider a travelling wave in one dimension for a linear system. If the wave is moving, then the wave function $\Psi(x,t)$ describing the shape of the wave, is a function of both x and t. The instantaneous amplitude of the wave $\Psi(x,t)$ could correspond to the transverse displacement of a wave on a string, the longitudinal amplitude of a wave on a spring, the pressure of a longitudinal sound wave, the transverse electric or magnetic fields in an electromagnetic wave, a matter wave, etc. If the wave train maintains its shape as it moves, then one can describe the wave train by the function $f(\phi)$ where the coordinate ϕ is measured relative to the shape of the wave, that is, it could correspond to the phase of a crest of the wave. Consider that $f(\phi = 0)$, corresponds to a constant phase, e.g. the peak of the travelling pulse, then assuming that the wave travels at a phase velocity v in the x direction and the peak is at $x = 0$ for $t = 0$, then it is at $x = vt$ at time t. That is, a point with phase ϕ fixed with respect to the waveform shape of the wave profile $f(\phi)$ moves in the $+x$ direction for $\phi = x - vt$ and in $-x$ direction for $\phi = x + vt$.

General wave motion can be described by solutions of a wave equation. The wave equation can be written in terms of the spatial and temporal derivatives of the wave function $\Psi(xt)$. Consider the first partial derivatives of $\Psi(xt) = f(x \mp vt) = f(\phi)$.

$$\frac{\partial \Psi}{\partial x} = \frac{d\Psi}{d\phi} \frac{\partial \phi}{\partial x} = \frac{d\Psi}{d\phi} \tag{3.90}$$

and

$$\frac{\partial \Psi}{\partial t} = \frac{d\Psi}{d\phi} \frac{\partial \phi}{\partial t} = \mp v \frac{d\Psi}{d\phi} \tag{3.91}$$

Factoring out $\frac{d\Psi}{d\phi}$ for the first derivatives gives

$$\frac{\partial \Psi}{\partial t} = \mp v \frac{\partial \Psi}{\partial x} \tag{3.92}$$

The sign in this equation depends on the sign of the wave velocity making it not a generally useful formula.

Consider the second derivatives

$$\frac{\partial^2 \Psi}{\partial x^2} = \frac{d^2 \Psi}{d\phi^2} \frac{\partial \phi}{\partial x} = \frac{d^2 \Psi}{d\phi^2} \tag{3.93}$$

and

$$\frac{\partial^2 \Psi}{\partial t^2} = \frac{d^2 \Psi}{d\phi^2} \frac{\partial \phi}{\partial t} = +v^2 \frac{d^2 \Psi}{d\phi^2} \tag{3.94}$$

Factoring out $\frac{d^2\Psi}{d\phi^2}$ gives

$$\frac{\partial^2 \Psi}{\partial x^2} = \frac{1}{v^2}\frac{\partial^2 \Psi}{\partial t^2} \tag{3.95}$$

This *wave equation in one dimension for a linear system* is independent of the sign of the velocity. There are an infinite number of possible shapes of waves both travelling and standing in one dimension, all of these must satisfy this one-dimensional wave equation. The converse is that any function that satisfies this one dimensional wave equation must be a wave in this one dimension.

The *Wave Equation in three dimensions* is

$$\nabla^2 \Psi \equiv \frac{\partial^2 \Psi}{\partial x^2} + \frac{\partial^2 \Psi}{\partial y^2} + \frac{\partial^2 \Psi}{\partial z^2} = \frac{1}{v^2}\frac{\partial^2 \Psi}{\partial t^2} \tag{3.96}$$

There are an unlimited number of possible solutions Ψ to this wave equation, any one of which corresponds to a wave motion with velocity v.

The Wave Equation is applicable to all manifestations of wave motion, both transverse and longitudinal, for linear systems. That is, it applies to waves on a string, water waves, seismic waves, sound waves, electromagnetic waves, matter waves, etc. If it can be shown that a wave equation can be derived for any system, discrete or continuous, then this is equivalent to proving the existence of waves of any waveform, frequency, or wavelength travelling with the phase velocity given by the wave equation.[Cra65]

3.8 Travelling and standing wave solutions of the wave equation

The wave equation can exhibit both travelling and standing-wave solutions. Consider a one-dimensional travelling wave with velocity v having a specific wavenumber $k \equiv \frac{2\pi}{\lambda}$. Then the travelling wave is best written in terms of the phase of the wave as

$$\Psi(x,t) = A(k)e^{i\frac{2\pi}{\lambda}(x\mp vt)} = A(k)e^{i(kx\mp\omega t)} \tag{3.97}$$

where the wave number $k \equiv \frac{2\pi}{\lambda}$, with λ being the wave length, and angular frequency $\omega \equiv kv$. This particular solution satisfies the wave equation and corresponds to a *travelling wave* with *phase velocity* $v = \frac{\omega_n}{k_n}$ in the positive or negative direction x depending on whether the sign is negative or positive. Assuming that the superposition principle applies, then the superposition of these two particular solutions of the wave equation can be written as

$$\Psi(x,t) = A(k)(e^{i(kx-\omega t)} + e^{i(kx+\omega t)}) = A(k)e^{ikx}(e^{-i\omega t} + e^{i\omega t}) = 2A(k)e^{ikx}\cos\omega t \tag{3.98}$$

Thus the superposition of two identical single wavelength *travelling waves* propagating in opposite directions can correspond to a *standing wave* solution. Note that a standing wave is identical to a stationary normal mode of the system discussed in chapter 14. This transformation between standing and travelling waves can be reversed, that is, the superposition of two standing waves, i.e. normal modes, can lead to a travelling wave solution of the wave equation.

Discussion of waveforms is simplified when using either of the following two limits.

1) The time dependence of the waveform at a given location $x = x_0$ which can be expressed using a Fourier decomposition, appendix *I.2*, of the time dependence as a function of angular frequency $\omega = n\omega_0$.

$$\Psi(x_0,t) = \sum_{n=-\infty}^{\infty} A_n e^{in(k_0 x_0 - \omega_0 t)} = \sum_{n=-\infty}^{\infty} B_n(x_0) e^{-in\omega_0 t} \tag{3.99}$$

2) The spatial dependence of the waveform at a given instant $t = t_0$ which can be expressed using a Fourier decomposition of the spatial dependence as a function of wavenumber $k = nk_0$

$$\Psi(x,t_0) = \sum_{n=-\infty}^{\infty} A_n e^{in(k_0 x - \omega_1 t_0)} = \sum_{n=-\infty}^{\infty} C_n(t_0) e^{ink_0 x} \tag{3.100}$$

The above is applicable both to discrete, or continuous linear oscillator systems, e.g. waves on a string.

In summary, stationary normal modes of a system are obtained by a superposition of travelling waves travelling in opposite directions, or equivalently, travelling waves can result from a superposition of stationary normal modes.

3.9   Waveform analysis

3.9.1   Harmonic decomposition

As described in appendix I, when superposition applies, then a Fourier series decomposition of the form 3.101 can be made of any periodic function where

$$F(t) = \sum_{n=1}^{N} \alpha_n \cos(n\omega_0 t + \phi_n) \qquad (3.101)$$

A more general Fourier Transform can be made for an aperiodic function where

$$F(t) = \int \alpha(\omega) \cos(\omega t + \phi(\omega)) dt \qquad (3.102)$$

Any linear system that is subject to the forcing function $F(t)$, has an output that can be expressed as a linear superposition of the solutions of the individual harmonic components of the forcing function. Fourier analysis of periodic waveforms in terms of harmonic trigonometric functions plays a key role in describing oscillatory motion in classical mechanics and signal processing for linear systems. Fourier's theorem states that any arbitrary forcing function $F(t)$ can be decomposed into a sum of harmonic terms. As a consequence two equivalent representations can be used to describe signals and waves; the first is in the time domain which describes the time dependence of the signal. The second is in the frequency domain which describes the frequency decomposition of the signal. Fourier analysis relates these equivalent representations.

For example, the superposition of two equal intensity harmonic oscillators in the time domain is given by

$$\begin{aligned} y(t) &= A\cos(\omega_1 t) + A\cos(\omega_2 t) \\ &= 2A\cos\left[\left(\frac{\omega_1 + \omega_2}{2}\right)t\right]\cos\left[\left(\frac{\omega_1 - \omega_2}{2}\right)t\right] \quad (3.103) \end{aligned}$$

which leads to the phenomenon of beats as illustrated for both the time domain and frequency domain in figure 3.11.

3.9.2   The free linearly-damped linear oscillator

The response of the free, linearly-damped, linear oscillator is one of the most frequently encountered waveforms in science and thus it is useful to investigate the Fourier transform of this waveform. The waveform amplitude for the underdamped case, shown in figure 3.5, is given by equation (3.35), that is

$$\begin{aligned} f(t) &= Ae^{-\frac{\Gamma}{2}t}\cos(\omega_1 t - \delta) \qquad & t \geq 0 \quad (3.104) \\ f(t) &= 0 \qquad & t < 0 \quad (3.105) \end{aligned}$$

where $\omega_1^2 = \omega_0^2 - \left(\frac{\Gamma}{2}\right)^2$ and where ω_0 is the angular frequency of the undamped system. The Fourier transform is given by

$$G(\omega) = \frac{\omega_0}{\left(\omega^2 - \omega_1^2\right)^2 + (\Gamma\omega)^2}\left[\left(\omega^2 - \omega_1^2\right) - i\Gamma\omega\right] \qquad (3.106)$$

which is complex and has the famous Lorentz form.

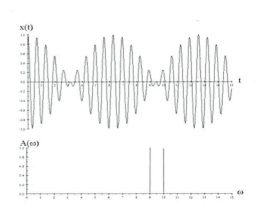

Figure 3.11: The time and frequency representations of a system exhibiting beats.

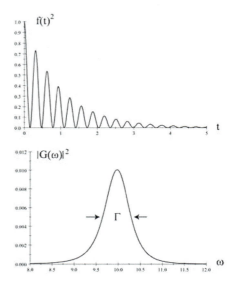

Figure 3.12:   The intensity $f(t)^2$ and Fourier transform $|G(\omega)|^2$ of the free linearly-underdamped harmonic oscillator with $\omega_0 = 10$ and damping $\Gamma = 1$.

The intensity of the wave gives

$$|f(t)|^2 = A^2 e^{-\Gamma t} \cos^2(\omega_1 t - \delta) \qquad (3.107)$$

$$|G(\omega)|^2 = \frac{\omega_0^2}{(\omega^2 - \omega_1^2)^2 + (\Gamma \omega)^2} \qquad (3.108)$$

Note that since the average over 2π of $\cos^2 = \frac{1}{2}$, then the average over the $\cos^2(\omega_1 t - \delta)$ term gives the intensity $I(t) = \frac{A^2}{2} e^{-\Gamma t}$ which has a mean lifetime for the decay of $\tau = \frac{1}{\Gamma}$. The $|G(\omega)|^2$ distribution has the classic Lorentzian shape, shown in figure 3.12, which has a full width at half-maximum, FWHM, equal to Γ. Note that $G(\omega)$ is complex and thus one also can determine the phase shift δ which is given by the ratio of the imaginary to real parts of equation 3.105, i.e. $\tan \delta = \frac{\Gamma \omega}{(\omega^2 - \omega_1^2)}$.

The mean lifetime of the exponential decay of the intensity can be determined either by measuring τ from the time dependence, or measuring the FWHM $\Gamma = \frac{1}{\tau}$ of the Fourier transform $|G(\omega)|^2$. In nuclear and atomic physics excited levels decay by photon emission with the wave form of the free linearly-damped, linear oscillator. Typically the mean lifetime τ usually can be measured when $\tau \gtrsim 10^{-12} s$ whereas for shorter lifetimes the radiation width Γ becomes sufficiently large to be measured. Thus the two experimental approaches are complementary.

3.9.3 Damped linear oscillator subject to an arbitrary periodic force

Fourier's theorem states that any arbitrary forcing function $F(t)$ can be decomposed into a sum of harmonic terms. Consider the response of a damped linear oscillator to an arbitrary periodic force.

$$F(t) = \sum_{n=0}^{N} \alpha_n F_0(\omega_n) \cos(\omega_n t + \delta_n) \qquad (3.109)$$

For each harmonic term ω_n the response of a linearly-damped linear oscillator to the forcing function $F(t) = F_0(\omega) \cos(\omega_n t)$ is given by equation $(3.65 - 67)$ to be

$$
\begin{aligned}
x(t)_{Total} &= x(t)_T + x(t)_S \\
&= \frac{F_0(\omega_n)}{m} \left[e^{-\frac{\Gamma}{2} t} \cos(\omega_1 t - \delta_n) + \frac{1}{\sqrt{(\omega_0^2 - \omega_n^2)^2 + (\Gamma \omega_n)^2}} \cos(\omega_n t - \delta_n) \right]
\end{aligned} \qquad (3.110)
$$

The amplitude is obtained by substituting into (3.110) the derived values $\frac{F_0(\omega_n)}{m}$ from the Fourier analysis.

3.2 Example: Vibration isolation

Frequently it is desired to isolate instrumentation from the influence of horizontal and vertical external vibrations that exist in the environment. One arrangement to achieve this isolation is to mount a heavy base of mass m on weak springs of spring constant k plus weak damping. The response of this system is given by equation 3.109 which exhibits a resonance at the angular frequency $\omega_R^2 = \omega_0^2 - 2\left(\frac{\Gamma}{2}\right)^2$ associated with each resonant frequency ω_0 of the system. For each resonant frequency the system amplifies the vibrational amplitude for angular frequencies close to resonance that is, below $\sqrt{2} \, \omega_0$, while it attenuates the vibration roughly by a factor of $\left(\frac{\omega_0}{\omega}\right)^2$ at higher frequencies. To avoid the amplification near the resonance it is necessary to make ω_0 very much smaller than the frequency range of the vibrational spectrum and have a moderately high Q value. This is achieved by use a very heavy base and weak spring constant so that ω_0 is very small. A typical table may have the resonance frequency at $0.5 Hz$ which is well below typical perturbing vibrational frequencies, and thus the table attenuates the vibration by 99% at $5 Hz$ and even more attenuation for higher frequency perturbations. This principle is used extensively in design of vibration-isolation tables for optics or microbalance equipment.

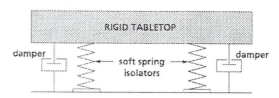

Seismic isolation of an optical bench.

3.10   Signal processing

It has been shown that the response of the linearly-damped linear oscillator, subject to any arbitrary periodic force, can be calculated using a frequency decomposition, (Fourier analysis), of the force, appendix I. The response also can be calculated using a time-ordered discrete-time sampling of the pulse shape; that is, the Green's function approach, appendix I. The linearly-damped, linear oscillator is the simplest example of a linear system that exhibits both resonance and frequency-dependent response. Typically physical linear systems exhibit far more complicated response functions having multiple resonances. For example, an automobile suspension system involves four wheels and associated springs plus dampers allowing the car to rock sideways, or forward and backward, in addition to the up-down motion, when subject to the forces produced by a rough road. Similarly a suspension bridge or aircraft wing can twist as well as bend due to air turbulence, or a building can undergo complicated oscillations due to seismic waves. An acoustic system exhibits similar complexity. Signal analysis and signal processing is of pivotal importance to elucidating the response of complicated linear systems to complicated periodic forcing functions. Signal processing is used extensively in engineering, acoustics, and science.

The response of a low-pass filter, such as an R-C circuit or a coaxial cable, to a input square wave, shown in figure 3.13, provides a simple example of the relative advantages of using the complementary Fourier analysis in the frequency domain, or the Green's discrete-function analysis in the time domain. The response of a repetitive square-wave input signal is shown in the time domain plus the Fourier transform to the frequency domain. The middle curves show the time dependence for the response of the low-pass filter to an impulse $I(t)$ and the corresponding Fourier transform $H(\omega)$. The output of the low-pass filter can be calculated by folding the input square wave and impulse time dependence in the time domain as shown on the left or by folding of their Fourier transforms shown on the right. Working in the frequency domain the response of linear mechanical systems, such as an automobile suspension or a musical instrument, as well as linear electronic signal processing systems such as amplifiers, loudspeakers and microphones, can be treated as black boxes having a certain **transfer function** $H(\omega, \phi)$ describing the gain and phase shift versus frequency. That is, the output wave frequency decomposition is

$$G(\omega)_{output} = H(\omega, \phi) \cdot G(\omega)_{input} \qquad (3.111)$$

Working in the time domain, the the low-pass system has an **impulse response** $I(t) = e^{-\frac{t}{\tau}}$, which is the Fourier transform of the transfer function $H(\omega, \phi)$. In the time domain

$$y(t)_{output} = \int_{-\infty}^{\infty} x(\tau) \cdot I(t - \tau) d\tau \qquad (3.112)$$

This is shown schematically in figure 3.13. The Fourier transformation connects the three quantities in the time domain with the corresponding three in the frequency domain. For example, the impulse response of the low-pass filter has a fall time of τ which is related by a Fourier transform to the width of the transfer function. Thus the time and frequency domain approaches are closely related and give the same result for the output signal for the low-pass filter to the applied square-wave input signal. The result is that the higher-frequency components are attenuated leading to slow rise and fall times in the time domain.

Analog signal processing and Fourier analysis were the primary tools to analyze and process all forms of periodic motion during the 20th century. For example, musical instruments, mechanical systems, electronic circuits, all employed resonant systems to enhance the desired frequencies and suppress the undesirable frequencies and the signals could be observed using analog oscilloscopes. The remarkable development of computing has enabled use of digital signal processing leading to a revolution in signal processing that has had a profound impact on both science and engineering. The digital oscilloscope, which can sample at frequencies above $10^9 Hz$, has replaced the analog oscilloscope because it allows sophisticated analysis of each individual signal that was not possible using analog signal processing. For example, the analog approach in nuclear physics used tiny analog electric signals, produced by many individual radiation detectors, that were transmitted hundreds of meters via carefully shielded and expensive coaxial cables to the data room where the signals were amplified and signal processed using analog filters to maximize the signal to noise in order to separate the signal from the background noise. Stray electromagnetic radiation picked up via the cables significantly degraded the signals. The performance and limitations of the analog electronics severely restricted the pulse processing capabilities. Digital signal processing has rapidly replaced analog signal processing.

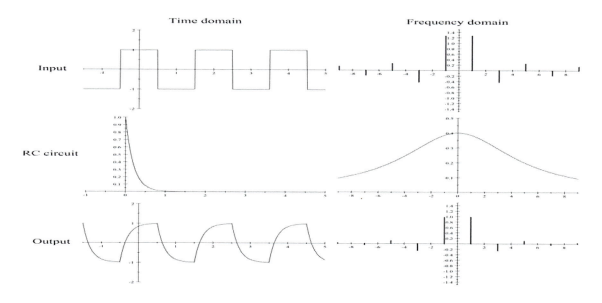

Figure 3.13: Response of an *RC* electrical circuit to an input square wave. The upper row shows the time and the exponential-form frequency representations of the square-wave input signal. The middle row gives the impulse response, and corresponding transfer function for the *RC* circuit. The bottom row shows the corresponding output properties in both the time and frequency domains

Analog to digital detector circuits are built directly into the electronics for each individual detector so that only digital information needs to be transmitted from each detector to the analysis computers. Computer processing provides unlimited and flexible processing capabilities for the digital signals greatly enhancing the response and sensitivity of our detector systems. Digital CD and DVD disks are common application of digital signal processing.

3.11 Wave propagation

Wave motion typically involves a packet of waves encompassing a finite number of wave cycles. Information in a wave only can be transmitted by starting, stopping, or modulating the amplitude of a wave train, which is equivalent to forming a wave packet. For example, a musician will play a note for a finite time, and this wave train propagates out as a wave packet of finite length. You have no information as to the frequency and amplitude of the sound prior to the wave packet reaching you, or after the wave packet has passed you. The velocity of the wavelets contained within the wave packet is called the **phase velocity**. For a dispersive system the phase velocity of the wavelets contained within the wave packet is frequency dependent and the shape of the wave packet travels at the **group velocity** which usually differs from the phase velocity. If the shape of the wave packet is time dependent, then neither the phase velocity, which is the velocity of the wavelets, nor the group velocity, which is the velocity of an instantaneous point fixed to the shape of the wave packet envelope, represent the actual velocity of the overall wavepacket.

A third wavepacket velocity, the **signal velocity**, is defined to be the velocity of the leading edge of the energy distribution, and corresponding information content, of the wave packet. For most linear systems the shape of the wave packet is not time dependent and then the group and signal velocities are identical. However, the group and signal velocities can be very different for non-linear systems as discussed in chapter 4.7. Note that even when the phase velocity of the waves within the wave packet travels faster than the group velocity of the shape, or the signal velocity of the energy content of the envelope of the wave packet, the information contained in a wave packet is only manifest when the wave packet envelope reaches the detector and this energy and information travel at the signal velocity. The modern ideas of wave propagation, including Hamilton's concept of group velocity, were developed by Lord Rayleigh when applied to the theory of sound[Ray1887]. The concept of phase, group, and signal velocities played a major role in discussion of electromagnetic waves as well as de Broglie's development of wave-particle duality in quantum mechanics.

3.11.1   Phase, group, and signal velocities of wave packets

The concepts of wave packets, as well as their phase, group, and signal velocities, are of considerable importance for propagation of information and other manifestations of wave motion in science and engineering. This importance warrants further discussion at this juncture.

Consider a particular k, ω, component of a one-dimensional wave,

$$q(x,t) = E e^{i(kx \pm \omega t)} \tag{3.113}$$

The argument of the exponential is called the **phase** ϕ of the wave where

$$\phi \equiv kx - \omega t \tag{3.114}$$

If we move along the x axis at a velocity such that the phase is constant then we perceive a stationary pattern in this moving frame. The velocity of this wave is called the **phase velocity.** To ensure constant phase requires that ϕ is constant, or assuming real k and ω

$$\omega dt = k dx \tag{3.115}$$

Therefore the **phase velocity** is defined to be

$$v_{phase} = \frac{\omega}{k} \tag{3.116}$$

The velocity discussed so far is just the phase velocity of the individual wavelets at the carrier frequency. If k or ω are complex then one must take the real parts to ensure that the velocity is real.

If the phase velocity of a wave is dependent on the wavelength, that is, $v_{phase}(k)$, then the system is said to be **dispersive** in that the wave is dispersed according the wavelength. The simplest illustration of dispersion is the refraction of light in glass prism which leads to dispersion of the light into the spectrum of wavelengths. Dispersion leads to development of wave packets that travel at group and signal velocities that usually differ from the phase velocity. To illustrate this behavior, consider two equal amplitude travelling waves having slightly different wave number k and angular frequency ω. Superposition of these waves gives

$$
\begin{aligned}
q(x,t) &= A(e^{i[kx - \omega t]} + e^{i[(k+\Delta k)x - (\omega + \Delta \omega)t]}) \\
&= A e^{i[(k+\frac{\Delta k}{2})x - (\omega + \frac{\Delta \omega}{2})t]} \cdot \{ e^{-i[\frac{\Delta k}{2}x - \frac{\Delta \omega}{2}t]} + e^{i[\frac{\Delta k}{2}x - \frac{\Delta \omega}{2}t]} \} \\
&= 2A e^{i[(k+\frac{\Delta k}{2})x - (\omega + \frac{\Delta \omega}{2})t]} \cos[\frac{\Delta k}{2}x - \frac{\Delta \omega}{2}t]
\end{aligned}
\tag{3.117}
$$

This corresponds to a wave with the average carrier frequency modulated by the cosine term which has a wavenumber of $\frac{\Delta k}{2}$ and angular frequency $\frac{\Delta \omega}{2}$, that is, this is the usual example of beats. The cosine term modulates the average wave producing wave packets as shown in figure 3.11. The velocity of these wave packets is called the **group velocity** given by requiring that the phase of the modulating term is constant, that is

$$\frac{\Delta k}{2} dx = \frac{\Delta \omega}{2} dt \tag{3.118}$$

Thus the **group velocity** is given by

$$v_{group} = \frac{dx}{dt} = \frac{\Delta \omega}{\Delta k} \tag{3.119}$$

If dispersion is present then the group velocity $v_{group} = \frac{\Delta \omega}{\Delta k}$ does not equal the phase velocity $v_{phase} = \frac{\omega}{k}$.

Expanding the above example to superposition of n waves gives

$$q(x,t) = \sum_{r=1}^{n} A_r e^{i(k_r x \pm \omega_r t)} \tag{3.120}$$

In the event that $n \to \infty$ and the frequencies are continuously distributed, then the summation is replaced by an integral

$$q(x,t) = \int_{-\infty}^{\infty} A(k) e^{i(kx \pm \omega t)} dk \tag{3.121}$$

where the factor $A(k)$ represents the distribution amplitudes of the component waves, that is the spectral decomposition of the wave. This is the usual Fourier decomposition of the spatial distribution of the wave.

Consider an extension of the linear superposition of two waves to a well defined wave packet where the amplitude is nonzero only for a small range of wavenumbers $k_0 \pm \Delta k$.

$$q(x,t) = \int_{k_0 - \Delta k}^{k_0 + \Delta k} A(k) e^{i(kx - \omega t)} dk \tag{3.122}$$

This functional shape is called a wave packet which only has meaning if $\Delta k << k_0$. The angular frequency can be expressed by making a Taylor expansion around k_0

$$\omega(k) = \omega(k_0) + \left(\frac{d\omega}{dk}\right)_{k_0} (k - k_0) + ... \tag{3.123}$$

For a linear system the phase then reduces to

$$kx - \omega t = (k_0 x - \omega_0 t) + (k - k_0)x - \left(\frac{d\omega}{dk}\right)_{k_0} (k - k_0)t \tag{3.124}$$

The summation of terms in the exponent given by 3.124 leads to the amplitude 3.122 having the form of a product where the integral becomes

$$q(x,t) = e^{i(k_0 x - \omega_0 t)} \int_{k_0 - \Delta k}^{k_0 + \Delta k} A(k) e^{i(k - k_0)[x - \left(\frac{d\omega}{dk}\right)_{k_0} t]} dk \tag{3.125}$$

The integral term modulates the $e^{i(k_0 x - \omega_0 t)}$ first term.

The group velocity is defined to be that for which the phase of the exponential term in the integral is constant. Thus

$$v_{group} = \left(\frac{d\omega}{dk}\right)_{k_0} \tag{3.126}$$

Since $\omega = k v_{phase}$ then

$$v_{group} = v_{phase} + k \frac{\partial v_{phase}}{\partial k} \tag{3.127}$$

For non-dispersive systems the phase velocity is independent of the wave number k or angular frequency ω and thus $v_{group} = v_{phase}$. The case discussed earlier, equation (3.103), for beating of two waves gives the same relation in the limit that $\Delta \omega$ and Δk are infinitessimal.

The group velocity of a wave packet is of physical significance for dispersive media where $v_{group} = \left(\frac{d\omega}{dk}\right)_{k_0} \neq \frac{\omega}{k} = v_{phase}$. Every wave train has a finite extent and thus we usually observe the motion of a group of waves rather than the wavelets moving within the wave packet. In general, for non-linear dispersive systems the derivative $\frac{\partial v_{phase}}{\partial k}$ can be either positive or negative and thus in principle the group velocity can either be greater than, or less than, the phase velocity. Moreover, if the group velocity is frequency dependent, that is, when group velocity dispersion occurs, then the overall shape of the wave packet is time dependent and thus the speed of a specific relative location defined by the shape of the envelope of the wave packet does not represent the signal velocity of the wave packet. Brillouin showed that the distribution of the energy, and corresponding information content, for any wave packet, travels at the signal velocity which can be different from the group velocity if the shape of the envelope of the wave packet is time dependent. For electromagnetic waves one has the possibility that the group velocity $v_{group} > v_{phase} = c$. In 1914 Brillouin[Bri14][Bri60] showed that the signal velocity of electromagnetic waves, defined by the leading edge of the time-dependent envelope of the wave packet, never exceeds c even though the group velocity corresponding to the velocity of the instantaneous shape of the wave packet may exceed c. Thus, there is no violation of Einstein's fundamental principle of relativity that the velocity of an electromagnetic wave cannot exceed c.

3.3 *Example: Water waves breaking on a beach*

The concepts of phase and group velocity are illustrated by the example of water waves moving at velocity v incident upon a straight beach at an angle α to the shoreline. Consider that the wavepacket comprises many wavelengths of wavelength λ. During the time it takes the wave to travel a distance λ, the point where the crest of one wave breaks on the beach travels a distance $\frac{\lambda}{\cos \alpha}$ along beach. Thus the phase velocity of the crest of the one wavelet in the wave packet is

$$v_{phase} = \frac{v}{\cos \alpha}$$

The velocity of the wave packet along the beach equals

$$v_{group} = v \cos \alpha$$

Note that for the wave moving parallel to the beach $\alpha = 0$ and $v_{phase} = v_{group} = v$. However, for $\alpha = \frac{\pi}{2}$ $v_{phase} \to \infty$ and $v_{group} \to 0$. In general for waves breaking on the beach

$$v_{phase} v_{group} = v^2$$

The same behavior is exhibited by surface waves bouncing off the sides of the Erie canal, sound waves in a trombone, and electromagnetic waves transmitted down a rectangular wave guide. In the latter case the phase velocity exceeds the velocity of light c in apparent violation of Einstein's theory of relativity. However, the information travels at the signal velocity which is less than c.

3.4 *Example: Surface waves for deep water*

In the "Theory of Sound"[Ray1887] Rayleigh discusses the example of surface waves for water. He derives a dispersion relation for the phase velocity v_{phase} and wavenumber k which are related to the density ρ, depth l, gravity g, and surface tension T, by

$$\omega^2 = gk + \frac{Tk^3}{\rho} \tanh(kl)$$

For deep water where the wavelength is short compared with the depth, that is $kl >> 1$, then $tanh(kl) \to 1$ and the dispersion relation is given approximately by

$$\omega^2 = gk + \frac{Tk^3}{\rho}$$

For long surface waves for deep water, that is, small k, then the gravitational first term in the dispersion relation dominates and the group velocity is given by

$$v_{group} = \left(\frac{d\omega}{dk} \right) = \frac{1}{2} \sqrt{\frac{g}{k}} = \frac{1}{2} \frac{\omega}{k} = \frac{v_{phase}}{2}$$

That is, the group velocity is half of the phase velocity. Here the wavelets are building at the back of the wave packet, progress through the wave packet and dissipate at the front. This can be demonstrated by dropping a pebble into a calm lake. It will be seen that the surface disturbance comprises a wave packet moving outwards at the group velocity with the individual waves within the wave packet expanding at twice the group velocity of the wavepacket, that is, they are created at the inner radius of the wave packet and disappear at the outer radius of the wave packet.

For small wavelength ripples, where k is large, then the surface tension term dominates and the dispersion relation is approximately given by

$$\omega^2 \simeq \frac{Tk^3}{\rho}$$

leading to a group velocity of

$$v_{group} = \left(\frac{d\omega}{dk} \right) = \frac{3}{2} v_{phase}$$

Here the group velocity exceeds the phase velocity and wavelets are building at the front of the wave packet and dissipate at the back. Note that for this linear system, the Brillion signal velocity equals the group velocity for both gravity and surface tension waves for deep water.

3.5 *Example: Electromagnetic waves in ionosphere*

The response to radio waves, incident upon a free electron plasma in the ionosphere, provides an excellent example that involves cut-off frequency, complex wavenumber k, as well as the phase, group, and signal velocities. Maxwell's equations give the most general wave equation for electromagnetic waves to be

$$\nabla^2 \mathbf{E} - \varepsilon\mu \frac{\partial^2 \mathbf{E}}{\partial t^2} = \mu \frac{\partial \mathbf{j}_{free}}{\partial t} + \nabla \cdot \left(\frac{\rho_{free}}{\varepsilon}\right)$$

$$\nabla^2 \mathbf{H} - \mu\varepsilon \frac{\partial^2 \mathbf{H}}{\partial t^2} = -\nabla \times \mathbf{j}_{free}$$

where ρ_{free} and $\mathbf{j}_{free}$ are the unbound charge and current densities. The effect of the bound charges and currents are absorbed into ε and μ. Ohm's Law can be written in terms of the electrical conductivity σ which is a constant

$$\mathbf{j} = \sigma \mathbf{E}$$

Assuming Ohm's Law plus assuming $\rho_{free} = 0$, in the plasma gives the relations

$$\nabla^2 \mathbf{E} - \varepsilon\mu \frac{\partial^2 \mathbf{E}}{\partial t^2} - \sigma\mu \frac{\partial \mathbf{E}}{\partial t} = 0$$

$$\nabla^2 \mathbf{H} - \mu\varepsilon \frac{\partial^2 \mathbf{H}}{\partial t^2} - \sigma\mu \frac{\partial \mathbf{H}}{\partial t} = 0$$

The third term in both of these wave equations is a damping term that leads to a damped solution of an electromagnetic wave in a good conductor.

The solution of these damped wave equations can be solved by considering an incident wave

$$\mathbf{E} = E_o \hat{\mathbf{x}} e^{i(\omega t - kz)}$$

Substituting for $\mathbf{E}$ in the first damped wave equation gives

$$-k^2 + \omega^2 \varepsilon\mu - i\omega\sigma\mu = 0$$

That is

$$k^2 = \omega^2 \varepsilon\mu \left[1 - \frac{i\sigma}{\omega\varepsilon}\right]$$

In general k is complex, that is, it has real k_R and imaginary k_I parts that lead to a solution of the form

$$\mathbf{E} = E_o e^{-k_I z} e^{i(\omega t - k_R z)}$$

The first exponential term is an exponential damping term while the second exponential term is the oscillating term.

Consider that the plasma involves the motion of a bound damped electron, of charge q of mass m, bound in a one dimensional atom or lattice subject to an oscillatory electric field of frequency ω. Assume that the electromagnetic wave is travelling in the $\hat{z}$ direction with the transverse electric field in the $\hat{x}$ direction. The equation of motion of an electron can be written as

$$\ddot{\mathbf{x}} + \Gamma\dot{\mathbf{x}} + \omega_0^2 x = \hat{\mathbf{x}} q E_0 e^{i(\omega t - kz)}$$

where Γ is the damping factor. The instantaneous displacement of the oscillating charge equals

$$\mathbf{x} = \frac{q}{m} \frac{1}{(\omega_0^2 - \omega^2) + i\Gamma\omega} \hat{\mathbf{x}} E_0 e^{i(\omega t - kz)}$$

and the velocity is

$$\dot{\mathbf{x}} = \frac{q}{m} \frac{i\omega}{(\omega_0^2 - \omega^2) + i\Gamma\omega} \hat{\mathbf{x}} E_0 e^{i(\omega t - kz)}$$

Thus the instantaneous current density is given by

$$\mathbf{j} = Nq\dot{\mathbf{x}} = \frac{Nq^2}{m} \frac{i\omega}{(\omega_0^2 - \omega^2) + i\Gamma\omega} \hat{\mathbf{x}} E_0 e^{i(\omega t - kz)}$$

Therefore the electrical conductivity is given by

$$\sigma = \frac{Nq^2}{m} \frac{i\omega}{(\omega_0^2 - \omega^2) + i\Gamma\omega}$$

Let us consider only unbound charges in the plasma, that is let $\omega_0 = 0$. Then the conductivity is given by

$$\sigma = \frac{Nq^2}{m} \frac{i\omega}{i\Gamma\omega - \omega^2}$$

For a low density ionized plasma $\omega \gg \Gamma$ thus the conductivity is given approximately by

$$\sigma \approx -i\frac{Nq^2}{m\omega}$$

Since σ is pure imaginary, then $\mathbf{j}$ and $\mathbf{E}$ have a phase difference of $\frac{\pi}{2}$ which implies that the average of the Joule heating over a complete period is $\langle \mathbf{j} \cdot \mathbf{E} \rangle = 0$. Thus there is no energy loss due to Joule heating implying that the electromagnetic energy is conserved.

Substitution of σ into the relation for k^2

$$k^2 = \omega^2 \varepsilon\mu \left[1 - \frac{i\sigma}{\omega\varepsilon} \right] = \omega^2 \varepsilon\mu \left[1 - \frac{Nq^2}{\varepsilon m\omega^2} \right]$$

Define the Plasma oscillation frequency ω_P to be

$$\omega_p \equiv \sqrt{\frac{Nq^2}{\varepsilon m}}$$

then k^2 can be written as

$$k^2 = \omega^2 \varepsilon\mu \left[1 - \left(\frac{\omega_P}{\omega}\right)^2 \right] \qquad (\alpha)$$

For a low density plasma the dielectric constant $\kappa_E \simeq 1$ and the relative permeability $\kappa_B \simeq 1$ and thus $\varepsilon = \kappa_E \varepsilon_0 \simeq \varepsilon_0$ and $\mu = \kappa_B \mu_0 \simeq \mu_0$. The velocity of light in vacuum $c = \frac{1}{\sqrt{\varepsilon_0\mu_0}}$. Thus for low density equation α can be written as

$$\omega^2 = \omega_p^2 + c^2 k^2 \qquad (\beta)$$

Differentiation of equation β with respect to k gives $2\omega\frac{d\omega}{dk} = 2c^2 k$. That is, $v_{phase}v_{group} = c^2$ and the phase velocity is

$$v_{phase} = \sqrt{c^2 + \frac{\omega_p^2}{k^2}}$$

There are three cases to consider.

1) $\omega > \omega_P$:   For this case $\left[1 - \left(\frac{\omega_P}{\omega}\right)^2 \right] > 1$ and thus k is a pure real number. Therefore the electromagnetic wave is transmitted with a phase velocity that exceeds c while the group velocity is less than c.

2) $\omega < \omega_P$:   For this case $\left[1 - \left(\frac{\omega_P}{\omega}\right)^2 \right] < 1$ and thus k is a pure imaginary number. Therefore the electromagnetic wave is not transmitted in the ionosphere and is attenuated rapidly as $e^{-\left(\frac{\omega_p}{c}\right)z}$. However, since there are no Joule heating losses, then the electromagnetic wave must be complete reflected. Thus the Plasma oscillation frequency serves as a cut-off frequency. For this example the signal and group velocities are identical.

For the ionosphere $N = 10^{-11} electrons/m^3$, which corresponds to a Plasma oscillation frequency of $v = \omega_P/2\pi = 3MHz$. Thus electromagnetic waves in the AM waveband ($< 1.6MHz$) are totally reflected by the ionosphere and bounce repeatedly around the Earth, whereas for VHF frequencies above $3MHz$, the waves are transmitted and refracted passing through the atmosphere. Thus light is transmitted by the ionosphere. By contrast, for a good conductor like silver, the Plasma oscillation frequency is around $10^{16}Hz$ which is in the far ultraviolet part of the spectrum. Thus, all lower frequencies, such as light, are totally reflected by such a good conductor, whereas X-rays have frequencies above the Plasma oscillation frequency and are transmitted.

3.11.2 Fourier transform of wave packets

The relation between the time distribution and the corresponding frequency distribution, or equivalently, the spatial distribution and the corresponding wave-number distribution, are of considerable importance in discussion of wave packets and signal processing. It directly relates to the uncertainty principle that is a characteristic of all forms of wave motion. The relation between the time and corresponding frequency distribution is given via the Fourier transform discussed in appendix *I*. The following are two examples of the Fourier transforms of typical but rather different wavepacket shapes that are encountered often in science and engineering.

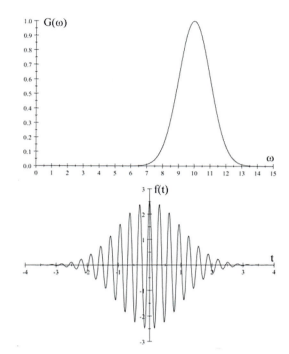

3.6 *Example: Fourier transform of a Gaussian wave packet:*

Assuming that the amplitude of the wave is a Gaussian wave packet shown in the adjacent figure where

$$G\left(\omega\right) = ce^{-\frac{(\omega-\omega_0)^2}{2\sigma_\omega^2}}$$

This *leads to the Fourier transform*

$$f\left(t\right) = c\sqrt{2\pi}\sigma_\omega e^{-\frac{\sigma_\omega^2 t^2}{2}}\cos\left(\omega_0 t\right)$$

Fourier transform of a Gaussian frequency distribution.

Note that the wavepacket has a standard deviation for the amplitude of the wavepacket of $\sigma_t = \frac{1}{\sigma_\omega}$, that is $\sigma_t \cdot \sigma_\omega = 1$. The Gaussian wavepacket results in the minimum product of the standard deviations of the frequency and time representations for a wavepacket. This has profound importance for all wave phenomena, and especially to quantum mechanics. Because matter exhibits wave-like behavior, the above property of wave packet leads to Heisenberg's Uncertainty Principle. For signal processing, it shows that if you truncate a wavepacket you will broaden the frequency distribution.

3.7 *Example: Fourier transform of a rectangular wave packet:*

Assume unity amplitude of the frequency distribution between $\omega_0 - \Delta\omega \leq \omega \leq \omega_0 + \Delta\omega$, that is, a single isolated square pulse of width τ that is described by the rectangular function Π defined as

$$\Pi(\omega) = \begin{cases} 1 & |\omega - \omega_0| < \Delta\omega \\ 0 & |\omega - \omega_0| > \Delta\omega \end{cases}$$

Then the Fourier transform us given by

$$f\left(t\right) = \left[\frac{\sin\Delta\omega t}{\Delta\omega t}\right]\cos\omega_0 t$$

That is, the transform of a rectangular wavepacket gives a cosine wave modulated by an unnormalized sinc function which is a nice example of a simple wave packet. That is, on the right hand side we have a wavepacket $\Delta t = \pm\frac{2\pi}{\Delta\omega}$ wide. Note that the product of the two measures of the widths $\Delta\omega \cdot \Delta t = \pm\pi$. Example I.2 considers a rectangular pulse of unity amplitude between $-\frac{\tau}{2} \leq t \leq \frac{\tau}{2}$ which resulted in a Fourier transform $G\left(\omega\right) = \tau\left(\frac{\sin\frac{\omega\tau}{2}}{\frac{\omega\tau}{2}}\right)$. That is, for a pulse of width $\Delta t = \pm\frac{\tau}{2}$ the frequency envelope has the first zero at $\Delta\omega = \pm\frac{\pi}{\tau}$. Note that this is the complementary system to the one considered here which has $\Delta\omega \cdot \Delta t = \pm\pi$ illustrating the symmetry of the Fourier transform and its inverse.

3.11.3   Wave-packet Uncertainty Principle

The Uncertainty Principle states that wavemotion exhibits a minimum product of the uncertainty in the simultaneously measured width in time of a wave packet, and the distribution width of the frequency decomposition of this wave packet. This was illustrated by the Fourier transforms of wave packets discussed above where it was shown the product of the widths is minimized for a Gaussian-shaped wave packet. The Uncertainty Principle implies that to make a precise measurement of the frequency of a sinusoidal wave requires that the wave packet be infinitely long. If the duration of the wave packet is reduced then the frequency distribution broadens. The crucial aspect needed for this discussion, is that, for the *amplitudes* of any wavepacket, the *standard deviations* $\sigma(t) = \sqrt{\langle t^2 \rangle - \langle t \rangle^2}$ characterizing the width of the spectral distribution in the angular frequency domain, $\sigma_A(\omega)$, and the width for the conjugate variable in time $\sigma_A(t)$ are related :

$$\sigma_A(t) \cdot \sigma_A(\omega) \geqslant 1 \qquad \text{(Relation between amplitude uncertainties.)}$$

This product of the *standard deviations equals unity only for the special case of Gaussian-shaped spectral distributions, and it is greater than unity for all other shaped spectral distributions.*

The *intensity* of the wave is the square of the amplitude leading to standard deviation widths for a Gaussian distribution where $\sigma_I(t)^2 = \frac{1}{2}\sigma_A(t)^2$, that is, $\sigma_I(t) = \frac{\sigma_A(t)}{\sqrt{2}}$. Thus the standard deviations for the spectral distribution and width of the *intensity* of the wavepacket are related by:

$$\sigma_I(t) \cdot \sigma_I(\omega) \geqslant \frac{1}{2} \qquad \text{(Uncertainty principle for frequency-time intensities)}$$

This states that the uncertainties with which you can simultaneously measure the time and frequency for the intensity of a given wavepacket are related. If you try to measure the frequency within a short time interval $\sigma_I(t)$ then the uncertainty in the frequency measurement $\sigma_I(\omega) \geqslant \frac{1}{2\sigma_I(t)}$. Accurate measurement of the frequency requires measurement times that encompass many cycles of oscillation, that is, a long wavepacket.

Exactly the same relations exist between the spectral distribution as a function of wavenumber k_x and the corresponding spatial dependence of a wave x which are conjugate representations. Thus the spectral distribution plotted versus k_x is directly related to the amplitude as a function of position x; the spectral distribution versus k_y is related to the amplitude as a function of y; and the k_z spectral distribution is related to the spatial dependence on z. Following the same arguments discussed above, the standard deviation, $\sigma_I(k_x)$ characterizing the width of the *spectral intensity* distribution of k_x, and the standard deviation $\sigma_I(x)$, characterizing the spatial width of the wave packet intensity as a function of x, are related by the Uncertainty Principle for position-wavenumber. Thus in summary the temporal and spatial uncertainty principles of the intensity of wave motion is,

$$\sigma_I(t) \cdot \sigma_I(\omega) \geqslant \frac{1}{2} \tag{3.128}$$

$$\sigma_I(x) \cdot \sigma_I(k_x) \geqslant \frac{1}{2} \qquad \sigma_I(y) \cdot \sigma_I(k_y) \geqslant \frac{1}{2} \qquad \sigma_I(z) \cdot \sigma_I(k_z) \geqslant \frac{1}{2}$$

This *applies to all forms of wave motion*, be they, sound waves, water waves, electromagnetic waves, or matter waves.

As discussed in chapter 18, the transition to quantum mechanics involves relating the matter-wave properties to the energy and momentum of the corresponding particle. That is, in the case of matter waves, multiplying both sides of equation 3.129 by $\hbar$ and using the de Broglie relations gives that the particle energy is related to the angular frequency by $E = \hbar\omega$ and the particle momentum is related to the wavenumber, that is $\overrightarrow{\mathbf{p}} = \hbar\overrightarrow{\mathbf{k}}$. These lead to the **Heisenberg Uncertainty Principle:**

$$\sigma_I(t) \cdot \sigma_I(E) \geqslant \frac{\hbar}{2} \tag{3.129}$$

$$\sigma_I(x) \cdot \sigma_I(p_x) \geqslant \frac{\hbar}{2} \qquad \sigma_I(y) \cdot \sigma_I(p_y) \geqslant \frac{\hbar}{2} \qquad \sigma_I(z) \cdot \sigma_I(p_z) \geqslant \frac{\hbar}{2}$$

This uncertainty principle applies equally to the wavefunction of the electron in the hydrogen atom, proton in a nucleus, as well as to a wavepacket describing a particle wave moving along some trajectory. This implies that, for a particle of given momentum, the wavefunction is spread out spatially. Planck's constant $\hbar = 1.05410^{-34} J \cdot s = 6.58210^{-16} eV \cdot s$ is extremely small compared with energies and times encountered in normal life, and thus the effects due to the Uncertainty Principle are not important for macroscopic dimensions.

Confinement of a particle, of mass m, within $\pm\sigma(x)$ of a fixed location implies that there is a corresponding uncertainty in the momentum

$$\sigma(p_x) \geq \frac{\hbar}{2\sigma(x)} \tag{3.130}$$

Now the variance in momentum $\mathbf{p}$ is given by the difference in the average of the square $\left\langle (\mathbf{p} \cdot \mathbf{p})^2 \right\rangle$, and the square of the average of $\langle \mathbf{p} \rangle^2$. That is

$$\sigma(\mathbf{p})^2 = \left\langle (\mathbf{p} \cdot \mathbf{p})^2 \right\rangle - \langle \mathbf{p} \rangle^2 \tag{3.131}$$

Assuming a fixed average location implies that $\langle \mathbf{p} \rangle = 0$, then

$$\left\langle (\mathbf{p} \cdot \mathbf{p})^2 \right\rangle = \sigma(p)^2 \geq \left(\frac{\hbar}{2\sigma(r)} \right)^2 \tag{3.132}$$

Since the kinetic energy is given by:

$$\text{Kinetic energy} = \frac{p^2}{2m} \geq \frac{\hbar^2}{8m\sigma(r)^2} \qquad \text{(Zero-point energy)}$$

This zero-point energy is the minimum kinetic energy that a particle of mass m can have if confined within a distance $\pm\sigma(r)$. This zero-point energy is a consequence of wave-particle duality and the uncertainty between the size and wavenumber for any wave packet. It is a quantal effect in that the classical limit has $\hbar \to 0$ for which the zero-point energy $\to 0$.

Inserting numbers for the zero-point energy gives that an electron confined to the radius of the atom, that is $\sigma(x) = 10^{-10}m$, has a zero-point kinetic energy of $\sim 1eV$. Confining this electron to $3 \times 10^{-15}m$, the size of a nucleus, gives a zero-point energy of $10^9 eV$ ($1GeV$). Confining a proton to the size of the nucleus gives a zero-point energy of $0.5MeV$. These values are typical of the level spacing observed in atomic and nuclear physics. If $\hbar$ was a large number, then a billiard ball confined to a billiard table would be a blur as it oscillated with the minimum zero-point kinetic energy. The smaller the spatial region that the ball was confined, the larger would be its zero-point energy and momentum causing it to rattle back and forth between the boundaries of the confined region. Life would be dramatically different if $\hbar$ was a large number.

In summary, Heisenberg's Uncertainty Principle is a well-known and crucially important aspect of quantum physics. What is less well known, is that the Uncertainty Principle applies for all forms of wave motion, that is, it is not restricted to matter waves. The following three examples illustrate application of the Uncertainty Principle to acoustics, the nuclear Mössbauer effect, and quantum mechanics.

3.8 *Example: Acoustic wave packet*

A violinist plays the note middle C (261.625Hz) with constant intensity for precisely 2 seconds. Using the fact that the velocity of sound in air is 343.2m/s calculate the following:

1) The wavelength of the sound wave in air: $\lambda = 343.2/261.625 = 1.312m$.

2) The length of the wavepacket in air: Wavepacket length $= 343.2 \times 2 = 686.4m$

3) The fractional frequency width of the note: Since the wave packet has a square pulse shape of length $\tau = 2s$, *then the Fourier transform is a sinc function having the first zeros when* $\sin \frac{\omega\tau}{2} = 0$, *that is,* $\Delta\nu = \frac{1}{\tau}$. *Therefore the fractional width is* $\frac{\Delta\nu}{\nu} = \frac{1}{\nu\tau} = 0.0019$. *Note that to achieve a purity of* $\frac{\Delta\nu}{\nu} = 10^{-6}$ *the violinist would have to play the note for 1.06hours.*

3.9 *Example: Gravitational red shift*

The Mössbauer effect in nuclear physics provides a wave packet that has an exceptionally small fractional width in frequency. For example, the ^{57}Fe nucleus emits a 14.4keV deexcitation-energy photon which corresponds to $\omega \approx 2 \times 10^{25} rad/s$ *with a decay time of* $\tau \approx 10^{-7}s$. *Thus the fractional width is* $\frac{\Delta\omega}{\omega} \approx 3 \times 10^{-18}$.

In 1959 Pound and Rebka used this to test Einstein's general theory of relativity by measurement of the gravitational red shift between the attic and basement of the 22.5m high physics building at Harvard. The magnitude of the predicted relativistic red shift is $\frac{\Delta E}{E} = 2.5 \times 10^{-15}$ which is what was observed with a fractional precision of about 1%.

3.10 *Example: Quantum baseball*

George Gamow, in his book "Mr. Tompkins in Wonderland", describes the strange world that would exist if $\hbar$ was a large number. As an example, consider you play baseball in a universe where $\hbar$ is a large number. The pitcher throws a 150g ball 20m to the batter at a speed of 40m/s. For a strike to be thrown, the ball's position must be pitched within the 30cm radius of the strike zone, that is, it is required that $\Delta x \leq 0.3m$. The uncertainty relation tells us that the transverse velocity of the ball cannot be less than $\Delta v = \frac{\hbar}{2m\Delta x}$. The time of flight of the ball from the mound to batter is $t = 0.5s$. Because of the transverse velocity uncertainty, Δv, the ball will deviate $t\Delta v$ transversely from the strike zone. This also must not exceed the size of the strike zone, that is;

$$t\Delta v = \frac{\hbar t}{2m\Delta x} \leq 0.3m \qquad \text{(Due to transverse velocity uncertainty)}$$

Combining both of these requirements gives

$$\hbar \leq \frac{2m\Delta x^2}{t} = 5.4 \ 10^{-2} J \cdot s.$$

This is 32 orders of magnitude larger than $\hbar$ so quantal effects are negligible. However, if $\hbar$ exceeded the above value, then the pitcher would have difficulty throwing a reliable strike.

3.12   Summary

Linear systems have the feature that the solutions obey the *Principle of Superposition*, that is, the amplitudes add linearly for the superposition of different oscillatory modes. Applicability of the Principle of Superposition to a system provides a tremendous advantage for handling and solving the equations of motion of oscillatory systems.

Geometric representations of the motion of dynamical systems provide sensitive probes of periodic motion. Configuration space $(\mathbf{q}, \mathbf{q}, t)$, state space $(\mathbf{q}, \dot{\mathbf{q}}, t)$ and phase space $(\mathbf{q}, \mathbf{p}, t)$, are powerful geometric representations that are used extensively for recognizing periodic motion where $\mathbf{q}, \dot{\mathbf{q}}$, and $\mathbf{p}$ are vectors in n-dimensional space.

Linearly-damped free linear oscillator   The free linearly-damped linear oscillator is characterized by the equation

$$\ddot{x} + \Gamma\dot{x} + \omega_0^2 x = 0 \tag{3.26}$$

The solutions of the linearly-damped free linear oscillator are of the form

$$z = e^{-\left(\frac{\Gamma}{2}\right)t}\left[z_1 e^{i\omega_1 t} + z_2 e^{-i\omega_1 t}\right] \qquad\qquad \omega_1 \equiv \sqrt{\omega_o^2 - \left(\frac{\Gamma}{2}\right)^2} \tag{3.33}$$

The solutions of the linearly-damped free linear oscillator have the following characteristic frequencies corresponding to the three levels of linear damping

$x(t) = Ae^{-\left(\frac{\Gamma}{2}\right)t}\cos(\omega_1 t - \beta)$	underdamped	$\omega_1 = \sqrt{\omega_o^2 - \left(\frac{\Gamma}{2}\right)^2} > 0$
$x(t) = [A_1 e^{-\omega_+ t} + A_2 e^{-\omega_- t}]$	overdamped	$\omega_\pm = -\left[-\frac{\Gamma}{2} \pm \sqrt{\left(\frac{\Gamma}{2}\right)^2 - \omega_o^2}\right]$
$x(t) = (A + Bt)e^{-\left(\frac{\Gamma}{2}\right)t}$	critically damped	$\omega_1 = \sqrt{\omega_o^2 - \left(\frac{\Gamma}{2}\right)^2} = 0$

The energy dissipation for the linearly-damped free linear oscillator time averaged over one period is given by

$$\langle E \rangle = E_0 e^{-\Gamma t} \tag{3.44}$$

The quality factor Q characterizing the damping of the free oscillator is defined to be

$$Q = \frac{E}{\Delta E} = \frac{\omega_1}{\Gamma} \tag{3.47}$$

where ΔE is the energy dissipated per radian.

Sinusoidally-driven, linearly-damped, linear oscillator   The linearly-damped linear oscillator, driven by a harmonic driving force, is of considerable importance to all branches of physics, and engineering. The equation of motion can be written as

$$\ddot{x} + \Gamma \dot{x} + \omega_0^2 x = \frac{F(t)}{m} \tag{3.49}$$

where $F(t)$ is the driving force. The complete solution of this second-order differential equation comprises two components, the complementary solution (*transient response*), and the particular solution (*steady-state response*). That is,

$$x(t)_{Total} = x(t)_T + x(t)_S \tag{3.65}$$

For the underdamped case, the transient solution is the complementary solution

$$x(t)_T = \frac{F_0}{m} e^{-\frac{\Gamma}{2}t} \cos(\omega_1 t - \delta) \tag{3.66}$$

and the steady-state solution is given by the particular solution

$$x(t)_S = \frac{\frac{F_0}{m}}{\sqrt{\left(\omega_0^2 - \omega^2\right)^2 + \left(\Gamma\omega\right)^2}} \cos(\omega t - \delta) \tag{3.67}$$

Resonance   A detailed discussion of resonance and energy absorption for the driven linearly-damped linear oscillator was given. For resonance of the linearly-damped linear oscillator the maximum amplitudes occur at the following resonant frequencies

Resonant system	Resonant frequency
undamped free linear oscillator	$\omega_0 = \sqrt{\frac{k}{m}}$
linearly-damped free linear oscillator	$\omega_1 = \sqrt{\omega_0^2 - \left(\frac{\Gamma}{2}\right)^2}$
driven linearly-damped linear oscillator	$\omega_R = \sqrt{\omega_0^2 - 2\left(\frac{\Gamma}{2}\right)^2}$

The energy absorption for the steady-state solution for resonance is given by

$$x(t)_S = A_{el} \cos\omega t + A_{abs} \sin\omega t \tag{3.73}$$

where the **elastic amplitude**

$$A_{el} = \frac{\frac{F_0}{m}}{\left(\omega_0^2 - \omega^2\right)^2 + \left(\Gamma\omega\right)^2} \left(\omega_0^2 - \omega^2\right) \tag{3.74}$$

while the **absorptive amplitude**

$$A_{abs} = \frac{\frac{F_0}{m}}{\left(\omega_0^2 - \omega^2\right)^2 + \left(\Gamma\omega\right)^2} \Gamma\omega \tag{3.75}$$

The time average power input is given by only the absorptive term

$$\langle P \rangle = \frac{1}{2} F_0 \omega A_{abs} = \frac{F_0^2}{2m} \frac{\Gamma\omega^2}{\left(\omega_0^2 - \omega^2\right)^2 + \left(\Gamma\omega\right)^2} \tag{3.133}$$

This power curve has the classic Lorentzian shape.

Wave propagation The wave equation was introduced and both travelling and standing wave solutions of the wave equation were discussed. Harmonic wave-form analysis, and the complementary time-sampled wave form analysis techniques, were introduced in this chapter and in appendix I. The relative merits of Fourier analysis and the digital Green's function waveform analysis were illustrated for signal processing.

The concepts of phase velocity, group velocity, and signal velocity were introduced. The phase velocity is given by

$$v_{phase} = \frac{\omega}{k} \tag{3.117}$$

and group velocity

$$v_{group} = \left(\frac{d\omega}{dk}\right)_{k_0} = v_{phase} + k\frac{\partial v_{phase}}{\partial k} \tag{3.128}$$

If the group velocity is frequency dependent then the information content of a wave packet travels at the signal velocity which can differ from the group velocity.

The Wave-packet Uncertainty Principle implies that making a precise measurement of the frequency of a sinusoidal wave requires that the wave packet be infinitely long. The *standard deviation* $\sigma\left(t\right) = \sqrt{\langle t^2 \rangle - \langle t \rangle^2}$ characterizing the width of the *amplitude* of the wavepacket spectral distribution in the angular frequency domain, $\sigma_A(\omega)$, and the corresponding width in time $\sigma_A(t)$, are related by :

$$\sigma_A(t) \cdot \sigma_A(\omega) \geqslant 1 \qquad \text{(Relation between amplitude uncertainties.)}$$

The standard deviations for the spectral distribution and width of the *intensity* of the wave packet are related by:

$$\sigma_I(t) \cdot \sigma_I(\omega) \geqslant \frac{1}{2} \tag{3.134}$$

$$\sigma_I(x) \cdot \sigma_I(k_x) \geqslant \frac{1}{2} \qquad \sigma_I(y) \cdot \sigma_I(k_y) \geqslant \frac{1}{2} \qquad \sigma_I(z) \cdot \sigma_I(k_z) \geqslant \frac{1}{2}$$

This *applies to all forms of wave motion*, including sound waves, water waves, electromagnetic waves, or matter waves.

Workshop exercises

1. Given below are a list of statements followed by a list of reasons related to harmonic motion. For each of the statements, determine the reason(s) that make that statement true. You may do this in small groups or as one large group–the teaching assistant will decide what works best for your workshop.
   Statements:

   - We can neglect the higher order terms in the Taylor expansion of $F(x)$.
   - The restoring force is a linear force.
   - F_0 must vanish.
   - $(dF/dx)_0$ is negative and k is positive.
   - We can write $F(x)$ as a Taylor series expansion.

   Reasons:

   - $F(x)$ depends only on x.
   - A position of stable equilibrium exists and we call this point the origin of our coordinate system.
   - $F(x)$ has continuous derivatives of all orders.
   - The restoring force is directed toward the equilibrium position.
   - We consider only small displacements.

2. Second-order ordinary differential equations are an important part of the physics of the harmonic oscillator.

   (a) What do each of the following terms mean with respect to differential equations?
      i. Ordinary
      ii. Second-order
      iii. Homogeneous
      iv. Linear

   (b) Give a mini-lesson on how to solve second-order differential equations by working through the following examples. Don't just provide a solution; explain the steps leading up to the solution.
      i. $y''+5y'+6y = 0$
      ii. $y''+y'+y = 0$
      iii. $y''+4y'+4y = 0$
      iv. $y''-3y'^{2x}$
      v. $y''-3y'-4y = 2\sin x$

3. Harmonic oscillations occur for many different types of systems and it is important to recognize when the equations for harmonic motion apply. Three different systems are described below. Each system can be approximately described using the equations for harmonic motion. Break up into three groups–one group per system. For your group's system, answer the following questions:

   (a) What approximations are necessary for this system to exhibit harmonic oscillations?

   (b) What is the differential equation that governs the motion of this system? Use Newton's second law to arrive at this equation.

   (c) What is the solution to the differential equation that you found in part (b)?

   (d) What is the natural frequency of oscillations?

   Here are the three systems:

   - A mass m is tied to a massless spring having a spring constant k. The system oscillates in one dimension along a horizontal frictionless surface.

- A particle of mass m is attached to a weightless, extensionless rod to form a pendulum. The length of the rod is L and the system oscillates in a single plane.

- A tube is bent into the shape of a U and is partially filled with a liquid of density ρ. The cross-sectional area of the tube is A and the length of the tube filled with liquid is L. The liquid is initially displaced so that it is higher on one side of the tube than the other.

Once each group has answered all of the questions, share the results with the entire class.

4. Consider a mass m attached to a spring of spring constant k. The spring is mounted horizontally so that the mass oscillates horizontally on a frictionless surface. The spring is attached to the wall on the right and the mass is initially moved to the right of its equilibrium position (compressing the spring) by a distance s and released. Working individually, determine how (if at all) the period of the motion would be affected by each of the changes below. Once you have answered each part on your own, compare your answers with a classmate.

   (a) The spring is replaced with a stiffer spring.

   (b) The mass is initially displaced a distance s to the left and released.

   (c) The mass is replaced with a heavier mass.

   (d) The mass is initially displaced a distance r $(r < s)$ to the right and released.

5. When you were first introduced to simple harmonic motion, you used the formula $m\ddot{x} = -kx$ to find the position of the oscillating mass as a function of time. This assumes that the origin is defined to be the equilibrium point. What happens if this is not the case? What would the equation of motion look like? How would the position of the oscillating mass as a function of time change?

6. For each of the situations described below, give a rough sketch of the state space diagram ($\dot{x}$ versus x) that represents the motion of each object. All of the motion takes place along the x-axis.

   (a) An eggplant is at rest at a point on the $+x$ axis.

   (b) A monkey on a skateboard skates with constant speed in the negative x direction.

   (c) A race car moving in the $+x$ direction undergoes constant acceleration until it abruptly stops.

   (d) A cantaloupe undergoes simple harmonic motion. The initial location of the cantaloupe is at a point on the $+x$ axis.

7. Consider a simple harmonic oscillator consisting of a mass m attached to a spring of spring constant k. For this oscillator $x(t) = A\sin(\omega_0 t - \delta)$.

   (a) Find an expression for $\dot{x}(t)$.

   (b) Eliminate t between $x(t)$ and $\dot{x}(t)$ to arrive at one equation similar to that for an ellipse.

   (c) Rewrite the equation in part (b) in terms of x, $\dot{x}$, k, m, and the total energy E.

   (d) Give a rough sketch of the phase space diagram ($\dot{x}$ versus x) for this oscillator. Also, on the same set of axes, sketch the phase space diagram for a similar oscillator with a total energy that is larger than the first oscillator.

   (e) What direction are the paths that you have sketched? Explain your answer.

   (f) Would different trajectories for the same oscillator ever cross paths? Why or why not?

8. Consider a damped, driven oscillator consisting of a mass m attached to a spring of spring constant k.

   (a) What is the equation of motion for this system?

   (b) Solve the equation in part (a). The solution consists of two parts, the complementary solution and the particular solution. When might it be possible to safely neglect one part of the solution?

   (c) What is the difference between amplitude resonance and kinetic energy resonance?

   (d) How might phase space diagrams look for this type of oscillator? What variables would affect the diagram?

9. A particle of mass m is subject to the following force

$$\mathbf{F} = A(x^3 - 4x^2 + 3x)\hat{\mathbf{x}}$$

where A is a constant.

(a) Determine the points when the particle is in equilibrium.

(b) Which of these points is stable and which are unstable?

(c) Is the motion bounded or unbounded?

10. A very long cylindrical shell has a mass density that depends upon the radial distance such that $\rho(r) = \frac{k}{r}$, where k is a constant. The inner radius of the shell is a and the outer radius is b.

(a) Determine the direction and the magnitude of the gravitational field for all regions of space.

(b) If the gravitational potential is zero at the origin, what is the difference between the gravitational potential at $r = b$ and $r = a$?

11. A mass m is constrained to move along one dimension. Two identical springs are attached to the mass, one on each side, and each spring is in turn attached to a wall. Both springs have the same spring constant k.

(a) Determine the frequency of the oscillation, assuming no damping.

(b) Now consider damping. It is observed that after n oscillations, the amplitude of the oscillation has dropped to one-half of its initial value. Find an expression for the damping constant.

(c) How long does it take for the amplitude to decrease to one-quarter of its initial value?

12. Discuss the motion of a continuous string when plucked at one third of the length of the string. That is, the initial condition is $\dot{q}(x,0) = 0$, and $q(x,0) = \left\{ \begin{array}{ll} \frac{3A}{L}x, & 0 \le x \le \frac{L}{3} \\ \frac{3A}{2L}(L - x), & \frac{L}{3} \le x \le L \end{array} \right\}$

13. When a particular driving force is applied to a stretched string it is observed that the string vibration in purely of the nth harmonic. Find the driving force.

14. Consider the two-mass system pivoted at its vertex where $M \ne m$. It undergoes oscillations of the angle θ with respect to the vertical in the plane of the triangle.

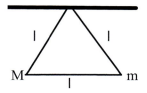

(a) Determine the angular frequency of small oscillations.

(b) Use your result from part (a) to show $\omega^2 \approx \frac{g}{l}$ for $M \gg m$.

(c) Show that your result from part (a) agrees with $\omega^2 = \frac{U''(\theta_e)}{I}$ where θ_e is the equilibrium angle and I is the moment of inertia.

(d) Assume the system has energy E. Setup an integral that determines the period of oscillation.

15. A cube of side a and mass m is immersed in water with density ρ past the point of equilibrium and then released. Assume there is no damping due to the water.

(a) Show that the cube's equation of motion is

$$\frac{d^2x}{dt^2} + Ax + B = 0.$$

where A and B are constants. Determine A and B.

(b) The solution to the equation of motion is

$$x(t) = \frac{B}{A} + C_1 \cos\left(\sqrt{A}t\right) + C_2 \sin\left(\sqrt{A}t\right)$$

where C_1 and C_2 are constants. If $x(0) = -a$, determine $x(t)$.

(c) Determine the period T of oscillation.

Problems

1. An unusual pendulum is made by fixing a string to a horizontal cylinder of radius R, wrapping the string several times around the cylinder, and then tying a mass m to the loose end. In equilibrium the mass hangs a distance l_0 vertically below the edge of the cylinder. Find the potential energy if the pendulum has swung to an angle ϕ from the vertical. Show that for small angles, it can be written in the Hooke's Law form $U = \frac{1}{2}k\phi^2$. Comment of the value of k.

2. Consider the two-dimensional anisotropic oscillator with motion with $\omega_x = p\omega$ and $\omega_y = q\omega$.

   a) Prove that if the ratio of the frequencies is rational (that is, $\frac{\omega_x}{\omega_y} = \frac{p}{q}$ where p and q are integers) then the motion is periodic. What is the period?

   b) Prove that if the same ratio is irrational, the motion never repeats itself.

3. A simple pendulum consists of a mass m suspended from a fixed point by a weight-less, extensionless rod of length l.

   a) Obtain the equation of motion, and in the approximation $\sin\theta \approx \theta$, show that the natural frequency is $\omega_0 = \sqrt{\frac{g}{l}}$, where g is the gravitational field strength.

   b) Discuss the motion in the event that the motion takes place in a viscous medium with retarding force $2m\sqrt{gl}\dot{\theta}$.

4. Derive the expression for the State Space paths of the plane pendulum if the total energy is $E > 2mgl$. Note that this is just the case of a particle moving in a periodic potential $U(\theta) = mgl(1-\cos\theta)$. Sketch the State Space diagram for both $E > 2mgl$ and $E < 2mgl$.

5. Consider the motion of a driven linearly-damped harmonic oscillator after the transient solution has died out, and suppose that it is being driven close to resonance, $\omega = \omega_o$.

   a) Show that the oscillator's total energy is $E = \frac{1}{2}m\omega^2 A^2$.

   b) Show that the energy ΔE_{dis} dissipated during one cycle by the damping force $\Gamma\dot{x}$ is $\pi\Gamma m\omega A^2$

6. Two masses m_1 and m_2 slide freely on a horizontal frictionless rail and are connected by a spring whose force constant is k. Find the frequency of oscillatory motion for this system.

7. A particle of mass m moves under the influence of a resistive force proportional to velocity and a potential U, that is l.

$$F(x, \dot{x}) = -b\dot{x} - \frac{\partial U}{\partial x}$$

   where $b > 0$ and $U(x) = (x^2 - a^2)^2$

   a) Find the points of stable and unstable equilibrium.

   b) Find the solution of the equations of motion for small oscillations around the stable equilibrium points

   c) Show that as $t \to \infty$ the particle approaches one of the stable equilibrium points for most choices of initial conditions. What are the exceptions? (Hint: You can prove this without finding the solutions explicitly.)

Chapter 4

Nonlinear systems and chaos

4.1 Introduction

In nature only a subset of systems have equations of motion that are linear. Contrary to the impression given by the analytic solutions presented in undergraduate physics courses, most dynamical systems in nature exhibit non-linear behavior that leads to complicated motion. The solutions of non-linear equations usually do not have analytic solutions, superposition does not apply, and they predict phenomena such as attractors, discontinuous period bifurcation, extreme sensitivity to initial conditions, rolling motion, and chaos. During the past four decades, exciting discoveries have been made in classical mechanics that are associated with the recognition that nonlinear systems can exhibit chaos. Chaotic phenomena have been observed in most fields of science and engineering such as, weather patterns, fluid flow, motion of planets in the solar system, epidemics, changing populations of animals, birds and insects, and the motion of electrons in atoms. The complicated dynamical behavior predicted by non-linear differential equations is not limited to classical mechanics, rather it is a manifestation of the mathematical properties of the solutions of the differential equations involved, and thus is generally applicable to solutions of first or second-order non-linear differential equations. It is important to understand that the systems discussed in this chapter follow a fully deterministic evolution predicted by the laws of classical mechanics, the evolution for which is based on the prior history. This behavior is completely different from a random walk where each step is based on a random process. The complicated motion of deterministic non-linear systems stems in part from sensitivity to the initial conditions.

The French mathematician Poincaré is credited with being the first to recognize the existence of chaos during his investigation of the gravitational three-body problem in celestial mechanics. At the end of the nineteenth century Poincaré noticed that such systems exhibit high sensitivity to initial conditions characteristic of chaotic motion, and the existence of nonlinearity which is required to produce chaos. Poincaré's work received little notice, in part it was overshadowed by the parallel development of the Theory of Relativity and quantum mechanics at the start of the 20th century. In addition, solving nonlinear equations of motion is difficult, which discouraged work on nonlinear mechanics and chaotic motion. The field blossomed during the $1960's$ when computers became sufficiently powerful to solve the nonlinear equations required to calculate the long-time histories necessary to document the evolution of chaotic behavior. Laplace, and many other scientists, believed in the deterministic view of nature which assumes that if the position and velocities of all particles are known, then one can unambiguously predict the future motion using Newtonian mechanics. Researchers in many fields of science now realize that this "clockwork universe" is invalid. That is, knowing the laws of nature can be insufficient to predict the evolution of nonlinear systems in that the time evolution can be extremely sensitive to the initial conditions even though they follow a completely deterministic development. There are two major classifications of nonlinear systems that lead to chaos in nature. The first classification encompasses nondissipative Hamiltonian systems such as Poincaré's three-body celestial mechanics system. The other main classification involves driven, damped, non-linear oscillatory systems.

Nonlinearity and chaos is a broad and active field and thus this chapter will focus only on a few examples that illustrate the general features of non-linear systems. Weak non-linearity is used to illustrate bifurcation and asymptotic attractor solutions for which the system evolves independent of the initial conditions. The common sinusoidally-driven linearly-damped plane pendulum illustrates several features characteristic of the evolution of a non-linear system from order to chaos. The impact of non-linearity on wavepacket propagation velocities and the existence of soliton solutions is discussed. The example of the three-body problem is discussed in chapter 11. The transition from laminar flow to turbulent flow is illustrated by fluid mechanics discussed in chapter 16.8. Analytic solutions of nonlinear systems usually are not available and thus one must resort to computer simulations. As a consequence the present discussion focusses on the main features of the solutions for these systems and ignores how the equations of motion are solved.

4.2   Weak nonlinearity

Most physical oscillators become non-linear with increase in amplitude of the oscillations. Consequences of non-linearity include breakdown of superposition, introduction of additional harmonics, and complicated chaotic motion that has great sensitivity to the initial conditions as illustrated in this chapter. Weak non-linearity is interesting since perturbation theory can be used to solve the non-linear equations of motion.

The potential energy function for a linear oscillator has a pure parabolic shape about the minimum location, that is, $U = \frac{1}{2}k(x - x_0)^2$ where x_0 is the location of the minimum. Weak non-linear systems have small amplitude oscillations Δx about the minimum allowing use of the Taylor expansion

$$U(\Delta x) = U(x_0) + \Delta x \frac{dU(x_0)}{dx} + \frac{\Delta x^2}{2!} \frac{d^2 U(x_0)}{dx^2} + \frac{\Delta x^3}{3!} \frac{d^3 U(x_0)}{dx^3} + \frac{\Delta x^4}{4!} \frac{d^4 U(x_0)}{dx^4} + ... \tag{4.1}$$

By definition, at the minimum $\frac{dU(x_0)}{dx} = 0$, and thus equation 4.1 can be written as

$$\Delta U = U(\Delta x) - U(x_0) = \frac{\Delta x^2}{2!} \frac{d^2 U(x_0)}{dx^2} + \frac{\Delta x^3}{3!} \frac{d^3 U(x_0)}{dx^3} + \frac{\Delta x^4}{4!} \frac{d^4 U(x_0)}{dx^4} + ... \tag{4.2}$$

For small amplitude oscillations the system is linear when only the second-order $\frac{\Delta x^2}{2!} \frac{d^2 U(x_0)}{dx^2}$ term in equation 4.2 is significant. The linearity for small amplitude oscillations greatly simplifies description of the oscillatory motion in that superposition applies, and complicated chaotic motion is avoided. For slightly larger amplitude motion, where the higher-order terms in the expansion are still much smaller than the second-order term, then perturbation theory can be used as illustrated by the simple plane pendulum which is non linear since the restoring force equals

$$mg \sin \theta \simeq mg(\theta - \frac{\theta^3}{3!} + \frac{\theta^5}{5!} - \frac{\theta^7}{7!} + ...) \tag{4.3}$$

This is linear only at very small angles where the higher-order terms in the expansion can be neglected. Consider the equation of motion at small amplitudes for the harmonically-driven, linearly-damped plane pendulum

$$\ddot{\theta} + \Gamma \dot{\theta} + \omega_0^2 \sin \theta = \ddot{\theta} + \Gamma \dot{\theta} + \omega_0^2 (\theta - \frac{\theta^3}{6}) = F_0 \cos(\omega t) \tag{4.4}$$

where only the first two terms in the expansion 4.3 have been included. It was shown in chapter 3 that when $\sin \theta \approx \theta$ then the steady-state solution of equation 4.4 is of the form

$$\theta(t) = A \cos(\omega t - \delta) \tag{4.5}$$

Insert this first-order solution into equation 4.4, then the cubic term in the expansion gives a term $cos^3 \omega t = \frac{1}{4}(\cos 3\omega t + 3 \cos \omega t)$. Thus the perturbation expansion to third order involves a solution of the form

$$\theta(t) = A \cos(\omega t - \delta) + B \cos 3(\omega t - \delta) \tag{4.6}$$

This perturbation solution shows that the non-linear term has distorted the signal by addition of the third harmonic of the driving frequency with an amplitude that depends sensitively on θ. This illustrates that the superposition principle is not obeyed for this non-linear system, but, if the non-linearity is weak, perturbation theory can be used to derive the solution of a non-linear equation of motion.

Figure 4.1 illustrates that for a potential $U(x) = 2x^2 + x^4$, the x^4 non-linear term are greatest at the maximum amplitude x, which makes the total energy contours in state-space more rectangular than the elliptical shape for the harmonic oscillator as shown in figure 3.3. The solution is of the form given in equation 4.6.

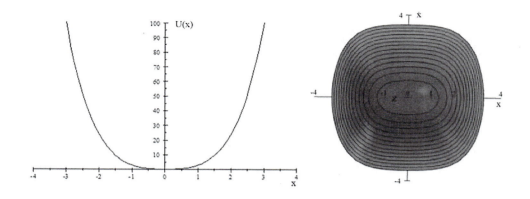

Figure 4.1: The left side shows the potential energy for a symmetric potential $U(x) = 2x^2 + x^4$. The right side shows the contours of constant total energy on a state-space diagram.

4.1 *Example: Non-linear oscillator*

Assume that a non-linear oscillator has a potential given by

$$U(x) = \frac{kx^2}{2} - \frac{m\lambda x^3}{3}$$

where λ is small. Find the solution of the equation of motion to first order in λ, assuming $x = 0$ at $t = 0$. The equation of motion for the nonlinear oscillator is

$$m\ddot{x} = -\frac{dU}{dx} = -kx + m\lambda x^2$$

If the $m\lambda x^2$ term is neglected, then the second-order equation of motion reduces to a normal linear oscillator with

$$x_0 = A\sin(\omega_0 t + \varphi)$$

where

$$\omega_0 = \sqrt{\frac{k}{m}}$$

Assume that the first-order solution has the form

$$x_1 = x_0 + \lambda x_1$$

Substituting this into the equation of motion, and neglecting terms of higher order than λ, gives

$$\ddot{x}_1 + \omega_0^2 x_1 = x_0^2 = \frac{A^2}{2}[1 - \cos(2\omega_0 t)]$$

To solve this try a particular integral

$$x_1 = B + C\cos(2\omega_0 t)$$

and substitute into the equation of motion gives

$$-3\omega_0^2 C\cos(2\omega_0 t) + \omega_0^2 B = \frac{A^2}{2} - \frac{A^2}{2}\cos(2\omega_0 t)$$

Comparison of the coefficients gives

$$B = \frac{A^2}{2\omega_0^2}$$

$$C = \frac{A^2}{6\omega_0^2}$$

The homogeneous equation is

$$\ddot{x}_1 + \omega_0^2 x_1 = 0$$

which has a solution of the form

$$x_1 = D_1 \sin(\omega_0 t) + D_2 \cos(\omega_0 t)$$

Thus combining the particular and homogeneous solutions gives

$$x_1 = (A + \lambda D_1) \sin(\omega_0 t) + \lambda \left[\frac{A^2}{2\omega_0^2} + D_2 \cos(\omega_0 t) + \frac{A^2}{6\omega_0^2} \cos(2\omega_0 t) \right]$$

The initial condition $x = 0$ at $t = 0$ then gives

$$D_2 = -\frac{2A^2}{3\omega^2}$$

and

$$x_1 = (A + \lambda D_1) \sin(\omega_0 t) + \frac{\lambda A^2}{\omega_0^2} \left[\frac{1}{2} - \frac{2}{3} \cos(\omega_0 t) + \frac{1}{6} \cos(2\omega_0 t) \right]$$

The constant $(A + \lambda D_1)$ is given by the initial amplitude and velocity.

*   This system is nonlinear in that the output amplitude is not proportional to the input amplitude. Secondly, a large amplitude second harmonic component is introduced in the output waveform; that is, for a non-linear system the gain and frequency decomposition of the output differs from the input. Note that the frequency composition is amplitude dependent. This particular example of a nonlinear system does not exhibit chaos. The Laboratory for Laser Energetics uses nonlinear crystals to double the frequency of laser light.*

4.3   Bifurcation, and point attractors

Interesting new phenomena, such as bifurcation, and attractors, occur when the non-linearity is large. In chapter 3 it was shown that the state-space diagram $(\dot{x}, x)$ for an undamped harmonic oscillator is an ellipse with dimensions defined by the total energy of the system. As shown in figure 3.5, for the damped harmonic oscillator, the state-space diagram spirals inwards to the origin due to dissipation of energy. Non-linearity distorts the shape of the ellipse or spiral on the state-space diagram, and thus the state-space, or corresponding phase-space, diagrams, provide useful representations of the motion of linear and non-linear periodic systems.

    The complicated motion of non-linear systems makes it necessary to distinguish between transient and asymptotic behavior. The damped harmonic oscillator executes a transient spiral motion that asymptotically approaches the origin. The transient behavior depends on the initial conditions, whereas the asymptotic limit of the steady-state solution is a specific location, that is called a **point attractor.** The point attractor for damped motion in the anharmonic potential well

$$U(x) = 2x^2 + x^4 \tag{4.7}$$

is at the minimum, which is the origin of the state-space diagram as shown in figure 4.1.

    The more complicated one-dimensional potential well

$$U(x) = 8 - 4x^2 + 0.5x^4 \tag{4.8}$$

shown in figure 4.2, has two minima that are symmetric about $x = 0$ with a saddle of height 8.

    The kinetic plus potential energies of a particle with mass $m = 2$, released in this potential, will be assumed to be given by

$$E(x, \dot{x}) = \dot{x}^2 + U(x) \tag{4.9}$$

The state-space plot in figure 4.2 shows contours of constant energy with the minima at $(x, \dot{x}) = (\pm 2, 0)$. At slightly higher total energy the contours are closed loops around either of the two minima at $x = \pm 2$. At total energies above the saddle energy of 8 the contours are peanut-shaped and are symmetric about the origin. Assuming that the motion is weakly damped, then a particle released with total energy E_{total} which is higher than E_{saddle} will follow a peanut-shaped spiral trajectory centered at $(x, \dot{x}) = (0, 0)$ in the

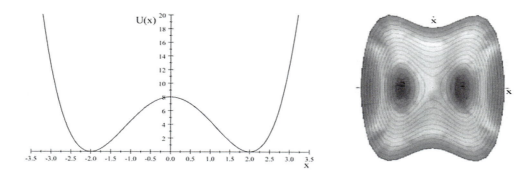

Figure 4.2: The left side shows the potential energy for a bimodal symmetric potential $U(x) = 8 - 4x^2 + 0.5x^4$. The right-hand figure shows contours of the sum of kinetic and potential energies on a state-space diagram. For total energies above the saddle point the particle follows peanut-shaped trajectories in state-space centered around $(x, \dot{x}) = (0, 0)$. For total energies below the saddle point the particle will have closed trajectories about either of the two symmetric minima located at $(x, \dot{x}) = (\pm 2, 0)$. Thus the system solution bifurcates when the total energy is below the saddle point.

state-space diagram for $E_{total} > E_{saddle}$. For $E_{total} < E_{saddle}$ there are two separate solutions for the two minimum centered at $x = \pm 2$ and $\dot{x} = 0$. This is an example of *bifurcation* where the one solution for $E_{total} > E_{saddle}$ bifurcates into either of the two solutions for $E_{total} < E_{saddle}$.

For an initial total energy $E_{total} > E_{saddle}$, damping will result in spiral trajectories of the particle that will be trapped in one of the two minima. For $E_{total} > E_{saddle}$ the particle trajectories are centered giving the impression that they will terminate at $(x, \dot{x}) = (0, 0)$ when the kinetic energy is dissipated. However, for $E_{total} < E_{saddle}$ the particle will be trapped in one of the two minimum and the trajectory will terminate at the bottom of that potential energy minimum occurring at $(x, \dot{x}) = (\pm 2, 0)$. These two possible terminal points of the trajectory are called *point attractors*. This example appears to have a single attractor for $E_{total} > E_{saddle}$ which bifurcates leading to two attractors at $(x, \dot{x}) = (\pm 2, 0)$ for $E_{total} < E_{saddle}$. The determination as to which minimum traps a given particle depends on exactly where the particle starts in state space and the damping etc. That is, for this case, where there is symmetry about the x-axis, the particle has an initial total energy $E_{total} > E_{saddle}$, then the initial conditions with π radians of state space will lead to trajectories that are trapped in the left minimum, and the other π radians of state space will be trapped in the right minimum. Trajectories starting near the split between these two halves of the starting state space will be sensitive to the exact starting phase. This is an example of sensitivity to initial conditions.

4.4 Limit cycles

4.4.1 Poincaré-Bendixson theorem

Coupled first-order differential equations in two dimensions of the form

$$\dot{x} = f(x, y) \qquad \dot{y} = g(x, y) \tag{4.10}$$

occur frequently in physics. The state-space paths do not cross for such two-dimensional autonomous systems, where an autonomous system is not explicitly dependent on time.

The Poincaré-Bendixson theorem states that, state-space, and phase-space, can have three possible paths:
(1) closed paths, like the elliptical paths for the undamped harmonic oscillator,
(2) terminate at an equilibrium point as $t \to \infty$, like the point attractor for a damped harmonic oscillator,
(3) tend to a limit cycle as $t \to \infty$.

The limit cycle is unusual in that the periodic motion tends asymptotically to the limit-cycle attractor independent of whether the initial values are inside or outside the limit cycle. The balance of dissipative forces and driving forces often leads to limit-cycle attractors, especially in biological applications. Identification of limit-cycle attractors, as well as the trajectories of the motion towards these limit-cycle attractors, is more complicated than for point attractors.

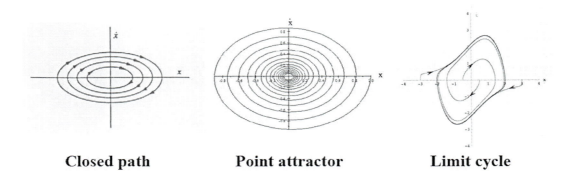

Closed path   **Point attractor**   **Limit cycle**

Figure 4.3: The Poincaré-Bendixson theorem allows the following three scenarios for two-dimensional autonomous systems. (1) Closed paths as illustrated by the undamped harmonic oscillator. (2) Terminate at an equilibrium point as $t \to \infty$, as illustrated by the damped harmonic oscillator, and (3) Tend to a limit cycle as $t \to \infty$ as illustrated by the van der Pol oscillator.

4.4.2   van der Pol damped harmonic oscillator:

The van der Pol damped harmonic oscillator illustrates a non-linear equation that leads to a well-studied, limit-cycle attractor that has important applications in diverse fields. The van der Pol oscillator has an equation of motion given by

$$\frac{d^2x}{dt^2} + \mu\left(x^2 - 1\right)\frac{dx}{dt} + \omega_0^2 x = 0 \tag{4.11}$$

The non-linear $\mu\left(x^2 - 1\right)\frac{dx}{dt}$ damping term is unusual in that the sign changes when $x = 1$ leading to positive damping for $x > 1$ and negative damping for $x < 1$. To simplify equation 4.11, assume that the term $\omega_0^2 x = x$, that is, $\omega_0^2 = 1$.

This equation was studied extensively during the 1920's and 1930's by the Dutch engineer, Balthazar van der Pol, for describing electronic circuits that incorporate feedback. The form of the solution can be simplified by defining a variable $y \equiv \frac{dx}{dt}$. Then the second-order equation 4.11 can be expressed as two coupled first-order equations.

$$y \quad \equiv \quad \frac{dx}{dt} \tag{4.12}$$

$$\frac{dy}{dt} \quad = \quad -x - \mu\left(x^2 - 1\right)y \tag{4.13}$$

It is advantageous to transform the $(\dot{x}, x)$ state space to polar coordinates by setting

$$x \quad = \quad r\cos\theta \tag{4.14}$$

$$y \quad = \quad r\sin\theta$$

and using the fact that $r^2 = x^2 + y^2$. Therefore

$$r\frac{dr}{dt} = x\frac{dx}{dt} + y\frac{dy}{dt} \tag{4.15}$$

Similarly for the angle coordinate

$$\frac{dx}{dt} \quad = \quad \frac{dr}{dt}\cos\theta - r\frac{d\theta}{dt}\sin\theta \tag{4.16}$$

$$\frac{dy}{dt} \quad = \quad \frac{dr}{dt}\sin\theta + r\frac{d\theta}{dt}\cos\theta \tag{4.17}$$

Multiply equation 4.16 by y and 4.17 by x and subtract gives

$$r^2\frac{d\theta}{dt} = x\frac{dy}{dt} - y\frac{dx}{dt} \tag{4.18}$$

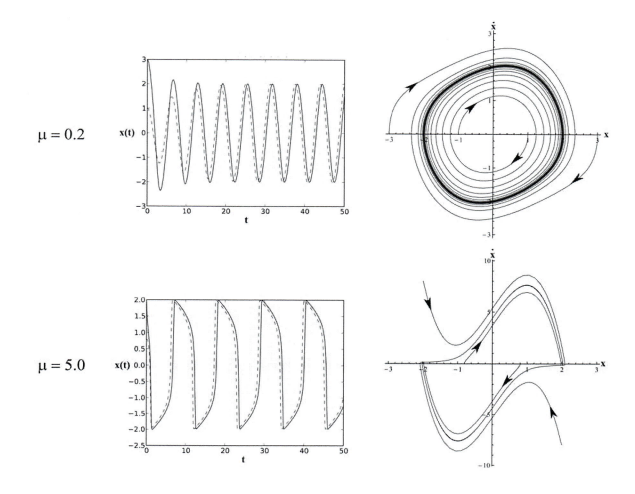

Figure 4.4: Solutions of the van der Pol system for $\mu = 0.2$ top row and $\mu = 5$ bottom row, assuming that $\omega_0^2 = 1$. The left column shows the time dependence $x(t)$. The right column shows the corresponding $(x, \dot{x})$ state space plots. **Upper: Weak nonlinearity,** $\mu= 0.2$; At large times the solution tends to one limit cycle for initial values inside or outside the limit cycle attractor. The amplitude $x(t)$ for two initial conditions approaches an approximately harmonic oscillation. **Lower: Strong nonlinearity,** $\mu = 5$; Solutions approach a common limit cycle attractor for initial values inside or outside the limit cycle attractor while the amplitude $x(t)$ approaches a common approximate square-wave oscillation.

Equations 4.15 and 4.18 allow the van der Pol equations of motion to be written in polar coordinates

$$\frac{dr}{dt} = -\mu \left(r^2 \cos^2 \theta - 1 \right) r \sin^2 \theta \tag{4.19}$$

$$\frac{d\theta}{dt} = -1 - \mu \left(r^2 \cos^2 \theta - 1 \right) \sin \theta \cos \theta \tag{4.20}$$

The non-linear terms on the right-hand side of equations $4.19 - 20$ have a complicated form.

Weak non-linearity: $\mu << 1$

In the limit that $\mu \to 0$, equations $4.19, 4.20$ correspond to a circular state-space trajectory similar to the harmonic oscillator. That is, the solution is of the form

$$x(t) = \rho \sin(t - t_0) \tag{4.21}$$

where ρ and t_0 are arbitrary parameters. For weak non-linearity, $\mu << 1$ the angular equation 4.20 has a rotational frequency that is unity since the $\sin \theta \cos \theta$ term changes sign twice per period, in addition to the small value of μ. For $\mu << 1$ and $r < 1$, the radial equation 4.19 has a sign of the $\left(r^2 \cos^2\theta - 1\right)$ term that is positive and thus the radius increases monotonically to unity. For $r > 1$, the bracket is predominantly negative resulting in a spiral decrease in the radius. Thus, for very weak non-linearity, this radial behavior results in the amplitude spiralling to a well defined limit-cycle attractor value of $\rho = 2$ as illustrated by the state-space plots in figure 4.4 for cases where the initial condition is inside or external to the circular attractor. The final amplitude for different initial conditions also approach the same asymptotic behavior.

Dominant non-linearity: $\mu >> 1$

For the case where the non-linearity is dominant, that is $\mu >> 1$, then as shown in figure 4.4, the system approaches a well defined attractor, but in this case it has a significantly skewed shape in state-space, while the amplitude approximates a square wave. The solution remains close to $x = +2$ until $y = \dot{x} \approx +7$ and then it relaxes quickly to $x = -2$ with $y = \dot{x} \approx 0$. This is followed by the mirror image. This behavior is called a relaxed vibration in that a tension builds up slowly then dissipates by a sudden relaxation process. The seesaw is an extreme example of a relaxation oscillator where the seesaw angle switches spontaneously from one solution to the other when the difference in their moment arms changes sign.

The study of feedback in electronic circuits was the stimulus for study of this equation by van der Pol. However, Lord Rayleigh first identified such *relaxation oscillator* behavior in 1880 during studies of vibrations of a stringed instrument excited by a bow, or the squeaking of a brake drum. In his discussion of non-linear effects in acoustics, he derived the equation

$$\ddot{x} - (a - b\dot{x}^2)\dot{x} + \omega_0^2 x \tag{4.22}$$

Differentiation of Rayleigh's equation 4.22 gives

$$\dddot{x} - (a - 3b\dot{x}^2)\ddot{x} + \omega_0^2 \dot{x} = 0 \tag{4.23}$$

Using the substitution of

$$y = y_0 \sqrt{\frac{3b}{a}} \dot{x} \tag{4.24}$$

leads to the relations

$$\dot{x} = \sqrt{\frac{a}{3b}} \frac{y}{y_0} \qquad\qquad \ddot{x} = \sqrt{\frac{a}{3b}} \frac{\dot{y}}{y_0} \qquad\qquad \dddot{x} = \sqrt{\frac{a}{3b}} \frac{\ddot{y}}{y_0} \tag{4.25}$$

Substituting these relations into equation 4.23 gives

$$\sqrt{\frac{a}{3b}} \frac{\ddot{y}}{y_0} - \sqrt{\frac{a}{3b}} \left[a - \frac{3ba}{b} \frac{\dot{y}^2}{y_0^2} \right] \frac{\dot{y}}{y_0} + \omega_0^2 \sqrt{\frac{a}{3b}} \frac{y}{y_0} = 0 \tag{4.26}$$

Multiplying by $y_0 \sqrt{\frac{3b}{a}}$ and rearranging leads to the van der Pol equation

$$\ddot{y} - \frac{a}{y_0^2}(y_0^2 - y^2)\dot{y} - \omega_0^2 y = 0 \tag{4.27}$$

The rhythm of a heartbeat driven by a pacemaker is an important application where the self-stabilization of the attractor is a desirable characteristic to stabilize an irregular heartbeat; the medical term is arrhythmia. The mechanism that leads to synchronization of the many pacemaker cells in the heart and human body due to the influence of an implanted pacemaker is discussed in chapter 14.12. Another biological application of limit cycles is the time variation of animal populations.

In summary the non-linear damping of the van der Pol oscillator leads to a self-stabilized, single limit-cycle attractor that is insensitive to the initial conditions. The van der Pol oscillator has many important applications such as bowed musical instruments, electrical circuits, and human anatomy as mentioned above. The van der Pol oscillator illustrates the complicated manifestations of the motion that can be exhibited by non-linear systems

4.5 Harmonically-driven, linearly-damped, plane pendulum

The harmonically-driven, linearly-damped, plane pendulum illustrates many of the phenomena exhibited by non-linear systems as they evolve from ordered to chaotic motion. It illustrates the remarkable fact that determinism does not imply either regular behavior or predictability. The well-known, harmonically-driven linearly-damped pendulum provides an ideal basis for an introduction to non-linear dynamics[1].

Consider a harmonically-driven linearly-damped plane pendulum of moment of inertia I and mass m in a gravitational field that is driven by a torque due to a force $F(t) = F_D \cos \omega t$ acting at a moment arm L. The damping term is b and the angular displacement of the pendulum, relative to the vertical, is θ. The equation of motion of the harmonically-driven linearly-damped simple pendulum can be written as

$$I\ddot{\theta} + b\dot{\theta} + mgL \sin \theta = LF_D \cos \omega t \tag{4.28}$$

Note that the sinusoidal restoring force for the plane pendulum is non-linear for large angles θ. The natural period of the free pendulum is

$$\omega_0 = \sqrt{\frac{mgL}{I}} \tag{4.29}$$

A dimensionless parameter γ, which is called the **drive strength,** is defined by

$$\gamma \equiv \frac{F_D}{mg} \tag{4.30}$$

The equation of motion 4.28 can be generalized by introducing dimensionless units for both time $\tilde{t}$ and relative drive frequency $\tilde{\omega}$ defined by

$$\tilde{t} \equiv \omega_0 t \qquad\qquad \tilde{\omega} \equiv \frac{\omega}{\omega_0} \tag{4.31}$$

In addition, define the inverse damping factor Q as

$$Q \equiv \frac{\omega_0 I}{b} \tag{4.32}$$

These definitions allow equation 4.28 to be written in the dimensionless form

$$\frac{d^2\theta}{d\tilde{t}^2} + \frac{1}{Q}\frac{d\theta}{d\tilde{t}} + \sin \theta = \gamma \cos \tilde{\omega}\tilde{t} \tag{4.33}$$

The behavior of the angle θ for the driven damped plane pendulum depends on the drive strength γ and the damping factor Q. Consider the case where equation 4.33 is evaluated assuming that the damping coefficient $Q = 2$, and that the relative angular frequency $\tilde{\omega} = \frac{2}{3}$, which is close to resonance where chaotic phenomena are manifest. The Runge-Kutta method is used to solve this non-linear equation of motion.

4.5.1 Close to linearity

For drive strength $\gamma = 0.2$ the amplitude is sufficiently small that $\sin \theta \simeq \theta$, superposition applies, and the solution is identical to that for the driven linearly-damped linear oscillator. As shown in figure 4.5, once the transient solution dies away, the steady-state solution asymptotically approaches one attractor that has an amplitude of ± 0.3 radians and a phase shift δ with respect to the driving force. The abscissa is given in units of the dimensionless time $\tilde{t} = \omega_0 t$. The transient solution depends on the initial conditions and dies away after about 5 periods, whereas the steady-state solution is independent of the initial conditions and has a state-space diagram that has an elliptical shape, characteristic of the harmonic oscillator. For all initial conditions, the time dependence and state space diagram for steady-state motion approaches a unique solution, called an "**attractor**", that is, the pendulum oscillates sinusoidally with a given amplitude at the frequency of the driving force and with a constant phase shift δ, i.e.

$$\theta(t) = A \cos(\omega t - \delta). \tag{4.34}$$

This solution is identical to that for the harmonically-driven, linearly-damped, linear oscillator discussed in chapter 3.6.

---

[1] A similar approach is used by the book *"Chaotic Dynamics"* by Baker and Gollub[Bak96].

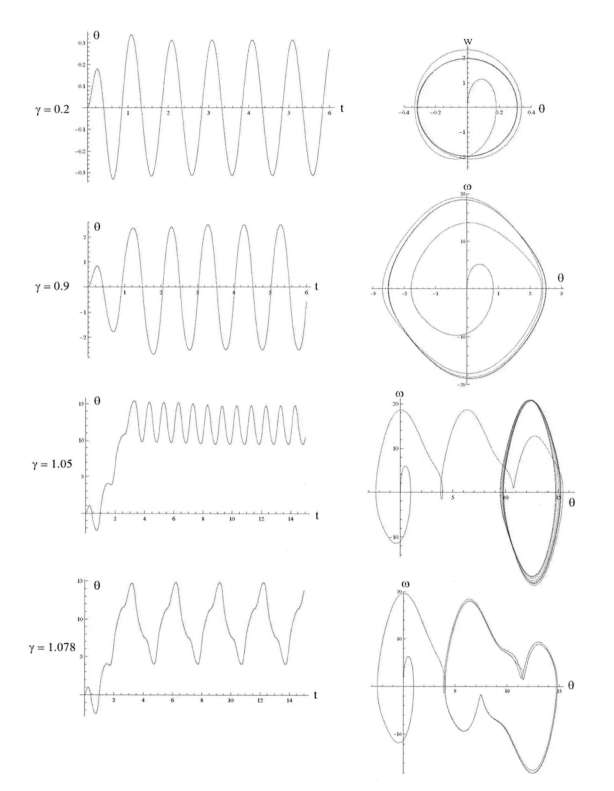

Figure 4.5: Motion of the driven damped pendulum for drive strengths of $\gamma = 0.2$, $\gamma = 0.9$, $\gamma = 1.05$, and $\gamma = 1.078$. The left side shows the time dependence of the deflection angle θ with the time axis expressed in dimensionless units $\tilde{t}$. The right side shows the corresponding state-space plots. These plots assume $\tilde{\omega} = \frac{\omega}{\omega_0} = \frac{2}{3}$, $Q = 2$, and the motion starts with $\theta = \omega = 0$.

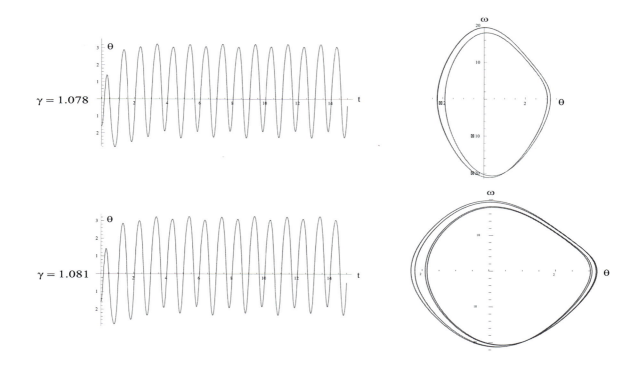

Figure 4.6: The driven damped pendulum assuming that $\tilde{\omega} = \frac{2}{3}$, $Q = 2$, with initial conditions $\theta(0) = -\frac{\pi}{2}$, $\omega(0) = 0$. The system exhibits period-two motion for drive strengths of $\gamma = 1.078$ as shown by the state space diagram for cycles $10 - 20$. For $\gamma = 1.081$ the system exhibits period-four motion shown for cycles $10 - 30$.

4.5.2 Weak nonlinearity

Figure 4.5 shows that for drive strength $\gamma = 0.9$, after the transient solution dies away, the steady-state solution settles down to one attractor that oscillates at the drive frequency with an amplitude of slightly more than $\frac{\pi}{2}$ radians for which the small angle approximation fails. The distortion due to the non-linearity is exhibited by the non-elliptical shape of the state-space diagram.

The observed behavior can be calculated using the successive approximation method discussed in chapter 4.2. That is, close to small angles the sine function can be approximated by replacing

$$\sin\theta \approx \theta - \frac{1}{6}\theta^3$$

in equation 4.33 to give

$$\ddot{\theta} + \frac{1}{Q}\dot{\theta} + \omega_0^2\left(\theta - \frac{1}{6}\theta^3\right) = \gamma\cos\tilde{\omega}\tilde{t} \tag{4.35}$$

As a first approximation assume that

$$\theta(\tilde{t}) \approx A\cos(\tilde{\omega}\tilde{t} - \delta)$$

then the small ϕ^3 term in equation 4.35 contributes a term proportional to $\cos^3(\tilde{\omega}\tilde{t} - \delta)$. But

$$\cos^3(\tilde{\omega}\tilde{t} - \delta) = \frac{1}{4}\left(\cos 3(\tilde{\omega}\tilde{t} - \delta) + 3\cos(\tilde{\omega}\tilde{t} - \delta)\right)$$

That is, the nonlinearity introduces a small term proportional to $\cos 3(\omega t - \delta)$. Since the right-hand side of equation 4.35 is a function of only $\cos\omega t$, then the terms in $\theta, \dot{\theta}$, and $\ddot{\theta}$ on the left hand side must contain the third harmonic $\cos 3(\omega t - \delta)$ term. Thus a better approximation to the solution is of the form

$$\theta(\tilde{t}) = A\left[\cos(\tilde{\omega}\tilde{t} - \delta) + \varepsilon\cos 3(\tilde{\omega}\tilde{t} - \delta)\right] \tag{4.36}$$

where the admixture coefficient $\varepsilon < 1$. This successive approximation method can be repeated to add additional terms proportional to $\cos n(\omega t - \delta)$ where n is an integer with $n \geq 3$. Thus the nonlinearity introduces progressively weaker n-fold harmonics to the solution. This successive approximation approach is viable only when the admixture coefficient $\varepsilon < 1$. Note that these harmonics are integer multiples of ω, thus the steady-state response is identical for each full period even though the state space contours deviate from an elliptical shape.

4.5.3 Onset of complication

Figure 4.5 shows that for $\gamma = 1.05$ the drive strength is sufficiently strong to cause the transient solution for the pendulum to rotate through two complete cycles before settling down to a single steady-state attractor solution at the drive frequency. However, this attractor solution is shifted two complete rotations relative to the initial condition. The state space diagram clearly shows the rolling motion of the transient solution for the first two periods prior to the system settling down to a single steady-state attractor. The successive approximation approach completely fails at this coupling strength since θ oscillates through large values that are multiples of π.

Figure 4.5 shows that for drive strength $\gamma = 1.078$ the motion evolves to a much more complicated periodic motion with a period that is three times the period of the driving force. Moreover the amplitude exceeds 2π corresponding to the pendulum oscillating over top dead center with the centroid of the motion offset by 3π from the initial condition. Both the state-space diagram, and the time dependence of the motion, illustrate the complexity of this motion which depends sensitively on the magnitude of the drive strength γ, in addition to the initial conditions, $(\theta(0), \omega(0))$ and damping factor Q as is shown in figure 4.6

4.5.4 Period doubling and bifurcation

For drive strength $\gamma = 1.078$, with the initial condition $(\theta(0), \omega(0)) = (0,0)$, the system exhibits a regular motion with a period that is three times the drive period. In contrast, if the initial condition is $[\theta(0) = -\frac{\pi}{2}, \omega(0) = 0]$ then, as shown in figure 4.6, the steady-state solution has the drive frequency with no offset in θ, that is, it exhibits period-one oscillation. This appearance of two separate and very different attractors for $\gamma = 1.078$, using different initial conditions, is called **bifurcation**.

An additional feature of the system response for $\gamma = 1.078$ is that changing the initial conditions to $[\theta(0) = -\frac{\pi}{2}, \omega(0) = 0]$ shows that the amplitude of the even and odd periods of oscillation differ slightly in shape and amplitude, that is, the system really has period-two oscillation. This period-two motion, i.e. **period doubling**, is clearly illustrated by the state space diagram in that, although the motion still is dominated by period-one oscillations, the even and odd cycles are slightly displaced. Thus, for different initial conditions, the system for $\gamma = 1.078$ bifurcates into either of two attractors that have very different waveforms, one of which exhibits period doubling.

The period doubling exhibited for $\gamma = 1.078$, is followed by a second period doubling when $\gamma = 1.081$ as shown in figure 4.6 . With increase in drive strength this period doubling keeps increasing in binary multiples to period 8, 16, 32, 64 etc. Numerically it is found that the threshold for period doubling is $\gamma_1 = 1.0663$, from two to four occurs at $\gamma_2 = 1.0793$ etc. Feigenbaum showed that this cascade increases with increase in drive strength according to the relation that obeys

$$(\gamma_{n+1} - \gamma_n) \simeq \frac{1}{\delta}(\gamma_n - \gamma_{n-1}) \tag{4.37}$$

where $\delta = 4.6692016$, δ is called a Feigenbaum number. As $n \to \infty$ this cascading sequence goes to a limit γ_c where

$$\gamma_c = 1.0829 \tag{4.38}$$

4.5.5 Rolling motion

It was shown that for $\gamma > 1.05$ the transient solution causes the pendulum to have angle excursions exceeding 2π, that is, the system rolls over top dead center. For drive strengths in the range $1.3 < \gamma < 1.4$, the steady-state solution for the system undergoes continuous rolling motion as illustrated in figure 4.7. The time dependence for the angle exhibits a periodic oscillatory motion superimposed upon a monotonic rolling motion, whereas the time dependence of the angular frequency $\omega = \frac{d\theta}{dt}$ is periodic. The state space plots

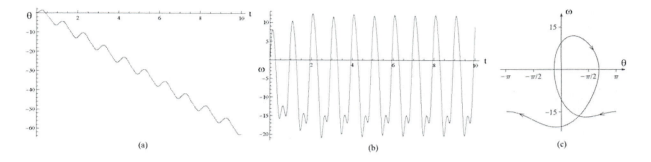

Figure 4.7: Rolling motion for the driven damped plane pendulum for $\gamma = 1.4$. (a) The time dependence of angle $\theta(t)$ increases by 2π per drive period whereas (b) the angular velocity $\omega(t)$ exhibits periodicity. (c) The state space plot for rolling motion is shown with the origin shifted by 2π per revolution to keep the plot within the bounds $-\pi < \theta < +\pi$

for rolling motion corresponds to a chain of loops with a spacing of 2π between each loop. The state space diagram for rolling motion is more compactly presented if the origin is shifted by 2π per revolution to keep the plot within bounds as illustrated in figure 4.7c.

4.5.6   Onset of chaos

When the drive strength is increased to $\gamma = 1.105$, then the system does not approach a unique attractor as illustrated by figure 4.8$left$ which shows state space orbits for cycles $25 - 200$. Note that these orbits do not repeat implying the onset of chaos. For drive strengths greater than $\gamma_c = 1.0829$ the driven damped plane pendulum starts to exhibit chaotic behavior. The onset of chaotic motion is illustrated by making a 3-dimensional plot which combines the time coordinate with the state-space coordinates as illustrated in figure 4.8$right$. This plot shows 16 trajectories starting at different initial values in the range $-0.15 < \theta < 0.15$ for $\gamma = 1.168$. Some solutions are erratic in that, while trying to oscillate at the drive frequency, they never settle down to a steady periodic motion which is characteristic of chaotic motion. Figure 4.8$right$ illustrates the considerable sensitivity of the motion to the initial conditions. That is, this deterministic system can exhibit either order, or chaos, dependent on miniscule differences in initial conditions.

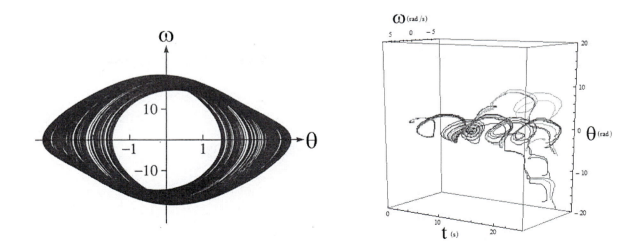

Figure 4.8: Left: Space-space orbits for the driven damped pendulum with $\gamma = 1.105$. Note that the orbits do not repeat for cycles 25 to 200. Right: Time-state-space diagram for $\gamma = 1.168$. The plot shows 16 trajectories starting with different initial values in the range $-0.15 < \theta < 0.15$.

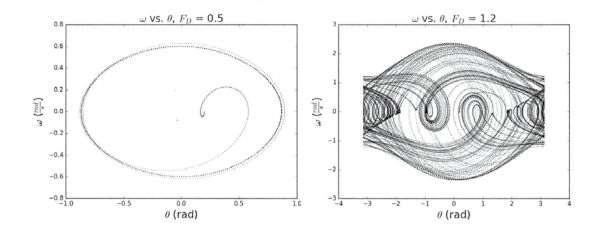

Figure 4.9: State-space plots for the harmonically-driven, linearly-damped, pendulum for driving amplitudes of $F_D = 0.5$ and $F_D = 1.2$. These calculations were performed using the Runge-Kutta method by E. Shah, (Private communication)

4.6   Differentiation between ordered and chaotic motion

Chapter 4.5 showed that motion in non-linear systems can exhibit both order and chaos. The transition between ordered motion and chaotic motion depends sensitively on both the initial conditions and the model parameters. It is surprisingly difficult to unambiguously distinguish between complicated ordered motion and chaotic motion. Moreover, the motion can fluctuate between order and chaos in an erratic manner depending on the initial conditions. The extremely sensitivity to initial conditions of the motion for non-linear systems, makes it essential to have quantitative measures that can characterize the degree of order, and interpret the complicated dynamical motion of systems. As an illustration, consider the harmonically-driven, linearly-damped, pendulum with $Q = 2$, and driving force $F(t) = F_D \sin \tilde{\omega} \tilde{t}$ where $\tilde{\omega} = \frac{2}{3}$. Figure 4.9 shows the state-space plots for two driving amplitudes, $F_D = 0.5$ which leads to ordered motion, and $F_D = 1.2$ which leads to possible chaotic motion. It can be seen that for $F_D = 0.5$ the state-space diagram converges to a single attractor once the transient solution has died away. This is in contrast to the case for $F_D = 1.2$, where the state-space diagram does not converge to a single attractor, but exhibits possible chaotic motion. Three quantitative measures can be used to differentiate ordered motion from chaotic motion for this system; namely, the Lyapunov exponent, the bifurcation diagram, and the Poincaré section, as illustrated below.

4.6.1   Lyapunov exponent

The Lyapunov exponent provides a quantitative and useful measure of the instability of trajectories, and how quickly nearby initial conditions diverge. It compares two identical systems that start with an infinitesimally small difference in the initial conditions in order to ascertain whether they converge to the same attractor at long times, corresponding to a stable system, or whether they diverge to very different attractors, characteristic of chaotic motion. If the initial separation between the trajectories in phase space at $t = 0$ is $|\delta Z_0|$, then to first order the time dependence of the difference can be assumed to depend exponentially on time. That is,

$$|\delta Z(t)| \sim e^{\lambda t} |Z_0| \tag{4.39}$$

where λ is the Lyapunov exponent. That is, the Lyapunov exponent is defined to be

$$\lambda = \lim_{t \to \infty} \lim_{\delta Z_0 \to 0} \frac{1}{t} \ln \frac{|\delta Z(t)|}{|Z_0|} \tag{4.40}$$

Systems for which the Lyapunov exponent $\lambda < 0$ (negative), converge exponentially to the same attractor solution at long times since $|\delta Z(t)| \to 0$ for $t \to \infty$. By contrast, systems for which $\lambda > 0$ (positive) diverge to completely different long-time solutions, that is, $|\delta Z(t)| \to \infty$ for $t \to \infty$. Even for infinitesimally

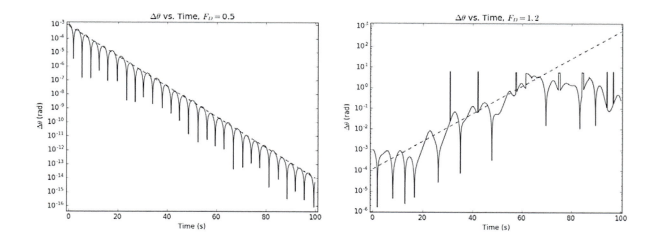

Figure 4.10: Lyapunov plots of $\Delta\theta$ versus time for two initial starting points differing by $\Delta\theta_0 = 0.001 rads$. The parameters are $Q = 2$, and $F(t) = F_D \sin(\frac{2}{3}t)$, and $\Delta t = 0.04s$. The Lyapunov exponent for $F_D = 0.5$ which is drawn as a dashed line, is convergent with $\lambda = -0.251$. For $F_D = 1.2$ the exponent is divergent as indicated by the dashed line which as a slope of $\lambda = 0.1538$. These calculations were performed using the Runge-Kutta method by E. Shah, (Private communication)

small differences in the initial conditions, systems having a positive Lyapunov exponent diverge to different attractors, whereas when the Lyapunov exponent $\lambda < 0$ they correspond to stable solutions.

Figure 4.10 illustrates Lyapunov plots for the harmonically-driven, linearly-damped, plane pendulum, with the same conditions discussed in chapter 4.5. Note that for the small driving amplitude $F_D = 0.5$, the Lyapunov plot converges to ordered motion with an exponent $\lambda = -0.251$, whereas for $F_D = 1.2$, the plot diverges characteristic of chaotic motion with an exponent $\lambda = 0.1538$. The Lyapunov exponent usually fluctuates widely at the local oscillator frequency, and thus the time average of the Lyapunov exponent must be taken over many periods of the oscillation to identify the general trend with time. Some systems near an order-to-chaos transition can exhibit positive Lyapunov exponents for short times, characteristic of chaos, and then converge to negative λ at longer time implying ordered motion. The Lyapunov exponents are used extensively to monitor the stability of the solutions for non-linear systems. For example the Lyapunov exponent is used to identify whether fluid flow is laminar or turbulent as discussed in chapter 16.8.

A dynamical system in n-dimensional phase space will have a set of n Lyapunov exponents $\{\lambda_1, \lambda_2, ..., \lambda_n\}$ associated with a set of attractors, the importance of which depend on the initial conditions. Typically one Lyapunov exponent dominates at one specific location in phase space, and thus it is usual to use the maximal Lyapunov exponent to identify chaos. The Lyapunov exponent is a very sensitive measure of the onset of chaos and provides an important test of the chaotic nature for the complicated motion exhibited by non-linear systems.

4.6.2   Bifurcation diagram

The bifurcation diagram simplifies the presentation of the dynamical motion by sampling the status of the system once per period, synchronized to the driving frequency, for many sets of initial conditions. The results are presented graphically as a function of one parameter of the system in the bifurcation diagram. For example, the wildly different behavior in the driven damped plane pendulum is represented on a bifurcation diagram in figure 4.11, which shows the observed angular velocity ω of the pendulum sampled once per drive cycle plotted versus drive strength. The bifurcation diagram is obtained by sampling either the angle θ, or angular velocity ω, once per drive cycle, that is, it represents the observables of the pendulum using a stroboscopic technique that samples the motion synchronous with the drive frequency. Bifurcation plots also can be created as a function of either the time $\tilde{t}$, the damping factor Q, the normalized frequency $\tilde{\omega} = \frac{\omega}{\omega_0}$, or the driving amplitude γ.

In the domain with drive strength $\gamma <$ 1.0663 there is one unique angle each drive cycle as illustrated by the bifurcation diagram. For slightly higher drive strength period-two bifurcation behavior results in two different angles per drive cycle. The Lyapunov exponent is negative for this region corresponding to ordered motion. The cascade of period doubling with increase in drive strength is readily apparent until chaos sets in at the critical drive strength γ_c when there is a random distribution of sampled angular velocities and the Lyapunov exponent becomes positive. Note that at $\gamma = 1.0845$ there is a brief interval of period-6 motion followed by another region of chaos. Around $\gamma = 1.1$ there is a region that is primarily chaotic which is reflected by chaotic values of the angular velocity on the bifurcation plot and large positive values of the Lyapunov exponent. The region around $\gamma = 1.12$ exhibits period three motion and negative Lyapunov exponent corresponding to ordered motion. The $1.15 < \gamma < 1.25$ region is mainly chaotic and has a large positive Lyapunov exponent. The region with $1.3 < \gamma < 1.4$ is striking in that this corresponds to rolling motion with reemergence of period one and negative Lyapunov exponent. This period-1 motion is due to a continuous rolling motion of the

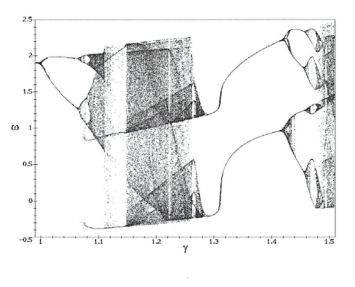

Figure 4.11: Bifurcation diagram samples the angular velocity ω once per period for the driven, linearly-damped, plane pendulum plotted as a function of the drive strength γ. Regions of period doubling, and chaos, as well as islands of stability all are manifest as the drive strength γ is changed. Note that the limited number of samples causes broadening of the lines adjacent to bifurcations.

plane pendulum as shown in figure 4.7 where it is seen that the average θ increases 2π per cycle, whereas the angular velocity ω exhibits a periodic motion. That is, on average the pendulum is rotating 2π per cycle. Above $\gamma = 1.4$ the system start to exhibit period doubling followed by chaos reminiscent of the behavior seen at lower γ values.

These results show that the bifurcation diagram nicely illustrates the order to chaos transitions for the harmonically-driven, linearly-damped, pendulum. Several transitions between order and chaos are seen to occur. The apparent ordered and chaotic regimes are confirmed by the corresponding Lyapunov exponents which alternate between negative and positive values for the ordered and chaotic regions respectively.

4.6.3   Poincaré Section

State-space plots are very useful for characterizing periodic motion, but they become too dense for useful interpretation when the system approaches chaos as illustrated in figure 4.11. Poincaré sections solve this difficulty by taking a stroboscopic sample once per cycle of the state-space diagram. That is, the point on the state space orbit is sampled once per drive frequency. For period-1 motion this corresponds to a single point (θ, ω). For period-2 motion this corresponds to two points etc. For chaotic systems the sequence of state-space sample points follow complicated trajectories. Figure 4.12 shows the Poincaré sections for the corresponding state space diagram shown in figure 4.9 for cycles 10 to 6000. Note the complicated curves do not cross or repeat. Enlargements of any part of this plot will show increasingly dense parallel trajectories, called **fractals**, that indicates the complexity of the chaotic cyclic motion. That is, zooming in on a small section of this Poincaré plot shows many closely parallel trajectories. The fractal attractors are surprisingly robust to large differences in initial conditions. Poincaré sections are a sensitive probe of periodic motion for systems where periodic motion is not readily apparent.

In summary, the behavior of the well-known, harmonically-driven, linearly-damped, plane pendulum becomes remarkably complicated at large driving amplitudes where non-linear effects dominate. That is,

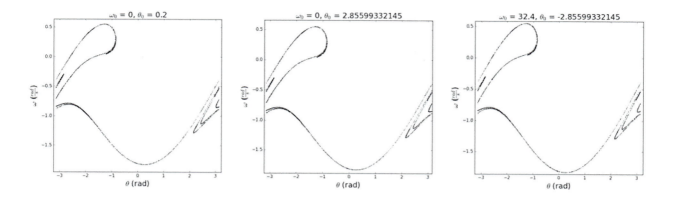

Figure 4.12: Three Poincaré section plots for the harmonically-driven, linearly-damped, pendulum for various initial conditions with $F_D = 1.2, \tilde{\omega} = \frac{2}{3}$, and $\Delta t = \frac{\pi}{100}$. These calculations used the Runge-Kutta method and were performed for $6000 cycles$ by E. Shah (Private communication).

when the restoring force is non-linear. The system exhibits bifurcation where it can evolve to multiple attractors that depend sensitively on the initial conditions. The system exhibits both oscillatory, and rolling, solutions depending on the amplitude of the motion. The system exhibits domains of simple ordered motion separated by domains of very complicated ordered motion as well as chaotic regions. The transitions between these dramatically different modes of motion are extremely sensitive to the amplitude and phase of the driver. Eventually the motion becomes completely chaotic. The Lyapunov exponent, bifurcation diagram, and Poincaré section plots, are sensitive measures of the order of the motion. These three sensitive measures of order and chaos are used extensively in many fields in classical mechanics. Considerable computing capabilities are required to elucidate the complicated motion involved in non-linear systems. Examples include laminar and turbulent flow in fluid dynamics and weather forecasting of hurricanes, where the motion can span a wide dynamic range in dimensions from 10^{-5} to $10^4 m$.

4.7 Wave propagation for non-linear systems

4.7.1 Phase, group, and signal velocities

Chapter 3 discussed the wave equation and solutions for linear systems. It was shown that, for linear systems, the wave motion obeys superposition and exhibits dispersion, that is, a frequency-dependent phase velocity, and, in some cases, attenuation. Nonlinear systems introduce intriguing new wave phenomena. For example for nonlinear systems, second, and higher terms must be included in the Taylor expansion given in equation 4.2. These second and higher order terms result in the group velocity being a function of ω, that is, group velocity dispersion occurs which leads to the shape of the envelope of the wave packet being time dependent. As a consequence the group velocity in the wave packet is not well defined, and does not equal the signal velocity of the wave packet or the phase velocity of the wavelets. Nonlinear optical systems have been studied experimentally where $v_{group} << c$, which is called slow light, while other systems have $v_{group} > c$ which is called superluminal light. The ability to control the velocity of light in such optical systems is of considerable current interest since it has signal transmission applications.

The dispersion relation for a nonlinear system can be expressed as a Taylor expansion of the form

$$k = k_0 + \left(\frac{\partial k}{\partial \omega}\right)_{\omega=\omega_0} (\omega - \omega_0) + \frac{1}{2}\left(\frac{\partial^2 k}{\partial \omega^2}\right)_{\omega=\omega_0} (\omega - \omega_0)^2 + .. \qquad (4.41)$$

where ω is used as the independent variable since it is invariant to phase transitions of the system. Note that the factor for the first derivative term is the reciprocal of the group velocity

$$\left(\frac{\partial k}{\partial \omega}\right)_{\omega=\omega_0} \equiv \frac{1}{v_{group}} \qquad (4.42)$$

while the factor for the second derivative term is

$$\left(\frac{\partial^2 k}{\partial \omega^2}\right)_{\omega=\omega_0} = \frac{\partial}{\partial \omega}\left[\frac{1}{v_{group}(\omega)}\right]_{\omega=\omega_0} = \left(-\frac{1}{v_{group}^2}\frac{\partial v_{group}}{\partial \omega}\right)_{\omega=\omega_0}$$

(4.43)

which gives the velocity dispersion for the system.

Since

$$k = \frac{\omega}{v_{phase}}$$

(4.44)

then

$$\frac{\partial k}{\partial \omega} \equiv \frac{1}{v_{group}} = \frac{1}{v_{phase}} + \omega \frac{\partial \frac{1}{v_{phase}}}{\partial \omega}$$

(4.45)

The inverse velocities for electromagnetic waves are best represented in terms of the corresponding refractive indices n, where

$$n \equiv \frac{c}{v_{phase}}$$

(4.46)

and the group refractive index

$$n_{group} \equiv \frac{c}{v_{group}}$$

(4.47)

Then equation 4.45 can be written in the more convenient form

$$n_{group} = n + \omega \frac{\partial n}{\partial \omega}$$

(4.48)

Wave propagation for an optical system that is subject to a single resonance gives one example of nonlinear frequency response that has applications to optics.

Figure 4.13 shows that the real n_R and imaginary n_I parts of the phase refractive index exhibit the characteristic resonance frequency dependence of the sinusoidally-driven, linear oscillator that was discussed in chapter 3.6 and as illustrated in figure 3.10. Figure 4.13 also shows the group refractive index n_{group} computed using equation 4.48.

Note that at resonance, n_{group} is reduced below the non-resonant value which corresponds to superluminal (fast) light, whereas in the wings of the resonance n_{group} is larger than the non-resonant value corresponding to slow light. Thus the nonlinear dependence of the refractive index n on angular frequency ω leads to fast or slow group velocities for isolated wave packets. Velocities of light as slow as $17m/\sec$ have been observed. Experimentally the energy absorption that occurs on resonance makes it difficult to observe the superluminal electromagnetic wave at resonance.

Note that Sommerfeld and Brillouin showed that even though the group velocity may exceed c, the signal velocity, which marks the arrival of the leading edge of the optical pulse, does not exceed c, the velocity of light in vacuum, as was postulated by Einstein.[Bri14]

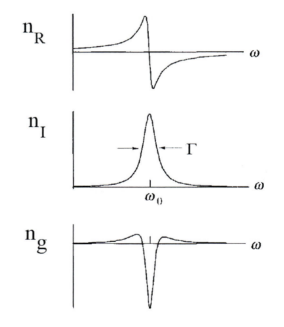

Figure 4.13: The real and imaginary parts of the phase refractive index n plus the real part of the group refractive index associated with an isolated atomic resonance.

4.7.2 Soliton wave propagation

The soliton is a fascinating and very special wave propagation phenomenon that occurs for certain non-linear systems. The soliton is a self-reinforcing solitary localized wave packet that maintains its shape while travelling long distances at a constant speed. Solitons are caused by a cancellation of phase modulation resulting from non-linear velocity dependence, and the group velocity dispersive effects in a medium. Solitons arise as solutions of a widespread class of weakly-nonlinear dispersive partial differential equations describing many physical systems. Figure 4.14 shows a soliton comprising a solitary water wave approaching the coast of Hawaii. While the soliton in Fig. 4.14 may appear like a normal wave, it is unique in that there are no other waves accompanying it. This wave was probably created far away from the shore when a normal wave was modulated by a geometrical change in the ocean depth, such as the rising sea floor, which forced it into the appropriate shape for a soliton. The wave then was able to travel to the coast intact,

Figure 4.14: A solitary wave approaches the coast of Hawaii. (Image: Robert Odom/University of Washington)

despite the apparently placid nature of the ocean near the beach. Solitons are notable in that they interact with each other in ways very different from normal waves. Normal waves are known for their complicated interference patterns that depend on the frequency and wavelength of the waves. Solitons, can pass right through each other without being a affected at all. This makes solitons very appealing to scientists because soliton waves are more sturdy than normal waves, and can therefore be used to transmit information in ways that are distinctly different than for normal wave motion. For example, optical solitons are used in optical fibers made of a dispersive, nonlinear optical medium, to transmit optical pulses with an invariant shape.

Solitons were first observed in 1834 by John Scott Russell (1808 − 1882). Russell was an engineer conducting experiments to increase the efficiency of canal boats. His experimental and theoretical investigations allowed him to recreate the phenomenon in wave tanks. Through his extensive studies, Scott Russell noticed that soliton propagation exhibited the following properties:

• The waves are stable and hold their shape for long periods of time.

• The waves can travel over long distances at uniform speed.

• The speed of propagation of the wave depends on the size of the wave, with larger waves traveling faster than smaller waves.

• The waves maintained their shape when they collided - seemingly passing right through each other.

Scott Russell's work was met with scepticism by the scientific community. The problem with the Wave of Translation was that it was an effect that depended on nonlinear effects, whereas previously existing theories of hydrodynamics (such as those of Newton and Bernoulli) only dealt with linear systems. George Biddell Airy, and George Gabriel Stokes, published papers attacking Scott Russell's observations because the observations could not be explained by their theories of wave propagation in water. Regardless, Scott Russell was convinced of the prime importance of the Wave of Translation, and history proved that he was correct. Scott Russell went on to develop the "wave line" system of hull construction that revolutionized nineteenth century naval architecture, along with a number of other great accomplishments leading him to fame and prominence. Despite all of the success in his career, he continued throughout his life to pursue his studies of the Wave of Translation.

In 1895 Korteweg and de Vries developed a wave equation for surface waves for shallow water.

$$\frac{\partial \phi}{\partial t} + \frac{\partial^3 \phi}{\partial x^3} + 6\phi\frac{\partial \phi}{\partial x} = 0 \tag{4.49}$$

A solution of this equation has the characteristics of a solitary wave with fixed shape. It is given by substituting the form $\phi(x,t) = f(x - vt)$ into the Korteweg-de Vries equation which gives

$$-v\frac{\partial f}{\partial x} + \frac{\partial^3 f}{\partial x^3} + 6f\frac{\partial f}{\partial x} = 0 \tag{4.50}$$

Integrating with respect to x gives

$$3f^2 + \frac{d^2 f}{dx^3} - cf = C \tag{4.51}$$

where C is a constant of integration. This non-linear equation has a solution

$$\phi(x,t) = \frac{1}{2}c\sec h^2\left[\frac{\sqrt{v}}{2}(x - vt - a)\right] \tag{4.52}$$

where a is a constant. Equation 4.52 is the equation of a solitary wave moving in the $+x$ direction at a velocity v.

Soliton behavior is observed in phenomena such as tsunamis, tidal bores that occur for some rivers, signals in optical fibres, plasmas, atmospheric waves, vortex filaments, superconductivity, and gravitational fields having cylindrical symmetry. Much work has been done on solitons for fibre optics applications. The soliton's inherent stability make long-distance transmission possible without the use of repeaters, and could potentially double the transmission capacity.

Before the discovery of solitons, mathematicians were under the impression that nonlinear partial differential equations could not be solved exactly. However, solitons led to the recognition that there are non-linear systems that can be solved analytically. This discovery has prompted much investigation into these so-called "integrable systems." Such systems are rare, as most non-linear differential equations admit chaotic behavior with no explicit solutions. Integrable systems nevertheless lead to very interesting mathematics ranging from differential geometry and complex analysis to quantum field theory and fluid dynamics.

Many of the fundamental equations in physics (Maxwell's, Schrödinger's) are linear equations. However, physicists have begun to recognize many areas of physics in which nonlinearity can result in qualitatively new phenomenon which cannot be constructed via perturbation theory starting from linearized equations. These include phenomena in magnetohydrodynamics, meteorology, oceanography, condensed matter physics, nonlinear optics, and elementary particle physics. For example, the European space mission Cluster detected a soliton-like electrical disturbances that travelled through the ionized gas surrounding the Earth starting about 50,000 kilometers from Earth and travelling towards the planet at about 8 km/s. It is thought that this soliton was generated by turbulence in the magnetosphere.

Efforts to understand the nonlinearity of solitons has led to much research in many areas of physics. In the context of solitons, their particle-like behavior (in that they are localized and preserved under collisions) leads to a number of experimental and theoretical applications. The technique known as bosonization allows viewing particles, such as electrons and positrons, as solitons in appropriate field equations. There are numerous macroscopic phenomena, such as internal waves on the ocean, spontaneous transparency, and the behavior of light in fiber optic cable, that are now understood in terms of solitons. These phenomena are being applied to modern technology.

4.8 Summary

The study of the dynamics of non-linear systems remains a vibrant and rapidly evolving field in classical mechanics as well as many other branches of science. This chapter has discussed examples of non-linear systems in classical mechanics.   It was shown that the superposition principle is broken even for weak nonlinearity. It was shown that increased nonlinearity leads to bifurcation, point attractors, limit-cycle attractors, and sensitivity to initial conditions.

Limit-cycle attractors: The Poincaré-Bendixson theorem for limit cycle attractors states that the paths, both in state-space and phase-space, can have three possible paths:

(1) closed paths, like the elliptical paths for the undamped harmonic oscillator,

(2) terminate at an equilibrium point as $t \to \infty$, like the point attractor for a damped harmonic oscillator,

(3) tend to a limit cycle as $t \to \infty$.

The limit cycle is unusual in that the periodic motion tends asymptotically to the limit-cycle attractor independent of whether the initial values are inside or outside the limit cycle. The balance of dissipative forces and driving forces often leads to limit-cycle attractors, especially in biological applications. Identification of limit-cycle attractors, as well as the trajectories of the motion towards these limit-cycle attractors, is more complicated than for point attractors.

The van der Pol oscillator is a common example of a limit-cycle system that has an equation of motion of the form

$$\frac{d^2x}{dt^2} + \mu \left(x^2 - 1\right) \frac{dx}{dt} + \omega_0^2 x = 0 \tag{4.11}$$

The van der Pol oscillator has a limit-cycle attractor that includes non-linear damping and exhibits periodic solutions that asymptotically approach one attractor solution independent of the initial conditions. There are many examples in nature that exhibit similar behavior.

Harmonically-driven, linearly-damped, plane pendulum: The non-linearity of the well-known driven linearly-damped plane pendulum was used as an example of the behavior of non-linear systems in nature. It was shown that non-linearity leads to discontinuous period bifurcation, extreme sensitivity to initial conditions, rolling motion and chaos.

Differentiation between ordered and chaotic motion: Lyapunov exponents, bifurcation diagrams, and Poincaré sections were used to identify the transition from order to chaos. Chapter 16.8 discusses the non-linear Navier-Stokes equations of viscous-fluid flow which leads to complicated transitions between laminar and turbulent flow. Fluid flow exhibits remarkable complexity that nicely illustrates the dominant role that non-linearity can have on the solutions of practical non-linear systems in classical mechanics.

Wave propagation for non-linear systems: Non-linear equations can lead to unexpected behavior for wave packet propagation such as fast or slow light as well as soliton solutions. Moreover, it is notable that some non-linear systems can lead to analytic solutions.

The complicated phenomena exhibited by the above non-linear systems is not restricted to classical mechanics, rather it is a manifestation of the mathematical behavior of the solutions of the differential equations involved. That is, this behavior is a general manifestation of the behavior of solutions for second-order differential equations. Exploration of this complex motion has only become feasible with the advent of powerful computer facilities during the past three decades. The breadth of phenomena exhibited by these examples is manifest in myriads of other nonlinear systems, ranging from many-body motion, weather patterns, growth of biological species, epidemics, motion of electrons in atoms, etc. Other examples of non-linear equations of motion not discussed here, are the three-body problem, which is mentioned in chapter 11, and turbulence in fluid flow which is discussed in chapter 16.

It is stressed that the behavior discussed in this chapter is very different from the random walk problem which is a stochastic process where each step is purely random and not deterministic. This chapter has assumed that the motion is fully deterministic and rigorously follows the laws of classical mechanics. Even though the motion is fully deterministic, and follows the laws of classical mechanics, the motion is extremely sensitive to the initial conditions and the non-linearities can lead to chaos. Computer modelling is the only viable approach for predicting the behavior of such non-linear systems. The complexity of solving non-linear equations is the reason that this book will continue to consider only linear systems. Fortunately, in nature, non-linear systems can be approximately linear when the small-amplitude assumption is applicable.

Workshop exercises

1. Consider the chaotic motion of the driven damped pendulum whose equation of motion is given by

$$\ddot{\phi} + \Gamma\dot{\phi} + \omega_0^2 \sin\phi = \gamma\omega_0^2 \cos\omega t$$

for which the Lyapunov exponent is $\lambda = 1$ with time measured in units of the drive period.

(a) Assume that you need to predict $\phi(t)$ with accuracy of $10^{-2} radians$, and that the initial value $\phi(0)$ is known to within $10^{-6} radians$. What is the maximum time horizon $t_{\max}$ for which you can predict $\phi(t)$ to within the required accuracy?

(b) Suppose that you manage to improve the accuracy of the initial value to $10^{-9} radians$ (that is, a thousand-fold improvement). What is the time horizon now for achieving the accuracy of $10^{-2} radians$?

(c) By what factor has $t_{\max}$ improved with the $1000 - fold$ improvement in initial measurement.

(d) What does this imply regarding long-term predictions of chaotic motion?

2. A non-linear oscillator satisfies the equation $\ddot{x} + \dot{x}^3 + x = 0$. Find the polar equations for the motion in the state-space diagram. Show that any trajectory that starts within the circle $r < 1$ encircle the origin infinitely many times in the clockwise direction. Show further that these trajectories in state space terminate at the origin.

3. Consider the system of a mass suspended between two identical springs as shown.

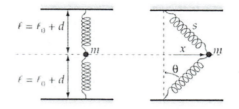

If each spring is stretched a distance d to attach the mass at the equilibrium position the mass is subject to two equal and oppositely directed forces of magnitude κd. Ignore gravity. Show that the potential in which the mass moves is approximately

$$U(x) = \left\{ \frac{\kappa d}{l} \right\} x^2 + \left\{ \frac{\kappa(l-d)}{4l^3} \right\} x^4$$

Construct a state-space diagram for this potential.

Problems

1. A non-linear oscillator satisfies the equation

$$\ddot{x} + (x^2 + \dot{x}^2 - 1)\dot{x} + x = 0$$

Find the polar equations for the motion in the state-space diagram. Show that any trajectory that starts in the domain $1 < r < \sqrt{3}$ spirals clockwise and tends to the limit cycle $r = 1$. [The same is true of trajectories that start in the domain $0 < r < 1$.] What is the period of the limit cycle?

2. A mass m moves in one direction and is subject to a constant force $+F_0$ when $x < 0$ and to a constant force $-F_0$ when $x > 0$. Describe the motion by constructing a state space diagram. Calculate the period of the motion in terms of m, F_0 and the amplitude A. Disregard damping.

3. Investigate the motion of an undamped mass subject to a force of the form

$$F(x) = (\begin{array}{ll} -kx & |x| < a \\ -(k+\delta)x + \delta a & |x| > a \end{array}$$

Chapter 5

Calculus of variations

5.1 Introduction

The prior chapters have focussed on the intuitive Newtonian approach to classical mechanics, which is based on vector quantities like force, momentum, and acceleration. Newtonian mechanics leads to second-order differential equations of motion. The calculus of variations underlies a powerful alternative approach to classical mechanics that is based on identifying the path that minimizes an integral quantity. This integral variational approach was first championed by Gottfried Wilhelm Leibniz, contemporaneously with Newton's development of the differential approach to classical mechanics.

During the 18th century, Bernoulli, who was a student of Leibniz, developed the field of variational calculus which underlies the integral variational approach to mechanics. He solved the brachistochrone problem which involves finding the path for which the transit time between two points is the shortest. The integral variational approach also underlies Fermat's principle in optics, which can be used to derive that the angle of reflection equals the angle of incidence, as well as derive Snell's law. Other applications of the calculus of variations include solving the catenary problem, finding the maximum and minimum distances between two points on a surface, polygon shapes having the maximum ratio of enclosed area to perimeter, or maximizing profit in economics. Bernoulli, developed the principle of virtual work used to describe equilibrium in static systems, and d'Alembert extended the principle of virtual work to dynamical systems. Euler, the preeminent Swiss mathematician of the 18th century and a student of Bernoulli, developed the calculus of variations with full mathematical rigor. The culmination of the development of the Lagrangian variational approach to classical mechanics is done by Lagrange (1736-1813), who was a student of Euler,.

The Euler-Lagrangian approach to classical mechanics stems from a deep philosophical belief that the laws of nature are based on the principle of economy.That is, the physical universe follows paths through space and time that are based on extrema principles. The standard **Lagrangian** L is defined as the difference between the kinetic and potential energy, that is

$$L = T - U \tag{5.1}$$

Chapters 6 through 9 will show that the laws of classical mechanics can be expressed in terms of **Hamilton's variational principle** which states that the motion of the system between the initial time t_1 and final time t_2 follows a path that minimizes the scalar **action integral** S defined as the time integral of the Lagrangian.

$$S = \int_{t_1}^{t_2} L dt \tag{5.2}$$

The calculus of variations provides the mathematics required to determine the path that minimizes the action integral. This variational approach is both elegant and beautiful, and has withstood the rigors of experimental confirmation. In fact, not only is it an exceedingly powerful alternative approach to the intuitive Newtonian approach in classical mechanics, but Hamilton's variational principle now is recognized to be more fundamental than Newton's Laws of Motion. The Lagrangian and Hamiltonian variational approaches to mechanics are the only approaches that can handle the Theory of Relativity, statistical mechanics, and the dichotomy of philosophical approaches to quantum physics.

5.2   Euler's differential equation

The calculus of variations, presented here, underlies the powerful variational approaches that were developed for classical mechanics. Variational calculus, developed for classical mechanics, now has become an essential approach to many other disciplines in science, engineering, economics, and medicine.

For the special case of one dimension, the calculus of variations reduces to varying the function $y(x)$ such that the scalar functional F is an extremum, that is, it is a maximum or minimum, where.

$$F = \int_{x_1}^{x_2} f\left[y(x), y'(x); x\right] dx \tag{5.3}$$

Here x is the independent variable, $y(x)$ the dependent variable, plus its first derivative $y' \equiv \frac{dy}{dx}$. The quantity $f\left[y(x), y'(x); x\right]$ has some given dependence on y, y' and x. The calculus of variations involves varying the function $y(x)$ until a stationary value of F is found, which is presumed to be an extremum. This means that if a function $y = y(x)$ gives a minimum value for the scalar functional F, then any neighboring function, no matter how close to $y(x)$, must increase F.   For all paths, the integral F is taken between two fixed points, x_1, y_1 and x_2, y_2. Possible paths between the initial and final points are illustrated in figure 5.1. Relative to any neighboring path, the functional F must have a stationary value which is presumed to be the correct extremum path.

Define a neighboring function using a parametric representation $y(\epsilon, x)$, such that for $\epsilon = 0$, $y = y(0, x) = y(x)$ is the function that yields the extremum for F. Assume that an infinitesimally small fraction ϵ of the neighboring function $\eta(x)$ is added to the extremum path $y(x)$. That is, assume

$$\begin{aligned} y(\epsilon, x) &= y(0, x) + \epsilon \eta(x) \\ y'(\epsilon, x) &\equiv \frac{dy(\epsilon, x)}{dx} = \frac{dy(0, x)}{dx} + \epsilon \frac{d\eta}{dx} \end{aligned} \tag{5.4}$$

where it is assumed that the extremum function $y(0, x)$ and the auxiliary function $\eta(x)$ are well behaved functions of x with continuous first derivatives, and where $\eta(x)$ vanishes at x_1 and x_2, because, for all possible paths, the function $y(\epsilon, x)$ must be identical with $y(x)$ at the end points of the path, i.e. $\eta(x_1) = \eta(x_2) = 0$. The situation is depicted in figure 5.1. It is possible to express any such parametric family of curves F as a function of ϵ

$$F(\epsilon) = \int_{x_1}^{x_2} f\left[y(\epsilon, x), y'(\epsilon, x); x\right] dx \tag{5.5}$$

The condition that the integral has a stationary (extremum) value is that F be independent of ϵ to first order along the path. That is, the extremum value occurs for $\epsilon = 0$ where

$$\left(\frac{dF}{d\epsilon}\right)_{\epsilon=0} = 0 \tag{5.6}$$

for all functions $\eta(x)$. This is illustrated on the right side of figure 5.1.

Applying condition (5.6) to equation (5.5), and since x is independent of ϵ, then

$$\frac{\partial F}{\partial \epsilon} = \int_{x_1}^{x_2} \left(\frac{\partial f}{\partial y}\frac{\partial y}{\partial \epsilon} + \frac{\partial f}{\partial y'}\frac{\partial y'}{\partial \epsilon}\right) dx = 0 \tag{5.7}$$

Since the limits of integration are fixed, the differential operation affects only the integrand. From equations (5.4),

$$\frac{\partial y}{\partial \epsilon} = \eta(x) \tag{5.8}$$

and

$$\frac{\partial y'}{\partial \epsilon} = \frac{d\eta}{dx} \tag{5.9}$$

Consider the second term in the integrand

$$\int_{x_1}^{x_2} \frac{\partial f}{\partial y'}\frac{\partial y'}{\partial \epsilon} dx = \int_{x_1}^{x_2} \frac{\partial f}{\partial y'}\frac{d\eta}{dx} dx \tag{5.10}$$

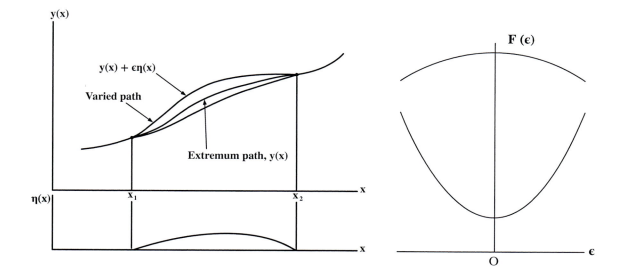

Figure 5.1: The left shows the extremum $y(x)$ and neighboring paths $y(\epsilon, x) = y(x) + \epsilon\eta(x)$ between (x_1, y_1) and (x_2, y_2) that minimizes the function $F = \int_{x_1}^{x_2} f\left[y(x), y'(x); x\right] dx$. The right shows the dependence of F as a function of the admixture coefficient ϵ for a maximum (upper) or a minimum (lower) at $\epsilon = 0$.

Integrate by parts

$$\int u\,dv = uv - \int v\,du \tag{5.11}$$

gives

$$\int_{x_1}^{x_2} \frac{\partial f}{\partial y'} \frac{d\eta}{dx} dx = \left[\frac{\partial f}{\partial y'}\eta(x)\right]_{x_1}^{x_2} - \int_{x_1}^{x_2} \eta(x)\frac{d}{dx}\left(\frac{\partial f}{\partial y'}\right) dx \tag{5.12}$$

Note that the first term on the right-hand side is zero since by definition $\frac{\partial y}{\partial \epsilon} = \eta(x) = 0$ at x_1 and x_2. Thus

$$\frac{\partial F}{\partial \epsilon} = \int_{x_1}^{x_2} \left(\frac{\partial f}{\partial y}\frac{\partial y}{\partial \epsilon} + \frac{\partial f}{\partial y'}\frac{\partial y'}{\partial \epsilon}\right) dx = \int_{x_1}^{x_2} \left(\frac{\partial f}{\partial y}\eta(x) - \eta(x)\frac{d}{dx}\left(\frac{\partial f}{\partial y'}\right)\right) dx$$

Thus equation 5.7 reduces to

$$\frac{\partial F}{\partial \epsilon} = \int_{x_1}^{x_2} \left(\frac{\partial f}{\partial y} - \frac{d}{dx}\frac{\partial f}{\partial y'}\right) \eta(x) dx \tag{5.13}$$

The function $\frac{\partial F}{\partial \epsilon}$ will be an extremum if it is stationary at $\epsilon = 0$. That is,

$$\frac{\partial F}{\partial \epsilon} = \int_{x_1}^{x_2} \left(\frac{\partial f}{\partial y} - \frac{d}{dx}\frac{\partial f}{\partial y'}\right) \eta(x) dx = 0 \tag{5.14}$$

This integral now appears to be independent of ϵ. However, the functions y and y' occurring in the derivatives are functions of ϵ. Since $\left(\frac{\partial F}{\partial \epsilon}\right)_{\epsilon=0}$ must vanish for a stationary value, and *because $\eta(x)$ is an arbitrary function subject to the conditions stated*, *then the above integrand must be zero*. This derivation that the integrand must be zero leads to **Euler's differential equation**

$$\frac{\partial f}{\partial y} - \frac{d}{dx}\frac{\partial f}{\partial y'} = 0 \tag{5.15}$$

where y and y' are the original functions, independent of ϵ. The basis of the calculus of variations is that the function $y(x)$ that satisfies Euler's equation is an stationary function. Note that the stationary value could be either a maximum or a minimum value. When Euler's equation is applied to mechanical systems using the Lagrangian as the functional, then Euler's differential equation is called the Euler-Lagrange equation.

5.3   Applications of Euler's equation

5.1 *Example: Shortest distance between two points*

Consider the path lies in the $x - y$ plane. The infinitessimal length of arc is

$$ds = \sqrt{dx^2 + dy^2} = \left[\sqrt{1 + \left(\frac{dy}{dx} \right)^2} \right] dx$$

Then the length of the arc is

$$J = \int_1^2 ds = \int_1^2 \left[\sqrt{1 + \left(\frac{dy}{dx} \right)^2} \right] dx$$

The function f is

$$f = \sqrt{1 + (y')^2}$$

Therefore

$$\frac{\partial f}{\partial y} = 0$$

and

$$\frac{\partial f}{\partial y'} = \frac{y'}{\sqrt{1 + (y')^2}}$$

Inserting these into Euler's equation 5.15 gives

$$0 + \frac{d}{dx} \left(\frac{y'}{\sqrt{1 + (y')^2}} \right) = 0$$

that is

$$\frac{y'}{\sqrt{1 + (y')^2}} = \text{constant} = C$$

This is valid if

$$y' = \frac{C}{\sqrt{1 - C^2}} = a$$

Therefore

$$y = ax + b$$

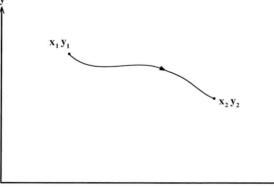

Shortest distance between two points in a plane.

which is the equation of a straight line in the plane. Thus the shortest path between two points in a plane is a straight line between these points, as is intuitively obvious. This stationary value obviously is a minimum.

This trivial example of the use of Euler's equation to determine an extremum value has given the obvious answer. It has been presented here because it provides a proof that a straight line is the shortest distance in a plane and illustrates the power of the calculus of variations to determine extremum paths.

5.2 *Example: Brachistochrone problem*

The Brachistochrone problem involves finding the path having the minimum transit time between two points. The Brachistochrone problem stimulated the development of the calculus of variations by John Bernoulli and Euler. For simplicity, take the case of frictionless motion in the $x - y$ plane with a uniform gravitational field acting in the $\widehat{\mathbf{y}}$ direction, as shown in the adjacent figure. The question is what constrained path will result in the minimum transit time between two points $(x_1 y_1)$ and $(x_2 y_2)$.

Consider that the particle of mass m starts at the origin $x_1 = 0, y_1 = 0$ with zero velocity. Since the problem conserves energy and assuming that initially $E = KE + PE = 0$ then

$$\frac{1}{2}mv^2 - mgy = 0$$

That is

$$v = \sqrt{2gy}$$

The transit time is given by

$$t = \int_{x_1}^{x_2} \frac{ds}{v} = \int_{x_1}^{x_2} \frac{\sqrt{dx^2 + dy^2}}{\sqrt{2gy}} = \int_{x_1}^{x_2} \sqrt{\frac{(1 + x'^2)}{2gy}}\, dy$$

where $x' \equiv \frac{dx}{dy}$. Note that, in this example, the independent variable has been chosen to be y and the dependent variable is $x(y)$.

The function f of the integral is

$$f = \frac{1}{\sqrt{2g}} \sqrt{\frac{(1 + x'^2)}{y}}$$

Factor out the constant $\sqrt{2g}$ term, which does not affect the final equation, and note that

$$\frac{\partial f}{\partial x} = 0$$

$$\frac{\partial f}{\partial x'} = \frac{x'}{\sqrt{y\left(1 + (x')^2\right)}}$$

Therefore Euler's equation gives

$$0 + \frac{d}{dy}\left(\frac{x'}{\sqrt{y\left(1 + (x')^2\right)}}\right) = 0$$

or

$$\frac{x'}{\sqrt{y\left(1 + (x')^2\right)}} = \text{constant} = \frac{1}{\sqrt{2a}}$$

That is

$$\frac{x'^2}{y\left(1 + (x')^2\right)} = \frac{1}{2a}$$

This may be rewritten as

$$x = \int_{y_1}^{y_2} \frac{y\,dy}{\sqrt{2ay - y^2}}$$

Change the variable to $y = a(1 - \cos\theta)$ gives that $dy = a\sin\theta d\theta$, leading to the integral

$$x = \int a\left(1 - \cos\theta\right) d\theta$$

or

$$x = a(\theta - \sin\theta) + \text{constant}$$

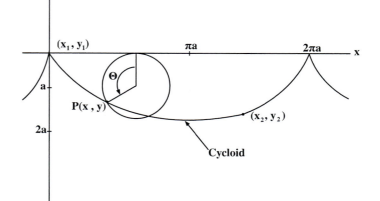

The Bachistochrone problem involves finding the path for the minimum transit time for constrained frictionless motion in a uniform gravitational field.

The parametric equations for a cycloid passing through the origin are

$$x = a(\theta - \sin\theta)$$
$$y = a(1 - \cos\theta)$$

which is the form of the solution found. That is, the shortest time between two points is obtained by constraining the motion of the mass to follow a cycloid shape. Thus the mass first accelerates rapidly by falling down steeply and then follows the curve and coasts upward at the end. The elapsed time is obtained by inserting the above parametric relations for x and y, in terms of θ, into the transit time integral giving $t = \sqrt{\frac{a}{g}}\theta$ where a and θ are fixed by the end point coordinates. Thus the time to fall from starting with zero velocity at the cusp to the minimum of the cycloid is $\pi\sqrt{\frac{a}{g}}$. If $y_2 = y_1 = 0$ then $x_2 = 2\pi a$ which defines the shape of the cycloid and the minimum time is $2\pi\sqrt{\frac{a}{g}} = \sqrt{\frac{2\pi x_2}{g}}$. If the mass starts with a non-zero initial velocity, then the starting point is not at the cusp of the cycloid, but down a distance d such that the kinetic energy equals the potential energy difference from the cusp.

A modern application of the Brachistochrone problem is determination of the optimum shape of the low-friction emergency chute that passengers slide down to evacuate a burning aircraft. Bernoulli solved the problem of rapid evacuation of an aircraft two centuries before the first flight of a powered aircraft.

5.3 *Example: Minimal travel cost*

Assume that the cost of flying an aircraft at height z is $e^{-\kappa z}$ per unit distance of flight-path, where κ is a positive constant. Consider that the aircraft flies in the (x, z)-plane from the point $(-a, 0)$ to the point $(a, 0)$ where $z = 0$ corresponds to ground level, and where the z-axis points vertically upwards. Find the extremal for the problem of minimizing the total cost of the journey.

The differential arc-length element of the flight path ds can be written as

$$ds = \sqrt{dx^2 + dz^2} = \sqrt{1 + z'^2}dx$$

where $z' \equiv \frac{dz}{dx}$. Thus the cost integral to be minimized is

$$C = \int_{-a}^{+a} e^{-\kappa z}ds = \int_{-a}^{+a} e^{-\kappa z}\sqrt{1 + z'^2}dx$$

The function of this integral is

$$f = e^{-\kappa z}\sqrt{1 + z'^2}$$

The partial differentials required for the Euler equations are

$$\frac{d}{dx}\frac{\partial f}{\partial z'} = \frac{z''e^{-\kappa z}}{\sqrt{1 + z'^2}} - \frac{\kappa z'^2 e^{-\kappa z}}{\sqrt{1 + z'^2}} - \frac{z''z'^2 e^{-\kappa z}}{(1 + z'^2)^{3/2}}$$

$$\frac{\partial f}{\partial z} = -\kappa e^{-\kappa z}\sqrt{1 + z'^2}$$

Therefore Euler's equation equals

$$\frac{\partial f}{\partial z} - \frac{d}{dx}\frac{\partial f}{\partial z'} = -\kappa e^{-\kappa z}\sqrt{1 + z'^2} - \frac{z''e^{-\kappa z}}{\sqrt{1 + z'^2}} + \frac{\kappa z'^2 e^{-\kappa z}}{\sqrt{1 + z'^2}} + \frac{z''z'^2 e^{-\kappa z}}{(1 + z'^2)^{3/2}} = 0$$

This can be simplified by multiplying the radical to give

$$-\kappa - 2\kappa z'^2 - \kappa z'^4 - z'' - z''z'^2 + \kappa z'^2 + \kappa z'^4 + z''z'^2 = 0$$

Cancelling terms gives

$$z'' + \kappa\left(1 + z'^2\right) = 0$$

Separating the variables leads to

$$\arctan z' = \int \frac{dz'}{z'^2 + 1} = -\int \kappa dx = -\kappa z + c_1$$

Integration gives

$$z(x) = \int_{-a}^{x} dz = \int_{-a}^{x} \tan(c_1 - \kappa x) dx = \frac{\ln(\cos(c_1 - \kappa x)) - \ln(\cos(c_1 + \kappa a))}{\kappa} + c_2 = \frac{\ln\left(\frac{\cos(c_1 - \kappa x)}{\cos(c_1 + \kappa a)}\right)}{\kappa} + c_2$$

Using the initial condition that $z(-a) = 0$ gives $c_2 = 0$. Similarly the final condition $z(a) = 0$ implies that $c_1 = 0$. Thus Euler's equation has determined that the optimal trajectory that minimizes the cost integral C is

$$z(x) = \frac{1}{\kappa} \ln\left(\frac{\cos(\kappa x)}{\cos(\kappa a)}\right)$$

This example is typical of problems encountered in economics.

5.4 Selection of the independent variable

A wide selection of variables can be chosen as the independent variable for variational calculus. The derivation of Euler's equation and example 5.1 both assumed that the independent variable is x, whereas example 5.2 used y as the independent variable, example 5.3 used z, and Lagrange mechanics uses time t as the independent variable. Selection of which variable to use as the independent variable does not change the physics of a problem, but some selections can simplify the mathematics for obtaining an analytic solution. The following example of a cylindrically-symmetric soap-bubble surface formed by blowing a soap bubble that stretches between two circular hoops, illustrates the importance when selecting the independent variable.

5.4 *Example: Surface area of a cylindrically-symmetric soap bubble*

Consider a cylindrically-symmetric soap-bubble surface formed by blowing a soap bubble that stretches between two circular hoops. The surface energy, that results from the surface tension of the soap bubble, is minimized when the surface area of the bubble is minimized. Assume that the axes of the two hoops lie along the z axis as shown in the adjacent figure. It is intuitively obvious that the soap bubble having the minimum surface area that is bounded by the two hoops will have a circular cross section that is concentric with the symmetry axis, and the radius will be smaller between the two hoops. Therefore, intuition can be used to simplify the problem to finding the shape of the contour of revolution around the axis of symmetry that defines the shape of the surface of minimum surface area. Use cylindrical coordinates (ρ, θ, z) and assume that hoop 1 at z_1 has radius ρ_1 and hoop 2 at z_2 has radius ρ_2. Consider the cases where either ρ, or z, are selected to be the independent variable.

The differential arc-length element of the circular annulus at constant θ between z and $z + dz$ is given by $ds = \sqrt{dz^2 + d\rho^2}$. Therefore the area of the infinitessimal circular annulus is $dS = 2\pi\rho ds$ which can be integrated to give the area of the surface S of the soap bubble bounded by the two circular hoops as

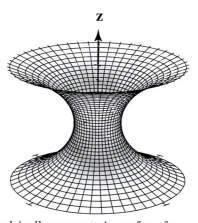

Cylindrically-symmetric surface formed by rotation about the z axis of a soap bubble suspended between two identical hoops centred on the z axis.

$$S = 2\pi \int_{1}^{2} \rho \sqrt{dz^2 + d\rho^2}$$

Independent variable z

Assuming that z is the independent variable, then the surface area can be written as

$$S = 2\pi \int_{1}^{2} \rho \sqrt{1 + \left(\frac{d\rho}{dz}\right)^2} \, dz = 2\pi \int_{1}^{2} \rho \sqrt{1 + \rho'^2} \, dz$$

where $\rho' \equiv \frac{d\rho}{dz}$. The function of the surface integral is $f = \rho\sqrt{1 + \rho'^2}$. The derivatives are

$$\frac{\partial f}{\partial \rho} = \sqrt{1 + \rho'^2}$$

and

$$\frac{\partial f}{\partial \rho'} = \frac{\rho\rho'}{\sqrt{1 + (\rho')^2}}$$

Therefore Euler's equation gives

$$\frac{d}{dz}\left(\frac{\rho\rho'}{\sqrt{1 + (\rho')^2}}\right) - \sqrt{1 + \rho'^2} = 0$$

This is not an easy equation to solve.

Independent variable ρ

Consider the case where the independent variable is chosen to be ρ, then the surface integral can be written as

$$S = 2\pi \int_1^2 \rho\sqrt{1 + \left(\frac{dz}{d\rho}\right)^2}\, d\rho = 2\pi \int \rho\sqrt{1 + z'^2}\, d\rho$$

where $z' \equiv \frac{dz}{d\rho}$. Thus the function of the surface integral is $f = \rho\sqrt{1 + z'^2}$. The derivatives are

$$\frac{\partial f}{\partial z} = 0$$

and

$$\frac{\partial f}{\partial z'} = \frac{\rho z'}{\sqrt{1 + (z')^2}}$$

Therefore Euler's equation gives

$$0 + \frac{d}{d\rho}\left(\frac{\rho z'}{\sqrt{1 + (z')^2}}\right) = 0$$

That is

$$\frac{\rho z'}{\sqrt{1 + (z')^2}} = a$$

where a is a constant. This can be rewritten as

$$z'^2\left(\rho^2 - a^2\right) = a^2$$

or

$$z' = \frac{dz}{d\rho} = \frac{a}{\sqrt{\rho^2 - a^2}}$$

The integral of this is

$$z = a\cosh^{-1}\left(\frac{\rho}{a}\right) + b$$

That is

$$\rho = a\cosh\frac{z - b}{a}$$

which is the equation of a catenary. The catenary is the shape of a uniform flexible cable hung in a uniform gravitational field. The constants a and b are given by the end points. The physics of the solution must be identical for either choice of independent variable. However, mathematically one case is easier to solve than the other because, in the latter case, one term in Euler's equation is zero.

5.5 Functions with several independent variables $y_i(x)$

The discussion has focussed on systems having only a single function $y(x)$ such that the functional is an extremum. It is more common to have a functional that is dependent upon several independent variables $f\left[y_1(x), y_1'(x), y_2(x), y_2'(x), ...; x\right]$ which can be written as

$$F = \int_{x_1}^{x_2} \sum_{i=1}^{N} f\left[y_i(x), y_i'(x); x\right] dx \tag{5.16}$$

where $i = 1, 2, 3, ..., N$.

By analogy with the one dimensional problem, define neighboring functions η_i for each variable. Then

$$
\begin{aligned}
y_i(\epsilon, x) &= y_i(0, x) + \epsilon \eta_i(x) \tag{5.17}\\
y_i'(\epsilon, x) &\equiv \frac{dy_i(\epsilon, x)}{dx} = \frac{dy_i(0, x)}{dx} + \epsilon \frac{d\eta_i}{dx}
\end{aligned}
$$

where η_i are independent functions of x that vanish at x_1 and x_2. Using equations 5.12 and 5.17 leads to the requirements for an extremum value to be

$$\frac{\partial F}{\partial \epsilon} = \int_{x_1}^{x_2} \sum_i^N \left(\frac{\partial f}{\partial y_i} \frac{\partial y_i}{\partial \epsilon} + \frac{\partial f}{\partial y_i'} \frac{\partial y_i'}{\partial \epsilon}\right) dx = \int_{x_1}^{x_2} \sum_i^N \left(\frac{\partial f}{\partial y_i} - \frac{d}{dx}\frac{\partial f}{\partial y_i'}\right) \eta_i(x) dx = 0 \tag{5.18}$$

If the variables $y_i(x)$ are *independent*, then the $\eta_i(x)$ are independent. Since the $\eta_i(x)$ are independent, then evaluating the above equation at $\epsilon = 0$ implies that each term in the bracket must vanish independently. That is, Euler's differential equation becomes a set of N *equations for the N independent variables*

$$\frac{\partial f}{\partial y_i} - \frac{d}{dx}\frac{\partial f}{\partial y_i'} = 0 \tag{5.19}$$

where $i = 1, 2, 3..N$. Thus, each of the N *equations can be solved independently when the N variables are independent*. Note that Euler's equation involves partial derivatives for the dependent variables y_i , y_i' and the total derivative for the independent variable x.

5.5 *Example: Fermat's Principle*

In 1662 Fermat's proposed that the propagation of light obeyed the generalized principle of least transit time. In optics, Fermat's principle, or the principle of least time, is the principle that the path taken between two points by a ray of light is the path that can be traversed in the least time. Historically, the proof of Fermat's principle by Johann Bernoulli was one of the first triumphs of the calculus of variations, and served as a guiding principle in the formulation of physical laws using variational calculus.

Consider the geometry shown in the figure, where the light travels from the point $P_1(0, y_1, 0)$ to the point $P_2(x_2, -y_2, 0)$. The light beam intersects a plane glass interface at the point $Q(x, 0, z)$.

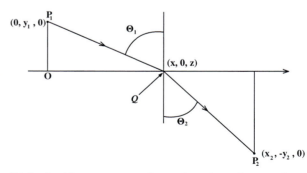

Light incident upon a plane glass interface in the (x, y) plane at $y = 0$.

The French mathematician Fermat discovered that the required path travelled by light is the path for which the travel time t is a minimum. That is, the transit time from the initial point P_1 to the final point P_2 is given by

$$t = \int_1^2 dt = \int_1^2 \frac{ds}{v} = \frac{1}{c}\int_1^2 n\,ds = \frac{1}{c}\int_1^2 n(x, y, z)\sqrt{1 + (x')^2 + (z')^2}\,dy$$

assuming that the velocity of light in any medium is given by $v = c/n$ where n is the refractive index of the medium and c is the velocity of light in vacuum.

This is a problem that has two dependent variables $x(y)$ and $z(y)$ with y chosen as the independent variable. The integral can be broken into two parts $y_1 \to 0$ and $0 \to -y_2$.

$$t = \frac{1}{c} \left[\int_{y_1}^{0} n_1 \sqrt{1 + (x')^2 + (z')^2} \, dy + \int_{0}^{-y_2} n_2 \sqrt{1 + (x')^2 + (z')^2} \, dy \right]$$

The functionals are functions of x' and z' but not x or z. Thus Euler's equation for z simplifies to

$$0 + \frac{d}{dy} \left(\frac{1}{c} \left(\frac{n_1 z'}{\sqrt{1 + x'^2 + z'^2}} + \frac{n_2 z'}{\sqrt{1 + x'^2 + z'^2}} \right) \right) = 0$$

This implies that $z' = 0$, therefore z is a constant. Since the initial and final values were chosen to be $z_1 = z_2 = 0$, therefore at the interface $z = 0$. Similarly Euler's equations for x are

$$0 + \frac{d}{dy} \left(\frac{1}{c} \left(\frac{n_1 x'}{\sqrt{1 + x'^2 + z'^2}} + \frac{n_2 x'}{\sqrt{1 + x'^2 + z'^2}} \right) \right) = 0$$

But $x' = \tan\theta_1$ for n_1 and $x' = -\tan\theta_2$ for n_2 and it was shown that $z' = 0$. Thus

$$0 + \frac{d}{dy} \left(\frac{1}{c} \left(\frac{n_1 \tan\theta_1}{\sqrt{1 + (\tan\theta_1)^2}} - \frac{n_2 \tan\theta_2}{\sqrt{1 + (\tan\theta_2)^2}} \right) \right) = \frac{d}{dy} \left(\frac{1}{c} (n_1 \sin\theta_1 - n_2 \sin\theta_2) \right) = 0$$

Therefore $\frac{1}{c}(n_1 \sin\theta_1 - n_2 \sin\theta_2) = $ constant which must be zero since when $n_1 = n_2$, then $\theta_1 = \theta_2$. Thus Fermat's principle leads to Snell's Law.

$$n_1 \sin\theta_1 = n_2 \sin\theta_2$$

The geometry of this problem is simple enough to directly minimize the path rather than using Euler's equations for the two parameters as performed above. The lengths of the paths $P_1 Q$ and $Q P_2$ are

$$P_1 Q = \sqrt{x^2 + y_1^2 + z^2}$$

$$Q P_2 = \sqrt{(x_2 - x)^2 + y_2^2 + z^2}$$

The total transit time is given by

$$t = \frac{1}{c} \left(n_1 \sqrt{x^2 + y_1^2 + z^2} + n_2 \sqrt{(x_2 - x)^2 + y_2^2 + z^2} \right)$$

This problem involves two dependent variables, $y(x)$ and $z(x)$. To find the minima, set the partial derivatives $\frac{\partial t}{\partial z} = 0$ and $\frac{\partial t}{\partial x} = 0$. That is,

$$\frac{\partial t}{\partial z} = \frac{1}{c} \left(\frac{n_1 z}{\sqrt{x^2 + y_1^2 + z^2}} + \frac{n_2 z}{\sqrt{(x_2 - x)^2 + y_2^2 + z^2}} \right) = 0$$

This is zero only if $z = 0$, that is the point Q lies in the plane containing P_1 and P_2. Similarly

$$\frac{\partial t}{\partial x} = \frac{1}{c} \left(\frac{n_1 x}{\sqrt{x^2 + y_1^2 + z^2}} - \frac{n_2 (x_2 - x)}{\sqrt{(x_2 - x)^2 + y_2^2 + z^2}} \right) = \frac{1}{c} (n_1 \sin\theta_1 - n_2 \sin\theta_2) = 0$$

This is zero only if Snell's law applies that is

$$n_1 \sin\theta_1 = n_2 \sin\theta_2$$

Fermat's principle has shown that the refracted light is given by Snell's Law, and is in a plane normal to the surface. The laws of reflection also are given since then $n_1 = n_2 = n$ and the angle of reflection equals the angle of incidence.

5.6 *Example: Minimum of* $(\nabla\phi)^2$ *in a volume*

Find the function $\phi(x_1, x_2, x_3)$ *that has the minimum value of* $(\nabla\phi)^2$ *per unit volume. For the volume* V *it is desired to minimize the following*

$$J = \frac{1}{V} \int \int \int (\nabla\phi)^2 \, dx_1 dx_2 dx_3 = \frac{1}{V} \int \int \int \left[\left(\frac{\partial\phi}{\partial x_1}\right)^2 + \left(\frac{\partial\phi}{\partial x_2}\right)^2 + \left(\frac{\partial\phi}{\partial x_3}\right)^2 \right] dx_1 dx_2 dx_3$$

Note that the variables x_1, x_2, x_3 *are independent, and thus Euler's equation for several independent variables can be used. To minimize the functional* J, *the function*

$$f = \left(\frac{\partial\phi}{\partial x_1}\right)^2 + \left(\frac{\partial\phi}{\partial x_2}\right)^2 + \left(\frac{\partial\phi}{\partial x_3}\right)^2 \tag{α}$$

must satisfy the Euler equation

$$\frac{\partial f}{\partial \phi} - \sum_{i=1}^{3} \frac{\partial}{\partial x_i} \left(\frac{\partial f}{\partial \phi'}\right) = 0$$

where $\phi' = \frac{\partial\phi}{\partial x_i}$. *Substitute* f *into Euler's equation gives*

$$\sum_{i=1}^{3} \frac{\partial}{\partial x_i} \left(\frac{\partial\phi}{\partial x_i}\right) = 0$$

This is just Laplace's equation

$$\nabla^2\phi = 0$$

Therefore ϕ *must satisfy Laplace's equation in order that the functional* J *be a minimum.*

5.6 Euler's integral equation

An integral form of the Euler differential equation can be written which is useful for cases when the function f does not depend explicitly on the independent variable x, that is, when $\frac{\partial f}{\partial x} = 0$. Note that

$$\frac{df}{dx} = \frac{\partial f}{\partial x} + \frac{\partial f}{\partial y}\frac{dy}{dx} + \frac{\partial f}{\partial y'}\frac{dy'}{dx} \tag{5.20}$$

But

$$\frac{d}{dx}\left(y'\frac{\partial f}{\partial y'}\right) = \frac{\partial f}{\partial y'}\frac{dy'}{dx} + y'\frac{d}{dx}\frac{\partial f}{\partial y'} \tag{5.21}$$

Combining these two equations gives

$$\frac{d}{dx}\left(y'\frac{\partial f}{\partial y'}\right) = \frac{df}{dx} - \frac{\partial f}{\partial x} - y'\frac{\partial f}{\partial y} + y'\frac{d}{dx}\frac{\partial f}{\partial y'} \tag{5.22}$$

The last two terms can be rewritten as

$$y'\left(\frac{d}{dx}\frac{\partial f}{\partial y'} - \frac{\partial f}{\partial y}\right) \tag{5.23}$$

which vanishes when the Euler equation is satisfied. Therefore the above equation simplifies to

$$\frac{\partial f}{\partial x} - \frac{d}{dx}\left(f - y'\frac{\partial f}{\partial y'}\right) = 0 \tag{5.24}$$

This integral form of Euler's equation is especially useful *when* $\frac{\partial f}{\partial x} = 0$, *that is, when* f *does not depend explicitly on the independent variable* x. Then the first integral of equation 5.24 is a constant, i.e.

$$f - y'\frac{\partial f}{\partial y'} = \text{constant} \tag{5.25}$$

This is Euler's integral variational equation. Note that the shortest distance between two points, the minimum surface of rotation, and the brachistochrone, described earlier, all are examples where $\frac{\partial f}{\partial x} = 0$ and thus the integral form of Euler's equation is useful for solving these cases.

5.7   Constrained variational systems

Imposing a constraint on a variational system implies:

1. The N constrained coordinates $y_i(x)$ are correlated which violates the assumption made in chapter 5.5 that the N variables are independent.

2. Constrained motion implies that constraint forces must be acting to account for the correlation of the variables. These constraint forces must be taken into account in the equations of motion.

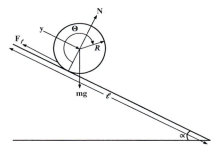

   For example, for a disk rolling down an inclined plane without slipping, there are three coordinates x [perpendicular to the wedge], y, [Along the surface of the wedge], and the rotation angle θ shown in figure 5.2. The constraint forces, $\mathbf{F}_f$ $\mathbf{N}$, lead to the correlation of the variables such that $x = R$, while $y = R\theta$. Basically there is only one independent variable, which can be either y or θ. The use of only one independent variable essentially buries the constraint forces under the rug, which is fine if you only need to know the equation of motion. If you need to determine the forces of constraint then it is necessary to include all coordinates explicitly in the equations of motion as discussed below.

Figure 5.2: A disk rolling down an inclined plane.

5.7.1   Holonomic constraints

Most systems involve restrictions or constraints that couple the coordinates. For example, the $y_i(x)$ may be confined to a surface in coordinate space. The constraints mean that the coordinates $y_i(x)$ are not independent, but are related by equations of constraint. A constraint is called **holonomic** if the equations of constraint can be expressed in the form of an algebraic equation that directly and unambiguously specifies the shape of the surface of constraint. A **non-holonomic** constraint does not provide an algebraic relation between the correlated coordinates. In addition to the holonomy of the constraints, the equations of constraint also can be grouped into the following three classifications depending on whether they are algebraic, differential, or integral. These three classifications for the constraints exhibit different holonomy relating the coupled coordinates. Fortunately the solution of constrained systems is greatly simplified if the equations of constraint are holonomic.

5.7.2   Geometric (algebraic) equations of constraint

Geometric constraints can be expressed in the form of algebraic relations that directly specify the shape of the surface of constraint in coordinate space $q_1, q_2, ..., q_j, ..q_n$.

$$g_k(q_1, q_2, ..q_j, ..q_n; t) = 0 \tag{5.26}$$

where $j = 1, 2, 3, ...n$. There can be m such equations of constraint where $0 \leq k \leq m$. An example of such a geometric constraint is when the motion is confined to the surface of a sphere of radius R in coordinate space which can be written in the form $g = x^2 + y^2 + z^2 - R^2 = 0$. Such algebraic constraint equations are called **Holonomic** which allows use of generalized coordinates as well as Lagrange multipliers to handle both the constraint forces and the correlation of the coordinates.

5.7.3   Kinematic (differential) equations of constraint

The m constraint equations also can be expressed in terms of the infinitessimal displacements of the form

$$\sum_{j=1}^{n} \frac{\partial g_k}{\partial q_j} dq_j + \frac{\partial g_k}{\partial t} dt = 0 \tag{5.27}$$

where $k = 1, 2, 3, ...m$, $j = 1, 2, 3, ...n$. If equation (5.27) represents the total differential of a function then it can be integrated to give a holonomic relation of the form of equation 5.26. However, if equation 5.27 is not the total differential, then it is non-holonomic and can be integrated only after having solved the full problem.

An example of differential constraint equations is for a wheel rolling on a plane without slipping which is non-holonomic and more complicated than might be expected. The wheel moving on a plane has five degrees of freedom since the height z is fixed. That is, the motion of the center of mass requires two coordinates (x, y) plus there are three angles (ϕ, θ, ψ) where ϕ is the rotation angle for the wheel, θ is the pivot angle of the axis, and ψ is the tilt angle of the wheel. If the wheel slides then all five degrees of freedom are active. If the axis of rotation of the wheel is horizontal, that is, the tilt angle $\psi = 0$ is constant, then this kinematic system leads to three differential constraint equations The wheel can roll with angular velocity $\dot{\phi}$, as well as pivot which corresponds to a change in θ. Combining these leads to two *differential* equations of constraint

$$dx - a\sin\theta d\phi = 0 \qquad\qquad dy + a\cos\theta d\phi = 0 \qquad\qquad (5.28)$$

These constraints are insufficient to provide finite relations between all the coordinates. That is, the constraints cannot be reduced by integration to the form of equation 5.26 because there is no functional relation between ϕ and the other three variables, x, y, θ. Many rolling trajectories are possible between any two points of contact on the plane that are related to different pivot angles. That is, the point of contact of the disk could pivot plus roll in a circle returning to the same point where x, y, θ are unchanged whereas the value of ϕ depends on the circumference of the circle. As a consequence the rolling constraint is non-holonomic except for the case where the disk rolls in a straight line and remains vertical.

5.7.4 Isoperimetric (integral) equations of constraint

Equations of constraint also can be expressed in terms of direct integrals. This situation is encountered for isoperimetric problems, such as finding the maximum volume bounded by a surface of fixed area, or the shape of a hanging rope of fixed length. Integral constraints occur in economics when minimizing some cost algorithm subject to a fixed total cost constraint.

A simple example of an isoperimetric problem involves finding the curve $y = y(x)$ such that the functional has an extremum where the curve $y(x)$ satisfies boundary conditions such that $y(x_1) = a$ and $y(x_2) = b$, that is

$$F(y) = \int_{x_1}^{x_2} f(y, y'; x) dx \qquad\qquad (5.29)$$

is an extremum such that the perimeter also is constrained to satisfy

$$G(y) = \int_{x_1}^{x_2} g(y, y'; x) dx = l \qquad\qquad (5.30)$$

where l is a fixed length. This integral constraint is geometric and holonomic. Another example is finding the minimum surface area of a closed surface subject to the enclosed volume being the constraint.

5.7.5 Properties of the constraint equations

Holonomic constraints   Geometric constraints can be expressed in the form of an algebraic equation that directly specifies the shape of the surface of constraint

$$g(y_1, y_2, y_3, ...; x) = 0 \qquad\qquad (5.31)$$

Such a system is called **holonomic** since there is a direct relation between the coupled variables. An example of such a holonomic geometric constraint is if the motion is confined to the surface of a sphere of radius R which can be written in the form

$$g = x^2 + y^2 + z^2 - R^2 = 0 \qquad\qquad (5.32)$$

Non-holonomic constraints   There are many classifications of non-holonomic constraints that exist if equation (5.31) is not satisfied. The algebraic approach is difficult to handle when the constraint is an inequality, such as the requirement that the location is restricted to lie inside a spherical shell of radius R which can be expressed as

$$g = x^2 + y^2 + z^2 - R^2 \leq 0 \qquad\qquad (5.33)$$

This non-holonomic constrained system has a one-sided constraint. Systems usually are non-holonomic if the constraint is kinematic as discussed above.

    Partial Holonomic constraints   Partial-holonomic constraints are holonomic for a restricted range of the constraint surface in coordinate space, and this range can be case specific. This can occur if the constraint force is one-sided and perpendicular to the path. An example is the pendulum with the mass attached to the fulcrum by a flexible string that provides tension but not compression. Then the pendulum length is constant only if the tension in the string is positive. Thus the pendulum will be holonomic if the gravitational plus centrifugal forces are such that the tension in the string is positive, but the system becomes non-hononomic if the tension is negative as can happen when the pendulum rotates to an upright angle where the centrifugal force outwards is insufficient to compensate for the vertical downward component of the gravitational force. There are many other examples where the motion of an object is holonomic when the object is pressed against the constraint surface, such as the surface of the Earth, but is unconstrained if the object leaves the surface.

Time dependence

A constraint is called *scleronomic* if the constraint is not explicitly time dependent. This ignores the time dependence contained within the solution of the equations of motion. Fortunately a major fraction of systems are scleronomic. The constraint is called *rheonomic* if the constraint is explicitly time dependent. An example of a rheonomic system is where the size or shape of the surface of constraint is explicitly time dependent such as a deflating pneumatic tire.

Energy conservation

The solution depends on whether the constraint is conservative or dissipative, that is, if friction or drag are acting. The system will be conservative if there are no drag forces, and the constraint forces are perpendicular to the trajectory of the path such as the motion of a charged particle in a magnetic field. Forces of constraint can result from sliding of two solid surfaces, rolling of solid objects, fluid flow in a liquid or gas, or result from electromagnetic forces. Energy dissipation can result from friction, drag in a fluid or gas, or finite resistance of electric conductors leading to dissipation of induced electric currents in a conductor, e.g. eddy currents.

    A rolling constraint is unusual in that friction between the rolling bodies is necessary to maintain rolling. A disk on a frictionless inclined plane will conserve it's angular momentum since there is no torque acting if the rolling contact is frictionless, that is, the disk will just slide. If the friction is sufficient to stop sliding, then the bodies will roll and not slide. A perfect rolling body does not dissipate energy since no work is done at the instantaneous point of contact where both bodies are in zero relative motion and the force is perpendicular to the motion. In real life, a rolling wheel can involve a very small energy dissipation due to deformation at the point of contact coupled with non-elastic properties of the material used to make the wheel and the plane surface. For example, a pneumatic tire can heat up and expand due to flexing of the tire.

5.7.6   Treatment of constraint forces in variational calculus

There are three major approaches to handle constraint forces in variational calculus. All three of them exploit the tremendous freedom and flexibility available when using generalized coordinates. The (1) **generalized coordinate** approach, described in chapter 5.8, exploits the correlation of the n coordinates due to the m constraint forces to reduce the dimension of the equations of motion to $s = n - m$ degrees of freedom. This approach embeds the m constraint forces, into the choice of generalized coordinates and does not determine the constraint forces, (2) **Lagrange multiplier** approach, described in chapter 5.9, exploits generalized coordinates but includes the m constraint forces into the Euler equations to determine both the constraint forces in addition to the n equations of motion. (3) **Generalized forces** approach, described in chapter 6.7.3, introduces constraint and other forces explicitly.

5.8 Generalized coordinates in variational calculus

Newtonian mechanics is based on a vectorial treatment of mechanics which can be difficult to apply when solving complicated problems in mechanics. Constraint forces acting on a system usually are unknown. In Newtonian mechanics constrained forces must be included explicitly so that they can be determined simultaneously with the solution of the dynamical equations of motion. The major advantage of the variational approaches is that solution of the dynamical equations of motion can be simplified by expressing the motion in terms of n independent **generalized coordinates.** These generalized coordinates can be any set of **independent variables**, q_i, where $1 \leq i \leq n$, plus the corresponding velocities $\dot{q}_i$ for Lagrangian mechanics, or the corresponding canonical variables, q_i, p_i for Hamiltonian mechanics. These generalized coordinates for the n variables are used to specify the scalar functional dependence on these generalized coordinates. The variational approach employs this scalar functional to determine the trajectory. The generalized coordinates used for the variational approach do not need to be orthogonal, they only need to be independent since they are used only to completely specify the *magnitude* of the scalar functional. This greatly expands the arsenal of possible generalized coordinates beyond what is available using Newtonian mechanics. For example, generalized coordinates can be the dimensionless amplitudes for the n normal modes of coupled oscillator systems, or action-angle variables. In addition, generalized coordinates having different dimensions can be used for each of the n variables. *Each generalized coordinate, q_i specifies an independent mode of the system, not a specific particle.* For example, each normal mode of coupled oscillators can involve correlated motion of several coupled particles. The major advantage of using generalized coordinates is that they can be chosen to be perpendicular to a corresponding constraint force, and therefore that specific constraint force does no work for motion along that generalized coordinate. Moreover, the constrained motion does no work in the direction of the constraint force for rigid constraints. Thus generalized coordinates allow specific constraint forces to be ignored in evaluation of the minimized functional. This freedom and flexibility of choice of generalized coordinates allows the correlated motion produced by the constraint forces to be embedded directly into the choice of the independent generalized coordinates, and the actual constraint forces can be ignored. Embedding of the constraint induced correlations into the generalized coordinates, effectively "sweeps the constraint forces under the rug" which greatly simplifies the equations of motion for any system that involve constraint forces. Selection of the appropriate generalized coordinates can be obvious, and often it is performed subconsciously by the user.

Three variational approaches are used that employ generalized coordinates to derive the equations of motion of a system that has n generalized coordinates subject to m constraints.

1) Minimal set of generalized coordinates: When the m equations of constraint are holonomic, then the m algebraic constraint relations can be used to transform the coordinates into $s = n - m$ independent *generalized coordinates q_i*. This approach reduces the number of unknowns, n, by the number of constraints m, to give a minimal set of $s = n - m$ independent generalized dynamical variables. The forces of constraint are not explicitly discussed, or determined, when this generalized coordinate approach is employed. This approach greatly simplifies solution of dynamical problems by avoiding the need for explicit treatment of the constraint forces. This approach is straight forward for holonomic constraints, since the n *spatial coordinates* $y_1(x), ...y_N(x)$, are coupled by m algebraic equations which can be used to make the transformation to generalized coordinates. Thus the n *coupled spatial coordinates* are transformed to $s = n - m$ *independent generalized dynamical coordinates* $q_1(x), ...q_s(x)$, and their generalized first derivatives $\dot{q}_1(x), ...\dot{q}_s(x)$. These *generalized coordinates are independent,* and thus it is possible to use Euler's equation for each independent parameter q_i

$$\frac{\partial f}{\partial q_i} - \frac{d}{dx}\frac{\partial f}{\partial q_i'} = 0 \tag{5.34}$$

where $i = 1, 2, 3..s$. There are $s = n - m$ such Euler equations. The freedom to choose generalized coordinates underlies the tremendous advantage of applying the variational approach.

2) Lagrange multipliers: The n Lagrange equations, plus the m equations of constraint, can be used to explicitly determine the n generalized coordinates plus the m constraint forces. That is, $n + m$ unknowns are determined. This approach is discussed in chapter 5.9.

3) Generalized forces: This approach introduces the constraint forces explicity. This approach, applied to Lagrangian mechanics, is discussed in chapter 6.6.3.

The above three approaches exploit generalized coordinates to handle constraint forces as described in chapter 6.

5.9   Lagrange multipliers for holonomic constraints

5.9.1   Algebraic equations of constraint

The Lagrange multiplier technique provides a powerful, and elegant, way to handle holonomic constraints using Euler's equations[1]. The general method of Lagrange multipliers for n variables, with m constraints, is best introduced using Bernoulli's ingenious exploitation of virtual infinitessimal displacements, which Lagrange signified by the symbol δ. The term "virtual" refers to an intentional variation of the generalized coordinates δq_i in order to elucidate the local sensitivity of a function $F(q_i, x)$ to variation of the variable. Contrary to the usual infinitessimal interval in differential calculus, where an actual displacement dq_i occurs during a time dt, a virtual displacement is imagined to be an instantaneous, infinitessimal, displacement of a coordinate, not an actual displacement, in order to elucidate the local dependence of F on the coordinate. The local dependence of any functional F, to virtual displacements of all n coordinates, is given by taking the partial differentials of F.

$$\delta F = \sum_{i}^{n} \frac{\partial F}{\partial q_i} \delta q_i \tag{5.35}$$

The function F is stationary, that is an extremum, if equation 5.35 equals zero. The extremum of the functional F, given by equation 5.16, can be expressed in a compact form using the virtual displacement formalism as

$$\delta F = \delta \int_{x_1}^{x_2} \sum_{i}^{n} f\left[q_i(x), q_i'(x); x\right] dx = \sum_{i}^{n} \frac{\partial F}{\partial q_i} \delta q_i = 0 \tag{5.36}$$

The auxiliary conditions, due to the m holonomic algebraic constraints for the n variables q_i, can be expressed by the m equations

$$g_k(\mathbf{q}) = 0 \tag{5.37}$$

where $1 \leq k \leq m$ and $1 \leq i \leq n$ with $m < n$. The variational problem for the m holonomic constraint equations also can be written in terms of m differential equations where $1 \leq k \leq m$

$$\delta g_k = \sum_{i=1}^{n} \frac{\partial g_k}{\partial q_i} \delta q_i = 0 \tag{5.38}$$

Since equations 5.36 and 5.38 both equal zero, the m equations 5.38 can be multiplied by arbitrary undetermined factors λ_k, and added to equations 5.36 to give.

$$\delta F(q_i, x) + \lambda_1 \delta g_1 + \lambda_2 \delta g_2 \cdots \lambda_k \delta g_k \cdots \lambda_m \delta g_m = 0 \tag{5.39}$$

Note that this is not trivial in that although the sum of the constraint equations for each y_i is zero; the individual terms of the sum are not zero.

Insert equations 5.36 plus 5.38 into 5.39, and collect all n terms, gives

$$\sum_{i}^{n} \left(\frac{\partial F}{\partial q_i} + \sum_{k=1}^{m} \lambda_k \frac{\partial g_k}{\partial q_i} \right) \delta q_i = 0 \tag{5.40}$$

Note that all the δq_i are free independent variations and thus the terms in the brackets, which are the coefficients of each δq_i, individually must equal zero. For each of the n values of i, the corresponding bracket implies

$$\frac{\partial F}{\partial q_i} + \sum_{k=1}^{m} \lambda_k \frac{\partial g_k}{\partial q_i} = 0 \tag{5.41}$$

This is equivalent to what would be obtained from the variational principle

$$\delta F + \sum_{k=1}^{m} \lambda_k \delta g_k = 0 \tag{5.42}$$

---

[1]This textbook uses the symbol q_i to designate a generalized coordinate, and q_i' to designate the corresponding first derivative with respect to the independent variable, in order to differentiate the spatial coordinates from the more powerful generalized coordinates.

Equation 5.42 is equivalent to a variational problem for finding the stationary value of F'

$$\delta\left(F'\right) = \delta\left(F + \sum_{k}^{m} \lambda_k g_k\right) = 0 \tag{5.43}$$

where F' is defined to be

$$F' \equiv \left(F + \sum_{k=1}^{m} \lambda_k g_k\right) \tag{5.44}$$

The solution to equation 5.43 can be found using Euler's differential equation 5.19 of variational calculus. At the extremum $\delta\left(F'\right) = 0$ corresponds to following contours of constant F' which are in the surface that is perpendicular to the gradients of the terms in F'. The Lagrange multiplier constants are required because, although these gradients are parallel at the extremum, the magnitudes of the gradients are not equal.

The beauty of the Lagrange multipliers approach is that the auxiliary conditions do not have to be handled explicitly, since they are handled automatically as m additional free variables during solution of Euler's equations for a variational problem with $n + m$ unknowns fit to $n + m$ equations. That is, the n variables q_i are determined by the variational procedure using the n variational equations

$$\frac{d}{dx}\left(\frac{\partial F'}{\partial q'_i}\right) - \left(\frac{\partial F'}{\partial q_i}\right) = \frac{d}{dx}\left(\frac{\partial F}{\partial q'_i}\right) - \left(\frac{\partial F}{\partial q_i}\right) - \sum_{k}^{m} \lambda_k \frac{\partial g_k}{\partial q_i} = 0 \tag{5.45}$$

simultaneously with the m variables λ_k which are determined by the m variational equations

$$\frac{d}{dx}\left(\frac{\partial F'}{\partial \lambda'_k}\right) - \left(\frac{\partial F'}{\partial \lambda_k}\right) = 0 \tag{5.46}$$

Equation 5.45 usually is expressed as

$$\left(\frac{\partial F}{\partial q_i}\right) - \frac{d}{dx}\left(\frac{\partial F}{\partial q'_i}\right) + \sum_{k}^{m} \lambda_k \frac{\partial g_k}{\partial q_i} = 0 \tag{5.47}$$

The elegance of Lagrange multipliers is that a single variational approach allows simultaneous determination of all $n + m$ unknowns. Chapter 6.2 shows that the forces of constraint are given directly by the $\lambda_k \frac{\partial g_k}{\partial q_i}$ terms.

5.7 Example: Two dependent variables coupled by one holonomic constraint

The powerful, and generally applicable, Lagrange multiplier technique is illustrated by considering the case of only two dependent variables, $y(x)$, and $z(x)$, with the function $f(y(x), y'(x), z(x), z(x)'; x)$ and with one holonomic equation of constraint coupling these two dependent variables. The extremum is given by requiring

$$\frac{\partial F}{\partial \epsilon} = \int_{x_1}^{x_2} \left[\left(\frac{\partial f}{\partial y} - \frac{d}{dx}\frac{\partial f}{\partial y'}\right)\frac{\partial y}{\partial \epsilon} + \left(\frac{\partial f}{\partial z} - \frac{d}{dx}\frac{\partial f}{\partial z'}\right)\frac{\partial z}{\partial \epsilon}\right] dx = 0 \tag{A}$$

with the constraint expressed by the auxiliary condition

$$g\left(y, z; x\right) = 0 \tag{B}$$

Note that the variations $\frac{\partial y}{\partial \epsilon}$ and $\frac{\partial z}{\partial \epsilon}$ are no longer independent because of the constraint equation, thus the the two terms in the brackets of equation A are not separately equal to zero at the extremum. However, differentiating the constraint equation B gives

$$\frac{dg}{d\epsilon} = \left(\frac{\partial g}{\partial y}\frac{\partial y}{\partial \epsilon} + \frac{\partial g}{\partial z}\frac{\partial z}{\partial \epsilon}\right) = 0 \tag{C}$$

No $\frac{\partial g}{\partial x}$ term applies because, for the independent variable, $\frac{\partial x}{\partial \epsilon} = 0$. Introduce the neighboring paths by adding the auxiliary functions

$$y(\epsilon, x) = y(x) + \epsilon \eta_1(x) \tag{D}$$
$$z(\epsilon, x) = z(x) + \epsilon \eta_2(x) \tag{E}$$

Insert the differentials of equations D and E, into C gives

$$\frac{dg}{d\epsilon} = \left(\frac{\partial g}{\partial y} \eta_1(x) + \frac{\partial g}{\partial z} \eta_2(x) \right) = 0 \tag{F}$$

implying that

$$\eta_2(x) = -\frac{\frac{\partial g}{\partial y}}{\frac{\partial g}{\partial z}} \eta_1(x)$$

Equation A can be rewritten as

$$\int_{x_1}^{x_2} \left[\left(\frac{\partial f}{\partial y} - \frac{d}{dx} \frac{\partial f}{\partial y'} \right) \eta_1(x) + \left(\frac{\partial f}{\partial z} - \frac{d}{dx} \frac{\partial f}{\partial z'} \right) \eta_2(x) \right] dx = 0$$

$$\int_{x_1}^{x_2} \left[\left(\frac{\partial f}{\partial y} - \frac{d}{dx} \frac{\partial f}{\partial y'} \right) - \left(\frac{\partial f}{\partial z} - \frac{d}{dx} \frac{\partial f}{\partial z'} \right) \frac{\frac{\partial g}{\partial y}}{\frac{\partial g}{\partial z}} \right] \eta_1(x) dx = 0 \tag{G}$$

Equation G now contains only a single arbitrary function $\eta_1(x)$ that is not restricted by the constraint. Thus the bracket in the integrand of equation G must equal zero for the extremum. That is

$$\left(\frac{\partial f}{\partial y} - \frac{d}{dx} \frac{\partial f}{\partial y'} \right) \left(\frac{\partial g}{\partial y} \right)^{-1} = \left(\frac{\partial f}{\partial z} - \frac{d}{dx} \frac{\partial f}{\partial z'} \right) \left(\frac{\partial g}{\partial z} \right)^{-1} \equiv -\lambda(x)$$

Now the left-hand side of this equation is only a function of f and g with respect to y and y' while the right-hand side is a function of f and g with respect to z and z'. Because both sides are functions of x then each side can be set equal to a function $-\lambda(x)$. Thus the above equations can be written as

$$\frac{d}{dx} \frac{\partial f}{\partial y'} - \frac{\partial f}{\partial y} = \lambda(x) \frac{\partial g}{\partial y} \qquad\qquad \frac{d}{dx} \frac{\partial f}{\partial z'} - \frac{\partial f}{\partial z} = \lambda(x) \frac{\partial g}{\partial z} \tag{H}$$

The complete solution of the three unknown functions. $y(x), z(x)$, and $\lambda(x)$. is obtained by solving the two equations, H, plus the equation of constraint F. The Lagrange multiplier $\lambda(x)$ is related to the force of constraint. This example of two variables coupled by one holonomic constraint conforms with the general relation for many variables and constraints given by equation 5.47.

5.9.2   Integral equations of constraint

The constraint equation also can be given in an integral form which is used frequently for isoperimetric problems. Consider a one dependent-variable isoperimetric problem, for finding the curve $q = q(x)$ such that the functional has an extremum, and the curve $q(x)$ satisfies boundary conditions such that $q(x_1) = a$ and $q(x_2) = b$. That is

$$F(y) = \int_{x_1}^{x_2} f(q, q'; x) dx \tag{5.48}$$

is an extremum such that the fixed length l of the perimeter satisfies the integral constraint

$$G(y) = \int_{x_1}^{x_2} g(q, q'; x) dx = l \tag{5.49}$$

Analogous to (5.44) these two functionals can be combined requiring that

$$\delta K(q, x, \lambda) \equiv \delta \left[F(q) + \lambda G(q) \right] = \delta \int_{x_1}^{x_2} [f + \lambda g] dx = 0 \tag{5.50}$$

That is, it is an extremum for both $q(x)$ and the Lagrange multiplier λ. This effectively involves finding the extremum path for the function $K(q, x, \lambda) = F(q, x) + \lambda G(q, x)$ where both $q(x)$ and λ are the minimized variables. Therefore the curve $q(x)$ must satisfy the differential equation

$$\frac{d}{dx} \frac{\partial f}{\partial q'_i} - \frac{\partial f}{\partial q_i} + \lambda \left[\frac{d}{dx} \frac{\partial g}{\partial q'_i} - \frac{\partial g}{\partial q_i} \right] = 0 \tag{5.51}$$

subject to the boundary conditions $q(x_1) = a$, $q(x_2) = b$, and $G(q) = l$.

5.8 *Example: Catenary*

One isoperimetric problem is the catenary which is the shape a uniform rope or chain of fixed length l that minimizes the gravitational potential energy. Let the rope have a uniform mass per unit length of σ kg/m.

The gravitational potential energy is

$$U = \sigma g \int_1^2 y \, ds = \sigma g \int_1^2 y \sqrt{dx^2 + dy^2} = \sigma g \int_1^2 y \sqrt{1 + y'^2} dx$$

The constraint is that the length be a constant l

$$l = \int_1^2 ds = \int_1^2 \sqrt{1 + y'^2} dx$$

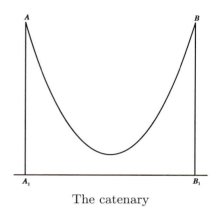

The catenary

Thus the function is $f(y, y'; x) = y\sqrt{1 + y'^2}$ while the integral constraint sets $g = \sqrt{1 + y'^2}$

These need to be inserted into the Euler equation (5.51) by defining

$$F = f + \lambda g = (y + \lambda)\sqrt{1 + y'^2}$$

Note that this case is one where $\frac{\partial F}{\partial x} = 0$ and λ is a constant; also defining $z = y + \lambda$ then $z' = y'$. Therefore the Euler's equations can be written in the integral form

$$F - z' \frac{\partial F}{\partial z'} = c = \text{constant}$$

Inserting the relation $F = z\sqrt{1 + z'^2}$ gives

$$z\sqrt{1 + z'^2} - z' \frac{zz'}{\sqrt{1 + z'^2}} = c$$

where c is an arbitrary constant. This simplifies to

$$z'^2 = \left(\frac{z}{c}\right)^2 - 1$$

The integral of this is

$$z = c \cosh\left(\frac{x + b}{c}\right)$$

where b and c are arbitrary constants fixed by the locations of the two fixed ends of the rope.

5.9 *Example: The Queen Dido problem*

A famous constrained isoperimetric legend is that of Dido, first Queen of Carthage. Legend says that, when Dido landed in North Africa, she persuaded the local chief to sell her as much land as an oxhide could contain. She cut an oxhide into narrow strips and joined them to make a continuous thread more than four kilometers in length which was sufficient to enclose the land adjoining the coast on which Carthage was built. Her problem was to enclose the maximum area for a given perimeter. Let us assume that the coast line is straight and the ends of the thread are at $\pm a$ on the coast line. The enclosed area is given by

$$A = \int_{-a}^{+a} y \, dx$$

The constraint equation is that the total perimeter equals l.

$$\int_{-a}^{a} \sqrt{1 + y'^2} dx = l$$

Thus we have that the functional $f(y, y', x) = y$ *and* $g(y, y', x) = \sqrt{1 + y'^2}$. *Then* $\frac{\partial f}{\partial y} = 1, \frac{\partial f}{\partial y'} = 0, \frac{\partial g}{\partial y} = 0$ *and* $\frac{\partial g}{\partial y'} = \frac{y'}{\sqrt{1+y'^2}}$. *Insert these into the Euler-Lagrange equation (5.51) gives*

$$1 - \lambda \frac{d}{dx}\left[\frac{y'}{\sqrt{1 + y'^2}}\right] = 0$$

That is

$$\frac{d}{dx}\left[\frac{y'}{\sqrt{1 + y'^2}}\right] = \frac{1}{\lambda}$$

Integrate with respect to x gives

$$\frac{\lambda y'}{\sqrt{1 + y'^2}} = x - b$$

where b is a constant of integration. This can be rearranged to give

$$y' = \frac{\pm (x - b)}{\sqrt{\lambda^2 - (x - b)^2}}$$

The integral of this is

$$y = \mp \sqrt{\lambda^2 - (x - b)^2} + c$$

Rearranging this gives

$$(x - b)^2 + (y - c)^2 = \lambda^2$$

This is the equation of a circle centered at (b, c). *Setting the bounds to be* $(-a, 0)$ *to* $(a, 0)$ *gives that* $b = c = 0$ *and the circle radius is* λ. *Thus the length of the thread must be* $l = \pi\lambda$. *Assuming that* $l = 4km$ *then* $\lambda = 1.27km$ *and Queen Dido could buy an area of* $2.53km^2$.

5.10   Geodesic

The geodesic is defined as the shortest path between two fixed points for motion that is constrained to lie on a surface. Variational calculus provides a powerful approach for determining the equations of motion constrained to follow a geodesic.

The use of variational calculus is illustrated by considering the geodesic constrained to follow the surface of a sphere of radius R. As discussed in appendix $C.2.3$, the element of path length on the surface of the sphere is given in spherical coordinates as $ds = R\sqrt{d\theta^2 + (\sin\theta d\phi)^2}$. Therefore the distance s between two points 1 and 2 is

$$s = R \int_1^2 \left[\sqrt{\left(\frac{d\theta}{d\phi}\right)^2 + \sin^2\theta}\right] d\phi \tag{5.52}$$

The function f for ensuring that s be an extremum value uses

$$f = \sqrt{\theta'^2 + \sin^2\theta} \tag{5.53}$$

where $\theta' = \frac{d\theta}{d\phi}$. This is a case where $\frac{\partial f}{\partial \phi} = 0$ and thus the integral form of Euler's equation can be used leading to the result that

$$\sqrt{\theta'^2 + \sin^2\theta} - \theta'\frac{\partial}{\partial \theta'}\sqrt{\theta'^2 + \sin^2\theta} = \text{constant} = a \tag{5.54}$$

This gives that

$$\sin^2\theta = a\sqrt{\theta'^2 + \sin^2\theta} \tag{5.55}$$

This can be rewritten as

$$\frac{d\phi}{d\theta} = \frac{1}{\theta'} = \frac{a\csc^2\theta}{\sqrt{1 - a^2\csc^2\theta}} \tag{5.56}$$

Solving for ϕ gives

$$\phi = \sin^{-1}\left(\frac{\cot\theta}{\beta}\right) + \alpha \tag{5.57}$$

where

$$\beta \equiv \frac{1-a^2}{a^2} \tag{5.58}$$

That is

$$\cot\theta = \beta\sin(\phi - \alpha) \tag{5.59}$$

Expanding the sine and cotangent gives

$$(\beta\cos\alpha)\,R\sin\theta\sin\phi - (\beta\sin\alpha)\,R\sin\theta\cos\phi = R\cos\theta \tag{5.60}$$

Since the brackets are constants, this can be written as

$$A\,(R\sin\theta\sin\phi) - B\,(R\sin\theta\cos\phi) = (R\cos\theta) \tag{5.61}$$

The terms in the brackets are just expressions for the rectangular coordinates x, y, z. That is,

$$Ay - Bx = z \tag{5.62}$$

This is the equation of a plane passing through the center of the sphere. Thus the geodesic on a sphere is the path where a plane through the center intersects the sphere as well as the initial and final locations. This geodesic is called a great circle. Euler's equation gives both the maximum and minimum extremum path lengths for motion on this great circle.

Chapter 17 discusses the geodesic in the four-dimensional space-time coordinates that underlie the General Theory of Relativity. As a consequence, the use of the calculus of variations to determine the equations of motion for geodesics plays a pivotal role in the General Theory of Relativity.

5.11   Variational approach to classical mechanics

This chapter has introduced the general principles of variational calculus needed for understanding the Lagrangian and Hamiltonian approaches to classical mechanics. Although variational calculus was developed originally for classical mechanics, now it has grown to be an important branch of mathematics with applications to many other fields outside of physics. The prologue of this book emphasized the dramatic differences between the differential vectorial approach of Newtonian mechanics, and the integral variational approaches of Lagrange and Hamiltonian mechanics. The Newtonian vectorial approach involves solving Newton's differential equations of motion that relate the force and momenta vectors. This requires knowledge of the time dependence of all the force vectors, including constraint forces, acting on the system which can be very complicated. Chapter 2 showed that the first-order time integrals, equations $2.10, 2.16$, relate the initial and final total momenta without requiring knowledge of the complicated instantaneous forces acting during the collision of two bodies. Similarly, for conservative systems, the first-order spatial integral, equation 2.21, relates the initial and final total energies to the net work done on the system without requiring knowledge of the instantaneous force vectors. The first-order spatial integral has the advantage that it is a scalar quantity, in contrast to time integrals which are vector quantities. These first-order integral relations are used frequently in Newtonian mechanics to derive solutions of the equations of motion that avoid having to solve complicated differential equations of motion.

This chapter has illustrated that variational principles provide a means of deriving more detailed information, such as the trajectories for the motion between given initial and final conditions, by requiring that scalar functionals have extrema values. For example, the solution of the brachistochrone problem determined the trajectory having the minimum transit time, based on only the magnitudes of the kinetic and gravitational potential energies. Similarly, the catenary shape of a suspended chain was derived by minimizing the gravitational potential energy. The calculus of variations uses Euler's equations to determine directly the differential equations of motion of the system that lead to the functional of interest being stationary at an extremum. The Lagrangian and Hamiltonian variational approaches to classical mechanics are discussed in chapters $6 - 16$. The broad range of applicability, the flexibility, and the power provided by variational approaches to classical mechanics and modern physics will be illustrated.

5.12   Summary

Euler's differential equation: The calculus of variations has been introduced and Euler's differential equation was derived. The calculus of variations reduces to varying the functions $y_i(x)$, where $i = 1, 2, 3, ...n$, such that the integral

$$F = \int_{x_1}^{x_2} f\left[y_i(x), y_i'(x); x\right] dx \tag{5.16}$$

is an extremum, that is, it is a maximum or minimum. Here x is the independent variable, $y_i(x)$ are the dependent variables plus their first derivatives $y_i' \equiv \frac{dy_i}{dx}$. The quantity $f\left[y(x), y'(x); x\right]$ has some given dependence on y_i, y_i' and x. The calculus of variations involves varying the functions $y_i(x)$ until a stationary value of F is found which is presumed to be an extremum. It was shown that *if the $y_i(x)$ are independent,* then the extremum value of F leads to n independent Euler equations

$$\frac{\partial f}{\partial y_i} - \frac{d}{dx}\frac{\partial f}{\partial y_i'} = 0 \tag{5.19}$$

where $i = 1, 2, 3..n$. This can be used to determine the functional form $y_i(x)$ that ensures that the integral $F = \int_{x_1}^{x_2} f\left[y(x), y'(x); x\right] dx$ is a stationary value, that is, presumably a maximum or minimum value.

Note that Euler's equation involves partial derivatives for the dependent variables y_i, y_i', and the total derivative for the independent variable x.

Euler's integral equation: It was shown that if the function $\int_{x_1}^{x_2} f\left[y_i(x), y_i'(x); x\right]$ does not depend on the independent variable, then Euler's differential equation can be written in an integral form. This integral form of Euler's equation is especially useful *when $\frac{\partial f}{\partial x} = 0$, that is, when f does not depend explicitly on x,* then the first integral of the Euler equation is a constant

$$f - y'\frac{\partial f}{\partial y'} = \text{constant} \tag{5.25}$$

Constrained variational systems: Most applications involve constraints on the motion. The equations of constraint can be classified according to whether the constraints are holonomic or non-holonomic, the time dependence of the constraints, and whether the constraint forces are conservative.

Generalized coordinates in variational calculus: Independent generalized coordinates can be chosen that are perpendicular to the rigid constraint forces and therefore the constraint does not contribute to the functional being minimized. That is, the constraints are embedded into the generalized coordinates and thus the constraints can be ignored when deriving the variational solution.

Minimal set of generalized coordinates: If the constraints are holonomic then the m holonomic equations of constraint can be used to transform the n coupled generalized coordinates to $s = n - m$ independent generalized variables q_i, q_i'. The generalized coordinate method then uses Euler's equations to determine these $s = n - m$ independent generalized coordinates.

$$\frac{\partial f}{\partial q_i} - \frac{d}{dx}\frac{\partial f}{\partial q_i'} = 0 \tag{5.35}$$

Lagrange multipliers for holonomic constraints: The Lagrange multipliers approach for n variables, plus m holonomic equations of constraint, determines all $N + m$ unknowns for the system. The holonomic forces of constraint acting on the N variables, are related to the Lagrange multiplier terms $\lambda_k(x)\frac{\partial g_k}{\partial y_i}$ that are introduced into the Euler equations. That is,

$$\frac{\partial f}{\partial y_i} - \frac{d}{dx}\frac{\partial f}{\partial y_i'} + \sum_{k}^{m} \lambda_k(x)\frac{\partial g_k}{\partial y_i} = 0 \tag{5.48}$$

where the holonomic equations of constraint are given by

$$g_k(y_i; x) = 0 \tag{5.38}$$

The advantage of using the Lagrange multiplier approach is that the variational procedure simultaneously determines both the equations of motion for the N variables plus the m constraint forces acting on the system.

Workshop exercises

1. Find the extremal of the functional

$$J(x) = \int\limits_1^2 \frac{\dot{x}^2}{t^3} dt$$

   that satisfies $x(1) = 3$ and $x(2) = 18$. Show that this extremal provides the global minimum of J.

2. Consider the use of equations of constraint.

   (a) A particle is constrained to move on the surface of a sphere. What are the equations of constraint for this system?

   (b) A disk of mass m and radius R rolls without slipping on the outside surface of a half-cylinder of radius $5R$. What are the equations of constraint for this system?

   (c) What are holonomic constraints? Which of the equations of constraint that you found above are holonomic?

   (d) Equations of constraint that do not explicitly contain time are said to be scleronomic. Moving constraints are rheonomic. Are the equations of constraint that you found above scleronomic or rheonomic?

3. For each of the following systems, describe the generalized coordinates that would work best. There may be more than one answer for each system.

   (a) An inclined plane of mass M is sliding on a smooth horizontal surface, while a particle of mass m is sliding on the smooth inclined surface.

   (b) A disk rolls without slipping across a horizontal plane. The plane of the disk remains vertical, but it is free to rotate about a vertical axis.

   (c) A double pendulum consisting of two simple pendula, with one pendulum suspended from the bob of the other. The two pendula have equal lengths and have bobs of equal mass. Both pendula are confined to move in the same plane.

   (d) A particle of mass m is constrained to move on a circle of radius R. The circle rotates in space about one point on the circle, which is fixed. The rotation takes place in the plane of the circle, with constant angular speed ω, in the absence of a gravitational force.

   (e) A particle of mass m is attracted toward a given point by a force of magnitude k/r^2, where k is a constant.

4. Looking back at the systems in problem 3, which ones could have equations of constraint? How would you classify the equations of constraint (holonomic, scleronomic, rheonomic, etc.)?

Problems

1. Find the extremal of the functional

$$J(x) = \int_0^\pi (2x \sin t - \dot{x}^2)dt$$

   that satisfies $x(o) = x(\pi) = 0$. Show that this extremal provides the global maximum of J.

2. Find and describe the path $y = y(x)$ for which the the integral $\int_{x_1}^{x_2} \sqrt{x}\sqrt{1 + (y')^2}dx$ is stationary.

3. Find the dimensions of the parallelepiped of maximum volume circumscribed by a sphere of radius R.

4. Consider a single loop of the cycloid having a fixed value of a as shown in the figure. A car released from rest at any point P_0 anywhere on the track between O and the lowest point P , that is, P_0 has a parameter $0 < \theta_0 < \pi$.

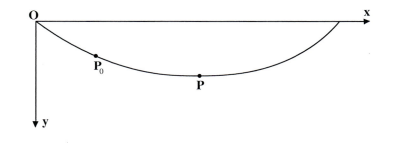

   (a) Show that the time T for the cart to slide from P_0 to P is given by the integral

$$T(P_0 \rightarrow P) = \sqrt{\frac{a}{g}} \int_{\theta_0}^\pi \sqrt{\frac{1 - \cos\theta}{\cos\theta_0 - \cos\theta}}d\theta$$

   (b) Prove that this time T is equal to $\pi\sqrt{a/g}$ which is independent of the position P_0.
   (c) Explain qualitatively how this surprising result can possibly be true.

5. Consider a medium for which the refractive index $n = \frac{a}{r^2}$ where a is a constant and r is the distance from the origin. Use Fermat's Principle to find the path of a ray of light travelling in a plane containing the origin. Hint, use two-dimensional polar coordinates with $\phi = \phi(r)$. Show that the resulting path is a circle through the origin.

6. Find the shortest path between the (x, y, z) points $(0, -1, 0)$ and $(0, 1, 0)$ on the conical surface

$$z = 1 - \sqrt{x^2 + y^2}$$

   What is the length of this path? Note that this is the shortest mountain path around a volcano.

7. Show that the geodesic on the surface of a right circular cylinder is a segment of a helix.

Chapter 6

Lagrangian dynamics

6.1   Introduction

Newtonian mechanics is based on vector observables such as momentum and force, and Newton's equations of motion can be derived if the forces are known. Newtonian mechanics becomes difficult to apply for many-body systems that involve constraint forces. The alternative algebraic Lagrangian mechanics approach is based on the concept of scalar energies which circumvent many of the difficulties in handling constraint forces and many-body systems.

The Lagrangian approach to classical dynamics is based on the calculus of variations introduced in chapter 5. It was shown that the calculus of variations determines the function $y_i(x)$ such that the scalar functional

$$F = \int_{x_1}^{x_2} \sum_i^n f\left[y_i(x), y_i'(x); x\right] dx \tag{6.1}$$

is an extremum, that is, a maximum or minimum. Here x is the independent variable, $y_i(x)$ are the n dependent variables, and their derivatives $y_i' \equiv \frac{dy_i}{dx}$, where $i = 1, 2, 3, ..n$. The function $f\left[y_i(x), y_i'(x); x\right]$ has an assumed dependence on y_i, y_i' and x. The calculus of variations determines the functional dependence of the dependent variables $y_i(x)$, on the independent variable x, that is needed to ensure that F is an extremum. For n *independent variables*, F has a stationary point, which is presumed to be an extremum, that is determined by solution of **Euler's differential equations**

$$\frac{d}{dx} \frac{\partial f}{\partial y_i'} - \frac{\partial f}{\partial y_i} = 0 \tag{6.2}$$

If the coordinates $y_i(x)$ are independent, then the Euler equations, (6.2), for each coordinate i are independent. However, for constrained motion, the constraints lead to auxiliary conditions that correlate the coordinates. As shown in chapter 5, a transformation to *independent generalized coordinates* can be made such that the correlations induced by the constraint forces are embedded into the choice of the independent generalized coordinates. The use of generalized coordinates in Lagrangian mechanics simplifies derivation of the equations of motion for constrained systems. For example, for a system of n coordinates, that involves m holonomic constraints, there are $s = n - m$ independent generalized coordinates. For such holonomic constrained motion, it will be shown that the Euler equations can be solved using either of the following three alternative ways.

1) The **minimal set of generalized coordinates** approach involves finding a set of $s = n - m$ independent generalized coordinates q_i that satisfy the assumptions underlying (6.2). These generalized coordinates can be determined if the m equations of constraint are holonomic, that is, related by algebraic equations of constraint

$$g_k(q_i; x) = 0 \tag{6.3}$$

where $k = 1, 2, 3, ...m$. These equations uniquely determine the relationship between the n correlated coordinates. This method has the advantage that it reduces the system of n coordinates, subject to m constraints, to $s = n - m$ independent generalized coordinates which reduces the dimension of the problem to be solved. However, it does not explicitly determine the forces of constraint which are effectively swept under the rug.

2) The **Lagrange multipliers** approach takes account of the correlation between the n coordinates and m holonomic constraints by introducing the Lagrange multipliers $\lambda_k(x)$. These n generalized coordinates q_i are correlated by the m holonomic constraints.

$$\frac{d}{dx}\frac{\partial f}{\partial q_i'} - \frac{\partial f}{\partial q_i} = \sum_k^m \lambda_k(x)\frac{\partial g_k}{\partial q_i} \tag{6.4}$$

where $i = 1, 2, 3, ...n$. The Lagrange multiplier approach has the advantage that Euler's calculus of variations automatically use the n Lagrange equations, plus the m equations of constraint, to explicitly determine both the n coordinates q_i and the m forces of constraint which are related to the Lagrange multipliers λ_k as given in equation (6.4). Chapter 6.2 shows that the $\sum_k^m \lambda_k(x)\frac{\partial g_k}{\partial y_i}$ terms are directly related to the holonomic forces of constraint.

3) The **generalized force** approach incorporates the forces of constraint explicitly as will be shown in chapter 6.5.4. Incorporating the constraint forces explicitly allows use of holonomic, non-holonomic, and non-conservative constraint forces.

Understanding the Lagrange formulation of classical mechanics is facilitated by use of a simple non-rigorous plausibility approach that is based on Newton's laws of motion. This introductory plausibility approach will be followed by two more rigorous derivations of the Lagrangian formulation developed using either d'Alembert Principle or Hamiltons Principle. These better elucidate the physics underlying the Lagrange and Hamiltonian analytic representations of classical mechanics. In 1788 Lagrange derived his equations of motion using the differential *d'Alembert Principle*, that extends to dynamical systems the Bernoulli Principle of infinitessimal virtual displacements and virtual work. The other approach, developed in 1834, uses the integral *Hamilton's Principle* to derive the Lagrange equations. Hamilton's Principle is discussed in more detail in chapter 9. Euler's variational calculus underlies d'Alembert's Principle and Hamilton's Principle since both are based on the philosophical belief that the laws of nature prefer economy of motion. Chapters $6.2 - 6.5$ show that both d'Alembert's Principle and Hamilton's Principle lead to the Euler-Lagrange equations. This will be followed by a series of examples that illustrate the use of Lagrangian mechanics in classical mechanics.

6.2   Newtonian plausibility argument for Lagrangian mechanics

Insight into the physics underlying Lagrange mechanics is given by showing the direct relationship between Newtonian and Lagrangian mechanics. The variational approaches to classical mechanics exploit the first-order spatial integral of the force, equation 2.17, which equals the work done between the initial and final conditions. The work done is a simple scalar quantity that depends on the initial and final location for conservative forces. Newton's equation of motion is

$$\mathbf{F} = \frac{d\mathbf{p}}{dt} \tag{6.5}$$

The kinetic energy is given by

$$T = \frac{1}{2}mv^2 = \frac{\mathbf{p}\cdot\mathbf{p}}{2m} = \frac{p_x^2}{2m} + \frac{p_y^2}{2m} + \frac{p_z^2}{2m}$$

It can be seen that

$$\frac{\partial T}{\partial \dot{x}} = p_x \tag{6.6}$$

and

$$\frac{d}{dt}\frac{\partial T}{\partial \dot{x}} = \frac{dp_x}{dt} = F_x \tag{6.7}$$

Consider that the force, acting on a mass m, is arbitrarily separated into two components, one part that is conservative, and thus can be written as the gradient of a scalar potential U, plus the excluded part of the force, F^{EX}. The excluded part of the force F^{EX} could include non-conservative frictional forces as well as forces of constraint which may be conservative or non-conservative. This separation allows the force to be written as

$$\mathbf{F} = -\boldsymbol{\nabla}U + \mathbf{F}^{EX} \tag{6.8}$$

Along each of the x_i axes,

$$\frac{d}{dt}\frac{\partial T}{\partial \dot{x}_i} = -\frac{\partial U}{\partial x_i} + F^{EX}_{x_i} \tag{6.9}$$

Equation (6.9) can be extended by transforming the cartesian coordinate x_i to the generalized coordinates q_i.

Define the standard Lagrangian to be the difference between the kinetic energy and the potential energy, which can be written in terms of the generalized coordinates q_i as

$$L(q_i, \dot{q}_i) \equiv T(\dot{q}_i) - U(q_i) \tag{6.10}$$

Assume that the potential is only a function of the generalized coordinates q_i, that is $\frac{\partial U}{\partial \dot{q}_i} = 0$, then

$$\frac{\partial L}{\partial \dot{q}_i} = \frac{\partial T}{\partial \dot{q}_i} + \frac{\partial U}{\partial \dot{q}_i} = \frac{\partial T}{\partial \dot{q}_i} \tag{6.11}$$

Using the above equations allows Newton's equation of motion (6.9) to be expressed as

$$\frac{d}{dt}\frac{\partial L}{\partial \dot{q}_i} - \frac{\partial L}{\partial q_i} = F^{EX}_{q_i} \tag{6.12}$$

The excluded force $F^{EX}_{q_i}$ can be partitioned into a holonomic constraint force $F^{HC}_{q_i}$, plus any remaining excluded forces F^{EXC}, as given by

$$F^{EX}_{q_i} = F^{HC}_{q_i} + F^{EXC} \tag{6.13}$$

A comparison of equations (6.12, 6.13) and (6.4) shows that the holonomic constraint forces $F^{HC}_{q_i}$, that are contained in the excluded force F^{EX}, can be identified with the Lagrange multiplier term in equation 6.4.

$$F^{HC}_{q_i} \equiv \sum_{k}^{m} \lambda_k(t)\frac{\partial g_k}{\partial q_i} \tag{6.14}$$

That is the Lagrange multiplier terms can be used to account for holonomic constraint forces $F^{HC}_{q_i}$. Thus equation 6.12 can be written as

$$\frac{d}{dt}\frac{\partial L}{\partial \dot{q}_i} - \frac{\partial L}{\partial q_i} = \sum_{k}^{m} \lambda_k(t)\frac{\partial g_k}{\partial q_i} + F^{EXC}_{q_i} \tag{6.15}$$

where the Lagrange multiplier term accounts for holonomic constraint forces, and $F^{EXC}_{q_i}$ includes all the remaining forces that are not accounted for by the scalar potential U, or the Lagrange multiplier terms $F^{HC}_{q_i}$.

For holonomic, conservative forces it is possible to absorb all the forces into the potential U plus the Lagrange multiplier term, that is $F^{EXC}_{q_i} = 0$. Moreover, the use of a minimal set of generalized coordinates allows the holonomic constraint forces to be ignored by explicitly reducing the number of coordinates from n dependent coordinates to $s = n - m$ independent generalized coordinates. That is, the correlations due to the constraint forces are embedded into the generalized coordinates. Then equation 6.15 reduces to the basic Euler differential equations.

$$\frac{d}{dt}\frac{\partial L}{\partial \dot{q}_i} - \frac{\partial L}{\partial q_i} = 0 \tag{6.16}$$

Note that equation 6.16 is identical to Euler's equation 5.34, if the independent variable x is replaced by time t. Thus Newton's equation of motion are equivalent to minimizing the action integral $S = \int_{t_1}^{t_2} Ldt$, that is

$$\delta S = \delta \int_{t_1}^{t_2} L(q_i, \dot{q}_i; t)dt = 0 \tag{6.17}$$

which is Hamilton's Principle. Hamilton's Principle underlies many aspects of physics and as discussed in chapter 9, and is used as the starting point for developing classical mechanics. Hamilton's Principle was postulated 46 years after Lagrange introduced Lagrangian mechanics.

The above plausibility argument, which is based on Newtonian mechanics, illustrates the close connection between the vectorial Newtonian mechanics and the algebraic Lagrangian mechanics approaches to classical mechanics.

6.3   Lagrange equations from d'Alembert's Principle

6.3.1   d'Alembert's Principle of Virtual Work

The Principle of Virtual Work provides a basis for a rigorous derivation of Lagrangian mechanics. Bernoulli introduced the concept of virtual infinitessimal displacement of a system mentioned in chapter 5.9.1. This refers to a change in the configuration of the system as a result of any arbitrary infinitessimal instantaneous change of the coordinates $\delta \mathbf{r}_i$, that is consistent with the forces and constraints imposed on the system at the instant t. Lagrange's symbol δ is used to designate a virtual displacement which is called "virtual" to imply that there is no change in time t, i.e. $\delta t = 0$. This distinguishes it from an actual displacement $d\mathbf{r}_i$ of body i during a time interval dt when the forces and constraints may change.

Suppose that the system of n particles is in equilibrium, that is, the total force on each particle i is zero. The virtual work done by the force $\mathbf{F}_i$ moving a distance $\delta \mathbf{r}_i$ is given by the dot product $\mathbf{F}_i \cdot \delta \mathbf{r}_i$. For equilibrium, the sum of all these products for the N bodies also must be zero

$$\sum_i^N \mathbf{F}_i \cdot \delta \mathbf{r}_i = 0 \tag{6.18}$$

Decomposing the force $\mathbf{F}_i$ on particle i into applied forces $\mathbf{F}_i^A$ and constraint forces $\mathbf{f}_i^C$ gives

$$\sum_i^N \mathbf{F}_i^A \cdot \delta \mathbf{r}_i + \sum_i^N \mathbf{f}_i^C \cdot \delta \mathbf{r}_i = 0 \tag{6.19}$$

The second term in equation 6.19 can be ignored if the virtual work due to the constraint forces is zero. This is rigorously true for rigid bodies and is valid for any forces of constraint where the constraint forces are perpendicular to the constraint surface and the virtual displacement is tangent to this surface. Thus if the constraint forces do no work, then (6.19) reduces to

$$\sum_i^N \mathbf{F}_i^A \cdot \delta \mathbf{r}_i = 0 \tag{6.20}$$

This relation is the Bernoulli's *Principle of Static Virtual Work* and is used to solve problems in statics.

Bernoulli introduced dynamics by using Newton's Law to related force and momentum.

$$\mathbf{F}_i = \dot{\mathbf{p}}_i \tag{6.21}$$

Equation (6.21) can be rewritten as

$$\mathbf{F}_i - \dot{\mathbf{p}}_i = 0 \tag{6.22}$$

In 1742, d'Alembert developed the *Principle of Dynamic Virtual Work* in the form

$$\sum_i^N (\mathbf{F}_i - \dot{\mathbf{p}}_i) \cdot \delta \mathbf{r}_i = 0 \tag{6.23}$$

Using equations (6.19) plus (6.23) gives

$$\sum_i^N (\mathbf{F}_i^A - \dot{\mathbf{p}}_i) \cdot \delta \mathbf{r}_i + \sum_i^N \mathbf{f}_i^C \cdot \delta \mathbf{r}_i = 0 \tag{6.24}$$

For the special case where the forces of constraint are zero, then equation 6.24 reduces to **d'Alembert's Principle**

$$\sum_i^N (\mathbf{F}_i^A - \dot{\mathbf{p}}_i) \cdot \delta \mathbf{r}_i = 0 \tag{6.25}$$

d'Alembert's Principle, by a stroke of genius, cleverly transforms the principle of virtual work from the realm of statics to dynamics. Application of virtual work to statics primarily leads to algebraic equations between the forces, whereas d'Alembert's principle applied to dynamics leads to differential equations.

6.3.2 Transformation to generalized coordinates

In classical mechanical systems the coordinates $\delta\mathbf{r}_i$ usually are not independent due to the forces of constraint and the constraint-force energy contributes to equation 6.24. These problems can be eliminated by expressing d'Alembert's Principle in terms of virtual displacements of n *independent generalized coordinates* q_i of the system for which the constraint force term $\sum_i^n \mathbf{f}_i^C \cdot \delta\mathbf{q}_i = 0$. Then the individual variational coefficients δq_i are independent and $(\mathbf{F}_i^A - \dot{\mathbf{p}}_i) \cdot \delta\mathbf{q}_i = 0$ can be equated to zero for each value of i.

The transformation of the N-body system to n independent generalized coordinates q_k can be expressed as

$$\mathbf{r}_i = \mathbf{r}_i(q_1, q_2, q_3..., q_n, t) \tag{6.26}$$

Assuming n independent coordinates, then the velocity $\mathbf{v}_i$ can be written in terms of general coordinates q_k using the chain rule for partial differentiation.

$$\mathbf{v}_i \equiv \frac{d\mathbf{r}_i}{dt} = \sum_j^n \frac{\partial \mathbf{r}_i}{\partial q_j} \dot{q}_j + \frac{\partial \mathbf{r}_i}{\partial t} \tag{6.27}$$

The arbitrary virtual displacement $\delta\mathbf{r}_i$ can be related to the virtual displacement of the generalized coordinate δq_j by

$$\delta\mathbf{r}_i = \sum_j^n \frac{\partial \mathbf{r}_i}{\partial q_j} \delta q_j \tag{6.28}$$

Note that by definition, a virtual displacement considers only displacements of the coordinates, and no time variation δt is involved.

The above transformations can be used to express d'Alembert's dynamical principle of virtual work in generalized coordinates. Thus the first term in d'Alembert's Dynamical Principle, (6.25) becomes

$$\sum_i^n \mathbf{F}_i^A \cdot \delta\mathbf{r}_i = \sum_{i,j}^n \mathbf{F}_i^A \cdot \frac{\partial \mathbf{r}_i}{\partial q_j} \delta q_j = \sum_j^n Q_j \delta q_j \tag{6.29}$$

where Q_j are called components of the *generalized force*,[1] defined as

$$Q_j \equiv \sum_i^n \mathbf{F}_i^A \cdot \frac{\partial \mathbf{r}_i}{\partial q_j} \tag{6.30}$$

Note that just as the generalized coordinates q_j need not have the dimensions of length, so the Q_j do not necessarily have the dimensions of force, but the product $Q_j \delta q_j$ must have the dimensions of work. For example, Q_j could be torque and δq_j could be the corresponding infinitessimal rotation angle.

The second term in d'Alembert's Principle (6.25) can be transformed using equation 6.28

$$\sum_i^n \dot{\mathbf{p}}_i \cdot \delta\mathbf{r}_i = \sum_i^n m_i \ddot{\mathbf{r}}_i \cdot \delta\mathbf{r}_i = \left(\sum_i^n m_i \ddot{\mathbf{r}}_i \cdot \frac{\partial \mathbf{r}_i}{\partial q_j} \right) \delta q_j \tag{6.31}$$

The right-hand side of (6.31) can be rewritten as

$$\left(\sum_i^n m_i \ddot{\mathbf{r}}_i \cdot \frac{\partial \mathbf{r}_i}{\partial q_j} \right) \delta q_j = \sum_i^n \left\{ \frac{d}{dt} \left(m_i \dot{\mathbf{r}}_i \cdot \frac{\partial \mathbf{r}_i}{\partial q_j} \right) - m_i \dot{\mathbf{r}}_i \cdot \frac{d}{dt} \left(\frac{\partial \mathbf{r}_i}{\partial q_j} \right) \right\} \delta q_j \tag{6.32}$$

Note that equation (6.27) gives that

$$\frac{\partial \mathbf{v}_i}{\partial \dot{q}_j} = \frac{\partial \mathbf{r}_i}{\partial q_j} \tag{6.33}$$

therefore the first right-hand term in (6.32) can be written as

$$\frac{d}{dt} \left(m_i \dot{\mathbf{r}}_i \cdot \frac{\partial \mathbf{r}_i}{\partial q_j} \right) = \frac{d}{dt} \left(m_i \mathbf{v}_i \cdot \frac{\partial \mathbf{v}_i}{\partial \dot{q}_j} \right) \tag{6.34}$$

---

[1] This proof, plus the notation, conform with that used by Goldstein [Go50] and by other texts on classical mechanics.

The second right-hand term in (6.32) can be rewritten by interchanging the order of the differentiation with respect to t and q_j

$$\frac{d}{dt}\left(\frac{\partial \mathbf{r}_i}{\partial q_j}\right) = \frac{\partial \mathbf{v}_i}{\partial q_j} \tag{6.35}$$

Substituting (6.34) and (6.35) into (6.32) gives

$$\sum_i^n \dot{\mathbf{p}}_i \cdot \delta \mathbf{r}_i = \left(\sum_i^n m_i \ddot{\mathbf{r}}_i \cdot \frac{\partial \mathbf{r}_i}{\partial q_j}\right)\delta q_j = \sum_i^N \left\{\frac{d}{dt}\left(m_i \mathbf{v}_i \cdot \frac{\partial \mathbf{v}_i}{\partial \dot{q}_j}\right) - m_i \mathbf{v}_i \cdot \frac{\partial \mathbf{v}_i}{\partial q_j}\right\}\delta q_j \tag{6.36}$$

Inserting (6.29) and (6.36) into d'Alembert's Principle (6.25) leads to the relation

$$\sum_i^n (\mathbf{F}_i^A - \dot{\mathbf{p}}_i) \cdot \delta \mathbf{r}_i = -\sum_j^N \left\{\frac{d}{dt}\left(\frac{\partial}{\partial \dot{q}_j}\left(\sum_i \frac{1}{2}m_i v_i^2\right)\right) - \frac{\partial}{\partial q_j}\left(\sum_i \frac{1}{2}m_i v_i^2\right) - Q_j\right\}\delta q_j = 0 \tag{6.37}$$

The $\sum_i^n \frac{1}{2}m_i v_i^2$ term can be identified with the system kinetic energy T. Thus d'Alembert Principle reduces to the relation

$$\sum_j^N \left[\left\{\frac{d}{dt}\left(\frac{\partial T}{\partial \dot{q}_j}\right) - \frac{\partial T}{\partial q_j}\right\} - Q_j\right]\delta q_j = 0 \tag{6.38}$$

For cartesian coordinates T is a function only of velocities $(\dot{x}, \dot{y}, \dot{z})$ and thus the term $\frac{\partial T}{\partial q_j} = 0$. However, as discussed in appendix $C.2.2$, for curvilinear coordinates $\frac{\partial T}{\partial q_j} \neq 0$ due to the curvature of the coordinates as is illustrated for polar coordinates where $\mathbf{v} = \dot{r}\hat{\mathbf{r}} + r\dot{\theta}\hat{\boldsymbol{\theta}}$.

If all the n generalized coordinates q_j are independent, then equation 6.38 implies that the term in the square brackets is zero for each individual value of j. This leads to the *basic Euler-Lagrange equations of motion for each of the independent generalized coordinates*

$$\left\{\frac{d}{dt}\left(\frac{\partial T}{\partial \dot{q}_j}\right) - \frac{\partial T}{\partial q_j}\right\} = Q_j \tag{6.39}$$

where $n \geq j \geq 1$. That is, this leads to n Euler-Lagrange equations of motion for the generalized forces Q_j. As discussed in chapter 5.8, when m holonomic constraint forces apply, it is possible to reduce the system to $s = n - m$ independent generalized coordinates for which equation 6.25 applies.

In 1687 Leibniz proposed minimizing the time integral of his "vis viva", which equals $2T$. That is,

$$\delta \int_{t_1}^{t_2} T dt = 0 \tag{6.40}$$

The variational equation 6.39 accomplishes the minimization of equation 6.40. It is remarkable that Leibniz anticipated the basic variational concept prior to the birth of the developers of Lagrangian mechanics, i.e., d'Alembert, Euler, Lagrange, and Hamilton.

6.3.3   Lagrangian

The handling of both conservative and non-conservative generalized forces Q_j is best achieved by assuming that the generalized force $Q_j = \sum_i^n \mathbf{F}_i^A \cdot \frac{\partial \bar{\mathbf{r}}_i}{\partial q_j}$ can be partitioned into a conservative velocity-independent term, that can be expressed in terms of the gradient of a scalar potential, $-\boldsymbol{\nabla}U_i$, plus an excluded generalized force Q_j^{EX} which contains the non-conservative, velocity-dependent, and all the constraint forces not explicitly included in the potential U_j. That is,

$$Q_j = -\boldsymbol{\nabla}U_j + Q_j^{EX} \tag{6.41}$$

Inserting (6.41) into (6.38), and *assuming that the potential U is velocity independent*, allows (6.38) to be rewritten as

$$\sum_j \left[\left\{\frac{d}{dt}\left(\frac{\partial(T-U)}{\partial \dot{q}_j}\right) - \frac{\partial(T-U)}{\partial q_j}\right\} - Q_j^{EX}\right]\delta q_j = 0 \tag{6.42}$$

The definition of the **Standard Lagrangian** is

$$L \equiv T - U \tag{6.43}$$

then (6.42) can be written as

$$\sum_{j}^{N} \left[\left\{ \frac{d}{dt} \left(\frac{\partial L}{\partial \dot{q}_j} \right) - \frac{\partial L}{\partial q_j} \right\} - Q_j^{EX} \right] \delta q_j = 0 \tag{6.44}$$

Note that equation (6.44) contains the basic Euler-Lagrange equation (6.38) as a special case when $U = 0$. In addition, note that *if all the generalized coordinates are independent,* then the square bracket terms are zero for each value of j, which leads to the *general Euler-Lagrange equations of motion*

$$\left\{ \frac{d}{dt} \left(\frac{\partial L}{\partial \dot{q}_j} \right) - \frac{\partial L}{\partial q_j} \right\} = Q_j^{EX} \tag{6.45}$$

where $n \geq j \geq 1$.

Chapter 6.5.3 will show that the holonomic constraint forces can be factored out of the generalized force term Q_j^{EX} which simplifies derivation of the equations of motion using Lagrangian mechanics. The general Euler-Lagrange equations of motion are used extensively in classical mechanics because conservative forces play a ubiquitous role in classical mechanics.

6.4   Lagrange equations from Hamilton's Action Principle

Hamilton published two papers in 1834 and 1835, announcing a fundamental new dynamical principle that underlies both Lagrangian and Hamiltonian mechanics. Hamilton was seeking a theory of optics when he developed Hamilton's Action Principle, plus the field of Hamiltonian mechanics, both of which play a crucial role in classical mechanics and modern physics. Hamilton's Action Principle states *" dynamical systems follow paths that minimize the time integral of the Lagrangian".* That is, the *action functional S*

$$S = \int_{t_1}^{t_2} L(\mathbf{q}, \dot{\mathbf{q}}, t) dt \tag{6.46}$$

has a minimum value for the correct path of motion. **Hamilton's Action Principle** can be written in terms of a virtual infinitessimal displacement δ, as

$$\delta S = \delta \int_{t_1}^{t_2} L dt = 0 \tag{6.47}$$

Variational calculus therefore implies that a system of s independent generalized coordinates must satisfy the basic Lagrange-Euler equations

$$\frac{d}{dt} \frac{\partial L}{\partial \dot{q}_j} - \frac{\partial L}{\partial q_j} = 0 \tag{6.48}$$

Note that for $Q_j^{EX} = 0$, this is the same as equation 6.45 which was derived using d'Alembert's Principle.

This discussion has shown that Euler's variational differential equation underlies both the differential variational d'Alembert Principle, and the more fundamental integral Hamilton's Action Principle. As discussed in chapter 9.2, Hamilton's Principle of Stationary Action adds a fundamental new dimension to classical mechanics which leads to derivation of both Lagrangian and Hamiltonian mechanics. That is, both Hamilton's Action Principle, and d'Alembert's Principle, can be used to derive Lagrangian mechanics leading to the most general Lagrange equations that are applicable to both holonomic and non-holonomic constraints, as well as conservative and non-conservative systems. In addition, Chapter 6.2 presented a plausibility argument showing that Lagrangian mechanics can be justified based on Newtonian mechanics. Hamilton's Action Principle, and d'Alembert's Principle, can be expressed in terms of generalized coordinates which is much broader in scope than the equations of motion implied using Newtonian mechanics.

6.5   Constrained systems

The motion for systems subject to constraints is difficult to calculate using Newtonian mechanics because all the unknown constraint forces must be included explicitly with the active forces in order to determine the equations of motion. Lagrangian mechanics avoids these difficulties by allowing selection of independent generalized coordinates that incorporate the correlated motion induced by the constraint forces. This allows the constraint forces acting on the system to be ignored by reducing the system to a minimal set of generalized coordinates. The holonomic constraint forces can be determined using the Lagrange multiplier approach, or all constraint forces can be determined by including them as generalized forces, as described below.

6.5.1   Choice of generalized coordinates

As discussed in chapter 5.8, the flexibility and freedom for selection of generalized coordinates is a considerable advantage of Lagrangian mechanics when handling constrained systems. The generalized coordinates can be any set of *independent* variables that completely specify the scalar action functional, equation 6.46. The generalized coordinates are not required to be orthogonal as is required when using the vectorial Newtonian approach. The secret to using generalized coordinates is to select coordinates that are perpendicular to the constraint forces so that the constraint forces do no work. Moreover, if the constraints are rigid, then the constraint forces do no work in the direction of the constraint force. As a consequence, the constraint forces do not contribute to the action integral and thus the $\sum_i^n \mathbf{f}_i^C \cdot \delta\mathbf{r}_i$ term in equation 6.19 can be omitted from the action integral. Generalized coordinates allow reducing the number of unknowns from n to $s = n - m$ when the system has m holonomic constraints. In addition, generalized coordinates facilitate using both the Lagrange multipliers, and the generalized forces, approaches for determining the constraint forces.

6.5.2   Minimal set of generalized coordinates

The set of n generalized coordinates q_i are used to describe the motion of the system. No restrictions have been placed on the nature of the constraints other than they are workless for a virtual displacement. *If the m constraints are holonomic,* then it is possible to find sets of $s = n - m$ *independent generalized coordinates* q_j that contain the m constraint conditions implicitly in the transformation equations

$$\mathbf{r}_i = \mathbf{r}_i(q_1, q_2, q_3..., q_s, t) \tag{6.49}$$

For the case of $s = n - m$ unknowns, *any virtual displacement* δq_j *is independent of* δq_k, therefore the only way for (6.44) to hold is for the term in brackets to vanish for each value of j, that is

$$\left\{ \frac{d}{dt}\left(\frac{\partial L}{\partial \dot{q}_j}\right) - \frac{\partial L}{\partial q_j} \right\} = Q_j^{EX} \tag{6.50}$$

where $j = 1, 2, 3,.. \ s$. These are the **Lagrange equations** for the minimal set of s *independent* generalized coordinates.

If all the generalized forces are conservative plus velocity independent, and are included in the potential U, and $Q_j^{EX} = 0$, then (6.50) simplifies to

$$\left\{ \frac{d}{dt}\left(\frac{\partial L}{\partial \dot{q}_j}\right) - \frac{\partial L}{\partial q_j} \right\} = 0 \tag{6.51}$$

This is Euler's differential equation, derived earlier using the calculus of variations. Thus d'Alembert's Principle leads to a solution that minimizes the action integral $\delta \int_{t_1}^{t_2} L dt = 0$ as stated by Hamilton's Principle.

6.5.3   Lagrange multipliers approach

Equation (6.44) sums over all n coordinates for N particles, providing n equations of motion. If the m constraints are holonomic they can be expressed by m algebraic equations of constraint

$$g_k(q_1, q_2, ..q_n, t) = 0 \tag{6.52}$$

where $k = 1, 2, 3, ...m$. Kinematic constraints can be expressed in terms of the infinitessimal displacements of the form

$$\sum_{j=1}^{n} \frac{\partial g_k}{\partial q_j}(\mathbf{q}, t) dq_j + \frac{\partial g_k}{\partial t} dt = 0 \tag{6.53}$$

where $k = 1, 2, 3, ...m$, $j = 1, 2, 3, ...n$, and where the $\frac{\partial g_k}{\partial q_j}$, and $\frac{\partial g_k}{\partial t}$ are functions of the generalized coordinates q_j, described by the vector $\mathbf{q}$, that are derived from the equations of constraint. As discussed in chapter 5.7, if (6.53) represents the total differential of a function, then it can be integrated to give a holonomic relation of the form of equation (6.52). However, if (6.53) is not the total differential, then it can be integrated only after having solved the full problem. If $\frac{\partial g_k}{\partial t} = 0$ then the kth constraint is scleronomic.

The discussion of Lagrange multipliers in chapter 5.9.1, showed that, for virtual displacements δq_j, the correlation of the generalized coordinates, due to the constraint forces, can be taken into account by multiplying (6.53) by unknown Lagrange multipliers λ_k and summing over all m constraints. Generalized forces can be partitioned into a Lagrange multiplier term plus a remainder force. That is

$$Q_j^{EX} = \sum_{k=1}^{m} \lambda_k \frac{\partial g_k}{\partial q_j}(\mathbf{q}, t) + Q_j^{EXC} \tag{6.54}$$

since by definition $\delta t = 0$ for virtual displacements.

Chapter 5.9.1 showed that holonomic forces of constraint can be taken into account by introducing the Lagrange undetermined multipliers approach, which is equivalent to defining an extended Lagrangian $L'(\mathbf{q}, \dot{\mathbf{q}}, \boldsymbol{\lambda}, t)$ where

$$L'(\mathbf{q}, \dot{\mathbf{q}}, \boldsymbol{\lambda}, t) = L(\mathbf{q}, \dot{\mathbf{q}}, t) + \sum_{k=1}^{m} \sum_{j=1}^{n} \lambda_k \frac{\partial g_k}{\partial q_j}(\mathbf{q}, t) \tag{6.55}$$

Finding the extremum for the extended Lagrangian $L'(\mathbf{q}, \dot{\mathbf{q}}, \boldsymbol{\lambda}, t)$ using (6.47) gives

$$\sum_{j}^{n} \left[\left\{ \frac{d}{dt} \left(\frac{\partial L}{\partial \dot{q}_j} \right) - \frac{\partial L}{\partial q_j} \right\} - \sum_{k=1}^{m} \lambda_k \frac{\partial g_k}{\partial q_j}(\mathbf{q}, t) - Q_j^{EXC} \right] \delta q_j = 0 \tag{6.56}$$

where Q_j^{EXC} is the remaining part of the generalized force Q_j after subtracting both the part of the force absorbed in the potential energy U, which is buried in the Lagrangian L, as well as the holonomic constraint forces which are included in the Lagrange multiplier terms $\sum_{k=1}^{m} \lambda_k \frac{\partial g_k}{\partial q_j}(\mathbf{q}, t)$. The m Lagrange multipliers λ_k can be chosen arbitrarily in (6.56). Utilizing the free choice of the m Lagrange multipliers λ_k allows them to be determined in such a way that the coefficients of the first m infinitesimals, i.e. the square brackets vanish. Therefore the expression in the square bracket must vanish for each value of $1 \leq j \leq m$. Thus it follows that

$$\left\{ \frac{d}{dt} \left(\frac{\partial L}{\partial \dot{q}_j} \right) - \frac{\partial L}{\partial q_j} \right\} - \sum_{k=1}^{m} \lambda_k \frac{\partial g_k}{\partial q_j}(\mathbf{q}, t) - Q_j^{EXC} = 0 \tag{6.57}$$

when $j = 1, 2, ..m$. Thus (6.56) reduces to a sum over the remaining coordinates between $m + 1 \leq j \leq n$

$$\sum_{j=m+1}^{n} \left[\left\{ \frac{d}{dt} \left(\frac{\partial L}{\partial \dot{q}_j} \right) - \frac{\partial L}{\partial q_j} \right\} - \sum_{k=1}^{m} \lambda_k \frac{\partial g_k}{\partial q_j}(\mathbf{q}, t) - Q_j^{EXC} \right] \delta q_j = 0 \tag{6.58}$$

In equation (6.58) the $s = n - m$ infinitesimals δq_j can be chosen freely since the $s = n - m$ degrees of freedom are *independent*. Therefore the expression in the square bracket must vanish for each value of $m + 1 \leq j \leq n$. Thus it follows that

$$\left\{ \frac{d}{dt} \left(\frac{\partial L}{\partial \dot{q}_j} \right) - \frac{\partial L}{\partial q_j} \right\} - \sum_{k=1}^{m} \lambda_k \frac{\partial g_k}{\partial q_j}(\mathbf{q}, t) - Q_j^{EXC} = 0 \tag{6.59}$$

where $j = m + 1, m + 2, ..n$. Combining equations (6.57) and (6.59) then gives the important general relation that for $1 \leq j \leq n$

$$\left\{ \frac{d}{dt} \left(\frac{\partial L}{\partial \dot{q}_j} \right) - \frac{\partial L}{\partial q_j} \right\} = \sum_{k=1}^{m} \lambda_k \frac{\partial g_k}{\partial q_j}(\mathbf{q}, t) + Q_j^{EXC} \tag{6.60}$$

To summarize, the Lagrange multiplier approach (6.60) automatically solves the n equations plus the m holonomic equations of constraint, which determines the $n + m$ unknowns, that is, the n coordinates plus the m forces of constraint. The beauty of the Lagrange multipliers is that all n variables, plus the m constraint forces, are found simultaneously by using the calculus of variations to determine the extremum for the expanded Lagrangian $L'(\mathbf{q}, \dot{\mathbf{q}}, \boldsymbol{\lambda}, t)$.

6.5.4 Generalized forces approach

The two right-hand terms in (6.60) can be understood to be those forces acting on the system that are not absorbed into the scalar potential U component of the Lagrangian L. The Lagrange multiplier terms $\sum_{k=1}^{m} \lambda_k \frac{\partial g_k}{\partial q_j}(\mathbf{q}, t)$ account for the holonomic forces of constraint that are not included in the conservative potential or in the generalized forces Q_j^{EXC}. The generalized force

$$Q_j^{EXC} = \sum_i^n \mathbf{F}_i^A \cdot \frac{\partial \mathbf{r}_i}{\partial q_j} \tag{6.17}$$

is the sum of the components in the q_j direction for all external forces that have not been taken into account by the scalar potential or the Lagrange multipliers. Thus the non-conservative generalized force Q_j^{EXC} contains non-holonomic constraint forces, including dissipative forces such as drag or friction, that are not included in U, or used in the Lagrange multiplier terms to account for the holonomic constraint forces.

The concept of generalized forces is illustrated by the case of spherical coordinate systems. The attached table gives the displacement elements δq_i, (taken from table $C4$) and the generalized force for the three coordinates. Note that Q_i has the dimensions of force and $Q_i . \delta q_i$ has the units of energy. By contrast equation 6.30 gives that $Q_\theta = F_\theta r$ and $Q_\phi = F_\phi r$ which have the dimensions of torque. However, $Q_\theta \delta\theta$ and $Q_\phi \delta\phi$ both have the dimensions of energy as is required in equation 6.30. This illustrates that the units used for generalized forces depend on the units of the corresponding generalized coordinate.

Unit vectors	δq_i	Q_i	$Q_i \cdot \delta q_i$
$\hat{r}$	$\hat{\mathbf{r}}dr$	$\hat{\mathbf{r}}F_r$	$F_r dr$
$\hat{\boldsymbol{\theta}}$	$\hat{\boldsymbol{\theta}}rd\theta$	$\hat{\boldsymbol{\theta}}F_\theta r$	$F_\theta rd\theta$
$\hat{\boldsymbol{\phi}}$	$\hat{\boldsymbol{\phi}}r\sin\theta d\phi$	$\hat{\boldsymbol{\phi}}F_\phi r\sin\theta$	$F_\phi r\sin\theta d\phi$

6.6 Applying the Euler-Lagrange equations to classical mechanics

d'Alembert's principle of virtual work has been used to derive the Euler-Lagrange equations, which also satisfy Hamilton's Principle, and the Newtonian plausibility argument. These imply that the actual path taken in configuration space $(q_i, \dot{q}_i, t)$ is the one that minimizes the action integral $\int_{t_1}^{t_2} L(q_j, \dot{q}_j; t)dt$. As a consequence, the Euler equations for the calculus of variations lead to the Lagrange equations of motion.

$$\left\{ \frac{d}{dt}\left(\frac{\partial L}{\partial \dot{q}_j} \right) - \frac{\partial L}{\partial q_j} \right\} = \sum_{k=1}^{m} \lambda_k \frac{\partial g_k}{\partial q_j}(\mathbf{q}, t) + Q_j^{EXC} \tag{6.60}$$

for n variables, with m equations of constraint. The generalized forces Q_j^{EXC} are not included in the conservative, potential energy U, or the Lagrange multipliers approach for holonomic equations of constraint.[2]

The following is a logical procedure for applying the Euler-Lagrange equations to classical mechanics.

1) Select a set of independent generalized coordinates:

Select an optimum set of independent generalized coordinates as described in chapter 6.5.1. Use of generalized coordinates is always advantageous since they incorporate the constraints, and can reduce the number of unknowns, both of which simplify use of Lagrangian mechanics

---

[2]Euler's differential equation is ubiquitous in Lagrangian mechanics. Thus, for brevity, it is convenient to define the concept of the **Lagrange linear operator** Λ_j, as described in appendix $F2$.

$$\Lambda_j \equiv \frac{d}{dt}\frac{\partial}{\partial \dot{q}_j} - \frac{\partial}{\partial q_j}$$

where Λ_j operates on the Lagrangian L. Then Euler's equations can be written compactly in the form $\Lambda_j L = 0$.

2) Partition of the active forces:

The active forces should be partitioned into the following three groups:

(i) **Conservative one-body forces plus the velocity-dependent electromagnetic force** which can be characterized by the scalar potential U, that is absorbed into the Lagrangian. The gravitational forces plus the velocity-dependent electromagnetic force can be absorbed into the potential U as discussed in chapter 6.10. This approach is by far the easiest way to account for such forces in Lagrangian mechanics.

(ii) **Holonomic constraint forces** provide algebraic relations that couple some of the generalized coordinates. This coupling can be used either to reduce the number of generalized coordinates, or to determine these holonomic constraint forces using the Lagrange multiplier approach.

(iii) **Generalized forces** provide a mechanism for introducing non-conservative and non-holonomic constraint forces into Lagrangian mechanics. Typically general forces are used to introduce dissipative forces.

Typical systems can involve a mixture of all three categories of active forces. For example, mechanical systems often include gravity, introduced as a potential, holonomic constraint forces are determined using Lagrange multipliers, and dissipative forces are included as generalized forces.

3) Minimal set of generalized coordinates:

The ability to embed constraint forces directly into the generalized coordinates is a tremendous advantage enjoyed by the Lagrangian and Hamiltonian variational approaches to classical mechanics. If the constraint forces are not required, then choice of a minimal set of generalized coordinates significantly reduces the number of equations of motion that need to be solved .

4) Derive the Lagrangian:

The Lagrangian is derived in terms of the generalized coordinates and including the conservative forces that are buried into the scalar potential U.

5) Derive the equations of motion:

Equation (6.60) is solved to determine the n generalized coordinates, plus the m Lagrange multipliers characterizing the holonomic constraint forces, plus any generalized forces that were included. The holonomic constraint forces then are given by evaluating the $\lambda_k \frac{\partial g_k}{\partial q_j}(\mathbf{q}, t)$ terms for the m holonomic forces.

In summary, Lagrangian mechanics is based on energies which are scalars in contrast to Newtonian mechanics which is based on vector forces and momentum. As a consequence, Lagrange mechanics allows use of any set of independent generalized coordinates, which do not have to be orthogonal, and they can have very different units for different variables. The generalized coordinates can incorporate the correlations introduced by constraint forces.

The active forces are split into the following three categories;

1. Velocity-independent conservative forces are taken into account using scalar potentials U_i.

2. Holonomic constraint forces can be determined using Lagrange multipliers.

3. Non-holonomic constraints require use of generalized forces Q_j^{EXC}.

Use of the concept of scalar potentials is a trivial and powerful way to incorporate conservative forces in Lagrangian mechanics. The Lagrange multipliers approach requires using the Euler-Lagrange equations for $n + m$ coordinates but determines both holonomic constraint forces and equations of motion simultaneously. Non-holonomic constraints and dissipative forces can be incorporated into Lagrangian mechanics via use of generalized forces which broadens the scope of Lagrangian mechanics.

Note that the equations of motion resulting from the Lagrange-Euler algebraic approach are the same equations of motion as obtained using Newtonian mechanics. However, the Lagrangian is a scalar which facilitates rotation into the most convenient frame of reference. This can greatly simplify determination of the equations of motion when constraint forces apply. As discussed in chapter 17, the Lagrangian and the Hamiltonian variational approaches to mechanics are the only viable way to handle relativistic, statistical, and quantum mechanics.

6.7   Applications to unconstrained systems

Although most dynamical systems involve constrained motion, it is useful to consider examples of systems subject to conservative forces with no constraints. For no constraints, the Lagrange-Euler equations (6.60) simplify to $\Lambda_j L = 0$ where $j = 1, 2, ..n$, and the transformation to generalized coordinates is of no consequence.

6.1 Example: Motion of a free particle, U=0

The Lagrangian in cartesian coordinates is $L = \frac{1}{2}m(\dot{x}^2 + \dot{y}^2 + \dot{z}^2)$. Then

$$\frac{\partial L}{\partial \dot{x}} = m\dot{x}$$

$$\frac{\partial L}{\partial \dot{y}} = m\dot{y}$$

$$\frac{\partial L}{\partial \dot{z}} = m\dot{z}$$

$$\frac{\partial L}{\partial x} = \frac{\partial L}{\partial y} = \frac{\partial L}{\partial z} = 0$$

Insert these in the Lagrange equation gives

$$\Lambda_x L = \frac{d}{dt}\frac{\partial L}{\partial \dot{x}} - \frac{\partial L}{\partial x} = \frac{d}{dt}m\dot{x} - 0 = 0$$

Thus

$$p_x = m\dot{x} = constant$$
$$p_y = m\dot{y} = constant$$
$$p_z = m\dot{z} = constant$$

That is, this shows that the linear momentum is conserved if U is a constant, that is, no forces apply. Note that momentum conservation has been derived without any direct reference to forces.

6.2 Example: Motion in a uniform gravitational field

Consider the motion is in the $x - y$ plane. The kinetic energy $T = \frac{1}{2}m\left(\dot{x}^2 + \dot{y}^2\right)$ while the potential energy is $U = mgy$ where $U(y = 0) = 0$. Thus

$$L = \frac{1}{2}m\left(\dot{x}^2 + \dot{y}^2\right) - mgy$$

Using the Lagrange equation for the x coordinate gives

$$\Lambda_x L = \frac{d}{dt}\frac{\partial L}{\partial \dot{x}} - \frac{\partial L}{\partial x} = \frac{d}{dt}m\dot{x} - 0 = 0$$

Thus the horizontal momentum $m\dot{x}$ is conserved and $\ddot{x} = 0$. The y coordinate gives

$$\Lambda_y L = \frac{d}{dt}\frac{\partial L}{\partial \dot{y}} - \frac{\partial L}{\partial y} = \frac{d}{dt}m\dot{y} + mg = 0$$

Thus the Lagrangian produces the same results as derived using Newton's Laws of Motion.

$$\ddot{x} = 0 \qquad \ddot{y} = -g$$

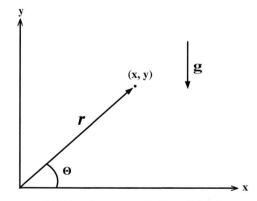

Motion in a gravitational field

The importance of selecting the most convenient generalized coordinates is nicely illustrated by trying to solve this problem using polar coordinates r, θ, where r is radial distance and θ the elevation angle from the x axis as shown in the adjacent figure. Then

$$T = \frac{1}{2}m\dot{r}^2 + \frac{1}{2}m\left(r\dot{\theta}\right)^2$$

$$U = mgr\sin\theta$$

Thus

$$L = \frac{1}{2}m\dot{r}^2 + \frac{1}{2}m\left(r\dot{\theta}\right)^2 - mgr\sin\theta$$

$\Lambda_r L = 0$ for the r coordinate

$$r\dot{\theta}^2 - g\sin\theta - \ddot{r} = 0$$

$\Lambda_\theta L = 0$ for the θ coordinate

$$-gr\cos\theta - 2r\dot{r}\dot{\theta} - r^2\ddot{\theta} = 0$$

These equations written in polar coordinates are more complicated than the result expressed in cartesian coordinates. This is because the potential energy depends directly on the y coordinate, whereas it is a function of both r, θ. This illustrates the freedom for using different generalized coordinates, plus the importance of choosing a sensible set of generalized coordinates.

6.3 Example: Central forces

Consider a mass m moving under the influence of a spherically-symmetric, conservative, attractive, inverse-square force. The potential then is

$$U = -\frac{k}{r}$$

It is natural to express the Lagrangian in spherical coordinates for this system. That is,

$$L = \frac{1}{2}m\dot{r}^2 + \frac{1}{2}m\left(r\dot{\theta}\right)^2 + \frac{1}{2}m(r\sin\theta\dot{\phi})^2 + \frac{k}{r}$$

$\Lambda_r L = 0$ for the r coordinate gives

$$m\ddot{r} - mr[\dot{\theta}^2 + \sin^2\theta\dot{\phi}^2] = \frac{k}{r^2}$$

where the $mr\sin^2\theta\dot{\phi}^2$ term comes from the centripetal acceleration.

$\Lambda_\phi L = 0$ for the ϕ coordinate gives

$$\frac{d}{dt}\left(mr^2\sin^2\theta\dot{\phi}\right) = 0$$

This implies that the derivative of the angular momentum about the ϕ axis, $\dot{p}_\phi = 0$ and thus $p_\phi = mr^2\sin^2\theta\dot{\phi}$ is a constant of motion.

$\Lambda_\theta L = 0$ for the θ coordinate gives

$$\frac{d}{dt}(mr^2\dot{\theta}) - mr^2\sin\theta\cos\theta\dot{\phi}^2 = 0$$

That is,

$$\dot{p}_\theta = mr^2\sin\theta\cos\theta\dot{\phi}^2 = \frac{p_\phi^2\cos\theta}{2mr^2\sin^3\theta}$$

Note that p_θ is a constant of motion if $p_\phi = 0$ and only the radial coordinate is influenced by the radial form of the central potential.

6.8   Applications to systems involving holonomic constraints

The equations of motion that result from the Lagrange-Euler algebraic approach are the same as those given by Newtonian mechanics. The solution of these equations of motion can be obtained mathematically using the chosen initial conditions. The following simple example of a disk rolling on an inclined plane, is useful for comparing the merits of the Newtonian method with Lagrange mechanics employing either minimal generalized coordinates, the Lagrange multipliers, or the generalized forces approaches.

6.4 *Example: Disk rolling on an inclined plane*

Consider a disk rolling down an inclined plane to compare the results obtained using Newton's laws with the results obtained using Lagrange's equations with either generalized coordinates, Lagrange multipliers, or generalized forces. All these cases assume that the friction is sufficient to ensure that the rolling equation of constraint applies and that the disk has a radius R and moment of inertia of I. Assume as generalized coordinates, distance along the inclined plane y which is perpendicular to the normal constraint force N, and perpendicular to the inclined plane x, plus the rolling angle θ. The constraint for rolling is holonomic

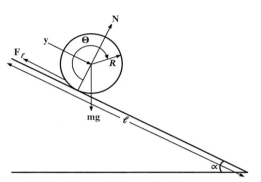

Disk rolling without slipping on an inclined plane.

$$y - R\theta = 0$$

The frictional force is F_f. The constraint that it rolls along the plane implies

$$x - R = 0$$

a) Newton's laws of motion

Newton's law for the components of the forces along the inclined plane gives

$$mg\sin\alpha - F_f = m\ddot{y} \qquad\qquad\qquad (a)$$

Perpendicular to the inclined plane, Newton's law gives

$$mg\cos\alpha = N \qquad\qquad\qquad (b)$$

The torque on the disk gives

$$F_f R = I\ddot{\theta} \qquad\qquad\qquad (c)$$

Assuming the disc rolls gives

$$\ddot{y} = R\ddot{\theta}$$

then

$$F_f = \frac{I}{R^2}\ddot{y}$$

Inserting this into equation (a) gives

$$\left(m + \frac{I}{R^2}\right)\ddot{y} - mg\sin\alpha = 0$$

The moment of inertia of a uniform solid circular disk is $I = \frac{1}{2}mR^2$
  Therefore

$$\ddot{y} = \frac{2}{3}g\sin\alpha$$

and the frictional force is

$$F_f = \frac{mg}{3}\sin\alpha$$

which is smaller than the gravitational force along the plane which is $mg\sin\alpha$.

b) Lagrange equations with a minimal set of generalized coordinates

Using the generalized coordinates defined above, the total kinetic energy is

$$T = \frac{1}{2}m\dot{y}^2 + \frac{1}{2}I\dot{\theta}^2$$

The conservative gravitational force can be absorbed into the potential energy

$$U = mg(l - y)\sin\alpha$$

Thus the Lagrangian is

$$L = \frac{1}{2}m\dot{y}^2 + \frac{1}{2}I\dot{\theta}^2 - mg(l - y)\sin\alpha$$

The holonomic equations of constraint are

$$g_1 = y - R\theta = 0$$
$$g_2 = x - R = 0$$

A holonomic constraint can be used to reduce the system to a single generalized coordinate y plus generalized velocity $\dot{y}$. Expressed in terms of this single generalized coordinate, the Lagrangian becomes

$$L = \frac{1}{2}\left(m + \frac{I}{R^2}\right)\dot{y}^2 - mg(l - y)\sin\alpha$$

The Lagrange equation $\Lambda_y L = 0$ gives

$$mg\sin\alpha = \left(m + \frac{I}{R^2}\right)\ddot{y}$$

Again if $I = \frac{1}{2}mR^2$ then

$$\ddot{y} = \frac{2}{3}g\sin\alpha$$

The solution for the x coordinate is trivial. This answer is identical to that obtained using Newton's laws of motion. Note that no forces have been determined using the single generalized coordinate.

c) Lagrange equation with Lagrange multipliers

Again the conservative gravitation force is absorbed into the scalar potential while the holonomic constraints are taken into account using Lagrange multipliers. Ignoring the trivial x dependence, the Lagrangian is given above to be

$$L = \frac{1}{2}m\dot{y}^2 + \frac{1}{2}I\dot{\theta}^2 - mg(l - y)\sin\alpha$$

The constraint equations are

$$g_1 = y - R\theta = 0$$
$$g_2 = x - R = 0$$

The Lagrange equation for the y coordinate

$$\frac{d}{dt}\frac{\partial L}{\partial \dot{y}} - \frac{\partial L}{\partial y} = \lambda_1\frac{\partial g_1}{\partial y} + \lambda_2 0$$

gives

$$m\ddot{y} - mg\sin\alpha = \lambda_1$$

The Lagrange equation for the θ coordinate

$$\frac{d}{dt}\frac{\partial L}{\partial \dot{\theta}} - \frac{\partial L}{\partial \theta} = \lambda_1\frac{\partial g_1}{\partial \theta} + \lambda_2 0$$

which gives

$$I\ddot{\theta} = -\lambda_1 R$$

The constraint can be written as

$$\ddot{y} = R\ddot{\theta}$$

Let $I = \frac{1}{2}MR^2$ *and solve for* y, θ *and* λ *gives*

$$\lambda_1 = -\frac{mg}{\left(1 + \frac{mR^2}{I}\right)}\sin\alpha = -\frac{mg}{3}\sin\alpha$$

The frictional force is given by

$$F_f = \lambda_1 \frac{\partial g_1}{\partial y} = \lambda_1 = -\frac{mg}{3}\sin\alpha$$

Also

$$m\ddot{y} = mg\sin\alpha + \lambda_1 = \frac{2}{3}mg\sin\alpha$$

and the torque is

$$-\lambda_1 R = F_f R = I\ddot{\theta}$$

d) Lagrange equation using a generalized force

Again the conservative gravitation force is absorbed into the scalar potential while the holonomic constraints are taken into account using generalized forces. Ignoring the trivial x *dependence, the Lagrangian was given above to be*

$$L = \frac{1}{2}m\dot{y}^2 + \frac{1}{2}I\dot{\theta}^2 - mg(l-y)\sin\alpha$$

The generalized forces (6.30) are

$$\begin{aligned} Q_y &= -F_f \\ Q_\theta &= F_f R \end{aligned}$$

The Euler-Lagrange equations are:

The $\Lambda_y L = Q_y$ *Lagrange equation for the* y *coordinate*

$$m\ddot{y} - mg\sin\alpha = Q_y = -F_f$$

The $\Lambda_\theta L = Q_\theta$ *Lagrange equation for the* θ *coordinate*

$$I\ddot{\theta} = Q_\theta = F_f R$$

The constraint equation gives that $y = R\theta$ *and assuming* $I = \frac{1}{2}mR^2$ *leads to the* Q_θ *relation*

$$\frac{Q_\theta}{R} = F_f = \frac{m}{2}\ddot{y}$$

Substitute this equation into the Q_y *relation gives that*

$$m\ddot{y} - mg\sin\alpha = Q_y = -F_f = \frac{m}{2}\ddot{y}$$

Thus

$$\ddot{y} = \frac{2}{3}g\sin\alpha$$

and

$$F_f = -\frac{mg}{3}\sin\alpha$$

The four methods for handling the equations of constraint all are equivalent and result in the same equations of motion. The scalar Lagrangian mechanics is able to calculate the vector forces acting in a direct and simple way. The Newton's law approach is more intuitive for this simple case and the ease and power of the Lagrangian approach is not apparent for this simple system.

The following series of examples will gradually increase in complexity, and will illustrate the power, elegance, plus superiority of the Lagrangian approach compared with the Newtonian approach.

6.5 Example: Two connected masses on frictionless inclined planes

Consider the system shown in the figure. This is a problem that has five constraints that will be solved using the method of generalized coordinates. The obvious generalized coordinates are x_1 and x_2 which are perpendicular to the normal constraint forces on the inclined planes. Another holonomic constraint is that the length of the rope connecting the masses is assumed to be constant. Thus the equation of constraint is that

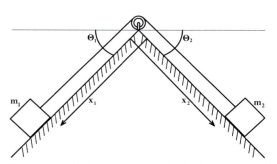

$$x_1 + x_2 - l = 0$$

The other four constraints ensure that the two masses slide directly down the inclined planes in the plane shown. This is assumed implicitly by using only the variables, x_1 and x_2. Let us chose x_1 as the primary generalized coordinate, thus

Two connected masses on frictionless inclined planes

$$\begin{aligned} x_2 &= l - x_1 \\ y_1 &= x_1 \sin \theta_1 \\ y_2 &= (l - x_1) \sin \theta_2 \end{aligned}$$

The conservative gravitational force is absorbed into the potential energy given by

$$U = -m_1 g x_1 \sin \theta_1 - m_2 g (l - x_1) \sin \theta_2$$

Since $\dot{x}_1 = -\dot{x}_2$ the kinetic energy is given by

$$T = \frac{1}{2} m_1 \dot{x}_1^2 + \frac{1}{2} m_2 \dot{x}_2^2 = \frac{1}{2} (m_1 + m_2) \dot{x}_1^2$$

The Lagrangian then gives that

$$L = \frac{1}{2} (m_1 + m_2) \dot{x}_1^2 + m_1 g x_1 \sin \theta_1 + m_2 g (l - x_1) \sin \theta_2$$

Therefore

$$\begin{aligned} \frac{\partial L}{\partial \dot{x}_1} &= (m_1 + m_2) \dot{x}_1 \\ \frac{\partial L}{\partial x_1} &= g (m_1 \sin \theta_1 - m_2 \sin \theta_2) \end{aligned}$$

Thus

$$\Lambda_{x_1} L = \frac{d}{dt} \frac{\partial L}{\partial \dot{x}_1} - \frac{\partial L}{\partial x_1} = 0 = (m_1 + m_2) \ddot{x}_1 - g (m_1 \sin \theta_1 - m_2 \sin \theta_2)$$

Note that the system acts as though the inertial mass is $(m_1 + m_2)$ while the driving force comes from the difference of the forces. The acceleration is zero if

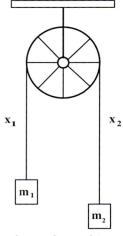

$$m_1 \sin \theta_1 = m_2 \sin \theta_2$$

A special case of this is the Atwood's machine with a massless pulley shown in the adjacent figure. For this case $\theta_1 = \theta_2 = 90^\circ$. Thus

Atwoods machine

$$(m_1 + m_2) \ddot{x}_1 = g (m_1 - m_2)$$

Note that this problem has been solved without any reference to the force in the rope or the normal constraint forces on the inclined planes.

6.6 Example: Two blocks connected by a frictionless bar

Two identical masses m are connected by a massless rigid bar of length l, and they are constrained to move in two frictionless slides, one vertical and the other horizontal as shown in the adjacent figure. Assume that the conservative gravitational force acts along the negative y axis and is incorporated into the scalar potential U. The generalized coordinate can be chosen to be the angle α corresponding to a single degree of freedom. The relative cartesian coordinates of the blocks are given by

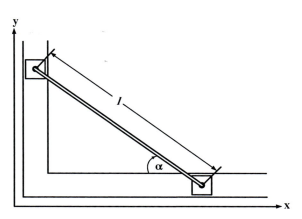

$$x = l\cos\alpha$$
$$y = l\sin\alpha$$

Thus

$$\dot{x} = -l(\sin\alpha)\dot{\alpha}$$
$$\dot{y} = l(\cos\alpha)\dot{\alpha}$$

Two frictionless masses that are connected by a bar and are constrained to slide in vertical and horizontal channels.

This constraint, that is absorbed into the generalized coordinate, is holonomic, scleronomic, and conservative. The kinetic energy is given by

$$T = \frac{1}{2}m\left(l^2(\sin\alpha)^2\dot{\alpha}^2 + l^2(\cos\alpha)^2\dot{\alpha}^2\right) = \frac{1}{2}ml^2\dot{\alpha}^2$$

The gravitational potential energy is given by

$$U = mgy = mgl\sin\alpha$$

Thus the Lagrangian is

$$L = \frac{1}{2}ml^2\dot{\alpha}^2 - mgl\sin\alpha$$

Using the Lagrange operator equation $\Lambda_\alpha L = 0$ gives

$$ml^2\ddot{\alpha} + mgl\cos\alpha = 0$$
$$\ddot{\alpha} + \frac{g}{l}\cos\alpha = 0$$

Multiply by $\dot{\alpha}$ *yields*

$$\ddot{\alpha}\dot{\alpha} + \frac{g}{l}\dot{\alpha}\cos\alpha = 0$$

This can be integrated to give

$$\frac{1}{2}\dot{\alpha}^2 + \frac{g}{l}\sin\alpha = c$$

where c *is a constant. That is*

$$\dot{\alpha} = \sqrt{2\left(c - \frac{g}{l}\sin\alpha\right)}$$

Separation of the variable gives

$$dt = \frac{d\alpha}{\sqrt{2\left(c - \frac{g}{l}\sin\alpha\right)}}$$

Integration of this gives

$$t - t_0 = \int_{\alpha_0}^{\alpha} \frac{d\alpha}{\sqrt{2\left(c - \frac{g}{l}\sin\alpha\right)}}$$

The constants c and t_0 are determined from the given initial conditions.

6.7 Example: Block sliding on a movable frictionless inclined plane

Consider a block of mass m free to slide on a smooth frictionless inclined plane of mass M that is free to slide horizontally as shown in the adjacent figure. The six degrees of freedom can be reduced to two independent generalized coordinates since the inclined plane and mass m are confined to slide along specific non-orthogonal directions. Choose x as the coordinate for movement of the inclined plane in the horizontal $\hat{\imath}$ direction and x' the position of the block with respect to the surface of the inclined plane in the $\hat{e}$ direction which is inclined downward at an angle θ. Thus the velocity of the inclined plane is

$$\mathbf{V} = \hat{\imath}\dot{x}$$

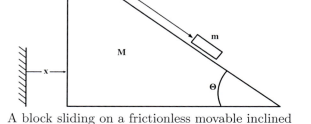

A block sliding on a frictionless movable inclined plane.

while the velocity of the small block on the inclined plane is

$$\mathbf{v} = \hat{\imath}\dot{x} + \hat{e}\dot{x}'$$

The kinetic energy is given by

$$T = \frac{1}{2}M\mathbf{V}\cdot\mathbf{V} + \frac{1}{2}m\mathbf{v}\cdot\mathbf{v} = \frac{1}{2}M\dot{x}^2 + \frac{1}{2}m[\dot{x}^2 + \dot{x}'^2 + 2\dot{x}\dot{x}'\cos\theta]$$

The conservative gravitational force is absorbed into the scalar potential energy which depends only on the vertical position of the block and is taken to be zero at the top of the wedge.

$$U = -mgx'\sin\theta$$

Thus the Lagrangian is

$$L = \frac{1}{2}M\dot{x}^2 + \frac{1}{2}m[\dot{x}^2 + \dot{x}'^2 + 2\dot{x}\dot{x}'\cos\theta] + mgx'\sin\theta$$

Consider the Lagrange-Euler equation for the x coordinate, $\Lambda_x L = 0$ which gives

$$\frac{d}{dt}[m(\dot{x} + \dot{x}'\cos\theta) + M\dot{x}] = 0 \qquad (a)$$

which states that $[m(\dot{x} + \dot{x}'\cos\theta) + M\dot{x}]$ is a constant of motion. This constant of motion is just the total linear momentum of the complete system in the x direction. That is, conservation of the linear momentum is satisfied automatically by the Lagrangian approach. The Newtonian approach also predicts conservation of the linear momentum since there are no external horizontal forces,

Consider the Lagrangian equation for the x' coordinate $\Lambda_{x'} L = 0$ which gives

$$\frac{d}{dt}[\dot{x}' + \dot{x}\cos\theta] = g\sin\theta \qquad (b)$$

Perform both of the time derivatives for equations a and b give

$$m[\ddot{x} + \ddot{x}'\cos\theta] + M\ddot{x} = 0$$
$$\ddot{x}' + \ddot{x}\cos\theta = g\sin\theta$$

Solving for $\ddot{x}$ and $\ddot{x}'$ gives

$$\ddot{x} = \frac{-g\sin\theta\cos\theta}{(m+M)/m - \cos^2\theta}$$

and.

$$\ddot{x}' = \frac{g\sin\theta}{1 - m\cos^2\theta/(m+M)}$$

This example illustrates the flexibility of being able to use non-orthogonal displacement vectors to specify the scalar Lagrangian energy. Newtonian mechanics would require more thought to solve this problem.

6.8 Example: Sphere rolling without slipping down an inclined plane on a frictionless floor.

A sphere of mass m and radius r rolls, without slipping, down an inclined plane, of mass M, sitting on a frictionless horizontal floor as shown in the adjacent figure. The velocity of the rolling sphere has horizontal and vertical components of

$$v_x = \dot{x} + R\dot{\theta}\cos\varphi$$
$$v_y = -R\dot{\theta}\sin\varphi$$

Assume initial conditions are $t = 0, \xi = 0, x = 0, \theta = 0, y = h, \dot{x} = \dot{\theta} = 0$. Choose the independent coordinates x and θ as generalized coordinates plus the holonomic constraint $\xi = R\theta$. Then the Lagrangian is

$$L = \frac{M}{2}\dot{x}^2 + \frac{m}{2}\left[\dot{x}^2 + r^2\dot{\theta}^2 + 2r\dot{x}\dot{\theta}\cos\varphi\right] + \frac{m}{5}r^2\dot{\theta}^2 - mg\left(h - r\theta\sin\varphi\right)$$

Lagrange's equations $\Lambda_x L = 0$ and $\Lambda_\theta L = 0$, give

$$(M + m)\ddot{x} + mr\ddot{\theta}\cos\varphi = 0$$
$$\ddot{x}\cos\varphi + \frac{7}{5}r\ddot{\theta} - g\sin\varphi = 0$$

Eliminating $\ddot{x}$ gives

$$\left(\frac{7}{5} - \frac{m\cos^2\varphi}{M+m}\right)\ddot{\theta} = g\frac{\sin\varphi}{r}$$

Integrate this equation assuming the initial conditions, results in

$$\theta = \frac{5(M+m)\sin\varphi}{2[7(M+m) - 5m\cos^2\varphi]}gt^2$$

Thus

$$x = -\frac{mr\cos\varphi}{M+m}\theta = \frac{5m\sin(2\varphi)}{4[7(M+m) - 5m\cos^2\varphi]}gt^2$$

Note that these equations predict conservation of linear momentum for the block plus sphere.

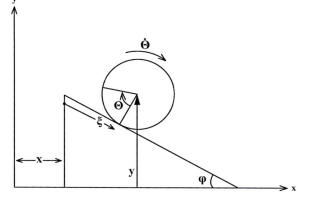

Solid sphere rolling without slipping on an inclined plane on a frictionless horizontal floor.

6.9 Example: Mass sliding on a rotating straight frictionless rod.

Consider a mass m sliding on a frictionless rod that rotates about one end of the rod with an angular velocity $\dot{\theta}$. Choose r and θ to be generalized coordinates. Then the kinetic energy is given by

$$T = \frac{1}{2}m\dot{r}^2 + \frac{1}{2}mr^2\dot{\theta}^2$$

and potential energy

$$U = 0$$

The Lagrange equation for θ gives

$$\Lambda_\theta L = \frac{d}{dt}\frac{\partial L}{\partial\dot{\theta}} - \frac{\partial L}{\partial\theta} = \frac{d}{dt}(mr^2\dot{\theta}) = 0$$

Thus the angular momentum is constant

$$mr^2\dot{\theta} = \text{constant} = p_\theta$$

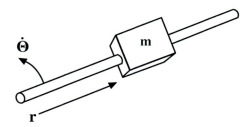

Mass sliding on a rotating straight frictionless rod.

The Lagrange equation for r gives

$$\Lambda_r L = \frac{d}{dt}\frac{\partial L}{\partial \dot{r}} - \frac{\partial L}{\partial r} = m\ddot{r} - mr\dot{\theta}^2 = 0$$

The θ equation states that the angular momentum is conserved for this case which is what we expect since there are no external torques acting on the system. The r equation states that the centrifugal acceleration is $\ddot{r} = r\omega^2$. These equations of motion were derived without reference to the forces between the rod and mass.

6.10 Example: Spherical pendulum

The spherical pendulum is a classic holonomic problem in mechanics that involves rotation plus oscillation where the pendulum is free to swing in any direction. This also applies to a particle constrained to slide in a smooth frictionless spherical bowl under gravity, such as a bar of soap in a wet hemispherical sink. Consider the equation of motion of the spherical pendulum of mass m and length b shown in the adjacent figure. The most convenient generalized coordinates are r, θ, ϕ with origin at the fulcrum, since the length is constrained to be $r = b$. The kinetic energy is

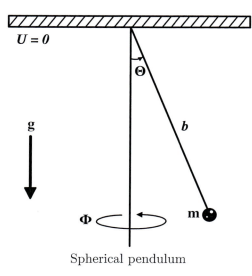

Spherical pendulum

$$T = \frac{1}{2}mb^2\dot{\theta}^2 + \frac{1}{2}mb^2 \sin^2\theta \dot{\phi}^2$$

The potential energy

$$U = -mgb\cos\theta$$

giving that

$$L = \frac{1}{2}mb^2\dot{\theta}^2 + \frac{1}{2}mb^2 \sin^2\theta \dot{\phi}^2 + mgb\cos\theta$$

The Lagrange equation for θ

$$\Lambda_\theta L = \frac{d}{dt}\frac{\partial L}{\partial \dot{\theta}} - \frac{\partial L}{\partial \theta} = 0$$

which gives

$$mb^2\ddot{\theta} = mb^2\dot{\phi}^2 \sin\theta\cos\theta - mgb\sin\theta$$

The Lagrange equation for ϕ

$$\Lambda_\phi L = \frac{d}{dt}\frac{\partial L}{\partial \dot{\phi}} - \frac{\partial L}{\partial \phi} = \frac{d}{dt}[mb^2 \sin^2\theta \dot{\phi}] = 0$$

which gives

$$mb^2 \sin^2\theta \dot{\phi} = p_\phi = \text{ constant}$$

This is just the angular momentum p_ϕ for the pendulum rotating in the ϕ direction. Automatically the Lagrange approach shows that the angular momentum p_ϕ is a conserved quantity. This is what is expected from Newton's Laws of Motion since there are no external torques applied about this vertical axis.

The equation of motion for θ can be simplified to

$$\ddot{\theta} + \frac{g}{b}\sin\theta - \frac{p_\phi^2 \cos\theta}{m^2 b^4 \sin^3\theta} = 0$$

There are many possible solutions depending on the initial conditions. The pendulum can just oscillate in the θ direction, or rotate in the ϕ direction or some combination of these. Note that if p_ϕ is zero, then the equation reduces to the simple harmonic pendulum, while the other extreme is when $\ddot{\theta} = 0$ for which the motion is that of a conical pendulum that rotates at a constant angle θ_0 to the vertical axis.

6.11 *Example: Spring plane pendulum*

A mass m is suspended by a spring with spring constant k in the gravitational field. Besides the longitudinal spring vibration, the spring performs a plane pendulum motion in the vertical plane, as illustrated in the adjacent figure. Find the Lagrangian, the equations of motion, and force in the spring.

The system is holonomic, conservative, and scleronomic. Introduce plane polar coordinates with radial length r and polar angle θ as generalized coordinates. The generalized coordinates are related to the cartesian coordinates by

$$y = r\cos\theta$$
$$x = r\sin\theta$$

Therefore the velocities are given by

$$\dot{y} = \dot{r}\cos\theta + r\dot{\theta}\sin\theta$$
$$\dot{x} = \dot{r}\sin\theta - r\dot{\theta}\cos\theta$$

The kinetic energy is given by

$$T = \frac{1}{2}m\left(\dot{x}^2 + \dot{y}^2\right) = \frac{1}{2}m\left(\dot{r}^2 + r^2\dot{\theta}^2\right)$$

The gravitational plus spring potential energies both can be absorbed into the potential U.

$$U = -mgr\cos\theta + \frac{k}{2}\left(r - r_0\right)^2$$

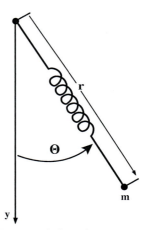

Spring pendulum having spring constant k and oscillating in a vertical plane.

where r_0 denotes the rest length of the spring. The Lagrangian thus equals

$$L = \frac{1}{2}m\left(\dot{r}^2 + r^2\dot{\theta}^2\right) + mgr\cos\theta - \frac{k}{2}\left(r - r_0\right)^2$$

For the polar angle θ, the Lagrange equation $\Lambda_\theta L = 0$ gives

$$\frac{d}{dt}\left(mr^2\dot{\theta}\right) = -mgr\sin\theta$$

The angular momentum $p_\theta = mr^2\dot{\theta}$, thus the equation of motion can be written as

$$\dot{p}_\theta = -mgr\sin\theta$$

Alternatively, evaluating $\frac{d}{dt}\left(mr^2\dot{\theta}\right)$ gives

$$mr^2\ddot{\theta} = -mgr\sin\theta - 2mr\dot{r}\dot{\theta}$$

The last term in the right-hand side is the Coriolis force caused by the time variation of the pendulum length. For the radial distance r, the Lagrange equation $\Lambda_r L = 0$ gives

$$m\ddot{r} = mr\dot{\theta}^2 + mg\cos\theta - k\left(r - r_0\right)$$

This equation just equals the tension in the spring, i.e. $F = m\ddot{r}$. The first term on the right-hand side represents the centrifugal radial acceleration, the second term is the component of the gravitational force, and the third term represents Hooke's Law for the spring. For small amplitudes of θ the motion appears as a superposition of harmonic oscillations in the r, θ plane.

In this example the orthogonal coordinate approach used gave the tension in the spring thus it is unnecessary to repeat this using the Lagrange multiplier approach.

6.12 *Example: The yo-yo*

Consider a yo-yo comprising a disc that has a string wrapped around it with one end attached to a fixed support. The disc is allowed to fall with the string unwinding as it falls as illustrated in the adjacent figure. Derive the equations of motion and the forces of constraint via use of Lagrange multipliers. Use y and ϕ as independent generalized coordinates.

The kinetic energy of the falling yo-yo is given by

$$T = \frac{1}{2}m\dot{y}^2 + \frac{1}{2}I\dot{\phi}^2 = \frac{1}{2}m\dot{y}^2 + \frac{1}{4}ma^2\dot{\phi}^2$$

where m is the mass of the disc, a the radius, and $I = \frac{1}{2}ma^2$ is the moment of inertia of the disc about its central axis. The potential energy of the disc is

$$U = -mgy$$

Thus the Lagrangian is

$$L = \frac{1}{2}m\dot{y}^2 + \frac{1}{4}ma^2\dot{\phi}^2 + mgy$$

The one equation of constraint is holonomic

$$g(y, \phi) = y - a\phi = 0$$

The two Lagrange equations are

$$\frac{\partial L}{\partial y} - \frac{d}{dt}\frac{\partial L}{\partial y'} + \lambda\frac{\partial g}{\partial y} = 0$$

$$\frac{\partial L}{\partial \phi} - \frac{d}{dt}\frac{\partial L}{\partial \phi'} + \lambda\frac{\partial g}{\partial \phi} = 0$$

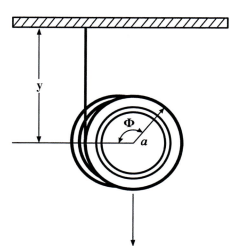

The yo-yo comprises a falling disc unrolling from a string attached to the disc at one end and a fixed support at the other end.

with only one Lagrange multiplier λ. Evaluating these two Euler-Lagrange equations leads to two equations of motion

$$mg - m\ddot{y} + \lambda = 0$$
$$-\frac{1}{2}ma^2\ddot{\phi} - \lambda a = 0$$

Differentiating the equation of constraint gives

$$\ddot{\phi} = \frac{\ddot{y}}{a}$$

Inserting this into the second equation and solving the two equations gives

$$\lambda = -\frac{1}{3}mg$$

Inserting λ into the two equations of motion gives

$$\ddot{y} = \frac{2}{3}g$$
$$\ddot{\phi} = \frac{2}{3}\frac{g}{a}$$

The generalized force of constraint

$$F_y = \lambda\frac{\partial g}{\partial y} = -\frac{1}{3}mg$$

and the constraint torque is

$$N_\phi = \lambda\frac{\partial g}{\partial \phi} = \frac{1}{3}mga$$

Thus the string reduces the acceleration of the disc in the gravitational field by a factor of $\frac{1}{3}$.

6.13 Example: Mass constrained to move on the inside of a frictionless paraboloid

A mass m moves on the frictionless inner surface of a paraboloid

$$x^2 + y^2 = \rho^2 = az$$

with a gravitational potential energy of $U = mgz$.

This system is holonomic, scleronomic, and conservative. Choose cylindrical coordinates ρ, ϕ, z with respect to the vertical axis of the paraboloid to be the generalized coordinates.

The Lagrangian is

$$L = \frac{1}{2}m\left(\dot{\rho}^2 + \rho^2\dot{\phi}^2 + \dot{z}^2\right) - mgz$$

The equation of constraint is

$$g(\rho, z) = \rho^2 - az = 0$$

The Lagrange multiplier approach will be used to determine the forces of constraint.

For $\Lambda_\rho L = \lambda\frac{\partial g}{\partial \rho}$

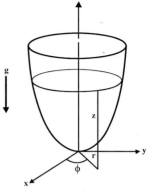

Mass constrained to slide on the inside of a frictionless paraboloid.

$$\frac{d}{dt}\frac{\partial L}{\partial \dot{r}} - \frac{\partial L}{\partial r} = \lambda_1 2\rho \qquad\qquad (a)$$

$$m\left(\ddot{\rho} - \rho\dot{\phi}^2\right) = \lambda_1 2\rho$$

For $\Lambda_\phi L = \lambda\frac{\partial g}{\partial \phi}$

$$\frac{d}{dt}\left(m\rho^2\dot{\phi}\right) = \dot{p}_\phi = 0 \qquad\qquad (b)$$

Thus the angular momentum p_ϕ is conserved, that is, it is a constant of motion.
For $\Lambda_z L = \lambda\frac{\partial g}{\partial z}$

$$m\ddot{z} = -mg - \lambda_1 a \qquad\qquad (c)$$

and the time differential of the constraint equation is

$$2\rho\dot{\rho} - a\dot{z} = 0 \qquad\qquad (d)$$

The above four equations of motion can be used to determine $r, \phi.z, \lambda_1$.

The radius of the circle at the intersection of the plane $z = h$, with the paraboloid $\rho^2 = az$, is given by $\rho_0 = \sqrt{ah}$. For a constant height $z = h$, then $\ddot{z} = 0$ and equation (c) reduces to

$$\lambda_1 = -\frac{mg}{a}$$

Therefore the constraint force F_c is given by

$$F_c = \lambda_1\frac{\partial g(\rho, z)}{\partial \rho} = -\frac{mg}{a}2\rho$$

Assuming that $\ddot{\rho} = 0$, then equation (a) for $\dot{\phi} = \omega$ and $\rho = \rho_0$ gives

$$m\left(0 - \rho_0\omega^2\right) = \lambda_1 2\rho_0 = -\frac{mg}{a}2\rho_0 = F_c$$

That is, the constraint force equals

$$F_c = -m\rho_0\omega^2$$

which is the usual centripetal force. These relations also give that the initial angular velocity required for such a stable trajectory with height h is

$$\dot{\phi} = \omega = \sqrt{\frac{2g}{a}}$$

6.14 *Example: Mass on a frictionless plane connected to a plane pendulum*

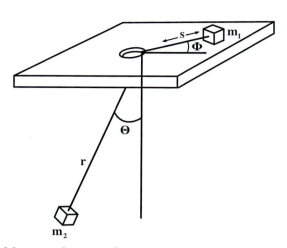

Two masses m_1 and m_2 are connected by a string of length l. Mass m_1 is on a horizontal frictionless table and it is assumed that mass m_2 moves in a vertical plane. This is another problem involving holonomic constrained motion. The constraints are:

1) m_1 moves in the horizontal plane
2) m_2 moves in the vertical plane
3) $r + s = l$. Therefore $\dot{r} = -\dot{s}$

There are $6 - 3 = 3$ remaining degrees of freedom after taking the constraints into account. Choose as a set of generalized coordinates, $r, \theta,$ and ϕ. In terms of these three generalized coordinates, the kinetic energy is

$$
\begin{aligned}
T &= \frac{1}{2}m_1\left(\dot{s}^2 + s^2\dot{\phi}^2\right) + \frac{1}{2}m_2\left(\dot{r}^2 + r^2\dot{\theta}^2\right)\\
&= \frac{1}{2}m_1\left(\dot{r}^2 + (l-r)^2\dot{\phi}^2\right) + \frac{1}{2}m_2\left(\dot{r}^2 + r^2\dot{\theta}^2\right)
\end{aligned}
$$

Mass m_2, hanging from a rope that is connected to m_1, which slides on a frictionless plane.

The potential energy in terms of the generalized coordinates relative to the horizontal plane, is

$$
U = 0 - m_2 gr\cos\theta
$$

Therefore the Lagrangian equals

$$
L = \frac{1}{2}m_1\left(\dot{r}^2 + (l-r)^2\dot{\phi}^2\right) + \frac{1}{2}m_2\left(\dot{r}^2 + r^2\dot{\theta}^2\right) + m_2 gr\cos\theta
$$

The differentials are

$$
\begin{aligned}
\frac{\partial L}{\partial r} &= -m(l-r)\dot{\phi}^2 + m_2 r\dot{\theta}^2 + mgr\cos\theta\\
\frac{\partial L}{\partial \dot{r}} &= (m_1 + m_2)\dot{r}\\
\frac{\partial L}{\partial \theta} &= -mgr\sin\theta\\
\frac{\partial L}{\partial \dot{\theta}} &= m_2 r^2\dot{\theta}\\
\frac{\partial L}{\partial \phi} &= 0\\
\frac{\partial L}{\partial \dot{\phi}} &= m_1(l-r)^2\dot{\phi}
\end{aligned}
$$

Thus the three Lagrange equations are

$$
\begin{aligned}
\Lambda_r L &= (m_1 + m_2)\ddot{r} + m_1(l-r)\dot{\phi}^2 - m_2 r\dot{\theta}^2 - m_2 g\cos\theta = 0\\
\Lambda_\theta L &= \frac{d}{dt}\left[m_2 r^2\dot{\theta}\right] + m_2 gr\sin\theta = 0
\end{aligned}
$$

that is

$$
2m_2\dot{r}\dot{\theta} + r^2 m_2\ddot{\theta} + m_2 gr\sin\theta = 0
$$

$$
\Lambda_\phi L = \frac{d}{dt}\left[m_1(l-r)^2\dot{\phi}\right] = 0
$$

This last equation is a statement of the conservation of angular momentum. These three differential equations of motion can be solved for known initial conditions.

6.15 Example: Two connected masses constrained to slide along a moving rod

Consider two identical masses m, constrained to move along the axis of a thin straight rod, of mass M and length l, which is free to both translate and rotate. Two identical springs link the two masses to the central point of the rod. Consider only motions of the system for which the extended lengths of the two springs are equal and opposite such that the two masses always are equal distances from the center of the rod keeping the center of mass at the center of the rod. Find the equations of motion for this system.

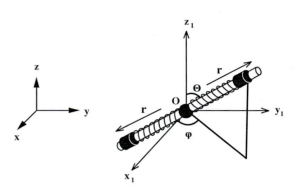

Use a fixed cartesian coordinate system (x, y, z) and a moving frame with the origin O at the center of the rod with its cartesian coordinates (x_1, y_1, z_1) being parallel to the fixed coordinate frame as shown in the figure. Let (r, θ, φ) be the spherical coordinates of a point referring to the center of the moving (x_1, y_1, z_1) frame as shown in the figure. Then the two masses m have spherical coordinates (r, θ, φ) and $(-r, \theta, \varphi)$ in the moving-rod fixed frame. The frictionless constraints are holonomic.

Two identical masses m constrained to slide on a moving rod of mass M. The masses are attached to the center of the rod by identical springs each having a spring constant K.

The kinetic energy of the system is equal to the kinetic energy for all the mass concentrated at the center of mass plus the kinetic energy about the center of mass. Since O is the center of mass then the kinetic energy can be separated into three terms

$$T = T_{cm} + T_{rot}^{masses} + T_{rot}^{rod}$$

Note that since the kinetic energy is a scalar quantity it is rotational invariant and thus can be evaluated in any rotated frame. Thus the kinetic energy of the center of mass is

$$T_{cm} = \frac{1}{2}(M + 2m)(\dot{x}^2 + \dot{y}^2 + \dot{z}^2)$$

The rotational kinetic energy of the two masses in the center of mass frame is

$$T_{rot}^{masses} = m(\dot{r}^2 + r^2\dot{\theta}^2 + r^2\dot{\varphi}^2 \sin^2 \theta)$$

The rotational kinetic energy of the rod T_{rot}^{rod} is a scalar and thus can be evaluated in any rotated frame of reference fixed with respect to the principal axis system of the rod. The angular velocity of the rod about O resolved along its principal axes is given by

$$\bar{\omega} = \dot{\varphi} \cos \theta \hat{\mathbf{e}}_r - \dot{\varphi} \sin \theta \hat{\mathbf{e}}_\theta - \dot{\theta} \hat{\mathbf{e}}_\varphi$$

The corresponding moments of inertia of the uniform infinitesimally-thin rod are $I_r = 0, I_\theta = \frac{1}{12}Ml^2, I_\varphi = \frac{1}{12}Ml^2$. Hence the rotational kinetic energy of the rod is

$$T_{rot}^{rod} = \frac{1}{2}(I_r\omega_r^2 + I_\theta\omega_\theta^2 + I_\varphi\omega_\varphi^2) = \frac{1}{24}Ml^2(\dot{\theta}^2 + \dot{\varphi}^2 \sin^2 \theta)$$

The only potential energy is due to the two extended springs which are assumed to have the same length r where r_0 is the unstretched length.

$$U = 2 \cdot \frac{1}{2}K(r - r_0)^2 = K(r - r_0)^2$$

Thus the Lagrangian is

$$L = \frac{1}{2}(M + 2m)(\dot{x}^2 + \dot{y}^2 + \dot{z}^2) + m(\dot{r}^2 + r^2\dot{\theta}^2 + r^2\dot{\varphi}^2 \sin^2 \theta) + \frac{1}{24}ML^2(\dot{\theta}^2 + \dot{\varphi}^2 \sin^2 \theta) - K(r - r_0)^2$$

Using Lagrange's equations $\Lambda_{q_i}L = 0$ for the generalized coordinates gives.

$$(M + 2m)\dot{x} = \text{constant} \qquad (\Lambda_x L = 0)$$

$$(M + 2m)\dot{y} = \text{constant} \qquad (\Lambda_y L = 0)$$

$$(M + 2m)\dot{z} = \text{constant} \qquad (\Lambda_z L = 0)$$

$$\left(2mr^2 + \frac{1}{12}Ml^2\right)\dot{\varphi}\sin^2\theta = \text{constant} \qquad (\Lambda_\varphi L = 0)$$

$$\ddot{r} - r\dot{\theta}^2 - r\dot{\varphi}^2\sin^2\theta + \frac{K}{m}(r - r_0) = 0 \qquad (\Lambda_r L = 0)$$

$$\left(r^2 + \frac{Ml^2}{24m}\right)\ddot{\theta} + 2r\dot{r}\dot{\theta} - \left(r^2 + \frac{ml^2}{24m}\right)\dot{\varphi}^2\sin\theta\cos\theta = 0 \qquad (\Lambda_\theta L = 0)$$

The first three equations show that the three components of the linear momentum of the center of mass are constants of motion. The fourth equation shows that the component of the angular momentum about the z' axis is a constant of motion. Since the z_1 axis has been arbitrarily chosen then the total angular momentum must be conserved. The fifth and sixth equations give the radial and angular equations of motion of the oscillating masses m.

6.9 Applications involving non-holonomic constraints

In general, non-holonomic constraints can be handled by use of generalized forces Q_j^{EXC} in the Lagrange-Euler equations 6.60. The following examples, $6.16 - 6.19$, involve one-sided constraints which exhibit holonomic behavior for restricted ranges of the constraint surface in coordinate space, and this range is case specific. When the forces of constraint press the object against the constraint surface, then the system is holonomic, but the holonomic range of coordinate space is limited to situations where the constraint forces are positive. When the constraint force is negative, the object flies free from the constraint surface. In addition, when the frictional force $F > N\mu_{static}$ where μ_{static} is the static coefficient of friction, then the object slides negating any rolling constraint that assumes static friction.

6.16 *Example: Mass sliding on a frictionless spherical shell*

Consider a mass starts from rest at the top of a frictionless fixed spherical shell of radius R. The questions are what is the force of constraint and determine the angle θ at which the mass leaves the surface of the spherical shell. The coordinates r, θ shown are the obvious generalized coordinates to use. The constraint will not apply if the force of constraint does not hold the mass against the surface of the spherical shell, that is, it is only holonomic in a restricted domain.

The Lagrangian is

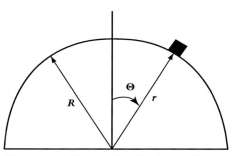

$$L = \frac{1}{2}m\left(\dot{r}^2 + r^2\dot{\theta}^2\right) - mgr\cos\theta$$

This Lagrangian is applicable irrespective of whether the constraint is obeyed, where the constraint is given by

Mass m sliding on frictionless cylinder of radius R.

$$g(r, \theta) = r - R = 0$$

For the restricted domain where this system is holonomic, it can be solved using generalized coordinates, generalized forces, Lagrange multipliers, or Newtonian mechanics as illustrated below.

Minimal generalized coordinates:

The minimal number of generalized coordinates reduces the system to one coordinate θ, which does not determine the constraint force that is needed to know if the constraint applies. Thus this approach is not useful for solving this partially-holonomic system.

Generalized forces:

The radial constraint has a corresponding generalized force Q_r. The Lagrange equation $\Lambda_r L = Q_r$ gives

$$m\ddot{r} + mg\cos\theta - mr\dot{\theta}^2 = Q_r \tag{a}$$

The Lagrange equation $\Lambda_\theta L = Q_\theta = 0$ since there is no tangential force for this frictionless system. Therefore

$$mr^2\ddot{\theta} - mgr\sin\theta + 2mr\dot{r}\dot{\theta} = 0 \tag{b}$$

When constrained to follow the surface of the spherical shell, the system is holonomic, i.e. $r = R$ and $\dot{r} = \ddot{r} = 0$. Thus the above two equations reduce to

$$\begin{aligned}
mg\cos\theta - mR\dot{\theta}^2 &= Q_r \\
mR^2\ddot{\theta} - mgR\sin\theta &= 0
\end{aligned} \tag{c}$$

That is

$$\ddot{\theta} = \frac{g}{R}\sin\theta$$

Integrate to get $\dot{\theta}$ using the fact that

$$\ddot{\theta} = \frac{d\dot{\theta}}{d\theta}\frac{d\theta}{dt} = \dot{\theta}\frac{d\dot{\theta}}{d\theta}$$

then

$$\int \ddot{\theta}d\theta = \int \dot{\theta}d\dot{\theta} = \frac{g}{R}\int \sin\theta d\theta$$

Therefore

$$\dot{\theta}^2 = \frac{2g}{R}\left(1 - \cos\theta\right) \tag{d}$$

assuming that $\dot{\theta} = 0$ at $\theta = 0$. Substituting equation (d) into equation (c) gives the constraint force, which is normal to the surface, to be

$$F = Q_r = mg(3\cos\theta - 2)$$

Note that $F = Q_r = 0$ when $\cos\theta = \frac{2}{3}$, that is $\theta = 48.2°$.

Lagrange multipliers:

For the holonomic regime, which obeys the constraint, $g(r,\theta) = r - R = 0$, the Lagrange equation for r is $\Lambda_r L = \lambda \frac{\partial g}{\partial r}$. Since $\frac{\partial g}{\partial r} = 1$, then

$$m\ddot{r} + mg\cos\theta - mr\dot{\theta}^2 = \lambda \tag{a}$$

The Lagrange equation for θ gives $\Delta_\theta L = \lambda\frac{\partial g}{\partial\theta} = 0$ since $\frac{\partial g}{\partial\theta} = 0$. Thus

$$mr^2\ddot{\theta} - mgr\sin\theta + 2mr\dot{r}\dot{\theta} = 0 \tag{b}$$

As above, when constrained to follow the surface of the spherical shell, the system is holonomic $r = R$, and $\dot{r} = \ddot{r} = 0$. Thus the above two equations reduce to

$$\begin{aligned}
mg\cos\theta - mR\dot{\theta}^2 &= \lambda \tag{c} \\
mR^2\ddot{\theta} - mgR\sin\theta &= 0 \tag{d}
\end{aligned}$$

That is, the answers are identical to that obtained using generalized forces, namely;

$$\dot{\theta}^2 = \frac{2g}{R}\left(1 - \cos\theta\right) \tag{d}$$

assuming that $\dot{\theta} = 0$ at $\theta = 0$.

The force of constraint applied by the surface is

$$F = \lambda\frac{\partial g}{\partial r} = \lambda$$

Substituting equation (d) into equation (c) gives

$$F = \lambda = mg(3\cos\theta - 2)$$

Note that $\lambda = 0$ when $\cos\theta = \frac{2}{3}$, that is $\theta = 48.2^o$.

  Both of the above methods give identical results and give that the force of constraint is negative when $\theta > 48.2^o$. Assuming that the surface cannot hold the mass against the surface, then the mass will fly off the spherical shell when $\theta > 48.2^o$ and the system reduces to an unconstrained object falling freely in a uniform gravitational field, which is holonomic, that is $Q_r = \lambda = 0$. Then the equations of motion (a) and (b) reduce to

$$m\ddot{r} + mg\cos\theta - mr\dot{\theta}^2 = 0 \tag{e}$$
$$mr^2\ddot{\theta} - mgr\sin\theta + 2mr\dot{r}\dot{\theta} = 0 \tag{f}$$

Energy conservation:
This problem can be solved using energy conservation

$$\frac{1}{2}mv^2 = mgR[1 - \cos\theta]$$

Thus the centripetal acceleration

$$\frac{v^2}{R} = 2g[1 - \cos\theta]$$

The normal force to the surface will cancel when the centripetal acceleration equals the gravitational acceleration, that is, when

$$\frac{v^2}{R} = 2g[1 - \cos\theta] = g\cos\theta$$

This occurs when $\cos\theta = \frac{2}{3}$. This is an unusual case where the Newtonian approach is the simplest.

6.17 *Example: Rolling solid sphere on a spherical shell*

  This is a similar problem to the prior one with the added complication of rolling which is assumed to move in a vertical plane making it holonomic. Here we would like to determine the forces of constraint to see when the solid sphere flies off the spherical shell and when the friction is insufficient to stop the rolling sphere from slipping.

  The best generalized coordinates are the distance of the center of the sphere from the center of the spherical shell, r, θ and ϕ. It is important to note that ϕ is measured with respect to the vertical, not the time-dependent vector $\mathbf{r}$. That is, the direction of the radius r is θ which is time dependent and thus is not a useful reference to use to define the angle ϕ. Let us assume that the sphere is uniform with a moment of inertia of $I = \frac{2}{5}ma^2$. If the tangential frictional force F is less than the limiting value $N\mu_{statics}$, with $N > 0$, then the sphere will roll without slipping on the surface of the cylinder and both constraints apply.

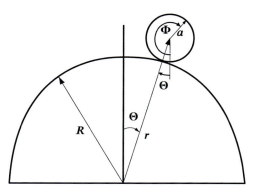

Disk of mass m, radius a, rolling on a cylindrical surface of radius R.

Under these conditions the system is holonomic and the solution is solved using Lagrange multipliers and the equations of constraint are the following:

  1) The center of the sphere follows the surface of the cylinder

$$g_1 = r - R - a = 0$$

  2) The sphere rolls without slipping

$$g_2 = a(\phi - \theta) - R\theta = 0$$

The kinetic energy is $T = \frac{1}{2}m\left(\dot{r}^2 + r^2\dot{\theta}^2\right) + \frac{1}{2}I\dot{\phi}^2$ and the potential energy is $U = mgr\cos\theta$. Thus the Lagrangian is

$$L = \frac{1}{2}m\left(\dot{r}^2 + r^2\dot{\theta}^2\right) + \frac{1}{2}I\dot{\phi}^2 - mgr\cos\theta$$

Consider the solution using Lagrange multipliers for the holonomic regime where both constraints are satisfied and lead to the following differential constraint relations

$$\frac{\partial g_1}{\partial r} = 1 \qquad \frac{\partial g_1}{\partial \phi} = 0 \qquad \frac{\partial g_1}{\partial \theta} = 0$$

$$\frac{\partial g_2}{\partial r} = 0 \qquad \frac{\partial g_2}{\partial \phi} = a \qquad \frac{\partial g_2}{\partial \theta} = -(R+a)$$

The Lagrange operator equation $\Lambda_r L$ gives,

$$\frac{d}{dt}\frac{\partial L}{\partial \dot{r}} - \frac{\partial L}{\partial r} = \lambda_1 \frac{\partial g_1}{\partial r} + \lambda_2 \frac{\partial g_2}{\partial r}$$

that is

$$m\ddot{r} + mg\cos\theta - mr\dot{\theta}^2 = \lambda_1 \tag{a}$$

$\Lambda_\theta L$ gives

$$mr^2\ddot{\theta} + 2mr\dot{r}\dot{\theta} - mgr\sin\theta = -\lambda_2(R+a) \tag{b}$$

$\Lambda_\phi L$ gives

$$I\ddot{\phi} = a\lambda_2 \tag{c}$$

Since the center of the sphere rolling on the spherical shell must have

$$r = R + a$$

then

$$\dot{r} = \ddot{r} = 0$$
$$\ddot{\phi} = \frac{r}{a}\ddot{\theta}$$

Substituting this into (c) gives

$$\ddot{\theta} = \frac{a^2}{rI}\lambda_2$$

Insert this into equation (b) gives

$$\lambda_2 = \frac{mgr\sin\theta}{\left(r + \frac{mr^2a^2}{rI}\right)}$$

The moment of inertia about the axis of a solid sphere is $I = \frac{2}{5}ma^2$. Then

$$\lambda_2 = \frac{2mg\sin\theta}{7}$$

But also

$$\ddot{\theta} = \dot{\theta}\frac{d\dot{\theta}}{d\theta} = \frac{a^2}{rI}\lambda_2 = \frac{5}{2mr}\lambda_2 = \frac{5g\sin\theta}{7r}$$

Integrating gives

$$\int \dot{\theta}d\dot{\theta} = \frac{5g}{7r}\int \sin\theta d\theta$$

That is

$$\dot{\theta}^2 = \frac{10g}{7r}(1 - \cos\theta)$$

assuming that $\dot{\theta} = 0$ at $\theta = 0$. Inserting this into equation (a) gives

$$-mr\frac{10g}{7r}[1 - \cos\theta] + mg\cos\theta = \lambda_1$$

That is

$$\lambda_1 = \frac{mg}{7}[17\cos\theta - 10]$$

Note that this equals zero when

$$\cos\theta = \frac{10}{17}$$

For larger angles λ_1 is negative implying that the solid sphere will fly off the surface of the spherical shell.

The sphere will leave the surface of the cylinder when $\cos\theta = \frac{10}{17}$ that is, $\theta = 53.97°$. This is a significantly larger angle than obtained for the similar problem where the mass is sliding on a frictionless cylinder because the energy stored in rotation implies that the linear velocity of the mass is lower at a given angle θ for the case of a rolling sphere.

The above discussion has omitted an important fact that, if $\mu_{static} < \infty$, the frictional force becomes insufficient to maintain the rolling constraint before $\theta = 53.97°$, that is, the frictional force will exceed the sliding limit $N\mu_{static}$. To determine when the rolling constraint fails it is necessary to determine the frictional torque

$$F_f R = -\lambda_2 R$$

Thus

$$F_f = -\lambda_2$$

It is in the negative direction because of the direction chosen for ϕ. The required coefficient of friction μ is given by the ratio of the frictional force to the normal force, that is

$$\mu = \frac{\lambda_2}{\lambda_1} = \frac{2\sin\theta}{[17\cos\theta - 10]}$$

For $\mu = 1$ the disk starts to slip when $\theta = 47.54°$. Note that the sphere starts slipping before it flies off the cylinder since a normal force is required to support a frictional force and the difference depends on the coefficient of friction. The no-slipping constraint is not satisfied once the sphere starts slipping and the frictional force should equal $\mu_{kinetic}\lambda_1$. Thus for the angles beyond $47.54°$ the problem needs to be solved with the rolling constraint changed to a sliding non-conservative frictional force. This is best handled by including the frictional force and normal forces as generalized forces. Fortunately this will be a small correction. The friction will slightly change the exact angle at which the normal force becomes zero and the system transitions to free motion of the sphere in a gravitational field.

6.18 *Example: Solid sphere rolling plus slipping on a spherical shell*

Consider the above case when the frictional force is insufficient to constrain the motion to rolling. Now the frictional force F is given by

$$F = N\mu_{sliding}$$

when N is positive.

This can be solved using generalized forces with the previous Lagrangian. Then

$$\frac{d}{dt}\frac{\partial L}{\partial \dot{r}} - \frac{\partial L}{\partial r} = Q_r = N$$

which gives

$$m\ddot{r} + mg\cos\theta - mr\dot{\theta}^2 = N$$

Similarly $\Lambda_\theta L = Q_\theta = -F(R+a)$ gives

$$mr^2\ddot{\theta} + 2mr\dot{r}\dot{\theta} - mgr\sin\theta = -F(R+a)$$

Similarly $\Lambda_\phi L = Q_\phi = aF$ gives

$$I\ddot{\phi} = aF$$

These can be solved by substituting the relation $F = N\mu_{sliding}$. The sphere flies off the spherical shell when $N \leq 0$ leading to free motion discussed in example 6.2. The problem of a solid uniform sphere rolling inside a hollow sphere can be solved the same way.

6.19 Example: Small body held by friction on the periphery of a rolling wheel

Assume that a small body of mass m is balanced on a rolling wheel of mass M and radius R as shown in the figure. The wheel rolls in a vertical plane without slipping on a horizontal surface. This example illustrates that it is possible to use simultaneously a mixture of holonomic constraints, partially-holonomic constraints, and generalized forces.[3]

Assume that at t = 0 the wheel touches the floor at x = y = 0 with the mass perched at the top of the wheel at x = 0. Let the frictional force acting on the mass m be F and the reaction force of the periphery of the wheel on the mass be N. Let $\dot\varphi$ be the angular velocity of the wheel, and $\dot x$ the horizontal velocity of the center of the wheel. The polar coordinates r, θ of the mass m are taken with r measured from the center of the wheel with θ measured with respect to the vertical. Thus the cartesian coordinates of the small mass m are $(x + r\sin\theta, R + r\cos\theta)$ with respect to the origin at x = y = 0.

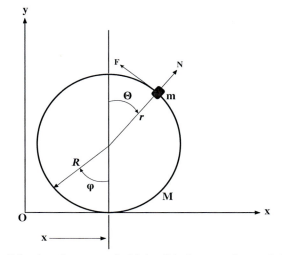

Small body of mass m held by friction on the periphery of a rolling wheel of mass M and radius R.

The kinetic energy is given by

$$T = \frac{1}{2}M\dot x^2 + \frac{1}{2}I\dot\varphi^2 + \frac{1}{2}m\left[\left(\dot x + r\dot\theta\cos\theta + \dot r\sin\theta\right)^2 + \left(\dot r\cos\theta - r\dot\theta\sin\theta\right)^2\right]$$

The gravitational force can be absorbed into the scalar potential term of the Lagrangian and includes only the potential energy of the mass m since the potential energy of the rolling wheel is constant.

$$U = +mg\left(R + r\cos\theta\right)$$

Thus the Lagrangian is

$$L = \frac{1}{2}\left(M + m\right)\dot x^2 + \frac{1}{2}I\dot\varphi^2 + \frac{1}{2}m\left[r^2\dot\theta^2 + 2r\dot x\dot\theta\cos\theta + 2\dot x\dot r\sin\theta + \dot r^2\right] - mg\left(R + r\cos\theta\right)$$

The equations of constraints are:

1) The wheel rolls without slipping on the ground plane leading to a holonomic constraint:

$$g_1 = x - R\varphi = \dot x - R\dot\varphi = 0$$

2) The mass m is touching the periphery of the wheel, that is, the normal force N > 0. This is a one-sided restricted holonomic constraint.

$$g_2 = R - r = 0$$

3) The mass m does not slip on the wheel if the frictional force $F < N\mu_{static}$. When this restricted holonomic constraint is satisfied, then

$$g_3 = \dot\theta - \dot\varphi = 0$$

The rolling constraint is holonomic, and can be accounted for using one Lagrange multiplier λ_x plus the differential constraint equations

---

[3]This problem is solved in detail in example 3.19 of " Classical Mechanics and Relativity". by Muller-Kirsten [*Mu06*].

$$\frac{\partial g_1}{\partial x} = 1$$

$$\frac{\partial g_1}{\partial \theta} = 0$$

$$\frac{\partial g_1}{\partial \varphi} = R$$

$$\frac{\partial g_1}{\partial r} = 0$$

The other two constraints are non-holonomic, and thus these constraint forces are expressed in terms of two generalized forces Q_θ, and Q_r that are related to the tangential force F and radial reaction force N. For simplicity, assume that the wheel is a thin-walled cylinder with a moment of inertia of

$$I = MR^2$$

The Euler-Lagrange equations for the four coordinates x, θ, φ, r are

$$-\frac{d}{dt}\left((M+m)\,\dot{x} + mr\dot{\theta}\cos\theta + \dot{r}\sin\theta\right) + \lambda_x + Q_x = 0 \qquad (\Lambda_x)$$

$$mr\dot{x}\dot{\theta}\sin\theta + \dot{x}\dot{r}\cos\theta - mgr\sin\theta - \frac{d}{dt}\left(mr^2\dot{\theta} + mr\dot{x}\cos\theta\right) + Q_\theta = 0 \qquad (\Lambda_\theta)$$

$$-\frac{d}{dt}\left(MR^2\dot{\varphi}\right) - R\lambda_x = 0 \qquad (\Lambda_\varphi)$$

$$-mg\cos\theta - \frac{d}{dt}(m\dot{x}\sin\theta + \dot{r}) + Q_r = 0 \qquad (\Lambda_r)$$

The generalized forces can be related to F and N using the definition

$$Q_{q_k} = \mathbf{F}(r)\cdot\frac{\partial \mathbf{r}}{\partial q_k}$$

where $F(r)$ is the vectorial sum of the forces acting at r. The components of vector $r = (x + r\sin\theta, R + r\cos\theta)$ and F, and N are in the directions defined in the figure which leads to the generalized forces

$$Q_x = -F\cos\theta + N\sin\theta$$

$$Q_\theta = (-F\cos\theta + N\sin\theta)(-R\cos\theta) - (F\sin\theta + N\cos\theta)\,R\sin\theta = -FR$$

$$Q_r = N$$

Solving the above 7 equations gives that

$$m\ddot{x}\sin\theta + mR\dot{\theta}^2 - mg\cos\theta + N = 0$$

This last equation can be derived by Newtonian mechanics from consideration of the forces acting.

The above equations of motion can be used to calculate the motion for the following conditions.

a) Mass not slipping:

This occurs if $\mu = \frac{F}{N} \leq \mu_{static}$ which also implies that $N > 0$, That is a situation where the system is holonomic with $r = R$, $\dot{x} = R\dot{\varphi}$, $\dot{\theta} = \dot{\varphi}$ which can be solved using the generalized coordinate approach with only one independent coordinate which can be taken to be θ.

b) Mass slipping:

Here the no-slip constraint is violated and thus one has to explicitly include the generalized forces Q_r, Q_φ, Q_θ and assume that sliding friction is given by $F = N\mu_{sliding}$.

c) Reaction force N is negative:

Here the mass is not subject to any constraints and it is in free fall.

The above example illustrates the flexibility provided by Lagrangian mechanics that allows simultaneous use of Lagrange multipliers, generalized forces, and scalar potential to handle combinations of several holonomic and nonholonomic constraints for a complicated problem.

6.10   Velocity-dependent Lorentz force

The Lorentz force in electromagnetism is unusual in that it is a velocity-dependent force, as well as being a conservative force that can be treated using the concept of potential. That is, the Lorentz force is

$$\mathbf{F} = q(\mathbf{E} + \mathbf{v} \times \mathbf{B}) \tag{6.61}$$

It is interesting to use Maxwell's equations and Lagrangian mechanics to show that the Lorentz force can be represented by a conservative potential in Lagrangian mechanics.

Maxwell's equations can be written as

$$\boldsymbol{\nabla} \cdot \mathbf{E} = \frac{\rho}{\varepsilon_0} \tag{6.62}$$

$$\boldsymbol{\nabla} \times \mathbf{E} + \frac{\partial \mathbf{B}}{\partial t} = 0$$

$$\boldsymbol{\nabla} \cdot \mathbf{B} = 0$$

$$\boldsymbol{\nabla} \times \mathbf{B} - \mu_0 \varepsilon_0 \frac{\partial \mathbf{E}}{\partial t} = \mathbf{J}$$

Since $\boldsymbol{\nabla} \cdot \mathbf{B} = 0$ then it follows from Appendix H that $\mathbf{B}$ can be represented by the curl of a vector potential, $\mathbf{A}$, that is

$$\mathbf{B} = \boldsymbol{\nabla} \times \mathbf{A} \tag{6.63}$$

Substituting this into $\boldsymbol{\nabla} \times \mathbf{E} + \frac{\partial \mathbf{B}}{\partial t} = 0$ gives that

$$\boldsymbol{\nabla} \times \mathbf{E} + \frac{\partial \boldsymbol{\nabla} \times \mathbf{A}}{\partial t} = 0 \tag{6.64}$$

$$\boldsymbol{\nabla} \times \left(\mathbf{E} + \frac{\partial \mathbf{A}}{\partial t} \right) = 0$$

Since this curl is zero it can be represented by the gradient of a scalar potential U

$$\mathbf{E} + \frac{\partial \mathbf{A}}{\partial t} = -\boldsymbol{\nabla} U \tag{6.65}$$

The following shows that this relation corresponds to taking the gradient of a potential U for the charge q where the potential U is given by the relation

$$U = q(\Phi - \mathbf{A} \cdot \mathbf{v}) \tag{6.66}$$

where Φ is the scalar electrostatic potential. This scalar potential U can be employed in the Lagrange equations using the Lagrangian

$$L = \frac{1}{2} m \mathbf{v} \cdot \mathbf{v} - q(\Phi - \mathbf{A} \cdot \mathbf{v}) \tag{6.67}$$

The Lorentz force can be derived from this Lagrangian by considering the Lagrange equation for the cartesian coordinate x

$$\frac{d}{dt} \frac{\partial L}{\partial \dot{x}} - \frac{\partial L}{\partial x} = 0 \tag{6.68}$$

Using the above Lagrangian (6.67) gives

$$m\ddot{x} + q \left[\frac{dA_x}{dt} + \frac{\partial \Phi}{\partial x} - \frac{\partial \mathbf{A}}{\partial x} \cdot \mathbf{v} \right] = 0 \tag{6.69}$$

But

$$\frac{dA_x}{dt} = \frac{\partial A_x}{\partial t} + \frac{\partial A_x}{\partial x} \dot{x} + \frac{\partial A_x}{\partial y} \dot{y} + \frac{\partial A_x}{\partial z} \dot{z} \tag{6.70}$$

and

$$\frac{\partial \mathbf{A}}{\partial x} \cdot \mathbf{v} = \frac{\partial A_x}{\partial x} \dot{x} + \frac{\partial A_y}{\partial x} \dot{y} + \frac{\partial A_z}{\partial x} \dot{z} \tag{6.71}$$

Inserting equations 6.70 and 6.71 into 6.69 gives

$$F_x = m\ddot{x} = q \left[\left(-\frac{\partial \Phi}{\partial x} - \frac{\partial A_x}{\partial t} \right) + \left(\frac{\partial A_y}{\partial x} - \frac{\partial A_x}{\partial y} \right) \dot{y} - \left(\frac{\partial A_x}{\partial z} - \frac{\partial A_z}{\partial x} \right) \dot{z} \right] = q \left[\mathbf{E} + \mathbf{v} \times \mathbf{B} \right]_x \qquad (6.72)$$

Corresponding expressions can be obtained for F_y and F_z. Thus the total force is the well-known Lorentz force

$$\mathbf{F} = q(\mathbf{E} + \mathbf{v} \times \mathbf{B}) \qquad (6.73)$$

This has demonstrated that the electromagnetic scalar potential

$$U = q(\Phi - \mathbf{A} \cdot \mathbf{v}) \qquad (6.74)$$

satisfies Maxwell's equations, gives the Lorentz force, and it can be absorbed into the Lagrangian. Note that the velocity-dependent Lorentz force is conservative since $\mathbf{E}$ is conservative, and because $(\mathbf{v} \times \mathbf{B} \times \mathbf{v})dt=0$, therefore the magnetic force does no work since it is perpendicular to the trajectory. The velocity-dependent conservative Lorentz force is an important and ubiquitous force that features prominently in many branches of science. It will be discussed further for the case of relativistic motion in example 17.6.

6.11 Time-dependent forces

All examples discussed in this chapter have assumed Lagrangians that are time independent. Mathematical systems where the ordinary differential equations do not depend explicitly on the independent variable, which in this case is time t, are called *autonomous* systems. Systems having differential equations governing the dynamical behavior that have time-dependent coefficients are called *non-autonomous* systems.

In principle it is trivial to incorporate time-dependent behavior into the equations of motion by introducing either a time dependent generalized force $Q(r,t)$, or allowing the Lagrangian to be time dependent. For example, in the rocket problem the mass is time dependent. In some cases the time dependent forces can be represented by a time-dependent potential energy rather than using a generalized force. Solutions for non-autonomous systems can be considerably more difficult to obtain, and can involve regions where the motion is stable and other regions where the motion is unstable or chaotic similar to the behavior discussed in chapter 4. The following case of a simple pendulum, whose support is undergoing vertical oscillatory motion, illustrates the complexities that can occur for systems involving time-dependent forces.

6.20 *Example: Plane pendulum hanging from a vertically-oscillating support*

Consider a plane pendulum having a mass M fastened to a massless rigid rod of length L that is at an angle $\theta(t)$ to the vertical gravitational field g. The pendulum is attached to a support that is subject to a vertical oscillatory force F such that the vertical position y of the support is

$$y = A \cos \omega t$$

The kinetic energy is

$$T = \frac{1}{2} M \left[\left(L\dot{\theta} \cos \theta \right)^2 + (\dot{y} + L\dot{\theta} \sin \theta)^2 \right] = \frac{1}{2} M \left[L^2 \dot{\theta}^2 + 2L\dot{\theta}\dot{y} \sin \theta + \dot{y}^2 \right]$$

and the potential energy is

$$U = Mg \left[L(1 - \cos \theta) + y \right]$$

Thus the Lagrangian is

$$L = \frac{1}{2} M \left[L^2 \dot{\theta}^2 + 2L\dot{\theta}\dot{y} \sin \theta + \dot{y}^2 \right] - Mg \left[L(1 - \cos \theta) + y \right]$$

The Euler-Lagrange equations lead to equations of motion for θ and y

$$ML^2 \ddot{\theta} + ML\ddot{y} \sin \theta + MgL \sin \theta = 0$$

$$Ml\ddot{\theta} \sin \theta + ML\dot{\theta}^2 \cos \theta + M\ddot{y} + Mg = F$$

Assume the small-angle approximation where $\theta \to 0$, then these two equations reduce to

$$\ddot{\theta} + \left(\frac{g}{L} + \frac{\ddot{y}}{L} \right) \theta = 0$$

$$\ddot{y} + g = \frac{F}{M}$$

Substitute $\ddot{y} = -A\omega^2 \cos \omega t$ into these equations gives

$$\ddot{\theta} + \left(\frac{g}{L} - \frac{A\omega^2}{L} \cos \omega t \right) \theta = 0$$

$$M \left(g - A\omega^2 \cos \omega t \right) = F$$

These correspond to stable harmonic oscillations about $\theta \approx 0$ if the bracket term is positive, and to unstable motion if the bracket is negative. Thus, for small amplitude oscillation about $\theta \approx 0$ the motion of the system can be unstable whenever the bracket is negative, that is, when the acceleration $A\omega^2 \cos \omega t > g$ and resonance behavior can occur coupling the pendulum period and the forcing frequency ω.

This discussion also applies to the inverted pendulum with a surprising result. It is well known that the pendulum is unstable near $\theta = \pi$. However, if the support is oscillating, then for $\theta \approx \pi$ the equations of motion become

$$\ddot{\theta} - \left(\frac{g}{L} - \frac{A\omega^2}{L} \cos \omega t \right) \theta = 0$$

$$m \left(g - A\omega^2 \cos \omega t \right) = F$$

The inverted pendulum has stable oscillations about $\theta \approx \pi$ if the bracket is negative, that is, if $A\omega^2 \cos \omega t > g$. This illustrates that nonautonomous dynamical systems can involve either stable or unstable motion.

6.12 Impulsive forces

Colliding bodies often involve large impulsive forces that act for a short time. As discussed in chapter 2.12.8, the treatment of impulsive forces or torques is greatly simplified if they act for a sufficiently short time that the displacement during the impact can be ignored, even though the instantaneous change in velocities may be large. The simplicity is achieved by taking the time integral of the Euler-Lagrange equations over the duration τ of the impulse and assuming $\tau \to 0$.

The impact of the impulse on a system can be handled two ways. The first approach is to use the Euler-Lagrange equation during the impulse to determine the equations of motion

$$\frac{d}{dt} \left(\frac{\partial L}{\partial \dot{q}_j} \right) - \frac{\partial L}{\partial q_j} = Q_j^{EXC} \tag{6.75}$$

where the impulsive force is introduced using the generalized force Q_j^{EXC}. Knowing the initial conditions at time t, the conditions at the time $t + \tau$ are given by integration of equation 6.75 over the duration τ of the impulse which gives

$$\int_t^{t+\tau} \frac{d}{dt} \left(\frac{\partial L}{\partial \dot{q}_j} \right) d\tau - \int_t^{t+\tau} \frac{\partial L}{\partial q_j} d\tau = \int_t^{t+\tau} Q_j^{EXC} d\tau \tag{6.76}$$

This integration determines the conditions at time $t + \tau$ which then are used as the initial conditions for the motion when the impulsive force Q_j^{EXC} is zero.

The second approach is to realize that equation 6.76 can be rewritten in the form

$$\lim_{\tau \to 0} \int_t^{t+\tau} \frac{d}{dt} \left(\frac{\partial L}{\partial \dot{q}_j} \right) dt = \lim_{\tau \to 0} \frac{\partial L}{\partial \dot{q}_j} \Big|_t^{t+\tau} = \Delta p_j = \lim_{\tau \to 0} \int_t^{t+\tau} \left(\left(\frac{\partial L}{\partial q_j} \right) + Q_j^{EXC} \right) d\tau \tag{6.77}$$

Note that in the limit that $\tau \to 0$ then the integral of the generalized momentum $p_j = \frac{\partial L}{\partial \dot{q}_j}$ simplifies to give the change in generalized momentum Δp_j. In addition, assuming that the non-impulsive forces $\left(\frac{\partial L}{\partial q_j} \right)$ are finite and independent of the instantaneous impulsive force during the infinitessimal duration τ, then the contribution of the non-impulsive forces $\int_t^{t+\tau} \left(\frac{\partial L}{\partial q_j} \right) d\tau$ during the impulse can be neglected relative to the large impulsive force term; $\lim_{\tau \to 0} \int_t^{t+\tau} Q_j^{EXC} d\tau$. Thus it can be assumed that

$$\Delta p_j = \lim_{\tau \to 0} \int_t^{t+\tau} Q_j^{EXC} d\tau = \tilde{Q}_j \tag{6.78}$$

where $\tilde{Q}_j$ is the generalized impulse associated with coordinate $j = 1, 2, 3, ..., n$. This generalized impulse can be derived from the time integral of the impulsive forces $\mathbf{P}_i$ given by equation 2.135 using the time integral of equation 6.77, that is

$$\Delta p_j = \tilde{Q}_j = \lim_{\tau \to 0} \int_t^{t+\tau} Q_j^{EXC} d\tau \equiv \lim_{\tau \to 0} \int_t^{t+\tau} \sum_i \mathbf{P}_i \cdot \frac{\partial \mathbf{r}_i}{\partial q_j} d\tau = \sum_i \tilde{\mathbf{P}}_i \cdot \frac{\partial \mathbf{r}_i}{\partial q_j} \tag{6.79}$$

Note that the generalized impulse $\tilde{Q}_j$ can be a translational impulse $\tilde{\mathbf{P}}_j$ with corresponding translational variable q_j, or an angular impulsive torque $\tilde{\tau}_j$ with corresponding angular variable ϕ_j.

Impulsive force problems usually are solved in two stages. Either equations 6.76 or 6.79 are used to determine the conditions of the system immediately following the impulse. If $\tau \to 0$ then impulse changes the generalized velocities $\dot{q}_j$ but not the generalized coordinates q_j. The subsequent motion then is determined using the Lagrangian equations of motion with the impulsive generalized force being zero, and assuming that the initial condition corresponds to the result of the impulse calculation.

6.21 *Example: Series-coupled double pendulum subject to impulsive force*

Consider a series-coupled double pendulum comprising two masses m_1 and m_2 connected by rigid massless rods of lengths L_1 and L_2 as shown in the figure. Initially the two pendula are at rest and hanging vertically when a horizontal impulse $\tilde{P}$ strikes the system at a distance D below the upper fulcrum where $L_1 < D < L_1 + L_2$. For this system the kinetic energy of the masses m_1 and m_2 are

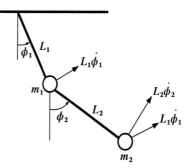

Two series-coupled plane pendula.

$$T_1 = \frac{1}{2} m_1 L_1^2 \dot{\phi}_1^2$$

$$T_2 = \frac{1}{2} m_2 [L_1^2 \dot{\phi}_1^2 + 2 L_1 L_2 \dot{\phi}_1 \dot{\phi}_2 \cos(\phi_1 - \phi_2) + L_2^2 \dot{\phi}_2^2]$$

Note the velocity of m_2 is the vector sum of the two velocities shown, separated by the angle $\phi_2 - \phi_1$. Thus the total kinetic energy is

$$T = \frac{1}{2}(m_1 + m_2) L_1^2 \dot{\phi}_1^2 + m_2 L_1 L_2 \dot{\phi}_1 \dot{\phi}_2 \cos(\phi_1 - \phi_2) + \frac{1}{2} m_2 L_2^2 \dot{\phi}_2^2$$

To first order in $\cos(\phi_1 - \phi_2)$

$$T = \frac{1}{2}(m_1 + m_2) L_1^2 \dot{\phi}_1^2 + m_2 L_1 L_2 \dot{\phi}_1 \dot{\phi}_2 + \frac{1}{2} m_2 L_2^2 \dot{\phi}_2^2$$

The total potential energy is

$$U = m_1 g L_1 (1 - \cos\phi_1) + m_2 g [L_1(1 - \cos\phi_1) + L_2(1 - \cos\phi_2)]$$

$$= (m_1 + m_2) g L_1 (1 - \cos\phi_1) + m_2 g L_2 (1 - \cos\phi_2)$$

Thus, assuming the small-angle approximation, the Lagrangian becomes

$$L = \frac{1}{2}(m_1 + m_2) L_1^2 \dot{\phi}_1^2 + m_2 L_1 L_2 \dot{\phi}_1 \dot{\phi}_2 + \frac{1}{2} m_2 L_2^2 \dot{\phi}_2^2 - \left(\frac{1}{2}(m_1 + m_2) g L_1 \phi_1^2 + \frac{1}{2} m_2 g L_2 \phi_2^2 \right)$$

Use equation 6.79 to transform to the generalized coordinates ϕ_1 and ϕ_2 with the corresponding generalized impulsive torques

$$\tilde{Q}_1 = \tilde{P}L_1$$
$$\tilde{Q}_2 = \tilde{P}(D - L_1)$$

Since the system starts at rest where $\phi_1 = \phi_2 = 0$, then using equation 6.77 gives the change in angular momentum immediately following the impulse to be

$$m_1 L_1^2 \dot{\phi}_1 + m_2 L_1 \left(L_1 \dot{\phi}_1 + L_2 \dot{\phi}_2 \right) = \tilde{P}L_1$$

$$m_2 L_2 \left(L_1 \dot{\phi}_1 + L_2 \dot{\phi}_2 \right) = \tilde{P}(D - L_1)$$

These two equations determine $\dot{\phi}_1$ and $\dot{\phi}_2$ immediately after the impulse; these can be used with $\phi_1 = \phi_2 = 0$ as initial conditions for solving the subsequent force-free motion when the generalized impulsive force is zero.

As described in example 14.5, the subsequent motion of this series coupled pendulum will be a superposition of the two normal modes with amplitudes determined by the result of the impulse calculation.

6.13   The Lagrangian versus the Newtonian approach to classical mechanics

It is useful to contrast the differences, and relative advantages, of the Newtonian and Lagrangian formulations of classical mechanics. The Newtonian force-momentum formulation is vectorial in nature, it has cause and effect embedded in it. The Lagrangian approach is cast in terms of kinetic and potential energies which involve only scalar functions and the equations of motion come from a single scalar function, i.e. Lagrangian. The directional properties of the equations of motion come from the requirement that the trajectory is specified by the principle of least action. The directional properties of the vectors in the Newtonian approach assist in our intuition when setting up a problem, but the Lagrangian method is simpler mathematically when the mechanical system is more complex.

The major advantage of the variational approaches to mechanics is that solution of the dynamical equations of motion can be simplified by expressing the motion in terms of independent **generalized coordinates**. For Lagrangian mechanics these generalized coordinates can be any set of **independent variables**, q_i, where $1 \leq i \leq n$, plus the corresponding velocities $\dot{q}_i$. These independent generalized coordinates completely specify the scalar potential and kinetic energies used in the Lagrangian or Hamiltonian. The variational approach allows for a much larger arsenal of possible generalized coordinates than the typical vector coordinates used in Newtonian mechanics. For example, the generalized coordinates can be dimensionless amplitudes for the N normal modes of coupled oscillator systems, or action-angle variables. Moreover, very different generalized coordinates can be used for each of the n variables. The tremendous freedom plus flexibility of the choice of generalized coordinates is important when constraint forces are acting on the system. Generalized coordinates allow the constraint forces to be ignored by including auxiliary conditions to account for the kinematic constraints that lead to correlated motion. The Lagrange method provides an incredibly consistent and mechanistic problem-solving strategy for many-body systems subject to constraints. Expressed in terms of generalized coordinates, the Lagrange's equations can be applied to a wide variety of physical problems including those involving fields. The manipulation of scalar quantities in a configuration space of generalized coordinates can greatly simplify problems compared with being confined to a rigid orthogonal coordinate system characterized by the Newtonian vector approach.

The use of generalized coordinates in Lagrange's equations of motion can be applied to a wide range of physical phenomena including field theory, such as for electromagnetic fields, which are beyond the applicability of Newton's equations of motion. The superiority of the Lagrangian approach compared to the Newtonian approach for solving problems in mechanics is apparent when dealing with holonomic constraint forces. Constraint forces must be known and included explicitly in the Newtonian equations of motion. Unfortunately, knowledge of the equations of motion is required to derive these constraint forces. For holonomic constrained systems, the equations of motion can be solved directly without calculating the constraint forces using the minimal set of generalized coordinate approach to Lagrangian mechanics. Moreover, the Lagrange approach has significant philosophical advantages compared to the Newtonian approach.

6.14 Summary

Newtonian plausibility argument for Lagrangian mechanics:

A justification for introducing the calculus of variations to classical mechanics becomes apparent when the concept of the Lagrangian $L \equiv T - U$ is used in the functional and time t is the independent variable. It was shown that Newton's equation of motion can be rewritten as

$$\frac{d}{dt} \frac{\partial L}{\partial \dot{q}_i} - \frac{\partial L}{\partial q_i} = F_{q_i}^{EX} \tag{6.12}$$

where $F_{y_i}^{EX}$ are the excluded forces of constraint plus any other conservative or non-conservative forces not included in the potential U. This corresponds to the Euler-Lagrange equation for determining the minimum of the time integral of the Lagrangian. Equation 6.12 can be written as

$$\frac{d}{dt} \frac{\partial L}{\partial \dot{q}_i} - \frac{\partial L}{\partial q_i} = \sum_{k}^{m} \lambda_k (t) \frac{\partial g_k}{\partial q_i} + F_{q_i}^{EXC} \tag{6.15}$$

where the Lagrange multiplier term accounts for holonomic constraint forces, and $F_{q_i}^{EXC}$ includes all additional forces not accounted for by the scalar potential U, or the Lagrange multiplier terms $F_{q_i}^{HC}$. The constraint forces can be included explicitly as generalized forces in the excluded term $F_{q_i}^{EXC}$ of equation 6.15.

d'Alembert's Principle

It was shown that d'Alembert's Principle

$$\sum_{i}^{N} (\mathbf{F}_i^A - \dot{\mathbf{p}}_i) \cdot \delta \mathbf{r}_i = 0 \tag{6.25}$$

cleverly transforms the principle of virtual work from the realm of statics to dynamics. Application of virtual work to statics primarily leads to algebraic equations between the forces, whereas d'Alembert's principle applied to dynamics leads to differential equations of motion.

Lagrange equations from d'Alembert's Principle

After transforming to generalized coordinates, d'Alembert's Principle leads to

$$\sum_{j}^{N} \left[\left\{ \frac{d}{dt} \left(\frac{\partial T}{\partial \dot{q}_j} \right) - \frac{\partial T}{\partial q_j} \right\} - Q_j \right] \delta q_j = 0 \tag{6.38}$$

If all the n generalized coordinates q_j are independent, then equation 6.38 implies that the term in the square brackets is zero for each individual value of j. That is, this implies the basic Euler-Lagrange equations of motion.

The handling of both conservative and non-conservative generalized forces Q_j is best achieved by assuming that the generalized force $Q_j = \sum_{i}^{n} \mathbf{F}_i^A \cdot \frac{\partial \mathbf{r}_i}{\partial q_j}$ can be partitioned into a conservative velocity-independent term, that can be expressed in terms of the gradient of a scalar potential, $-\boldsymbol{\nabla} U_i$, plus an excluded generalized force Q_j^{EX} which contains the non-conservative, velocity-dependent, and all the constraint forces not explicitly included in the potential U_j. That is,

$$Q_j = -\boldsymbol{\nabla} U_j + Q_j^{EX} \tag{6.41}$$

Inserting (6.41) into (6.38), and *assuming that the potential U is velocity independent*, allows (6.38) to be rewritten as

$$\sum_{j} \left[\left\{ \frac{d}{dt} \left(\frac{\partial (T - U)}{\partial \dot{q}_j} \right) - \frac{\partial (T - U)}{\partial q_j} \right\} - Q_j^{EX} \right] \delta q_j = 0 \tag{6.42}$$

Expressed in terms of the standard Lagrangian $L = T - U$ this gives

$$\sum_{j}^{N} \left[\left\{ \frac{d}{dt} \left(\frac{\partial L}{\partial \dot{q}_j} \right) - \frac{\partial L}{\partial q_j} \right\} - Q_j^{EX} \right] \delta q_j = 0 \tag{6.44}$$

Note that equation (6.44) contains the basic Euler-Lagrange equation (6.38) for the special case when $U = 0$. In addition, note that *if all the generalized coordinates are independent,* then the square bracket terms are zero for each value of j, which leads to the n *general Euler-Lagrange equations of motion*

$$\left\{ \frac{d}{dt} \left(\frac{\partial L}{\partial \dot{q}_j} \right) - \frac{\partial L}{\partial q_j} \right\} = Q_j^{EX} \tag{6.45}$$

where $n \geq j \geq 1$.

Newtonian mechanics has trouble handling constraint forces because they lead to coupling of the degrees of freedom. Lagrangian mechanics is more powerful since it provides the following three ways to handle such correlated motion.

1) Minimal set of generalized coordinates

If the n coordinates q_j are independent, then the square bracket equals zero for each value of j in equation 6.44, which corresponds to Euler's equation for each of the n independent coordinates. If the n generalized coordinates are coupled by m constraints, then the coordinates can be transformed to a minimal set of $s = n - m$ independent coordinates which then can be solved by applying equation 6.45 to the minimal set of s independent coordinates.

2) Lagrange multipliers approach

The Lagrangian method concentrates solely on active forces, completely ignoring all other internal forces. In Lagrangian mechanics the generalized forces, corresponding to each generalized coordinate, can be partitioned three ways

$$Q_j = -\nabla U + \sum_{k=1}^{m} \lambda_k \frac{\partial g_k}{\partial q_j}(\mathbf{q}, t) + Q_j^{EXC}$$

where the velocity-independent conservative forces can be absorbed into a scalar potential U, the holonomic constraint forces can be handled using the Lagrange multiplier term $\sum_{k=1}^{m} \lambda_k \frac{\partial g_k}{\partial q_j}(\mathbf{q}, t)$, and the remaining part of the active forces can be absorbed into the generalized force Q_j^{EXC}. The scalar potential energy U is handled by absorbing it into the standard Lagrangian $L = T - U$. If the constraint forces are holonomic then these forces are easily and elegantly handled by use of Lagrange multipliers. All remaining forces, including dissipative forces, can be handled by including them explicitly in the the generalized force Q_j^{EXC}.

Combining the above two equations gives

$$\sum_{j}^{N} \left[\left\{ \frac{d}{dt} \left(\frac{\partial L}{\partial \dot{q}_j} \right) - \frac{\partial L}{\partial q_j} \right\} - Q_j^{EXC} - \sum_{k=1}^{m} \lambda_k \frac{\partial g_k}{\partial q_j}(\mathbf{q}, t) \right] \delta q_j = 0 \tag{6.56}$$

Use of the Lagrange multipliers to handle the m constraint forces ensures that all n infinitessimals δq_j are independent implying that the expression in the square bracket must be zero for each of the n values of j. This leads to n Lagrange equations plus m constraint relations

$$\left\{ \frac{d}{dt} \left(\frac{\partial L}{\partial \dot{q}_j} \right) - \frac{\partial L}{\partial q_j} \right\} = Q_j^{EXC} + \sum_{k=1}^{m} \lambda_k \frac{\partial g_k}{\partial q_j}(\mathbf{q}, t) \tag{6.60}$$

where $j = 1, 2, 3, ...n$.

3) Generalized forces approach

The two right-hand terms in (6.60) can be understood to be those forces acting on the system that are not absorbed into the scalar potential U component of the Lagrangian L. The Lagrange multiplier terms $\sum_{k=1}^{m} \lambda_k \frac{\partial g_k}{\partial q_j}(\mathbf{q}, t)$ account for the holonomic forces of constraint that are not included in the conservative potential or in the generalized forces Q_j^{EXC}. The generalized force

$$Q_j^{EXC} = \sum_{i}^{n} \mathbf{F}_i^A \cdot \frac{\partial \mathbf{r}_i}{\partial q_j} \tag{6.17}$$

is the sum of the components in the q_j direction for all external forces that have not been taken into account by the scalar potential or the Lagrange multipliers. Thus the non-conservative generalized force Q_j^{EXC} contains non-holonomic constraint forces, including dissipative forces such as drag or friction, that are not included in U, or used in the Lagrange multiplier terms to account for the holonomic constraint forces.

Applying the Euler-Lagrange equations in mechanics:
The optimal way to exploit Lagrangian mechanics is as follows:

1. Select a set of independent generalized coordinates.

2. Partition the active forces into three groups:

    (a) Conservative one-body forces
    (b) Holonomic constraint forces
    (c) Generalized forces

3. Minimize the number of generalized coordinates.

4. Derive the Lagrangian

5. Derive the equations of motion

Velocity-dependent Lorentz force:
Usually velocity-dependent forces are non-holonomic. However, electromagnetism is a special case where the velocity-dependent Lorentz force $\mathbf{F} = q(\mathbf{E} + \mathbf{v} \times \mathbf{B})$ can be obtained from a velocity-dependent potential function $U(q, \dot{q}, t)$. It was shown that the velocity-dependent potential

$$U = q\Phi - q\mathbf{v} \cdot \mathbf{A} \tag{6.74}$$

leads to the Lorentz force where Φ is the scalar electric potential and $\mathbf{A}$ the vector potential.

Time-dependent forces:
It was shown that time-dependent forces can lead to complicated motion having both stable regions and unstable regions of motion that can exhibit chaos.

Impulsive forces:
A generalized impulse $\tilde{Q}_j$ can be derived for an instantaneous impulsive force from the time integral of the impulsive forces $\mathbf{P}_i$ given by equation 2.135 using the time integral of equation 6.78, that is

$$\Delta p_j = \tilde{Q}_j = \lim_{\tau \to 0} \int_t^{t+\tau} Q_j^{EXC} d\tau \equiv \lim_{\tau \to 0} \int_t^{t+\tau} \sum_i \mathbf{F}_i \cdot \frac{\partial \mathbf{r}_i}{\partial q_j} d\tau = \sum_i \tilde{\mathbf{P}}_i \cdot \frac{\partial \mathbf{r}_i}{\partial q_j} \tag{6.79}$$

Note that the generalized impulse $\tilde{Q}_j$ can be a translational impulse $\tilde{\mathbf{P}}_j$ with corresponding translational variable q_j or an angular impulsive torque $\tilde{\mathbf{T}}_j$ with corresponding angular variable ϕ_j.

Comparison of Newtonian and Lagrangian mechanics:
In contrast to Newtonian mechanics, which is based on knowing all the vector forces acting on a system, Lagrangian mechanics can derive the equations of motion using generalized coordinates without requiring knowledge of the constraint forces acting on the system. Lagrangian mechanics provides a remarkably powerful, and incredibly consistent approach to solving for the equations of motion in classical mechanics, and is especially powerful for handling systems that are subject to holonomic constraints.

Workshop exercises

1. A disk of mass M and radius R rolls without slipping down a plane inclined from the horizontal by an angle α. The disk has a short weightless axle of negligible radius. From this axis is suspended a simple pendulum of length $l < R$ and whose bob has a mass m. Assume that the motion of the pendulum takes place in the plane of the disk.

   (a) What generalized coordinates would be appropriate for this situation?

   (b) Are there any equations of constraint? If so, what are they?

   (c) Find Lagrange's equations for this system.

2. A Lagrangian for a particular system can be written as

$$L = \frac{m}{2}(a\dot{x}^2 + 2b\dot{x}\dot{y} + c\dot{y}^2) - \frac{K}{2}(ax^2 + 2bxy + cy^2)$$

   where a, b, and c are arbitrary constants, but subject to the condition that $b^2 - 4ac \neq 0$.

   (a) What are the equations of motion?

   (b) Examine the case $a = 0 = c$. What physical system does this represent?

   (c) Examine the case $b = 0$ and $a = -c$. What physical system does this represent?

   (d) Based on your answers to (b) and (c), determine the physical system represented by the Lagrangian given above.

3. Consider a particle of mass m moving in a plane and subject to an inverse square attractive force.

   (a) Obtain the equations of motion.

   (b) Is the angular momentum about the origin conserved?

   (c) Obtain expressions for the generalized forces. Recall that the generalized forces are defined by

$$Q_j = \sum_i F_i \frac{\partial x_i}{\partial q_j}.$$

4. Consider a Lagrangian function of the form $L(q_i, \dot{q}_i, \ddot{q}_i, t)$. Here the Lagrangian contains a time derivative of the generalized coordinates that is higher than the first. When working with such Lagrangians, the term "generalized mechanics" is used.

   (a) Consider a system with one degree of freedom. By applying the methods of the calculus of variations, and assuming that Hamilton's principle holds with respect to variations which keep both q and $\dot{q}$ fixed at the end points, show that the corresponding Lagrange equation is

$$\frac{d^2}{dt^2}\left(\frac{\partial L}{\partial \ddot{q}}\right) - \frac{d}{dt}\left(\frac{\partial L}{\partial \dot{q}}\right) + \frac{\partial L}{\partial q} = 0.$$

   Such equations of motion have interesting applications in chaos theory.

   (b) Apply this result to the Lagrangian

$$L = -\frac{m}{2}q\ddot{q} - \frac{k}{2}q^2.$$

   Do you recognize the equations of motion?

5. A bead of mass m slides under gravity along a smooth wire bent in the shape of a parabola $x^2 = az$ in the vertical (x, z) plane.

   (a) What kind (holonomic, nonholonomic, scleronomic, rheonomic) of constraint acts on m?

   (b) Set up Lagrange's equation of motion for x with the constraint embedded.

(c) Set up Lagrange's equations of motion for both x and z with the constraint adjoined and a Lagrangian multiplier λ introduced.

(d) Show that the same equation of motion for x results from either of the methods used in part (b) or part (c).

(e) Express λ in terms of x and $\dot{x}$.

(f) What are the x and z components of the force of constraint in terms of x and $\dot{x}$?

6. Consider the two Lagrangians

$$L(q, \dot{q}; t) \quad \text{and} \quad L'(q, \dot{q}; t) = L(q, \dot{q}; t) + \frac{dF(q, t)}{dt}$$

where $F(q, t)$ is an arbitrary function of the generalized coordinates $q(t)$. Show that these two Lagrangians yield the same Euler-Lagrange equations. As a consequence two Lagrangians that differ only by an exact time derivative are said to be equivalent.

7. Consider the double pendulum comprising masses m_1 and m_2 connected by inextensible strings as shown in the figure. Assume that the motion of the pendulum takes place in a vertical plane.

(a) Are there any equations of constraint? If so, what are they?

(b) Find Lagrange's equations for this system.

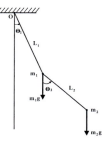

8 Consider the system shown in the figure which consists of a mass m suspended via a constrained massless link of length L where the point A is acted upon by a spring of spring constant k. The spring is unstretched when the massless link is horizontal. Assume that the holonomic constraints at A and B are frictionless.

a Derive the equations of motion for the system using the method of Lagrange multipliers.

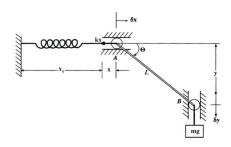

9 Consider a pendulum, with mass m, connected to a (horizontally) moveable support of mass M.

(a) Determine the Lagrangian of the system.

(b) Determine the equations of motion for $\theta \ll 1$.

(c) Find an equation of motion in θ alone. What is the frequency of oscillation?

(d) What is the frequency of oscillation for $M \gg m$? Does this make sense?

Problems

1. A sphere of radius ρ is constrained to roll without slipping on the lower half of the inner surface of a hollow cylinder of radius R. Determine the Lagrangian function, the equation of constraint, and the Lagrange equations of motion. Find the frequency of small oscillations.

2. A particle moves in a plane under the influence of a force $f = -Ar^{\alpha-1}$ directed toward the origin; A and $\alpha\,(>0)$ are constants. Choose generalized coordinates with the potential energy zero at the origin.

   a) Find the Lagrangian equations of motion.

   b) Is the angular momentum about the origin conserved?

   c) Is the total energy conserved?

3. Two blocks, each of mass M, are connected by an extensionless, uniform string of length l. One block is placed on a frictionless horizontal surface, and the other block hangs over the side, the string passing over a frictionless pulley. Describe the motion of the system:

   a) when the mass of the string is negligible

   b) when the string has mass m.

4. Two masses m_1 and m_2 $(m_1 \neq m_2)$ are connected by a rigid rod of length d and of negligible mass. An extensionless string of length l_1 is attached to m_1 and connected to a fixed point of the support P. Similarly a string of length l_2 $(l_1 \neq l_2)$ connects m_2 and P. Obtain the equation of motion describing the motion in the plane of m_1, m_2, and P, and find the frequency of small oscillation around the equilibrium position.

5. A thin uniform rigid rod of length $2L$ and mass M is suspended by a massless string of length l. Initially the system is hanging vertically downwards in the gravitational field g. Use as generalized coordinates the angles given in the diagram.

   a) Derive the Lagrangian for the system.

   b) Use the Lagrangian to derive the equations of motion.

   c) A horizontal impulsive force F_x in the x direction strikes the bottom end of the rod for an infinitessimal time τ. Derive the initial conditions for the system immediately after the impulse has occurred.

   d) Draw a diagram showing the geometry of the pendulum shortly after the impulse when the displacement angles are significant.

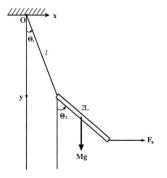

Chapter 7

Symmetries, Invariance and the Hamiltonian

7.1 Introduction

The chapter 7 discussion of Lagrangian dynamics illustrates the power of Lagrangian mechanics for deriving the equations of motion. In contrast to Newtonian mechanics, which is expressed in terms of force vectors acting on a system, the Lagrangian method, based on d'Alembert's Principle or Hamilton's Principle, is expressed in terms of the scalar kinetic and potential energies of the system. The Lagrangian approach is a sophisticated alternative to Newton's laws of motion, that provides a simpler derivation of the equations of motion that allows constraint forces to be ignored. In addition, the use of Lagrange multipliers or generalized forces allows the Lagrangian approach to determine the constraint forces when these forces are of interest. The equations of motion, derived either from Newton's Laws or Lagrangian dynamics, can be non-trivial to solve mathematically. It is necessary to integrate second-order differential equations, which for n degrees of freedom, imply $2n$ constants of integration.

Chapter 7 will explore the remarkable connection between symmetry and invariance of a system under transformation, and the related conservation laws that imply the existence of constants of motion. Even when the equations of motion cannot be solved easily, it is possible to derive important physical principles regarding the first-order integrals of motion of the system directly from the Lagrange equation, as well as for elucidating the underlying symmetries plus invariance. This property is contained in **Noether's theorem** which states that conservation laws are associated with differentiable symmetries of a physical system.

7.2 Generalized momentum

Consider a holonomic system of N masses under the influence of conservative forces that depend on position q_j but not velocity $\dot{q}_j$, that is, the potential is velocity independent. Then for the x coordinate of particle i for N particles

$$
\begin{aligned}
\frac{\partial L}{\partial \dot{x}_i} &= \frac{\partial T}{\partial \dot{x}_i} - \frac{\partial U}{\partial \dot{x}_i} = \frac{\partial T}{\partial \dot{x}_i} \\
&= \frac{\partial}{\partial \dot{x}_i} \sum_{i=1}^{N} \frac{1}{2} m_i \left(\dot{x}_i^2 + \dot{y}_i^2 + \dot{z}_i^2 \right) \\
&= m_i \dot{x}_i = p_{i,x}
\end{aligned}
\tag{7.1}
$$

Thus for a holonomic, conservative, velocity-independent potential we have

$$
\frac{\partial L}{\partial \dot{x}_i} = p_{i,x}
\tag{7.2}
$$

which is the x component of the linear momentum for the ith particle.

This result suggests an obvious extension of the concept of momentum to generalized coordinates. The **generalized momentum** associated with the coordinate q_j is defined to be

$$\frac{\partial L}{\partial \dot{q}_j} \equiv p_j \tag{7.3}$$

Note that p_j also is called the **conjugate momentum** or **canonical momentum** to q_j where q_j, p_j are conjugate, or canonical, variables. Remember that the linear momentum p_j is the first-order time integral given by equation 2.10. If q_j is not a spatial coordinate, then p_j is the generalized momentum, not the kinematic linear momentum. For example, if q_j is an angle, then p_j will be angular momentum. That is, the generalized momentum may differ from the usual linear or angular momentum since the definition (7.3) is more general than the usual $p_x = m\dot{x}$ definition of linear momentum in classical mechanics. This is illustrated by the case of a moving charged particles m_j, e_j in an electromagnetic field. Chapter 6 showed that electromagnetic forces on a charge e_j can be described in terms of a scalar potential U_j where

$$U_j = e_j(\Phi - \mathbf{A} \cdot \mathbf{v}_j) \tag{7.4}$$

Thus the Lagrangian for the electromagnetic force can be written as

$$L = \sum_{j=1}^{N} \left[\frac{1}{2} m_j \mathbf{v}_j \cdot \mathbf{v}_j - e_j(\Phi - \mathbf{A} \cdot \mathbf{v}_j) \right] \tag{7.5}$$

The generalized momentum to the coordinate x_j for charge e_j, and mass m_j, is given by the above Lagrangian

$$p_{j,x} = \frac{\partial L}{\partial \dot{x}_j} = m_j \dot{x}_j + e_j A_x \tag{7.6}$$

Note that this includes both the mechanical linear momentum plus the correct electromagnetic momentum. The fact that the electromagnetic field carries momentum should not be a surprise since electromagnetic waves also carry energy as is illustrated by the transmission of radiant energy from the sun.

7.1 *Example: Feynman's angular-momentum paradox*

Feynman[Fey84] posed the following paradox. A circular insulating disk, mounted on frictionless bearings, has a circular ring of total charge q uniformly distributed around the perimeter of the circular disk at the radius R. A superconducting long solenoid of radius s, where $s < R$, is fixed to the disk and is mounted coaxial with the bearings. The moment of inertia of the system about the rotation axis is I. Initially the disk plus superconducting solenoid are stationary with a steady current producing a uniform magnetic field B_0 inside the solenoid. Assume that a rise in temperature of the solenoid destroys the superconductivity leading to a rapid dissipation of the electric current and resultant magnetic field. Assume that the system is free to rotate, no other forces or torques are acting on the system, and that the charge carriers in the solenoid have zero mass and thus do not contribute to the angular momentum. Does the system rotate when the current in the solenoid stops?

Initially the system is stationary with zero mechanical angular momentum. Faraday's Law states that, when the magnetic field dissipates from B_0 to zero, there will be a torque $\mathbf{N}$ acting on the circumferential charge q at radius R due to the change in magnetic flux Φ.

$$\mathbf{N}(t) = -qR\frac{d\Phi}{dt}$$

Since $\frac{d\Phi}{dt} < 0$, this torque leads to an angular impulse which will equal the final mechanical angular momentum.

$$\mathbf{L}_{final}^{MECH} = \mathbf{T} = \int_t \mathbf{N}(t)dt = qR\mathbf{\Phi}$$

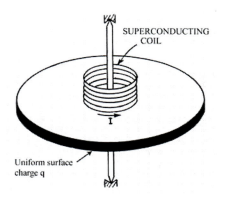

SUPERCONDUCTING COIL

I

Uniform surface charge q

The initial angular momentum in the electromagnetic field can be derived using equation 7.6, plus Stoke's theorem (Appendix H3). Equation 2.142 gives that the final angular momentum equals the angular impulse

$$\mathbf{L}^{EM}_{initial} = R \int_t \oint r\dot{p}_\phi \, dl \, dt = R \oint r p_\phi \, dl = qR \oint A_\phi \, dl = qR \int \mathbf{B} \cdot d\mathbf{S} = qR\Phi$$

where $\Phi = \oint A_\phi \, dl = \int \mathbf{B} \cdot d\mathbf{S}$ is the initial total magnetic flux through the solenoid. Thus the total initial angular momentum is given by

$$\mathbf{L}^{TOTAL}_{initial} = 0 + \mathbf{L}^{EM}_{initial} = qR\Phi$$

Since the final electromagnetic field is zero the final total angular momentum is given by

$$\mathbf{L}^{TOTAL}_{final} = \mathbf{L}^{MECH}_{final} + 0 = qR\Phi$$

Note that the total angular momentum is conserved. That is, initially all the angular momentum is stored in the electromagnetic field, whereas the final angular momentum is all mechanical. This explains the paradox that the mechanical angular momentum is not conserved, only the total angular momentum of the system is conserved, that is, the sum of the mechanical and electromagnetic angular momenta.

7.3   Invariant transformations and Noether's Theorem

One of the great advantages of Lagrangian mechanics is the freedom it allows in choice of generalized coordinates which can simplify derivation of the equations of motion. For example, for any set of coordinates, q_j, a reversible point transformation can define another set of coordinates q'_j such that

$$q'_j = q'_j(q_1, q_2, ..q_n; t) \tag{7.7}$$

The new set of generalized coordinates satisfies Lagrange's equations of motion with the new Lagrangian

$$L(q', \dot{q}', t) = L(q, \dot{q}, t) \tag{7.8}$$

The Lagrangian is a scalar, with units of energy, which does not change if the coordinate representation is changed. Thus $L(q', \dot{q}', t)$ can be derived from $L(q, \dot{q}, t)$ by substituting the inverse relation $q_i = q_i(q'_1, q'_2, ..q'_n; t)$ into $L(q, \dot{q}, t)$. That is, the value of the Lagrangian L is independent of which coordinate representation is used. Although the general form of Lagrange's equations of motion is preserved in any point transformation, the explicit equations of motion for the new variables usually look different from those with the old variables. A typical example is the transformation from cartesian to spherical coordinates. For a given system, there can be particular transformations for which the explicit equations of motion are the same for both the old and new variables. Transformations for which the equations of motion are invariant, are called *invariant transformations*. It will be shown that if the Lagrangian does not explicitly contain a particular coordinate of displacement q_i, then the corresponding conjugate momentum, p_i, is conserved. This relation is called Noether's theorem which states *"For each symmetry of the Lagrangian, there is a conserved quantity"*.

Noether's Theorem will be used to consider invariant transformations for two dependent variables, $x(t)$, and $\theta(t)$, plus their conjugate momenta p_x and p_θ. For a closed system, these provide up to six possible conservation laws for the three axes. Then we will discuss the independent variable t, and its relation to the Generalized Energy Theorem, which provides another possible conservation law. For simplicity, these discussions will assume that the systems are holonomic and conservative.

The Lagrange equations using generalized coordinates for holonomic systems, was given by equation 6.60 to be

$$\left\{ \frac{d}{dt}\left(\frac{\partial L}{\partial \dot{q}_j} \right) - \frac{\partial L}{\partial q_j} \right\} = \sum_{k=1}^{m} \lambda_k \frac{\partial g_k}{\partial q_j}(\mathbf{q}, t) + Q_j^{EXC} \tag{7.9}$$

This can be written in terms of the generalized momentum as

$$\left\{ \frac{d}{dt} p_j - \frac{\partial L}{\partial q_j} \right\} = \sum_{k=1}^{m} \lambda_k \frac{\partial g_k}{\partial q_j}(\mathbf{q}, t) + Q_j^{EXC} \tag{7.10}$$

or equivalently as

$$\dot{p}_j = \frac{\partial L}{\partial q_j} + \left[\sum_{k=1}^{m} \lambda_k \frac{\partial g_k}{\partial q_j}(\mathbf{q}, t) + Q_j^{EXC}\right] \tag{7.11}$$

Note that if the Lagrangian L does not contain q_i explicitly, that is, the Lagrangian is invariant to a linear translation, or equivalently, is **spatially homogeneous**, and if the Lagrange multiplier constraint force and generalized force terms are zero, then

$$\frac{\partial L}{\partial q_j} + \left[\sum_{k=1}^{m} \lambda_k \frac{\partial g_k}{\partial q_j}(\mathbf{q}, t) + Q_j^{EXC}\right] = 0 \tag{7.12}$$

In this case the Lagrange equation reduces to

$$\dot{p}_j = \frac{dp_j}{dt} = 0 \tag{7.13}$$

Equation 7.13 corresponds to p_j being a constant of motion. Stated in words, *the generalized momentum p_i is a constant of motion if the Lagrangian is invariant to a spatial translation of q_i, and the constraint plus generalized force terms are zero.* Expressed another way, if the Lagrangian does not contain a given coordinate q_i and the corresponding constraint plus generalized forces are zero, then the generalized momentum associated with this coordinate is conserved. Note that this example of Noether's theorem applies to any component of $\mathbf{q}$. For example, in the uniform gravitational field at the surface of the earth, the Lagrangian does not depend on the x and y coordinates in the horizontal plane, thus p_x and p_y are conserved, whereas, due to the gravitational force, the Lagrangian does depend on the vertical z axis and thus p_z is not conserved.

7.2 Example: Atwoods machine

Assume that the linear momentum is conserved for the Atwood's machine shown in the figure below. Let the left mass rise a distance x and the right mass rise a distance y. Then the middle mass must drop by $x + y$ to conserve the length of the string. The Lagrangian of the system is

$$L = \frac{1}{2}(4m)\dot{x}^2 + \frac{1}{2}(3m)(-\dot{x} - \dot{y})^2 + \frac{1}{2}m\dot{y}^2 - (4mgx + 3mg(-x - y) + mgy) = \frac{7}{2}m\dot{x}^2 + 3m\dot{x}\dot{y} + 2m\dot{y}^2 - mg(x - 2y)$$

Note that the transformation

$$x = x_0 + 2\epsilon$$
$$y = y_0 + \epsilon$$

results in the potential energy term $mg(x - 2y) = mg(x_0 - 2y_0)$ which is a constant of motion. As a result the Lagrangian is independent of ϵ, which means that it is invariant to the small perturbation ϵ, and thus $\frac{dL}{d\epsilon} = 0$. Therefore, according to Noether's theorem, the corresponding linear momentum $P_\epsilon = \frac{dL}{d\dot{\epsilon}}$ is conserved. This conserved linear momentum then is given by

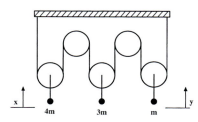

Example of an Atwood's machine

$$P_\epsilon = \frac{dL}{d\dot{\epsilon}} = \frac{\partial L}{\partial \dot{x}}\frac{\partial \dot{x}}{\partial \dot{\epsilon}} + \frac{\partial L}{\partial \dot{y}}\frac{\partial \dot{y}}{\partial \dot{\epsilon}} = m(7\dot{x} + 3\dot{y})(2) + m(3\dot{x} + 4\dot{y}) = m(17\dot{x} + 10\dot{y})$$

Thus, if the system starts at rest with $P_\epsilon = 0$, then $\dot{x}$ always equals $-\frac{10}{17}\dot{y}$ since P_ϵ is constant.

Note that this also can be shown using the Euler-Lagrange equations in that $\Lambda_x L = 0$ and $\Lambda_y L = 0$ give

$$7m\ddot{x} + 3m\ddot{y} = -mg$$
$$3m\ddot{x} + 4m\ddot{y} = 2mg$$

Adding the second equation to twice the first gives

$$17m\ddot{x} + 10m\ddot{y} = \frac{d}{dt}(17m\dot{x} + 10m\dot{y}) = 0$$

This is the result obtained directly using Noether's theorem.

7.4 Rotational invariance and conservation of angular momentum

The arguments, used above, apply equally well to conjugate momenta p_θ and θ for rotation about any axis. The Lagrange equation is

$$\left\{ \frac{d}{dt} p_\theta - \frac{\partial L}{\partial \theta} \right\} = \sum_{k=1}^{m} \lambda_k \frac{\partial g_k}{\partial \theta}(\mathbf{q}, t) + Q_\theta^{EXC} \tag{7.14}$$

If no constraint or generalized torques act on the system, then the right-hand side of equation 7.14 is zero. Moreover if the Lagrangian in not an explicit function of θ, then $\frac{\partial L}{\partial \theta} = 0$, and assuming that the constraint plus generalized torques are zero, then p_θ is a constant of motion.

Noether's Theorem illustrates this general result which can be stated as, *if the Lagrangian is rotationally invariant about some axis, then the component of the angular momentum along that axis is conserved.* Also this is true for the more general case where the Lagrangian is invariant to rotation about any axis, which leads to conservation of the total angular momentum.

7.3 Example: Conservation of angular momentum for rotational invariance:

The Noether theorem result for rotational-invariance about an axis also can be derived using cartesian coordinates as shown below. As discussed in appendix D, it is necessary to limit discussion of rotation to infinitessimal rotation angles in order to represent the rotation by a vector. Consider an infinitessimal rotation $\delta\theta$ about some axis, which is a vector. As illustrated in the adjacent figure, this can be expressed as

$$\delta \mathbf{r} = \delta\boldsymbol{\theta} \times \mathbf{r}$$

The velocity vectors also change on rotation of the system obeying the transformation equation which is common to all vectors, that is,

$$\delta \dot{\mathbf{r}} = \delta\boldsymbol{\theta} \times \dot{\mathbf{r}}$$

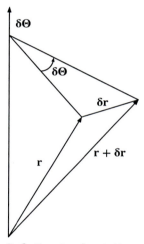

Infinitessimal rotation

If the Lagrangian is unaffected by the orientation of the system, that is, it is rotationally invariant, then it can be shown that the angular momentum is conserved. For example, consider that the Lagrangian is invariant to rotation about some axis q_i. Since the Lagrangian is a function

$$L = L(q_i, \dot{q}_i; t)$$

then the expression that the Lagrangian does not change due to an infinitesimal rotation $\delta\theta$ about this axis can be expressed as

$$\delta L = \sum_i \frac{\partial L}{\partial x_i} \delta x_i + \sum_i \frac{\partial L}{\partial \dot{x}_i} \delta \dot{x}_i = 0 \tag{A}$$

where cartesian coordinates have been used.
    Using the generalized momentum

$$\frac{\partial L}{\partial \dot{x}_i} = p_i$$

then, Lagrange's equation gives

$$\frac{d}{dt} p_i - \frac{\partial L}{\partial x_i} = 0$$

that is

$$\dot{p}_i = \frac{\partial L}{\partial x_i}$$

Inserting this into equation A gives

$$\delta L = \sum_i^3 \dot{p} \delta x_i + \sum_i^3 p_i \delta \dot{x}_i = 0$$

This is equivalent to the scalar products

$$\dot{\mathbf{p}} \cdot \delta\mathbf{r} + \mathbf{p} \cdot \delta\dot{\mathbf{r}} = 0$$

For an infinitessimal rotation $\delta\theta$, then $\delta r = \delta\theta \times r$ and $\delta\dot{r} = \delta\theta \times \dot{r}$. Therefore

$$\dot{\mathbf{p}} \cdot (\delta\boldsymbol{\theta} \times \mathbf{r}) + \mathbf{p} \cdot (\delta\boldsymbol{\theta} \times \dot{\mathbf{r}}) = 0$$

The cyclic order can be permuted giving

$$\begin{aligned}
\delta\boldsymbol{\theta} \cdot (\mathbf{r} \times \dot{\mathbf{p}}) + \delta\boldsymbol{\theta} \cdot (\dot{\mathbf{r}} \times \mathbf{p}) &= 0 \\
\delta\boldsymbol{\theta} \cdot [(\mathbf{r} \times \dot{\mathbf{p}}) + (\dot{\mathbf{r}} \times \mathbf{p})] &= 0 \\
\delta\boldsymbol{\theta} \cdot \frac{d}{dt}(\mathbf{r} \times \mathbf{p}) &= 0
\end{aligned}$$

Because the infinitessimal angle $\delta\theta$ is arbitrary, then the time derivative

$$\frac{d}{dt}(\mathbf{r} \times \mathbf{p}) = 0$$

about the axis of rotation $\delta\theta$. But the bracket $(\mathbf{r} \times \mathbf{p})$ equals the angular momentum. That is;

$$\text{Angular momentum} = (\mathbf{r} \times \mathbf{p}) = \text{constant}$$

This proves the Noether' theorem that the angular momentum about any axis is conserved if the Lagrangian is rotationally invariant about that axis.

7.4 *Example: Diatomic molecules and axially-symmetric nuclei*

An interesting example of Noether's theorem applies to diatomic molecules such as H_2, N_2, F_2, O_2, Cl_2 and Br_2. The electric field produced by the two charged nuclei of the diatomic molecule has cylindrical symmetry about the axis through the two nuclei. Electrons are bound to this dumbbell arrangement of the two nuclear charges which may be rotating and vibrating in free space. Assuming that there are no external torques acting on the diatomic molecule in free space, then the angular momentum about any fixed axis in free space must be conserved according to Noether's theorem. If no external torques are applied, then the component of the angular momentum about any fixed axis is conserved, that is, the total angular momentum is conserved. What is especially interesting is that since the electrostatic potential, and thus the Lagrangian, of the diatomic molecule has cylindrical symmetry, that is $\frac{\partial L}{\partial \phi} = 0$, then the component of the angular momentum with respect to this symmetry axis also is conserved irrespective of how the diatomic molecule rotates or vibrates in free space. That is, an additional symmetry has been identified that leads to an additional conservation law that applies to the angular momentum.

An example of Noether's theorem is in nuclear physics where some nuclei have a spheroidal shape similar to an american football or a rugby ball. This spheroidal shape has an axis of symmetry along the long axis. The Lagrangian is rotationally invariant about the symmetry axis resulting in the angular momentum about the symmetry axis being conserved in addition to conservation of the total angular momentum.

7.5   Cyclic coordinates

Translational and rotational invariance occurs when a system has a cyclic coordinate q_k. *A cyclic coordinate is one that does not explicitly appear in the Lagrangian.* The term cyclic is a natural name when one has cylindrical or spherical symmetry. In Hamiltonian mechanics a cyclic coordinate often is called an *ignorable coordinate*. By virtue of Lagrange's equations

$$\frac{d}{dt}\frac{\partial L}{\partial \dot{q}_k} - \frac{\partial L}{\partial q_k} = 0 \tag{7.15}$$

then a cyclic coordinate q_k, is one for which $\frac{\partial L}{\partial q_k} = 0$. Thus

$$\frac{d}{dt}\frac{\partial L}{\partial \dot{q}_k} = \dot{p}_k = 0 \tag{7.16}$$

that is, p_k *is a constant of motion if the conjugate coordinate q_k is cyclic.* This is just Noether's Theorem.

7.6 Kinetic energy in generalized coordinates

Application of Noether's theorem to the conservation of energy requires the kinetic energy to be expressed in generalized coordinates. In terms of fixed rectangular coordinates, the kinetic energy for N bodies, each having three degrees of freedom, is expressed as

$$T = \frac{1}{2} \sum_{\alpha=1}^{N} \sum_{i=1}^{3} m_\alpha \dot{x}_{\alpha,i}^2 \tag{7.17}$$

These can be expressed in terms of generalized coordinates as $x_{\alpha,i} = x_{\alpha,i}(q_j, t)$ and in terms of generalized velocities

$$\dot{x}_{\alpha,i} = \sum_{j=1}^{s} \frac{\partial x_{\alpha,i}}{\partial q_j} \dot{q}_j + \frac{\partial x_{\alpha,i}}{\partial t} \tag{7.18}$$

Taking the square of $\dot{x}_{\alpha,i}$ and inserting into the kinetic energy relation gives

$$T(\mathbf{q}, \dot{\mathbf{q}}, t) = \sum_\alpha \sum_{i,j,k} \frac{1}{2} m_\alpha \frac{\partial x_{\alpha,i}}{\partial q_j} \frac{\partial x_{\alpha,i}}{\partial q_k} \dot{q}_j \dot{q}_k + \sum_\alpha \sum_{i,j} m_\alpha \frac{\partial x_{\alpha,i}}{\partial q_j} \frac{\partial x_{\alpha,i}}{\partial t} \dot{q}_j + \sum_\alpha \sum_i \frac{1}{2} m_\alpha \left(\frac{\partial x_{\alpha,i}}{\partial t} \right)^2 \tag{7.19}$$

This can be abbreviated as

$$T(\mathbf{q}, \dot{\mathbf{q}}, t) = T_2(\mathbf{q}, \dot{\mathbf{q}}, t) + T_1(\mathbf{q}, \dot{\mathbf{q}}, t) + T_0(\mathbf{q}, t) \tag{7.20}$$

where

$$T_2(\mathbf{q}, \dot{\mathbf{q}}, t) = \sum_\alpha \sum_{i,j,k} \frac{1}{2} m_\alpha \frac{\partial x_{\alpha,i}}{\partial q_j} \frac{\partial x_{\alpha,i}}{\partial q_k} \dot{q}_j \dot{q}_k = \sum_{j,k} a_{jk} \dot{q}_j \dot{q}_k \tag{7.21}$$

$$T_1(\mathbf{q}, \dot{\mathbf{q}}, t) = \sum_\alpha \sum_{i,j} m_\alpha \frac{\partial x_{\alpha,i}}{\partial q_j} \frac{\partial x_{\alpha,i}}{\partial t} \dot{q}_j = \sum_{j,k} b_j \dot{q}_j \tag{7.22}$$

$$T_0(\mathbf{q}, t) = \sum_\alpha \sum_i \frac{1}{2} m_\alpha \left(\frac{\partial x_{\alpha,i}}{\partial t} \right)^2 \tag{7.23}$$

where

$$a_{jk} \equiv \sum_{\alpha=1}^{n} \sum_{i,=1}^{3} \frac{1}{2} m_\alpha \frac{\partial x_{\alpha,i}}{\partial q_j} \frac{\partial x_{\alpha,i}}{\partial q_k} \tag{7.24}$$

When the transformed system is scleronomic, time does not appear explicitly in the transformation equations to generalized coordinates since $\frac{\partial x_{\alpha,i}}{\partial t} = 0$. Then $T_1 = T_0 = 0$, and the kinetic energy reduces to a homogeneous quadratic function of the generalized velocities

$$T(\mathbf{q}, \dot{\mathbf{q}}, t) = T_2(\mathbf{q}, \dot{\mathbf{q}}, t) \tag{7.25}$$

A useful relation can be derived by taking the differential of equation 7.21 with respect to $\dot{q}_l$. That is

$$\frac{\partial T_2(\mathbf{q}, \dot{\mathbf{q}}, t)}{\partial \dot{q}_l} = \sum_k a_{lk} \dot{q}_k + \sum_j a_{jl} \dot{q}_j \tag{7.26}$$

Multiply this by $\dot{q}_l$ and sum over l gives

$$\sum_l \dot{q}_l \frac{\partial T_2(\mathbf{q}, \dot{\mathbf{q}}, t)}{\partial \dot{q}_l} = \sum_{k,l} a_{lk} \dot{q}_k \dot{q}_l + \sum_{j,l} a_{jl} \dot{q}_j \dot{q}_l = 2 \sum_{j,k} a_{lk} \dot{q}_k \dot{q}_l = 2T_2$$

Similarly, the products of the generalized velocities $\dot{q}$, with the corresponding derivatives of T_1 and T_0 give

$$\sum_l \dot{q}_l \frac{\partial T_2}{\partial \dot{q}_l} = 2T_2 \tag{7.27}$$

$$\sum_l \dot{q}_l \frac{\partial T_1(\mathbf{q}, \dot{\mathbf{q}}, t)}{\partial \dot{q}_l} = T_1(\mathbf{q}, \dot{\mathbf{q}}, t) \tag{7.28}$$

$$\sum_l \dot{q}_l \frac{\partial T_0(\mathbf{q}, t)}{\partial \dot{q}_l} = 0 \tag{7.29}$$

Equation 7.25 gives that $T = T_2$ when the transformed system is scleronomic, i.e. $\frac{\partial x_{\alpha,i}}{\partial t} = 0$, and then the kinetic energy is a quadratic function of the generalized velocities $\dot{q}_j$. Using the definition of the generalized momentum equation 7.3, assuming $T = T_2$, and that the potential U is velocity independent, gives that

$$p_l \equiv \frac{\partial L}{\partial \dot{q}_l} = \frac{\partial T}{\partial \dot{q}_l} - \frac{\partial U}{\partial \dot{q}_l} = \frac{\partial T_2}{\partial \dot{q}_l} \tag{7.30}$$

Then equation 7.27 reduces to the useful relation that

$$T_2 = \frac{1}{2} \sum_l \dot{q}_l p_l = \frac{1}{2} \dot{\mathbf{q}} \cdot \mathbf{p} \tag{7.31}$$

where, for compactness, the summation is abbreviated as a scalar product.

7.7   Generalized energy and the Hamiltonian function

Consider the time derivative of the Lagrangian, plus the fact that time is the independent variable in the Lagrangian. Then the total time derivative is

$$\frac{dL}{dt} = \sum_j \frac{\partial L}{\partial q_j} \dot{q}_j + \sum_j \frac{\partial L}{\partial \dot{q}_j} \ddot{q}_j + \frac{\partial L}{\partial t} \tag{7.32}$$

The Lagrange equations for a conservative force are given by equation 6.60 to be

$$\frac{d}{dt} \frac{\partial L}{\partial \dot{q}_j} - \frac{\partial L}{\partial q_j} = Q_j^{EXC} + \sum_{k=1}^m \lambda_k \frac{\partial g_k}{\partial q_j}(\mathbf{q}, t) \tag{7.33}$$

The holonomic constraints can be accounted for using the Lagrange multiplier terms while the generalized force Q_j^{EXC} includes non-holonomic forces or other forces not included in the potential energy term of the Lagrangian, or holonomic forces not accounted for by the Lagrange multiplier terms.

Substituting equation 7.33 into equation 7.32 gives

$$\begin{aligned}
\frac{dL}{dt} &= \sum_j \dot{q}_j \frac{d}{dt} \frac{\partial L}{\partial \dot{q}_j} - \sum_j \dot{q}_j \left[Q_j^{EXC} + \sum_{k=1}^m \lambda_k \frac{\partial g_k}{\partial q_j}(\mathbf{q}, t) \right] + \sum_j \frac{\partial L}{\partial \dot{q}_j} \ddot{q}_j + \frac{\partial L}{\partial t} \\
&= \sum_j \frac{d}{dt} \left(\dot{q}_j \frac{\partial L}{\partial \dot{q}_j} \right) - \sum_j \dot{q}_j \left[Q_j^{EXC} + \sum_{k=1}^m \lambda_k \frac{\partial g_k}{\partial q_j}(\mathbf{q}, t) \right] + \frac{\partial L}{\partial t}
\end{aligned} \tag{7.34}$$

This can be written in the form

$$\frac{d}{dt} \left[\sum_j \left(\dot{q}_j \frac{\partial L}{\partial \dot{q}_j} \right) - L \right] = \sum_j \dot{q}_j \left[Q_j^{EXC} + \sum_{k=1}^m \lambda_k \frac{\partial g_k}{\partial q_j}(\mathbf{q}, t) \right] - \frac{\partial L}{\partial t} \tag{7.35}$$

Define Jacobi's **Generalized Energy**[1] $h(\mathbf{q}, \dot{\mathbf{q}}, t)$ by

$$h(\mathbf{q}, \dot{\mathbf{q}}, t) \equiv \sum_j \left(\dot{q}_j \frac{\partial L}{\partial \dot{q}_j} \right) - L(\mathbf{q}, \dot{\mathbf{q}}, t) \tag{7.36}$$

Jacobi's generalized momentum, equation 7.3, can be used to express the generalized energy $h(q, \dot{q}, t)$ in terms of the canonical coordinates $\dot{q}_i$ and p_i, plus time t. Define the **Hamiltonian function** to equal the generalized energy expressed in terms of the conjugate variables (q_j, p_j), that is,

$$H(\mathbf{q}, \mathbf{p}, t) \equiv h(\mathbf{q}, \dot{\mathbf{q}}, t) \equiv \sum_j \left(\dot{q}_j \frac{\partial L}{\partial \dot{q}_j} \right) - L(\mathbf{q}, \dot{\mathbf{q}}, t) = \sum_j (\dot{q}_j p_j) - L(\mathbf{q}, \dot{\mathbf{q}}, t) \tag{7.37}$$

This Hamiltonian $H(\mathbf{q}, \mathbf{p}, t)$ underlies Hamiltonian mechanics which plays a profoundly important role in most branches of physics as illustrated in chapters 8, 15 and 18.

---

[1] Most textbooks call the function $h(\mathbf{q}, \dot{\mathbf{q}}, t)$ *Jacobi's energy integral*. This book adopts the more descriptive name *Generalized energy* in analogy with use of generalized coordinates $\mathbf{q}$ and generalized momentum $\mathbf{p}$.

7.8 Generalized energy theorem

The Hamilton function, 7.37 plus equation 7.35 lead to the *generalized energy theorem*

$$\frac{dH\left(\mathbf{q},\mathbf{p},t\right)}{dt} = \frac{dh(\mathbf{q},\dot{\mathbf{q}},t)}{dt} = \sum_j \dot{q}_j \left[Q_j^{EXC} + \sum_{k=1}^m \lambda_k \frac{\partial g_k}{\partial q_j}(\mathbf{q},t)\right] - \frac{\partial L(\mathbf{q},\dot{\mathbf{q}},t)}{\partial t} \qquad (7.38)$$

Note that for the special case where all the external forces $\left[Q_j^{EXC} + \sum_{k=1}^m \lambda_k \frac{\partial g_k}{\partial q_j}(\mathbf{q},t)\right] = 0$, then

$$\frac{dH}{dt} = -\frac{\partial L}{\partial t} \qquad (7.39)$$

Thus the Hamiltonian is time independent if both $\left[Q_j^{EXC} + \sum_{k=1}^m \lambda_k \frac{\partial g_k}{\partial q_j}(\mathbf{q},t)\right] = 0$ and the Lagrangian are time-independent. For an isolated closed system having no external forces acting, then the Lagrangian is time independent because the velocities are constant, and there is no external potential energy. That is, the Lagrangian is time-independent, and

$$\frac{d}{dt}\left[\sum_j \left(\dot{q}_j \frac{\partial L}{\partial \dot{q}_j}\right) - L\right] = \frac{dH}{dt} = -\frac{\partial L}{\partial t} = 0 \qquad (7.40)$$

As a consequence, the Hamiltonian $H\left(\mathbf{q},\mathbf{p},t\right)$, and generalized energy $h(\mathbf{q},\dot{\mathbf{q}},t)$, both are constants of motion if the Lagrangian is a constant of motion, and if the external non-potential forces are zero. This is an example of Noether's theorem, where the symmetry of time independence leads to conservation of the conjugate variable, which is the Hamiltonian or Generalized energy.

7.9 Generalized energy and total energy

The generalized kinetic energy, equation 7.20, can be used to write the generalized Lagrangian as

$$L(\mathbf{q},\dot{\mathbf{q}},t) = T_2(\mathbf{q},\dot{\mathbf{q}},t) + T_1(\mathbf{q},\dot{\mathbf{q}},t) + T_0(\mathbf{q},t) - U(\mathbf{q},t) \qquad (7.41)$$

If the potential energy U does not depend explicitly on velocities $\dot{q}_i$ or time, then

$$p_j = \frac{\partial L}{\partial \dot{q}_j} = \frac{\partial (T-U)}{\partial \dot{q}_j} = \frac{\partial T}{\partial \dot{q}_j} \qquad (7.42)$$

Equation 7.42 can be used to write the **Hamiltonian**, equation 7.37, as

$$H\left(\mathbf{q},\mathbf{p},t\right) = \sum_i \left(\dot{q}_j \frac{\partial T_2}{\partial \dot{q}_j}\right) + \sum_i \left(\dot{q}_j \frac{\partial T_1}{\partial \dot{q}_j}\right) + \sum_i \left(\dot{q}_j \frac{\partial T_0}{\partial \dot{q}_j}\right) - L(\mathbf{q},\dot{\mathbf{q}},t) \qquad (7.43)$$

Using equations $7.27, 7.28, 7.29$ gives that the total generalized Hamiltonian $H\left(\mathbf{q},\mathbf{p},t\right)$ equals

$$H\left(\mathbf{q},\mathbf{p},t\right) = 2T_2 + T_1 - (T_2 + T_1 + T_0 - U) = T_2 - T_0 + U \qquad (7.44)$$

But the sum of the kinetic and potential energies equals the total energy. Thus equation 7.44 can be rewritten in the form

$$H\left(\mathbf{q},\mathbf{p},t\right) = (T+U) - (T_1 + 2T_0) = E - (T_1 + 2T_0) \qquad (7.45)$$

Note that Jacobi's generalized energy and the Hamiltonian do not equal the total energy E. However, in the special case where the transformation is scleronomic, then $T_1 = T_0 = 0$, and if the potential energy U does not depend explicitly of $\dot{q}_i$, then the generalized energy (Hamiltonian) equals the total energy, that is, $H = E$. Recognition of the relation between the Hamiltonian and the total energy facilitates determining the equations of motion.

7.10   Hamiltonian invariance

Chapters 7.8, 7.9 addressed two important and independent features of the Hamiltonian regarding: *a)* when H is conserved, and *b)* when H equals the total mechanical energy. These important results are summarized below with a discussion of the assumptions made in deriving the Hamiltonian, as well as the implications.

a) Conservation of generalized energy

The generalized energy theorem (7.38) was given as

$$\frac{dH\left(\mathbf{q},\mathbf{p},t\right)}{dt} = \frac{dh(\mathbf{q},\dot{\mathbf{q}},t)}{dt} = \sum_j \dot{q}_j \left[Q_j^{EXC} + \sum_{k=1}^m \lambda_k \frac{\partial g_k}{\partial q_j}(\mathbf{q},t) \right] - \frac{\partial L(\mathbf{q},\dot{\mathbf{q}},t)}{\partial t} \tag{7.46}$$

Note that when $\sum_j \dot{q}_j \left[Q_j^{EXC} + \sum_{k=1}^m \lambda_k \frac{\partial g_k}{\partial q_j}(\mathbf{q},t) \right] = 0$, then equation 7.46 reduces to

$$\frac{dH}{dt} = -\frac{\partial L}{\partial t} \tag{7.47}$$

Also, when $\sum_j \dot{q}_j \left[Q_j^{EXC} + \sum_{k=1}^m \lambda_k \frac{\partial g_k}{\partial q_j}(\mathbf{q},t) \right] = 0$, and if the Lagrangian is not an explicit function of time, then the Hamiltonian is a constant of motion. That is, H is conserved if, and only if, the Lagrangian, and consequently the Hamiltonian, are not explicit functions of time, and if the external forces are zero.

b) The generalized energy and total energy

If the following two requirements are satisfied
  1) The kinetic energy has a homogeneous quadratic dependence on the generalized velocities, that is, the transformation to generalized coordinates is independent of time, $\frac{\partial x_{\alpha,i}}{\partial t} = 0$.
  2) The potential energy is not velocity dependent, thus the terms $\frac{\partial U}{\partial \dot{q}_i} = 0$.
  Then equation 7.45 implies that the Hamiltonian equals the total mechanical energy, that is,

$$H = T + U = E \tag{7.48}$$

Expressed in words, the generalized energy (Hamiltonian) equals the total energy if the constraints are time independent and the potential energy is velocity independent. This is equivalent to stating that, *if the constraints, or generalized coordinates, for the system are time independent, then $H = E$.*
  The four combinations of the above two independent conditions, assuming that the external forces term in equation 7.46 is zero, are summarized in table 7.1.

Table 7.1: Hamiltonian and total energy

Hamiltonian	Constraints and coordinate transformation	
Time behavior	Time independent	Time dependent
$\frac{dH}{dt} = -\frac{\partial L}{\partial t} = 0$	H conserved, $H = E$	H conserved, $H \neq E$
$\frac{dH}{dt} = -\frac{\partial L}{\partial t} \neq 0$	H not conserved, $H = E$	H not conserved, $H \neq E$

Note the following general facts regarding the Lagrangian and the Hamiltonian.
  (1) the Lagrangian is indefinite with respect to addition of a constant to the scalar potential,
  (2) the Lagrangian is indefinite with respect to addition of a constant velocity,
  (3) there is no unique choice of generalized coordinates.
  (4) the Hamiltonian is a scalar function that is derived from the Lagrangian scalar function.
  (5) the generalized momentum is derived from the Lagrangian.
  These facts, plus the ability to recognize the conditions under which H is conserved, and when $H = E$, can greatly facilitate solving problems as shown by the following two examples.

7.5 *Example: Linear harmonic oscillator on a cart moving at constant velocity*

Consider a linear harmonic oscillator located on a cart that is moving with constant velocity v_0 in the x direction, as shown in the adjacent figure. Let the laboratory frame be the unprimed frame, and the cart frame be designated the primed frame. Assume that $x = x'$ at $t = 0$. Then

$$x' = x - v_0 t \qquad \dot{x}' = \dot{x} - v_0 \qquad \ddot{x}' = \ddot{x}$$

The harmonic oscillator will have a potential energy of

$$U = \frac{1}{2} k x'^2 = \frac{1}{2} k \left(x - v_0 t \right)^2$$

Laboratory frame: *The Lagrangian is*

$$L(x, \dot{x}, t) = \frac{m\dot{x}^2}{2} - \frac{1}{2} k \left(x - v_0 t \right)^2$$

Lagrange equation $\Lambda_x L = 0$ gives the equation of motion to be

$$m\ddot{x} = -k(x - v_0 t)$$

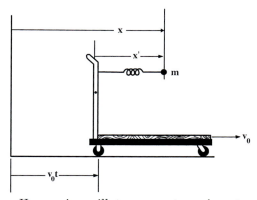

Harmonic oscillator on cart moving at uniform velocity v_0.

The definition of generalized momentum gives

$$p = \frac{\partial L}{\partial \dot{x}} = m\dot{x}$$

The Hamiltonian is

$$H(x, p, t) = \sum_i \dot{q}_i \frac{\partial L}{\partial \dot{q}_i} - L = \frac{p^2}{2m} + \frac{1}{2} k \left(x - v_0 t \right)^2$$

The Hamiltonian is the sum of the kinetic and potential energies and equals the total energy of the system, but it is not conserved since L and H are both explicit functions of time, that is $\frac{dH}{dt} = \frac{\partial H}{\partial t} = -\frac{\partial L}{\partial t} \neq 0$. Physically this is understood in that energy must flow into and out of the external constraint keeping the cart moving uniformly at a constant velocity v_0 against the reaction to the oscillating mass. That is, assuming a uniform velocity for the moving cart constitutes a time-dependent constraint on the mass, and the force of constraint does work in actual displacement of the complete system. If the constraint did not exist, then the cart momentum would oscillate such that the total momentum of cart plus spring system is conserved.

Cart frame: *Transform the Lagrangian to the primed coordinates in the moving frame of reference, which also is an inertial frame. Then the Lagrangian L, in terms of the moving cart frame coordinates, is*

$$L(x', \dot{x}', t) = \frac{m}{2} \left(\dot{x}'^2 + 2\dot{x}' v_0 + v_0^2 \right) - \frac{1}{2} k x'^2$$

The Lagrange equation of motion $\Lambda_{x'} L = 0$ gives the equation of motion to be

$$m\ddot{x}' = -k x'$$

where x' is the displacement of the mass with respect to the cart. This implies that an observer on the cart will observe simple harmonic motion as is to be expected from the principle of equivalence in Galilean relativity.

The definition of the generalized momentum gives the linear momentum in the primed frame coordinates to be

$$p' = \frac{\partial L}{\partial \dot{x}'} = m\dot{x}' + mv_0$$

The cart-frame Hamiltonian also can be expressed in terms of the coordinates in the moving frame to be

$$H(x', p', t) = \dot{x}' \frac{\partial L}{\partial \dot{x}'} - L = \frac{(p' - mv_0)^2}{2m} + \frac{1}{2} k x'^2 - \frac{m}{2} v_0^2$$

Note that the Lagrangian and Hamiltonian expressed in terms of the coordinates in the cart frame of reference are not explicitly time dependent, therefore H is conserved. However, the cart-frame Hamiltonian does not equal the total energy since the coordinate transformation is time dependent. Actually the first two terms in the above Hamiltonian are the energy of the harmonic oscillator in the cart frame. This example shows that the Hamiltonians differ when expressed in terms of either the laboratory or cart frames of reference

7.6 Example: Isotropic central force in a rotating frame

Consider a mass subject to a central isotropic radial force $U(r)$ as shown in the adjacent figure. Compare the Hamiltonian H in the fixed frame of reference S, with the Hamiltonian H' in a frame of reference S' that is rotating about the center of the force with constant angular velocity ω. Restrict this case to rotation about one axis so that only two polar coordinates r and ϕ need to be considered. The transformations are

$$\begin{aligned} r' &= r \\ \phi' &= \phi - \omega t \end{aligned}$$

Also

$$U(r) = U(r')$$

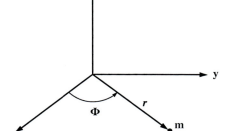

Mass subject to radial force

Fixed frame of reference S:

$$L = T - U = \frac{m}{2}\left(\dot{r}^2 + r^2\dot{\phi}^2\right) - U(r)$$

Since the Lagrangian is not explicitly time dependent, then the Hamiltonian is conserved. For this fixed-frame Hamiltonian the generalized momenta are

$$p_\phi = \frac{\partial L}{\partial \dot{\phi}} = m\dot{r}^2\dot{\phi}$$

$$p_r = \frac{\partial L}{\partial \dot{r}} = m\dot{r}$$

The Hamiltonian equals

$$H(p_r, p_\phi, r, \phi) = \sum_i \dot{q}_i \frac{\partial L}{\partial \dot{q}_i} - L = \frac{1}{2m}\left(p_r^2 + \frac{p_\phi^2}{r^2}\right) + U(r) = E$$

The Hamiltonian in the fixed frame is conserved and equals the total energy, that is $H = T + U$.

Rotating frame of reference S'

The above inertial fixed-frame Lagrangian can be written in terms of the primed (non-inertial rotating frame) coordinates as

$$L = T - U = \frac{m}{2}\left(\dot{r}^2 + r^2\dot{\phi}^2\right) - U(r) = \frac{m}{2}\left(\dot{r}'^2 + r'^2\left(\dot{\phi}' + \omega\right)^2\right) - U(r')$$

The generalized momenta derived from this Lagrangian are

$$p'_\phi = \frac{\partial L}{\partial \dot{\phi}'} = m\dot{r}'^2\left(\dot{\phi}' + \omega\right) = p'_{\phi'} + mr'^2\omega$$

$$p'_r = \frac{\partial L}{\partial \dot{r}'} = m\dot{r}'^2 = p_r$$

The Hamiltonian expressed in terms of the non-inertial rotating frame coordinates is

$$H'(p'_r, p'_\phi, r', \phi') = \frac{\partial L}{\partial \dot{r}'}\dot{r}' + \frac{\partial L}{\partial \dot{\phi}'}\dot{\phi}' - L = \frac{1}{2m}\left(p_r'^2 + \frac{\left(p'_{\phi'} + mr^2\omega\right)}{r^2}\right) + U(r')$$

Note that $H'(p'_r, p'_\phi, r', \phi')$ is time independent and therefore is conserved, but $H(p'_r, p'_\phi, r', \phi') \neq E$ because the generalized coordinates are time dependent. In addition, $p'_{\phi'}$ is conserved since

$$\dot{p}'_\phi = \frac{\partial H}{\partial \phi'} = -\frac{\partial L}{\partial \phi'} = 0$$

7.7 *Example: The plane pendulum*

The simple plane pendulum in a uniform gravitational field g is an example that illustrates Hamiltonian invariance. There is only one generalized coordinate, θ and the Lagrangian for this system is

$$L = \frac{1}{2}ml^2\dot{\theta}^2 + mgl\cos\theta$$

The momentum conjugate to θ is

$$p_\theta = \frac{\partial L}{\partial\dot{\theta}} = ml^2\dot{\theta}$$

which is the angular momentum about the pivot point. Using the Lagrange-Euler equation this gives that

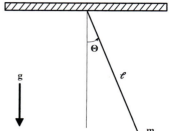

The plane pendulum constrained to oscillate in a vertical plane in a uniform gravitational field.

$$\frac{d}{dt}p_\theta = \dot{p}_\theta = \frac{\partial L}{\partial\theta} = -mgl\sin\theta$$

Note that the angular momentum p_θ is not a constant of motion since it explicitly depends on θ. The Hamiltonian is

$$H = \sum_i p_i\dot{q}_i - L = p_\theta\dot{\theta} - L = \frac{1}{2}ml^2\dot{\theta}^2 - mgl\cos\theta = \frac{p_\theta^2}{2ml^2} - mgl\cos\theta$$

Note that the Lagrangian and Hamiltonian are not explicit functions of time, therefore they are conserved. Also the potential is velocity independent and there is no coordinate transformation, thus the Hamiltonian equals the total energy E, which is a constant of motion.

$$H = \frac{p_\theta^2}{2ml^2} - mgl\cos\theta = E$$

7.8 *Example: Oscillating cylinder in a cylindrical bowl*

It is important to correctly account for constraint forces when using Noether's theorem for constrained systems. Noether's theorem assumes the variables are independent. This is illustrated by considering the example of a solid cylinder rolling in a fixed cylindrical bowl. Assume that a uniform cylinder of radius ρ and mass m is constrained to roll without slipping on the inner surface of the lower half of a hollow cylinder of radius R. The motion is constrained to ensure that the axes of both cylinders remain parallel and ρ < R.

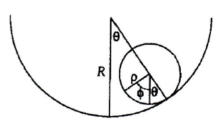

The generalized coordinates are taken to be the angles θ and φ which are measured with respect to a fixed vertical axis. Then the kinetic energy and potential energy are

$$T = \frac{1}{2}m\left[(R-\rho)\dot{\theta}\right]^2 + \frac{1}{2}I\dot{\phi}^2 \qquad\qquad U = [R - (R-\rho)\cos\theta]\,mg$$

where m is the mass of the small cylinder and where $U = 0$ at the lowest position of the sphere. The moment of inertia of a uniform cylinder is $I = \frac{1}{2}m\rho^2$.

The Lagrangian is

$$L - T - U = \frac{1}{2}m\left[(R-\rho)\dot{\theta}\right]^2 + \frac{1}{4}m\rho^2\dot{\phi}^2 - [R - (R-\rho)\cos\theta]\,mg$$

Since the solid cylinder rotates without slipping inside the cylindrical shell, then the equation of constraint is

$$g(\theta\phi) = R\theta - \rho(\phi + \theta) = 0$$

Using the Lagrangian, plus the one equation of constraint, requires one Lagrange multiplier. Then the Lagrange equations of motion for θ and φ are

$$\frac{\partial L}{\partial \theta} - \frac{d}{dt}\left[\frac{\partial L}{\partial \dot{\theta}}\right] + \lambda \frac{\partial g}{\partial \theta} = 0$$

$$\frac{\partial L}{\partial \phi} - \frac{d}{dt}\left[\frac{\partial L}{\partial \dot{\phi}}\right] + \lambda \frac{\partial g}{\partial \phi} = 0$$

Substitute the Lagrangian and the equation of constraint gives two equations of motion

$$-(R - \rho)\, mg \sin\theta - m\,(R-\rho)^2\, \ddot{\theta} + \lambda\,(R-\rho) = 0$$

$$-\frac{1}{2}m\rho^2 \ddot{\phi} - \lambda\rho = 0$$

The lower equation of motion gives that

$$\lambda = -\frac{1}{2}m\rho\ddot{\phi}$$

Substitute this into the equation of constraint gives

$$\lambda = -\frac{1}{2}m\,(R-\rho)\,\ddot{\theta}$$

Substitute this into the first equation of motion gives the equation of motion for θ to be

$$\ddot{\theta} = \frac{2g}{3\,(R-\rho)}\sin\theta$$

that is

$$\lambda = -\frac{mg}{3}\sin\theta$$

The torque acting on the small cylinder due to the frictional force is

$$F\rho = \frac{1}{2}m\rho^2 \ddot{\phi} = -\lambda\rho$$

Thus the frictional force is

$$F = -\lambda = \frac{mg}{3}\sin\theta$$

Noether's theorem can be used to ascertain if the angular momentum p_θ is a constant of motion. The derivative of the Lagrangian

$$\frac{\partial L}{\partial \theta} = (R-\rho)\,mg\sin\theta$$

and thus the Lagrange equations tells us that $\dot{p}_\theta = (R-\rho)\,mg\sin\theta$. Therefore p_θ is not a constant of motion.

The Lagrangian is not an explicit function of φ, which would suggest that p_ϕ is a constant of motion. But this is incorrect because the constraint equation $\phi = \frac{(R-\rho)}{\rho}\theta$ couples θ and φ, that is, they are not independent variables, and thus p_θ and p_ϕ are coupled by the constraint equation. As a result p_ϕ is not a constant of motion because it is directly coupled to $p_\theta = (R-\rho)\,mg\sin\theta$ which is not a constant of motion. Thus neither p_θ nor p_ϕ are constants of motion. This illustrates that one must account carefully for equations of constraint, and the concomitant constraint forces, when applying Noether's theorem which tacitly assumes independent variables.

The Hamiltonian can be derived using the generalized momenta

$$p_\theta = \frac{\partial L}{\partial \dot{\theta}} = m\,(R-\rho)^2\,\dot{\theta}$$

$$p_\phi = \frac{\partial L}{\partial \dot{\phi}} = \frac{1}{2}m\rho^2\,\dot{\phi}$$

Then the Hamiltonian is given by

$$H = p_\theta\dot{\theta} + p_\phi\dot{\phi} - L = \frac{p_\theta^2}{2m\,(R-\rho)^2} + \frac{p_\phi^2}{m\rho^2} + [R - (R-\rho)\cos\theta]\,mg$$

Note that the transformation to generalized coordinates is time independent and the potential is not velocity dependent, thus the Hamiltonian also equals the total energy. Also the Hamiltonian is conserved since $\frac{dH}{dt} = 0$.

7.11 Hamiltonian for cyclic coordinates

It is interesting to discuss the properties of the Hamiltonian for cyclic coordinates q_k for which $\frac{\partial L}{\partial q_k} = 0$. Ignoring the external and Lagrange multiplier terms,

$$\dot{p}_k = \frac{\partial L}{\partial q_k} = -\frac{\partial H}{\partial q_k} = 0 \tag{7.49}$$

That is, a cyclic coordinate has a constant corresponding momentum p_k for the Hamiltonian as well as for the Lagrangian. Conversely, if a generalized coordinate does not occur in the Hamiltonian, then the corresponding generalized momentum is conserved. Cyclic coordinates were discussed earlier when discussing symmetries and conservation-law aspects of the Lagrangian. For example, if the Lagrangian, or Hamiltonian do not depend on a linear coordinate x, then p_x is conserved. Similarly for θ and p_θ. An extension of this principle has been derived for the relationship between time independence and total energy of a system, that is, the Hamiltonian equals the total energy if the transformation to generalized coordinates is time independent and the potential is velocity independent.

A valuable feature of the Hamiltonian formulation is that it allows elimination of cyclic variables which reduces the number of degrees of freedom to be handled. As a consequence, cyclic variables are called *ignorable* variables in Hamiltonian mechanics. For example, consider that the Lagrangian has one cyclic variable q_n. As a consequence, the Lagrangian does not depend on q_n, and thus it can be written as $L = L(q_1, ..., q_{n-1}; \dot{q}_1, ..., \dot{q}_n; t)$. The Lagrangian still contains n generalized velocities, thus one still has to treat n degrees of freedom even though one degree of freedom q_n is cyclic. However, in the Hamiltonian formulation, only $n-1$ degrees of freedom are required since the momentum for the cyclic degree of freedom is a constant $p_n = \alpha$. Thus the Hamiltonian can be written as $H = H(q_1, ..., q_{n-1}; p_1, ..., p_{n-1}; \alpha; t)$, that is, the Hamiltonian includes only $n-1$ degrees of freedom. Thus the dimension of the problem has been reduced by one since the conjugate cyclic (ignorable) variables (q_n, p_n) are eliminated. Hamiltonian mechanics can significantly reduce the dimension of the problem when the system involves several cyclic variables. This is in contrast to the situation for the Lagrangian approach as discussed in chapters 8 and 15.

7.12 Symmetries and invariance

This chapter has shown that the *symmetries* of a system lead to *invariance* of physical quantities as was proposed by Noether. The symmetry properties of the Lagrangian can lead to the conservation laws summarized in table 7.2.

Table 7.2: Symmetries and conservation laws in classical mechanics

Symmetry	Lagrange property	Conserved quantity
Spatial invariance	Translational invariance	Linear momentum
Spatial homogeneous	Rotational invariance	Angular momentum
Time invariance	Time independence	Total energy

The importance of the relations between invariance and symmetry cannot be overemphasized. It extends beyond classical mechanics to quantum physics and field theory. For a three-dimensional closed system, there are three possible constants for linear momentum, three for angular momentum, and one for energy. It is especially interesting in that these, and only these, seven integrals have the property that they are *additive* for the particles comprising a system, and this occurs independent of whether there is an interaction among the particles. That is, this behavior is obeyed by the whole assemble of particles for finite systems. Because of its profound importance to physics, these relations between symmetry and invariance are used extensively.

7.13 Hamiltonian in classical mechanics

The Hamiltonian was defined by equation 7.37 during the discussion of time invariance and energy conservation. The Hamiltonian is of much more profound importance to physics than implied by the ad hoc definition given by equation 7.37. This relates to the fact that the Hamiltonian is written in terms of the fundamental coordinate q_i and its generalized momentum p_i defined by equation 7.3.

It is more convenient to write the n generalized coordinates q_i, plus their generalized momentum p_i, as vectors, e.g. $\mathbf{q} \equiv (q_1, q_2, ..q_n)$, $\mathbf{p} \equiv (p_1, p_2, ..p_n)$. The generalized momenta conjugate to the coordinate q_i, defined by 7.3, then can be written in the form

$$p_i = \frac{\partial L(\mathbf{q}, \dot{\mathbf{q}}, \mathbf{t})}{\partial \dot{q}_i} \tag{7.50}$$

Substituting this definition of the generalized momentum into the Hamiltonian defined in (7.37), and expressing it in terms of the coordinate $\mathbf{q}$ and its conjugate generalized momenta $\mathbf{p}$, leads to

$$H(\mathbf{q}, \mathbf{p}, t) = \sum_i p_i \dot{q}_i - L(\mathbf{q}, \dot{\mathbf{q}}, t) \tag{7.51}$$

$$= \mathbf{p} \cdot \dot{\mathbf{q}} - L(\mathbf{q}, \dot{\mathbf{q}}, t) \tag{7.52}$$

Note that the scalar product $\mathbf{p} \cdot \dot{\mathbf{q}} = \sum_i p_i \dot{q}_i$ equals $2T$ for systems that are scleronomic and when the potential is velocity independent.

The crucial feature of the Hamiltonian is that it is expressed as $H(\mathbf{q}, \mathbf{p}, t)$, that is, it is a function of the n generalized coordinates $\mathbf{q}$ and their conjugate momenta $\mathbf{p}$, *which are taken to be independent*, in addition to the independent variable, t. This is in contrast to the Lagrangian $L(\mathbf{q}, \dot{\mathbf{q}}, t)$ which is a function of the n generalized coordinates q_j, the corresponding velocities $\dot{q}_j$, and time t. The velocities $\dot{\mathbf{q}}$ are the time derivatives of the coordinates $\mathbf{q}$ and thus these are related. In physics, the fundamental conjugate coordinates are $(\mathbf{q}, \mathbf{p})$, which are the coordinates underlying the Hamiltonian. This is in contrast to $(\mathbf{q}, \dot{\mathbf{q}})$ which are the coordinates that underlie the Lagrangian. Thus the Hamiltonian is more fundamental than the Lagrangian and is a reason why the Hamiltonian mechanics, rather than the Lagrangian mechanics, was used as the foundation for development of quantum and statistical mechanics.

Hamiltonian mechanics will be derived two other ways. Chapter 8 uses the Legendre transformation between the conjugate variables $(\mathbf{q}, \dot{\mathbf{q}}, t)$ and $(\mathbf{q}, \mathbf{p}, t)$ where the generalized coordinate $\mathbf{q}$ and its conjugate generalized momentum, $\mathbf{p}$ are *independent*. This shows that Hamiltonian mechanics is based on the same variational principles as those used to derive Lagrangian mechanics. Chapter 9 derives Hamiltonian mechanics directly from Hamilton's Principle of Least action. Chapter 8 will introduce the algebraic Hamiltonian mechanics, that is based on the Hamiltonian. The powerful capabilities provided by Hamiltonian mechanics will be described in chapter 15.

7.14   Summary

This chapter has explored the importance of symmetries and invariance in Lagrangian mechanics and has introduced the Hamiltonian. The following summarizes the important conclusions derived in this chapter.

Noether's theorem:

Noether's theorem explores the remarkable connection between symmetry, plus the invariance of a system under transformation, and related conservation laws which imply the existence of important physical principles, and constants of motion. Transformations where the equations of motion are invariant are called *invariant transformations*. Variables that are invariant to a transformation are called cyclic variables. It was shown that if the Lagrangian does not explicitly contain a particular coordinate of displacement, q_i then the corresponding conjugate momentum, $\dot{p}_i$ is conserved. This is Noether's theorem which states *"For each symmetry of the Lagrangian, there is a conserved quantity"*. In particular it was shown that translational invariance in a given direction leads to the conservation of linear momentum in that direction, and rotational invariance about an axis leads to conservation of angular momentum about that axis. These are the first-order spatial and angular integrals of the equations of motion. Noether's theorem also relates the properties of the Hamiltonian to time invariance of the Lagrangian, namely;

(1) H is conserved if, and only if, the Lagrangian, and consequently the Hamiltonian, are not explicit functions of time.

(2) The Hamiltonian gives the total energy if the constraints and coordinate transformations are time independent and the potential energy is velocity independent. This is equivalent to stating that $H = E$ *if the constraints, or generalized coordinates, for the system are time independent.*

Noether's theorem is of importance since it underlies the relation between symmetries, and invariance in all of physics; that is, its applicability extends beyond classical mechanics.

Generalized momentum:

The generalized momentum associated with the coordinate q_j is defined to be

$$\frac{\partial L}{\partial \dot{q}_j} \equiv p_j \tag{7.3}$$

where p_j is also called the **conjugate momentum** (or **canonical momentum**) to q_j where q_j, p_j are conjugate, or canonical, variables. Remember that the linear momentum p_j is the first-order time integral given by equation 2.10. Note that if q_j is not a spatial coordinate, then p_j is not linear momentum, but is the conjugate momentum. For example, if q_j is an angle, then p_j will be angular momentum.

Kinetic energy in generalized coordinates:

It was shown that the kinetic energy can be expressed in terms of generalized coordinates by

$$
\begin{aligned}
T(\mathbf{q}, \dot{\mathbf{q}}, t) &= \sum_\alpha \sum_{i,j,k} \frac{1}{2} m_\alpha \frac{\partial x_{\alpha,i}}{\partial q_j} \frac{\partial x_{\alpha,i}}{\partial q_k} \dot{q}_j \dot{q}_k + \sum_\alpha \sum_{i,j} m_\alpha \frac{\partial x_{\alpha,i}}{\partial q_j} \frac{\partial x_{\alpha,i}}{\partial t} \dot{q}_j + \sum_\alpha \sum_i \frac{1}{2} m_\alpha \left(\frac{\partial x_{\alpha,i}}{\partial t} \right)^2 \tag{7.19} \\
&= T_2(\mathbf{q}, \dot{\mathbf{q}}, t) + T_1(\mathbf{q}, \dot{\mathbf{q}}, t) + T_0(\mathbf{q}, t) \tag{7.53}
\end{aligned}
$$

For scleronomic systems with a potential that is velocity independent, then the kinetic energy can be expressed as

$$T = T_2 = \frac{1}{2} \sum_l \dot{q}_l p_l = \frac{1}{2} \dot{\mathbf{q}} \cdot \mathbf{p} \tag{7.31}$$

Generalized energy

Jacobi's **Generalized Energy** $h(\mathbf{q}, \dot{q}, t)$ was defined as

$$h(\mathbf{q}, \dot{\mathbf{q}}, t) \equiv \sum_j \left(\dot{q}_j \frac{\partial L}{\partial \dot{q}_j} \right) - L(\mathbf{q}, \dot{\mathbf{q}}, t) \tag{7.36}$$

Hamiltonian function

The Hamiltonian $H(\mathbf{q}, \mathbf{p}, t)$ was defined in terms of the generalized energy $h(\mathbf{q}, \dot{\mathbf{q}}, t)$ and by introducing the generalized momentum. That is

$$H(\mathbf{q}, \mathbf{p}, t) \equiv h(\mathbf{q}, \dot{\mathbf{q}}, t) = \sum_j p_j \dot{q}_j - L(\mathbf{q}, \dot{\mathbf{q}}, t) = \mathbf{p} \cdot \dot{\mathbf{q}} - L(\mathbf{q}, \dot{\mathbf{q}}, t) \tag{7.37}$$

Generalized energy theorem

The equations of motion lead to the generalized energy theorem which states that the time dependence of the Hamiltonian is related to the time dependence of the Lagrangian.

$$\frac{dH(\mathbf{q}, \mathbf{p}, t)}{dt} = \sum_j \dot{q}_j \left[Q_j^{EXC} + \sum_{k=1}^m \lambda_k \frac{\partial g_k}{\partial q_j}(\mathbf{q}, t) \right] - \frac{\partial L(\mathbf{q}, \dot{\mathbf{q}}, t)}{\partial t} \tag{7.38}$$

Note that if all the generalized non-potential forces are zero, then the bracket in equation 7.38 is zero, and if the Lagrangian is not an explicit function of time, then the Hamiltonian is a constant of motion.

Generalized energy and total energy:

The generalized energy, and corresponding Hamiltonian, equal the total energy if:

1) The kinetic energy has a homogeneous quadratic dependence on the generalized velocities and the transformation to generalized coordinates is independent of time, $\frac{\partial x_{\alpha,i}}{\partial t} = 0$.

2) The potential energy is not velocity dependent, thus the terms $\frac{\partial U}{\partial \dot{q}_i} = 0$.

Chapter 8 will introduce Hamiltonian mechanics that is built on the Hamiltonian, and chapter 15 will explore applications of Hamiltonian mechanics.

Workshop exercises

1. Consider a particle of mass m moving in a plane and subject to an inverse square attractive force.

    (a) Obtain the equations of motion.

    (b) Is the angular momentum about the origin conserved?

    (c) Obtain expressions for the generalized forces.

2. Consider a Lagrangian function of the form $L(q_i, \dot{q}_i, \ddot{q}_i, t)$. Here the Lagrangian contains a time derivative of the generalized coordinates that is higher than the first. When working with such Lagrangians, the term "generalized mechanics" is used.

    (a) Consider a system with one degree of freedom. By applying the methods of the calculus of variations, and assuming that Hamilton's principle holds with respect to variations which keep both q and $\dot{q}$ fixed at the end points, show that the corresponding Lagrange equation is

    $$\frac{d^2}{dt^2}\left(\frac{\partial L}{\partial \ddot{q}}\right) - \frac{d}{dt}\left(\frac{\partial L}{\partial \dot{q}}\right) + \frac{\partial L}{\partial q} = 0.$$

    Such equations of motion have interesting applications in chaos theory.

    (b) Apply this result to the Lagrangian

    $$L = -\frac{m}{2}q\ddot{q} - \frac{k}{2}q^2.$$

    Do you recognize the equations of motion?

3. A uniform solid cylinder of radius R and mass M rests on a horizontal plane and an identical cylinder rests on it touching along the top of the first cylinder with the axes of both cylinders parallel. The upper cylinder is given an infinitessimal displacement so that both cylinders roll without slipping in the directions shown by the arrows.

    (a) Find Lagrangian for this system

    (b) What are the constants of motion?

    (c) Show that as long as the cylinders remain in contact then

    $$\dot{\theta}^2 = \frac{12g(1 - \cos\theta)}{R(17 + 4\cos\theta - 4\cos^2\theta)}$$

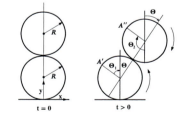

4. Consider a diatomic molecule which has a symmetry axis along the line through the center of the two atoms comprising the molecule. Consider that this molecule is rotating about an axis perpendicular to the symmetry axis and that there are no external forces acting on the molecule. Use Noether's Theorem to answer the following questions:

    a) Is the total angular momentum conserved?

    b) Is the projection of the total angular momentum along a space-fixed z axis conserved?

    c) Is the projection of the angular momentum along the symmetry axis of the rotating molecule conserved?

    d) Is the projection of the angular momentum perpendicular to the rotating symmetry axis conserved?

5. A bead of mass m slides under gravity along a smooth wire bent in the shape of a parabola $x^2 = az$ in the vertical (x, z) plane.

   (a) What kind (holonomic, nonholonomic, scleronomic, rheonomic) of constraint acts on m?
   (b) Set up Lagrange's equation of motion for x with the constraint embedded.
   (c) Set up Lagrange's equations of motion for both x and z with the constraint adjoined and a Lagrangian multiplier λ introduced.
   (d) Show that the same equation of motion for x results from either of the methods used in part (b) or part (c).
   (e) Express λ in terms of x and $\dot{x}$.
   (f) What are the x and z components of the force of constraint in terms of x and $\dot{x}$?

Problems

1. Let the horizontal plane be the $x - y$ plane. A bead of mass m is constrained to slide with speed v along a curve described by the function $y = f(x)$. What force does the curve apply to the bead? (Ignore gravity)

2. Consider the Atwoods machine shown. The masses are $4m$, $5m$, and $3m$. Let x and y be the heights of the right two masses relative to their initial positions.

   a) Solve this problem using the Euler-Lagrange equations
   b) Use Noether's theorem to find the conserved momentum.

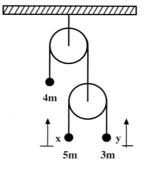

3. A cube of side $2b$ and center of mass C, is placed on a fixed horizontal cylinder of radius r and center O as shown in the figure. Originally the cube is placed such that C is centered above O but it can roll from side to side without slipping. (a) Assuming that $b < r$ use the Lagrangian approach to to find the frequency for small oscillations about the top of the cylinder. For simplicity make the small angle approximation for L before using the Lagrange-Euler equations. (b) What will be the motion if $b > r$? Note that the moment of inertia of the cube about the center of mass is $\frac{2}{3}mb^2$.

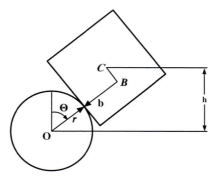

4. Two equal masses of mass m are glued to a massless hoop of radius R is free to rotate about its center in a vertical plane. The angle between the masses is 2θ, as shown. Find the frequency of oscillations.

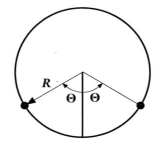

5. Three massless sticks each of length $2r$, and mass m with the center of mass at the center of each stick, are hinged at their ends as shown. The bottom end of the lower stick is hinged at the ground. They are held so that the lower two sticks are vertical, and the upper one is tilted at a small angle ε with respect to the vertical. They are then released. At the instant of release what are the three equations of motion derived from the Lagrangian derived assuming that ε is small? Use these to determine the initial angular accelerations of the three sticks.

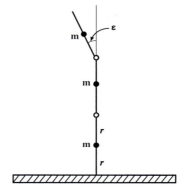

Chapter 8

Hamiltonian mechanics

8.1 Introduction

The three major formulations of classical mechanics are

1. **Newtonian mechanics** which is the most intuitive vector formulation used in classical mechanics.

2. **Lagrangian mechanics** is a powerful algebraic formulation of classical mechanics derived using either d'Alembert's Principle, or Hamilton's Principle. The latter states "*A dynamical system follows a path that minimizes the time integral of the difference between the kinetic and potential energies*".

3. **Hamiltonian mechanics** has a beautiful superstructure that, like Lagrangian mechanics, is built upon variational calculus, Hamilton's principle, and Lagrangian mechanics.

Hamiltonian mechanics is introduced at this juncture since it is closely interwoven with Lagrange mechanics. Hamiltonian mechanics plays a fundamental role in modern physics, but the discussion of the important role it plays in modern physics will be deferred until chapters 15 and 18 where applications to modern physics are addressed.

The following important concepts were introduced in chapter 7:

The **generalized momentum** was defined to be given by

$$p_i \equiv \frac{\partial L(\mathbf{q}, \dot{\mathbf{q}}, t)}{\partial \dot{q}_i} \tag{8.1}$$

Note that, as discussed in chapter 7.2, if the potential is velocity dependent, such as the Lorentz force, then the generalized momentum includes terms in addition to the usual mechanical momentum.

Jacobi's **generalized energy function** $h(\mathbf{q}, \dot{\mathbf{q}}, t)$ was introduced where

$$h(\mathbf{q}, \dot{\mathbf{q}}, t) = \sum_i^n \left(\dot{q}_i \frac{\partial L}{\partial \dot{q}_i} \right) - L(\mathbf{q}, \dot{\mathbf{q}}, t) \tag{8.2}$$

The **Hamiltonian function** was defined to be given by expressing the generalized energy function, equation 8.2, in terms of the generalized momentum. That is, the Hamiltonian $H(\mathbf{q}, \mathbf{p}, t)$ is expressed as

$$H(\mathbf{q}, \mathbf{p}, t) = \sum_i^n p_i \dot{q}_i - L(\mathbf{q}, \dot{\mathbf{q}}, t) \tag{8.3}$$

The symbols $\mathbf{q}$, $\mathbf{p}$, designate vectors of n generalized coordinates, $\mathbf{q} \equiv (q_1, q_2, ..q_n)$, $\mathbf{p} \equiv (p_1, p_2, ..p_n)$. Equation 8.3 can be written compactly in a symmetric form using the scalar product $\mathbf{p} \cdot \dot{\mathbf{q}} = \sum_i p_i \dot{q}_i$.

$$H(\mathbf{q}, \mathbf{p}, t) + L(\mathbf{q}, \dot{\mathbf{q}}, t) = \mathbf{p} \cdot \dot{\mathbf{q}} \tag{8.4}$$

A crucial feature of Hamiltonian mechanics is that the Hamiltonian is expressed as $H(\mathbf{q}, \mathbf{p}, t)$, that is, *it is a function of the n generalized coordinates and their conjugate momenta, which are taken to be independent*, plus the independent variable, time. This contrasts with the Lagrangian $L(\mathbf{q}, \dot{\mathbf{q}}, t)$ which is a function of the n generalized coordinates q_j, and the corresponding velocities $\dot{q}_j$, that is the time derivatives of the coordinates q_i, plus the independent variable, time.

8.2   Legendre Transformation between Lagrangian and Hamiltonian mechanics

Hamiltonian mechanics can be derived directly from Lagrange mechanics by considering the Legendre transformation between the conjugate variables $(\mathbf{q}, \dot{\mathbf{q}}, t)$ and $(\mathbf{q}, \mathbf{p}, t)$. Such a derivation is of considerable importance in that it shows that Hamiltonian mechanics is based on the same variational principles as those used to derive Lagrangian mechanics; that is d'Alembert's Principle and Hamilton's Principle. The general problem of converting Lagrange's equations into the Hamiltonian form hinges on the inversion of equation (8.1) that defines the generalized momentum $\mathbf{p}$. This inversion is simplified by the fact that (8.1) is the first partial derivative of the Lagrangian scalar function $L(\mathbf{q}, \dot{\mathbf{q}}, \mathbf{t})$.

As described in appendix $F4$, consider transformations between two functions $F(\mathbf{u}, \mathbf{w})$ and $G(\mathbf{v}, \mathbf{w})$, where $\mathbf{u}$ and $\mathbf{v}$ are the active variables related by the functional form

$$\mathbf{v} = \boldsymbol{\nabla}_{\mathbf{u}} F(\mathbf{u}, \mathbf{w}) \tag{8.5}$$

and where $\mathbf{w}$ designates passive variables. The function $\boldsymbol{\nabla}_{\mathbf{u}} F(\mathbf{u}, \mathbf{w})$ is the first-order derivative, (gradient) of $F(\mathbf{u}, \mathbf{w})$ with respect to the components of the vector $\mathbf{u}$. The Legendre transform states that the inverse formula can always be written as a first-order derivative

$$\mathbf{u} = \boldsymbol{\nabla}_{\mathbf{v}} G(\mathbf{v}, \mathbf{w}) \tag{8.6}$$

The function $G(\mathbf{v}, \mathbf{w})$ is related to $F(\mathbf{u}, \mathbf{w})$ by the symmetric relation

$$G(\mathbf{v}, \mathbf{w}) + F(\mathbf{u}, \mathbf{w}) = \mathbf{u} \cdot \mathbf{v} \tag{8.7}$$

where the scalar product $\mathbf{u} \cdot \mathbf{v} = \sum_{i=1}^{N} u_i v_i$.

Furthermore the first-order derivatives with respect to all the passive variables w_i are related by

$$\boldsymbol{\nabla}_{\mathbf{w}} F(\mathbf{u}, \mathbf{w}) = -\boldsymbol{\nabla}_{\mathbf{w}} G(\mathbf{v}, \mathbf{w}) \tag{8.8}$$

The relationship between the functions $F(\mathbf{u}, \mathbf{w})$ and $G(\mathbf{v}, \mathbf{w})$ is symmetrical and each is said to be the Legendre transform of the other.

The general Legendre transform can be used to relate the Lagrangian and Hamiltonian by identifying the active variables $\mathbf{v}$ with $\mathbf{p}$, and $\mathbf{u}$ with $\dot{\mathbf{q}}$, the passive variable $\mathbf{w}$ with $\mathbf{q}, t$, and the corresponding functions $F(\mathbf{u}, \mathbf{w}) = L(\mathbf{q}, \dot{\mathbf{q}}, t)$ and $G(\mathbf{v}, \mathbf{w}) = H(\mathbf{q}, \mathbf{p}, t)$. Thus the generalized momentum (8.1) corresponds to

$$\mathbf{p} = \boldsymbol{\nabla}_{\dot{\mathbf{q}}} L(\mathbf{q}, \dot{\mathbf{q}}, t) \tag{8.9}$$

where $(\mathbf{q}, t)$ are the passive variables. Then the Legendre transform states that the transformed variable $\dot{\mathbf{q}}$ is given by the relation

$$\dot{\mathbf{q}} = \boldsymbol{\nabla}_{\mathbf{p}} H(\mathbf{q}, \mathbf{p}, t) \tag{8.10}$$

Since the functions $L(\mathbf{q}, \dot{\mathbf{q}}, t)$ and $H(\mathbf{q}, \mathbf{p}, t)$ are the Legendre transforms of each other, they satisfy the relation

$$H(\mathbf{q}, \mathbf{p}, t) + L(\mathbf{q}, \dot{\mathbf{q}}, t) = \mathbf{p} \cdot \dot{\mathbf{q}} \tag{8.11}$$

The function $H(\mathbf{q}, \mathbf{p}, t)$, which is the Legendre transform of the Lagrangian $L(\mathbf{q}, \dot{\mathbf{q}}, t)$, is called the **Hamiltonian function** and equation (8.11) is identical to our original definition of the Hamiltonian given by equation (8.3). The variables $\mathbf{q}$ and t are passive variables thus equation (8.8) gives that

$$\boldsymbol{\nabla}_{\mathbf{q}} L(\dot{\mathbf{q}}, \mathbf{q}, t) = -\boldsymbol{\nabla}_{\mathbf{q}} H(\mathbf{p}, \mathbf{q}, t) \tag{8.12}$$

Written in component form equation 8.12 gives the partial derivative relations

$$\frac{\partial L(\dot{\mathbf{q}}, \mathbf{q}, t)}{\partial q_i} = -\frac{\partial H(\mathbf{p}, \mathbf{q}, t)}{\partial q_i} \tag{8.13}$$

$$\frac{\partial L(\dot{\mathbf{q}}, \mathbf{q}, t)}{\partial t} = -\frac{\partial H(\mathbf{p}, \mathbf{q}, t)}{\partial t} \tag{8.14}$$

Note that equations 8.13 and 8.14 are strictly a result of the Legendre transformation. To complete the transformation from Lagrangian to Hamiltonian mechanics it is necessary to invoke the calculus of variations via the Lagrange-Euler equations. The symmetry of the Legendre transform is illustrated by equation 8.11.

Equation 7.31 gives that the scalar product $\mathbf{p} \cdot \dot{\mathbf{q}} = 2T_2$. For scleronomic systems, with velocity independent potentials U, the standard Lagrangian $L = T - U$ and $H = 2T - T + U = T + U$. Thus, for this simple case, equation 8.11 reduces to an identity $H + L = 2T$.

8.3 Hamilton's equations of motion

The explicit form of the Legendre transform 8.10 gives that the time derivative of the generalized coordinate q_j is

$$\dot{q}_j = \frac{\partial H(\mathbf{q}, \mathbf{p}, t)}{\partial p_j} \qquad (8.15)$$

The Euler-Lagrange equation 6.60 is

$$\frac{d}{dt}\frac{\partial L}{\partial \dot{q}_j} - \frac{\partial L}{\partial q_j} = \sum_{k=1}^{m} \lambda_k \frac{\partial g_k}{\partial q_j} + Q_j^{EXC} \qquad (8.16)$$

This gives the corresponding Hamilton equation for the time derivative of p_i to be

$$\frac{d}{dt}\frac{\partial L}{\partial \dot{q}_j} = \dot{p}_j = \frac{\partial L}{\partial q_j} + \sum_{k=1}^{m} \lambda_k \frac{\partial g_k}{\partial q_j} + Q_j^{EXC} \qquad (8.17)$$

Substitute equation 8.13 into equation 8.17 leads to the second Hamilton equation of motion

$$\dot{p}_j = -\frac{\partial H(\mathbf{q}, \mathbf{p}, t)}{\partial q_j} + \sum_{k=1}^{m} \lambda_k \frac{\partial g_k}{\partial q_j} + Q_j^{EXC} \qquad (8.18)$$

One can explore further the implications of Hamiltonian mechanics by taking the time differential of (8.3) giving.

$$\frac{dH(\mathbf{q}, \mathbf{p}, t)}{dt} = \sum_j \left(\dot{q}_j \frac{dp_j}{dt} + p_j \frac{d\dot{q}_j}{dt} - \frac{\partial L}{\partial q_j}\frac{dq_j}{dt} - \frac{\partial L}{\partial \dot{q}_j}\frac{d\dot{q}_j}{dt} \right) - \frac{\partial L}{\partial t} \qquad (8.19)$$

Inserting the conjugate momenta $p_i \equiv \frac{\partial L}{\partial \dot{q}_i}$ and equation 8.17 into equation 8.19 results in

$$\frac{dH(\mathbf{q}, \mathbf{p}, t)}{dt} = \sum_j \left(\dot{q}_j \dot{p}_j + p_j \frac{d\dot{q}_j}{dt} - \left[\dot{p}_j - \sum_{k=1}^{m} \lambda_k \frac{\partial g_k}{\partial q_j} - Q_j^{EXC} \right] \dot{q}_j - p_j \frac{d\dot{q}_j}{dt} \right) - \frac{\partial L}{\partial t} \qquad (8.20)$$

The second and fourth terms cancel as well as the $\dot{q}_j \dot{p}_j$ terms, leaving

$$\frac{dH(\mathbf{q}, \mathbf{p}, t)}{dt} = \sum_j \left(\left[\sum_{k=1}^{m} \lambda_k \frac{\partial g_k}{\partial q_j} + Q_j^{EXC} \right] \dot{q}_j \right) - \frac{\partial L}{\partial t} \qquad (8.21)$$

This is the **generalized energy theorem** given by equation 7.38.

The total differential of the Hamiltonian also can be written as

$$\frac{dH(\mathbf{q}, \mathbf{p}, t)}{dt} = \sum_j \left(\frac{\partial H}{\partial p_j}\dot{p}_j + \frac{\partial H}{\partial q_j}\dot{q}_j \right) + \frac{\partial H}{\partial t} \qquad (8.22)$$

Use equations 8.15 and 8.18 to substitute for $\frac{\partial H}{\partial p_j}$ and $\frac{\partial H}{\partial q_j}$ in equation 8.22 gives

$$\frac{dH(\mathbf{q}, \mathbf{p}, t)}{dt} = \sum_j \left(\left[\sum_{k=1}^{m} \lambda_k \frac{\partial g_k}{\partial q_j} + Q_j^{EXC} \right] \dot{q}_j \right) + \frac{\partial H(\mathbf{q}, \mathbf{p}, t)}{\partial t} \qquad (8.23)$$

Note that equation 8.23 must equal the generalized energy theorem, i.e. equation 8.21. Therefore,

$$\frac{\partial H}{\partial t} = -\frac{\partial L}{\partial t} \tag{8.24}$$

In summary, **Hamilton's equations of motion** are given by

$$\dot{q}_j = \frac{\partial H(\mathbf{q}, \mathbf{p}, t)}{\partial p_j} \tag{8.25}$$

$$\dot{p}_j = -\frac{\partial H(\mathbf{q}, \mathbf{p}, t)}{\partial q_j} + \left[\sum_{k=1}^{m} \lambda_k \frac{\partial g_k}{\partial q_j} + Q_j^{EXC}\right] \tag{8.26}$$

$$\frac{dH(\mathbf{q}, \mathbf{p}, t)}{dt} = \sum_j \left(\left[\sum_{k=1}^{m} \lambda_k \frac{\partial g_k}{\partial q_j} + Q_j^{EXC}\right]\dot{q}_j\right) - \frac{\partial L(\mathbf{q}, \dot{\mathbf{q}}, t)}{\partial t} \tag{8.27}$$

The symmetry of Hamilton's equations of motion is illustrated when the Lagrange multiplier and generalized forces are zero. Then

$$\dot{q}_j = \frac{\partial H(\mathbf{q}, \mathbf{p}, t)}{\partial p_j} \tag{8.28}$$

$$\dot{p}_j = -\frac{\partial H(\mathbf{p}, \mathbf{q}, t)}{\partial q_j} \tag{8.29}$$

$$\frac{dH(\mathbf{p}, \mathbf{q}, t)}{dt} = \frac{\partial H(\mathbf{p}, \mathbf{q}, t)}{\partial t} = -\frac{\partial L(\dot{\mathbf{q}}, \mathbf{q}, t)}{\partial t} \tag{8.30}$$

This simplified form illustrates the symmetry of Hamilton's equations of motion. Many books present the Hamiltonian only for this special simplified case where it is holonomic, conservative, and generalized coordinates are used.

8.3.1   Canonical equations of motion

Hamilton's equations of motion, summarized in equations $8.25 - 27$, use either a minimal set of generalized coordinates, or the Lagrange multiplier terms, to account for holonomic constraints, or generalized forces Q_j^{EXC} to account for non-holonomic or other forces. Hamilton's equations of motion usually are called the **canonical equations of motion**. Note that the term "canonical" has nothing to do with religion or canon law; the reason for this name has bewildered many generations of students of classical mechanics. The term was introduced by Jacobi in 1837 to designate a simple and fundamental set of conjugate variables and equations. Note the symmetry of Hamilton's two canonical equations, plus the fact that the canonical variables p_k, q_k are treated as independent canonical variables. The Lagrange mechanics coordinates $(\mathbf{q}, \dot{\mathbf{q}}, t)$ are replaced by the Hamiltonian mechanics coordinates $(\mathbf{q}, \mathbf{p}, t)$, *where the conjugate momenta* $\mathbf{p}$ *are taken to be independent of the coordinate* $\mathbf{q}$.

Lagrange was the first to derive the canonical equations but he did not recognize them as a basic set of equations of motion. Hamilton derived the canonical equations of motion from his fundamental variational principle, chapter 9.2, and made them the basis for a far-reaching theory of dynamics. Hamilton's equations give $2s$ first-order differential equations for p_k, q_k for each of the $s = n - m$ degrees of freedom. Lagrange's equations give s second-order differential equations for the s independent generalized coordinates $q_k, \dot{q}_k$.

It has been shown that $H(\mathbf{p}, \mathbf{q}, t)$ and $L(\dot{\mathbf{q}}, \mathbf{q}, t)$ are the Legendre transforms of each other. Although the Lagrangian formulation is ideal for solving numerical problems in classical mechanics, the Hamiltonian formulation provides a better framework for conceptual extensions to other fields of physics since it is written in terms of the fundamental conjugate coordinates, $\mathbf{q}, \mathbf{p}$. The Hamiltonian is used extensively in modern physics, including quantum physics, as discussed in chapters 15 and 18. For example, in quantum mechanics there is a straightforward relation between the classical and quantal representations of momenta; this does not exist for the velocities.

The concept of state space, introduced in chapter 3.3.2, applies naturally to Lagrangian mechanics since $(\dot{q}, q)$ are the generalized coordinates used in Lagrangian mechanics. The concept of Phase Space, introduced in chapter 3.3.3, naturally applies to Hamiltonian phase space since (p, q) are the generalized coordinates used in Hamiltonian mechanics.

8.4 Hamiltonian in different coordinate systems

Prior to solving problems using Hamiltonian mechanics, it is useful to express the Hamiltonian in cylindrical and spherical coordinates for the special case of conservative forces since these are encountered frequently in physics.

8.4.1 Cylindrical coordinates ρ, z, ϕ

Consider cylindrical coordinates ρ, z, ϕ. Expressed in cartesian coordinate

$$
\begin{aligned}
x &= \rho \cos \phi \\
y &= \rho \sin \phi \\
z &= z
\end{aligned}
\tag{8.31}
$$

Using appendix table $C.3$, the Lagrangian can be written in cylindrical coordinates as

$$
L = T - U = \frac{m}{2} \left(\dot{\rho}^2 + \rho^2 \dot{\phi}^2 + \dot{z}^2 \right) - U(\rho, z, \phi)
\tag{8.32}
$$

The conjugate momenta are

$$
p_\rho = \frac{\partial L}{\partial \dot{\rho}} = m \dot{\rho}
\tag{8.33}
$$

$$
p_\phi = \frac{\partial L}{\partial \dot{\phi}} = m \rho^2 \dot{\phi}
\tag{8.34}
$$

$$
p_z = \frac{\partial L}{\partial \dot{z}} = m \dot{z}
\tag{8.35}
$$

Assume a conservative force, then H is conserved. Since the transformation from cartesian to non-rotating generalized cylindrical coordinates is time independent, then $H = E$. Then using $(8.32 - 8.35)$ gives the Hamiltonian in cylindrical coordinates to be

$$
\begin{aligned}
H(\mathbf{q}, \mathbf{p}, t) &= \sum_i p_i \dot{q}_i - L(\mathbf{q}, \dot{\mathbf{q}}, t) \\
&= \left(p_\rho \dot{\rho} + p_\phi \dot{\phi} + p_z \dot{z} \right) - \frac{m}{2} \left(\dot{\rho}^2 + \rho^2 \dot{\phi}^2 + \dot{z}^2 \right) + U(\rho, z, \phi) \\
&= \frac{1}{2m} \left(p_\rho^2 + \frac{p_\phi^2}{\rho^2} + p_z^2 \right) + U(\rho, z, \phi)
\end{aligned}
\tag{8.36}
$$

$$
\tag{8.37}
$$

The canonical equations of motion in cylindrical coordinates can be written as

$$
\dot{p}_\rho = -\frac{\partial H}{\partial \rho} = \frac{p_\phi^2}{m \rho^3} - \frac{\partial U}{\partial \rho}
\tag{8.38}
$$

$$
\dot{p}_\phi = -\frac{\partial H}{\partial \phi} = -\frac{\partial U}{\partial \phi}
\tag{8.39}
$$

$$
\dot{p}_z = -\frac{\partial H}{\partial z} = -\frac{\partial U}{\partial z}
\tag{8.40}
$$

$$
\dot{\rho} = \frac{\partial H}{\partial p_\rho} = \frac{p_\rho}{m}
\tag{8.41}
$$

$$
\dot{\phi} = \frac{\partial H}{\partial p_\phi} = \frac{p_\phi}{m \rho^2}
\tag{8.42}
$$

$$
\dot{z} = \frac{\partial H}{\partial p_z} = \frac{p_z}{m}
\tag{8.43}
$$

Note that if ϕ is cyclic, that is $\frac{\partial U}{\partial \phi} = 0$, then the angular momentum about the z axis, p_ϕ, is a constant of motion. Similarly, if z is cyclic, then p_z is a constant of motion.

8.4.2   Spherical coordinates, r, θ, ϕ

Appendix table $C.4$ shows that the spherical coordinates are related to the cartesian coordinates by

$$
\begin{aligned}
x &= r \sin \theta \cos \phi \\
y &= r \sin \theta \sin \phi \\
z &= r \cos \theta
\end{aligned} \tag{8.44}
$$

The Lagrangian is

$$
L = T_i - U = \frac{m}{2} \left(\dot{r}^2 + r^2 \dot{\theta}^2 + r^2 \sin^2 \theta \dot{\phi}^2 \right) - U(r\theta\phi) \tag{8.45}
$$

The conjugate momenta are

$$
p_r = \frac{\partial L}{\partial \dot{r}} = m\dot{r} \tag{8.46}
$$

$$
p_\theta = \frac{\partial L}{\partial \dot{\theta}} = mr^2 \dot{\theta} \tag{8.47}
$$

$$
p_\phi = \frac{\partial L}{\partial \dot{\phi}} = mr^2 \sin^2 \theta \dot{\phi} \tag{8.48}
$$

Assuming a conservative force then H is conserved. Since the transformation from cartesian to generalized spherical coordinates is time independent, then $H = E$. Thus using $(8.46 - 8.48)$ the Hamiltonian is given in spherical coordinates by

$$
\begin{aligned}
H(\mathbf{q}, \mathbf{p}, t) &= \sum_i p_i \dot{q}_i - L(\mathbf{q}, \dot{\mathbf{q}}, t) \tag{8.49} \\
&= \left(p_r \dot{r} + p_\theta \dot{\theta} + p_\phi \dot{\phi} \right) - \frac{m}{2} \left(\dot{r}^2 + r^2 \dot{\theta}^2 + r^2 \sin^2 \theta \dot{\phi}^2 \right) + U(r, \theta, \phi) \tag{8.50} \\
&= \frac{1}{2m} \left(p_r^2 + \frac{p_\theta^2}{r^2} + \frac{p_\phi^2}{r^2 \sin^2 \theta} \right) + U(r, \theta, \phi) \tag{8.51}
\end{aligned}
$$

Then the canonical equations of motion in spherical coordinates are

$$
\dot{p}_r = -\frac{\partial H}{\partial r} = \frac{1}{mr^3} \left(p_\theta^2 + \frac{p_\phi^2}{\sin^2 \theta} \right) - \frac{\partial U}{\partial r} \tag{8.52}
$$

$$
\dot{p}_\theta = -\frac{\partial H}{\partial \theta} = \frac{1}{mr^2} \left(\frac{p_\phi^2 \cos \theta}{\sin^3 \theta} \right) - \frac{\partial U}{\partial \theta} \tag{8.53}
$$

$$
\dot{p}_\phi = -\frac{\partial H}{\partial \phi} = -\frac{\partial U}{\partial \phi} \tag{8.54}
$$

$$
\dot{r} = \frac{\partial H}{\partial p_r} = \frac{p_r}{m} \tag{8.55}
$$

$$
\dot{\theta} = \frac{\partial H}{\partial p_\theta} = \frac{p_\theta}{mr^2} \tag{8.56}
$$

$$
\dot{\phi} = \frac{\partial H}{\partial p_\phi} = \frac{p_\phi}{mr^2 \sin^2 \theta} \tag{8.57}
$$

Note that if the coordinate ϕ is cyclic, that is $\frac{\partial U}{\partial \phi} = 0$ then the angular momentum p_ϕ is conserved. Also if the θ coordinate is cyclic, and $p_\phi = 0$, that is, there is no change in the angular momentum perpendicular to the z axis, then p_θ is conserved.

An especially important spherically-symmetric Hamiltonian is that for a central field. Central fields, such as the gravitational or Coulomb fields of a uniform spherical mass, or charge, distributions, are spherically symmetric and then both θ and ϕ are cyclic. Thus the projection of the angular momentum p_ϕ about the z axis is conserved for these spherically symmetric potentials. In addition, since both p_θ and p_ϕ, are conserved, then the total angular momentum also must be conserved as is predicted by Noether's theorem.

8.5 Applications of Hamiltonian Dynamics

The equations of motion of a system can be derived using the Hamiltonian coupled with Hamilton's equations of motion, that is, equations $8.25 - 8.27$.

Formally the Hamiltonian is constructed from the Lagrangian. That is

1) Select a set of independent generalized coordinates q_i
2) Partition the active forces.
3) Construct the Lagrangian $L(q_i, \dot{q}_i, t)$
4) Derive the conjugate generalized momenta via $p_i = \frac{\partial L}{\partial \dot{q}_i}$
5) Knowing $L, \dot{q}_i, p_i$ derive $H = \sum_i p_i \dot{q}_i - L$
6) Derive $\dot{q}_k = \frac{\partial H}{\partial p_k}$ and $\dot{p}_j = -\frac{\partial H(\mathbf{q}, \mathbf{p}, t)}{\partial q_j} + \sum_{k=1}^{m} \lambda_k \frac{\partial g_k}{\partial q_j} + Q_j^{EXC}$.

This procedure appears to be unnecessarily complicated compared to just using the Lagrangian plus Lagrangian mechanics to derive the equations of motion. Fortunately the above lengthy procedure often can be bypassed for conservative systems. That is, if the following conditions are satisfied;

i) $L = T(\dot{q}) - U(q)$, that is, $U(q)$ is independent of the velocity $\dot{q}$.
ii) the generalized coordinates are time independent.
then it is possible to use the fact that $H = T + U = E$.

The following five examples illustrate the use of Hamiltonian mechanics to derive the equations of motion.

8.1 Example: Motion in a uniform gravitational field

Consider a mass m in a uniform gravitational field acting in the $-\mathbf{z}$ direction. The Lagrangian for this simple case is

$$L = \frac{1}{2}m\left(\dot{x}^2 + \dot{y}^2 + \dot{z}^2\right) - mgz$$

Therefore the generalized momenta are $p_x = \frac{\partial L}{\partial \dot{x}} = m\dot{x}$, $p_y = \frac{\partial L}{\partial \dot{y}} = m\dot{y}$, $p_z = \frac{\partial L}{\partial \dot{z}} = m\dot{z}$. The corresponding Hamiltonian H is

$$
\begin{aligned}
H &= \sum_i p_i \dot{q}_i - L = p_x \dot{x} + p_y \dot{y} + p_z \dot{z} - L \\
&= \frac{p_x^2}{m} + \frac{p_y^2}{m} + \frac{p_z^2}{m} - \frac{1}{2}\left(\frac{p_x^2}{m} + \frac{p_y^2}{m} + \frac{p_z^2}{m}\right) + mgz = \frac{1}{2}\left(\frac{p_x^2}{m} + \frac{p_y^2}{m} + \frac{p_z^2}{m}\right) + mgz
\end{aligned}
$$

Note that the Lagrangian is not explicitly time dependent, thus the Hamiltonian is a constant of motion. Hamilton's equations give that

$$
\begin{aligned}
\dot{x} &= \frac{\partial H}{\partial p_x} = \frac{p_x}{m} & -\dot{p}_x &= \frac{\partial H}{\partial x} = 0 \\
\dot{y} &= \frac{\partial H}{\partial p_y} = \frac{p_y}{m} & -\dot{p}_y &= \frac{\partial H}{\partial y} = 0 \\
\dot{z} &= \frac{\partial H}{\partial p_z} = \frac{p_z}{m} & -\dot{p}_z &= \frac{\partial H}{\partial z} = mg
\end{aligned}
$$

Combining these gives that $\ddot{x} = 0$, $\ddot{y} = 0$, $\ddot{z} = -g$. Note that the linear momenta p_x and p_y are constants of motion whereas the rate of change of p_z is given by the gravitational force mg. Note also that $H = T + U$ for this conservative system.

8.2 Example: One-dimensional harmonic oscillator

Consider a mass m subject to a linear restoring force with spring constant k. The Lagrangian $L = T - U$ equals

$$L = \frac{1}{2}m\dot{x}^2 - \frac{1}{2}kx^2$$

Therefore the generalized momentum is

$$p_x = \frac{\partial L}{\partial \dot{x}} = m\dot{x}$$

The Hamiltonian H is

$$H = \sum_i p_i \dot{q}_i - L = p_x \dot{x} - L$$

$$= \frac{p_x p_x}{m} - \frac{1}{2}\frac{p_x^2}{m} + \frac{1}{2}kx^2 = \frac{1}{2}\frac{p_x^2}{m} + \frac{1}{2}kx^2$$

Note that the Lagrangian is not explicitly time dependent, thus the Hamiltonian will be a constant of motion.
Hamilton's equations give that

$$\dot{x} = \frac{\partial H}{\partial p_x} = \frac{p_x}{m}$$

or

$$p_x = m\dot{x}$$

In addition

$$-\dot{p}_x = \frac{\partial H}{\partial x} = \frac{\partial U}{\partial x} = kx$$

Combining these gives that

$$\ddot{x} + \frac{k}{m}x = 0$$

which is the equation of motion for the harmonic oscillator.

8.3 *Example: Plane pendulum*

The plane pendulum, in a uniform gravitational field g, is an interesting system to consider. There is
only one generalized coordinate, θ and the Lagrangian for this system is

$$L = \frac{1}{2}ml^2\dot{\theta}^2 + mgl\cos\theta$$

The momentum conjugate to θ is

$$p_\theta = \frac{\partial L}{\partial \dot{\theta}} = ml^2\dot{\theta}$$

which is the angular momentum about the pivot point.
The Hamiltonian is

$$H = \sum_i p_i \dot{q}_i - L = p_\theta\dot{\theta} - L = \frac{1}{2}ml^2\dot{\theta}^2 - mgl\cos\theta = \frac{p_\theta^2}{2ml^2} - mgl\cos\theta$$

Hamilton's equations of motion give

$$\dot{\theta} = \frac{\partial H}{\partial p_\theta} = \frac{p_\theta}{ml^2}$$

$$\dot{p}_\theta = -\frac{\partial H}{\partial \theta} = -mgl\sin\theta$$

Note that the Lagrangian and Hamiltonian are not explicit functions of time, therefore they are conserved.
Also the potential is velocity independent and there is no coordinate transformation, thus the Hamiltonian
equals the total energy, that is

$$H = \frac{p_\theta^2}{2ml^2} - mgl\cos\theta = E$$

where E is a constant of motion. Note that the angular momentum p_θ is not a constant of motion since $\dot{p}_\theta$
explicitly depends on θ.

The solutions for the plane pendulum on a (θ, p_θ) phase diagram, shown in the adjacent figure, illustrate the motion. The upper phase-space plot shows the range $(\theta = \pm\pi, p_\theta)$. Note that the $\theta = +\pi$ and $-\pi$ correspond to the same physical point, that is the phase diagram should be rolled into a cylinder connected along the dashed lines. The lower phase space plot shows two cycles for θ to better illustrate the cyclic nature of the phase diagram. The corresponding state-space diagram is shown in figure 3.4. The trajectories are ellipses for low energy $-mgl < E < mgl$ corresponding to oscillations of the pendulum about $\theta = 0$. The center of the ellipse $(0,0)$ is a stable equilibrium point for the oscillation. However, there is a phase change to rotational motion about the horizontal axis when $|E| > mgl$, that is, the pendulum swings around a circle continuously, i.e. it rotates continuously in one direction about the horizontal axis. The phase change occurs at $E = mgl$. and is designated by the separatrix trajectory.

The plot of p_θ versus θ for the plane pendulum is better presented on a cylindrical phase space representation since θ is a cyclic variable that cycles around the cylinder, whereas p_θ oscillates equally about zero having both positive and negative values. When wrapped around a cylinder then the unstable and stable equilibrium points will be at diametrically opposite locations on the surface of the cylinder at $p_\theta = 0$. For small oscillations about equilibrium, also called librations, the correlation between p_θ and θ is given by the clockwise closed ellipses wrapped on the cylindrical surface, whereas for energies $|E| > mgl$ the positive p_θ corresponds to counterclockwise rotations while the negative p_θ corresponds to clockwise rotations.

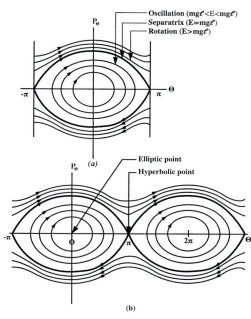

Phase-space diagrams for the plane pendulum. The separatrix (bold line) separates the oscillatory solutions from the rolling solutions. The upper (a) shows one complete cycle while the lower (b) shows two complete cycles.

8.4 Example: Hooke's law force constrained to the surface of a cylinder

Consider the case where a mass m is attracted by a force directed toward the origin and proportional to the distance from the origin. Determine the Hamiltonian if the mass is constrained to move on the surface of a cylinder defined by

$$x^2 + y^2 = R^2$$

It is natural to transform this problem to cylindrical coordinates ρ, z, θ. Since the force is just Hooke's law

$$\mathbf{F} = -k\mathbf{r}$$

the potential is the same as for the harmonic oscillator, that is

$$U = \frac{1}{2}kr^2 = \frac{1}{2}k(\rho^2 + z^2)$$

This is independent of θ, and thus θ is cyclic.

In cylindrical coordinates the velocity is

$$v^2 = \dot{\rho}^2 + \rho^2\dot{\theta}^2 + \dot{z}^2$$

Confined to the surface of the cylinder means that

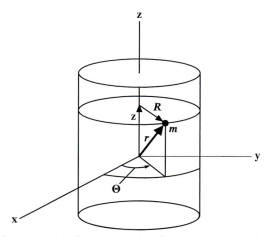

Mass attracted to origin by force proportional to distance from origin with the motion constrained to the surface of a cylinder.

$$\begin{aligned} \rho &= R \\ \dot{\rho} &= 0 \end{aligned}$$

Then the Lagrangian simplifies to

$$L = T - U = \frac{1}{2}m\left(R^2\dot{\theta}^2 + \dot{z}^2\right) - \frac{1}{2}k(R^2 + z^2)$$

The generalized coordinates are θ, z and the corresponding generalized momenta are

$$p_\theta = \frac{\partial L}{\partial \dot{\theta}} = mR^2\dot{\theta} \tag{a}$$

$$p_z = \frac{\partial L}{\partial \dot{z}} = m\dot{z} \tag{b}$$

The system is conservative, and the transformation from rectangular to cylindrical coordinates does not depend explicitly on time. Therefore the Hamiltonian is conserved and equals the total energy. That is

$$H = \sum_i p_i\dot{q}_i - L = \frac{p_\theta^2}{2mR^2} + \frac{p_z^2}{2m} + \frac{1}{2}k(R^2 + z^2) = E$$

The equations of motion then are given by the canonical equations

$$\dot{p}_\theta = -\frac{\partial H}{\partial \theta} = 0 \qquad\qquad \dot{\theta} = \frac{\partial H}{\partial p_\theta} = \frac{p_\theta}{mR^2} \tag{c}$$

$$\dot{p}_z = -\frac{\partial H}{\partial z} = -kz \qquad\qquad \dot{z} = \frac{\partial H}{\partial p_z} = \frac{p_z}{m} \tag{d}$$

Equation (a) and (c) imply that

$$p_\theta = \frac{\partial L}{\partial \dot{\theta}} = mR^2\dot{\theta} = \text{constant}$$

Thus the angular momentum about the axis of the cylinder is conserved, that is, it is a cyclic variable. Combining equations (b) and (d) implies that

$$\ddot{z} + \frac{k}{m}z = 0$$

This is the equation for simple harmonic motion with angular frequency $\omega = \sqrt{\frac{k}{m}}$. The symmetries imply that this problem has the same solutions for the z coordinate as the harmonic oscillator, while the θ coordinate moves with constant angular velocity.

8.5 Example: Electron motion in a cylindrical magnetron

A magnetron comprises a hot cylindrical wire cathode that emits electrons and is at a high negative voltage. It is surrounded by a larger diameter concentric cylindrical anode at ground potential. A uniform magnetic field runs parallel to the cylindrical axis of the magnetron. The electron beam excites a multiple set of microwave cavities located around the circumference of the cylindrical wall of the anode. The magnetron was invented in England during World War 2 to generate microwaves required for the development of radar.

Consider a non-relativistic electron of mass m and charge $-e$ in a cylindrical magnetron moving between the central cathode wire, of radius a at a negative electric potential $-\phi_0$, and a concentric cylindrical anode conductor of radius R which has zero electric potential. There is a uniform constant magnetic field B parallel to the cylindrical axis of the magnetron.

Using SI units and cylindrical coordinates (r, θ, z) aligned with the axis of the magnetron, the electromagnetic force Lagrangian, given in chapter 6.10, equals

$$L = \frac{1}{2}m\dot{\mathbf{r}}^2 + e(\phi - \dot{\mathbf{r}} \cdot \mathbf{A})$$

The electric and vector potentials for the magnetron geometry are

$$\phi = -\phi_0\frac{\ln(\frac{r}{R})}{\ln(\frac{a}{R})}$$

$$\mathbf{A} = \frac{1}{2}Br\hat{e}_\theta$$

Thus expressed in cylindrical coordinates the Lagrangian equals

$$L = \frac{1}{2}m\left(\dot{r}^2 + r^2\dot{\theta}^2 + \dot{z}^2\right) + e\phi - \frac{1}{2}eBr^2\dot{\theta}$$

The generalized momenta are

$$p_r = \frac{\partial L}{\partial \dot{r}} = m\dot{r}$$

$$p_\theta = \frac{\partial L}{\partial \dot{\theta}} = mr^2\dot{\theta} - \frac{1}{2}eBr^2$$

$$p_z = \frac{\partial L}{\partial \dot{z}} = m\dot{z}$$

Note that the vector potential A contributes an additional term to the angular momentum p_θ.
Using the above generalized momenta leads to the Hamiltonian

$$
\begin{aligned}
H &= p_r\dot{r} + p_\theta\dot{\theta} + p_z\dot{z} - L \\
&= \frac{1}{2}m\left(\dot{r}^2 + r^2\dot{\theta}^2 + \dot{z}^2\right) - e\phi + \frac{1}{2}eBr^2\dot{\theta} \\
&= \frac{p_r^2}{2m} + \frac{1}{2mr^2}\left(p_\theta + \frac{1}{2}eBr^2\right)^2 + \frac{p_z^2}{2m} - e\phi \\
&= \frac{1}{2m}\left[p_r^2 + \left(\frac{p_\theta}{r} + \frac{1}{2}eBr\right)^2 + p_z^2\right] - e\phi
\end{aligned}
$$

Note that the Hamiltonian is not an explicit function of time, therefore it is a constant of motion which equals the total energy.

$$H = \frac{1}{2m}\left[p_r^2 + \left(\frac{p_\theta}{r} + \frac{1}{2}eBr\right)^2 + p_z^2\right] - e\phi = E$$

Since $\dot{p}_i = -\frac{\partial H}{\partial q_i}$, and if H is not an explicit function of q_i, then $\dot{p}_i = 0$, that is, p_i is a constant of motion. Thus p_θ and p_z are constants of motion.

Consider the initial conditions $r = a, \dot{r} = \dot{\theta} = \dot{z} = 0$. Then

$$
\begin{aligned}
p_\theta &= \frac{\partial L}{\partial \dot{\theta}} = mr^2\dot{\theta} - \frac{1}{2}eBr^2 = -\frac{1}{2}eBa^2 \\
p_z &= 0 \\
H &= \frac{1}{2m}\left[p_r^2 + \left(\frac{p_\theta}{r} + \frac{1}{2}eBr\right)^2 + p_z^2\right] + e\phi_0\frac{\ln(\frac{r}{R})}{\ln(\frac{a}{R})} = e\phi_0
\end{aligned}
$$

Note that at $r = R$, then p_r is given by the last equation since the Hamiltonian equals a constant $e\phi_0$. That is, assuming that $a << R$ then

$$p_r^2 = 2me\phi_0 - (\frac{1}{2}eBR)^2$$

Define a critical magnetic field by

$$B_c \equiv \frac{2}{R}\sqrt{\frac{2m\phi_0}{e}}$$

then

$$\left(p_r^2\right)_{r=R} = \left(B_c^2 - B^2\right)\left(\frac{1}{2}eR\right)^2$$

Note that if $B < B_c$ then p_r is real at $r = R$. However, if $B > B_c$ then p_r is imaginary at $r = R$ implying that there must be a maximum orbit radius r_0 for the electron where $r_0 < R$. That is, the electron trajectories are confined spatially to coaxial cylindrical orbits concentric with the magnetron electromagnetic fields. These closed electron trajectories excite the microwave cavities located in the nearby outer cylindrical wall of the anode.

8.6   Routhian reduction

Noether's theorem states that if the coordinate q_j is cyclic, and if the Lagrange multiplier plus generalized force contributions for the jth coordinates are zero, then the canonical momentum of the cyclic variable, p_j, is a constant of motion as is discussed in chapter 7.3. Therefore, both (q_j, p_j) are constants of motion for cyclic variables, and these constant (q_j, p_j) coordinates can be factored out of the Hamiltonian $H(\mathbf{p}, \mathbf{q}, t)$. This reduces the number of degrees of freedom included in the Hamiltonian. For this reason, cyclic variables are called *ignorable* variables in Hamiltonian mechanics. This advantage does not apply to the $(q_j, \dot{q}_j)$ variables used in Lagrangian mechanics since $\dot{q}$ is not a constant of motion for a cyclic coordinate. The ability to eliminate the cyclic variables as unknowns in the Hamiltonian is a valuable advantage of Hamiltonian mechanics that is exploited extensively for solving problems, as is described in chapter 15.

It is advantageous to have the ability to exploit both the Lagrangian and Hamiltonian formulations simultaneously when handling systems that involve a mixture of cyclic and non-cyclic coordinates. The equations of motion for each *independent generalized coordinate* can be derived independently of the remaining generalized coordinates. Thus it is possible to select either the Hamiltonian or the Lagrangian formulations for each generalized coordinate, independent of what is used for the other generalized coordinates. Routh[Rou1860] devised an elegant, and useful, hybrid technique that separates the cyclic and non-cyclic generalized coordinates in order to simultaneously exploit the differing advantages of both the Hamiltonian and Lagrangian formulations of classical mechanics. The Routhian reduction approach partitions the $\sum_{i=1}^{n} p_i \dot{q}_i$ kinetic energy term in the Hamiltonian into a cyclic group, plus a non-cyclic group, i.e.

$$H(q_1, ..., q_n; p_1, ..., p_n; t) = \sum_{i=1}^{n} p_i \dot{q}_i - L = \sum_{cyclic}^{s} p_i \dot{q}_i + \sum_{noncyclic}^{n-s} p_i \dot{q}_i - L \tag{8.58}$$

Routh's clever idea was to define a new function, called the **Routhian**, that include only one of the two partitions of the kinetic energy terms. This makes the Routhian a Hamiltonian for the coordinates for which the kinetic energy terms are included, while the Routhian acts like a negative Lagrangian for the coordinates where the kinetic energy term is omitted. This book defines two Routhians.

$$R_{cyclic}(q_1, ..., q_n; \dot{q}_1, ..., \dot{q}_s; p_{s+1}, ..., p_n; t) \equiv \sum_{cyclic}^{m} p_i \dot{q}_i - L \tag{8.59}$$

$$R_{noncyclic}(q_1, ..., q_n; p_1, ..., p_s; \dot{q}_{s+1}, ..., \dot{q}_n; t) \equiv \sum_{noncyclic}^{s} p_i \dot{q}_i - L \tag{8.60}$$

The first, Routhian, called R_{cyclic}, includes the kinetic energy terms only for the cyclic variables, and behaves like a Hamiltonian for the cyclic variables, and behaves like a Lagrangian for the non-cyclic variables. The second Routhian, called $R_{non-cyclic}$, includes the kinetic energy terms for only the non-cyclic variables, and behaves like a Hamiltonian for the non-cyclic variables, and behaves like a negative Lagrangian for the cyclic variables. These two Routhians complement each other in that they make the Routhian either a Hamiltonian for the cyclic variables, or the converse where the Routhian is a Hamiltonian for the non-cyclic variables. The Routhians use $(q_i, \dot{q}_i)$ to denote those coordinates for which the Routhian behaves like a Lagrangian, and (q_i, p_i) for those coordinates where the Routhian behaves like a Hamiltonian. For uniformity, it is assumed that the degrees of freedom between $1 \leq i \leq s$ are non-cyclic, while those between $s+1 \leq i \leq n$ are ignorable cyclic coordinates.

The Routhian is a hybrid of Lagrangian and Hamiltonian mechanics. Some textbooks minimize discussion of the Routhian on the grounds that this hybrid approach is not fundamental. However, the Routhian is used extensively in engineering in order to derive the equations of motion for rotating systems. In addition it is used when dealing with rotating nuclei in nuclear physics, rotating molecules in molecular physics, and rotating galaxies in astrophysics. The Routhian reduction technique provides a powerful way to calculate the intrinsic properties for a rotating system in the rotating frame of reference. The Routhian approach is included in this textbook because it plays an important role in practical applications of rotating systems, plus it nicely illustrates the relative advantages of the Lagrangian and Hamiltonian formulations in mechanics.

8.6.1 R_{cyclic} - Routhian is a Hamiltonian for the cyclic variables

The cyclic Routhian R_{cyclic} is defined assuming that the variables between $1 \leq i \leq s$ are non-cyclic, where $s = n - m$, while the m variables between $s+1 \leq i \leq n$ are ignorable cyclic coordinates. The cyclic Routhian R_{cyclic} expresses the cyclic coordinates in terms of (q, p) which are required for use by Hamilton's equations, while the non-cyclic variables are expressed in terms of $(q, \dot{q})$ for use by the Lagrange equations. That is, the cyclic Routhian R_{cyclic} is defined to be

$$R_{cyclic}(q_1, ..., q_n; \dot{q}_1, ..., \dot{q}_s; p_{s+1}, ..., p_n; t) \equiv \sum_{cyclic}^{m} p_i \dot{q}_i - L \tag{8.61}$$

where the summation $\sum_{cyclic} p_i \dot{q}_i$ is over only the m cyclic variables $s+1 \leq i \leq n$. Note that the Lagrangian can be split into the cyclic and the non-cyclic parts

$$R_{cyclic}(q_1, ..., q_n; \dot{q}_1, ..., \dot{q}_s; p_{s+1}, ..., p_n; t) = \sum_{cyclic}^{m} p_i \dot{q}_i - L_{cyclic} - L_{noncyclic} \tag{8.62}$$

The first two terms on the right can be combined to give the Hamiltonian H_{cyclic} for only the m cyclic variables, $i = s + 1, s + 2, .., n$, that is

$$R_{cyclic}(q_1, ..., q_n; \dot{q}_1, ..., \dot{q}_s; p_{s+1}, ..., p_n; t) = H_{cyclic} - L_{noncyclic} \tag{8.63}$$

The Routhian $R_{cyclic}(q_1, ..., q_n; \dot{q}_1, ..., \dot{q}_s; p_{s+1}, ..., p_n; t)$ also can be written in an alternate form

$$R_{cyclic}(q_1, ..., q_n; \dot{q}_1, ..., \dot{q}_s; p_{s+1}, ..., p_n; t) \equiv \sum_{cyclic}^{m} p_i \dot{q}_i - L = \sum_{i=1}^{n} p_i \dot{q}_i - L - \sum_{noncyclic}^{s} p_i \dot{q} \tag{8.64}$$

$$= H - \sum_{noncyclic}^{s} p_i \dot{q}_i \tag{8.65}$$

which is expressed as the complete Hamiltonian minus the kinetic energy term for the noncyclic coordinates. The Routhian R_{cyclic} behaves like a Hamiltonian for the m cyclic coordinates and behaves like a negative Lagrangian $L_{noncyclic}$ for all the $s = n - m$ noncyclic coordinates $i = 1, 2, ..., s$. Thus the equations of motion for the s non-cyclic variables are given using Lagrange's equations of motion, while the Routhian behaves like a Hamiltonian H_{cyclic} for the m ignorable cyclic variables $i = s + 1, ..., n$.

Ignoring both the Lagrange multiplier and generalized forces, then the partitioned equations of motion for the non-cyclic and cyclic generalized coordinates are given in Table 8.1.

Table 8.1; Equations of motion for the Routhian R_{cyclic}

	Lagrange equations	Hamilton equations
Coordinates	Noncyclic: $1 \leq i \leq s$	Cyclic: $(s + 1) \leq i \leq n$
Equations of motion	$\dfrac{\partial R_{cyclic}}{\partial q_i} = -\dfrac{\partial L_{noncyclic}}{\partial q_i}$	$\dfrac{\partial R_{cyclic}}{\partial q_i} = -\dot{p}_i$
	$\dfrac{\partial R_{cyclic}}{\partial \dot{q}_i} = -\dfrac{\partial L_{noncyclic}}{\partial \dot{q}_i}$	$\dfrac{\partial R_{cyclic}}{\partial p_i} = \dot{q}_i$

Thus there are m cyclic (ignorable) coordinates $(q, p)_{s+1}, ..., (q, p)_n$ which obey Hamilton's equations of motion, while the the first $s = n - m$ non-cyclic (non-ignorable) coordinates $(q, \dot{q})_1, ..., (q, \dot{q})_s$ for $i = 1, 2, ..., s$ obey Lagrange equations. The solution for the cyclic variables is trivial since they are constants of motion and thus the Routhian R_{cyclic} has reduced the number of equations of motion that must be solved from n to the $s = n - m$ non-cyclic variables. This Routhian provides an especially useful way to reduce the number of equations of motion for rotating systems.

Note that there are several definitions used to define the Routhian, for example some books define this Routhian as being the negative of the definition used here so that it corresponds to a positive Lagrangian. However, this sign usually cancels when deriving the equations of motion, thus the sign convention is unimportant if a consistent sign convention is used.

8.6.2 $R_{noncyclic}$ - Routhian is a Hamiltonian for the non-cyclic variables

The non-cyclic Routhian $R_{noncyclic}$ complements R_{cyclic}. Again the generalized coordinates between $1 \le i \le s$ are assumed to be non-cyclic, while those between $s+1 \le i \le n$ are ignorable cyclic coordinates. However, the expression in terms of (q, p) and $(q, \dot{q})$ are interchanged, that is, the cyclic variables are expressed in terms of $(q, \dot{q})$ and the non-cyclic variables are expressed in terms of (q, p) which is opposite of what was used for R_{cyclic}.

$$R_{noncyclic}(q_1, ..., q_n; p_1, ..., p_s; \dot{q}_{s+1}, ..., \dot{q}_n; t) \quad = \quad \sum_{noncyclic}^{s} p_i \dot{q}_i - L_{noncyclic} - L_{cyclic} \qquad (8.66)$$

$$= \quad H_{noncyclic} - L_{cyclic} \qquad (8.67)$$

It can be written in a frequently used form

$$R_{noncyclic}(q_1, ..., q_n; p_1, ..., p_s; \dot{q}_{s+1}, ..., \dot{q}_n; t) \quad \equiv \quad \sum_{noncyclic}^{s} p_i \dot{q}_i - L = \sum_{i=1}^{n} p_i \dot{q}_i - L - \sum_{cyclic}^{m} p_i \dot{q}_i$$

$$= \quad H - \sum_{cyclic}^{m} p_i \dot{q}_i \qquad (8.68)$$

This Routhian behaves like a Hamiltonian for the s non-cyclic variables which are expressed in terms of q and p appropriate for a Hamiltonian. This Routhian writes the m cyclic coordinates in terms of q, and $\dot{q}$, appropriate for a Lagrangian, which are treated assuming the Routhian R_{cyclic} is a negative Lagrangian for these cyclic variables as summarized in table 8.2.

Table 8.2; Equations of motion for the Routhian $R_{noncyclic}$

	Hamilton equations	Lagrange equations
Coordinates	Noncyclic: $1 \le i \le s$	Cyclic: $(s+1) \le i \le n$
Equations of motion	$\frac{\partial R_{noncyclic}}{\partial q_i} = -\dot{p}_i$	$\frac{\partial R_{noncyclic}}{\partial q_i} = -\frac{\partial L_{cyclic}}{\partial q_i}$
	$\frac{\partial R_{noncyclic}}{\partial p_i} = \dot{q}_i$	$\frac{\partial R_{noncyclic}}{\partial \dot{q}_i} = -\frac{\partial L_{cyclic}}{\partial \dot{q}_i}$

This non-cyclic Routhian $R_{noncyclic}$ is especially useful since it equals the Hamiltonian for the non-cyclic variables, that is, the kinetic energy for motion of the cyclic variables has been removed. Note that since the cyclic variables are constants of motion, then $R_{noncyclic}$ is a constant of motion if H is a constant of motion. However, $R_{noncyclic}$ does not equal the total energy since the coordinate transformation is time dependent, that is, $R_{noncyclic}$ corresponds to the energy of the non-cyclic parts of the motion. For example, when used to describe rotational motion, $R_{noncyclic}$ corresponds to the energy in the non-inertial rotating body-fixed frame of reference. This is especially useful in treating rotating systems such as rotating galaxies, rotating machinery, molecules, or rotating strongly-deformed nuclei as discussed in chapter 12.9.

The Lagrangian and Hamiltonian are the fundamental algebraic approaches to classical mechanics. The Routhian reduction method is a valuable hybrid technique that exploits a trick to reduce the number of variables that have to be solved for complicated problems encountered in science and engineering. The Routhian $R_{noncyclic}$ provides the most useful approach for solving the equations of motion for rotating molecules, deformed nuclei, or astrophysical objects in that it gives the Hamiltonian in the non-inertial body-fixed rotating frame of reference ignoring the rotational energy of the frame. By contrast, the cyclic Routhian R_{cyclic} is especially useful to exploit Lagrangian mechanics for solving problems in rigid-body rotation such as the Tippe Top described in example 13.13.

Note that the Lagrangian, Hamiltonian, plus both the $R_{noncyclic}$ and $R_{noncyclic}$ Routhian's, all are scalars under rotation, that is, they are rotationally invariant. However, they may be expressed in terms of the coordinates in either the stationary or a rotating frame. The major difference is that the Routhian includes only subsets of the kinetic energy term $\sum_j p_j \dot{q}_j$. The relative merits of using Lagrangian, Hamiltonian, and both the $R_{noncyclic}$ and $R_{noncyclic}$ Routhian reduction methods, are illustrated by the following examples.

8.6 *Example: Spherical pendulum using Hamiltonian mechanics*

The spherical pendulum provides a simple test case for comparison of the use of Lagrangian mechanics, Hamiltonian mechanics, and both approaches to Routhian reduction. The Lagrangian mechanics solution of the spherical pendulum is described in example 6.10. The solution using Hamiltonian mechanics is given in this example followed by solutions using both of the Routhian reduction approaches.

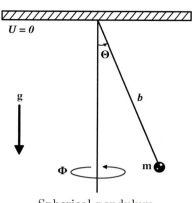

Spherical pendulum

Consider the equations of motion of a spherical pendulum of mass m and length b. The generalized coordinates are θ, ϕ since the length is fixed at $r = b$. The kinetic energy is

$$T = \frac{1}{2}mb^2\dot{\theta}^2 + \frac{1}{2}mb^2\sin^2\theta\dot{\phi}^2$$

The potential energy $U = -mgb\cos\theta$ giving that

$$L(r, \theta, \phi, \dot{r}, \dot{\theta}, \dot{\phi}) = \frac{1}{2}mb^2\dot{\theta}^2 + \frac{1}{2}mb^2\sin^2\theta\dot{\phi}^2 + mgb\cos\theta$$

The generalized momenta are

$$p_\theta = \frac{\partial L}{\partial \dot{\theta}} = mb^2\dot{\theta} \qquad\qquad p_\phi = \frac{\partial L}{\partial \dot{\phi}} = mb^2\sin^2\theta\dot{\phi}$$

Since the system is conservative, and the transformation from rectangular to spherical coordinates does not depend explicitly on time, then the Hamiltonian is conserved and equals the total energy. The generalized momenta allow the Hamiltonian to be written as

$$H(r, \theta, \phi, p_r, p_\theta, p_\phi) = \frac{p_\theta^2}{2mb^2} + \frac{p_\phi^2}{2mb^2\sin^2\theta} - mgb\cos\theta$$

The equations of motion are

$$\dot{p}_\theta = -\frac{\partial H}{\partial \theta} = \frac{p_\phi^2\cos\theta}{2mb^2\sin^3\theta} - mgb\sin\theta \tag{a}$$

$$\dot{p}_\phi = -\frac{\partial H}{\partial \phi} = 0 \tag{b}$$

$$\dot{\theta} = \frac{\partial H}{\partial p_\theta} = \frac{p_\theta}{mb^2} \tag{c}$$

$$\dot{\phi} = \frac{\partial H}{\partial p_\phi} = \frac{p_\phi}{mb^2\sin^2\theta} \tag{d}$$

Take the time derivative of equation (c) and use (a) to substitute for $\dot{p}_\theta$ gives that

$$\ddot{\theta} - \frac{p_\phi^2\cos\theta}{m^2b^4\sin^3\theta} + \frac{g}{b}\sin\theta = 0 \tag{e}$$

Note that equation (b) shows that ϕ is a cyclic coordinate. Thus

$$p_\phi = mb^2\sin^2\theta\dot{\phi} = \text{constant}$$

that is the angular momentum about the vertical axis is conserved. Note that although p_ϕ is a constant of motion, $\dot{\phi} = \frac{p_\phi}{mb^2\sin^2\theta}$ is a function of θ, and thus in general it is not conserved. There are various solutions depending on the initial conditions. If $p_\phi = 0$ then the pendulum is just the simple pendulum discussed previously that can oscillate, or rotate in the θ direction. The opposite extreme is where $p_\theta = 0$ where the pendulum rotates in the ϕ direction with constant θ. In general the motion is a complicated coupling of the θ and ϕ motions.

8.7 Example: Spherical pendulum using $R_{cyclic}(r, \theta, \phi, \dot{r}, \dot{\theta}, p_\phi)$

The Lagrangian for the spherical pendulum is

$$L(r, \theta, \phi, \dot{r}, \dot{\theta}, \dot{\phi}) = \frac{1}{2}mb^2\dot{\theta}^2 + \frac{1}{2}mb^2\sin^2\theta\dot{\phi}^2 + mgb\cos\theta$$

Note that the Lagrangian is independent of ϕ, therefore ϕ is an ignorable variable with

$$\dot{p}_\phi = \frac{\partial L}{\partial \phi} = -\frac{\partial H}{\partial \phi} = 0$$

Therefore p_ϕ is a constant of motion equal to

$$p_\phi = \frac{\partial L}{\partial \dot{\phi}} = mb^2\sin^2\theta\dot{\phi}$$

The Routhian $R_{cyclic}(r, \theta, \phi, \dot{r}, \dot{\theta}, p_\phi)$ equals

$$
\begin{aligned}
R_{cyclic}(r, \theta, \phi, \dot{r}, \dot{\theta}, p_\phi) &= p_\phi\dot{\phi} - L \\
&= -\left[\frac{1}{2}mb^2\dot{\theta}^2 + \frac{1}{2}mb^2\sin^2\theta\dot{\phi}^2 + mgb\cos\theta - mb^2\sin^2\theta\dot{\phi}^2\right] \\
&= -\frac{1}{2}mb^2\dot{\theta}^2 + \frac{1}{2}\frac{p_\phi^2}{mb^2\sin^2\theta} + mgb\cos\theta
\end{aligned}
$$

The Routhian $R_{cyclic}(r, \theta, \phi, \dot{r}, \dot{\theta}, p_\phi)$ behaves like a Hamiltonian for ϕ, and like a Lagrangian $L' = -R_{cyclic}$ for θ. Use of Hamilton's canonical equations for ϕ give

$$
\begin{aligned}
\dot{\phi} &= \frac{\partial R_{cyclic}}{\partial p_\phi} = \frac{p_\phi}{mb^2\sin^2\theta} \\
-\dot{p}_\phi &= \frac{\partial R_{cyclic}}{\partial \phi} = 0
\end{aligned}
$$

These two equations show that p_ϕ is a constant of motion given by

$$mb^2\sin^2\theta\dot{\phi} = p_\phi = \text{ constant} \tag{α}$$

Note that the Hamiltonian only includes the kinetic energy for the ϕ motion which is a constant of motion, but this energy does not equal the total energy. This solution is what is predicted by Noether's theorem due to the symmetry of the Lagrangian about the vertical ϕ axis.

Since $R_{cyclic}(r, \theta, \phi, \dot{r}, \dot{\theta}, p_\phi)$ behaves like a Lagrangian for θ then the Lagrange equation for θ is

$$\Lambda_\theta L = \frac{d}{dt}\frac{\partial R_{cyclic}}{\partial \dot{\theta}} - \frac{\partial R_{cyclic}}{\partial \theta} = 0$$

where the negative sign of the Lagrangian in $R_{cyclic}(r, \theta, \phi, \dot{r}, \dot{\theta}, p_\phi)$ cancels. This leads to

$$mb^2\ddot{\theta} = \frac{p_\phi^2\cos\theta}{mb^2\sin^3\theta} - mgb\sin\theta$$

that is

$$\ddot{\theta} - \frac{p_\phi^2\cos\theta}{m^2b^4\sin^3\theta} + \frac{g}{b}\sin\theta = 0 \tag{β}$$

This result is identical to the one obtained using Lagrangian mechanics in example 6.10 and Hamiltonian mechanics given in example 8.6. The Routhian R_{cyclic} simplified the problem to one degree of freedom θ by absorbing into the Hamiltonian the ignorable cyclic ϕ coordinate and its conserved conjugate momentum p_ϕ. Note that the central term in equation β is the centrifugal term which is due to rotation about the vertical axis. This term is zero for plane pendulum motion when $p_\phi = 0$.

8.8 *Example: Spherical pendulum using* $R_{noncyclic}(r, \theta, \phi, p_r, p_\theta, \dot{\phi})$

For a rotational system the Routhian $R_{noncyclic}(r, \theta, \phi, p_r, p_\theta, \dot{\phi})$ also can be used to project out the Hamiltonian for the active variables in the rotating body-fixed frame of reference. Consider the spherical pendulum where the rotating frame is rotating with angular velocity $\dot{\phi}$. The Lagrangian for the spherical pendulum is

$$L(r, \theta, \phi, \dot{r}, \dot{\theta}, \dot{\phi}) = \frac{1}{2}mb^2\dot{\theta}^2 + \frac{1}{2}mb^2\sin^2\theta\dot{\phi}^2 + mgb\cos\theta$$

Note that the Lagrangian is independent of ϕ, therefore ϕ is an ignorable variable with

$$\dot{p}_\phi = \frac{\partial L}{\partial \phi} = -\frac{\partial H}{\partial \phi} = 0$$

Therefore p_ϕ is a constant of motion equal to

$$p_\phi = \frac{\partial L}{\partial \dot{\phi}} = mb^2\sin^2\theta\dot{\phi}$$

The total Hamiltonian is given by

$$H(r, \theta, \phi, p_r, p_\theta, p_\phi) = \sum_i p_i\dot{q}_i - L = \frac{p_\theta^2}{2mb^2} + \frac{p_\phi^2}{2mb^2\sin^2\theta} - mgb\cos\theta$$

The Routhian for the rotating frame of reference H_{rot} is given by equation 8.68, that is

$$
\begin{aligned}
R_{noncyclic}(r, \theta, \phi, p_r, p_\theta, \dot{\phi}) &= \sum_{i=1}^{n} p_i\dot{q}_i - p_\phi\dot{\phi} - L = H - p_\phi\dot{\phi} \\
&= \frac{p_\theta^2}{2mb^2} + \frac{p_\phi^2}{2mb^2\sin^2\theta} - mgb\cos\theta - p_\phi\dot{\phi} \\
&= \frac{p_\theta^2}{2mb^2} - \frac{1}{2}mb^2\sin^2\theta\dot{\phi}^2 - mgb\cos\theta \quad\quad (\gamma)
\end{aligned}
$$

This behaves like a negative Lagrangian for ϕ and a Hamiltonian for θ. The conjugate momenta are

$$
\begin{aligned}
p_\phi &= \frac{\partial L}{\partial \dot{\phi}} = -\frac{\partial R_{noncyclic}}{\partial \dot{\phi}} = mb^2\sin^2\theta\dot{\phi} \\
\dot{p}_\phi &= \frac{\partial L}{\partial \phi} = -\frac{\partial R_{noncyclic}}{\partial \phi} = 0
\end{aligned}
$$

that is, p_ϕ is a constant of motion.

Hamilton's equations of motion give

$$\dot{\theta} = \frac{\partial R_{noncyclic}}{\partial p_\theta} = \frac{p_\theta}{mb^2} \quad\quad (\delta)$$

$$-\dot{p}_\theta = \frac{\partial R_{noncyclic}}{\partial \theta} = -\frac{p_\phi^2\cos\theta}{mb^2\sin^3\theta} + mgb\sin\theta \quad\quad (\epsilon)$$

Equation δ gives that

$$\frac{\partial}{\partial t}\dot{\theta} = \ddot{\theta} = \frac{\dot{p}_\theta}{mb^2}$$

Inserting this into equation ϵ gives

$$\ddot{\theta} - \frac{p_\phi^2\cos\theta}{m^2b^4\sin^3\theta} + \frac{g}{b}\sin\theta = 0$$

which is identical to the equation of motion α derived using R_{cyclic}. The Hamiltonian in the rotating frame is a constant of motion given by γ, but it does not include the total energy.

Note that these examples show that both forms of the Routhian, as well as the complete Lagrangian formalism, shown in example 6.10, and complete Hamiltonian formalism, shown in example 8.6, all give the same equations of motion. This illustrates that the Lagrangian, Hamiltonian, and Routhian mechanics all give the same equations of motion and this applies both in the static inertial frame as well as a rotating frame since the Lagrangian, Hamiltonian and Routhian all are scalars under rotation, that is, they are rotationally invariant.

8.9 Example: Single particle moving in a vertical plane under the influence of an inverse-square central force

The Lagrangian for a single particle of mass m, moving in a vertical plane and subject to a central inverse square central force, is specified by two generalized coordinates, r, and θ.

$$L = \frac{m}{2}(\dot{r}^2 + r^2\dot{\theta}^2) + \frac{k}{r}$$

The ignorable coordinate is θ, since it is cyclic. Let the constant conjugate momentum be denoted by $p_\theta = \frac{\partial L}{\partial \dot{\theta}} = mr^2\dot{\theta}$. Then the corresponding cyclic Routhian is

$$R_{cyclic}(r,\theta,\dot{r},p_\theta) = p_\theta\dot{\theta} - L = \frac{p_\theta^2}{2mr^2} - \frac{1}{2}mr^2 - \frac{k}{r}$$

This Routhian is the equivalent one-dimensional potential $U(r)$ minus the kinetic energy of radial motion. Applying Hamilton's equation to the cyclic coordinate θ gives

$$\dot{p}_\theta = 0 \qquad\qquad \frac{p_\theta}{mr^2} = \dot{\theta}$$

implying a solution

$$p_\theta = mr^2\dot{\theta} = l$$

where the angular momentum l is a constant.

The Lagrange-Euler equation can be applied to the non-cyclic coordinate r

$$\Lambda_r L = \frac{d}{dt}\frac{\partial R_{cyclic}}{\partial \dot{r}} - \frac{\partial R_{cyclic}}{\partial r} = 0$$

where the negative sign of R_{cyclic} cancels. This leads to the radial solution

$$m\ddot{r} - \frac{p_\theta^2}{mr^3} + \frac{k}{r^2} = 0$$

where $p_\theta = l$ which is a constant of motion in the centrifugal term. Thus the problem has been reduced to a one-dimensional problem in radius r that is in a rotating frame of reference.

8.7   Variable-mass systems

Lagrangian and Hamiltonian mechanics assume that the total mass and energy of the system are conserved. Variable-mass systems involve transferring mass and energy between donor and receptor bodies. However, such systems still can be conservative if the Lagrangian or Hamiltonian include all the active degrees of freedom for the combined donor-receptor system. The following examples of variable mass systems illustrate subtle complications that occur handling such problems using algebraic mechanics.

8.7.1   Rocket propulsion:

Newtonian mechanics was used to solve the rocket problem in chapter 2.12.6.   The equation of motion (2.113) relating the rocket thrust F_{ex} to the rate of change of the momentum separated into two terms,

$$F_{ex} = \dot{p}_y = m\ddot{y} + \dot{m}\dot{y} \tag{8.69}$$

The first term is the usual mass times acceleration, while the second term arises from the rate of change of mass times the velocity. The equation of motion for rocket motion is easily derived using either Lagrangian or Hamiltonian mechanics by relating the rocket thrust to the generalized force Q_j^{EXC}.

8.7.2 Moving chains:

The motion of a flexible, frictionless, heavy chain that is falling in a gravitational field, often can be split into two coupled variable-mass partitions that have different chain-link velocities. These partitions are coupled at the moving intersection between the chain partitions. That is, these partitions share time-dependent fractions of the total chain mass. Moving chains were discussed first by Caley in 1857[Cay1857] and since then the moving chain problem has had a controversial history due to the frequent erroneous assumption that, in the gravitational field, the chain partitions fall with acceleration g rather than applying the correct energy conservation assumption for this conservative system. The following two examples of conservative falling-chain systems illustrate solutions obtained using variational principles applied to a single chain that is partitioned into two variable length sections.[1]

Consider the following two possible scenarios for motion of a flexible, heavy, frictionless, chain located in a uniform gravitational field g. The first scenario is the "folded chain" system which assumes that one end of the chain is held fixed, while the adjacent free end is released at the same altitude as the top of the fixed arm, and this free end is allowed to fall in the constant gravitational field g. The second "falling chain", scenario assumes that one end of the chain is hanging down through a hole in a frictionless, smooth, rigid, horizontal table, with the stationary partition of the chain sitting on the table surrounding the hole. The falling section of this chain is being pulled out of the stationary pile by the hanging partition. Both of these systems are conservative since it is assumed that the total mass of the chain is fixed, and no dissipative forces are acting. The chains are assumed to be inextensible, flexible, and frictionless, and subject to a uniform gravitational field g in the vertical y direction. In both examples, the chain, with mass M and length L, is partitioned into a stationary segment, plus a moving segment, where the mass per unit length of the chain is $\mu = \frac{M}{L}$. These partitions are strongly coupled at their intersection which propagates downward with time for the "folded chain" and propagates upward, relative to the lower end of the falling chain, for the "falling chain". For the "folded chain", the chain links are transferred from the moving segment to the stationary segment as the moving section falls. By contrast, for the "falling system", the chain links are transferred from the stationary upper section to the moving lower segment of the chain.

8.10 *Example: Folded chain*

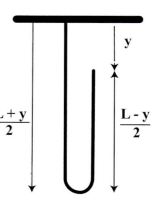

The folded chain of length L and mass-per-unit-length $\mu = \frac{M}{L}$ hangs vertically downwards in a gravitational field g with both ends held initially at the same height. The fixed end is attached to a fixed support while the free end of the chain is dropped at time $t = 0$ with the free end at the same height and adjacent to the fixed end. Let y be the distance the falling free end is below the fixed end. Using an idealized one-dimensional assumption, the Lagrangian $\mathcal{L}$ is given by

$$\mathcal{L}(y, \dot{y}) = \frac{M}{4L}(L - y)\dot{y}^2 + Mg\frac{1}{4L}(L^2 + 2Ly - y^2) \qquad (8.70)$$

where the bracket in the second term is the height of the center of mass of the folded chain with respect to the fixed upper end of the chain.

The Hamiltonian is given by

$$H(y, p_R) = p_R\dot{y} - \mathcal{L}(y, \dot{y}) = \frac{p_R}{\mu(L - y)} - Mg\frac{(L^2 + 2Ly - y^2)}{4L} \qquad (8.71)$$

where p_R is the linear momentum of the right-hand arm of the folded chain.

As shown in the discussion of the Generalized Energy Theorem, (chapters 7.8 and 7.9), when all the active forces are included in the Lagrangian and the Hamiltonian, then the total mechanical energy E is given by $E = H$. Moreover, both the Lagrangian and the Hamiltonian are time independent, since

$$\frac{dE}{dt} = \frac{dH}{dt} = -\frac{\partial \mathcal{L}}{\partial t} = 0 \qquad (8.72)$$

Therefore the "folded chain" Hamiltonian equals the total energy, which is a constant of motion. Energy conservation for this system can be used to give

---

[1] Discussions with Professor Frank Wolfs stimulated inclusion of these two examples of moving chains.

$$\frac{\mu}{4}\left(L-y\right)\dot{y}^2 - \frac{1}{4}\mu g(L^2 + 2Ly - y^2) = -\frac{1}{4}\mu g L^2 \tag{8.73}$$

Solve for $\dot{y}^2$ gives

$$\dot{y}^2 = g\frac{(2Ly - y^2)}{L - y} \tag{8.74}$$

The acceleration of the falling arm, $\ddot{y}$, is given by taking the time derivative of equation 8.74

$$\ddot{y} = g + \frac{g\left(2Ly - y^2\right)}{2\left(L - y\right)} \tag{8.75}$$

The rate of change in linear momentum for the moving right side of the chain, $\dot{p}_R$, is given by

$$\dot{p}_R = m_R\ddot{y} + \dot{m}_R\dot{y} = m_R g + m_R g\frac{(2Ly - y^2)}{2\left(L - y\right)} \tag{8.76}$$

For this energy-conserving chain, the tension in the chain T_0 at the fixed end of the chain is given by

$$T_0 = \frac{\mu g}{2}\left(L + y\right) + \frac{1}{4}\mu\dot{y}^2 \tag{8.77}$$

Equations 8.74 and 8.76, imply that the tension T_o diverges to infinity when $y \to L$. Calkin and March measured the y dependence of the chain tension at the support for the folded chain and observed the predicted y dependence. The maximum tension was $\simeq 25Mg$, which is consistent with that predicted using equation 8.77 after taking into account the finite size and mass of individual links in the chain. This result is very different from that obtained using the erroneous assumption that the right arm falls with the free-fall acceleration g, which implies a maximum tension $T_0 = 2Mg$. Thus the free-fall assumption disagrees with the experimental results, in addition to violating energy conservation and the tenets of Lagrangian and Hamiltonian mechanics. That is, the experimental result demonstrates unambiguously that the energy conservation predictions apply in contradiction with the erroneous free-fall assumption.

The unusual feature of variable mass problems, such as the folded chain problem, is that the rate of change of momentum in equation 8.76 includes two contributions to the force and rate of change of momentum, that is, it includes both the acceleration term $m_R\ddot{y}$ plus the variable mass term $\dot{m}_R\dot{y}$ that accounts for the transfer of matter at the intersection of the moving and stationary partitions of the chain. At the transition point of the chain, moving links are transferred from the moving section and are added to the stationary subsection. Since this moving section is falling downwards, and the stationary section is stationary, then the transferred momentum is in a downward direction corresponding to an increased effective downward force. Thus the measured acceleration of the moving arm actually is faster than g. A related phenomenon is the loud cracking sound heard when cracking a whip.

8.11 Example: Falling chain

 The "falling chain", scenario assumes that one end of the chain is hanging down through a hole in a frictionless, smooth, rigid, horizontal table, with the stationary partition of the chain lying on the frictionless table surrounding the hole. The falling section of this chain is being pulled out of the stationary pile by the hanging partition. The analysis for the problem of the falling chain behaves differently from the folded chain. For the "falling-chain" let y be the falling distance of the lower end of the chain measured with respect to the table top. The Lagrangian and Hamiltonian are given by

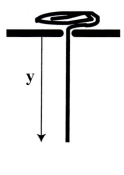

$$\mathcal{L}(y, \dot{y}) = \frac{\mu}{2}y\dot{y}^2 + \mu g\frac{y^2}{2} \tag{8.78}$$

$$p_y = \frac{\partial \mathcal{L}}{\partial \dot{y}} = \mu y\dot{y} \tag{8.79}$$

$$H = \frac{p_y^2}{2\mu y} - \frac{\mu g y^2}{2} = E \tag{8.80}$$

The Lagrangian and Hamiltonian are not explicitly time dependent, and the Hamiltonian equals the initial total energy, E_0. Thus energy conservation can be used to give that

$$E = \frac{1}{2}\mu y(\dot{y}^2 - gy) = E_0 \tag{8.81}$$

Lagrange's equation of motion gives

$$\dot{p}_y = m_y\ddot{y} + \dot{m}_y\dot{y} = m_yg + \frac{1}{2}\mu\dot{y}^2 = Mg - T_0 \tag{8.82}$$

The important difference between the folded chain and falling chain is that the moving component of the falling chain is gaining mass with time rather than losing mass. Also the tension in the chain T_0 reduces the acceleration of the falling chain making it less than the free-fall value g. This is in contrast to that for the folded chain system where the acceleration exceeds g.

The above discussion shows that Lagrangian and Hamiltonian can be applied to variable-mass systems if both the donor and receptor degrees of freedom are included to ensure that the total mass is conserved.

8.8 Summary

Hamilton's equations of motion

Inserting the generalized momentum into Jacobi's generalized energy relation was used to define the Hamiltonian function to be

$$H(\mathbf{q}, \mathbf{p}, t) = \mathbf{p} \cdot \dot{\mathbf{q}} - L(\mathbf{q}, \dot{\mathbf{q}}, t) \tag{8.3}$$

The Legendre transform of the Lagrange-Euler equations, led to Hamilton's equations of motion.

$$\dot{q}_j = \frac{\partial H}{\partial p_j} \tag{8.25}$$

$$\dot{p}_j = -\frac{\partial H}{\partial q_j} + \left[\sum_{k=1}^{m} \lambda_k \frac{\partial g_k}{\partial q_j} + Q_j^{EXC}\right] \tag{8.26}$$

The generalized energy equation 7.38 gives the time dependence

$$\frac{dH(\mathbf{q}, \mathbf{p}, t)}{dt} = \sum_j \left(\left[\sum_{k=1}^{m} \lambda_k \frac{\partial g_k}{\partial q_j} + Q_j^{EXC}\right]\dot{q}_j\right) - \frac{\partial L(\mathbf{q}, \dot{\mathbf{q}}, t)}{\partial t} \tag{8.27}$$

where

$$\frac{\partial H}{\partial t} = -\frac{\partial L}{\partial t} \tag{8.24}$$

The p_k, q_k are treated as independent canonical variables. Lagrange was the first to derive the canonical equations but he did not recognize them as a basic set of equations of motion. Hamilton derived the canonical equations of motion from his fundamental variational principle and made them the basis for a far-reaching theory of dynamics. Hamilton's equations give $2s$ first-order differential equations for p_k, q_k for each of the s degrees of freedom. Lagrange's equations give s second-order differential equations for the variables $q_k, \dot{q}_k$.

Routhian reduction technique

The Routhian reduction technique is a hybrid of Lagrangian and Hamiltonian mechanics that exploits the advantages of both approaches for solving problems involving cyclic variables. It is especially useful for solving motion in rotating systems in science and engineering. Two Routhians are used frequently for solving the equations of motion of rotating systems. Assuming that the variables between $1 \leq i \leq s$ are non-cyclic, while the m variables between $s + 1 \leq i \leq n$ are ignorable cyclic coordinates, then the two Routhians are:

$$R_{cyclic}(q_1, ..., q_n; \dot{q}_1, ..., \dot{q}_s; p_{s+1}, ..., p_n; t) = \sum_{cyclic}^{m} p_i\dot{q}_i - L = H - \sum_{noncyclic}^{s} p_i\dot{q}_i \tag{8.65}$$

$$R_{noncyclic}(q_1, ..., q_n; p_1, ..., p_s; \dot{q}_{s+1}, ..., \dot{q}_n; t) = \sum_{noncyclic}^{s} p_i\dot{q}_i - L = H - \sum_{cyclic}^{m} p_i\dot{q}_i \tag{8.68}$$

The Routhian R_{cyclic} is a negative Lagrangian for the non-cyclic variables between $1 \leq i \leq s$, where $s = n - m$, and is a Hamiltonian for the m cyclic variables between $s + 1 \leq i \leq n$. Since the cyclic variables are constants of the Hamiltonian, their solution is trivial, and the number of variables included in the Lagrangian is reduced from n to $s = n - m$. The Routhian R_{cyclic} is useful for solving some problems in classical mechanics. The Routhian $R_{noncyclic}$ is a Hamiltonian for the non-cyclic variables between $1 \leq i \leq s$, and is a negative Lagrangian for the m cyclic variables between $s + 1 \leq i \leq n$. Since the cyclic variables are constants of motion, the Routhian $R_{noncyclic}$ also is a constant of motion but it does not equal the total energy since the coordinate transformation is time dependent. The Routhian $R_{noncyclic}$ is especially valuable for solving rotating many-body systems such as galaxies, molecules, or nuclei, since the Routhian $R_{noncyclic}$ is the Hamiltonian in the rotating body-fixed coordinate frame.

Variable mass systems:

Two examples of heavy flexible chains falling in a uniform gravitational field were used to illustrate how variable mass systems can be handled using Lagrangian and Hamiltonian mechanics. The falling-mass system is conservative assuming that both the donor plus the receptor body systems are included.

Comparison of Lagrangian and Hamiltonian mechanics

Lagrangian and the Hamiltonian dynamics are two powerful and related variational algebraic formulations of mechanics that are based on Hamilton's action principle. They can be applied to any conservative degrees of freedom as discussed in chapters $6, 8$, and 15. Lagrangian and Hamiltonian mechanics both concentrate solely on active forces and can ignore internal forces. They can handle many-body systems and allow convenient generalized coordinates of choice. This ability is impractical or impossible using Newtonian mechanics. Thus it is natural to compare the relative advantages of these two algebraic formalisms in order to decide which should be used for a specific problem.

For a system with n generalized coordinates, plus m constraint forces that are not required to be known, then the Lagrangian approach, using a minimal set of generalized coordinates, reduces to only $s = n - m$ *second-order* differential equations and unknowns compared to the Newtonian approach where there are $n + m$ unknowns. Alternatively, use of Lagrange multipliers allows determination of the constraint forces resulting in $n + m$ second order equations and unknowns. The Lagrangian potential function is limited to conservative forces, Lagrange multipliers can be used to handle holonomic forces of constraint, while generalized forces can be used to handle non-conservative and non-holonomic forces. The advantage of the Lagrange equations of motion is that they can deal with any type of force, conservative or non-conservative, and they directly determine $q, \dot{q}$ rather than q, p which then requires relating p to $\dot{q}$.

For a system with n generalized coordinates, the Hamiltonian approach determines $2n$ *first-order* differential equations which are easier to solve than second-order equations. However, the $2n$ solutions must be combined to determine the equations of motion. The Hamiltonian approach is superior to the Lagrange approach in its ability to obtain an analytical solution of the integrals of the motion. Hamiltonian dynamics also has a means of determining the unknown variables for which the solution assumes a soluble form. Important applications of Hamiltonian mechanics are to quantum mechanics and statistical mechanics, where quantum analogs of q_i and p_i, can be used to relate to the fundamental variables of Hamiltonian mechanics. This does not apply for the variables q_i and $\dot{q}_i$ of Lagrangian mechanics. The Hamiltonian approach is especially powerful when the system has m cyclic variables, then the m conjugate momenta p_i are constants. Thus the m conjugate variables (q_i, p_i) can be factored out of the Hamiltonian, which reduces the number of conjugate variables required to $n - m$. This is not possible using the Lagrangian approach since, even though the m coordinates q_i can be factored out, the velocities $\dot{q}_i$ still must be included, thus the n conjugate variables must be included. The Lagrange approach is advantageous for obtaining a numerical solution of systems in classical mechanics. However, Hamiltonian mechanics expresses the variables in terms of the fundamental canonical variables $(\mathbf{q}, \mathbf{p})$ which provides a more fundamental insight into the underlying physics.[2]

---

[2]**Recommended reading:** *"Classical Mechanics"* H. Goldstein, Addison-Wesley, Reading (1950). The present chapter closely follows the notation used by Goldstein to facilitate cross-referencing and reading the many other textbooks that have adopted this notation.

Workshop exercises

1. A block of mass m rests on an inclined plane making an angle θ with the horizontal. The inclined plane (a triangular block of mass M) is free to slide horizontally without friction. The block of mass m is also free to slide on the larger block of mass M without friction.

   (a) Construct the Lagrangian function.

   (b) Derive the equations of motion for this system.

   (c) Calculate the canonical momenta.

   (d) Construct the Hamiltonian function.

   (e) Find which of the two momenta found in part (c) is a constant of motion and discuss why it is so. If the two blocks start from rest, what is the value of this constant of motion?

2. Discuss among yourselves the following four conditions that can exist for the Hamiltonian and give several examples of systems exhibiting each of the four conditions.

   (a) The Hamiltonian is conserved and equals the total mechanical energy

   (b) The Hamiltonian is conserved but does not equal the total mechanical energy

   (c) The Hamiltonian is not conserved but does equal the total mechanical energy

   (d) The Hamiltonian is not conserved and does not equal the mechanical total energy.

3. A block of mass m rests on an inclined plane making an angle θ with the horizontal. The inclined plane (a triangular block of mass M) is free to slide horizontally without friction. The block of mass m is also free to slide on the larger block of mass M without friction.

   (a) Construct the Lagrangian function.

   (b) Derive the equations of motion for this system.

   (c) Calculate the canonical momenta.

   (d) Construct the Hamiltonian function.

   (e) Find which of the two momenta found in part (c) is a constant of motion and discuss why it is so. If the two blocks start from rest, what is the value of this constant of motion?

4. Discuss among yourselves the following four conditions that can exist for the Hamiltonian and give several examples of systems exhibiting each of the four conditions.

   a) The Hamiltonian is conserved and equals the total mechanical energy

   b) The Hamiltonian is conserved but does not equal the total mechanical energy

   c) The Hamiltonian is not conserved but does equal the total mechanical energy

   d) The Hamiltonian is not conserved and does not equal the mechanical total energy

5. Compare the Lagrangian formalism and the Hamiltonian formalism by creating a two-column chart. Label one side "Lagrangian" and the other side "Hamiltonian" and discuss the similarities and differences. Here are some ideas to get you started:

   • What are the basic variables in each formalism?

   • What are the form and number of the equations of motion derived in each case?

   • How does the Lagrangian "state space" compare to the Hamiltonian "phase space"?

6. It can be shown that if $L(q, \dot{q}, t)$ is the Lagrangian of a particle moving in one dimension, then $L = L'$ where $L'(q, \dot{q}, t) = L(q, \dot{q}, t) + \frac{df}{dt}$ and $f(q, t)$ is an arbitrary function. This problem explores the consequences of this on the Hamiltonian formalism.

(a) Relate the new canonical momentum p', for L', to the old canonical momentum p, for L.

(b) Express the new Hamiltonian $H'(q', p', t)$ for L' in terms of the old Hamiltonian $H(q, p, t)$ and f.

(c) Explicitly show that the new Hamilton's equations for H' are equivalent to the old Hamilton's equations for H.

7. A massless hoop of radius R is rotating about an axis perpendicular to its central axis at constant angular velocity ω. A mass m can freely slide around the hoop.

(a) Determine the Lagrangian of the system.

(b) Determine the Hamiltonian of the system. Does it equal the total mechanical energy?

(c) Determine the Lagrangian of the system with respect to a coordinate frame in which $H = T + V_{\text{eff}}$. What is V_{eff}? What force generates the additional term in V_{eff}?

8. Consider a pendulum of length L attached to the end of rod of length R. The rod is rotating at constant angular velocity ω in the plane. Assume the pendulum is always taut.

(a) Determine equations of motion.

(b) For what value of $\omega^2 R$ is this system the same as a plane pendulum in a constant gravitational field?

(c) Show $H \neq E$. What is the reason?

Problems

1) A particle of mass m in a gravitational field slides on the inside of a smooth parabola of revolution whose axis is vertical. Using the distance from the axis r, and the azimuthal angle φ as generalized coordinates, find the following.
   a) The Lagrangian of the system.
   b) The generalized momenta and the corresponding Hamiltonian
   c) The equation of motion for the coordinate r as a function of time.
   d) If $\frac{d\varphi}{dt} = 0$, show that the particle can execute small oscillations about the lowest point of the paraboloid and find the frequency of these oscillations.

2) Consider a particle of mass m which is constrained to move on the surface of a sphere of radius R. There are no external forces of any kind acting on the particle.
   a) What is the number of generalized coordinates necessary to describe the problem?
   b) Choose a set of generalized coordinates and write the Lagrangian of the system.
   c) What is the Hamiltonian of the system? Is it conserved?
   d) Prove that the motion of the particle is along a great circle of the sphere.

3. A block of mass m is attached to a wedge of mass M by a spring with spring constant k. The inclined frictionless surface of the wedge makes an angle α to the horizontal. The wedge is free to slide on a horizontal frictionless surface as shown in the figure.
   a) Given that the relaxed length of the spring is d, find the values s_0 when both book and wedge are stationary.
   b) Find the Lagrangian for the system as a function of the x coordinate of the wedge and the length of spring s. Write down the equations of motion.
   c) What is the natural frequency of vibration?

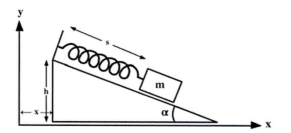

4. A fly-ball governor comprises two masses m connected by 4 hinged arms of length l to a vertical shaft and to a mass M which can slide up or down the shaft without friction in a uniform vertical gravitational field as shown in the figure. The assembly is constrained to rotate around the axis of the vertical shaft with same angular velocity as that of the vertical shaft. Neglect the mass of the arms, air friction, and assume that the mass M has a negligible moment of inertia. Assume that the whole system is constrained to rotate with a constant angular velocity ω_0.

a) Choose suitable coordinates and use the Lagrangian to derive equations of motion of the system around the equilibrium position.

b) Determine the height z of the mass M above its lowest position as a function of ω_0.

c) Find the frequency of small oscillations about this steady motion.

d) Derive a Routhian that provides the Hamiltonian in the rotating system.

e) Is the total energy of the fly-ball governor in the rotating frame of reference constant in time?

f) Suppose that the shaft and assembly are not constrained to rotate at a constant angular velocity ω_0, that is, it is allowed to rotate freely at angular velocity $\dot{\varphi}$. What is the difference in the overall motion?

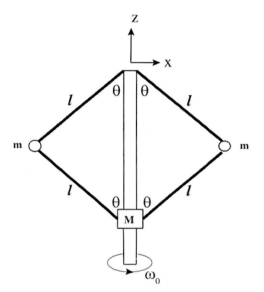

5. A rigid straight, frictionless, massless, rod rotates about the z axis at an angular velocity $\dot{\theta}$. A mass m slides along the frictionless rod and is attached to the rod by a massless spring of spring constant κ.

a; Derive the Lagrangian and the Hamiltonian

b; Derive the equations of motion in the stationary frame using Hamiltonian mechanics.

c; What are the constants of motion?

d; If the rotation is constrained to have a constant angular velocity $\dot{\theta} = \omega$ then is the non-cyclic Routhian $R_{noncyclic} = H - p_\theta \dot{\theta}$ a constant of motion, and does it equal the total energy?

e; Use the non-cyclic Routhian $R_{noncyclic}$ to derive the radial equation of motion in the rotating frame of reference for the cranked system with $\dot{\theta} = \omega$.

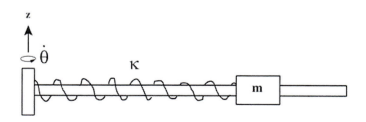

6. A thin uniform rod of length $2L$ and mass M is suspended from a massless string of length l tied to a nail. Initially the rod hangs vertically. A weak horizontal force F is applied to the rod's free end.

a) Write the Lagrangian for this system.

b) For very short times such that all angles are small, determine the angles that string and the rod make with the vertical. Start from rest at $t = 0$.

c) Draw a diagram to illustrate the initial motion of the rod.

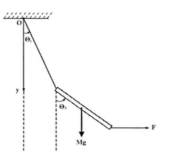

7. A uniform ladder of mass M and length $2L$ is leaning against a frictionless vertical wall with its feet on a frictionless horizontal floor. Initially the stationary ladder is released at an angle $\theta_0 = 60°$ to the floor. Assume that gravitation field $g = 9.81 m/s^2$ acts vertically downward and that the moment of inertia of the ladder about its midpoint is $I = \frac{1}{3}ML^2$.

a) Derive the Lagrangian

b) Derive the Hamiltonian

c) Explain if the Hamiltonian is conserved and/or if it equals the total energy

d) Use the Lagrangian to derive the equations of motion

e) Derive the angle θ at which the ladder loses contact with the vertical wall?

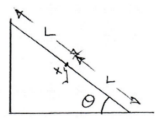

8. The classical mechanics exam induces Jacob to try his hand at bungee jumping. Assume Jacob's mass m is suspended in a gravitational field by the bungee of unstretched length b and spring constant k. Besides the longitudinal oscillations due to the bungee jump, Jacob also swings with plane pendulum motion in a vertical plane. Use polar coordinates r, ϕ, neglect air drag, and assume that the bungee always is under tension.

a; Derive the Lagrangian

b; Determine Lagrange's equation of motion for angular motion and identify by name the forces contributing to the angular motion.

c; Determine Lagrange's equation of motion for radial oscillation and identify by name the forces contributing to the tension in the spring.

d; Derive the generalized momenta

e; Determine the Hamiltonian and give all of Hamilton's equations of motion.

Chapter 9

Hamilton's Action Principle

9.1 Introduction

Hamilton's principle of stationary action was introduced in two papers published by Hamilton in 1834 and 1835. As mentioned in the Prologue, Hamilton's Action Principle is the foundation of the hierarchy of three philosophical stages that are used in applying analytical mechanics. The first stage is to use Hamilton's Action Principle to derive either the Hamiltonian and Lagrangian for the system. The second stage is to use either Lagrangian mechanics, or Hamiltonian mechanics, to derive the equations of motion for the system. The third stage is to solve these equations of motion for the assumed initial conditions. Lagrange had pioneered Lagrangian mechanics in 1788 based on d'Alembert's Principle. Hamilton's Action Principle now underlies theoretical physics, and many other disciplines in mathematics and economics. In 1834 Hamilton was seeking a theory of optics when he developed both his Action Principle, and the field of Hamiltonian mechanics.

Hamilton's Action Principle is based on defining the **action functional**[1] S for n generalized coordinates which are expressed by the vector $\mathbf{q}$, and their corresponding velocity vector $\dot{\mathbf{q}}$.

$$S = \int_{t_i}^{t_f} L(\mathbf{q}, \dot{\mathbf{q}}, t) dt \tag{9.1}$$

The scalar action S, is a functional of the Lagrangian $L(\mathbf{q}, \dot{\mathbf{q}}, t)$, integrated between an initial time t_i and final time t_f. In principle, higher order time derivatives of the generalized coordinates could be included, but most systems in classical mechanics are described adequately by including only the generalized coordinates, plus their velocities. The definition of the action functional allows for more general Lagrangians than the simple Standard Lagrangian $L(\mathbf{q}, \dot{\mathbf{q}}, t) = T(\dot{\mathbf{q}}, t) - U(\mathbf{q}, t)$ that has been used throughout chapters $5 - 8$. Hamilton stated that the actual trajectory of a mechanical system is that given by requiring that the action functional is stationary with respect to change of the variables. The action functional is stationary when the variational principle can be written in terms of a virtual infinitessimal displacement, δ, to be

$$\delta S = \delta \int_{t_i}^{t_f} L(\mathbf{q}, \dot{\mathbf{q}}, t) dt = 0 \tag{9.2}$$

Typically the stationary point corresponds to a minimum of the action functional. Applying variational calculus to the action functional leads to the same Lagrange equations of motion for systems as the equations derived using d'Alembert's Principle, if the additional generalized force terms, $\sum_{k=1}^{m} \lambda_k \frac{\partial g_k}{\partial q_j}(\mathbf{q}, t) + Q_j^{EXC}$, are omitted in the corresponding equations of motion.

These are used to derive the equations of motion, which then are solved for an assumed set of initial conditions. Prior to Hamilton's Action Principle, Lagrange developed Lagrangian mechanics based on d'Alembert's Principle while the Newtonian equations of motion are defined in terms of Newton's Laws of Motion.

---

[1] The term "action functional" was named "Hamilton's Principal Function" in older texts. The name usually is abbreviated to "action" in modern mechanics.

9.2   Hamilton's Principle of Stationary Action

Hamilton's crowning achievement was his use of the general form of Hamilton's principle of stationary action S, equation 9.2, to derive both Lagrangian mechanics, and Hamiltonian mechanics. Consider the action S_A for the extremum path of a system in configuration space, that is, along path A for $j = 1, 2, ..., n$ coordinates $q_j(t_i)$ at initial time t_i to $q_j(t_f)$ at a final time t_f as shown in figure 9.1. Then the action S_A is given by

$$S_A = \int_{t_i}^{t_f} L(\mathbf{q}(t), \dot{\mathbf{q}}(t), t)dt \qquad (9.3)$$

As used in chapter 5.2, a family of neighboring paths is defined by adding an infinitessimal fraction ϵ of a continuous, well-behaved neighboring function η_j where $\epsilon = 0$ for the extremum path. That is,

$$q_j(t, \epsilon) = q_j(t, 0) + \epsilon \eta_j(t) \qquad (9.4)$$

In contrast to the variational case discussed when deriving Lagrangian mechanics, the variational path used here does not assume that the functions $\eta_i(t)$ vanish at the end points. Assume that the neighboring path B has an action S_B where

$$S_B = \int_{t_i + \Delta t}^{t_f + \Delta t} L(\mathbf{q}(t) + \boldsymbol{\delta}\mathbf{q}(t), \dot{\mathbf{q}}(t) + \boldsymbol{\delta}\dot{\mathbf{q}}(t))dt \qquad (9.5)$$

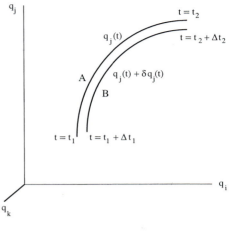

Figure 9.1: Extremum path A, plus the neighboring path B, shown in configuration space.

Expanding the integrand of S_B in equation 9.5 gives that, relative to the extremum path A, the incremental change in action is

$$\delta S = S_B - S_A = \int_{t_i}^{t_f} \sum_j \left(\frac{\partial L}{\partial q_j} \delta q_j + \frac{\partial L}{\partial \dot{q}_j} \delta \dot{q}_j \right) dt + [L\Delta t]_{t_i}^{t_f} \qquad (9.6)$$

The second term in the integral can be integrated by parts since $\delta \dot{q}_j = d\left(\frac{\delta q_j}{dt} \right)$ leading to

$$\delta S = \int_{t_i}^{t_f} \sum_j \left(\frac{\partial L}{\partial q_j} - \frac{d}{dt}\frac{\partial L}{\partial \dot{q}_j} \right) \delta q_j dt + \left[\sum_j \frac{\partial L}{\partial \dot{q}_j} \delta q_j + L\Delta t \right]_{t_i}^{t_f} \qquad (9.7)$$

Note that equation 9.7 includes contributions from the entire path of the integral as well as the variations at the ends of the curve and the Δt terms. Equation 9.7 leads to the following two pioneering principles of least action in variational mechanics that were developed by Hamilton.

9.2.1   Stationary-action principle in Lagrangian mechanics

Derivation of Lagrangian mechanics in chapter 6 was based on the extremum path for neighboring paths *between two given locations* $\mathbf{q}(t_i)$ and $\mathbf{q}(t_f)$ that the system occupies at the initial and final times t_i and t_f respectively. For the special case, where the end points do not vary, that is, when $\delta q_i(t_i) = \delta q_i(t_f) = 0$, and $\Delta t_i = \Delta t_f = 0$, then the least action δS for the stationary path (9.8) reduces to

$$\delta S = \int_{t_i}^{t_f} \sum_j \left(\frac{\partial L}{\partial q_j} - \frac{d}{dt}\frac{\partial L}{\partial \dot{q}_j} \right) \delta q_j dt = 0 \qquad (9.8)$$

For independent generalized coordinates δq_j, the integrand in brackets vanishes leading to the Euler-Lagrange equations. Conversely, if the Euler-Lagrange equations in 9.8 are satisfied, then, $\delta S = 0$, that is, the path is stationary. This leads to the statement that *the path in configuration space between two configurations* $\mathbf{q}(t_i)$ *and* $\mathbf{q}(t_f)$ *that the system occupies at times* t_i *and* t_f *respectively, is that for which the action* S *is stationary*. This is a statement of Hamilton's Principle.

9.2.2 Stationary-action principle in Hamiltonian mechanics

Hamilton used the general variation of the least-action path to derive of the basic equations of Hamiltonian mechanics. For the general path, the integral term in equation 9.7 vanishes because the Euler-Lagrange equations are obeyed for the stationary path. Thus the only remaining non-zero contributions are due to the end point terms, which can be written by defining the total variation of each end point to be

$$\Delta q_j = \delta q_j + \dot{q}_j \Delta t \tag{9.9}$$

where δq_i and $\dot{q}_i$ are evaluated at t_i and t_f. Then equation 9.7 reduces to

$$\delta S = \left[\sum_j \frac{\partial L}{\partial \dot{q}_j} \delta q_j + L \Delta t \right]_{t_i}^{t_f} = \left[\sum_j \frac{\partial L}{\partial \dot{q}_j} \Delta q_j + \left(-\sum_j \frac{\partial L}{\partial \dot{q}_j} \dot{q}_j + L \right) \Delta t \right]_{t_i}^{t_f} \tag{9.10}$$

Since the generalized momentum $p_j = \frac{\partial L}{\partial \dot{q}_j}$, then equation 9.10 can be expressed in terms of the Hamiltonian and generalized momentum as

$$\delta S = \left[\sum_j p_j \Delta q_j - H \Delta t \right]_{t_i}^{t_f} = [\mathbf{p} \cdot \Delta \mathbf{q} - H \Delta t]_{t_i}^{t_f} \tag{9.11}$$

$$\frac{\partial S}{\partial q_j} = \frac{\partial L}{\partial \dot{q}_j} = p_j \tag{9.12}$$

Equation 9.11 contains Hamilton's Principle of Least-action. Equation 9.12 gives an alternative relation of the generalized momentum p_j that is expressed in terms of the action functional S. Note that equations 9.11 and 9.12, were derived directly without invoking reference to the Lagrangian.

Integrating the action δS, equation 9.10, between the end points gives the action for the path between $t = t_i$ and $t = t_f$, that is, $S(q_j(t_i), t_1, q_j(t_f), t_2)$ to be

$$S(q_j(t_i), t_i, q_j(t_f), t_f) = \int_i^f [\mathbf{p} \cdot \dot{\mathbf{q}} - H(\mathbf{q}, \mathbf{p}, t)] \, dt \tag{9.13}$$

The stationary path is obtained by using the variational principle

$$\delta S = \delta \int_i^f [\mathbf{p} \cdot \dot{\mathbf{q}} - H(\mathbf{q}, \mathbf{p}, t)] \, dt = 0 \tag{9.14}$$

The integrand, $I = [\mathbf{p} \cdot \dot{\mathbf{q}} - H(\mathbf{q}, \mathbf{p}, t)]$, in this modified Hamilton's principle, can be used in the n Euler-Lagrange equations for $j = 1, 2, 3, ..., n$ to give

$$\frac{d}{dt}\left(\frac{\partial I}{\partial \dot{q}_j}\right) - \frac{\partial I}{\partial q_j} = \dot{p}_j + \frac{\partial H}{\partial q_j} = 0 \tag{9.15}$$

Similarly, the other n Euler-Lagrange equations give

$$\frac{d}{dt}\left(\frac{\partial I}{\partial \dot{p}_j}\right) - \frac{\partial I}{\partial p_j} = -\dot{q}_j + \frac{\partial H}{\partial p_j} = 0 \tag{9.16}$$

Thus Hamilton's principle of least-action leads to Hamilton's equations of motion, that is equations 9.15, and 9.16.

The total time derivative of the action S, which is a function of the coordinates and time, is

$$\frac{dS}{dt} = \frac{\partial S}{\partial t} + \sum_j^n \frac{\partial S}{\partial q_j} \dot{q}_j = \frac{\partial S}{\partial t} + \mathbf{p} \cdot \dot{\mathbf{q}}_j \tag{9.17}$$

But the total time derivative of equation 9.14 equals

$$\frac{dS}{dt} = \mathbf{p} \cdot \dot{\mathbf{q}} - H(\mathbf{q}, \mathbf{p}, t) \tag{9.18}$$

Combining equations 9.17 and 9.18 gives the *Hamilton-Jacobi equation* which is discussed in chapter 15.4.

$$\frac{\partial S}{\partial t} + H(\mathbf{q}, \mathbf{p}, t) = 0 \tag{9.19}$$

In summary, Hamilton's principle of least action leads directly to Hamilton's equations of motion (9.15, 9.16) plus the Hamilton-Jacobi equation (9.19). Note that the above discussion has derived both Hamilton's Action Principle (9.8), and Hamilton's equations of motion (9.15, 9.16), directly from Hamilton's variational concept of stationary action, S, without explicitly invoking the Lagrangian.

9.2.3   Abbreviated action

Hamilton's Action Principle determines completely the path of the motion and the position on the path as a function of time. If the Lagrangian and the Hamiltonian are time independent, that is, conservative, then $H = E$ and equation 9.13 equals

$$S(q_j(t_1), t_1, q_j(t_2), t_2) = \int_i^f \left[\mathbf{p} \cdot \dot{\mathbf{q}} - E\right] dt = \int_i^f \mathbf{p} \cdot \delta\mathbf{q} - E(t_f - t_i) \tag{9.20}$$

The $\int_1^2 \mathbf{p} \cdot \boldsymbol{\delta}\dot{\mathbf{q}}$ term in equation 9.20, is called the **abbreviated action** which is defined as

$$S_0 \equiv \int_i^f \mathbf{p} \cdot \delta\dot{\mathbf{q}} \, dt = \int_i^f \mathbf{p} \cdot \delta\mathbf{q} \tag{9.21}$$

The abbreviated action can be simplified assuming use of the standard Lagrangian $L = T - U$ with a velocity-independent potential U, then equation 8.4 gives.

$$S_0 \equiv \int_{t_i}^{t_f} \sum_j^n p_j \dot{q}_j \, dt = \int_{t_i}^{t_f} (L + H) \, dt = \int_{t_i}^{t_f} 2T \, dt = \int_{t_i}^{t_f} \mathbf{p} \cdot \delta\mathbf{q} \tag{9.22}$$

Abbreviated action provides for use of a simplified form of the principle of least action that is based on the kinetic energy, and not potential energy. For conservative systems it determines the path of the motion, but not the time dependence of the motion. Consider virtual motions where the path satisfies energy conservation, and where the end points are held fixed, that is $\delta q_i = 0$, but allow for a variation δt in the final time. Then using the Hamilton-Jacobi equation, 9.19

$$\delta S = -H \delta t = -E \delta t \tag{9.23}$$

However, equation 9.21 gives that

$$\delta S = \delta S_0 - E \delta t \tag{9.24}$$

Therefore

$$\delta S_0 = 0 \tag{9.25}$$

That is, the abbreviated action has a minimum with respect to all paths that satisfy the conservation of energy which can be written as

$$\delta S_0 = \delta \int_{t_i}^{t_f} 2T \, dt = 0 \tag{9.26}$$

Equation 9.26 is called the *Maupertuis' least-action principle* which he proposed in 1744 based on Fermat's Principle in optics. Credit for the formulation of least action commonly is given to Maupertuis; however, the Maupertuis principle is similar to the use of least action applied to the "vis viva", as was proposed by Leibniz four decades earlier. Maupertuis used teleological arguments, rather than scientific rigor, because of his limited mathematical capabilities. In 1744 Euler provided a scientifically rigorous argument, presented above, that underlies the Maupertuis principle. Euler derived the correct variational relation for the abbreviated action to be

$$\delta S_0 = \int_{t_i}^{t_f} \sum_j^n p_j \delta q_j = 0 \tag{9.27}$$

Hamilton's use of the principle of least action to derive both Lagrangian and Hamiltonian mechanics is a remarkable accomplishment. It underlies both Lagrangian and Hamiltonian mechanics and confirmed the conjecture of Maupertuis.

9.2.4 Hamilton's Principle applied using initial boundary conditions

Galley[Gal13] identified a subtle inconsistency in the applications of Hamilton's Principle of Stationary Action to both Lagrangian and Hamiltonian mechanics. The inconsistency involves the fact that *Hamilton's Principle is defined as the action integral between the initial time t_i and the final time t_f as boundary conditions, that is, it is assumed to be time symmetric. However, most applications in Lagrangian and Hamiltonian mechanics assume that the action integral is evaluated based on the initial values as the boundary conditions*, rather than the initial t_i and final times t_f. That is, typical applications require use of a time-asymmetric version of Hamilton's principle. Galley[Gal13][Gal14] proposed a framework for transforming Hamilton's Principle to a time-asymmetric form in order to handle problems where the boundary conditions are based on using only the initial values at the initial time t_i, rather than the initial plus final times (t_i, t_f) that is assumed in the time-symmetric definition of the action in Hamilton's Principle.

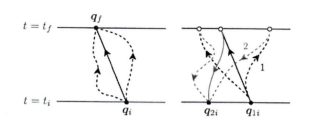

Figure 9.2: The left schematic shows paths between the initial $\mathbf{q}(t_i)$ and final $\mathbf{q}(t_f)$ times for conservative mechanics. The solid line designates the path for which the action is stationary, while the dashed lines represent the varied paths. The right schematic shows the paths applied to the doubled degrees of freedom with two initial boundary conditions, that is, $\mathbf{q}_1(t_i)$ and $\mathbf{q}_2(t_i)$ plus assuming that both paths are identical at their intersection and that they intersect at the same final time, that is, $\mathbf{q}_1(t_f) = \mathbf{q}_2(t_f)$.

The following describes the framework proposed by Galley for transforming Hamilton's Principle to a time-asymmetric form. Let $\mathbf{q}$ and $\dot{\mathbf{q}}$ designate sets of N generalized coordinates, plus their velocities, where $\mathbf{q}$ and $\dot{\mathbf{q}}$ are the fundamental variables assumed in the definition of the Lagrangian used by Hamilton's Principle. As illustrated schematically in figure 9.2, Galley proposed doubling the number of degrees of freedom for the system considered, that is, let $\mathbf{q} \rightarrow (\mathbf{q}_1, \mathbf{q}_2)$ and $\dot{\mathbf{q}} \rightarrow (\dot{\mathbf{q}}_1, \dot{\mathbf{q}}_2)$. In addition he defines two identical variational paths 1 and 2, where path 2 is the time reverse of path 1. That is, path 1 starts at the initial time t_i, and ends at t_f, whereas path 2 starts at t_f and ends at t_i. That is, he assumes that $\mathbf{q}$ and $\dot{\mathbf{q}}$ specify the two paths in the space of the doubled degrees of freedom that are identical, and that they intersect at the final time t_f. The arrows shown on the paths in figure 9.2 designate the assumed direction of the time integration along these paths.

For the doubled system of degrees of freedom, the total action for the sum of the two paths is given by the time integral of the doubled variables, $S(\mathbf{q}_1, \mathbf{q}_2)$ which can be written as

$$S(\mathbf{q}_1, \mathbf{q}_2) = \int_{t_i}^{t_f} L(\mathbf{q}_1, \dot{\mathbf{q}}_1, t)\, dt + \int_{t_f}^{t_i} L(\mathbf{q}_2, \dot{\mathbf{q}}_2, t)\, dt = \int_{t_i}^{t_f} [L(\mathbf{q}_1, \dot{\mathbf{q}}_1, t)\, dt - L(\mathbf{q}_2, \dot{\mathbf{q}}_2, t)]\, dt \tag{9.28}$$

The above relation assumes that the doubled variables $(\mathbf{q}_1, \dot{\mathbf{q}}_1)$ and $(\mathbf{q}_2, \dot{\mathbf{q}}_2)$ are decoupled from each other. More generally one can assume that the two sets of variables are coupled by some arbitrary function $K(\mathbf{q}_1, \dot{\mathbf{q}}_1, \mathbf{q}_2, \dot{\mathbf{q}}_2, t)$. Then the action can be written as

$$S(\mathbf{q}_1, \mathbf{q}_2) = \int_{t_i}^{t_f} [L(\mathbf{q}_1, \dot{\mathbf{q}}_1, \mathbf{t})\, dt - L(\mathbf{q}_2, \dot{\mathbf{q}}_2, \mathbf{t}) + K(\mathbf{q}_1, \dot{\mathbf{q}}_1, \mathbf{q}_2, \dot{\mathbf{q}}_2, t)]\, dt \tag{9.29}$$

The effective Lagrangian for this doubled system then can be defined as

$$\Lambda(\mathbf{q}_1, \mathbf{q}_2, \dot{\mathbf{q}}_1, \dot{\mathbf{q}}_2, t) \equiv [L(\mathbf{q}_1, \dot{\mathbf{q}}_1, t)\, dt - L(\mathbf{q}_2, \dot{\mathbf{q}}_2, t) + K(\mathbf{q}_1, \dot{\mathbf{q}}_1, \mathbf{q}_2, \dot{\mathbf{q}}_2, t)] \tag{9.30}$$

and the action can be written as

$$S(\mathbf{q}_1, \mathbf{q}_2) = \int_{t_i}^{t_f} \Lambda(\mathbf{q}_1, \dot{\mathbf{q}}_1, \mathbf{q}_2, \dot{\mathbf{q}}_2, t)\, dt \tag{9.31}$$

The coupling term $K(\mathbf{q}_1, \dot{\mathbf{q}}_1, \mathbf{q}_2, \dot{\mathbf{q}}_2, t)$ for the doubled system of degrees of freedom must satisfy the following two properties.

(a) If it can be expressed as the difference of two scalar potentials, $\Delta U\left(\mathbf{q}_1, \mathbf{q}_2\right) = U\left(\mathbf{q}_1\right) - U\left(\mathbf{q}_2\right)$, then it can be absorbed into the potential term for each of the doubled variables in the Lagrangian. This implies that $K = 0$, and there is no reason to double the number of degrees of freedom because the system is conservative. Thus K describes generalized forces that are not derivable from potential energy, that is, not conservative.

(b) A second property of the coupling term $K\left(\mathbf{q}_1, \dot{\mathbf{q}}_1, \mathbf{q}_2, \dot{\mathbf{q}}_2, t\right)$ is that it must be antisymmetric under interchange of the arbitrary labels $1 \leftrightarrow 2$. That is,

$$K\left(\mathbf{q}_2, \dot{\mathbf{q}}_2, \mathbf{q}_1, \dot{\mathbf{q}}_1, t\right) = -K\left(\mathbf{q}_1, \dot{\mathbf{q}}_1, \mathbf{q}_2, \dot{\mathbf{q}}_2, t\right) \tag{9.32}$$

Therefore the antisymmetric function $K\left(\mathbf{q}_1, \dot{\mathbf{q}}_1, \mathbf{q}_2, \dot{\mathbf{q}}_2, t\right)$ vanishes when $\mathbf{q}_2 = \mathbf{q}_1$.

The variational condition requires that the action $S\left(\mathbf{q}_1, \mathbf{q}_2\right)$ has a well defined stationary point for the doubled system. This is achieved by parametrizing both coordinate paths as

$$\mathbf{q}_{1,2}(t, \epsilon) = \mathbf{q}_{1,2}(t, 0) + \epsilon \eta_{1,2}(t) \tag{9.33}$$

where $\mathbf{q}_{1,2}(t, 0)$ are the coordinates for which the action is stationary, $\epsilon \ll 1$. and where $\eta_{1,2}(t)$ are arbitrary functions of time denoting virtual displacements of the paths. The doubled system has two independent paths connecting the two initial boundary conditions at t_i, and it requires that these paths intersect at t_f. The variational system for the two intersecting paths requires specifying four conditions, two per path. Two of the four conditions are determined by requiring that at t_i the initial boundary conditions satisfies that $\eta_{1,2}(t_i) = 0$. The remaining two conditions are derived by requiring that the variation of the action $S\left(\mathbf{q}_1, \mathbf{q}_2\right)$ satisfies

$$\left[\frac{dS}{d\epsilon}\right]_{\epsilon=0} = 0 = \int_{t_i}^{t_f} dt \left\{\eta_1 \left[\frac{\partial \Lambda}{\partial q_1} - \frac{d\pi_1}{dt}\right]_{\epsilon=0} - \eta_2 \left[\frac{\partial \Lambda}{\partial q_2} - \frac{d\pi_2}{dt}\right]_{\epsilon=0}\right\} + \left[\eta_1 \pi_1 - \eta_2 \pi_2\right]_{t=t_f} \tag{9.34}$$

The canonical momenta $\pi_{1,2}$ conjugate to the doubled coordinates $\mathbf{q}_{1,2}$ are defined using the nonconservative Lagrangian Λ to be

$$\pi_1^I\left(\mathbf{q}_{1,2}, \dot{\mathbf{q}}_{1,2}\right) \equiv \frac{\partial \Lambda}{\partial \dot{q}_1^I(t)} = \frac{\partial L\left(\mathbf{q}_1, \dot{\mathbf{q}}_1, t\right)}{\partial \dot{q}_1^I(t)} + \frac{\partial K\left(\mathbf{q}_1, \dot{\mathbf{q}}_1, \mathbf{q}_2, \dot{\mathbf{q}}_2, t\right)}{\partial \dot{q}_1^I(t)} \tag{9.35}$$

where the superscript I designates the solution based on the initial conditions. Note that the conjugate momentum $p_1^I = \frac{\partial L\left(\mathbf{q}_1, \dot{\mathbf{q}}_1, t\right)}{\partial \dot{q}_1^I(t)}$ while the $\frac{\partial K\left(\mathbf{q}_1, \dot{\mathbf{q}}_1, \mathbf{q}_2, \dot{\mathbf{q}}_2, t\right)}{\partial \dot{q}_1^I(t)}$ term is part of the total momentum due to the nonconservative interaction. Similarly the momentum for the second path is

$$\pi_2^I\left(\mathbf{q}_{1,2}, \dot{\mathbf{q}}_{1,2}\right) \equiv \frac{\partial \Lambda}{\partial \dot{q}_2^I(t)} = \frac{\partial L\left(\mathbf{q}_1, \dot{\mathbf{q}}_1, t\right)}{\partial \dot{q}_2^I(t)} + \frac{\partial K\left(\mathbf{q}_1, \dot{\mathbf{q}}_1, \mathbf{q}_2, \dot{\mathbf{q}}_2, t\right)}{\partial \dot{q}_2^I(t)} \tag{9.36}$$

The last term in equation 9.34, that is, the term $\left[\eta_1 \pi_1 - \eta_2 \pi_2\right]_{t=t_f}$ results from integration by parts, which will vanish if

$$\eta_1^I(t_f)\pi_1^I(t_f) = \eta_2^I(t_f)\pi_2^I(t_f) \tag{9.37}$$

The equality condition at the intersection of the two paths at t_f requires that

$$\eta_1^I(t_f) = \eta_2^I(t_f) \tag{9.38}$$

Therefore equations 9.37 and 9.38 imply that

$$\pi_1^I(t_f) = \pi_2^I(t_f) \tag{9.39}$$

Therefore equations 9.38 and 9.39 constitute the equality condition that must be satisfied when the two paths intersect at t_f. The equality condition ensures that the boundary term for integration by parts in equation 9.34 will vanish for arbitrary variations provided that the two unspecified paths agree at the final time t_f. Similarly the conjugate momenta $\pi_1^I(t_f), \pi_2^I(t_f)$ must agree, but otherwise are unspecified. As a consequence, the equality condition ensures that the variational principle is consistent with the final state at

t_f not being specified. That is, the equations of motion are only specified by the initial boundary conditions of the time asymmetric action for the doubled system.

More physics insight is provided by using a more convenient parametrization of the coordinates in terms of their average and difference. That is, let

$$q_+^I \equiv \frac{q_1^I + q_2^I}{2} \qquad\qquad q_-^I \equiv q_1^I - q_2^I \tag{9.40}$$

Then the physical limit is

$$q_+^I \to q^I \qquad\qquad q_-^I \to 0 \tag{9.41}$$

That is, the average history is the relevant physical history, while the difference coordinate simply vanishes. For these coordinates, the nonconservative Lagrangian is $\Lambda\left(\mathbf{q}_+, \mathbf{q}_-, \dot{\mathbf{q}}_+, \dot{\mathbf{q}}_-, t\right)$ and the equality conditions reduce to

$$\pi_-(t_f) = 0 \tag{9.42}$$
$$\eta_-(t_f) = 0 \tag{9.43}$$

which implies that the physically relevant average $(+)$ quantities are not specified at the final time t_f in order to have a well-defined variational principle.

The canonical momenta are given by

$$\pi_+^I = \frac{\pi_1^I + \pi_2^I}{2} = \frac{\partial \Lambda}{\partial \dot{q}_-^I} \tag{9.44}$$

$$\pi_-^I = \pi_1^I - \pi_2^I = \frac{\partial \Lambda}{\partial \dot{q}_+^I} \tag{9.45}$$

The equations of motion can be written as.

$$\frac{d}{dt}\frac{\partial \Lambda}{\partial \dot{q}_\pm^I} = \frac{\partial \Lambda}{\partial q_\pm^I} \tag{9.46}$$

Equation 9.46 is identically zero for the $+$ subscript, while, in the physical limit (PL), the negative subscript gives that

$$\left[\frac{d}{dt}\frac{\partial \Lambda}{\partial \dot{q}_-^I} - \frac{\partial \Lambda}{\partial q_-^I}\right]_{PL} = 0 \tag{9.47}$$

Substituting for the Lagrangian Λ gives that

$$\frac{d}{dt}\frac{\partial L}{\partial \dot{q}_-^I} - \frac{\partial L}{\partial q_-^I} = \left[\frac{\partial K}{\partial q_-^I} - \frac{d}{dt}\frac{\partial K}{\partial \dot{q}_-^I}\right]_{PL} \equiv Q^I\left(\mathbf{q}_1, \dot{\mathbf{q}}_1, t\right) \tag{9.48}$$

where Q^I is a generalized nonconservative force derived from K.

Note that equation 9.46 can be derived equally well by taking the direct functional derivative with respect to $q_-^I(t)$, that is,

$$0 = \left[\frac{\delta S}{\delta q_-^I(t)}\right]_{PL} \tag{9.49}$$

The above time-asymmetric formalism applies Hamilton's action principle to systems that involve initial boundary conditions while the second path corresponds to the final boundary conditions. This framework, proposed recently by Galley[Gal13], provides a remarkable advance for the handling of nonconservative action in Lagrangian and Hamiltonian mechanics.[2] This formalism directly incorporates the variational principle for initial boundary conditions and causal dynamics that are usually required for applications of Lagrangian and Hamiltonian mechanics. Currently, there is limited exploitation of this new formalism because there has been insufficient time for it to become well known, for full recognition of its importance, and for the development and publication of applications. Chapter 10 discusses an application of this formalism to nonconservative systems in classical mechanics.

---

[2] This topic goes beyond the planned scope of this book. It is recommended that the reader refer to the work of Galley, Tsang, and Stein[Gal13, Gal14] for further discussion plus examples of applying this formalism to nonconservative systems in classical mechanics, electromagnetic radiation, RLC circuits, fluid dynamics, and field theory.

9.3   Lagrangian

9.3.1   Standard Lagrangian

Lagrangian mechanics, as introduced in chapter 6, was based on the concepts of kinetic energy and potential energy. d'Alembert's principle of virtual work was used to derive Lagrangian mechanics in chapter 6 and this led to the definition of the *standard Lagrangian*. That is, the *standard Lagrangian* was defined in chapter 6.2 to be the difference between the kinetic and potential energies.

$$L(\mathbf{q}, \dot{\mathbf{q}}, t) = T(\dot{\mathbf{q}}, t) - U(\mathbf{q}, t) \tag{9.50}$$

Hamilton extended Lagrangian mechanics by defining Hamilton's Principle, equation 9.2, which states that *a dynamical system follows a path for which the action functional is stationary, that is, the time integral of the Lagrangian.* Chapter 6 showed that using the standard Lagrangian for defining the action functional leads to the Euler-Lagrange variational equations

$$\left\{ \frac{d}{dt} \left(\frac{\partial L}{\partial \dot{q}_j} \right) - \frac{\partial L}{\partial q_j} \right\} = Q_j^{EXC} + \sum_{k=1}^{m} \lambda_k \frac{\partial g_k}{\partial q_j}(\mathbf{q}, t) \tag{9.51}$$

The Lagrange multiplier terms handle the holonomic constraint forces and Q_j^{EXC} handles the remaining excluded generalized forces. Chapters $6-8$ showed that the use of the standard Lagrangian, with the Euler-Lagrange equations (9.51), provides a remarkably powerful and flexible way to derive second-order equations of motion for dynamical systems in classical mechanics.

Note that the Euler-Lagrange equations, expressed solely in terms of the standard Lagrangian (9.51), that is, excluding the $Q_j^{EXC} + \sum_{k=1}^{m} \lambda_k \frac{\partial g_k}{\partial q_j}(\mathbf{q}, t)$ terms, are valid only under the following conditions:

1. The forces acting on the system, apart from any forces of constraint, must be derivable from scalar potentials.

2. The equations of constraint must be relations that connect the coordinates of the particles and may be functions of time, that is, the constraints are holonomic.

The $Q_j^{EXC} + \sum_{k=1}^{m} \lambda_k \frac{\partial g_k}{\partial q_j}(\mathbf{q}, t)$ terms extend the range of validity of using the standard Lagrangian in the Lagrange-Euler equations by introducing constraint and omitted forces explicitly.

Chapters $6-8$ exploited Lagrangian mechanics based on use of the standard definition of the Lagrangian. The present chapter will show that the powerful Lagrangian formulation, using the standard Lagrangian, can be extended to include alternative non-standard Lagrangians that may be applied to dynamical systems where use of the standard definition of the Lagrangian is inapplicable. If these non-standard Lagrangians satisfy Hamilton's Action Principle, 9.2, then they can be used with the Euler-Lagrange equations to generate the correct equations of motion, even though the Lagrangian may not have the simple relation to the kinetic and potential energies adopted by the standard Lagrangian. Currently, the development and exploitation of non-standard Lagrangians is an active field of Lagrangian mechanics.

9.3.2   Gauge invariance of the standard Lagrangian

Note that the standard Lagrangian is not unique in that there is a continuous spectrum of equivalent standard Lagrangians that all lead to identical equations of motion. This is because the Lagrangian L is a scalar quantity that is invariant with respect to coordinate transformations. The following transformations change the standard Lagrangian, but leave the equations of motion unchanged.

1. The Lagrangian is indefinite with respect to addition of a constant to the scalar potential which cancels out when the derivatives in the Euler-Lagrange differential equations are applied.

2. The Lagrangian is indefinite with respect to addition of a constant kinetic energy.

3. The Lagrangian is indefinite with respect to addition of a total time derivative of the form $L_2 \rightarrow L_1 + \frac{d}{dt}[\Lambda(q_i, t)]$, for any differentiable function $\Lambda(q_i t)$ of the generalized coordinates plus time, that has continuous second derivatives.

This last statement can be proved by considering a transformation between two related standard Lagrangians of the form

$$L_2(\mathbf{q}, \dot{q}, t) = L_1(\mathbf{q}, \dot{q}, t) + \frac{d\Lambda(\mathbf{q}, t)}{dt} = L_1(\mathbf{q}, \dot{q}, t) + \left(\frac{\partial \Lambda(\mathbf{q}, t)}{\partial q_j} \dot{q}_j + \frac{\partial \Lambda(\mathbf{q}, t)}{\partial t} \right) \tag{9.52}$$

This leads to a standard Lagrangian L_2 that has the same equations of motion as L_1 as is shown by substituting equation 9.52 into the Euler-Lagrange equations. That is,

$$\frac{d}{dt}\left(\frac{\partial L_2}{\partial \dot{q}_j} \right) - \frac{\partial L_2}{\partial q_j} = \frac{d}{dt}\left(\frac{\partial L_1}{\partial \dot{q}_j} \right) - \frac{\partial L_1}{\partial q_j} + \frac{\partial^2 \Lambda(\mathbf{q}, t)}{\partial t \partial q_j} - \frac{\partial^2 \Lambda(\mathbf{q}, t)}{\partial t \partial q_j} = \frac{d}{dt}\left(\frac{\partial L_1}{\partial \dot{q}_j} \right) - \frac{\partial L_1}{\partial q_j} \tag{9.53}$$

Thus even though the related Lagrangians L_1 and L_2 are different, they are completely equivalent in that they generate identical equations of motion.

There is an unlimited range of equivalent standard Lagrangians that all lead to the same equations of motion and satisfy the requirements of the Lagrangian. That is, there is no unique choice among the wide range of equivalent standard Lagrangians expressed in terms of generalized coordinates. This discussion is an example of gauge invariance in physics.

Modern theories in physics describe reality in terms of potential fields. Gauge invariance, which also is called gauge symmetry, is a property of field theory for which different underlying fields lead to identical observable quantities. Well-known examples are the static electric potential field and the gravitational potential field where any arbitrary constant can be added to these scalar potentials with zero impact on the observed static electric field or the observed gravitational field. Gauge theories constrain the laws of physics in that the impact of gauge transformations must cancel out when expressed in terms of the observables. Gauge symmetry plays a crucial role in both classical and quantal manifestations of field theory, e.g. it is the basis of the Standard Model of electroweak and strong interactions.

Equivalent Lagrangians are a clear manifestation of gauge invariance as illustrated by equations 9.52, 9.53 which show that adding any total time derivative of a scalar function $\Lambda(\mathbf{q}, t)$ to the Lagrangian has no observable consequences on the equations of motion. That is, although addition of the total time derivative of the scalar function $\Lambda(\mathbf{q}, t)$ changes the value of the Lagrangian, it does not change the equations of motion for the observables derived using equivalent standard Lagrangians.

For Lagrangian formulations of classical mechanics, the gauge invariance is readily apparent by direct inspection of the Lagrangian.

9.1 *Example: Gauge invariance in electromagnetism*

The scalar electric potential Φ and the vector potential A fields in electromagnetism are examples of gauge-invariant fields. These electromagnetic-potential fields are not directly observable, that is, the electromagnetic observable quantities are the electric field E and magnetic field B which can be derived from the scalar and vector potential fields Φ and A. An advantage of using the potential fields is that they reduce the problem from 6 components, 3 each for E and B, to 4 components, one for the scalar field Φ and 3 for the vector potential A. The Lagrangian for the velocity-dependent Lorentz force, given by equation 6.67, provides an example of gauge invariance. Equations 6.63 and 6.65 showed that the electric and magnetic fields can be expressed in terms of scalar and vector potentials Φ and $\mathbf{A}$ by the relations

$$\mathbf{B} = \nabla \times \mathbf{A}$$

$$\mathbf{E} = -\nabla\Phi - \frac{\partial \mathbf{A}}{\partial t}$$

The equations of motion for a charge q in an electromagnetic field can be obtained by using the Lagrangian

$$L = \frac{1}{2}m\mathbf{v} \cdot \mathbf{v} - q(\Phi - \mathbf{A} \cdot \mathbf{v})$$

Consider the transformations $(\mathbf{A}, \Phi) \rightarrow (\mathbf{A}', \Phi')$ in the transformed Lagrangian L' where

$$\mathbf{A}' = \mathbf{A} + \nabla\Lambda(\mathbf{r}, t)$$

$$\Phi' = \Phi - \frac{\partial\Lambda(\mathbf{r}, t)}{\partial t}$$

The transformed Lorentz-force Lagrangian L' is related to the original Lorentz-force Lagrangian L by

$$L' = L + q \left[\dot{\mathbf{r}} \cdot \nabla \Lambda(\mathbf{r},t) + \frac{\partial \Lambda(\mathbf{r},t)}{\partial t} \right] = L + q \frac{d}{dt} \Lambda(\mathbf{r},t)$$

Note that the additive term $q\frac{d}{dt}\Lambda(\mathbf{r},t)$ is an exact time differential. Thus the Lagrangian L' is gauge invariant implying identical equations of motion are obtained using either of these equivalent Lagrangians.

The force fields $\mathbf{E}$ and $\mathbf{B}$ can be used to show that the above transformation is gauge-invariant. That is,

$$\mathbf{E}' = -\boldsymbol{\nabla}\Phi' - \frac{\partial \mathbf{A}'}{\partial t} = -\boldsymbol{\nabla}\Phi - \frac{\partial \mathbf{A}}{\partial t} = \mathbf{E}$$

$$\mathbf{B}' = \boldsymbol{\nabla} \times \mathbf{A}' = \boldsymbol{\nabla} \times \mathbf{A} = \mathbf{B}$$

That is, the additive terms due to the scalar field $\Lambda(\mathbf{r},t)$ cancel. Thus the electromagnetic force fields following a gauge-invariant transformation are shown to be identical in agreement with what is inferred directly by inspection of the Lagrangian.

9.3.3   Non-standard Lagrangians

The definition of the standard Lagrangian was based on d'Alembert's differential variational principle. The flexibility and power of Lagrangian mechanics can be extended to a broader range of dynamical systems by employing an extended definition of the Lagrangian that is based on Hamilton's Principle, equation 9.2. Note that Hamilton's Principle was introduced 46 years after development of the standard formulation of Lagrangian mechanics. Hamilton's Principle provides a general definition of the Lagrangian that applies to standard Lagrangians, which are expressed as the difference between the kinetic and potential energies, as well as to non-standard Lagrangians where there may be no clear separation into kinetic and potential energy terms. These non-standard Lagrangians can be used with the Euler-Lagrange equations to generate the correct equations of motion, even though they may have no relation to the kinetic and potential energies. The extended definition of the Lagrangian based on Hamilton's action functional 9.1 can be exploited for developing non-standard definitions of the Lagrangian that may be applied to dynamical systems where use of the standard definition is inapplicable. Non-standard Lagrangians can be equally as useful as the standard Lagrangian for deriving equations of motion for a system. Secondly, non-standard Lagrangians, that have no energy interpretation, are available for deriving the equations of motion for many nonconservative systems. Thirdly, Lagrangians are useful irrespective of how they were derived. For example, they can be used to derive conservation laws or the equations of motion. Coordinate transformations of the Lagrangian is much simpler than that required for transforming the equations of motion. The relativistic Lagrangian defined in chapter 17.6 is a well-known example of a non-standard Lagrangian.

9.3.4   Inverse variational calculus

Non-standard Lagrangians and Hamiltonians are not based on the concept of kinetic and potential energies. Therefore, development of non-standard Lagrangians and Hamiltonians require an alternative approach that ensures that they satisfy Hamilton's Principle, equation 9.2, which underlies the Lagrangian and Hamiltonian formulations. One useful alternative approach is to derive the Lagrangian or Hamiltonian via an inverse variational process based on the assumption that the equations of motion are known. Helmholtz developed the field of inverse variational calculus which plays an important role in development of non-standard Lagrangians. An example of this approach is use of the well-known Lorentz force as the basis for deriving a corresponding Lagrangian to handle systems involving electromagnetic forces. Inverse variational calculus is a branch of mathematics that is beyond the scope of this textbook. The Douglas theorem[Dou41] states that, if the three Helmholtz conditions are satisfied, then there exists a Lagrangian that, when used with the Euler-Lagrange differential equations, leads to the given set of equations of motion. Thus, it will be assumed that the inverse variational calculus technique can be used to derive a Lagrangian from known equations of motion.

9.4 Application of Hamilton's Action Principle to mechanics

Knowledge of the equations of motion is required to predict the response of a system to any set of initial conditions. Hamilton's action principle, that is built into Lagrangian and Hamiltonian mechanics, coupled with the availability of a wide arsenal of variational principles and techniques, provides a remarkably powerful and broad approach to deriving the equations of motions required to determine the system response.

As mentioned in the Prologue, derivation of the equations of motion for any system, based on Hamilton's Action Principle, separates naturally into a hierarchical set of three stages that differ in both sophistication and understanding, as described below.

1. **Action stage:** The primary "action stage" employs Hamilton's Action functional, $S = \int_{t_i}^{t_f} L(\mathbf{q}, \dot{\mathbf{q}}, t) dt$ to derive the Lagrangian and Hamiltonian functionals. This action stage provides the most fundamental and sophisticated level of understanding. It involves specifying all the active degrees of freedom, as well as the interactions involved. Symmetries incorporated at this primary action stage can simplify subsequent use of the Hamiltonian and Lagrangian functionals.

2. **Hamiltonian/Lagrangian stage:** The "Hamiltonian/Lagrangian stage" uses the Lagrangian or Hamiltonian functionals, that were derived at the action stage, in order to derive the equations of motion for the system of interest. Symmetries, not already incorporated at the primary action stage, may be included at this secondary stage.

3. **Equations of motion stage:** The "equations-of-motion stage" uses the derived equations of motion to solve for the motion of the system subject to a given set of initial boundary conditions. Nonconservative forces, such as dissipative forces, that were not included at the primary and secondary stages, may be added at the equations of motion stage.

Lagrange omitted the action stage when he used d'Alembert's Principle to derive Lagrangian mechanics. The Newtonian mechanics approach omits both the primary "action" stage, as well as the secondary "Hamiltonian/Lagrangian" stage, since Newton's Laws of Motion directly specify the "equations-of-motion stage". Thus these did not allow exploiting the considerable advantages provided by use of action, the Lagrangian, and the Hamiltonian. Newtonian mechanics requires that all the active forces be included when deriving the equations of motion, which involves dealing with vector quantities. In Newtonian mechanics, symmetries must be incorporated directly at the equations of motion stage, which is more difficult than when done at the primary "action" stage, or the secondary "Lagrangian/Hamiltonian" stage. The "action" and "Hamiltonian/Lagrangian" stages allow for use of the powerful arsenal of mathematical techniques that have been developed for applying variational principles.

There are considerable advantages to deriving the equations of motion based on Hamilton's Principle, rather than derive them using Newtonian mechanics. It is significantly easier to use variational principles to handle the scalar functionals, action, Lagrangian, and Hamiltonian, rather than starting at the equations-of-motion stage. For example, utilizing all three stages of algebraic mechanics facilitates accommodating extra degrees of freedom, symmetries, and interactions. The symmetries identified by Noether's theorem are more easily recognized during the primary "action" and secondary "Hamiltonian/Lagrangian" stages rather than at the subsequent "equations of motion" stage. Approximations made at the "action" stage are easier to implement than at the "equations-of-motion" stage. Constrained motion is much more easily handled at the primary "action", or secondary "Hamilton/Lagrangian" stages, than at the equations-of-motion stage. An important advantage of using Hamilton's Action Principle, is that there is a close relationship between action in classical and quantal mechanics, as discussed in chapters 15 and 18. Algebraic principles, that underly analytical mechanics, naturally encompass applications to many branches of modern physics, such as relativistic mechanics, fluid motion, and field theory.

In summary, the use of the single fundamental invariant quantity, action, as described above, provides a powerful and elegant framework, that was developed first for classical mechanics, but now is exploited in a wide range of science, engineering, and economics. An important feature of using the algebraic approach to classical mechanics is the tremendous arsenal of powerful mathematical techniques that have been developed for use of variational calculus applied to Lagrangian and Hamiltonian mechanics. Some of these variational techniques were presented in chapters 6, 7, 8, and 9, while others will be introduced in chapter 15.

9.5   Summary

The Hamilton's 1834 publication, introducing both Hamilton's Principle of Stationary Action and Hamiltonian mechanics, marked the crowning achievements for the development of variational principles in classical mechanics. A fundamental advantage of Hamiltonian mechanics is that it uses the conjugate coordinates $\mathbf{q}, \mathbf{p}$, plus time t, which is a considerable advantage in most branches of physics and engineering. Compared to Lagrangian mechanics, Hamiltonian mechanics has a significantly broader arsenal of powerful techniques that can be exploited to obtain an analytical solution of the integrals of the motion for complicated systems, as described in chapter 15. In addition, Hamiltonian dynamics provides a means of determining the unknown variables for which the solution assumes a soluble form, and is ideal for study of the fundamental underlying physics in applications to fields such as quantum or statistical physics. As a consequence, Hamiltonian mechanics has become the preeminent variational approach used in modern physics.

This chapter has introduced and discussed Hamilton's Principle of Stationary Action, which underlies the elegant and remarkably powerful Lagrangian and Hamiltonian representations of algebraic mechanics. The basic concepts employed in algebraic mechanics are summarized below.

Hamilton's Action Principle: As discussed in chapter 9.2, Hamiltonian mechanics is built upon Hamilton's action functional

$$S(\mathbf{q}, \mathbf{p}, t) = \int_{t_i}^{t_f} L(\mathbf{q}, \dot{\mathbf{q}}, t)dt \tag{9.1}$$

Hamilton's Principle of least action states that

$$\delta S(\mathbf{q}, \mathbf{p}, t) = \delta \int_{t_i}^{t_f} L(\mathbf{q}, \dot{\mathbf{q}}, t)dt = 0 \tag{9.2}$$

Generalized momentum p: In chapter 7.2, the generalized (canonical) momentum was defined in terms of the Lagrangian L to be

$$p_i \equiv \frac{\partial L(\mathbf{q}, \dot{\mathbf{q}}, t)}{\partial \dot{q}_i} \tag{7.3}$$

Chapter 9.2.2 defined the generalized momentum in terms of the action functional S to be

$$p_j = \frac{\partial S(\mathbf{q}, \mathbf{p}, t)}{\partial q_j} \tag{9.12}$$

Generalized energy $h(\mathbf{q}, \dot{q}, t)$:   Jacobi's Generalized Energy $h(\mathbf{q}, \dot{q}, t)$ was defined in equation 7.37 as

$$h(\mathbf{q}, \dot{\mathbf{q}}, t) \equiv \sum_j \left(\dot{q}_j \frac{\partial L(\mathbf{q}, \dot{\mathbf{q}}, t)}{\partial \dot{q}_j} \right) - L(\mathbf{q}, \dot{\mathbf{q}}, t) \tag{7.37}$$

Hamiltonian function: $H(\mathbf{q}, \mathbf{p}, t)$ The Hamiltonian $H(\mathbf{q}, \mathbf{p}, t)$ was defined in terms of the generalized energy $h(\mathbf{q}, \dot{\mathbf{q}}, t)$ plus the generalized momentum. That is

$$H(\mathbf{q}, \mathbf{p}, t) \equiv h(\mathbf{q}, \dot{\mathbf{q}}, t) = \sum_j p_j \dot{q}_j - L(\mathbf{q}, \dot{\mathbf{q}}, t) = \mathbf{p} \cdot \dot{\mathbf{q}} - L(\mathbf{q}, \dot{\mathbf{q}}, t) \tag{7.37}$$

where $\mathbf{p}, \mathbf{q}$ correspond to n-dimensional vectors, e.g. $\mathbf{q} \equiv (q_1, q_2, ..., q_n)$ and the scalar product $\mathbf{p} \cdot \dot{\mathbf{q}} = \sum_i p_i \dot{q}_i$. Chapter 8.2 used a Legendre transformation to derive this relation between the Hamiltonian and Lagrangian functions. Note that whereas the Lagrangian $L(\mathbf{q}, \dot{\mathbf{q}}, t)$ is expressed in terms of the coordinates $\mathbf{q}$, plus conjugate velocities $\dot{\mathbf{q}}$, the Hamiltonian $H(\mathbf{q}, \mathbf{p}, t)$ is expressed in terms of the coordinates $\mathbf{q}$ plus their conjugate momenta $\mathbf{p}$. For scleronomic systems, using the standard Lagrangian, in equations 7.44 and 7.29, shows that the Hamiltonian simplifies to be equal to the total mechanical energy, that is, $H = T + U$.

Generalized energy theorem: The equations of motion lead to the generalized energy theorem which states that the time dependence of the Hamiltonian is related to the time dependence of the Lagrangian.

$$\frac{dH(\mathbf{q},\mathbf{p},t)}{dt} = \sum_j \dot{q}_j \left[Q_j^{EXC} + \sum_{k=1}^m \lambda_k \frac{\partial g_k}{\partial q_j}(\mathbf{q},t) \right] - \frac{\partial L(\mathbf{q},\dot{\mathbf{q}},t)}{\partial t} \tag{7.38}$$

Note that if all the generalized non-potential forces and Lagrange multiplier terms are zero, and if the Lagrangian is not an explicit function of time, then the Hamiltonian is a constant of motion.

Lagrange equations of motion: Equation 6.60 gives that the N Lagrange equations of motion are

$$\left\{ \frac{d}{dt}\left(\frac{\partial L}{\partial \dot{q}_j} \right) - \frac{\partial L}{\partial q_j} \right\} = \sum_{k=1}^m \lambda_k \frac{\partial g_k}{\partial q_j}(\mathbf{q},t) + Q_j^{EXC} \tag{6.60}$$

where $j = 1, 2, 3, ...N$.

Hamilton's equations of motion: Chapter 8.3 showed that a Legendre transform, plus the Lagrange-Euler equations, (9.64, 9.65) lead to Hamilton's equations of motion. Hamilton derived these equations of motion directly from the action functional, as shown in chapter 9.2.

$$\dot{q}_j = \frac{\partial H(\mathbf{q},\mathbf{p},t)}{\partial p_j} \tag{8.25}$$

$$\dot{p}_j = -\frac{\partial H}{\partial q_j}(\mathbf{q},\mathbf{p},t) + \left[\sum_{k=1}^m \lambda_k \frac{\partial g_k}{\partial q_j} + Q_j^{EXC} \right] \tag{8.26}$$

$$\frac{\partial H(\mathbf{q},\mathbf{p},t)}{\partial t} = -\frac{\partial L(\mathbf{q},\dot{\mathbf{q}},t)}{\partial t} \tag{8.24}$$

Note the symmetry of Hamilton's two canonical equations. The canonical variables p_k, q_k are treated as independent canonical variables. Lagrange was the first to derive the canonical equations but he did not recognize them as a basic set of equations of motion. Hamilton derived the canonical equations of motion from his fundamental variational principle and made them the basis for a far-reaching theory of dynamics. Hamilton's equations give $2s$ first-order differential equations for p_k, q_k for each of the s degrees of freedom. Lagrange's equations give s second-order differential equations for the variables $q_k, \dot{q}_k$.

Hamilton-Jacobi equation: Hamilton used Hamilton's Principle plus equation 9.19 to derive the Hamilton-Jacobi equation.

$$\frac{\partial S}{\partial t} + H(\mathbf{q},\mathbf{p},t) = 0 \tag{9.19}$$

The solution of Hamilton's equations is trivial if the Hamiltonian is a constant of motion, or when a set of generalized coordinate can be identified for which all the coordinates q_i are constant, or are cyclic (also called *ignorable* coordinates). Jacobi developed the mathematical framework of canonical transformation required to exploit the Hamilton-Jacobi equation.

Hamilton's Principle applied using initial boundary conditions: The definition of Hamilton's Principle assumes integration between the initial time t_i and final time t_f. A recent development has extended applications of Hamilton's Principle to apply to systems that are defined in terms of only the initial boundary conditions. This method doubles the number of degrees of freedom and uses a coupling Lagrangian $K(\mathbf{q}_2, \dot{\mathbf{q}}_2, \mathbf{q}_1, \dot{\mathbf{q}}_1, t)$ between the corresponding $\mathbf{q}_1$ and $\mathbf{q}_2$ doubled degrees of freedom

$$\frac{d}{dt}\frac{\partial L}{\partial \dot{q}_-^I} - \frac{\partial L}{\partial q_-^I} = \left[\frac{\partial K}{\partial q_-^I} - \frac{d}{dt}\frac{\partial K}{\partial \dot{q}_-^I} \right]_{PL} \equiv Q^I(\mathbf{q}_1, \dot{\mathbf{q}}_1, t) \tag{9.50}$$

and where Q^I is a generalized nonconservative force derived from K.

Standard Lagrangians: Derivation of Lagrangian mechanics, using d'Alembert's principle of virtual work, assumed that the Lagrangian is defined by equation 9.52

$$L(\mathbf{q}, \dot{\mathbf{q}},t) = T(\dot{\mathbf{q}},t) - U(\mathbf{q}, t) \tag{9.52}$$

This was used in equation 9.3 to derive the action in terms of the fundamental Lagrangian defined by equation 9.52. The assumption that the action S is the fundamental property inverts this procedure and now equation 9.3 is used to derived the Lagrangian. That is, the assumption that Hamilton's Principle is the foundation of algebraic mechanics defines the Lagrangian in terms of the fundamental action S.

Non-standard Lagrangians: The flexibility and power of Lagrangian mechanics can be extended to a broader range of dynamical systems by employing an extended definition of the Lagrangian that assumes that the action is the fundamental property, and then the Lagrangian is defined in terms of Hamilton's variational action principle using equation 9.2. It was illustrated that the inverse variational calculus formalism can be used to identify non-standard Lagrangians that generate the required equations of motion. These non-standard Lagrangians can be very different from the standard Lagrangian and do not separate into kinetic and potential energy components. These alternative Lagrangians can be used to handle dissipative systems which are beyond the range of validity when using standard Lagrangians. That is, it was shown that several very different Lagrangians and Hamiltonians can be equivalent for generating useful equations of motion of a system. Currently the use of non-standard Lagrangians is a narrow, but active, frontier of classical mechanics with important applications to relativistic mechanics.

Gauge invariance of the standard Lagrangian: It was shown that there is a continuum of equivalent standard Lagrangians that lead to the same set of equations of motion for a system. This feature is related to gauge invariance in mechanics. The following transformations change the standard Lagrangian, but leave the equations of motion unchanged.

1. The Lagrangian is indefinite with respect to addition of a constant to the scalar potential which cancels out when the derivatives in the Euler-Lagrange differential equations are applied.

2. Similarly the Lagrangian is indefinite with respect to addition of a constant kinetic energy.

3. The Lagrangian is indefinite with respect to addition of a total time derivative of the form $L \rightarrow L + \frac{d}{dt}[\Lambda(q_i, t)]$ for any differentiable function $\Lambda(q_i t)$ of the generalized coordinates, plus time, that has continuous second derivatives.

Application of Hamilton's Action Principle to mechanics: The derivation of the equations of motion for any system can be separated into a hierarchical set of three stages in both sophistication and understanding. Variational principles are employed during the primary "action" stage and secondary "Hamilton/Lagrangian" stage to derive the required equations of motion, which then are solved during the third "equations-of-motion stage". Hamilton's Action Principle, is a scalar function that is the basis for deriving the Lagrangian and Hamiltonian functions. The primary "action stage" uses Hamilton's Action functional, $S = \int_{t_i}^{t_f} L(\mathbf{q}, \dot{\mathbf{q}},t)dt$ to derive the Lagrangian and Hamiltonian functionals that are based on Hamilton's action functional and provide the most fundamental and sophisticated level of understanding. The second "Hamiltonian/Lagrangian stage" involves using the Lagrangian and Hamiltonian functionals to derive the equations of motion. The third "equations-of-motion stage" uses the derived equations of motion to solve for the motion subject to a given set of initial boundary conditions. The Newtonian mechanics approach bypasses the primary "action" stage, as well as the secondary "Hamiltonian/Lagrangian" stage. That is, Newtonian mechanics starts at the third "equations-of-motion" stage, which does not allow exploiting the considerable advantages provided by use of action, the Lagrangian, and the Hamiltonian. Newtonian mechanics requires that all the active forces be included when deriving the equations of motion, which involves dealing with vector quantities. This is in contrast to the action, Lagrangian, and Hamiltonian which are scalar functionals. Both the primary "action" stage, and the secondary "Lagrangian/Hamiltonian" stage, exploit the powerful arsenal of mathematical techniques that have been developed for exploiting variational principles.

Chapter 10

Nonconservative systems

10.1 Introduction

Hamilton's action principle, Lagrangian mechanics, and Hamiltonian mechanics, all exploit the concept of action which is a single, invariant, quantity. These algebraic formulations of mechanics all are based on energy, which is a scalar quantity, and thus these formulations are easier to handle than the vector concept of force employed in Newtonian mechanics. Algebraic formulations provide a powerful and elegant approach to understand and develop the equations of motion of systems in nature. Chapters $6 - 9$ applied variational principles to Hamilton's action principle which led to the Lagrangian, and Hamiltonian formulations that simplify determination of the equations of motion for systems in classical mechanics.

A conservative force has the property that the total work done moving between two points is independent of the taken path. That is, a conservative force is time symmetric and can be expressed in terms of the gradient of a scalar potential V. *Hamilton's action principle implicitly assumes that the system is conservative for those degrees of freedom that are built into the definition of the action, and the related Lagrangian, and Hamiltonian.* The focus of this chapter is to discuss the origins of nonconservative motion and how it can be handled in algebraic mechanics.

10.2 Origins of nonconservative motion

Nonconservative degrees of freedom involve irreversible processes, such as dissipation, damping, and also can result from course-graining, or ignoring coupling to active degrees of freedom. The nonconservative role of ignored active degrees of freedom is illustrated by the weakly-coupled double harmonic oscillator system discussed below. Let the two harmonic oscillators have masses (m_1, m_2), uncoupled angular frequencies (ω_1, ω_2), and oscillation amplitudes (q_1, q_2). Assume that the coupling potential energy is $U = \lambda q_1 q_2$. The Lagrangian for this weakly-coupled double oscillator is

$$L(q_1, q_2, \dot{q}_1, \dot{q}_2, t) = \frac{m_1}{2}\left(\dot{q}_1^2 - \omega_1^2 q_1^2\right) + \lambda q_1 q_2 + \frac{m_2}{2}\left(\dot{q}_2^2 - \omega_2^2 q_2^2\right) \tag{10.1}$$

Note that the total Lagrangian is conservative since the Lagrangian is explicitly time independent. As shown in chapter 14.2, the solution for the amplitudes of the oscillation for the coupled system are given by

$$q_1(t) = D \sin\left[\left(\frac{\omega_1 + \omega_2}{2}\right)t\right]\sin\left[\left(\frac{\omega_1 - \omega_2}{2}\right)t\right] \tag{10.2}$$

$$q_2(t) = D \cos\left[\left(\frac{\omega_1 + \omega_2}{2}\right)t\right]\cos\left[\left(\frac{\omega_1 - \omega_2}{2}\right)t\right] \tag{10.3}$$

The system exhibits the common "beats" behavior where the coupled harmonic oscillators have an angular frequency that is the average oscillator frequency $\omega_{average} = \left(\frac{\omega_1 + \omega_2}{2}\right)$, and the oscillation intensities are modulated at the difference frequency, $\omega_{difference} = \left(\frac{\omega_1 - \omega_2}{2}\right)$. Although the total energy is conserved for this conservative system, this shared energy flows back and forth between the two coupled harmonic oscillators at the difference frequency. If the equations of motion for oscillator 1 ignore the coupling to the motion of oscillator 2, that is, assume a constant average value $q_2 = \langle q_2 \rangle$ is used, then the intensity $|q_1|^2$ and energy of the first oscillator still is modulated by the $\left| \sin\left(\frac{\omega_1 - \omega_2}{2} \right) t \right|^2$ term. Thus the total energy for this truncated coupled-oscillator system is no longer conserved due to neglect of the energy flowing into and out of oscillator 1 due to its coupling to oscillator 2. That is, the solution for the truncated system of oscillator 1 is not conservative since it is exchanging energy with the coupled, but ignored, second oscillator. This elementary example illustrates that ignoring active degrees of freedom can transform a conservative system into a nonconservative system, for which the equations of motion derived using the truncated Lagrangian is incorrect.

The above example illustrates the importance of including all active degrees of freedom when deriving the equations of motion, in order to ensure that the total system is conservative. Unfortunately, nonconservative systems due to viscous or frictional dissipation typically result from weak thermal interactions with an enormous number of nearby atoms, which makes inclusion of all of these degrees of freedom impractical. Even though the detailed behavior of such dissipative degrees of freedom may not be of direct interest, all the active degrees of freedom must be included when applying Lagrangian or Hamiltonian mechanics.

10.3   Algebraic mechanics for nonconservative systems

Since Lagrangian and Hamiltonian formulations are invalid for the nonconservative degrees of freedom, the following three approaches are used to include nonconservative degrees of freedom directly in the Lagrangian and Hamiltonian formulations of mechanics.

1. Expand the number of degrees of freedom used to include all active degrees of freedom for the system, so that the expanded system is conservative. This is the preferred approach when it is viable. Hamilton's action principle based on initial conditions, introduced in chapter 9.2.4, doubles the number of degrees of freedom, which can be used to account for the dissipative forces providing one approach to solve nonconservative systems. However, this approach typically is impractical for handling dissipated processes because of the large number of degrees of freedom that are involved in thermal dissipation.

2. Nonconservative forces can be introduced directly at the equations of motion stage as generalized forces Q_j^{EXC}. This approach is used extensively. For the case of linear velocity dependence, the Rayleigh's dissipation function provides an elegant and powerful way to express the generalized forces in terms of scalar potential energies.

3. New degrees of freedom or effective forces can be postulated that are then incorporated into the Lagrangian or the Hamiltonian in order to mimic the effects of the nonconservative forces.

Examples that exploit the above three ways to introduce nonconservative dissipative forces in algebraic formulations are given below.

10.4   Rayleigh's dissipation function

As mentioned above, nonconservative systems involving viscous or frictional dissipation, typically result from weak thermal interactions with many nearby atoms, making it impractical to include a complete set of active degrees of freedom. In addition, dissipative systems usually involve complicated dependences on the velocity and surface properties that are best handled by including the dissipative drag force explicitly as a generalized drag force in the Euler-Lagrange equations. The drag force can have any functional dependence on velocity, position, or time.

$$\mathbf{F}^{drag} = -f(\dot{\mathbf{q}}, \mathbf{q}, t)\hat{\mathbf{v}} \tag{10.4}$$

Note that since the drag force is dissipative the dominant component of the drag force must point in the opposite direction to the velocity vector.

In 1881 Lord Rayleigh[Ray1881, Ray1887] showed that if a dissipative force $\mathbf{F}$ depends linearly on velocity, it can be expressed in terms of a scalar potential functional of the generalized coordinates called the *Rayleigh dissipation function* $\mathcal{R}(\dot{\mathbf{q}})$. The Rayleigh dissipation function is an elegant way to include linear velocity-dependent dissipative forces in both Lagrangian and Hamiltonian mechanics, as is illustrated below for both Lagrangian and Hamiltonian mechanics.

10.4.1 Generalized dissipative forces for linear velocity dependence

Consider n equations of motion for the n degrees of freedom, and assume that the dissipation depends linearly on velocity. Then, allowing all possible cross coupling of the equations of motion for q_j, the equations of motion can be written in the form

$$\sum_{i=1}^{n} [m_{ij}\ddot{q}_j + b_{ij}\dot{q}_j + c_{ij}q_j - Q_i(t)] = 0 \tag{10.5}$$

Multiplying equation 10.5 by $\dot{q}_i$, take the time integral, and sum over i, j, gives the following energy equation

$$\sum_{i=1}^{n}\sum_{j=1}^{n}\int_0^t m_{ij}\ddot{q}_j\dot{q}_i dt + \sum_{i=1}^{n}\sum_{j=1}^{n}\int_0^t b_{ij}\dot{q}_j\dot{q}_i dt + \sum_{i=1}^{n}\sum_{j=1}^{n}\int_0^t c_{ij}q_j\dot{q}_i dt = \sum_{i}^{n}\int_0^t Q_i(t)\dot{q}_i dt \tag{10.6}$$

The right-hand term is the total energy supplied to the system by the external generalized forces $Q_i(t)$ at the time t. The first time-integral term on the left-hand side is the total kinetic energy, while the third time-integral term equals the potential energy. The second integral term on the left is defined to equal $2\mathcal{R}(\dot{\mathbf{q}})$ where Rayeigh's dissipation function $\mathcal{R}(\dot{\mathbf{q}})$ is defined as

$$\mathcal{R}(\dot{\mathbf{q}}) \equiv \frac{1}{2}\sum_{i=1}^{n}\sum_{j=1}^{n} b_{ij}\dot{q}_i\dot{q}_j \tag{10.7}$$

and the summations are over all n particles of the system. This definition allows for complicated cross-coupling effects between the n particles.

The particle-particle coupling effects usually can be neglected allowing use of the simpler definition that includes only the diagonal terms. Then the diagonal form of the Rayleigh dissipation function simplifies to

$$\mathcal{R}(\dot{\mathbf{q}}) \equiv \frac{1}{2}\sum_{i=1}^{n} b_i\dot{q}_i^2 \tag{10.8}$$

Therefore the frictional force in the q_i direction depends linearly on velocity $\dot{q}_i$, that is

$$F_{q_i}^f = -\frac{\partial\mathcal{R}(\dot{\mathbf{q}})}{\partial\dot{q}_i} = -b_i\dot{q}_i \tag{10.9}$$

In general, the dissipative force is the velocity gradient of the Rayleigh dissipation function,

$$\mathbf{F}^f = -\nabla_{\dot{\mathbf{q}}}\mathcal{R}(\dot{\mathbf{q}}) \tag{10.10}$$

The physical significance of the Rayleigh dissipation function is illustrated by calculating the work done by one particle i *against* friction, which is

$$dW_i^f = -\mathbf{F}_i^f \cdot d\mathbf{r} = -\mathbf{F}_i^f \cdot \dot{\mathbf{q}}_i dt = b_i\dot{q}_i^2 dt \tag{10.11}$$

Therefore

$$2\mathcal{R}(\dot{\mathbf{q}}) = \frac{dW^f}{dt} \tag{10.12}$$

which is the rate of energy (power) loss due to the dissipative forces involved. The same relation is obtained after summing over all the particles involved.

Transforming the frictional force into generalized coordinates requires equation 6.27

$$\dot{\mathbf{r}}_i = \sum_{k}\frac{\partial\mathbf{r}_i}{\partial q_k}\dot{q}_k + \frac{\partial\mathbf{r}_i}{\partial t} \tag{10.13}$$

Note that the derivative with respect to $\dot{q}_k$ equals

$$\frac{\partial\dot{\mathbf{r}}_i}{\partial\dot{q}_j} = \frac{\partial\mathbf{r}_i}{\partial q_j} \tag{10.14}$$

Using equations 6.28 and 6.29, the j component of the generalized frictional force Q_j^f is given by

$$Q_j^f = \sum_{i=1}^{n} \mathbf{F}_i^f \cdot \frac{\partial \mathbf{r}_i}{\partial q_j} = \sum_{i=1}^{n} \mathbf{F}_i^f \cdot \frac{\partial \dot{\mathbf{r}}_i}{\partial \dot{q}_j} = -\sum_{i=1}^{n} \nabla_{v_i} \mathcal{R}(\dot{\mathbf{q}}) \cdot \frac{\partial \dot{\mathbf{r}}_i}{\partial \dot{q}_j} = -\frac{\partial \mathcal{R}(\dot{\mathbf{q}})}{\partial \dot{q}_j} \qquad (10.15)$$

Equation 10.15 provides an elegant expression for the generalized dissipative force Q_j^f in terms of the Rayleigh's scalar dissipation potential $\mathcal{R}$.

10.4.2   Generalized dissipative forces for nonlinear velocity dependence

The above discussion of the Rayleigh dissipation function was restricted to the special case of linear velocity-dependent dissipation. Virga[Vir15] proposed that the scope of the classical Rayleigh-Lagrange formalism can be extended to include nonlinear velocity dependent dissipation by assuming that the nonconservative dissipative forces are defined by

$$\mathbf{F}_i^f = -\frac{\partial R(\mathbf{q}, \dot{\mathbf{q}})}{\partial \dot{\mathbf{q}}} \qquad (10.16)$$

where the generalized Rayleigh dissipation function $\mathcal{R}(\mathbf{q}, \dot{\mathbf{q}})$ satisfies the general Lagrange mechanics relation

$$\frac{\delta L}{\delta q} - \frac{\partial R}{\partial \dot{q}} = 0 \qquad (10.17)$$

This generalized Rayleigh's dissipation function eliminates the prior restriction to linear dissipation processes, which greatly expands the range of validity for using Rayleigh's dissipation function.

10.4.3   Lagrange equations of motion

Linear dissipative forces can be directly, and elegantly, included in Lagrangian mechanics by using Rayleigh's dissipation function as a generalized force Q_j^f. Inserting Rayleigh dissipation function 10.15 in the generalized Lagrange equations of motion 6.60 gives

$$\left\{ \frac{d}{dt} \left(\frac{\partial L}{\partial \dot{q}_j} \right) - \frac{\partial L}{\partial q_j} \right\} = \left[\sum_{k=1}^{m} \lambda_k \frac{\partial g_k}{\partial q_j}(\mathbf{q}, t) + Q_j^{EXC} \right] - \frac{\partial \mathcal{R}(\mathbf{q}, \dot{\mathbf{q}})}{\partial \dot{q}_j} \qquad (10.18)$$

Where Q_j^{EXC} corresponds to the generalized forces remaining after removal of the generalized linear, velocity-dependent, frictional force Q_j^f. The holonomic forces of constraint are absorbed into the Lagrange multiplier term.

10.4.4   Hamiltonian mechanics

If the nonconservative forces depend linearly on velocity, and are derivable from Rayleigh's dissipation function according to equation 10.15, then using the definition of generalized momentum gives

$$\dot{p}_i = \frac{d}{dt} \frac{\partial L}{\partial \dot{q}_j} = \frac{\partial L}{\partial q_i} + \left[\sum_{k=1}^{m} \lambda_k \frac{\partial g_k}{\partial q_j}(\mathbf{q}, t) + Q_j^{EXC} \right] - \frac{\partial \mathcal{R}(\mathbf{q}, \dot{\mathbf{q}})}{\partial \dot{q}_j} \qquad (10.19)$$

$$\dot{p}_i = -\frac{\partial H(\mathbf{p}, \mathbf{q}, t)}{\partial q_i} + \left[\sum_{k=1}^{m} \lambda_k \frac{\partial g_k}{\partial q_j}(\mathbf{q}, t) + Q_j^{EXC} \right] - \frac{\partial \mathcal{R}(\mathbf{q}, \dot{\mathbf{q}})}{\partial \dot{q}_j} \qquad (10.20)$$

Thus Hamilton's equations become

$$\dot{q}_i = \frac{\partial H}{\partial p_i} \qquad (10.21)$$

$$\dot{p}_i = -\frac{\partial H}{\partial q_i} + \left[\sum_{k=1}^{m} \lambda_k \frac{\partial g_k}{\partial q_j}(\mathbf{q}, t) + Q_j^{EXC} \right] - \frac{\partial \mathcal{R}(\mathbf{q}, \dot{\mathbf{q}})}{\partial \dot{q}_j} \qquad (10.22)$$

The Rayleigh dissipation function $\mathcal{R}(\mathbf{q}, \dot{\mathbf{q}})$ provides an elegant and convenient way to account for dissipative forces in both Lagrangian and Hamiltonian mechanics.

10.1 *Example: Driven, linearly-damped, coupled linear oscillators*

Consider the two identical, linearly damped, coupled oscillators (damping constant β) shown in the figure. A periodic force $F = F_0 \cos(\omega t)$ is applied to the left-hand mass m. The kinetic energy of the system is

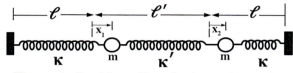

$$T = \frac{1}{2}m(\dot{x}_1^2 + \dot{x}_2^2)$$

The potential energy is

Harmonically-driven, linearly-damped, coupled linear oscillators.

$$U = \frac{1}{2}\kappa x_1^2 + \frac{1}{2}\kappa x_2^2 + \frac{1}{2}\kappa'(x_2 - x_1)^2 = \frac{1}{2}(\kappa + \kappa')x_1^2 + \frac{1}{2}(\kappa + \kappa')x_2^2 - \kappa' x_1 x_2$$

Thus the Lagrangian equals

$$L = \frac{1}{2}m(\dot{x}_1^2 + \dot{x}^2) - \left[\frac{1}{2}(\kappa + \kappa')x_1^2 + \frac{1}{2}(\kappa + \kappa')x_2^2 - \kappa' x_1 x_2\right]$$

Since the damping is linear, it is possible to use the Rayleigh dissipation function

$$\mathcal{R} = \frac{1}{2}\beta(\dot{x}_1^2 + \dot{x}_2^2)$$

The applied generalized forces are

$$Q_1' = F_o \cos(\omega t) \qquad\qquad Q_2' = 0$$

Use the Euler-Lagrange equations 10.18 to derive the equations of motion

$$\left\{\frac{d}{dt}\left(\frac{\partial L}{\partial \dot{q}_j}\right) - \frac{\partial L}{\partial q_j}\right\} + \frac{\partial \mathcal{F}}{\partial \dot{q}_j} = Q_j' + \sum_{k=1}^{m} \lambda_k \frac{\partial g_k}{\partial q_j}(\mathbf{q}, t)$$

gives

$$m\ddot{x}_1 + \beta\dot{x}_1 + (\kappa + \kappa')x_1 - \kappa' x_2 = F_0 \cos(\omega t)$$
$$m\ddot{x}_2 + \beta\dot{x}_2 + (\kappa + \kappa')x_2 - \kappa' x_1 = 0$$

These two coupled equations can be decoupled and simplified by making a transformation to normal coordinates, η_1, η_2 where

$$\eta_1 = x_1 - x_2 \qquad\qquad \eta_2 = x_1 + x_2$$

Thus

$$x_1 = \frac{1}{2}(\eta_1 + \eta_2) \qquad\qquad x_2 = \frac{1}{2}(\eta_2 - \eta_1)$$

Insert these into the equations of motion gives

$$m(\ddot{\eta}_1 + \ddot{\eta}_2) + \beta(\dot{\eta}_1 + \dot{\eta}_2) + (\kappa + \kappa')(\eta_1 + \eta_2) - \kappa'(\eta_2 - \eta_1) = 2F_0 \cos(\omega t)$$
$$m(\eta_2 - \eta_1) + \beta(\eta_2 - \eta_1) + (\kappa + \kappa')(\eta_2 - \eta_1) - \kappa'(\eta_1 + \eta_2) = 0$$

Add and subtract these two equations gives the following two decoupled equations

$$\ddot{\eta}_1 + \frac{\beta}{m}\dot{\eta}_1 + \frac{(\kappa + 2\kappa')}{m}\eta_1 = \frac{F_0}{m}\cos(\omega t)$$
$$\ddot{\eta}_2 + \frac{\beta}{m}\dot{\eta}_2 + \frac{\kappa}{m}\eta_2 = \frac{F_0}{m}\cos(\omega t)$$

Define $\Gamma = \frac{\beta}{m}, \omega_1 = \sqrt{\frac{(\kappa + 2\kappa')}{m}}, \omega_2 = \sqrt{\frac{\kappa}{m}}, A = \frac{F_0}{m}$. Then the two independent equations of motion become

$$\ddot{\eta}_1 + \Gamma\dot{\eta}_1 + \omega_1^2\eta_1 = A\cos(\omega t) \qquad\qquad \ddot{\eta}_2 + \Gamma\dot{\eta}_2 + \omega_2^2\eta_2 = A\cos(\omega t)$$

This solution is a superposition of two independent, linearly-damped, driven normal modes η_1 and η_2 that have different natural frequencies ω_1 and ω_2. For weak damping these two driven normal modes each undergo damped oscillatory motion with the η_1 and η_2 normal modes exhibiting resonances at $\omega_1' = \sqrt{\omega_1^2 - 2\left(\frac{\Gamma}{2}\right)^2}$ and $\omega_2' = \sqrt{\omega_2^2 - 2\left(\frac{\Gamma}{2}\right)^2}$

10.2 *Example: Kirchhoff's rules for electrical circuits*

The mathematical equations governing the behavior of mechanical systems and LRC electrical circuits have a close similarity. Thus variational methods can be used to derive the analogous behavior for electrical circuits. For example, for a system of n separate circuits, the magnetic flux Φ_{ik} through circuit i, due to electrical current $I_k = \dot{q}_k$ flowing in circuit k, is given by

$$\Phi_{ik} = M_{ik}\dot{q}_k$$

where M_{ik} is the mutual inductance. The diagonal term $M_{ii} = L_i$ corresponds to the self inductance of circuit i. The net magnetic flux Φ_i through circuit i, due to all n circuits, is the sum

$$\Phi_i = \sum_{k=1}^{n} M_{ik}\dot{q}_k$$

Thus the total magnetic energy W_{mag}, which is analogous to kinetic energy T, is given by summing over all n circuits to be

$$W_{mag} = T = \frac{1}{2}\sum_{i=1}^{n}\sum_{k=1}^{n} M_{ik}\dot{q}_i\dot{q}_k$$

Similarly the electrical energy W_{elect} stored in the mutual capacitance C_{ik} between the n circuits, which is analogous to potential energy, U, is given by

$$W_{elect} = U = \frac{1}{2}\sum_{i=1}^{n}\sum_{k=1}^{n} \frac{q_i q_k}{C_{ik}}$$

Thus the standard Lagrangian for this electric system is given by

$$L = T - U = \frac{1}{2}\sum_{i=1}^{n}\sum_{k=1}^{n}\left[M_{ik}\dot{q}_i\dot{q}_k - \frac{q_i q_k}{C_{ik}}\right] \tag{α}$$

Assuming that Ohm's Law is obeyed, that is, the dissipation force depends linearly on velocity, then the Rayleigh dissipation function can be written in the form

$$\mathcal{R} \equiv \frac{1}{2}\sum_{i=1}^{n}\sum_{k=1}^{n} R_{ik}\dot{q}_i\dot{q}_k \tag{β}$$

where R_{ik} is the resistance matrix. Thus the dissipation force, expressed in volts, is given by

$$F_i = -\frac{\partial\mathcal{R}}{\partial\dot{q}_j} = \frac{1}{2}\sum_{k=1}^{n} R_{ik}\dot{q}_k \tag{γ}$$

Inserting equations α, β, and γ into equation 10.18, plus making the assumption that an additional generalized electrical force $Q_i = \xi_i(t)$ volts is acting on circuit i, then the Euler-Lagrange equations give the following equations of motion.

$$\sum_{k=1}^{n}\left[M_{ik}\ddot{q}_k + R_{ik}\dot{q}_k + \frac{q_k}{C_{ik}}\right] = \xi_i(t)$$

This is a generalized version of Kirchhoff's loop rule which can be seen by considering the case where the diagonal term $i = k$ is the only non-zero term. Then

$$\left[M_{ii}\ddot{q}_i + R_{ii}\dot{q}_i + \frac{q_i}{C_{ii}}\right] = \xi_i(t)$$

This sum of the voltages is identical to the usual expression for Kirchhoff's loop rule. This example illustrates the power of variational methods when applied to fields beyond classical mechanics.

10.5 Dissipative Lagrangians

The prior discussion of nonconservative systems mentioned the following three ways to incorporate dissipative processes into Lagrangian or Hamiltonian mechanics. (1) Expand the number of degrees of freedom to include all the active dissipative active degrees of freedom as well as the conservative ones. (2) Use generalized forces to incorporate dissipative processes. (3) Add dissipative terms to the Lagrangian or Hamiltonian to mimic dissipation. The following illustrates the use of dissipative Lagrangians.

Bateman[Bat31] pointed out that an isolated dissipative system is physically incomplete, that is, a complete system must comprise at least two coupled subsystems where energy is transferred from a dissipating subsystem to an absorbing subsystem. A complete system should comprise both the dissipating and absorbing systems to ensure that the total system Lagrangian and Hamiltonian are conserved, as is assumed in conventional Lagrangian and Hamiltonian mechanics. Both Bateman and Dekker[Dek75] have illustrated that the equations of motion for a linearly-damped, free, one-dimensional harmonic oscillator are derivable using the Hamilton variational principle via introduction of a fictitious complementary subsystem that mimics dissipative processes. The following example illustrate that deriving the equations of motion for the linearly-damped, linear oscillator may be handled by three alternative equivalent non-standard Lagrangians that assume either: (1) a multidimensional system, (2) explicit time dependent Lagrangians and Hamiltonians, or (3) complex non-standard Lagrangians.

10.3 *Example: The linearly-damped, linear oscillator:*

Three toy dynamical models have been used to describe the linearly-damped, linear oscillator employing very different non-standard Lagrangians to generate the required Hamiltonians, and to derive the correct equations of motion.

1: Dual-component Lagrangian: L_{Dual}

Bateman proposed a dual system comprising a mass m subject to two coupled one-dimensional variables (x, y) where x is the observed variable and y is the mirror variable for the subsystem that absorbs the energy dissipated by the subsystem x.

Assume a non-standard Lagrangian of the form

$$L_{Dual} = \frac{m}{2}\left[\dot{x}\dot{y} - \frac{\Gamma}{2}[y\dot{x} - x\dot{y}] - \omega_0^2 xy\right] \tag{a}$$

where $\Gamma = \frac{\lambda}{m}$ is the damping coefficient. Minimizing by variation of the auxiliary variable y, that is, $\Lambda_y L = 0$, leads to the uncoupled equation of motion for x

$$\frac{m}{2}\left[\ddot{x} + \Gamma\dot{x} + \omega_0^2 x\right] = 0 \tag{b}$$

Similarly minimizing by variation of the primary variable x, that is $\Lambda_x L = 0$, leads to the uncoupled equation of motion for y

$$\frac{m}{2}\left[\ddot{y} - \Gamma\dot{y} + \omega_0^2 y\right] = 0 \tag{c}$$

Note that equation of motion (b), which was obtained by variation of the auxiliary variable y, corresponds to that for the usual free, linearly-damped, one-dimensional harmonic oscillator for the x variable which dissipates energy as is discussed in chapter 3.5. The equation of motion (c) is obtained by variation of the primary variable x and corresponds to a free linear, one-dimensional, oscillator for the y variable that is absorbing the energy dissipated by the dissipating x system.

The generalized momenta,

$$p_i \equiv \frac{\partial L}{\partial \dot{q}_i}$$

can be used to derive the corresponding Hamiltonian

$$H_{Dual}(x, p_x, y, p_y) = [p_x\dot{x} + p_y\dot{y} - L] = \frac{p_x p_y}{2m} - \frac{\Gamma}{2}[xp_x - yp_y] + \frac{m}{2}\left(\omega_0^2 - \left(\frac{\Gamma}{2}\right)^2\right)xy \tag{d}$$

Note that this Hamiltonian is time independent, and thus is conserved for this complete dual-variable system. Using Hamilton's equations of motion gives the same two uncoupled equations of motion as obtained using the Lagrangian, i.e. (b) and (c).

2: Time-dependent Lagrangian: L_{Damped}

The complementary subsystem of the above dual-component Lagrangian, that is added to the primary dissipative subsystem, is the adjoint to the equations for the primary subsystem of interest. In some cases, a set of the solutions of the complementary equations can be expressed in terms of the solutions of the primary subsystem allowing the equations of motion to be expressed solely in terms of the variables of the primary subsystem. Inspection of the solutions of the damped harmonic oscillator, presented in chapter 3.5, implies that x and y must be related by the function

$$y = xe^{\Gamma t} \qquad (e)$$

Therefore Bateman proposed a time-dependent, non-standard Lagrangian L_{Damped} of the form

$$L_{Damped} = \frac{m}{2}e^{\Gamma t}\left[\dot{x}^2 - \omega_0^2 x^2\right] \qquad (f)$$

This Lagrangian L_{Dampes} corresponds to a harmonic oscillator for which the mass $m = m_0 e^{\Gamma t}$ is accreting exponentially with time in order to mimic the exponential energy dissipation. Use of this Lagrangian in the Euler-Lagrange equations gives the solution

$$me^{\Gamma t}\left[\ddot{x} + \Gamma\dot{x} + \omega_0^2 x\right] = 0 \qquad (g)$$

If the factor outside of the bracket is non-zero, then the equation in the bracket must be zero. The expression in the bracket is the required equation of motion for the linearly-damped linear oscillator. This Lagrangian generates a generalized momentum of

$$p_x = me^{\Gamma t}\dot{x}$$

and the Hamiltonian is

$$H_{Damped} = p_x\dot{x} - L_2 = \frac{p_x^2}{2m}e^{-\Gamma t} + \frac{m}{2}\omega_0^2 e^{\Gamma t}x^2 \qquad (h)$$

The Hamiltonian is time dependent as expected. This leads to Hamilton's equations of motion

$$\dot{x} = \frac{\partial H_{Damped}}{\partial p_x} = \frac{p_x}{m}e^{-\Gamma t} \qquad (i)$$

$$-\dot{p}_x = \frac{\partial H_{Damped}}{\partial x} = m\omega_0^2 e^{\Gamma t}x \qquad (j)$$

Take the total time derivative of equation h and use equation i to substitute for $\dot{p}_x$ gives

$$me^{\Gamma t}\left[\ddot{x} + \Gamma\dot{x} + \omega_0^2 x\right] = 0 \qquad (k)$$

If the term $me^{\Gamma t}$ is non-zero, then the term in brackets is zero. The term in the bracket is the usual equation of motion for the linearly-damped harmonic oscillator.

3: Complex Lagrangian: $L_{Complex}$

Dekker proposed use of complex dynamical variables for solving the linearly-damped harmonic oscillator. It exploits the fact that, in principle, each second order differential equation can be expressed in terms of a set of first-order differential equations. This feature is the essential difference between Lagrangian and Hamiltonian mechanics. Let q be complex and assume it can be expressed in the form of a real variable x as

$$q = \dot{x} - \left(i\omega + \frac{\Gamma}{2}\right)x \qquad (l)$$

Substituting this complex variable into the relation

$$\dot{q} + \left[i\omega + \frac{\Gamma}{2}\right]q = 0 \qquad (m)$$

leads to the second-order equation for the real variable x of

$$\ddot{x} + \Gamma\dot{x} + \omega_0^2 = 0 \qquad (n)$$

This is the desired equation of motion for the linearly-damped harmonic oscillator. This result also can be shown by taking the time derivative of equation (m) and taking only the real part, i.e.

$$\ddot{q} + i\omega\dot{q} + \frac{\Gamma}{2}\dot{q} = \ddot{q} + \left(i\omega - \frac{\Gamma}{2}\right)\dot{q} + \Gamma\dot{q} = \ddot{q} + \Gamma\dot{q} + \omega_0^2 x = 0 \tag{o}$$

This feature is exploited using the following Lagrangian

$$L_{Complex} = \frac{i}{2}\left(q^*\dot{q} - q\dot{q}^*\right) - \left[\omega - i\frac{\Gamma}{2}\right]q^*q \tag{p}$$

where $\omega^2 \equiv \omega_0^2 - \left(\frac{\Gamma}{2}\right)^2$. The Lagrangian $L_{Complex}$ is real for a conservative system and complex for a dissipative system. Using the Lagrange-Euler equation for variation of q^*, that is, $\Lambda_{q^*}L_{Complex} = 0$, gives equation (m) which leads to the required equation of motion (n).

The canonical conjugate momenta are given by

$$p = \frac{\partial L_{Complex}}{\partial \dot{q}} \qquad \tilde{p} = \frac{\partial L_{Complex}}{\partial \dot{q}^*} \tag{q}$$

The above Lagrangian plus canonically conjugate momenta lead to the complimentary Hamiltonians

$$H_{Complex}(p, q, \tilde{p}, q^*) = \left(i\omega + \frac{\Gamma}{2}\right)(\tilde{p}_* q^* - pq) \tag{s}$$

$$\tilde{H}_{Complex}(p, q, \tilde{p}, q^*) = \left(i\omega - \frac{\Gamma}{2}\right)(\tilde{p}_* q^* - pq) \tag{r}$$

These Hamiltonians give Hamilton equations of motion that lead to the correct equations of motion for q and q^*

The above examples have shown that three very different, non-standard, Lagrangians, plus their corresponding Hamiltonians, all lead to the correct equation of motion for the linearly-damped harmonic oscillator. This illustrates the power of using non-standard Lagrangians to describe dissipative motion in classical mechanics. However, postulating non-standard Lagrangians to produce the required equations of motion appears to be of questionable usefulness. A fundamental approach is needed to build a firm foundation upon which non-standard Lagrangian mechanics can be based. Non-standard Lagrangian mechanics remains an active, albeit narrow, frontier of classical mechanics

10.6 Summary

Dissipative drag forces are non-conservative and usually are velocity dependent. Chapter 4 showed that the motion of non-linear dissipative dynamical systems can be highly sensitive to the initial conditions and can lead to chaotic motion.

Algebraic mechanics for nonconservative systems Since Lagrangian and Hamiltonian formulations are invalid for the nonconservative degrees of freedom, the following three approaches are used to include nonconservative degrees of freedom directly in the Lagrangian and Hamiltonian formulations of mechanics.

1. Expand the number of degrees of freedom used to include all active degrees of freedom for the system, so that the expanded system is conservative. This is the preferred approach when it is viable. Unfortunately this approach typically is impractical for handling dissipated processes because of the large number of degrees of freedom that are involved in thermal dissipation.

2. Nonconservative forces can be introduced directly at the equations of motion stage as generalized forces Q_j^{EXC}. This approach is used extensively. For the case of linear velocity dependence, the Rayleigh's dissipation function provides an elegant and powerful way to express the generalized forces in terms of scalar potential energies.

3. New degrees of freedom or effective forces can be postulated that are then incorporated into the Lagrangian or the Hamiltonian in order to mimic the effects of the nonconservative forces.

Rayleigh's dissipation function Generalized dissipative forces that have a linear velocity dependence can be easily handled in Lagrangian or Hamiltonian mechanics by introducing the powerful Rayleigh's dissipation function $\mathcal{R}(\dot{\mathbf{q}})$ where

$$\mathcal{R}(\dot{\mathbf{q}})\equiv\frac{1}{2}\sum_{i=1}^{n}\sum_{j=1}^{n}b_{ij}\dot{q}_i\dot{q}_j \tag{10.7}$$

This approach is used extensively in physics. This approach has been generalized by defining a linear velocity dependent Rayleigh dissipation function

$$\mathbf{F}_i^f = -\frac{\partial R(\mathbf{q},\dot{\mathbf{q}})}{\partial\dot{\mathbf{q}}} \tag{10.16}$$

where the generalized Rayleigh dissipation function $\mathcal{R}(\mathbf{q},\dot{\mathbf{q}})$ satisfies the general Lagrange mechanics relation

$$\frac{\delta L}{\delta q} - \frac{\partial R}{\partial\dot{q}} = 0 \tag{10.17}$$

This generalized Rayleigh's dissipation function eliminates the prior restriction to linear dissipation processes, which greatly expands the range of validity for using Rayleigh's dissipation function.

Rayleigh dissipation in Lagrange equations of motion Linear dissipative forces can be directly, and elegantly, included in Lagrangian mechanics by using Rayleigh's dissipation function as a generalized force Q_j^f. Inserting Rayleigh dissipation function 10.15 in the generalized Lagrange equations of motion 6.60 gives

$$\left\{\frac{d}{dt}\left(\frac{\partial L}{\partial\dot{q}_j}\right) - \frac{\partial L}{\partial q_j}\right\} = \left[\sum_{k=1}^{m}\lambda_k\frac{\partial g_k}{\partial q_j}(\mathbf{q},t) + Q_j^{EXC}\right] - \frac{\partial\mathcal{R}(\mathbf{q},\dot{\mathbf{q}})}{\partial\dot{q}_j} \tag{10.18}$$

Where Q_j^{EXC} corresponds to the generalized forces remaining after removal of the generalized linear, velocity-dependent, frictional force Q_j^f. The holonomic forces of constraint are absorbed into the Lagrange multiplier term.

Rayleigh dissipation in Hamiltonian mechanics If the nonconservative forces depend linearly on velocity, and are derivable from Rayleigh's dissipation function according to equation 10.15, then using the definition of generalized momentum gives

$$\dot{p}_i = \frac{d}{dt}\frac{\partial L}{\partial\dot{q}_j} = \frac{\partial L}{\partial q_i} + \left[\sum_{k=1}^{m}\lambda_k\frac{\partial g_k}{\partial q_j}(\mathbf{q},t) + Q_j^{EXC}\right] - \frac{\partial\mathcal{R}(\mathbf{q},\dot{\mathbf{q}})}{\partial\dot{q}_j} \tag{10.19}$$

$$\dot{p}_i = -\frac{\partial H(\mathbf{p},\mathbf{q},t)}{\partial q_i} + \left[\sum_{k=1}^{m}\lambda_k\frac{\partial g_k}{\partial q_j}(\mathbf{q},t) + Q_j^{EXC}\right] - \frac{\partial\mathcal{R}(\mathbf{q},\dot{\mathbf{q}})}{\partial\dot{q}_j} \tag{10.20}$$

Thus Hamilton's equations become

$$\dot{q}_i = \frac{\partial H}{\partial p_i} \tag{10.21}$$

$$\dot{p}_i = -\frac{\partial H}{\partial q_i} + \left[\sum_{k=1}^{m}\lambda_k\frac{\partial g_k}{\partial q_j}(\mathbf{q},t) + Q_j^{EXC}\right] - \frac{\partial\mathcal{R}(\mathbf{q},\dot{\mathbf{q}})}{\partial\dot{q}_j} \tag{10.22}$$

The Rayleigh dissipation function $\mathcal{R}(\mathbf{q},\dot{\mathbf{q}})$ provides an elegant and convenient way to account for dissipative forces in both Lagrangian and Hamiltonian mechanics.

Dissipative Lagrangians or Hamiltonians New degrees of freedom or effective forces can be postulated that are then incorporated into the Lagrangian or the Hamiltonian in order to mimic the effects of the nonconservative forces. This approach has been used for special cases.

Chapter 11

Conservative two-body central forces

11.1 Introduction

Conservative two-body central forces are important in physics because of the pivotal role that the Coulomb and the gravitational forces play in nature. The Coulomb force plays a role in electrodynamics, molecular, atomic, and nuclear physics, while the gravitational force plays an analogous role in celestial mechanics. Therefore this chapter focusses on the physics of systems involving conservative two-body central forces because of the importance and ubiquity of these conservative two-body central forces in nature.

A conservative two-body central force has the following three important attributes.

1. **Conservative:** A conservative force depends only on the particle position, that is, the force is not time dependent. Moreover the work done by the force moving a body between any two points 1 and 2 is path independent. Conservative fields are discussed in chapter 2.10.

2. **Two-body:** A two-body force between two bodies depends only on the relative locations of the two interacting bodies and is not influenced by the proximity of additional bodies. For two-body forces acting between n bodies, the force on body 1 is the vector superposition of the two-body forces due to the interactions with each of the other $n-1$ bodies. This differs from three-body forces where the force between any two bodies is influenced by the proximity of a third body.

3. **Central:** A central force field depends on the distance r_{12} from the origin of the force at point 1, to the body location at point 2, and the force is directed along the line joining them, that is, $\hat{\mathbf{r}}_{12}$.

A conservative, two-body, central force combines the above three attributes and can be expressed as,

$$\mathbf{F}_{21} = f(r_{12})\hat{\mathbf{r}}_{12} \qquad (11.1)$$

The force field $\mathbf{F}_{21}$ has a magnitude $f(r_{12})$ that depends only on the magnitude of the relative separation vector $\mathbf{r}_{12} = \mathbf{r}_2 - \mathbf{r}_1$ between the origin of the force at point 1 and point 2 where the force acts, and the force is directed along the line joining them, that is, $\hat{\mathbf{r}}_{12}$.

Chapter 2.10 showed that if a two-body central force is conservative, then it can be written as the gradient of a scalar potential energy $U(r)$ which is a function of the distance from the center of the force field.

$$\mathbf{F}_{21} = -\boldsymbol{\nabla} U(r_{12}) \qquad (11.2)$$

As discussed in chapter 2, the ability to represent the conservative central force by a scalar function $U(r)$ greatly simplifies the treatment of central forces.

The Coulomb and gravitational forces both are true conservative, two-body, central forces whereas the nuclear force between nucleons in the nucleus has three-body components. Two bodies interacting via a two-body central force is the simplest possible system to consider, but equation 11.1 is applicable equally for n bodies interacting via two-body central forces because the superposition principle applies for two-body central forces. This chapter will focus first on the motion of two bodies interacting via conservative two-body central forces followed by a brief discussion of the motion for $n > 2$ interacting bodies.

11.2   Equivalent one-body representation for two-body motion

The motion of two bodies, 1 and 2, interacting via two-body central forces, requires 6 spatial coordinates, that is, three each for $\mathbf{r}_1$ and $\mathbf{r}_2$. Since the two-body central force only depends on the relative separation $\mathbf{r} = \mathbf{r}_1 - \mathbf{r}_2$ of the two bodies, it is more convenient to separate the 6 degrees of freedom into 3 spatial coordinates of relative motion $\mathbf{r}$, plus 3 spatial coordinates for the center-of-mass location $\mathbf{R}$ as described in chapter 2.7. It will be shown here that the equation of motion for relative motion of the two-bodies in the center of mass can be represented by an equivalent one-body problem which simplifies the mathematics.

Consider two bodies acted upon by a conservative two-body central force, where the position vectors $\mathbf{r}_1$ and $\mathbf{r}_2$ specify the location of each particle as illustrated in figure 11.1. An alternate set of six variables would be the three components of the center of mass position vector $\mathbf{R}$ and the three components specifying the difference vector $\mathbf{r}$ defined by figure 11.1. Define the vectors $\mathbf{r}'_1$ and $\mathbf{r}'_2$ as the position vectors of the masses m_1 and m_2 with respect to the center of mass. Then

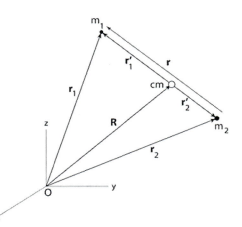

$$\begin{aligned} \mathbf{r}_1 &= \mathbf{R} + \mathbf{r}'_1 \\ \mathbf{r}_2 &= \mathbf{R} + \mathbf{r}'_2 \end{aligned} \tag{11.3}$$

Figure 11.1: Center of mass cordinates for the two-body system.

By the definition of the center of mass

$$\mathbf{R} = \frac{m_1 \mathbf{r}_1 + m_2 \mathbf{r}_2}{m_1 + m_2} \tag{11.4}$$

and

$$m_1 \mathbf{r}'_1 + m_2 \mathbf{r}'_2 = 0 \tag{11.5}$$

so that

$$-\frac{m_1}{m_2}\mathbf{r}'_1 = \mathbf{r}'_2 \tag{11.6}$$

Therefore

$$\mathbf{r} = \mathbf{r}'_1 - \mathbf{r}'_2 = \frac{m_1 + m_2}{m_2}\mathbf{r}'_1 \tag{11.7}$$

that is,

$$\mathbf{r}'_1 = \frac{m_2}{m_1 + m_2}\mathbf{r} \tag{11.8}$$

Similarly;

$$\mathbf{r}'_2 = -\frac{m_1}{m_1 + m_2}\mathbf{r} \tag{11.9}$$

Substituting these into equation 11.3 gives

$$\begin{aligned} \mathbf{r}_1 &= \cdot\ \mathbf{R} + \mathbf{r}'_1 = \mathbf{R} + \frac{m_2}{m_1 + m_2}\mathbf{r} \\ \mathbf{r}_2 &= \mathbf{R} + \mathbf{r}'_2 = \mathbf{R} - \frac{m_1}{m_1 + m_2}\mathbf{r} \end{aligned} \tag{11.10}$$

That is, the two vectors $\mathbf{r}_1, \mathbf{r}_2$ are written in terms of the position vector for the center of mass $\mathbf{R}$ and the position vector $\mathbf{r}$ for relative motion in the center of mass.

Assuming that the two-body central force is conservative and represented by $U(r)$, then the Lagrangian of the two-body system can be written as

$$L = \frac{1}{2}m_1 |\dot{\mathbf{r}}_1|^2 + \frac{1}{2}m_2 |\dot{\mathbf{r}}_2|^2 - U(r) \tag{11.11}$$

Differentiating equations 11.10, with respect to time, and inserting them into the Lagrangian, gives

$$L = \frac{1}{2}M\left|\dot{\mathbf{R}}\right|^2 + \frac{1}{2}\mu\left|\dot{\mathbf{r}}\right|^2 - U(r) \tag{11.12}$$

where the total mass M is defined as

$$M = m_1 + m_2 \tag{11.13}$$

and the **reduced mass** μ is defined by

$$\mu \equiv \frac{m_1 m_2}{m_1 + m_2} \tag{11.14}$$

or equivalently

$$\frac{1}{\mu} = \frac{1}{m_1} + \frac{1}{m_2} \tag{11.15}$$

The total Lagrangian can be separated into two independent parts

$$L = \frac{1}{2}M\left|\dot{\mathbf{R}}\right|^2 + L_{cm} \tag{11.16}$$

where

$$L_{cm} = \frac{1}{2}\mu\left|\dot{\mathbf{r}}\right|^2 - U(r) \tag{11.17}$$

Assuming that no external forces are acting, then $\frac{\partial L}{\partial \mathbf{R}} = 0$ and the three Lagrange equations for each of the three coordinates of the $\mathbf{R}$ coordinate can be written as

$$\frac{d}{dt}\frac{\partial L}{\partial \dot{\mathbf{R}}} = \frac{d\mathbf{P}_{cm}}{dt} = 0 \tag{11.18}$$

That is, for a pure central force, the center-of-mass momentum $\mathbf{P}_{cm}$ is a constant of motion where

$$\mathbf{P}_{cm} = \frac{\partial L}{\partial \dot{\mathbf{R}}} = M\dot{\mathbf{R}} \tag{11.19}$$

It is convenient to work in the center-of-mass frame using the effective Lagrangian L_{cm}. In the center-of-mass frame of reference, the translational kinetic energy $\frac{1}{2}M\left|\dot{\mathbf{R}}\right|^2$ associated with center-of-mass motion is ignored, and only the energy in the center-of-mass is considered. This center-of-mass energy is the energy involved in the interaction between the colliding bodies. *Thus, in the center-of-mass, the problem has been reduced to an equivalent one-body problem of a mass μ moving about a fixed force center with a path given by $\mathbf{r}$ which is the separation vector between the two bodies, as shown in figure 11.2.* In reality, both masses revolve around their center of mass, also called the barycenter, in the center-of-mass frame as shown in figure 11.2. Knowing $\mathbf{r}$ allows the trajectory of each mass about the center of mass $\mathbf{r}'_1$ and $\mathbf{r}'_2$ to be calculated. Of course the true path in the laboratory frame of reference must take into account both the translational motion of the center of mass, in addition to the motion of the equivalent one-body representation relative to the barycenter. Be careful to remember the difference between the actual trajectories of each body, and the effective trajectory assumed when using the reduced mass which only determines the *relative* separation $\mathbf{r}$ of the two bodies. This reduction to an equivalent one-body problem greatly simplifies the solution of the motion, but it misrepresents the actual trajectories and the spatial locations of each mass in space. The equivalent one-body representation will be used extensively throughout this chapter.

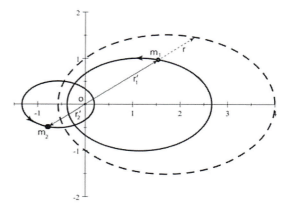

Figure 11.2: Orbits of a two-body system with mass ratio of 2 rotating about the center-of-mass, O. The dashed ellipse is the equivalent one-body orbit with the center of force at the focus O.

11.3   Angular momentum L

The notation used for the angular momentum vector is **L** where the magnitude is designated by $|\mathbf{L}| = l$. Be careful not to confuse the angular momentum vector **L** with the Lagrangian L_{cm}. Note that the angular momentum for two-body rotation about the center of mass with angular velocity ω is identical when evaluated in either the laboratory or equivalent two-body representation. That is, using equations 11.8 and 11.9

$$\mathbf{L} = \mathbf{m}_1 r_1'^2 \boldsymbol{\omega} + \mathbf{m}_2 r_2'^2 \boldsymbol{\omega} = \mu r^2 \boldsymbol{\omega} \qquad (11.20)$$

The center-of-mass Lagrangian leads to the following two general properties regarding the angular momentum vector **L**.

1) The motion lies entirely in a plane perpendicular to the fixed direction of the total angular momentum vector. This is because

$$\mathbf{L} \cdot \mathbf{r} = \mathbf{r} \times \mathbf{p} \cdot \mathbf{r} = 0 \qquad (11.21)$$

that is, *the radius vector is in the plane perpendicular to the total angular momentum vector.* Thus, it is possible to express the Lagrangian in polar coordinates, (r, ψ) rather than spherical coordinates. In polar coordinates the center-of-mass Lagrangian becomes

$$L_{cm} = \frac{1}{2}\mu\left(\dot{r}^2 + r^2\dot{\psi}^2\right) - U(r) \qquad (11.22)$$

2) If the potential is spherically symmetric, then the polar angle ψ is cyclic and therefore Noether's theorem gives that the angular momentum $\mathbf{p}_\psi \equiv \mathbf{L} = \mathbf{r} \times \mathbf{p}$ is a constant of motion. That is, since $\frac{\partial L_{cm}}{\partial \psi} = 0$, then the Lagrange equations imply that

$$\dot{\mathbf{p}}_\psi = \frac{d}{dt}\frac{\partial L_{cm}}{\partial \dot{\psi}} = 0 \qquad (11.23)$$

where the vectors $\dot{\mathbf{p}}_\psi$ and $\dot{\boldsymbol{\psi}}$ imply that equation 11.23 refers to three independent equations corresponding to the three components of these vectors. *Thus the angular momentum $\mathbf{p}_\psi$, conjugate to $\boldsymbol{\psi}$, is a constant of motion.* The generalized momentum $\mathbf{p}_\psi$ is a first integral of the motion which equals

$$\mathbf{p}_\psi = \frac{\partial L_{cm}}{\partial \dot{\psi}} = \mu r^2 \dot{\boldsymbol{\psi}} = \hat{\mathbf{p}}_\psi l \qquad (11.24)$$

where the magnitude of the angular momentum l, and the direction $\hat{\mathbf{p}}_\psi$, both are constants of motion.

A simple geometric interpretation of equation 11.24 is illustrated in figure 11.3. The radius vector sweeps out an area $d\mathbf{A}$ in time dt where

$$d\mathbf{A} = \frac{1}{2}\mathbf{r} \times \mathbf{v}dt \qquad (11.25)$$

and the vector **A** is perpendicular to the $x - y$ plane. The rate of change of area is

$$\frac{d\mathbf{A}}{dt} = \frac{1}{2}\mathbf{r} \times \mathbf{v} \qquad (11.26)$$

But the angular momentum is

$$\mathbf{L} = \mathbf{r} \times \mathbf{p} = \mu\mathbf{r} \times \mathbf{v} = 2\mu\frac{d\mathbf{A}}{dt} \qquad (11.27)$$

Thus the conservation of angular momentum implies that the areal velocity $\frac{dA}{dt}$ also is a constant of motion. This fact is called Kepler's second law of planetary motion which he deduced in 1609 based on Tycho Brahe's 55 years of observational records of the motion of Mars. Kepler's second law implies that a planet moves fastest when closest to the sun and slowest when farthest from the sun. Note that Kepler's second law is a statement of *the conservation of angular momentum which is independent of the radial form of the central potential.*

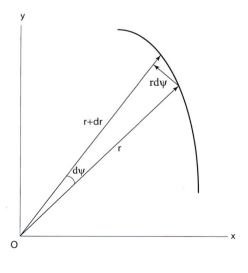

Figure 11.3: Area swept out by the radius vector in the time dt.

11.4   Equations of motion

The equations of motion for two bodies interacting via a conservative two-body central force can be determined using the center of mass Lagrangian, L_{cm}, given by equation 11.22. For the radial coordinate, the operator equation $\Lambda_r L_{cm} = 0$ for Lagrangian mechanics leads to

$$\frac{d}{dt}(\mu\dot{r}) - \mu r\dot{\psi}^2 + \frac{\partial U}{\partial r} = 0 \tag{11.28}$$

But

$$\dot{\psi} = \frac{l}{\mu r^2} \tag{11.29}$$

therefore the radial equation of motion is

$$\mu\ddot{r} = -\frac{\partial U}{\partial r} + \frac{l^2}{\mu r^3} \tag{11.30}$$

Similarly, for the angular coordinate, the operator equation $\Lambda_\psi L_{cm} = 0$ leads to equation 11.24. That is, the angular equation of motion for the magnitude of p_ψ is

$$p_\psi = \frac{\partial L}{\partial \dot{\psi}} = \mu r^2 \dot{\psi} = l \tag{11.31}$$

Lagrange's equations have given two equations of motion, one dependent on radius r and the other on the polar angle ψ. Note that the radial acceleration is just a statement of Newton's Laws of motion for the radial force F_r in the center-of-mass system of

$$F_r = -\frac{\partial U}{\partial r} + \frac{l^2}{\mu r^3} \tag{11.32}$$

This can be written in terms of an effective potential

$$U_{eff}(r) \equiv U(r) + \frac{l^2}{2\mu r^2} \tag{11.33}$$

which leads to an equation of motion

$$F_r = \mu\ddot{r} = -\frac{\partial U_{eff}(r)}{\partial r} \tag{11.34}$$

Since $\frac{l^2}{\mu r^3} = \mu r\dot{\psi}^2$, the second term in equation (11.33) is the usual centrifugal force that originates because the variable r is in a non-inertial, rotating frame of reference. Note that the angular equation of motion is independent of the radial dependence of the conservative two-body central force.

Figure 11.4 shows, by dashed lines, the radial dependence of the potential corresponding to the attractive inverse square law force, that is $U = -\frac{k}{r}$, and the potential corresponding to the centrifugal term $\frac{l^2}{2\mu r^2}$ corresponding to a repulsive centrifugal force. The sum of these two potentials $U_{eff}(r)$, shown by the solid line, has a minimum U_{min} value at a certain radius similar to that manifest by the diatomic molecule discussed in example 2.7.

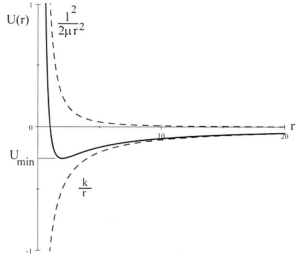

Figure 11.4: The attractive inverse-square law potential ($\frac{k}{r}$), the centrifugal potential ($\frac{l^2}{2\mu r^2}$), and the combined effective bound potential.

It is remarkable that the six-dimensional equations of motion, for two bodies interacting via a two-body central force, has been reduced to trivial center-of-mass translational motion, plus a *one-dimensional one-body problem* given by (11.34) in terms of the relative separation r and an effective potential $U_{eff}(r)$.

11.5   Differential orbit equation:

The differential orbit equation relates the shape of the orbital motion, in plane polar coordinates, to the radial dependence of the two-body central force. A Binet coordinate transformation, which depends on the functional form of $\mathbf{F}(\mathbf{r})$, can simplify the differential orbit equation. For the inverse-square law force, the best Binet transformed variable is u which is defined to be

$$u \equiv \frac{1}{r} \tag{11.35}$$

Inserting the transformed variable u into equation 11.29 gives

$$\dot{\psi} = \frac{lu^2}{\mu} \tag{11.36}$$

From the definition of the new variable

$$\frac{dr}{dt} = -u^{-2}\frac{du}{dt} = -u^{-2}\frac{du}{d\psi}\dot{\psi} = -\frac{l}{\mu}\frac{du}{d\psi} \tag{11.37}$$

Differentiating again gives

$$\frac{d^2r}{dt^2} = -\frac{l}{\mu}\frac{d}{dt}\left(\frac{du}{d\psi}\right) = -\left(\frac{lu}{\mu}\right)^2\frac{d^2u}{d\psi^2} \tag{11.38}$$

Substituting these into Lagrange's radial equation of motion gives

$$\frac{d^2u}{d\psi^2} + u = -\frac{\mu}{l^2}\frac{1}{u^2}F(\frac{1}{u}) \tag{11.39}$$

Binet's *differential orbit equation* directly relates ψ and r which determines the overall shape of the orbit trajectory. This shape is crucial for understanding the orbital motion of two bodies interacting via a two-body central force. Note that for the special case of an inverse square-law force, that is where $F(\frac{1}{u}) = ku^2$, then the right-hand side of equation 11.39 equals a constant $-\frac{\mu k}{l^2}$ since the orbital angular momentum is a conserved quantity.

11.1 *Example: Central force leading to a circular orbit* $r = 2R\cos\theta$

Binet's differential orbit equation can be used to derive the central potential that leads to the assumed circular trajectory of $r = 2R\cos\theta$ where R is the radius of the circular orbit. Note that this circular orbit passes through the origin of the central force when $r = 2R\cos\theta = 0$

Inserting this trajectory into Binet's differential orbit equation 11.39 gives

$$\frac{1}{2R}\frac{d^2\left(\cos\theta\right)^{-1}}{d\theta^2} + \frac{1}{2R}\left(\cos\theta\right)^{-1} = -\frac{\mu}{l^2}4R^2\left(\cos\theta\right)^2 F(\frac{1}{u}) \quad (\alpha)$$

Note that the differential is given by

$$\frac{d^2\left(\cos\theta\right)^{-1}}{d\theta^2} = \frac{d}{d\theta}\left(\frac{\sin\theta}{\cos^3\theta}\right) = \frac{2\sin^2\theta}{\cos^3\theta} + \frac{1}{\cos\theta}$$

Inserting this differential into equation α gives

$$\frac{2\sin^2\theta}{\cos^3\theta} + \frac{1}{\cos\theta} + \frac{1}{\cos\theta} = \frac{2}{\cos^3\theta} = -\frac{\mu}{l^2}8R^3\left(\cos\theta\right)^2 F(\frac{1}{u})$$

Thus the radial dependence of the required central force is

$$F = -\frac{l^2}{8R^3\mu}\frac{2}{\cos^5\theta} = -\frac{8R^2l^2}{\mu}\frac{1}{r^5} = -\frac{k}{r^5}$$

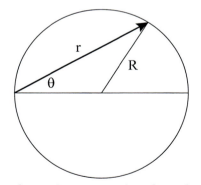

Circular trajectory passing through the origin of the central force.

This corresponds to an attractive central force that depends to the fifth power on the inverse radius r. Note that this example is unrealistic since the assumed orbit implies that the potential and kinetic energies are infinite when $r \to 0$ at $\theta \to \frac{\pi}{2}$.

11.6 Hamiltonian

Since the center-of-mass Lagrangian is not an explicit function of time, then

$$\frac{dH_{cm}}{dt} = -\frac{\partial L_{cm}}{\partial t} = 0 \tag{11.40}$$

Thus *the center-of mass Hamiltonian H_{cm} is a constant of motion.* However, since the transformation to center of mass can be time dependent, then $H_{cm} \neq E$, that is, it does not include the total energy because the kinetic energy of the center-of-mass motion has been omitted from H_{cm}. Also, since no transformation is involved, then

$$H_{cm} = T_{cm} + U = E_{cm} \tag{11.41}$$

That is, the center-of-mass Hamiltonian H_{cm} equals the center-of-mass total energy. The center-of-mass Hamiltonian then can be written using the effective potential (11.33) in the form

$$H_{cm} = \frac{p_r^2}{2\mu} + \frac{p_\theta^2}{2\mu r^2} + U(r) = \frac{p_r^2}{2\mu} + \frac{l^2}{2\mu r^2} + U(r) = \frac{p_r^2}{2\mu} + U_{eff}(r) = E_{cm} \tag{11.42}$$

It is convenient to express the center-of-mass Hamiltonian H_{cm} in terms of the energy equation for the orbit in a central field using the transformed variable $u = \frac{1}{r}$. Substituting equations 11.33 and 11.37 into the Hamiltonian equation 11.42 gives the *energy equation of the orbit*

$$\frac{l^2}{2\mu}\left[\left(\frac{du}{d\psi}\right)^2 + u^2\right] + U\left(u^{-1}\right) = E_{cm} \tag{11.43}$$

Energy conservation allows the Hamiltonian to be used to solve problems directly. That is, since

$$H_{cm} = \frac{\mu \dot{r}^2}{2} + \frac{l^2}{2\mu r^2} + U(r) = E_{cm} \tag{11.44}$$

then

$$\dot{r} = \frac{dr}{dt} = \pm\sqrt{\frac{2}{\mu}\left(E_{cm} - U - \frac{l^2}{2\mu r^2}\right)} \tag{11.45}$$

The time dependence can be obtained by integration

$$t = \int \frac{\pm dr}{\sqrt{\frac{2}{\mu}\left(E_{cm} - U - \frac{l^2}{2\mu r^2}\right)}} + \text{ constant} \tag{11.46}$$

An inversion of this gives the solution in the standard form $r = r(t)$. However, it is more interesting to find the relation between r and θ. From relation 11.46 for $\frac{dr}{dt}$ then

$$dt = \frac{\pm dr}{\sqrt{\frac{2}{\mu}\left(E_{cm} - U - \frac{l^2}{2\mu r^2}\right)}} \tag{11.47}$$

while equation 11.29 gives

$$d\psi = \frac{ldt}{\mu r^2} = \frac{\pm l dr}{r^2\sqrt{2\mu\left(E_{cm} - U - \frac{l^2}{2\mu r^2}\right)}} \tag{11.48}$$

Therefore

$$\psi = \int \frac{\pm l dr}{r^2\sqrt{2\mu\left(E_{cm} - U - \frac{l^2}{2\mu r^2}\right)}} + \text{ constant} \tag{11.49}$$

which can be used to calculate the angular coordinate. This gives the relation between the radial and angular coordinates which specifies the trajectory.

Although equations (11.45) and (11.49) formally give the solution, the actual solution can be derived analytically only for certain specific forms of the force law and these solutions differ for attractive versus repulsive interactions.

11.7   General features of the orbit solutions

It is useful to look at the general features of the solutions of the equations of motion given by the equivalent one-body representation of the two-body motion. These orbits depend on the net center of mass energy E_{cm}. There are five possible situations depending on the center-of-mass total energy E_{cm}.

1) $\mathbf{E}_{cm} > \mathbf{0}$: The trajectory is hyperbolic and has a minimum distance, but no maximum. The distance of closest approach is given when $\dot{r} = 0$. At the turning point $E_{cm} = U + \frac{l^2}{2\mu r^2}$

2) $\mathbf{E}_{cm} = \mathbf{0}$: It can be shown that the orbit for this case is parabolic.

3) $\mathbf{0} > \mathbf{E}_{cm} > \mathbf{U}_{\min}$: For this case the equivalent orbit has both a maximum and minimum radial distance at which $\dot{r} = 0$. At the turning points the radial kinetic energy term is zero so $E_{cm} = U + \frac{l^2}{2\mu r^2}$. For the attractive inverse square law force the path is an ellipse with the focus at the center of attraction (Figure 11.5), which is Kepler's First Law. During the time that the radius ranges from $r_{\min}$ to $r_{\max}$ and back the radius vector turns through an angle $\Delta\psi$ which is given by

$$\Delta\psi = 2 \int_{r_{\min}}^{r_{\max}} \frac{\pm l\,dr}{r^2 \sqrt{2\mu \left(E_{cm} - U - \frac{l^2}{2\mu r^2} \right)}} \tag{11.50}$$

The general path prescribes a rosette shape which is a closed curve only if $\Delta\psi$ is a rational fraction of 2π.

4) $\mathbf{E}_{cm} = \mathbf{U}_{\min}$: In this case r is a constant implying that the path is circular since

$$\dot{r} = \frac{dr}{dt} = \pm \sqrt{\frac{2}{\mu} \left(E_{cm} - U - \frac{l^2}{2\mu r^2} \right)} = 0 \tag{11.51}$$

5) $\mathbf{E}_{cm} < \mathbf{U}_{\min}$: For this case the square root is imaginary and there is no real solution.

In general the orbit is not closed, and such open orbits do not repeat. Bertrand's Theorem states that the inverse-square central force, and the linear harmonic oscillator, are the only radial dependences of the central force that lead to stable closed orbits.

11.2 Example: Orbit equation of motion for a free body

It is illustrative to use the differential orbit equation 11.39 to show that a body in free motion travels in a straight line. Assume that a line through the origin O intersects perpendicular to the instantaneous trajectory at the point Q which has polar coordinates (r_0, ϕ) relative to the origin. The point P, with polar coordinates (r, ϕ), lies on straight line through Q that is perpendicular to OQ if, and only if, $r\cos(\phi - \delta) = r_0$. Since the force is zero then the differential orbit equation simplifies to

$$\frac{d^2 u(\phi)}{d\phi^2} + u(\phi) = 0$$

A solution of this is

$$u(\phi) = \frac{1}{r_0}\cos(\phi - \delta)$$

where r_0 and δ are arbitrary constants. This can be rewritten as

$$r(\phi) = \frac{r_0}{\cos(\phi - \delta)}$$

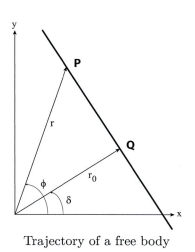

Trajectory of a free body

This is the equation of a straight line in polar coordinates as illustrated in the adjacent figure. This shows that a free body moves in a straight line if no forces are acting on the body.

11.8 Inverse-square, two-body, central force

The most important conservative, two-body, central interaction is the attractive inverse-square law force, which is encountered in both gravitational attraction and the Coulomb force. This force $\mathbf{F}(r)$ can be written in the form

$$\mathbf{F}(r) = \frac{k}{r^2}\widehat{\mathbf{r}} \tag{11.52}$$

The force constant k is defined to be negative for an attractive force and positive for a repulsive force. In S.I. units the force constant $k = -Gm_1m_2$ for the gravitational force and $k = +\frac{q_1 q_2}{4\pi\epsilon_0}$ for the Coulomb force. Note that this sign convention is the opposite of what is used in many books which use a negative sign in equation 11.52 and assume k to be positive for an attractive force and negative for a repulsive force.

The conservative, inverse-square, two-body, central force is unique in that the underlying symmetries lead to four conservation laws, all of which are of pivotal importance in nature.

1. **Conservation of angular momentum:** Like all conservative central forces, the inverse-square central two-body force conserves angular momentum as proven in chapter 11.3.

2. **Conservation of energy:** This conservative central force can be represented in terms of a scalar potential energy $U(r)$ as given by equation 11.2, where for this central force

$$U(r) = \frac{k}{r} \tag{11.53}$$

   Moreover, equation 11.42 showed that the center-of-mass Hamiltonian is conserved, that is, $H_{cm} = E_{cm}$

3. **Gauss' Law:** For a conservative, inverse-square, two-body, central force, the flux of the force field out of any closed surface is proportional to the algebraic sum of the sources and sinks of this field that are located inside the closed surface. The net flux is independent of the distribution of the sources and sinks inside the closed surface, as well as the size and shape of the closed surface. Chapter 2.14.5 proved this for the gravitational force field.

4. **Closed orbits:** Two bodies interacting via the conservative, inverse-square, two-body, central force follow closed (degenerate) orbits as stated by Bertrand's Theorem. The first consequence of this symmetry is that Kepler's laws of planetary motion have stable, single-valued orbits. The second consequence of this symmetry is the conservation of the eccentricity vector discussed in chapter 11.84.

Observables that depend on Gauss's Law, or on closed planetary orbits, are extremely sensitive to addition of even a miniscule incremental exponent ξ to the radial dependence $r^{-(2\pm\xi)}$ of the force. The statement that the inverse-square, two-body, central force leads to closed orbits can be proven by inserting equation 11.52 into the orbit differential equation,

$$\frac{d^2u}{d\psi^2} + u = -\frac{\mu}{l^2}\frac{1}{u^2}ku^2 = -\frac{\mu k}{l^2} \tag{11.54}$$

Using the transformation

$$y \equiv u + \frac{\mu k}{l^2} \tag{11.55}$$

the orbit equation becomes

$$\frac{d^2y}{d\psi^2} + y = 0 \tag{11.56}$$

A solution of this equation is

$$y = B\cos(\psi - \psi_0) \tag{11.57}$$

Therefore

$$u = \frac{1}{r} = -\frac{\mu k}{l^2}\left[1 + \epsilon\cos(\psi - \psi_0)\right] \tag{11.58}$$

This the equation of a conic section. For an attractive, inverse-square, central force, equation 11.58 is the equation for an ellipse with the *origin of r at one of the foci of the ellipse* that has eccentricity ϵ, defined as

$$\epsilon \equiv B\frac{l^2}{\mu k} \tag{11.59}$$

Equation 11.58 is the polar equation of a conic section. Equation 11.58 also can be derived with the origin at a focus by inserting the inverse square law potential into equation 11.49 which gives

$$\psi = \int \frac{\pm du}{\sqrt{\frac{2\mu E_{cm}}{l^2} + \frac{2\mu k}{l^2}u - u^2}} + \text{constant} \tag{11.60}$$

The solution of this gives

$$u = \frac{1}{r} = -\frac{\mu k}{l^2}\left[1 + \sqrt{1 + \frac{2E_{cm}l^2}{\mu k^2}}\cos\left(\psi - \psi_0\right)\right] \tag{11.61}$$

Equations 11.58 and 11.61 are identical if the eccentricity ϵ equals

$$\epsilon = \sqrt{1 + \frac{2E_{cm}l^2}{\mu k^2}} \tag{11.62}$$

The value of ψ_0 merely determines the orientation of the major axis of the equivalent orbit. Without loss of generality, it is possible to assume that the angle ψ is measured with respect to the major axis of the orbit, that is $\psi_0 = 0$. Then the equation can be written as

$$u = \frac{1}{r} = -\frac{\mu k}{l^2}\left[1 + \epsilon\cos\left(\psi\right)\right] = -\frac{\mu k}{l^2}\left[1 + \sqrt{1 + \frac{2E_{cm}l^2}{\mu k^2}}\cos\left(\psi\right)\right] \tag{11.63}$$

This is the equation of a conic section where ϵ is the eccentricity of the conic section. The conic section is a hyperbola if $\epsilon > 1$, parabola if $\epsilon = 1$, ellipse if $\epsilon < 1$, and a circle if $\epsilon = 0$. All the equivalent one-body orbits for an attractive force have the origin of the force at a focus of the conic section. The orbits depend on whether the force is attractive or repulsive, on the conserved angular momentum l, and on the center-of-mass energy E_{cm}.

11.8.1   Bound orbits

Closed bound orbits occur only if the following requirements are satisfied.

1. The force must be attractive, $(k < 0)$ then equation 11.63 ensures that r is positive.

2. For a closed elliptical orbit. the eccentricity $\epsilon < 1$ of the equivalent one-body representation of the orbit implies that the total center-of-mass energy $E_{cm} < 0$, that is, the closed orbit is bound.

Bound elliptical orbits have the center-of-force at one interior focus F_1 of the elliptical one-body representation of the orbit as shown in figure 11.5.

The minimum value of the orbit $r = r_{\min}$ occurs when $\psi = 0$, where

$$r_{\min} = -\frac{l^2}{\mu k\left[1 + \epsilon\right]} \tag{11.64}$$

This minimum distance is called the *periapsis*[1].

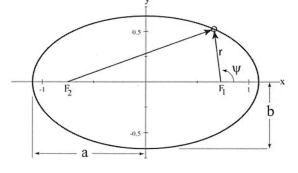

Figure 11.5: Bound elliptical orbit.

---

[1]The greek term apsis refers to the points of greatest or least distance of approach for an orbiting body from one of the foci of the elliptical orbit. The term periapsis or pericenter both are used to designate the closest distance of approach, while apoapsis or apocenter are used to designate the farthest distance of approach. Attaching the terms "perí-" and "apo-" to the general term "-apsis" is preferred over having different names for each object in the solar system. For example, frequently used terms are "-helion" for orbits of the sun, "-gee" for orbits around the earth, and "-cynthion" for orbits around the moon.

The maximum distance, $r = r_{\max}$, which is called the *apoapsis*, occurs when $\psi = 180^o$

$$r_{\max} = -\frac{l^2}{\mu k \, [1 - \epsilon]} \tag{11.65}$$

Remember that since $k < 0$ for bound orbits, the negative signs in equations 11.64 and 11.65 lead to $r > 0$. The most bound orbit is a circle having $\epsilon = 0$ which implies that $E_{cm} = -\frac{\mu k^2}{l^2}$.

The shape of the elliptical orbit also can be described with respect to the center of the elliptical equivalent orbit by deriving the lengths of the semi-major axis a and the semi-minor axis b shown in figure 11.5.

$$a = \frac{1}{2} (r_{\min} + r_{\max}) = \frac{1}{2} \left(\frac{l^2}{\mu k \, [1 + \epsilon]} + \frac{l^2}{\mu k \, [1 - \epsilon]} \right) = \frac{l^2}{\mu k \, [1 - \epsilon^2]} \tag{11.66}$$

$$b = a\sqrt{1 - \epsilon^2} = \frac{l^2}{\mu k \sqrt{[1 - \epsilon^2]}} \tag{11.67}$$

Remember that the predicted bound elliptical orbit corresponds to the equivalent one-body representation for the two-body motion as illustrated in figure 11.2. This can be transformed to the individual spatial trajectories of the each of the two bodies in an inertial frame.

11.8.2 Kepler's laws for bound planetary motion

Kepler's three laws of motion apply to the motion of two bodies in a bound orbit due to the attractive gravitational force for which $k = -Gm_1 m_2$.

1) Each planet moves in an elliptical orbit with the sun at one focus
2) The radius vector, drawn from the sun to a planet, describes equal areas in equal times
3) The square of the period of revolution about the sun is proportional to the cube of the major axis of the orbit.

Two bodies interacting via the gravitational force, which is a conservative, inverse-square, two-body central force, is best handled using the equivalent orbit representation. The first and second laws were proved in chapters 11.8 and 11.3. That is, the second law is equivalent to the statement that the angular momentum is conserved. The third law can be derived using the fact that the area of an ellipse is

$$A = \pi a b = \pi a^2 \sqrt{1 - \epsilon^2} = \frac{\pi l}{\sqrt{-\mu k}} a^{\frac{3}{2}} \tag{11.68}$$

Equations 11.26 and 11.27 give that the rate of change of area swept out by the radius vector is

$$\frac{dA}{dt} = \frac{1}{2} r^2 \dot{\psi} = \frac{l}{2\mu} \tag{11.69}$$

Therefore the period for one revolution τ is given by the time to sweep out one complete ellipse

$$\tau = \frac{A}{\left(\frac{dA}{dt} \right)} = 2\pi \left(\frac{\mu}{-k} \right)^{\frac{1}{2}} a^{\frac{3}{2}} \tag{11.70}$$

This leads to **Kepler's 3rd law**

$$\tau^2 = 4\pi^2 \frac{\mu}{-k} a^3 \tag{11.71}$$

Bound orbits occur only for attractive forces for which the force constant k is negative, and thus cancel the negative sign in equation 11.71. For example, for the gravitational force $k = -Gm_1 m_2$.

Note that the reduced mass $\mu = \frac{m_1 m_2}{m_1 + m_2}$ occurs in Kepler's 3rd law. That is, Kepler's third law can be written in terms of the actual masses of the bodies to be

$$\tau^2 = \frac{4\pi^2}{G \, (m_1 + m_2)} a^3 \tag{11.72}$$

In relating the relative periods of the different planets Kepler made the approximation that the mass of the planet m_1 is negligible relative to the mass of the sun m_2.

The eccentricity of the major planets ranges from $\epsilon = 0.2056$ for Mercury, to $\epsilon = 0.0068$ for Venus. The Earth has an eccentricity of $\epsilon = 0.0167$ with $r_{\min} = 91 \cdot 10^6$ miles and $r_{\max} = 95 \cdot 10^6$ miles. On the other hand, $\epsilon = 0.967$ for Halley's comet, that is, the radius vector ranges from 0.6 to 18 times the radius of the orbit of the Earth.

The orbit energy can be derived by substituting the eccentricity, given by equation 11.62, into the semi-major axis length a, given by equation 11.66, which leads to the center-of-mass energy of

$$E_{cm} = -\frac{k}{2a} \tag{11.73}$$

However, the Hamiltonian, given by equation 11.42, implies that E_{cm} is

$$E_{cm} = \frac{1}{2}\mu v^2 + \left(-\frac{k}{r}\right) = -\frac{k}{2a} \tag{11.74}$$

For the simple case of a circular orbit, $a = r$ then the velocity v equals

$$v = \sqrt{\frac{k}{\mu r}} \tag{11.75}$$

For a circular orbit, the drag on a satellite lowers the total energy resulting in a decrease in the radius of the orbit and a concomitant increase in velocity. That is, when the orbit radius is decreased, part of the gain in potential energy accounts for the work done against the drag, and the remaining part goes towards increase of the kinetic energy. Also note that, as predicted by the Virial Theorem, the kinetic energy always is half the potential energy for the inverse square law force.

11.8.3   Unbound orbits

Attractive inverse-square central forces lead to hyperbolic orbits for $\epsilon > 1$ for which $E_{cm} > 0$, that is, the orbit is unbound. In addition, the orbits always are unbound for a repulsive force since $U = \frac{k}{r}$ is positive as is the kinetic energy T_{cm}, thus $E_{cm} = T_{cm} + U_{cm} > 0$. The radial orbit equation for either an attractive or a repulsive force is

$$r = -\frac{l^2}{\mu k \left[1 + \epsilon \cos \psi\right]} \tag{11.76}$$

For a repulsive force k is positive and l^2 always is positive. Therefore to ensure that r remain positive the bracket term must be negative. That is

$$[1 + \epsilon \cos \psi] < 0 \qquad\qquad k > 0 \tag{11.77}$$

For an attractive force k is negative and since l^2 is positive then the bracket term must be positive to ensure that r is positive. That is,

$$[1 + \epsilon \cos \psi] > 0 \qquad\qquad k < 0 \tag{11.78}$$

Figure 11.6 shows both branches of the hyperbola for a given angle ψ for the equivalent two-body orbits where the center of force is at the origin. For an attractive force, $k < 0$, the center of force is at the interior focus of the hyperbola, whereas for a repulsive force the center of force is at the exterior focus. For a given value of $|\psi|$ the asymptotes of the orbits both are displaced by the same **impact parameter** b from parallel lines passing through the center of force. The scattering angle, between the outgoing direction of the scattered body and the incident direction, is designated to be θ, which is related to the angle ψ by $\theta = 180° - 2\psi$.

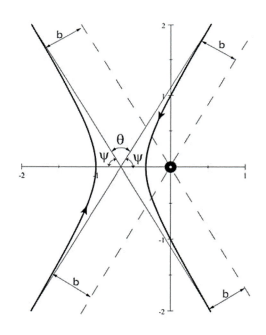

Figure 11.6: Hyperbolic two-body orbits for a repulsive (left) and attractive (right) inverse-square, central two-body forces. Both orbits have the angular momentum vector pointing upwards out of the plane of the orbit

11.8.4   Eccentricity vector

Two-bodies interacting via a conservative two-body central force have two invariant first-order integrals, namely the conservation of energy and the conservation of angular momentum. For the special case of the inverse-square law, there is a third invariant of the motion, which Hamilton called the **eccentricity vector**[2], that unambiguously defines the orientation and direction of the major axis of the elliptical orbit. It will be shown that the angular momentum plus the eccentricity vector completely define the plane and orientation of the orbit for a conservative inverse-square law central force.

Newton's second law for a central force can be written in the form

$$\dot{\mathbf{p}} = f(r)\hat{\mathbf{r}} \tag{11.79}$$

Note that the angular moment $\mathbf{L} = \mathbf{r} \times \mathbf{p}$ is conserved for a central force, that is $\dot{\mathbf{L}} = 0$. Therefore the time derivative of the product $\mathbf{p} \times \mathbf{L}$ reduces to

$$\frac{d}{dt}\left(\mathbf{p} \times \mathbf{L}\right) = \dot{\mathbf{p}} \times \mathbf{L} = f(r)\hat{\mathbf{r}} \times (\mathbf{r} \times \mu\dot{\mathbf{r}}) = f(r)\frac{\mu}{r}\left[\mathbf{r}\left(\mathbf{r} \cdot \dot{\mathbf{r}}\right) - r^2\dot{\mathbf{r}}\right] \tag{11.80}$$

This can be simplified using the fact that

$$\mathbf{r} \cdot \dot{\mathbf{r}} = \frac{1}{2}\frac{d}{dt}\left(\mathbf{r} \cdot \mathbf{r}\right) = r\dot{r} \tag{11.81}$$

thus

$$f(r)\frac{\mu}{r}\left[\mathbf{r}\left(\mathbf{r} \cdot \dot{\mathbf{r}}\right) - r^2\dot{\mathbf{r}}\right] = -\mu f(r)r^2\left[\frac{\dot{\mathbf{r}}}{r} - \frac{\mathbf{r}\dot{r}}{r^2}\right] = -\mu f(r)r^2\frac{d}{dt}\left(\frac{\mathbf{r}}{r}\right) \tag{11.82}$$

This allows equation 11.80 to be reduced to

$$\frac{d}{dt}\left(\mathbf{p} \times \mathbf{L}\right) = -\mu f(r)r^2\frac{d}{dt}\left(\frac{\mathbf{r}}{r}\right) \tag{11.83}$$

Assume the special case of the inverse-square law, equation 11.52, then the central force equation 11.83 reduces to

$$\frac{d}{dt}\left(\mathbf{p} \times \mathbf{L}\right) = -\frac{d}{dt}\left(\mu k\hat{\mathbf{r}}\right) \tag{11.84}$$

or

$$\frac{d}{dt}\left[\left(\mathbf{p} \times \mathbf{L}\right) + \left(\mu k\hat{\mathbf{r}}\right)\right] = 0 \tag{11.85}$$

Define the eccentricity vector $\mathbf{A}$ as

$$\mathbf{A} \equiv \left(\mathbf{p} \times \mathbf{L}\right) + \left(\mu k\hat{\mathbf{r}}\right) \tag{11.86}$$

then equation 11.85 corresponds to

$$\frac{d\mathbf{A}}{dt} = 0 \tag{11.87}$$

This is a statement that *the eccentricity vector A is a constant of motion for an inverse-square, central force.*

The definition of the eccentricity vector $\mathbf{A}$ and angular momentum vector $\mathbf{L}$ implies a zero scalar product,

$$\mathbf{A} \cdot \mathbf{L} = 0 \tag{11.88}$$

Thus the eccentricity vector $\mathbf{A}$ and angular momentum $\mathbf{L}$ are mutually perpendicular, that is, $\mathbf{A}$ is in the plane of the orbit while $\mathbf{L}$ is perpendicular to the plane of the orbit. The eccentricity vector $\mathbf{A}$, *always points along the major axis of the ellipse from the focus to the periapsis* as illustrated on the left side in figure 11.7.

---

[2]The symmetry underlying the eccentricity vector is less intuitive than the energy or angular momentum invariants leading to it being discovered independently several times during the past three centuries. Jakob Hermann was the first to indentify this invariant for the special case of the inverse-square central force. Bernoulli generalized his proof in 1710. Laplace derived the invariant at the end of the 18th century using analytical mechanics. Hamilton derived the connection between the invariant and the orbit eccentricity. Gibbs derived the invariant using vector analysis. Runge published the Gibb's derivation in his textbook which was referenced by Lenz in a 1924 paper on the quantal model of the hydrogen atom. Goldstein named this invariant the "Laplace-Runge-Lenz vector", while others have named it the "Runge-Lenz vector" or the "Lenz vector". This book uses Hamilton's more intuitive name of "eccentricity vector".

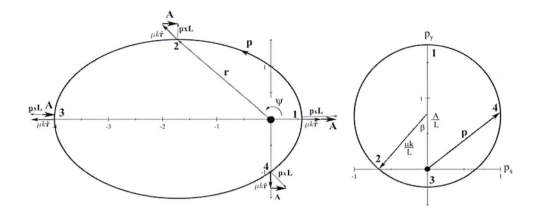

Figure 11.7: The elliptical trajectory and eccentricity vector **A** for two bodies interacting via the inverse-square, central force for eccentricity $\epsilon = 0.75$. The left plot shows the elliptical spatial trajectory where the semi-major axis is assumed to be on the x-axis and the angular momentum $\mathbf{L} = l\hat{\mathbf{z}}$, is out of the page. The force centre is at one foci of the ellipse. The vector coupling relation $\mathbf{A} \equiv (\mathbf{p} \times \mathbf{L}) + (\mu k \hat{\mathbf{r}})$ is illustrated at four points on the spatial trajectory. The right plot is a hodograph of the linear momentum **p** for this trajectory. The periapsis is denoted by the number **1** and the apoapsis is marked as **3** on both plots. Note that the eccentricity vector **A** is a constant that points parallel to the major axis towards the perapsis.

As a consequence, the two orthogonal vectors **A** and **L** completely define the plane of the orbit, plus the orientation of the major axis of the Kepler orbit, in this plane. The three vectors $\mathbf{A}$, $\mathbf{p} \times \mathbf{L}$, and $(\mu k \hat{\mathbf{r}})$ obey the triangle rule as illustrated in the left side of figure 11.7.

Hamilton noted the direct connection between the eccentricity vector **A** and the eccentricity ϵ of the conic section orbit. This can be shown by considering the scalar product

$$\mathbf{A} \cdot \mathbf{r} = Ar \cos \psi = \mathbf{r} \cdot (\mathbf{p} \times \mathbf{L}) + \mu k r \qquad (11.89)$$

Note that the triple scalar product can be permuted to give

$$\mathbf{r} \cdot (\mathbf{p} \times \mathbf{L}) = (\mathbf{r} \times \mathbf{p}) \cdot \mathbf{L} = \mathbf{L} \cdot \mathbf{L} = l^2 \qquad (11.90)$$

Inserting equation 11.90 into 11.89 gives

$$\frac{1}{r} = -\frac{\mu k}{l^2} \left(1 - \frac{A}{\mu k} \cos \psi \right) \qquad (11.91)$$

Note that equations 11.63 and 11.91 are identical if $\psi_0 = 0$. This implies that the eccentricity ϵ and A are related by

$$\epsilon = -\frac{A}{\mu k} \qquad (11.92)$$

where k is defined to be negative for an attractive force. The relation between the eccentricity and total center-of-mass energy can be used to rewrite equation 11.62 in the form

$$A^2 = \mu^2 k^2 + 2\mu E_{cm} l^2 \qquad (11.93)$$

The combination of the eccentricity vector **A** and the angular momentum vector **L** completely specifies the orbit for an inverse square-law central force. The trajectory is in the plane perpendicular to the angular momentum vector **L**, while the eccentricity, plus the orientation of the orbit, both are defined by the eccentricity vector **A**. The eccentricity vector and angular momentum vector each have three independent coordinates, that is, these two vector invariants provide six constraints, while the scalar invariant energy E, adds one additional constraint. The exact location of the particle moving along the trajectory is not defined and thus there are only five independent coordinates governed by the above seven constraints. Thus the eccentricity vector, angular momentum, and center-of-mass energy are related by the two equations 11.88 and 11.93.

Noether's theorem states that each conservation law is a manifestation of an underlying symmetry. Identification of the underlying symmetry responsible for the conservation of the eccentricity vector $\mathbf{A}$ is elucidated using equation 11.86 to give

$$(\mu k \hat{\mathbf{r}}) = \mathbf{A} - (\mathbf{p} \times \mathbf{L}) \tag{11.94}$$

Take the scalar product

$$(\mu k \hat{\mathbf{r}}) \cdot (\mu k \hat{\mathbf{r}}) = (\mu k)^2 = p^2 L^2 + A^2 - 2L \cdot (\mathbf{p} \times \mathbf{L}) \tag{11.95}$$

Choose the angular momentum to be along the z-axis, that is, $\mathbf{L} = l\hat{\mathbf{z}}$, and, since $\mathbf{p}$ and $\mathbf{A}$ are perpendicular to $\mathbf{L}$, then $\mathbf{p}$ and $\mathbf{A}$ are in the $\hat{\mathbf{x}} - \hat{\mathbf{y}}$ plane. Assume that the semimajor axis of the elliptical orbit is along the $\mathbf{x}$-axis, then the locus of the momentum vector on a momentum hodograph has the equation

$$p_x^2 + \left(p_y - \frac{A}{L} \right)^2 = \left(\frac{\mu k}{L} \right)^2 \tag{11.96}$$

Equation 11.96 implies that the locus of the momentum vector is a circle of radius $\left| \frac{\mu k}{L} \right|$ with the center displaced from the origin at coordinates $\left(0, \frac{A}{L} \right)$ as shown by the momentum hodograph on the right side of an figure 11.7. The angle β and eccentricity ϵ are related by,

$$\cos \beta = -\frac{A/L}{\mu k/L} = -\frac{A}{\mu k} = \epsilon \tag{11.97}$$

The circular orbit is centered at the origin for $\epsilon = -\frac{A}{\mu k} = 0$, and thus the magnitude $|\mathbf{p}|$ is a constant around the whole trajectory.

The inverse-square, central, two-body, force is unusual in that it leads to stable closed bound orbits because the radial and angular frequencies are degenerate, i.e. $\omega_r = \omega_\psi$. In momentum space, the locus of the linear momentum vector $\mathbf{p}$ is a perfect circle which is the underlying symmetry responsible for both the fact that the orbits are closed, and the invariance of the eccentricity vector. Mathematically this symmetry for the Kepler problem corresponds to the body moving freely on the boundary of a four-dimensional sphere in space and momentum. The invariance of the eccentricity vector is a manifestation of the special property of the inverse-square, central force under certain rotations in this four-dimensional space; this $O(4)$ symmetry is an example of a hidden symmetry.

11.9 Isotropic, linear, two-body, central force

Closed orbits occur for the two-dimensional linear oscillator when $\frac{\omega_x}{\omega_y}$ is a rational fraction as discussed in chapter 3.3. **Bertrand's Theorem** states that *the linear oscillator, and the inverse-square law (Kepler problem), are the only two-body central forces that have single-valued, stable, closed orbits of the coupled radial and angular motion.* The invariance of the eccentricity vector was the underlying symmetry leading to single-valued, stable, closed orbits for the Kepler problem. It is interesting to explore the symmetry that leads to stable closed orbits for the harmonic oscillator. For simplicity, this discussion will restrict discussion to the isotropic, harmonic, two-body, central force where $\omega_x = \omega_y = \omega$, for which the two-body, central force is linear

$$\mathbf{F}(r) = k\mathbf{r} \tag{11.98}$$

where $k > 0$ corresponds to a repulsive force and $k < 0$ to an attractive force. This isotropic harmonic force can be expressed in terms of a spherical potential $U(r)$ where

$$U(r) = -\frac{1}{2} k r^2 \tag{11.99}$$

Since this is a central two-body force, both the equivalent one-body representation, and the conservation of angular momentum, are equally applicable to the harmonic two-body force. As discussed in section 11.3, since the two-body force is central, the motion is confined to a plane, and thus the Lagrangian can be expressed in polar coordinates. In addition, since the force is spherically symmetric, then the angular momentum is conserved. The orbit solutions are conic sections as described in chapter 11.7. The shape of the orbit for the harmonic two-body central force can be derived using either polar or cartesian coordinates as illustrated below.

11.9.1   Polar coordinates

The origin of the equivalent orbit for the harmonic force will be found to be at the center of an ellipse, rather than the foci of the ellipse as found for the inverse square law. The shape of the orbit can be defined using a Binet differential orbit equation that employs the transformation

$$u' \equiv \frac{1}{r^2} \tag{11.100}$$

Then

$$\frac{du'}{d\psi} = -\frac{2}{r^3}\frac{dr}{d\psi} \tag{11.101}$$

The chain rule gives that

$$\dot{r} = \frac{dr}{d\psi}\dot{\psi} = -\frac{r^3}{2}\dot{\psi}\frac{du'}{d\psi} = -\frac{r}{2}\frac{p_\psi}{\mu}\frac{du'}{d\psi} \tag{11.102}$$

Substitute this into the Hamiltonian H_{cm}, equation 11.42, gives

$$\frac{1}{2}\mu\dot{r}^2 = \frac{1}{8}\frac{p_\psi^2}{u'\mu}\left(\frac{du'}{d\psi}\right)^2 = E - \frac{p_\psi^2}{2\mu}u' + \frac{k}{2u'} \tag{11.103}$$

Rearranging this equation gives

$$\left(\frac{du'}{d\psi}\right)^2 + 4u'^2 - \frac{8E\mu}{p_\psi^2}u' = \frac{4k\mu}{p_\psi^2} \tag{11.104}$$

Addition of a constant to both sides of the equation completes the square

$$\left[\frac{d}{d\psi}\left(u' - \frac{E\mu}{p_\psi^2}\right)\right]^2 + 4\left(u' - \frac{E\mu}{p_\psi^2}\right)^2 = +\frac{4k\mu}{p_\psi^2} + 4\left(\frac{E\mu}{p_\psi^2}\right)^2 \tag{11.105}$$

The right-hand side of equation 11.105 is a constant. The solution of 11.105 must be a sine or cosine function with polar angle $\psi = \omega t$. That is

$$\left(u' - \frac{E\mu}{p_\psi^2}\right) = \left[\left(\frac{E\mu}{p_\psi^2}\right)^2 + \frac{k\mu}{p_\psi^2}\right]^{\frac{1}{2}}\cos 2\left(\psi - \psi_0\right) \tag{11.106}$$

That is,

$$u' = \frac{1}{r^2} = \frac{E\mu}{p_\psi^2}\left(1 + \left(1 + \frac{kp_\psi^2}{E^2\mu}\right)^{\frac{1}{2}}\cos 2(\psi - \psi_0)\right) \tag{11.107}$$

Equation 11.107 corresponds to a closed orbit centered at the origin of the elliptical orbit as illustrated in figure 11.8. The eccentricity ϵ of this closed orbit is given by

$$\left(1 + \frac{kp_\psi^2}{E^2\mu}\right)^{\frac{1}{2}} = \frac{\epsilon^2}{2 - \epsilon^2} \tag{11.108}$$

Equations 11.66, 11.67 give that the eccentricity is related to the semi-major a and semi-minor b axes by

$$\epsilon^2 = 1 - \left(\frac{b}{a}\right)^2 \tag{11.109}$$

Note that for a repulsive force $k > 0$, then $\epsilon \geq 1$ leading to unbound hyperbolic or parabolic orbits centered on the origin. An attractive force, $k < 0$, allows for bound elliptical, as well as unbound parabolic and hyperbolic orbits.

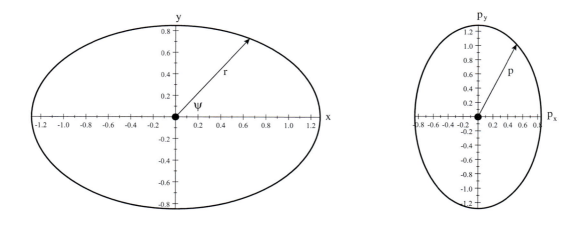

Figure 11.8: The elliptical equivalent trajectory for two bodies interacting via the linear, central force for eccentricity $\epsilon = 0.75$. The left plot shows the elliptical spatial trajectory where the semi-major axis is assumed to be on the x-axis and the angular momentum $\mathbf{L} = l\hat{\mathbf{z}}$, is out of the page. The force center is at the center of the ellipse. The right plot is a hodograph of the linear momentum $\mathbf{p}$ for this trajectory.

11.9.2 Cartesian coordinates

The isotropic harmonic oscillator, expressed in terms of cartesian coordinates in the (x, y) plane of the orbit, is separable because there is no direct coupling term between the x and y motion. That is. the center-of-mass Lagrangian in the (x, y) plane separates into independent motion for x and y.

$$L = \frac{1}{2}\mu\dot{\mathbf{r}} \cdot \dot{\mathbf{r}} + \frac{1}{2}k\mathbf{r} \cdot \mathbf{r} = \left[\frac{1}{2}\mu\dot{x}^2 + \frac{1}{2}kx^2\right] + \left[\frac{1}{2}\mu\dot{y}^2 + \frac{1}{2}ky^2\right] \tag{11.110}$$

Solutions for the independent coordinates, and their corresponding momenta, are

$$\mathbf{r} = \hat{\imath}A\cos(\omega t + \alpha) + \hat{\jmath}B\cos(\omega t + \beta) \tag{11.111}$$
$$\mathbf{p} = -\hat{\imath}A\mu\omega\sin(\omega t + \alpha) - \hat{\jmath}B\mu\omega\sin(\omega t + \beta) \tag{11.112}$$

where $\omega = \sqrt{\frac{k}{\mu}}$. Therefore

$$r^2 = x^2 + y^2 = [A\cos(\omega t + \alpha)]^2 + [B\cos(\omega t + \beta)]^2 \tag{11.113}$$
$$= \frac{A^2 + B^2}{2} + \frac{\sqrt{A^4 + B^4 + 2AB^2\cos(\alpha - \beta)}}{2}\cos(2\omega t + \psi_0)$$

where

$$\cos\psi_0 = \frac{A^2\cos\alpha + B^2\cos\beta}{\sqrt{A^4 + B^4 + 2AB^2\cos(\alpha - \beta)}} \tag{11.114}$$

For a phase difference $\alpha - \beta = \pm\frac{\pi}{2}$, this equation describes an ellipse centered at the origin which agrees with equation 11.107 that was derived using polar coordinates.

The two normal modes of the isotropic harmonic oscillator are degenerate, therefore x, y are equally good normal modes with two corresponding total energies, E_1, E_2, while the corresponding angular momentum J points in the z direction.

$$E_1 = \frac{p_x^2}{2\mu} + \frac{1}{2}kx^2 \tag{11.115}$$
$$E_2 = \frac{p_y^2}{2\mu} + \frac{1}{2}ky^2 \tag{11.116}$$
$$J = \mu(xp_y - yp_x) \tag{11.117}$$

Figure 11.8 shows the closed elliptical equivalent orbit plus the corresponding momentum hodograph for the isotropic harmonic two-body central force. Figures 11.7 and 11.8 contrast the differences between the elliptical orbits for the inverse-square force, and those for the harmonic two-body central force. Although the orbits for bound systems with the harmonic two-body force, and the inverse-square force, both lead to elliptical bound orbits, there are important differences. Both the radial motion and momentum are two valued per cycle for the reflection-symmetric harmonic oscillator, whereas the radius and momentum have only one maximum and one minimum per revolution for the inverse-square law. Although the inverse-square, and the isotropic, harmonic, two-body central forces both lead to closed bound elliptical orbits for which the angular momentum is conserved and the orbits are planar, there is another important difference between the orbits for these two interactions. The orbit equation for the Kepler problem is *expressed with respect to a foci of the elliptical equivalent orbit,* as illustrated in figure 11.7, whereas the orbit equation for the isotropic harmonic oscillator orbit is *expressed with respect to the center of the ellipse* as illustrated in figure 11.8.

11.9.3   Symmetry tensor $\mathbf{A}'$

The invariant vectors $\mathbf{L}$ and $\mathbf{A}$ provide a complete specification of the geometry of the bound orbits for the inverse square-law Kepler system. It is interesting to search for a similar invariant that fully specifies the orbits for the isotropic harmonic central force. In contrast to the Kepler problem, the harmonic force center is at the center of the elliptical orbit, and the orbit is reflection symmetric with the radial and angular frequencies related by $\omega_r = 2\omega_\psi$. Since the orbit is reflection-symmetric, the orientation of the major axis of the orbit cannot be uniquely specified by a vector. Therefore, for the harmonic interaction it is necessary to specify the orientation of the principal axis by the **symmetry tensor**. The symmetry of the isotropic harmonic, two-body, central force leads to the symmetry tensor $\mathbf{A}'$, which is an invariant of the motion analogous to the eccentricity vector $\mathbf{A}$. Like a rotation matrix, the symmetry tensor defines the orientation, but not direction, of the major principal axis of the elliptical orbit. In the plane of the polar orbit the 3×3 symmetry tensor $\mathbf{A}'$ reduces to a 2×2 matrix having matrix elements defined to be,

$$A'_{ij} = \frac{p_i p_j}{2\mu} + \frac{1}{2} k x_i x_j \tag{11.118}$$

The diagonal matrix elements $A'_{11} = E_1$, and $A'_{22} = E_2$ which are constants of motion. The off-diagonal term is given by

$$A'^2_{12} \equiv \left(\frac{p_x p_y}{2\mu} + \frac{1}{2} k x y \right)^2 = \left(\frac{p_x^2}{2\mu} + \frac{1}{2} k x^2 \right) \left(\frac{p_y^2}{2\mu} + \frac{1}{2} k y^2 \right) - 4\mu \left(x p_y - y p_x \right)^2 = E_1 E_2 - \frac{k J^2}{4\mu^3} \tag{11.119}$$

The terms on the right-hand side of equation 11.119 all are constants of motion, therefore A'^2_{12} also is a constant of motion. Thus the 3×3 symmetry tensor $\mathbf{A}'$ can be reduced to a 2×2 symmetry tensor for which all the matrix elements are constants of motion, and the trace of the symmetry tensor is equal to the total energy.

In summary, the inverse-square, and harmonic oscillator two-body central interactions both lead to closed, elliptical equivalent orbits, the plane of which is perpendicular to the conserved angular momentum vector. However, for the inverse-square force, the origin of the equivalent orbit is at the focus of the ellipse and $\omega_r = \omega_\phi$, whereas the origin is at the center of the ellipse and $\omega_r = 2\omega_\phi$ for the harmonic force. As a consequence, the elliptical orbit is reflection symmetric for the harmonic force but not for the inverse square force. The eccentricity vector and symmetry tensor both specify the major axes of these elliptical orbits, the plane of which are perpendicular to the angular momentum vector. The eccentricity vector, and the symmetry tensor, both are directly related to the eccentricity of the orbit and the total energy of the two-body system. Noether's theorem states that the invariance of the eccentricity vector and symmetry tensor, plus the corresponding closed orbits, are manifestations of underlying symmetries. The dynamical $SU3$ symmetry underlies the invariance of the symmetry tensor, whereas the dynamical $O4$ symmetry underlies the invariance of the eccentricity vector. These symmetries lead to stable closed elliptical bound orbits only for these two specific two-body central forces, and not for other two-body central forces.

11.10 Closed-orbit stability

Bertrand's theorem states that the linear oscillator and the inverse-square law are the only two-body, central forces for which all bound orbits are single-valued, and stable closed orbits. The stability of closed orbits can be illustrated by studying their response to perturbations. For simplicity, the following discussion of stability will focus on circular orbits, but the general principles are the same for elliptical orbits.

A circular orbit occurs whenever the attractive force just balances the effective "centrifugal force" in the rotating frame. This can occur for any radial functional form for the central force. The effective potential, equation 11.33 will have a stationary point when

$$\left(\frac{\partial U_{eff}}{\partial r}\right)_{r=r_0} = 0 \qquad (11.120)$$

that is, when

$$\left(\frac{\partial U}{\partial r}\right)_{r=r_0} - \frac{l^2}{\mu r_0^3} = 0 \qquad (11.121)$$

This is equivalent to the statement that the net force is zero. Since the central attractive force is given by

$$F(r) = -\frac{\partial U_{eff}}{\partial r} \qquad (11.122)$$

then the stationary point occurs when

$$F(r_0) = -\frac{l^2}{\mu r_0^3} = -\mu r_0 \dot{\psi}^2 \qquad (11.123)$$

This is the so-called centrifugal force in the rotating frame. The Hamiltonian, equation 11.44, gives that

$$\dot{r} = \pm\sqrt{\frac{2}{\mu}\left(E_{cm} - U - \frac{l^2}{2\mu r^2}\right)} \qquad (11.124)$$

For a circular orbit $\dot{r} = 0$ that is

$$E_{cm} = U - \frac{l^2}{2\mu r^2} \qquad (11.125)$$

A stable circular orbit is possible if both equations (11.121) and (11.125) are satisfied. Such a circular orbit will be a **stable orbit** at the minimum when

$$\left(\frac{d^2 U_{eff}}{dr^2}\right)_{r=r_0} > 0 \qquad (11.126)$$

Examples of stable and unstable orbits are shown in figure 11.9.

Stability of a circular orbit requires that

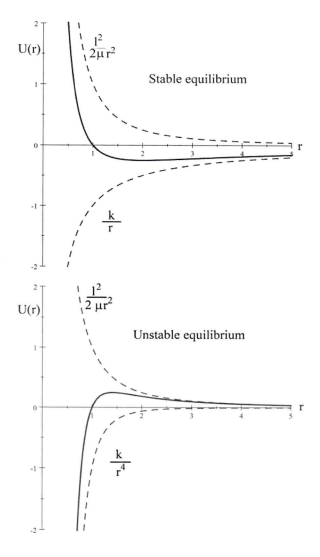

Figure 11.9: Stable and unstable effective central potentials. The repulsive centrifugal and the attractive potentials (k<0) are shown dashed. The solid curve is the effective potential.

$$\left(\frac{\partial^2 U}{\partial r^2}\right)_{r=r_0} + \frac{3l^2}{\mu r_0^4} > 0 \qquad (11.127)$$

which can be written in terms of the central force for a **stable orbit** as

$$-\left(\frac{\partial F}{\partial r}\right)_{r_0} + \frac{3F(r_0)}{r_0} > 0 \tag{11.128}$$

If the attractive central force can be expressed as a power law

$$F(r) = -kr^n \tag{11.129}$$

then stability requires

$$kr_0^{n-1}(3+n) > 0 \tag{11.130}$$

or

$$n > -3 \tag{11.131}$$

Stable equivalent orbits will undergo oscillations about the stable orbit if perturbed. To first order, the restoring force on a bound reduced mass μ is given by

$$F_{restore} = -\left(\frac{d^2U_{eff}}{dr^2}\right)_{r=r_0}(r - r_0) = \mu\ddot{r} \tag{11.132}$$

To the extent that this linear restoring force dominates over higher-order terms, then a perturbation of the stable orbit will undergo simple harmonic oscillations about the stable orbit with angular frequency

$$\omega = \sqrt{\frac{\left(\frac{d^2U_{eff}}{dr^2}\right)_{r=r_0}}{\mu}} \tag{11.133}$$

The above discussion shows that a small amplitude radial oscillation about the stable orbit with amplitude ξ will be of the form

$$\xi = A\sin(2\pi\omega t + \delta)$$

The orbit will be closed if the product of the oscillation frequency ω, and the orbit period τ is an integer value.

The fact that planetary orbits in the gravitational field are observed to be closed is strong evidence that the gravitational force field must obey the inverse square law. Actually there are small precessions of planetary orbits due to perturbations of the gravitational field by bodies other than the sun, and due to relativistic effects. Also the gravitational field near the earth departs slightly from the inverse square law because the earth is not a perfect sphere, and the field does not have perfect spherical symmetry. The study of the precession of satellites around the earth has been used to determine the oblate quadrupole and slight octupole (pear shape) distortion of the shape of the earth.

The most famous test of the inverse square law for gravitation is the precession of the perihelion of Mercury. If the attractive force experienced by Mercury is of the form

$$\mathbf{F}(r) = -G\frac{m_s m_m}{r^{2+\alpha}}\hat{\mathbf{r}}$$

where $|\alpha|$ is small, then it can be shown that, for approximate circular orbitals, the perihelion will advance by a small angle $\pi\alpha$ per orbit period. That is, the precession is zero if $\alpha = 0$, corresponding to an inverse square law dependence which agrees with Bertrand's theorem. The position of the perihelion of Mercury has been measured with great accuracy showing that, after correcting for all known perturbations, the perihelion advances by $43(\pm 5)$ seconds of arc per century, that is 5×10^{-7} radians per revolution. This corresponds to $\alpha = 1.6 \times 10^{-7}$ which is small but still significant. This precession remained a puzzle for many years until 1915 when Einstein predicted that one consequence of his general theory of relativity is that the planetary orbit of Mercury should precess at 43 seconds of arc per century, which is in remarkable agreement with observations.

11.3 *Example: Linear two-body restoring force*

The effective potential for a linear two-body restoring force $F = -kr$ is

$$U_{eff} = \frac{1}{2}kr^2 + \frac{l^2}{2\mu r^2}$$

At the minimum

$$\left(\frac{\partial U_{eff}}{\partial r}\right)_{r=r_0} = kr - \frac{l^2}{\mu r^3} = 0$$

Thus

$$r_0 = \left(\frac{l^2}{\mu k}\right)^{\frac{1}{4}}$$

and

$$\left(\frac{d^2 U_{eff}}{dr^2}\right)_{r=r_0} = \frac{3l^2}{\mu r_0^4} + k = 4k > 0$$

which is a stable orbit. Small perturbations of such a stable circular orbit will have an angular frequency

$$\omega = \sqrt{\frac{\left(\frac{d^2 U_{eff}}{dr^2}\right)_{r=r_0}}{\mu}} = 2\sqrt{\frac{k}{\mu}}$$

Note that this is twice the frequency for the planar harmonic oscillator with the same restoring coefficient. This is due to the central repulsion, the effective potential well for this rotating oscillator example has about half the width for the corresponding planar harmonic oscillator. Note that the kinetic energy for the rotational motion, which is $\frac{l^2}{2\mu r^2}$, equals the potential energy $\frac{1}{2}kr^2$ at the minimum as predicted by the Virial Theorem for a linear two-body restoring force.

11.4 *Example: Inverse square law attractive force*

The effective potential for an inverse square law restoring force $F = -\frac{k}{r^2}\hat{r}$, where k is assumed to be positive,

$$U_{eff} = -\frac{k}{r} + \frac{l^2}{2\mu r^2}$$

At the minimum

$$\left(\frac{\partial U_{eff}}{\partial r}\right)_{r=r_0} = \frac{k}{r^2} - \frac{l^2}{\mu r^3} = 0$$

Thus

$$r_0 = \frac{l^2}{\mu k}$$

and

$$\left(\frac{d^2 U_{eff}}{dr^2}\right)_{r=r_0} = \frac{3l^2}{\mu r_0^4} - \frac{2k}{r_0^3} = \frac{k}{r_0^3} > 0$$

which is a stable orbit. Small perturbations about such a stable circular orbit will have an angular frequency

$$\omega = \sqrt{\frac{\left(\frac{d^2 U_{eff}}{dr^2}\right)_{r=r_0}}{\mu}} = \frac{\mu k^2}{l^3}$$

The kinetic energy for oscillations about this stable circular orbit, which is $\frac{l^2}{2\mu r^2}$, equals half the magnitude of the potential energy $-\frac{k}{r}$ at the minimum as predicted by the Virial Theorem.

11.5 *Example: Attractive inverse cubic central force*

The inverse cubic force is an interesting example to investigate the stability of the orbit equations. One solution of the inverse cubic central force, for a reduced mass μ, is a spiral orbit

$$r = r_0 e^{\alpha\psi}$$

That this is true can be shown by inserting this orbit into the differential orbit equation.

Using a Binet transformation of the variable r to u gives

$$
\begin{aligned}
u &= \frac{1}{r} = \frac{1}{r_0} e^{-\alpha\psi} \\
\frac{du}{d\psi} &= -\frac{\alpha}{r_0} e^{-\alpha\psi} \\
\frac{d^2u}{d\psi^2} &= \frac{\alpha^2}{r_0} e^{-\alpha\psi}
\end{aligned}
$$

Substituting these into the differential equation of the orbit

$$\frac{d^2u}{d\psi^2} + u = -\frac{\mu}{l^2}\frac{1}{u^2}F\left(\frac{1}{u}\right)$$

gives

$$\frac{\alpha^2}{r_0}e^{-\alpha\psi} + \frac{1}{r_0}e^{-\alpha\psi} = -\frac{\mu}{l^2}r_0^2 e^{2\alpha\psi}F\left(\frac{1}{u}\right)$$

That is

$$F\left(\frac{1}{u}\right) = -\frac{\left(\alpha^2 + 1\right)l^2}{\mu}r_0^{-3}e^{-3\alpha\psi} = -\frac{\left(\alpha^2 + 1\right)l^2}{\mu r^3}$$

which is a central attractive inverse cubic force.

The time dependence of the spiral orbit can be derived since the angular momentum gives

$$\dot{\psi} = \frac{l}{\mu r^2} = \frac{l}{\mu r_0^2 e^{2\alpha\psi}}$$

This can be written as

$$e^{2\alpha\psi}d\psi = \frac{l}{\mu r_0^2}dt$$

Integrating gives

$$\frac{e^{2\alpha\psi}}{2\alpha} = \frac{lt}{\mu r_0^2} + \beta$$

where β is a constant. But the orbit gives

$$r^2 = r_0^2 e^{2\alpha\psi} = \frac{2\alpha lt}{\mu} + 2\alpha\beta$$

Thus the radius increases or decreases as the square root of the time. That is, an attractive cubic central force does not have a stable orbit which is what is expected since there is no minimum in the effective potential energy. Note that it is obvious that there will be no minimum or maximum for the summation of effective potential energy since, if the force is $F = -\frac{k}{r^3}$, then the effective potential energy is

$$U_{eff} = -\frac{k}{2r^2} + \frac{l^2}{2\mu r^2} = \left(\frac{l^2}{\mu} - k\right)\frac{1}{2r^2}$$

which has no stable minimum or maximum.

11.6 *Example: Spiralling mass attached by a string to a hanging mass*

An example of an application of orbit stability is the case shown in the adjacent figure. A particle of mass m moves on a horizontal frictionless table. This mass is attached by a light string of fixed length b and rotates about a hole in the table. The string is attached to a second equal mass m that is hanging vertically downwards with no angular motion.

The equations are most conveniently expressed in cylindrical coordinates (r, θ, z) with the origin at the hole in the table, and z vertically upward. The fixed length of the string requires $z = r - b$. The potential energy is

$$U = mgz = mg(r - b)$$

The system is central and conservative, thus the Hamiltonian can be written as

$$H = \frac{m}{2}\left(\dot{r}^2 + r^2\dot{\theta}^2\right) + \frac{m}{2}\dot{r}^2 + mg(r - b) = E$$

The Lagrangian is independent of θ, that is, θ is cyclic, thus the angular momentum $mr^2\dot{\theta} = l$ is a constant of motion. Substituting this into the Hamiltonian equation gives

$$m\dot{r}^2 + \frac{l^2}{2mr^2} + mg(r - b) = E$$

The effective potential is

$$U_{eff} = \frac{l^2}{2mr^2} + mg(r - b)$$

which is shown in the adjacent figure. The stationary value occurs when

$$\left(\frac{\partial U_{eff}}{\partial r}\right)_{r_0} = -\frac{l^2}{mr_0^3} + mg = 0$$

That is, when the angular momentum is related to the radius by

$$l^2 = m^2 g r_0^3$$

Note that $r_0 = 0$ if $l = 0$.

The stability of the solution is given by the second derivative

$$\left(\frac{\partial^2 U_{eff}}{\partial r^2}\right)_{r_0} = \frac{3l^2}{mr_0^4} = \frac{3mg}{r_0} > 0$$

Therefore the stationary point is stable.

Note that the equation of motion for the minimum can be expressed in terms of the restoring force on the two masses

$$2m\ddot{r} = -\left(\frac{\partial^2 U_{eff}}{\partial r^2}\right)_{r_0}(r - r_0)$$

Thus the system undergoes harmonic oscillation with frequency

$$\omega = \sqrt{\frac{\frac{3mg}{r_0}}{2m}} = \sqrt{\frac{3g}{2r_0}}$$

The solution of this system is stable and undergoes simple harmonic motion.

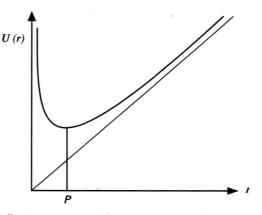

Rotating mass m on a frictionless horizontal table connected to a suspended mass m.

Effective potential for two connected masses.

11.11   The three-body problem

Two bodies interacting via conservative central forces can be solved analytically for the inverse square law and the Hooke's law radial dependences as already discussed. Central forces that have other radial dependences for the equations of motion may not be expressible in terms of simple functions, nevertheless the motion always can be given in terms of an integral. For a gravitational system comprising $n \geq 3$ bodies that are interacting via the two-body central gravitational force, then the equations of motion can be written as

$$m_j \ddot{\mathbf{q}} = \mathbf{G} \sum_{\substack{k \\ k \neq j}}^{n} m_j m_k \frac{(\mathbf{q}_k - \mathbf{q}_j)}{|\mathbf{q}_k - \mathbf{q}_j|^3} \qquad (j = 1, 2, .., n)$$

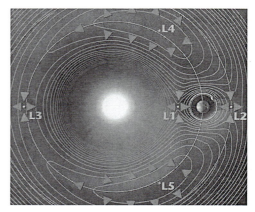

Figure 11.10: A contour plot of the effective potential for the Sun-Earth gravitational system in the rotating frame where the Sun and Earth are stationary. The 5 Lagrange points L_i are saddle points where the net force is zero. (Figure created by NASA)

Even when all the n bodies are interacting via two-body central forces, the problem usually is insoluble in terms of known analytic integrals. Newton first posed the difficulty of the three-body Kepler problem which has been studied extensively by mathematicians and physicists. No known general analytic integral solution has been found. Each body for the n-body system has 6 degrees of freedom, that is, 3 for position and 3 for momentum. The center-of-mass motion can be factored out, therefore the center-of-mass system for the n-body system has $6n - 10$ degrees of freedom after subtraction of 3 degrees for location of the center of mass, 3 for the linear momentum of the center of mass, 3 for rotation of the center of mass, and 1 for the total energy of the system. Thus for $n = 2$ there are $12 - 10 = 2$ degrees of freedom for the two-body system for which the Kepler approach takes to be $\mathbf{r}$ and θ. For $n = 3$ there are 8 degrees of freedom in the center of mass system that have to be determined.

Numerical solutions to the three-body problem can be obtained using successive approximation or perturbation methods in computer calculations. The problem can be simplified by restricting the motion to either of following two approximations:

1) Planar approximation

This approximation assumes that the three masses move in the same plane, that is, the number of degrees of freedom are reduced from 8 to 6 which simplifies the numerical solution.

2) Restricted three-body approximation

The restricted three-body approximation assumes that two of the masses are large and bound while the third mass is negligible such that the perturbation of the motion of the larger two by the third body is negligible. This approximation essentially reduces the system to a two body problem in order to calculate the gravitational fields that act on the third much lighter mass.

Euler and Lagrange showed that the restricted three-body system has five points at which the combined gravitational attraction plus centripetal force of the two large bodies cancel. These are called the Lagrange points and are used for parking satellites in stable orbits with respect to the Earth-Moon system, or with respect to the Sun-Earth system. Figure 11.10 illustrates the five Lagrange points for the Earth-Sun system. Only two of the Lagrange points, L_4 and L_5 lead to stable orbits. Note that these Lagrange points are fixed with respect to the Earth-Sun system which rotates with respect to inertial coordinate frames. The 1900's discovery of the Trojan asteroids at the L_4 and L_5 Lagrange points of the Sun-Jupiter system confirmed the Lagrange predictions.

Poincaré showed that the motion of a light mass bound to two heavy bodies can exhibit extreme sensitivity to initial conditions as well as characteristics of chaos. Solution of the three-body problem has remained a largely unsolved problem since Newton identified the difficulties involved.

11.12 Two-body scattering

Two moving bodies, that are interacting via a central force, scatter when the force is repulsive, or when an attractive system is unbound. Two-body scattering of bodies is encountered extensively in the fields of astronomy, atomic, nuclear, and particle physics. The probability of such scattering is most conveniently expressed in terms of scattering cross sections defined below.

11.12.1 Total two-body scattering cross section

The concept of scattering cross section for two-body scattering is most easily described for the total two-body cross section. The probability P that a beam of n_B incident point particles/second, distributed over a cross sectional area A_B, will hit a single solid object, having a cross sectional area σ, is given by the ratio of the areas as illustrated in figure 11.11. That is,

$$P = \frac{\sigma}{A_B} \qquad (11.134)$$

where it is assumed that $A_B >> \sigma$. For a spherical target body of radius r, the cross section $\sigma = \pi r^2$. The scattering probability P is proportional to the cross section σ which is the cross section of the target body perpendicular to the beam; thus σ has the units of area.

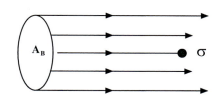

Figure 11.11: Scattering probability for an incident beam of cross sectional area A by a target body of cross sectional area σ.

Since the incident beam of n_B incident point particles/second, has a cross sectional area A_B, then it will have an areal density I given by

$$I = \frac{n_B}{A_B} \text{ beam particles}/m^2/\sec \qquad (11.135)$$

The number of beam particles scattered per second N_S by this single target scatterer equals

$$N_S = P n_B = \frac{\sigma}{A_B} I A_B = \sigma I \qquad (11.136)$$

Thus the cross section for scattering by this single target body is

$$\sigma = \frac{N_S}{I} = \frac{\text{Scattered particles/sec}}{\text{incident beam/m}^2/\text{sec}}$$

Realistically one will have many target scatterers in the target and the total scattering probability increases proportionally to the number of target scatterers. That is, for a target comprising an areal density of η_T target bodies per unit area of the incident beam, then the number scattered will increase proportional to the target areal density η_T. That is, there will be $\eta_T A_B$ scattering bodies that interact with the beam assuming that the target has a larger area than the beam. Thus the total number scattered per second N_S by a target that comprises multiple scatterers is

$$N_S = \sigma \frac{n_B}{A_B} \eta_T A_B = \sigma n_B \eta_T \qquad (11.137)$$

Note that this is independent of the cross sectional area of the beam assuming that the target area is larger than that of the beam. That is, the number scattered per second is proportional to the cross section σ times the product of the number of incident particles per second, n_B, and the areal density of target scatterers, η_T. Typical cross sections encountered in astrophysics are $\sigma \approx 10^{14} m^2$, in atomic physics: $\sigma \approx 10^{-20} m^2$, and in nuclear physics; $\sigma \approx 10^{-28} m^2 = barns$.[3]

N. B., the above proof assumed that the target size is larger than the cross sectional area of the incident beam. If the size of the target is smaller than the beam, then n_B is replaced by the areal density/sec of the beam η_B and η_T is replaced by the number of target particles n_T and the cross-sectional size of the target cancels.

---

[3]The term "barn" was chosen because nuclear physicists joked that the cross sections for neutron scattering by nuclei were as large as a barn door.

11.12.2   Differential two-body scattering cross section

The differential two-body scattering cross section gives much more detailed information of the scattering force than does the total cross section because of the correlation between the impact parameter and the scattering angle. That is, a measurement of the number of beam particles scattered into a given solid angle as a function of scattering angles θ, ϕ probes the radial form of the scattering force.

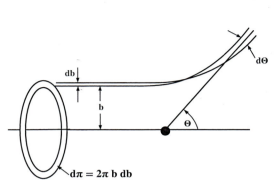

The differential cross section for scattering of an incident beam by a single target body into a solid angle $d\Omega$ at scattering angles θ, ϕ is defined to be

$$\frac{d\sigma}{d\Omega}(\theta\phi) \equiv \frac{1}{I}\frac{dN_S(\theta,\phi)}{d\Omega} \qquad (11.138)$$

where the right-hand side is the ratio of the number scattered per target nucleus into solid angle $d\Omega(\theta, \phi)$, to the incident beam intensity $I \; particles/m^2/sec$.

Figure 11.12: The equivalent one-body problem for scattering of a reduced mass μ by a force centre in the centre of mass system.

Similar reasoning used to derive equation 11.137 leads to the number of beam particles scattered into a solid angle $d\Omega$ for n_B beam particles incident upon a target with areal density η_T is

$$\frac{dN_S(\theta,\phi)}{d\Omega} = n_B\eta_T\frac{d\sigma}{d\Omega}(\theta\phi) \qquad (11.139)$$

Consider the equivalent one-body system for scattering of one body by a scattering force center in the center of mass. As shown in figures 11.6 and 11.12, the perpendicular distance between the center of force of the two body system and trajectory of the incoming body at infinite distance is called the *impact parameter b*. For a central force the scattering system has cylindrical symmetry, therefore the solid angle $d\Omega(\theta\phi) = \sin\theta d\theta d\phi$ can be integrated over the azimuthal angle ϕ to give $d\Omega(\theta) = 2\pi\sin\theta d\theta$.

For the inverse-square, two-body, central force there is a one-to-one correspondence between impact parameter b and scattering angle θ for a given bombarding energy. In this case, assuming conservation of flux means that the incident beam particles passing through the impact-parameter annulus between b and $b + db$ must equal the the number passing between the corresponding angles θ and $\theta + d\theta$. That is, for an incident beam flux of $I \; particles/m^2/sec$ the number of particles per second passing through the annulus is

$$I2\pi b\,|db| = 2\pi\frac{d\sigma}{d\Omega}I\sin\theta\,|d\theta| \qquad (11.140)$$

The modulus is used to ensure that the number of particles is always positive. Thus

$$\frac{d\sigma}{d\Omega} = \frac{b}{\sin\theta}\left|\frac{db}{d\theta}\right| \qquad (11.141)$$

11.12.3   Impact parameter dependence on scattering angle

If the function $b = f(\theta, E_{cm})$ is known, then it is possible to evaluate $\left|\frac{db}{d\theta}\right|$ which can be used in equation 11.141 to calculate the differential cross section. A simple and important case to consider is two-body elastic scattering for the inverse-square law force such as the Coulomb or gravitational forces. To avoid confusion in the following discussion, the center-of-mass scattering angle will be called θ, while the angle used to define the hyperbolic orbits in the discussion of trajectories for the inverse square law, will be called ψ.

In chapter 11.8 the equivalent one-body representation gave that the radial distance for a trajectory for the inverse square law is given by

$$\frac{1}{r} = -\frac{\mu k}{l^2}[1 + \epsilon\cos\psi] \qquad (11.142)$$

Note that closest approach occurs when $\psi = 0$ while for $r \to \infty$ the bracket must equal zero, that is

$$\cos\psi_\infty = \pm\left|\frac{1}{\epsilon}\right| \qquad (11.143)$$

The polar angle ψ is measured with respect to the symmetry axis of the two-body system which is along the line of distance of closest approach as shown in figure 11.6. The geometry and symmetry show that the scattering angle θ is related to the trajectory angle ψ_∞ by

$$\theta = \pi - 2\psi_\infty \tag{11.144}$$

Equation 11.50 gives that

$$\psi_\infty = \int_{r_{min}}^{\infty} \frac{\pm l\, dr}{r^2 \sqrt{2\mu\left(E_{cm} - U - \frac{l^2}{2\mu r^2}\right)}} \tag{11.145}$$

Since

$$l^2 = b^2 p^2 = b^2 2\mu E_{cm} \tag{11.146}$$

then the scattering angle can be written as.

$$\psi_\infty = \frac{\pi - \theta}{2} = \int_{r_{min}}^{\infty} \frac{b\, dr}{r^2 \sqrt{\left(1 - \frac{U}{E_{cm}} - \frac{b^2}{r^2}\right)}} \tag{11.147}$$

Let $u = \frac{1}{r}$, then

$$\psi_\infty = \frac{\pi - \theta}{2} = \int_{r_{min}}^{\infty} \frac{b\, du}{\sqrt{\left(1 - \frac{U}{E_{cm}} - b^2 u^2\right)}} \tag{11.148}$$

For the repulsive inverse square law

$$U = -\frac{k}{r} = -ku \tag{11.149}$$

where k is taken to be positive for a repulsive force. Thus the scattering angle relation becomes

$$\psi_\infty = \frac{\pi - \theta}{2} = \int_{r_{min}}^{\infty} \frac{b\, du}{\sqrt{\left(1 + \frac{ku}{E_{cm}} - b^2 u^2\right)}} \tag{11.150}$$

The solution of this equation is given by equation 11.63 to be

$$u = \frac{1}{r} = -\frac{\mu k}{l^2}\left[1 + \epsilon \cos\psi\right] \tag{11.151}$$

where the eccentricity

$$\epsilon = \sqrt{1 + \frac{2E_{cm}l^2}{\mu k^2}} \tag{11.152}$$

For $r \to \infty$, $u = 0$ then, as shown previously,

$$\left|\frac{1}{\epsilon}\right| = \cos\psi_\infty = \cos\frac{\pi - \theta}{2} = \sin\frac{\theta}{2} \tag{11.153}$$

Therefore

$$\frac{2E_{cm}b}{k} = \sqrt{\epsilon^2 - 1} = \cot\frac{\theta}{2} \tag{11.154}$$

that is, the impact parameter b is given by the relation

$$b = \frac{k}{2E_{cm}} \cot\frac{\theta}{2} \tag{11.155}$$

Thus, for an inverse-square law force, the two-body scattering has a one-to-one correspondence between impact parameter b and scattering angle θ as shown schematically in figure 11.13.

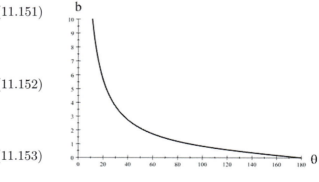

Figure 11.13: Impact parameter dependence on scattering angle for Rutherford scattering.

If k is negative, which corresponds to an attractive inverse square law, then one gets the same relation between impact parameter and scattering angle except that the sign of the impact parameter b is opposite. This means that the hyperbolic trajectory has an interior rather than exterior focus. That is, the trajectory partially orbits around the center of force rather than being repelled away.

Note that the **distance of closest approach** is related to the eccentricity ϵ by equation 11.151, therefore

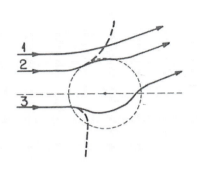

$$r_{\min} = \frac{k}{2E_{cm}} (1 + \epsilon) \qquad (11.156)$$

$$r_{\min} = \frac{k}{2E_{cm}} \left(1 + \frac{1}{\sin \frac{\theta}{2}} \right) \qquad (11.157)$$

Note that for $\theta = 180^o$ then

$$E_{cm} = \frac{k}{r_{\min}} = U(r_{\min}) \qquad (11.158)$$

which is what you would expect from equating the incident kinetic energy to the potential energy at the distance of closest approach.

For scattering of two nuclei by the repulsive Coulomb force, if the impact parameter becomes small enough, the attractive nuclear force also acts leading to impact-parameter dependent effective potentials illustrated in figure 11.14. Trajectory 1 does not overlap the nuclear force and thus is pure Coulomb. Trajectory 2 interacts at the periph-

Figure 11.14: Classical trajectories for scattering to a given angle by the repulsive Coulomb field plus the attractive nuclear field for three different impact parameters. Path 1 is pure Coulomb. Paths 2 and 3 include Coulomb plus nuclear interactions. The dashed parts of trajectories 2 and 3 correspond to only the Coulomb force acting, i.e. zero nuclear force

ery of the nuclear potential and the trajectory deviates from pure Coulomb shown dashed. Trajectory 3 passes through the interior of the nuclear potential. These three trajectories all can lead to the same scattering angle and thus there no longer is a one-to-one correspondence between scattering angle and impact parameter.

11.12.4   Rutherford scattering

Two models of the nucleus evolved in the 1900's, the Rutherford model assumed electrons orbiting around a small nucleus like planets around the sun, while J.J. Thomson's "plum-pudding" model assumed the electrons were embedded in a uniform sphere of positive charge the size of the atom. When Rutherford derived his classical formula in 1911 he realized that it can be used to determine the size of the nucleus since the electric field obeys the inverse square law only when outside of the charged spherical nucleus. Inside a uniform sphere of charge the electric field is $\mathbf{E} \propto \mathbf{r}$ and thus the scattering cross section will not obey the Rutherford relation for distances of closest approach that are less than the radius of the sphere of negative charge. Observation of the angle beyond which the Rutherford formula breaks down immediately determines the radius of the nucleus.

For pure Coulomb scattering, equation 11.155 can be used to evaluate $\left| \frac{db}{d\theta} \right|$, which when used in equation 11.141, gives the center-of-mass **Rutherford scattering cross section**

$$\frac{d\sigma}{d\Omega} = \frac{1}{4} \left(\frac{k}{2E_{cm}} \right)^2 \frac{1}{\sin^4 \frac{\theta}{2}} \qquad (11.159)$$

This cross section assumes elastic scattering by a repulsive two-body inverse-square central force. For scattering of nuclei in the Coulomb potential, the constant k is given to be

$$k = \frac{Z_p Z_T e^2}{4\pi\varepsilon_o} \qquad (11.160)$$

The cross section, scattering angle and E_{cm} of equation 11.159 are evaluated in the center-of-mass coordinate system, whereas usually two-body elastic scattering data involve scattering of the projectiles by a stationary target as discussed in chapter 11.13.

Gieger and Marsden performed scattering of 7.7 MeV α particles from a thin gold foil and proved that the differential scattering cross section obeyed the Rutherford formula back to angles corresponding to a distance of closest approach of $10^{-14}m$ which is much smaller that the $10^{-10}m$ size of the atom. This validated the Rutherford model of the atom and immediately led to the Bohr model of the atom which played such a crucial role in the development of quantum mechanics. Bohr showed that the agreement with the Rutherford formula implies the Coulomb field obeys the inverse square law to small distances. This work was performed at Manchester University, England between 1908 and 1913. It is fortunate that the classical result is identical to the quantal cross section for scattering, otherwise the development of modern physics could have been delayed for many years.

Scattering of very heavy ions, such as ^{208}Pb, can electromagnetically excite target nuclei. For the Coulomb force the impact parameter b and the distance of closest approach, r_{min} are directly related to the scattering angle θ by equation 11.155. Thus observing the angle of the scattered projectile unambiguously determines the hyperbolic trajectory and thus the electromagnetic impulse given to the colliding nuclei. This process, called Coulomb excitation, uses the measured angular distribution of the scattered ions for inelastic excitation of the nuclei to precisely and unambiguously determine the Coulomb excitation cross section as a function of impact parameter. This unambiguously determines the shape of the nuclear charge distribution.

11.7 Example: Two-body scattering by an inverse cubic force

Assume two-body scattering by a potential $U = \frac{k}{r^2}$ where $k > 0$. This corresponds to a repulsive two-body force $\mathbf{F} = \frac{2k}{r^3}\hat{\mathbf{r}}$. Insert this force into Binet's differential orbit, equation 11.39, gives

$$\frac{d^2 u}{d\phi^2} + u\left(1 + \frac{2k\mu}{l^2}\right) = 0$$

The solution is of the form $u = A\sin(\omega\psi + \beta)$ where A and β are constants of integration, $l = \mu r^2 \dot{\psi}$, and

$$\omega^2 = \left(1 + \frac{2k\mu}{l^2}\right)$$

Initially $r = \infty$, $u = 0$, and therefore $\beta = 0$. Also at $r = \infty$, $E = \frac{1}{2}\mu \dot{r}_\infty^2$, that is $|\dot{r}_\infty| = \sqrt{\frac{2E}{\mu}}$. Then

$$\dot{r} = \frac{dr}{d\psi}\dot{\psi} = \frac{dr}{d\psi}\frac{l}{\mu r^2} = -\frac{l}{\mu}\frac{du}{d\psi} = -A\frac{l}{\mu}\omega\cos(\omega\psi)$$

The initial energy gives that $A = \frac{1}{l\omega}\sqrt{2\mu E}$. Hence the orbit equation is

$$u = \frac{1}{r} = \frac{\sqrt{2\mu E}}{l\omega}\sin(\omega\psi)$$

The above trajectory has a distance of closest approach, r_{min}, when $\psi_{min} = \frac{\pi}{2\omega}$. Moreover, due to the symmetry of the orbit, the scattering angle θ is given by

$$\theta = \pi - 2\psi_0 = \pi\left(1 - \frac{1}{\omega}\right)$$

Since $l^2 = \mu^2 b^2 \dot{r}_\infty^2 = 2b^2\mu E$ then

$$1 - \frac{\theta}{\pi} = \left(1 + \frac{2k\mu}{l^2}\right)^{-\frac{1}{2}} = \left(1 + \frac{k}{b^2 E}\right)^{-\frac{1}{2}}$$

This gives that the impact parameter b is related to scattering angle by

$$b^2 = \frac{k}{E}\frac{(\pi - \theta)^2}{(2\pi - \theta)\theta}$$

This impact parameter relation can be used in equation 11.141 to give the differential cross section

$$\frac{d\sigma}{d\Omega} = \frac{b}{\sin\theta}\left|\frac{db}{d\theta}\right| = \frac{k}{E\sin\theta}\frac{\pi^2(\pi - \theta)}{(2\pi - \theta)^2\theta^2}$$

These orbits are called Cotes spirals.

11.13   Two-body kinematics

So far the discussion has been restricted to the center-of-momentum system. Actual scattering measurements are performed in the laboratory frame, and thus it is necessary to transform the scattering angle, energies and cross sections between the laboratory and center-of-momentum coordinate frame. In principle the transformation between the center-of-momentum and laboratory frames is straightforward, using the vector addition of the center-of-mass velocity vector and the center-of-momentum velocity vectors of the two bodies. The following discussion assumes non-relativistic kinematics apply.

In chapter 2.8 it was shown that, for Newtonian mechanics, the center-of-mass and center-of-momentum frames of reference are identical. By definition, in the center-of-momentum frame the vector sum of the linear momentum of the incoming projectile, $p_P^{Initial}$ and target, $p_T^{Initial}$ are equal and opposite. That is

$$\mathbf{p}_P^{Initial} + \mathbf{p}_T^{Initial} = 0 \tag{11.161}$$

Using the center-of-momentum frame, coupled with the conservation of linear momentum, implies that the vector sum of the final momenta of the N reaction products, p_i^{Final}, also is zero. That is

$$\sum_{i=1}^{N} \mathbf{p}_i^{Final} = 0 \tag{11.162}$$

An additional constraint is that energy conservation relates the initial and final kinetic energies by

$$\frac{\left(p_P^{Initial}\right)^2}{2m_P} + \frac{\left(p_T^{Initial}\right)^2}{2m_T} + Q = \frac{\left(p_P^{Final}\right)^2}{2m_P} + \frac{\left(p_T^{Final}\right)^2}{2m_T} \tag{11.163}$$

where the Q value is the energy contributed to the final total kinetic energy by the reaction between the incoming projectile and target. For exothermic reactions, $Q > 0$, the summed kinetic of the reaction products exceeds the sum of the incoming kinetic energies, while for endothermic reactions, $Q < 0$, the summed kinetic energy of the reaction products is less than that of the incoming channel.

For two-body kinematics, the following are three advantages to working in the center-of-momentum frame of reference.

1. The two incident colliding bodies are colinear as are the two final bodies.

2. The linear momenta for the two colliding bodies are identical in both the incident channel and the outgoing channel.

3. The total energy in the center-of-momentum coordinate frame is the energy available to the reaction during the collision. The trivial kinetic energy of the center-of-momentum frame relative to the laboratory frame is handled separately.

The kinematics for two-body reactions is easily determined using the conservation of linear momentum along and perpendicular to the beam direction plus the conservation of energy, $11.161 - 11.163$. Note that it is common practice to use the term "center-of-mass" rather than "center-of-momentum" in spite of the fact that, for relativistic mechanics, only the center-of-momentum is a meaningful concept.

General features of the transformation between the center-of-momentum and laboratory frames of reference are best illustrated by elastic or inelastic scattering of nuclei where the two reaction products in the final channel are identical to the incident bodies. Inelastic excitation of an excited state energy of ΔE_{ex} in either reaction product corresponds to $Q = -\Delta E_{exc}$, while elastic scattering corresponds to $Q = -\Delta E_{exc} = 0$.

For inelastic scattering, the conservation of linear momenta for the outgoing channel in the center-of-momentum simplifies to

$$\mathbf{p}_P^{Final} + \mathbf{p}_T^{Final} = 0 \tag{11.164}$$

that is, the linear momenta of the two reaction products are equal and opposite.

Assume that the center-of-momentum direction of the scattered projectile is at an angle $\vartheta_{cm}^P = \vartheta$ relative to the direction of the incoming projectile and that the scattered target nucleus is scattered at a center-of-momentum direction $\vartheta_{cm}^T = \pi - \vartheta$. Elastic scattering corresponds to simple scattering for which the magnitudes of the incoming and outgoing projectile momenta are equal, that is, $\left|p_P^{Final}\right| = \left|p_P^{Initial}\right|$.

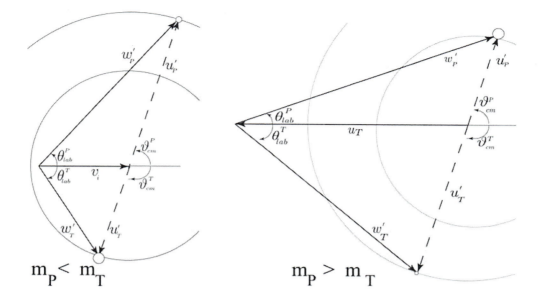

Figure 11.15: Vector hodograph of the scattered projectile and target velocities for a projectile, with incident velocity v_i, that is elastically scattered by a stationary target body. The circles show the magnitude of the projectile and target body final velocities in the center of mass. The center-of-mass velocity vectors are shown as dashed lines while the laboratory vectors are shown as solid lines. The left hodograph shows normal kinematics where the projectile mass is less than the target mass. The right hodograph shows inverse kinematics where the projectile mass is greater than the target mass. For elastic scattering $u_T = u'_T$.

Velocities

The transformation between the center-of-momentum and laboratory frames requires knowledge of the particle velocities which can be derived from the linear momenta since the particle masses are known. Assume that a projectile, mass m_P, with incident energy E_P in the laboratory frame bombards a stationary target with mass m_T. The incident projectile velocity v_i is given by

$$v_i = \sqrt{\frac{2E_P}{m_P}} \tag{11.165}$$

The initial velocities in the laboratory frame are taken to be

$$
\begin{aligned}
w_P &= v_i \\
w_T &= 0
\end{aligned}
\qquad \text{(Initial Lab velocities)}
$$

The final velocities in the laboratory frame after the inelastic collision are

$$
\begin{aligned}
w'_P \\
w'_T
\end{aligned}
\qquad \text{(Final Lab velocities)}
$$

In the center-of-momentum coordinate system, equation 11.10 implies that the initial center-of-momentum velocities are

$$
\begin{aligned}
u_P &= v_i \frac{m_T}{m_P + m_T} \\
u_T &= v_i \frac{m_P}{m_P + m_T}
\end{aligned}
\tag{11.166}
$$

It is simple to derive that the final center-of-momentum velocities after the inelastic collision are given by

$$u'_P = \frac{m_T}{m_P + m_T}\sqrt{\frac{2}{m_P}\tilde{E}}$$

$$u'_T = \frac{m_P}{m_P + m_T}\sqrt{\frac{2}{m_P}\tilde{E}} \tag{11.167}$$

The energy $\tilde{E}$ is defined to be given by

$$\tilde{E} = E_P + Q(1 + \frac{m_P}{m_T}) \tag{11.168}$$

where $Q = -\Delta E$ which is the excitation energy of the final excited states in the outgoing channel.

Angles

The angles of the scattered recoils are written as

$$\theta_{lab}^P \qquad \qquad \text{(Final laboratory angles)}$$
$$\theta_{lab}^T$$

and

$$\vartheta_{cm}^P = \vartheta \qquad \qquad \text{(Final CM angles)}$$
$$\vartheta_{cm}^T = \pi - \vartheta$$

where ϑ is the center-of-mass (center-of-momentum) scattering angle.

Figure 11.15 shows that the angle relations between the laboratory and center of momentum frames for the *scattered projectile* are connected by

$$\frac{\sin(\vartheta_{cm}^P - \theta_{lab}^P)}{\sin\theta_{lab}^P} = \frac{m_P}{m_T}\sqrt{\frac{E_P}{\tilde{E}}} \equiv \tau \tag{11.169}$$

where

$$\tau = \frac{m_P}{m_T}\frac{1}{\sqrt{1 + \frac{Q}{E_P}(1 + \frac{m_P}{m_T})}} = \frac{m_P}{m_T}\frac{1}{\sqrt{1 + \frac{Q}{E_P/m_P}(\frac{m_P+m_T}{m_P m_T})}} \tag{11.170}$$

and $\frac{E_P}{m_P}$ is the energy per nucleon on the incident projectile.

Equation 11.169 can be rewritten as

$$\tan\theta_{lab}^P = \frac{\sin\vartheta_{cm}^P}{\cos\vartheta_{cm}^P + \tau} \tag{11.171}$$

Another useful relation from equation 11.169 gives the center-of-momentum scattering angle in terms of the laboratory scattering angle.

$$\vartheta_{cm}^P = \sin^{-1}(\tau\sin\theta_{lab}^P) + \theta_{lab}^P \tag{11.172}$$

This gives the difference in angle between the lab scattering angle and the center-of-momentum scattering angle. Be careful with this relation since ϑ_{lab}^P is two-valued for inverse kinematics corresponding to the two possible signs for the solution.

The angle relations between the lab and center-of-momentum for the *recoiling target nucleus* are connected by

$$\frac{\sin(\vartheta_{cm}^T - \theta_{lab}^T)}{\sin\theta_{lab}^T} = \sqrt{\frac{E_P}{\tilde{E}}} \equiv \tilde{\tau} \tag{11.173}$$

That is

$$\vartheta_{cm}^T = \sin^{-1}(\tilde{\tau}\sin\theta_{lab}^T) + \theta_{lab}^T \tag{11.174}$$

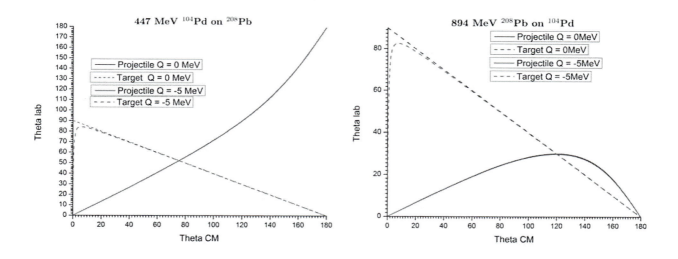

Figure 11.16: The kinematic correlation of the laboratory and center-of-mass scattering angles of the recoiling projectile and target nuclei for scattering for $4.3 MeV$/nucleon ^{104}Pd on ^{208}Pb (left) and for the inverse $4.3 MeV$/nucleon ^{208}Pb on ^{104}Pd (right). The projectile scattering angles are shown by solid lines while the recoiling target angles are shown by dashed lines. The blue curves correspond to elastic scattering, that is $Q = 0$, while the red curves correspond to inelastic scattering with $Q = -5 MeV$.

where

$$\tilde{\tau} = \frac{1}{\sqrt{1 + \frac{Q}{E_P}\left(1 + \frac{m_P}{m_T}\right)}} = \frac{1}{\sqrt{1 + \frac{Q}{E_P/m_P}\left(\frac{m_P + m_T}{m_P m_T}\right)}} \tag{11.175}$$

Note that $\tilde{\tau}$ is the same under interchange of the two nuclei at the same incident energy/nucleon, and that $\tilde{\tau}$ is always larger than or equal to unity since Q is negative. For elastic scattering $\tilde{\tau} = 1$ which gives

$$\theta_{lab}^T = \frac{1}{2}(\pi - \vartheta) \qquad \text{(Recoil lab angle for elastic scattering)}$$

For the target recoil equation 11.173 can be rewritten as

$$\tan\theta_{lab}^T = \frac{\sin\vartheta_{cm}^T}{\cos\vartheta_{cm}^T + \tilde{\tau}} \qquad \text{(Target lab to CM angle conversion)}$$

Velocity vector hodographs provide useful insight into the behavior of the kinematic solutions. As shown in figure 11.15, in the center-of-momentum frame the scattered projectile has a fixed final velocity u'_P, that is, the velocity vector describes a circle as a function of ϑ. The vector addition of this vector and the velocity of the center-of-mass vector $-u_T$ gives the laboratory frame velocity w'_P. Note that for normal kinematics, where $m_P < m_T$, then $|u_T| < |u'_P|$ leading to a monotonic one-to-one mapping of the center-of-momentum angle ϑ_P and θ_{lab}^P. However, for inverse kinematics, where $m_P > m_T$, then $|u_T| > |u'_P|$ leading to two valued ϑ solutions at any fixed laboratory scattering angle θ.

Billiard ball collisions are an especially simple example where the two masses are identical and the collision is essentially elastic. Then essentially $\tau = \tilde{\tau} = 1$, $\theta_{lab}^P = \frac{\vartheta_{cm}^P}{2}$, and $\theta_{lab}^T = \frac{1}{2}\left(\pi - \vartheta_{cm}^P\right)$, that is, the angle between the scattered billiard balls is $\frac{\pi}{2}$.

Both normal and inverse kinematics are illustrated in figure 11.16 which shows the dependence of the projectile and target scattering angles in the laboratory frame as a function of center-of-momentum scattering angle for the Coulomb scattering of ^{104}Pd by ^{208}Pb, that is, for a mass ratio of 2 : 1. Both normal and inverse kinematics are shown for the same bombarding energy of $4.3 MeV/nucleon$ for elastic scattering and for inelastic scattering with a Q-value of $-5 MeV$.

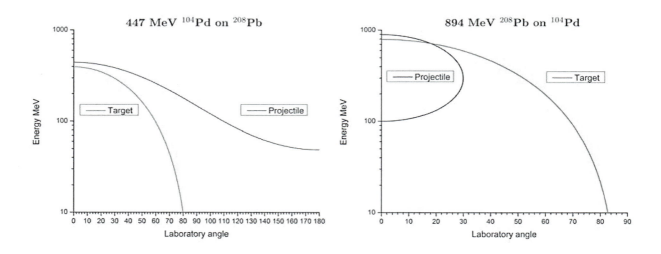

Figure 11.17: Recoil energies, in MeV, versus laboratory scattering angle, shown on the left for scattering of $447MeV$ ^{104}Pd by ^{208}Pb with $Q = -5.0MeV$, and shown on the right for scattering of $894MeV$ ^{208}Pb on ^{104}Pd with $Q = -5.0MeV$.

Since $\sin(\vartheta_{cm}^T - \theta_{lab}^T) \leq 1$ then equation 11.173 implies that $\tilde{\tau} \sin \theta_{lab}^T \leq 1$. Since $\tilde{\tau}$ is always larger than or equal to unity there is a maximum scattering angle in the laboratory frame for the recoiling target nucleus given by

$$\sin \theta_{\max}^T = \frac{1}{\tilde{\tau}} \qquad (11.176)$$

For elastic scattering $\theta_{lab}^T = \sin^{-1}(\frac{1}{\tilde{\tau}}) = 90°$ since $\tilde{\tau} = 1$ for both $894MeV$ ^{208}Pb bombarding ^{104}Pd, and the inverse reaction using a $447MeV$ ^{104}Pd beam scattered by a ^{208}Pb target. A Q-value of $-5MeV$ gives $\tilde{\tau} = 1.002808$ which implies a maximum scattering angle of $\theta_{lab}^T = 85.71°$ for both $894MeV$ ^{208}Pb bombarding ^{104}Pd, and the inverse reaction of a $447MeV$ ^{104}Pd beam scattered by a ^{208}Pb target. As a consequence there are two solutions for ϑ_{cm}^T for any allowed value of θ_{lab}^T as illustrated in figure 11.16.

Since $\sin(\vartheta_{cm}^P - \theta_{lab}^P) \leq 1$ then equation 11.150 implies that $\tau \sin \theta_{lab}^P \leq 1$. For a $447MeV$ ^{104}Pd beam scattered by a ^{208}Pb target $\frac{m_P}{m_T} = 0.50$, thus $\tau = 0.5$ for elastic scattering which implies that there is no upper bound to θ_{lab}^P. This leads to a one-to-one correspondence between θ_{lab}^P and ϑ_{cm}^P for normal kinematics. In contrast, the projectile has a maximum scattering angle in the laboratory frame for inverse kinematics since $\frac{m_P}{m_T} = 2.0$ leading to an upper bound to θ_{lab}^P given by

$$\sin \theta_{\max}^P = \frac{1}{\tau} \qquad (11.177)$$

For elastic scattering $\tau = 2$ implying $\theta_{\max}^P = 30°$. In addition to having a maximum value for θ_{lab}^P, when $\tau > 1$, also there are two solutions for ϑ_{cm}^P for any allowed value of θ_{lab}^P. For the example of $894MeV$ ^{208}Pb bombarding ^{178}Hf leads to a maximum projectile scattering angle of $\theta_{lab}^P = 30.0°$ for elastic scattering and $\theta_{lab}^P = 29.907°$ for $Q = -5MeV$.

Kinetic energies

The initial total kinetic energy in the center-of-momentum frame is

$$E_{cm}^{Initial} = E_P \frac{m_T}{m_P + m_T} \qquad (11.178)$$

The final total kinetic energy in the center-of-momentum frame is

$$E_{cm}^{Final} = E_{cm}^{Initial} + Q = \tilde{E} \frac{m_T}{m_P + m_T} \qquad (11.179)$$

In the laboratory frame the kinetic energies of the scattered projectile and recoiling target nucleus are given by

$$E_P^{Lab} = \left(\frac{m_T}{m_P + m_T}\right)^2 \left(1 + \tau^2 + 2\tau \cos \vartheta_{cm}^P\right) \tilde{E} \tag{11.180}$$

$$E_T^{Lab} = \frac{m_P m_T}{(m_P + m_T)^2} \left(1 + \tilde{\tau}^2 + 2\tilde{\tau} \cos \vartheta_{cm}^T\right) \tilde{E} \tag{11.181}$$

where ϑ_{cm}^P and ϑ_{cm}^T are the center-of-mass scattering angles respectively for the scattered projectile and target nuclei.

For the chosen incident energies the normal and inverse reactions give the same center-of-momentum energy of $298 MeV$ which is the energy available to the interaction between the colliding nuclei. However, the kinetic energy of the center-of-momentum is $447 - 298 = 149 MeV$ for normal kinematics and $894 - 298 = 596 MeV$ for inverse kinematics. This trivial center-of-momentum kinetic energy does not contribute to the reaction. Note that inverse kinematics focusses all the scattered nuclei into the forward hemisphere which reduces the required solid angle for recoil-particle detection.

Solid angles

The laboratory-frame solid angles for the scattered projectile and target are taken to be $d\omega_P$ and $d\omega_T$ respectively, while the center-of-momentum solid angles are $d\Omega_P$ and $d\Omega_T$ respectively. The Jacobian relating the solid angles is

$$\frac{d\omega_P}{d\Omega_P} = \left(\frac{\sin \theta_{lab}^P}{\sin \vartheta_{cm}^P}\right)^2 \left|\cos(\vartheta_{cm}^P - \theta_{lab}^P)\right| \tag{11.182}$$

$$\frac{d\omega_T}{d\Omega_T} = \left(\frac{\sin \theta_{lab}^T}{\sin \vartheta_{cm}^T}\right)^2 \left|\cos(\vartheta_{cm}^T - \theta_{lab}^T)\right| \tag{11.183}$$

These can be used to transform the calculated center-of-momentum differential cross sections to the laboratory frame for comparison with measured values. Note that relative to the center-of-momentum frame, the forward focussing increases the observed differential cross sections in the forward laboratory frame and decreases them in the backward hemisphere.

Exploitation of two-body kinematics

Computing the above non-trivial transform relations between the center-of-mass and laboratory coordinate frames for two-body scattering is used extensively in many fields of physics. This discussion has assumed non-relativistic two-body kinematics. Relativistic two-body kinematics encompasses non-relativistic kinematics as discussed in chapter 17.4. Many computer codes are available that can be used for making either non-relativistic or relativistic transformations.

It is stressed that the underlying physics for two interacting bodies is identical irrespective of whether the reaction is observed in the center-of-mass or the laboratory coordinate frames. That is, no new physics is involved in the kinematic transformation. However, the transformation between these frames can dramatically alter the angles and velocities of the observed scattered bodies which can be beneficial for experimental detection. For example, in heavy-ion nuclear physics the projectile and target nuclei can be interchanged leading to very different velocities and scattering angles in the laboratory frame of reference. This can greatly facilitate identification and observation of the velocities vectors of the scattered nuclei. In high-energy physics it is advantageous to collide beams having identical, but opposite, linear momentum vectors, since then the laboratory frame is the center-of-mass frame, and the energy required to accelerate the colliding bodies is minimized.

11.14   Summary

This chapter has focussed on the classical mechanics of bodies interacting via conservative, two-body, central interactions. The following are the main topics presented in this chapter.

Equivalent one-body representation for two bodies interacting via a central interaction   The equivalent one-body representation of the motion of two bodies interacting via a two-body central interaction greatly simplifies solution of the equations of motion. The position vectors $\mathbf{r}_1$ and $\mathbf{r}_2$ are expressed in terms of the center-of-mass vector $\mathbf{R}$ plus total mass $M = m_1 + m_2$ while the position vector $\mathbf{r}$, plus associated reduced mass $\mu = \frac{m_1 m_2}{m_1 + m_2}$, describe the relative motion of the two bodies in the center of mass. The total Lagrangian then separates into two independent parts

$$L = \frac{1}{2} M \left| \dot{\mathbf{R}} \right|^2 + L_{cm} \tag{11.16}$$

where the center-of-mass Lagrangian is

$$L_{cm} = \frac{1}{2} \mu \left| \dot{\mathbf{r}} \right|^2 - U(r) \tag{11.17}$$

Equations 11.10, and 11.11 can be used to derive the actual spatial trajectories of the two bodies expressed in terms of $\mathbf{r}_1$ and $\mathbf{r}_2$, from the relative equations of motion, written in terms of $\mathbf{R}$ and $\mathbf{r}$, for the equivalent one-body solution..

Angular momentum   Noether's theorem shows that the angular momentum is conserved if only a spherically-symmetric two-body central force acts between the interacting two bodies. The plane of motion is perpendicular to the angular momentum vector and thus the Lagrangian can be expressed in polar coordinates as

$$L_{cm} = \frac{1}{2} \mu \left(\dot{r}^2 + r^2 \dot{\psi}^2 \right) - U(r) \tag{11.22}$$

Differential orbit equation of motion   The Binet transformation $u = \frac{1}{r}$ allows the center-of-mass Lagrangian L_{cm} for a central force $\mathbf{F} = f(r)\hat{\mathbf{r}}$ to be used to express the differential orbit equation for the radial motion as

$$\frac{d^2 u}{d\psi^2} + u = -\frac{\mu}{l^2} \frac{1}{u^2} F(\frac{1}{u}) \tag{11.39}$$

The Lagrangian, and the Hamiltonian all were used to derive the equations of motion for two bodies interacting via a two-body, conservative, central interaction. The general features of the conservation of angular momentum and conservation of energy for a two-body, central potential were presented.

Inverse-square, two-body, central force   The inverse-square, two-body, central force is of pivotal importance in nature since it is applies to both the gravitational force and the Coulomb force. The underlying symmetries of the inverse-square, two-body, central interaction, lead to conservation of angular momentum, conservation of energy, Gauss's law, and that the two-body orbits follow closed, degenerate, orbits that are conic sections, for which the eccentricity vector is conserved. The radial dependence, relative to the force center lying at one focus of the conic section, is given by

$$\frac{1}{r} = -\frac{\mu k}{l^2} \left[1 + \epsilon \cos \left(\psi - \psi_0 \right) \right] \tag{11.58}$$

where the orbit eccentricity ϵ equals

$$\epsilon = \sqrt{1 + \frac{2 E_{cm} l^2}{\mu k^2}} \tag{11.62}$$

These lead to Kepler's three laws of motion for two bodies in a bound orbit due to the attractive gravitational force for which $k = -G m_1 m_2$. The inverse-square law is special in that the eccentricity vector $\mathbf{A}$ is a third invariant of the motion, where

$$\mathbf{A} \equiv (\mathbf{p} \times \mathbf{L}) + (\mu k \hat{\mathbf{r}}) \tag{11.86}$$

The eccentricity vector unambiguously defines the orientation and direction of the major axis of the elliptical orbit. The invariance of the eccentricity vector, and the existence of stable closed orbits, are manifestations of the dynamical 04 symmetry.

Isotropic, harmonic, two-body, central force The isotropic, harmonic, two-body, central interaction is of interest since, like the inverse-square law force, it leads to closed elliptical orbits described by

$$\frac{1}{r^2} = \frac{E\mu}{p_\psi^2}\left(1 + \left(1 + \frac{kp_\psi^2}{E^2\mu}\right)^{\frac{1}{2}}\cos 2(\psi - \psi_0)\right) \tag{11.107}$$

where the eccentricity ϵ is given by

$$\left(1 + \frac{kp_\psi^2}{E^2\mu}\right)^{\frac{1}{2}} = \frac{\epsilon^2}{2 - \epsilon^2} \tag{11.108}$$

The harmonic force orbits are distinctly different from those for the inverse-square law in that the force center is at the center of the ellipse, rather than at the focus for the inverse-square law force. This elliptical orbit is reflection symmetric for the harmonic force, but not for the inverse square force. The isotropic harmonic two-body force leads to invariance of the symmetry tensor, $\mathbf{A}'$ which is an invariant of the motion analogous to the eccentricity vector $\mathbf{A}$. This leads to stable closed orbits, which are manifestations of the dynamical $SU3$ symmetry.

Orbit stability Bertrand's theorem states that only the inverse square law and the linear radial dependences of the central forces lead to stable closed bound orbits that do not precess. These are manifestation of the dynamical symmetries that occur for these two specific radial forms of two-body forces.

The three-body problem The difficulties encountered in solving the equations of motion for three bodies, that are interacting via two-body central forces, was discussed. The three-body motion can include the existence of chaotic motion. It was shown that solution of the three-body problem is simplified if either the planar approximation, or the restricted three-body approximation, are applicable.

Two-body scattering The total and differential two-body scattering cross sections were introduced. It was shown that for the inverse-square law force there is a simple relation between the impact parameter b and scattering angle θ given by

$$b = \frac{k}{2E_{cm}}\cot\frac{\theta}{2} \tag{11.155}$$

This led to the solution for the differential scattering cross-section for Rutherford scattering due to the Coulomb interaction.

$$\frac{d\sigma}{d\Omega} = \frac{1}{4}\left(\frac{k}{2E_{cm}}\right)^2\frac{1}{\sin^4\frac{\theta}{2}} \tag{11.159}$$

This cross section assumes elastic scattering by a repulsive two-body inverse-square central force. For scattering of nuclei in the Coulomb potential the constant k is given to be

$$k = \frac{Z_p Z_T e^2}{4\pi\varepsilon_o} \tag{11.160}$$

Two-body kinematics The transformation from the center-of-momentum frame to laboratory frames of reference was introduced. Such transformations are used extensively in many fields of physics for theoretical modelling of scattering, and for analysis of experiment data.

Workshop exercises

1. Listed below are several statements concerning central force motion. For each statement, give the reason for why the statement is true. If a statement is only true in certain situations, then explain when it holds and when it doesn't. The system referred to below consists of mass m_1 located at r_1 and mass m_2 located at r_2.

   • The potential energy of the system depends only on the difference $r_1 - r_2$, not on r_1 and r_2 separately.

   • The potential energy of the system depends only on the magnitude of $r_1 - r_2$, not the direction.

   • It is possible to choose an inertial reference frame in which the center of mass of the system is at rest.

   • The total energy of the system is conserved.

   • The total angular momentum of the system is conserved.

2. A particle of mass m moves in a potential $U(r) = -U_0 e^{-\lambda^2 r^2}$.

   (a) Given the constant l, find an implicit equation for the radius of the circular orbit. A circular orbit at $r = \rho$ is possible if

   $$\left(\frac{\partial V}{\partial r}\right)\bigg|_{r=\rho} = 0$$

   where V is the effective potential.

   (b) What is the largest value of l for which a circular orbit exists? What is the value of the effective potential at this critical orbit?

3. A particle of mass m is observed to move in a spiral orbit given by the equation $r = k\theta$, where k is a constant. Is it possible to have such an orbit in a central force field? If so, determine the form of the force function.

4. The interaction energy between two atoms of mass m is given by the Lennard-Jones potential, $U(r) = \epsilon\left[(r_0/r)^{12} - 2(r_0/r)^6\right]$

   (a) Determine the Lagrangian of the system where r_1 and r_2 are the positions of the first and second mass, respectively.

   (b) Rewrite the Lagrangian as a one-body problem in which the center-of-mass is stationary.

   (c) Determine the equilibrium point and show that it is stable.

   (d) Determine the frequency of small oscillations about the stable point.

5. Consider two bodies of mass m in circular orbit of radius $r_0/2$, attracted to each other by a force $F(r)$, where r is the distance between the masses.

   (a) Determine the Lagrangian of the system in the center-of-mass frame (Hint: a one-body problem subject to a central force).

   (b) Determine the angular momentum. Is it conserved?

   (c) Determine the equation of motion in r in terms of the angular momentum and $|\mathbf{F}(r)|$.

   (d) Expand your result in (c) about an equilibrium radius r_0 and show that the condition for stability is, $\frac{F'(r_0)}{F(r_0)} + \frac{3}{r_0} > 0$

6. Consider two charges of equal magnitude q connected by a spring of spring constant k' in circular orbit. Can the charges oscillate about some equilibrium? If so, what condition must be satisfied?

7. Consider a mass m in orbit around a mass M, which is subject to a force $F = -\frac{k}{r^2}\hat{r}$, where r is the distance between the masses. Show that the eccentricity vector $A = p \times L - \mu k\,\hat{r}$ is conserved.

Problems

1. Show that the areal velocity is constant for a particle moving under the influence of an attractive force given by $F(r) = -kr$. Calculate the time averages of the kinetic and potential energies and compare with the the results of the virial theorem.

2. Assume that the Earth's orbit is circular and that the Sun's mass suddenly decreases by a factor of two. (a) What orbit will the earth then have? (b) Will the Earth escape the solar system?

3. Discuss the motion of a particle in a central inverse-square-law force field for a superimposed force whose magnitude is inversely proportional to the cube of the distance from the particle to force center; that is

$$F(r) = -\frac{k}{r^2} - \frac{\lambda}{r^3} \qquad (\text{k, } \lambda > 0)$$

Show that the motion is described by a precessing ellipse. Consider the cases

a) $\lambda < \frac{l^2}{\mu}$, b) $\lambda = \frac{l^2}{\mu}$, c) $\lambda > \frac{l^2}{\mu}$ where l is the angular momentum and μ the reduced mass.

4. A communications satellite is in a circular orbit around the earth at a radius R and velocity v. A rocket accidentally fires quite suddenly, giving the rocket an outward velocity v in addition to its original tangential velocity v.

a) Calculate the ratio of the new energy and angular momentum to the old.

b) Describe the subsequent motion of the satellite and plot $T(r)$, $U(r)$, the net effective potential, and $E(r)$ after the rocket fires.

5. Two identical point objects, each of mass m are bound by a linear two-body force $F = -kr$ where r is the vector distance between the two point objects. The two point objects each slide on a horizontal frictionless plane subject to a vertical gravitational field g. The two-body system is free to translate, rotate and oscillate on the surface of the frictionless plane.

a) Derive the Lagrangian for the complete system including translation and relative motion.
b) Use Noether's theorem to identify all constants of motion.
c) Use the Lagrangian to derive the equations of motion for the system.
d) Derive the generalized momenta and the corresponding Hamiltonian.
e) Derive the period for small amplitude oscillations of the relative motion of the two masses.

6. A bound binary star system comprises two spherical stars of mass m_1 and m_2 bound by their mutual gravitational attraction. Assume that the only force acting on the stars is their mutual gravitation attraction and let r be the instantaneous separation distance between the centers of the two stars where r is much larger than the sum of the radii of the stars.

a) Show that the two-body motion of the binary star system can be represented by an equivalent one-body system and derive the Lagrangian for this system.

b) Show that the motion for the equivalent one-body system in the center of mass frame lies entirely in a plane and derive the angle between the normal to the plane and the angular momentum vector.

c) Show whether H_{cm} is a constant of motion and whether it equals the total energy.

d) It is known that a solution to the equation of motion for the equivalent one-body orbit for this gravitational force has the form

$$\frac{1}{r} = -\frac{\mu k}{l^2} [1 + \epsilon \cos \theta]$$

and that the angular momentum is a constant of motion $L = l$. Use these to prove that the attractive force leading to this bound orbit is

$$\mathbf{F} = \frac{k}{r^2} \hat{\mathbf{r}}$$

where k must be negative.

7 When performing the Rutherford experiment, Gieger and Marsden scattered $7.7 MeV$ ^{4}He particles (alpha particles) from ^{238}U at a scattering angle in the laboratory frame of $\theta = 90^{0}$. Derive the following observables as measured in the laboratory frame.

   (a) The recoil scattering angle of the ^{238}U in the laboratory frame.

   (b) The scattering angles of the ^{4}He and ^{238}U in the center-of-mass frame

   (c) The kinetic energies of the ^{4}He and ^{238}U in the laboratory frame

   (d) The impact parameter

   (e) The distance of closest approach $r_{\min}$

Chapter 12

Non-inertial reference frames

12.1 Introduction

Newton's Laws of motion apply only to inertial frames of reference. Inertial frames of reference make it possible to use either Newton's laws of motion, or Lagrangian, or Hamiltonian mechanics, to develop the necessary equations of motion. There are certain situations where it is much more convenient to treat the motion in a non-inertial frame of reference. Examples are motion in frames of reference undergoing translational acceleration, rotating frames of reference, or frames undergoing both translational and rotational motion. This chapter will analyze the behavior of dynamical systems in accelerated frames of reference, especially rotating frames such as on the surface of the Earth. Newtonian mechanics, as well as the Lagrangian and Hamiltonian approaches, will be used to handle motion in non-inertial reference frames by introducing extra inertial forces that correct for the fact that the motion is being treated with respect to a non-inertial reference frame. These inertial forces are often called fictitious even though they appear real in the non-inertial frame. The underlying reasons for each of the inertial forces will be discussed followed by a presentation of important applications.

12.2 Translational acceleration of a reference frame

Consider an inertial system $(x_{fix}, y_{fix}, z_{fix})$ which is fixed in space, and a non-inertial system $(x'_{mov}, y'_{mov}, z'_{mov})$ that is moving in a direction relative to the fixed frame such as to maintain constant orientations of the axes relative to the fixed frame, as illustrated in figure 12.1. The fixed frame is designated to be the unprimed frame and, to avoid confusion the subscript fix is attached to the fixed coordinates taken with respect to the fixed coordinate frame. Similarly, the translating reference frame, which is undergoing translational acceleration, has the subscript mov attached to the coordinates taken with respect to the translating frame of reference. Newton's Laws of motion are obeyed only in the inertial (unprimed) reference frame. The respective position vectors are related by

$$\mathbf{r}_{fix} = \mathbf{R}_{fix} + \mathbf{r}'_{mov} \qquad (12.1)$$

where $\mathbf{r}_{fix}$ is the vector relative to the fixed frame, $\mathbf{r}'_{mov}$ is the vector relative to the translationally accelerating frame and $\mathbf{R}_{fix}$ is the vector from the origin of the fixed frame to the origin of the accelerating frame. Differentiating equation 12.1 gives the velocity vector relation

$$\mathbf{v}_{fix} = \mathbf{V}_{fix} + \mathbf{v}'_{mov} \qquad (12.2)$$

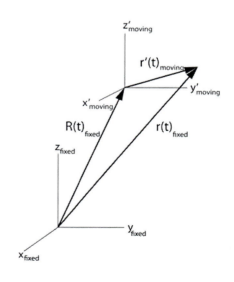

Figure 12.1: Inertial reference frame (unprimed), and translational accelerating frame (primed).

289

where $\mathbf{v}_{fix} = \frac{d\mathbf{r}_{fix}}{dt}$, $\mathbf{v}'_{mov} = \frac{d\mathbf{r}'_{mov}}{dt}$ and $\mathbf{V}_{fix} = \frac{d\mathbf{R}_{fix}}{dt}$. Similarly the acceleration vector relation is

$$\mathbf{a}_{fix} = \mathbf{A}_{fix} + \mathbf{a}'_{mov} \tag{12.3}$$

where $\mathbf{a}_{fix} = \frac{d^2\mathbf{r}_{fix}}{dt^2}$, $\mathbf{a}'_{mov} = \frac{d^2\mathbf{r}'_{mov}}{dt^2}$ and $\mathbf{A}_{fix} = \frac{d^2\mathbf{R}_{fix}}{dt^2}$.
  In the fixed frame, Newton's laws give that

$$\mathbf{F}_{fix} = m\mathbf{a}_{fix} \tag{12.4}$$

The force in the fixed frame can be separated into two terms, the acceleration of the accelerating frame of reference $\mathbf{A}_{fix}$ plus the acceleration with respect to the accelerating frame $\mathbf{a}'_{mov}$.

$$\mathbf{F}_{fix} = m\mathbf{A}_{fix} + m\mathbf{a}'_{mov} \tag{12.5}$$

Relative to the accelerating reference frame the acceleration is given by

$$m\mathbf{a}'_{mov} = \mathbf{F}_{fix} - m\mathbf{A}_{fix} \tag{12.6}$$

The accelerating frame of reference can exploit Newton's Laws of motion using an effective translational force $\mathbf{F}'_{tran} \equiv \mathbf{F}_{fix} - m\mathbf{A}_{fix}$. The additional $-m\mathbf{A}_{fix}$ term is called an inertial force; it can be altered by choosing a different non-inertial frame of reference, that is, it is dependent on the frame of reference in which the observer is situated.

12.3   Rotating reference frame

Consider a rotating frame of reference which will be designated as the double-primed (rotating) frame to differentiate it from the non-rotating primed (moving) frame, since both of which may be undergoing translational acceleration relative to the inertial fixed unprimed frame as described above.

12.3.1   Spatial time derivatives in a rotating, non-translating, reference frame

For simplicity assume that $\mathbf{R}_{fix} = \mathbf{V}_{fix} = 0$, that is, the primed reference frame is stationary and identical to the fixed stationary unprimed frame. The double-primed (rotating) frame is a non-inertial frame rotating with respect to the origin of the fixed primed frame. Appendix $D.2.3$ shows that an infinitessimal rotation $d\theta$ about an instantaneous axis of rotation leads to an infinitessimal displacement $d\mathbf{r}^R$ where

$$d\mathbf{r}^R = d\boldsymbol{\theta} \times \mathbf{r}'_{mov} \tag{12.7}$$

Consider that during a time dt, the position vector in the fixed primed reference frame moves by an arbitrary infinitessimal distance $d\mathbf{r}'_{mov}$. As illustrated in figure 12.2, this infinitessimal distance in the primed non-rotating frame can be split into two parts:

a) $d\mathbf{r}^R = d\boldsymbol{\theta} \times \mathbf{r}'_{mov}$ which is due to rotation of the rotating frame with respect to the translating primed frame.

b) $(d\mathbf{r}''_{rot})$ which is the motion *with respect to the rotating (double-primed) frame.*

That is, the motion has been arbitrarily divided into a part that is due to the rotation of the double-primed frame, plus the vector displacement measured in this rotating (double-primed) frame. It is always possible to make such a decomposition of the displacement as long as the vector sum can be written as

$$d\mathbf{r}'_{mov} = d\mathbf{r}''_{rot} + d\boldsymbol{\theta} \times \mathbf{r}'_{mov} \tag{12.8}$$

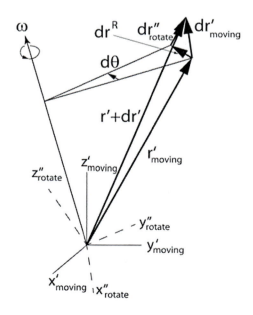

Figure 12.2: Infinitessimal displacement in the non rotating primed frame and in the rotating double-primed reference frame frame.

Since $d\boldsymbol{\theta} = \boldsymbol{\omega} dt$ then the time differential of the displacement, equation 12.8, can be written as

$$\left(\frac{d\mathbf{r}'}{dt}\right)_{mov} = \left(\frac{d\mathbf{r}''}{dt}\right)_{rot} + \boldsymbol{\omega} \times \mathbf{r}'_{mov} \tag{12.9}$$

The important conclusion is that a velocity measured in a non-rotating reference frame $\left(\frac{d\mathbf{r}'}{dt}\right)_{mov}$ can be expressed as the sum of the velocity $\left(\frac{d\mathbf{r}''}{dt}\right)_{rot}$, measured relative to a rotating frame, plus the term $\boldsymbol{\omega} \times \mathbf{r}'_{mov}$ which accounts for the rotation of the frame. The division of the $d\mathbf{r}'_{rot}$ vector into two parts, a part due to rotation of the frame plus a part with respect to the rotating frame, is valid for any vector as shown below.

12.3.2 General vector in a rotating, non-translating, reference frame

Consider an arbitrary vector $\mathbf{G}$ which can be expressed in terms of components along the three unit vector basis $\hat{\mathbf{e}}_i^{fix}$ in the fixed inertial frame as

$$\mathbf{G} = \sum_{i=1}^{3} G_i^{fix} \hat{\mathbf{e}}_i^{fix} \tag{12.10}$$

Neglecting translational motion, then it can be expressed in terms of the three unit vectors in the non-inertial rotating frame unit vector basis $\hat{\mathbf{e}}_i^{rot}$ as

$$\mathbf{G} = \sum_{i=1}^{3} (G_i)_{rot} \hat{\mathbf{e}}_i^{rot} \tag{12.11}$$

Since the unit basis vectors $\hat{\mathbf{e}}_i^{rot}$ are constant in the rotating frame, that is,

$$\left(\frac{d\hat{\mathbf{e}}_i^{rot}}{dt}\right)_{rot} = 0 \tag{12.12}$$

then the time derivatives of $\mathbf{G}$ in the rotating coordinate system $\hat{\mathbf{e}}_i^{rot}$ can be written as

$$\left(\frac{d\mathbf{G}}{dt}\right)_{rot} = \sum_{i=1}^{3} \left(\frac{dG_i}{dt}\right)_{rot} \hat{\mathbf{e}}_i^{rot} \tag{12.13}$$

The inertial-frame time derivative taken with components along the rotating coordinate basis $\hat{\mathbf{e}}_i^{rot}$, equation 12.11, is

$$\left(\frac{d\mathbf{G}}{dt}\right)_{fix} = \sum_{i=1}^{3} \left(\frac{dG_i}{dt}\right)_{rot} \hat{\mathbf{e}}_i^{rot} + \sum_{i=1}^{3} (G_i)_{rot} \frac{d\hat{\mathbf{e}}_i^{rot}}{dt} \tag{12.14}$$

Substitute the unit vector $\hat{\mathbf{e}}^{rot}$ for $\mathbf{r}'_{mov}$ in equation 12.9, plus using equation 12.12, gives that

$$\left(\frac{d\hat{\mathbf{e}}^{rot}}{dt}\right)_{fix} = \boldsymbol{\omega} \times \hat{\mathbf{e}}^{rot} \tag{12.15}$$

Substitute this into the second term of equation 12.14 gives

$$\left(\frac{d\mathbf{G}}{dt}\right)_{fix} = \left(\frac{d\mathbf{G}}{dt}\right)_{rot} + \boldsymbol{\omega} \times \mathbf{G} \tag{12.16}$$

This important identity relates the time derivatives of any vector expressed in both the inertial frame and the rotating non-inertial frame bases. Note that the $\boldsymbol{\omega} \times \mathbf{G}$ term originates from the fact that the unit basis vectors of the rotating reference frame are time dependent with respect to the non-rotating frame basis vectors as given by equation (12.15). Equation (12.16) is used extensively for problems involving rotating frames. For example, for the special case where $\mathbf{G} = \mathbf{r}'$, then equation (12.16) relates the velocity vectors in the fixed and rotating frames as given in equation (12.9).

Another example is the vector $\dot{\boldsymbol{\omega}}$

$$\dot{\boldsymbol{\omega}} = \left(\frac{d\boldsymbol{\omega}}{dt}\right)_{fix} = \left(\frac{d\boldsymbol{\omega}}{dt}\right)_{rot} + \boldsymbol{\omega} \times \boldsymbol{\omega} = \left(\frac{d\boldsymbol{\omega}}{dt}\right)_{rot} = \dot{\boldsymbol{\omega}} \tag{12.17}$$

That is, the angular acceleration $\dot{\boldsymbol{\omega}}$ has the same value in both the fixed and rotating frames of reference.

12.4   Reference frame undergoing rotation plus translation

Consider the case where the system is accelerating in translation as well as rotating, that is, the primed frame is the non-rotating translating frame. The position vector $\mathbf{r}_{fix}$ is taken with respect to the inertial fixed unprimed frame which can be written in terms of the fixed unit basis vectors $(\widehat{\mathbf{i}}_{fix}, \widehat{\mathbf{j}}_{fix}, \widehat{\mathbf{k}}_{fix})$. This $\mathbf{r}_{fix}$ vector can be written as the vector sum of the translational motion $\mathbf{R}_{fix}$ of the origin of the rotating system with respect to the fixed frame, plus the position $\mathbf{r}'_{mov}$ with respect to this translating primed frame basis

$$\mathbf{r}_{fix} = \mathbf{R}_{fix} + \mathbf{r}'_{mov} \tag{12.18}$$

The time differential is

$$\left(\frac{d\mathbf{r}}{dt}\right)_{fix} = \left(\frac{d\mathbf{R}}{dt}\right)_{fix} + \left(\frac{d\mathbf{r}'_{mov}}{dt}\right) \tag{12.19}$$

The vector $d\mathbf{r}'$ is the position with respect to the translating frame of reference which can be expressed in terms of the unit vectors $(\widehat{\mathbf{i}}'_{mov}, \widehat{\mathbf{j}}'_{mov}, \widehat{\mathbf{k}}'_{mov})$.

Equation 12.19 takes into account the translational motion of the moving primed frame basis. Now, assuming that the double primed frame rotates about the origin of the moving primed frame, then the net displacement with respect to the original inertial frame basis can be combined with equation 12.9 leading to the relation

$$\left(\frac{d\mathbf{r}}{dt}\right)_{fix} = \left(\frac{d\mathbf{R}}{dt}\right)_{fix} + \left(\frac{d\mathbf{r}''}{dt}\right)_{rot} + \boldsymbol{\omega} \times \mathbf{r}'_{mov} \tag{12.20}$$

Here the double-primed frame is both rotating and translating. Vectors in this frame are expressed in terms of the unit basis vectors $(\widehat{\mathbf{i}}''_{rot}, \widehat{\mathbf{j}}''_{rot}, \widehat{\mathbf{k}}''_{rot})$.

Expressed as velocities, equation 12.20 can be written as

$$\mathbf{v}_{fix} = \mathbf{V}_{fix} + \mathbf{v}''_{rot} + \boldsymbol{\omega} \times \mathbf{r}'_{mov} \tag{12.21}$$

where:

$\mathbf{v}_{fix}$ is the velocity measured with respect to the inertial (unprimed) frame basis.

$\mathbf{V}_{fix}$ is the velocity of the origin of the non-inertial translating (primed) frame basis with respect to the origin of the inertial (unprimed) frame basis.

$\mathbf{v}''_{rot}$ is the velocity of the particle with respect to the non-inertial rotating (double-primed) frame basis the origin of which is both translating and rotating.

$\boldsymbol{\omega} \times \mathbf{r}'_{mov}$ is the motion of the rotating (double-primed) frame with respect to the linearly-translating (primed) frame basis.

Thus this relation takes into account both the translational velocity plus rotation of the reference coordinate frame basis vectors.

12.5   Newton's law of motion in a non-inertial frame

The acceleration of the system in the rotating inertial frame can be derived by differentiating the general velocity relation for $\mathbf{v}$, equation 12.21, in the fixed frame basis which gives

$$\mathbf{a}_{fix} = \left(\frac{d\mathbf{v}_{fix}}{dt}\right)_{fixed} = \left(\frac{d\mathbf{V}_{fix}}{dt}\right)_{fixed} + \left(\frac{d\mathbf{v}''_{rot}}{dt}\right)_{fixed} + \left(\frac{d\boldsymbol{\omega}}{dt}\right)_{fixed} \times \mathbf{r}'_{mov} + \boldsymbol{\omega} \times \left(\frac{d\mathbf{r}'_{mov}}{dt}\right)_{fixed} \tag{12.22}$$

Now we wish to use the general transformation to a rotating frame basis which requires inclusion of the time dependence of the unit vectors in the rotating frame, that is,

$$\left(\frac{d\mathbf{v}''_{rot}}{dt}\right)_{fixed} = \left(\frac{d\mathbf{v}''_{rot}}{dt}\right)_{rotating} + \boldsymbol{\omega} \times \mathbf{v}''_{rot} \tag{12.23}$$

$$\left(\frac{d\boldsymbol{\omega}}{dt}\right)_{fixed} \times \mathbf{r}'_{mov} = \left(\frac{d\boldsymbol{\omega}}{dt}\right)_{rot} \times \mathbf{r}'_{mov} \tag{12.24}$$

$$\boldsymbol{\omega} \times \left(\frac{d\mathbf{r}'_{mov}}{dt}\right)_{fixed} = \boldsymbol{\omega} \times \mathbf{v}''_{rot} + \boldsymbol{\omega} \times (\boldsymbol{\omega} \times \mathbf{r}'_{mov}) \tag{12.25}$$

Using equations 12.23, 12.24, 12.25 gives

$$\mathbf{a}_{fix} = \mathbf{A}_{fix} + \mathbf{a}''_{rot} + 2\boldsymbol{\omega} \times \mathbf{v}''_{rot} + \boldsymbol{\omega} \times (\boldsymbol{\omega} \times \mathbf{r}'_{mov}) + \dot{\boldsymbol{\omega}} \times \mathbf{r}'_{mov} \tag{12.26}$$

where the acceleration in the rotating frame is $\mathbf{a}''_{rot} = \left(\frac{d\mathbf{v}''_{rot}}{dt}\right)_{rot}$ while the velocity is $\mathbf{v}''_{rot} = \left(\frac{d\mathbf{r}''_{rot}}{dt}\right)_{rot}$ and $\mathbf{A}_{fix}$ is with respect to the fixed frame.

Newton's laws of motion are obeyed in the inertial frame, that is

$$\mathbf{F}_{fix} = m\mathbf{a}_{fix} = m\left(\mathbf{A}_{fix} + \mathbf{a}''_{rot} + 2\boldsymbol{\omega} \times \mathbf{v}''_{rot} + \boldsymbol{\omega} \times (\boldsymbol{\omega} \times \mathbf{r}'_{mov}) + \dot{\boldsymbol{\omega}} \times \mathbf{r}'_{mov}\right) \tag{12.27}$$

In the double-primed frame, which may be both rotating and accelerating in translation, one can ascribe an effective force $\mathbf{F}^{eff}_{rot}$ that obeys an effective Newton's law for the acceleration $\mathbf{a}''_{rot}$ in the rotating frame

$$\mathbf{F}^{eff}_{rot} = m\mathbf{a}''_{rot} = \mathbf{F}_{fix} - m\left(\mathbf{A}_{fix} + 2\boldsymbol{\omega} \times \mathbf{v}''_{rot} + \boldsymbol{\omega} \times (\boldsymbol{\omega} \times \mathbf{r}'_{mov}) + \dot{\boldsymbol{\omega}} \times \mathbf{r}'_{mov}\right) \tag{12.28}$$

Note that the effective force $\mathbf{F}^{eff}_{rot}$ comprises the physical force $\mathbf{F}_{fixed}$ minus four non-inertial forces that are introduced to correct for the fact that the rotating reference frame is a non-inertial frame.

12.6 Lagrangian mechanics in a non-inertial frame

The above derivation of the equations of motion in the rotating frame is based on Newtonian mechanics. Lagrangian mechanics provides another derivation of these equations of motion for a rotating frame of reference by exploiting the fact that the Lagrangian is a scalar which is frame independent, that is, it is invariant to rotation of the frame of reference.

The Lagrangian in any frame is given by

$$L = \frac{1}{2}m\mathbf{v} \cdot \mathbf{v} - U(r) \tag{12.29}$$

The scalar product $\mathbf{v} \cdot \mathbf{v}$ is the same in any rotated frame and can be evaluated in terms of the rotating frame variables using the same decomposition of the translational plus rotational motion as used previously and given in equation 12.21.

Equation (12.21) decomposes the velocity in the fixed inertial frame $\mathbf{v}_{fix}$ into four vector terms, the translational velocity $\mathbf{V}_{fix}$ of the translating frame, the velocity in the rotating-translating frame $\mathbf{v}''_{rot}$, and rotational velocity $(\boldsymbol{\omega} \times \mathbf{r}'_{mov})$. Using equations 12.29 and 12.21, plus appendix equation $B.21$ for the triple products, gives that the Lagrangian evaluated using $\mathbf{v}_{fix} \cdot \mathbf{v}_{fix}$ equals

$$L = \frac{1}{2}m\left[\mathbf{V}_{fix} \cdot \mathbf{V}_{fix} + \mathbf{v}''_{rot} \cdot \mathbf{v}''_{rot} + 2\mathbf{V}_{fix} \cdot \mathbf{v}''_{rot} + 2\mathbf{V}_{fix} \cdot (\boldsymbol{\omega} \times \mathbf{r}'_{mov}) + 2\mathbf{v}''_{rot} \cdot (\boldsymbol{\omega} \times \mathbf{r}'_{mov}) + (\boldsymbol{\omega} \times \mathbf{r}'_{mov})^2\right] - U(r) \tag{12.30}$$

This can be used to derive the canonical momentum in the rotating frame

$$\mathbf{p}''_{rot} = \frac{\partial L}{\partial \mathbf{v}''_{rot}} = m\left[\mathbf{V}_{fix} + \mathbf{v}''_{rot} + \boldsymbol{\omega} \times \mathbf{r}'_{mov}\right] \tag{12.31}$$

The Lagrange equations can be used to derive the equations of motion in terms of the variables evaluated in the rotating reference frame. The required Lagrange derivatives are

$$\frac{d}{dt}\frac{\partial L}{\partial \mathbf{v}''_{rot}} = m\left[\mathbf{A}_{fix} + \mathbf{a}''_{rot} + (\boldsymbol{\omega} \times \mathbf{v}''_{rot}) + (\dot{\boldsymbol{\omega}} \times \mathbf{r}'_{mov})\right]_{rot} \tag{12.32}$$

and

$$\frac{\partial L}{\partial \mathbf{r}'} = -m\left[(\boldsymbol{\omega} \times \mathbf{V}_{fix}) - (\boldsymbol{\omega} \times \mathbf{v}''_{rot}) - \boldsymbol{\omega} \times (\boldsymbol{\omega} \times \mathbf{r}'_{mov})\right]_{rot} - \boldsymbol{\nabla}U \tag{12.33}$$

where the scalar triple product, equation $B.21$, has been used. Thus the Lagrange equations give for the rotating frame basis that

$$m\mathbf{a}''_{rot} = -\boldsymbol{\nabla}U - m[\mathbf{A}_{fix} + (\boldsymbol{\omega} \times \mathbf{V}_{fix}) + 2(\boldsymbol{\omega} \times \mathbf{v}''_{rot}) + \boldsymbol{\omega} \times (\boldsymbol{\omega} \times \mathbf{r}'_{mov}) + (\dot{\boldsymbol{\omega}} \times \mathbf{r}'_{mov})]_{rot} \tag{12.34}$$

The external force is identified as $\mathbf{F}_{fixed} = -\boldsymbol{\nabla}U$. Equation 12.16 can be used to transform between the fixed and the rotating bases.

$$\mathbf{A}_{fix} = \left[\mathbf{A}_{fix} + (\boldsymbol{\omega} \times \mathbf{V})_{fix}\right]_{rot} \tag{12.35}$$

This leads to an effective force in the non-inertial translating plus rotating frame that corresponds to an effective Newtonian force of

$$\mathbf{F}_{rot}^{eff} = m\mathbf{a}_{rot}'' = \mathbf{F} - m[\mathbf{A}_{fix} + 2\boldsymbol{\omega} \times \mathbf{v}_{rot}'' + \boldsymbol{\omega} \times (\boldsymbol{\omega} \times \mathbf{r}_{mov}') + (\dot{\boldsymbol{\omega}} \times \mathbf{r}_{mov}')] \tag{12.36}$$

where $\mathbf{A}_{fix}$ is expressed in the fixed frame. The derivation of equation 12.36 using Lagrangian mechanics, confirms the identical formula 12.29 derived using Newtonian mechanics.

The four correction terms for the non-inertial frame basis correspond to the following effective forces.

Translational acceleration: $\mathbf{F}_{mov}^{eff} = -m\mathbf{A}_{fix}$ is the usual inertial force experienced in a linearly accelerating frame of reference, and where $\mathbf{A}_{fix}$ is with respect to the fixed frame .

Coriolis force; $\mathbf{F}_{cor}^{eff} = -2m\boldsymbol{\omega} \times \mathbf{v}_{rot}''$ This is a new type of inertial force that is present only when a particle is moving in the rotating frame. This force is proportional to the velocity in the rotating frame and is independent of the position in the rotating frame

Centrifugal force: $\mathbf{F}_{cf}^{eff} = -m\boldsymbol{\omega} \times (\boldsymbol{\omega} \times \mathbf{r}_{mov}')$ This is due to the centripetal acceleration of the particle owing to the rotation of the moving axis about the axis of rotation.

Transverse (azimuthal) force: $\mathbf{F}_{az}^{eff} = -m\dot{\boldsymbol{\omega}} \times \mathbf{r}_{mov}'$ This is a straightforward term due to acceleration of the particle due to the angular acceleration of the rotating axes.

The above inertial forces are correction terms arising from trying to extend Newton's laws of motion to a non-inertial frame involving both translation and rotation. These correction forces are often referred to as "fictitious" forces. However, these non-inertial forces are very real when located in the non-inertial frame. Since the centrifugal and Coriolis terms are unusual they are discussed below.

12.7   Centrifugal force

The centrifugal force was defined as

$$\mathbf{F}_{cf} = -m\boldsymbol{\omega} \times (\boldsymbol{\omega} \times \mathbf{r}_{mov}') \tag{12.37}$$

Note that

$$\boldsymbol{\omega} \cdot \mathbf{F}_{cf} = 0 \tag{12.38}$$

therefore the centrifugal force is perpendicular to the axis of rotation.

Using the vector identity, equation $B.24$, allows the centrifugal force to be written as

$$\mathbf{F}_{cf} = -m\left[(\boldsymbol{\omega} \cdot \mathbf{r}_{mov}')\boldsymbol{\omega} - \omega^2 \mathbf{r}_{mov}'\right] \tag{12.39}$$

For the case where the radius $\mathbf{r}'$ is perpendicular to $\boldsymbol{\omega}$ then $\boldsymbol{\omega} \cdot \mathbf{r}' = 0$ and thus for this special case

$$\mathbf{F}_{cf} = m\omega^2 \mathbf{r}_{mov}' \tag{12.40}$$

The centrifugal force is experienced when riding in a car driven rapidly around a bend. The passenger experiences an apparent centrifugal (center fleeing) force that thrusts them to the outside of the bend relative to the inside of the turning car. In reality, relative to the fixed inertial frame, i.e. the road, the friction between the car tires and the road is changing the direction of the car towards the inside of the bend and the car seat is causing the centripetal (center seeking) acceleration of the passenger. A bucket of water attached to a rope can be swung around in a vertical plane without spilling any water if the centrifugal force exceeds the gravitation force at the top of the trajectory.

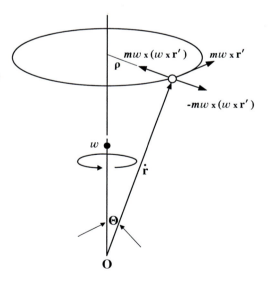

Figure 12.3: Centrifugal force.

12.8 Coriolis force

The Coriolis force was defined to be

$$\mathbf{F}_{cor} = -2m\boldsymbol{\omega} \times \mathbf{v}''_{rot} \qquad (12.41)$$

where $\mathbf{v}''$ is the velocity measured in the rotating (double-primed) frame. The Coriolis force is an interesting force; it is perpendicular to both the axis of rotation and the velocity vector in the rotating frame, that is, it is analogous to the $q\mathbf{v} \times \mathbf{B}$ Lorentz magnetic force .

The understanding of the Coriolis effect is facilitated by considering the physics of a hockey puck sliding on a rotating frictionless table. Assume that the table rotates with constant angular frequency $\boldsymbol{\omega} = \omega\widehat{\mathbf{k}}$ about the z axis. For this system the origin of the

Figure 12.4: Free-force motion of a hockey puck sliding on a rotating frictionless table of radius R that is rotating with constant angular frequency ω out of the page.

rotating system is fixed, and the angular frequency is constant, thus $\mathbf{A}$ and $\dot{\omega} \times \mathbf{r}'$ are zero. Also it is assumed that there are no external forces acting on the hockey puck, thus the net acceleration of the puck sliding on the table, as seen in the rotating frame, simplifies to

$$\mathbf{a}''_{rot} = -2\boldsymbol{\omega} \times \mathbf{v}''_{rot} - \boldsymbol{\omega} \times (\boldsymbol{\omega} \times \mathbf{r}'_{mov}) = -2\omega\widehat{\mathbf{k}} \times \mathbf{v}''_{rot} + \omega^2\mathbf{r}'_{mov} \qquad (12.42)$$

The centrifugal acceleration $+\omega^2\mathbf{r}'_{mov}$ is radially outwards while the Coriolis acceleration $-2\omega\widehat{\mathbf{k}} \times \mathbf{v}''_{rot}$ is to the right. Integration of the equations of motion can be used to calculate the trajectories in the rotating frame of reference.

Figure 12.4 illustrates trajectories of the hockey puck in the rotating reference frame when no external forces are acting, that is, in the inertial frame the puck moves in a straight line with constant velocity $\mathbf{v}_0$. In the rotating reference frame the Coriolis force accelerates the puck to the right leading to trajectories that exhibit spiral motion. The apparent complicated trajectories are a result of the observer being in the rotating frame for which that the straight inertial-frame trajectories of the moving puck exhibit a spiralling trajectory in the rotating-frame.

The Coriolis force is the reason that winds circulate in an anticlockwise direction about low-pressure regions in the Earth's northern hemisphere. It also has important consequences in many activities on earth such as ballet dancing, ice skating, acrobatics, nuclear and molecular rotation, and the motion of missiles.

12.1 *Example: Accelerating spring plane pendulum*

Comparison of the relative merits of using a non-inertial frame versus an inertial frame is given by a spring pendulum attached to an accelerating fulcrum. As shown in the figure, the spring pendulum comprises a mass m attached to a massless spring that has a rest length r_0 and spring constant k. The system is in a vertical gravitational field g and the fulcrum of the pendulum is accelerating vertically upwards with a constant acceleration a. Assume that the spring pendulum oscillates only in the vertical θ plane.

Inertial frame:

This problem can be solved in the fixed inertial coordinate system with coordinates (x, y). These coordinates, and their time derivatives, are given in terms of r and θ by

$$
\begin{aligned}
x &= r\sin\theta & \dot{x} &= \dot{r}\sin\theta + r\dot{\theta}\cos\theta \\
y &= -r\cos\theta + \frac{1}{2}at^2 & \dot{y} &= r\dot{\theta}\sin\theta - \dot{r}\cos\theta + at
\end{aligned}
$$

Thus

$$
\begin{aligned}
L &= \frac{1}{2}m\left(\dot{x}^2 + \dot{y}^2\right) - mgy - \frac{1}{2}k(r - r_0)^2 \\
&= \frac{1}{2}m\left[\dot{r}^2 + r^2\dot{\theta}^2 + a^2t^2 + 2at\left(r\dot{\theta}\sin\theta - \dot{r}\cos\theta\right)\right] + mg\left(r\cos\theta - \frac{1}{2}at^2\right) - \frac{1}{2}k(r - r_0)^2
\end{aligned}
$$

The Lagrange equations of motion are given by
$$\Lambda_r L = 0$$

$$\ddot{r} - r\dot{\theta}^2 - (a + g)\cos\theta + \frac{k}{m}(r - r_0) = 0$$

$$\Lambda_\theta L = 0$$

$$\ddot{\theta} + \frac{2}{r}\dot{r}\dot{\theta} + \frac{(a + g)}{r}\sin\theta = 0$$

The generalized momenta are

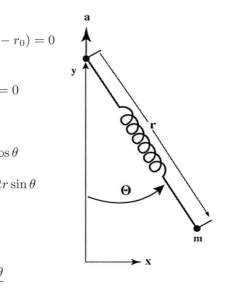

$$p_r = \frac{\partial L}{\partial \dot{r}} = m\dot{r} - mat\cos\theta$$

$$p_\theta = \frac{\partial L}{\partial \dot{\theta}} = mr^2\dot{\theta} + matr\sin\theta$$

These lead to the corresponding velocities of

$$\dot{r} = \frac{p_r}{m} + at\cos\theta$$

$$\dot{\theta} = \frac{p_\theta}{mr^2} - \frac{at\sin\theta}{r}$$

and thus the Hamiltonian is given by

$$H = p_r\dot{r} + p_\theta\dot{\theta} - L$$
$$= \frac{p_r^2}{2m} + \frac{p_\theta}{2mr^2} - \frac{at}{r}p_\theta\sin\theta + atp_r\cos\theta + \frac{1}{2}k(r - r_0)^2 + \frac{1}{2}mgat^2 - mgr\cos\theta$$

The Hamilton equations of motion give that

$$\dot{r} = \frac{\partial H}{\partial p_r} = \frac{p_r}{m} + at\cos\theta$$

$$\dot{\theta} = \frac{\partial H}{\partial p_\theta} = \frac{p_\theta}{mr^2} - \frac{at\sin\theta}{r}$$

These radial and angular velocities are the same as obtained using Lagrangian mechanics.
   The Hamilton equations for $\dot{p}_r$ and $\dot{p}_\theta$ are given by

$$\dot{p}_r = -\frac{\partial H}{\partial r} = -\frac{at}{r^2}p_\theta\sin\theta - k(r - r_0) + mg\cos\theta + \frac{p_\theta^2}{mr^3}$$

Similarly

$$\dot{p}_\theta = -\frac{\partial H}{\partial \theta} = \frac{at}{r}p_\theta\cos\theta + atp_r\sin\theta - mgr\sin\theta$$

   The transformation equations relating the generalized coordinates r, θ are time dependent so the Hamiltonian H does not equal the total energy E. In addition neither the Lagrangian nor the Hamiltonian are conserved since they both are time dependent. The fact that the Hamiltonian is not conserved is obvious since the whole system is accelerating upwards leading to increasing kinetic and potential energies. Moreover, the time derivative of the angular momentum $\dot{p}_\theta$ is non-zero so the angular momentum p_θ is not conserved.

Non-inertial fulcrum frame:
   This system also can be addressed in the accelerating non-inertial fulcrum frame of reference which is fixed to the fulcrum of the spring of the pendulum. In this non-inertial frame of reference, the acceleration of the frame can be taken into account using an effective acceleration a which is added to the gravitational force; that is, g is replaced by an effective gravitational force $(g + a)$. Then the Lagrangian in the fulcrum frame simplifies to

$$L_{fulcrum} = \frac{1}{2}m\dot{r}^2 + r^2\dot{\theta}^2 + m(g + a)(r\cos\theta) - \frac{1}{2}k(r - r_0)^2$$

   The Lagrange equations of motion in the fulcrum frame are given by

$$\Lambda_r L_{fulcrum} = 0$$

$$\ddot{r} - r\dot{\theta}^2 - (a+g)\cos\theta + \frac{k}{m}(r - r_0) = 0$$

$$\Lambda_\theta L_{fulcrum} = 0$$

$$\ddot{\theta} + \frac{2}{r}\dot{r}\dot{\theta} + \frac{(a+g)}{r}\sin\theta = 0$$

These are identical to the Lagrange equations of motion derived in the inertial frame. The $L_{fulcrum}$ can be used to derive the momenta in the non-inertial fulcrum frame

$$\tilde{p}_r = \frac{\partial L_{fulcrum}}{\partial \dot{r}} = m\dot{r}$$

$$\tilde{p}_\theta = \frac{\partial L_{fulcrumr}}{\partial \dot{\theta}} = mr^2\dot{\theta}$$

which comprise only a part of the momenta derived in the inertial frame. These partial fulcrum momenta lead to a Hamiltonian for the fulcum-frame of

$$H_{fulcrum} = \tilde{p}_r \dot{r} + \tilde{p}_\theta \dot{\theta} - L_{fulcrum} = \frac{\tilde{p}_r^2}{2m} + \frac{\tilde{p}_\theta}{2mr^2} + \frac{1}{2}k(r - r_0)^2 - m(g+a)r\cos\theta$$

Both $L_{fulcrum}$ and $H_{fulcrum}$ are time independent and thus the fulcrum Hamiltonian $H_{fulcrum}$ is a constant of motion in the fulcrum frame. However, $H_{fulcrum}$ does not equal the total energy which is increasing with time due to the acceleration of the fulcrum frame relative to the inertial frame. This example illustrates that use of non-inertial frames can simplify solution of accelerating systems.

12.2 Example: Surface of rotating liquid

Find the shape of the surface of liquid in a bucket that rotates with angular speed ω as shown in the adjacent figure. Assume that the liquid is at rest in the frame of the bucket. Therefore, in the coordinate system rotating with the bucket of liquid, the centrifugal force is important whereas the Coriolis, translational, and transverse forces are zero. The external force

$$\mathbf{F} = \mathbf{F}' - m\mathbf{g}$$

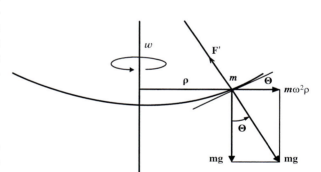

where $\mathbf{F}'$ is the pressure which is perpendicular to the surface. At equilibrium the acceleration of the surface is zero that is

$$m\mathbf{a}'' = 0 = \mathbf{F}' + m(\mathbf{g} - \boldsymbol{\omega} \times (\boldsymbol{\omega} \times \mathbf{r}'))$$

The effective gravitational force is

$$\mathbf{g}_{eff} = (\mathbf{g} - \boldsymbol{\omega} \times (\boldsymbol{\omega} \times \mathbf{r}'))$$

which must be perpendicular to the surface of the liquid since $\mathbf{F}'$ is perpendicular to the surface of a fluid, and the net force is zero. In cylindrical coordinates this can be written as

$$\mathbf{g}_{eff} = -g\widehat{\mathbf{z}} + \rho\omega^2\widehat{\boldsymbol{\rho}}$$

From the figure it can be deduced that

$$\tan\theta = \frac{dz}{d\rho} = \frac{\rho\omega^2}{g}$$

By integration

$$z = \frac{\omega^2}{2g}\rho^2 + \text{constant}$$

This is the equation of a paraboloid and corresponds to a parabolic gravitational equipotential energy surface. Astrophysicists build large parabolic mirrors for telescopes by continuously spinning a large vat of glass while it solidifies. This is much easier than grinding a large cylindrical block of glass into a parabolic shape.

12.3 *Example: The pirouette*

An interesting application of the Coriolis force is the problem of a spinning ice skater or ballet dancer. Her angular frequency increases when she draws in her arms. The conventional explanation is that angular momentum is conserved in the absence of any external forces which is correct. Thus since her moment of inertia decreases when she retracts her arms, her angular velocity must increase to maintain a constant angular momentum $\mathbf{L} = I \ \boldsymbol{\omega}$. *But this explanation does not address the question as to what are the forces that cause the angular frequency to increase? The real radial forces the skater feels when she retracts her arms cannot directly lead to angular acceleration since radial forces are perpendicular to the rotation. The following derivation shows that the Coriolis force* $-2m\boldsymbol{\omega} \times \mathbf{v}''_{rot}$ *acts tangentially to the radial retraction velocity of her arms leading to the angular acceleration required to maintain constant angular momentum.*

Consider that a mass m is moving radially at a velocity $\dot{r}''_{rot}$ *then the Coriolis force in the rotating frame is*

$$\mathbf{F}_{cor} = -2m\boldsymbol{\omega} \times \dot{\mathbf{r}}''_{rot}$$

This Coriolis force leads to an angular acceleration of the mass of

$$\dot{\boldsymbol{\omega}} = -\frac{2\boldsymbol{\omega} \times \dot{\mathbf{r}}''_{rot}}{r"} \qquad (\alpha)$$

that is, the rotational frequency decreases if the radius is increased. Note that, as shown in equation 12.17, $\dot{\omega} = \dot{\omega}''$. *This nonzero value of* $\dot{\omega}$ *obviously leads to an azimuthal force in addition to the Coriolis force. Consider the rate of change of angular momentum for the rotating mass m assuming that the angular momentum comes purely from the rotation* ω. *Then in the rotating frame*

$$\dot{\mathbf{p}}_{\theta''} = \frac{d}{dt}(mr''^2\boldsymbol{\omega}) = 2mr''\dot{r}''\boldsymbol{\omega} + mr''^2\dot{\boldsymbol{\omega}}$$

Substituting equation α *for* $\dot{\omega}$ *in the second term gives*

$$\dot{\mathbf{p}}_{\theta"} = 2mr''\dot{r}''\boldsymbol{\omega} - 2mr''\dot{r}''\boldsymbol{\omega} = 0$$

That is, the two terms cancel. Thus the angular momentum is conserved for this case where the velocity is radial. Note that, since p_θ*" is assumed to be colinear with* ω*, then it is the same in both the stationary and rotating frames of reference and thus angular momentum is conserved in both frames. In addition, in the fixed frame, the angular momentum is conserved if no external torques are acting as assumed above.*

Note that the rotational energy is

$$E_{rot} = \frac{1}{2}I\omega^2$$

Also the angular momentum is conserved, that is

$$\mathbf{p}_\theta = I\boldsymbol{\omega} = l\hat{\boldsymbol{\omega}}$$

Substituting $\boldsymbol{\omega} = \frac{\mathbf{p}_\theta}{I}$ *in the rotational energy gives*

$$E_{rot} = \frac{p_\theta^2}{2I} = \frac{l^2}{2I}$$

Therefore the rotational energy actually increases as the moment of inertia decreases when the ice skater pulls her arms close to her body. This increase in rotational energy is provided by the work done as the dancer pulls her arms inward against the centrifugal force.

12.9 Routhian reduction for rotating systems

The Routhian reduction technique, that was introduced in chapter 8.6, is a hybrid variational approach. It was devised by Routh to handle the cyclic and non-cyclic variables separately in order to simultaneously exploit the differing advantages of the Hamiltonian and Lagrangian formulations. The Routhian reduction technique is a powerful method for handling rotating systems ranging from galaxies to molecules, or deformed nuclei, as well as rotating machinery in engineering. A valuable feature of the Hamiltonian formulation is that it allows elimination of cyclic variables which reduces the number of degrees of freedom to be handled. As a consequence, cyclic variables are called *ignorable* variables in Hamiltonian mechanics. The Lagrangian, the Hamiltonian and the Routhian all are scalars under rotation and thus are invariant to rotation of the frame of reference. Note that often there are only two cyclic variables for a rotating system, that is, $\dot{\boldsymbol{\theta}} = \boldsymbol{\omega}$ and the corresponding canonical total angular momentum $\mathbf{p}_\theta = \mathbf{J}$.

As mentioned in chapter 8.6, there are two possible Routhians that are useful for handling rotation frames of reference. For rotating systems the cyclic Routhian R_{cyclic} simplifies to

$$R_{cyclic}(q_1, ..., q_n; \dot{q}_1, ..., \dot{q}_s; p_{s+1}, ..., p_n; t) = H_{cyclic} - L_{noncyclic} = \boldsymbol{\omega} \cdot \mathbf{J} - L \tag{12.43}$$

This Routhian behaves like a Hamiltonian for the ignorable cyclic coordinates $\boldsymbol{\omega}, \mathbf{J}$. Simultaneously it behaves like a negative Lagrangian $L_{noncyclic}$ for all the other coordinates.

The non-cyclic Routhian $R_{noncyclic}$ complements R_{cyclic} in that it is defined as

$$R_{noncyclic}(q_1, ..., q_n; p_1, ..., p_s; \dot{q}_{s+1}, ..., \dot{q}_n; t) = H_{noncyclic} - L_{cyclic} = H - \boldsymbol{\omega} \cdot \mathbf{J} \tag{12.44}$$

This non-cyclic Routhian behaves like a Hamiltonian for all the non-cyclic variables and behaves like a negative Lagrangian for the two cyclic variables ω, p_ω. Since the cyclic variables are constants of motion, then $R_{noncyclic}$ is a constant of motion that equals the energy in the rotating frame if H is a constant of motion. However, $R_{noncyclic}$ does not equal the total energy since the coordinate transformation is time dependent, that is, the Routhian $R_{noncyclic}$ corresponds to the energy of the non-cyclic parts of the motion.

For example, the Routhian $R_{noncyclic}$ for a system that is being cranked about the ϕ axis at some fixed angular frequency $\dot{\phi} = \omega$, with corresponding total angular momentum $\mathbf{p}_\phi = \mathbf{J}$, can be written as[1]

$$
\begin{aligned}
R_{noncyclic} &= H - \boldsymbol{\omega} \cdot \mathbf{J} \\
&= \frac{1}{2}m \left[\mathbf{V} \cdot \mathbf{V} + \mathbf{v}'' \cdot \mathbf{v}'' + 2\mathbf{V} \cdot \mathbf{v}'' + 2\mathbf{V} \cdot (\boldsymbol{\omega} \times \mathbf{r}') + 2\mathbf{v}'' \cdot (\boldsymbol{\omega} \times \mathbf{r}') + (\boldsymbol{\omega} \times \mathbf{r}')^2 \right] - \boldsymbol{\omega} \cdot \mathbf{J} + U(r)
\end{aligned}
\tag{12.45}
$$

Note that $R_{noncyclic}$ is a constant of motion if $\frac{\partial L}{\partial t} = 0$, which is the case when the system is being cranked at a constant angular frequency. However the Hamiltonian in the rotating frame $H_{rot} = H - \boldsymbol{\omega} \cdot \mathbf{J}$ is given by $R_{noncyclic} = H_{rot} \neq E$ since the coordinate transformation is time dependent. The canonical Hamilton equations for the fourth and fifth terms in the bracket can be identified with the Coriolis force $2m\boldsymbol{\omega} \times \mathbf{v}''$, while the last term in the bracket is identified with the centrifugal force. That is, define

$$U_{cf} \equiv -\frac{1}{2}m \left(\boldsymbol{\omega} \times \mathbf{r}' \right)^2 \tag{12.46}$$

where the gradient of U_{cf} gives the usual centrifugal force.

$$\mathbf{F}_{cf} = -\boldsymbol{\nabla} U_{cf} = \frac{m}{2} \boldsymbol{\nabla} \left[\omega^2 r'^2 - (\boldsymbol{\omega} \cdot \mathbf{r}')^2 \right] = m \left[\omega^2 \mathbf{r}' - (\boldsymbol{\omega} \cdot \mathbf{r}') \boldsymbol{\omega} \right] = -m\boldsymbol{\omega} \times (\boldsymbol{\omega} \times \mathbf{r}') \tag{12.47}$$

The Routhian reduction method is used extensively in science and engineering to describe rotational motion of rigid bodies, molecules, deformed nuclei, and astrophysical objects. The cyclic variables describe the rotation of the frame and thus the Routhian $R_{noncyclic} = H_{rot}$ corresponds to the Hamiltonian for the non-cyclic variables in the rotating frame.

---

[1] For clarity sections 10.1 to 10.8 of this chapter adopted a naming convention that uses unprimed coordinates with the subscript *fix* for the inertial frame of reference, primed coordinates with the subscript *mov* for the translating coordinates, and double-primed coordinates with the subscript *rot* for the translating plus rotating frame. For brevity the subsequent discussion omits the redundant subscripts *fix, mov, rot* since the single and double prime superscripts completely define the moving and rotating frames of reference.

12.4 *Example: Cranked plane pendulum*

The cranked plane pendulum, which is also called the rotating plane pendulum, comprises a plane pendulum that is cranked around a vertical axis at a constant angular velocity $\dot{\phi} = \omega$ as determined by some external drive mechanism. The parameters are illustrated in the adjacent figure. The cranked pendulum nicely illustrates the advantages of working in a non-inertial rotating frame for a driven rotating system. Although the cranked plane pendulum looks similar to the spherical pendulum, there is one very important difference; for the spherical pendulum $p_\phi = ml^2 \sin^2 \theta \dot{\phi}$ is a constant of motion and thus the angular velocity varies with θ, i.e. $\dot{\phi} = \frac{p_\phi}{ml^2 \sin^2 \theta}$, whereas for the cranked plane pendulum, the constant of motion is $\dot{\phi} = \omega$ and thus the angular momentum varies with θ, i.e. $p_\phi = l \sin^2 \theta \omega$. For the cranked plane pendulum, the energy must flow into and out of the cranking drive system that is providing the constraint force to satisfy the equation of constraint

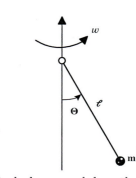

Cranked plane pendulum that is cranked around the vertical axis with angular velocity $\dot{\phi} = \omega$.

$$g_\phi = \dot{\phi} - \omega = 0$$

The easiest way to solve the equations of motion for the cranked plane pendulum is to use generalized coordinates to absorb the equation of constraint and applied constraint torque. This is done by incorporating the $\dot{\phi} = \omega$ constraint explicitly in the Lagrangian or Hamiltonian and solving for just θ in the rotating frame.

Assuming that $\dot{\phi} = \omega$, and using generalized coordinates to absorb the cranking constraint forces, then the Lagrangian for the cranked pendulum can be written as.

$$L = \frac{1}{2} ml^2(\dot{\theta}^2 + \sin^2 \theta \omega^2) + mgl \cos \theta$$

The momentum conjugate to θ is

$$p_\theta = \frac{\partial L}{\partial \dot{\theta}} = ml^2 \dot{\theta}$$

Consider the Routhian $R_{noncyclic} = p_\theta \dot{\theta} - L = H - p_\phi \dot{\phi}$ which acts as a Hamiltonian H_{rot} in the rotating frame

$$R_{noncyclic} = p_\theta \dot{\theta} - L = H - p_\phi \dot{\phi} = \frac{p_\theta^2}{2ml^2} - \frac{1}{2} ml^2 \omega^2 \sin^2 \theta - mgl \cos \theta$$

Note that if $\dot{\phi} = \omega$ is constant, then $R_{noncyclic}$ is a constant of motion for rotation about the ϕ axis since it is independent of ϕ. Also $\frac{dR_{noncyclic}}{dt} = -\frac{\partial L}{\partial t} = 0$ thus the energy in the rotating non-inertial frame of the pendulum $R_{noncyclic} = H_{rot} = H - p_\phi \dot{\phi}$ is a constant of motion, but it does not equal the total energy since the rotating coordinate transformation is time dependent. The driver that cranks the system at a constant ω provides or absorbs the energy $dW = dE = \omega dp_\phi$ as θ changes in order to maintain a constant ω.

The Routhian $R_{noncyclic}$ can be used to derive the equations of motion using Hamiltonian mechanics.

$$\dot{\theta} = \frac{\partial R_{noncyclic}}{\partial p_\theta} = \frac{p_\theta}{ml^2}$$

$$\dot{p}_\theta = -\frac{\partial R_{noncyclic}}{\partial \theta} = -mgl \sin \theta \left[1 - \frac{l}{g} \cos \theta \omega^2\right]$$

Since $\dot{p}_\theta = ml^2 \ddot{\theta}$, then the equation of motion is

$$\ddot{\theta} + \frac{g}{l} \sin \theta \left[1 - \frac{l}{g} \cos \theta \omega^2\right] = 0 \qquad (\alpha)$$

Assuming that $\sin \theta \approx \theta$, then equation α leads to linear harmonic oscillator solutions about a minimum at $\theta = 0$ if the term in brackets is positive. That is, when the bracket $\left[1 - \frac{l}{g} \cos \theta \omega^2\right] > 0$ then equation α corresponds to a harmonic oscillator with angular velocity Ω given by

$$\Omega^2 = \frac{g}{l} \left[1 - \frac{l}{g} \cos \theta \omega^2\right]$$

The adjacent figure shows the phase-space diagrams for a plane pendulum rotating about a vertical axis at angular velocity ω for (a) $\omega < \sqrt{\frac{g}{l}}$ and (b) $\omega > \sqrt{\frac{g}{l}}$. The upper phase plot shows small ω when the square bracket of equation α is positive and the the phase space trajectories are ellipses around the stable equilibrium point $(0,0)$. As ω increases the bracket becomes smaller and changes sign when $\omega^2 \cos\theta = \frac{g}{l}$. For larger ω the bracket is negative leading to hyperbolic phase space trajectories around the $(\theta, p_\theta) = (0,0)$ equilibrium point, that is, an unstable equilibrium point. However, new stable equilibrium points now occur at angles $(\theta, p_\theta) = (\pm\theta_0, 0)$ where $\cos\theta_0 = \frac{g}{l\omega^2}$. That is, the equilibrium point $(0,0)$ undergoes bifurcation as illustrated in the lower figure. These new equilibrium points are stable as illustrated by the elliptical trajectories around these points. It is interesting that these new equilibrium points $\pm\theta_0$ move to larger angles given by $\cos\theta_0 = \frac{g}{l\omega^2}$ beyond the bifurcation point at $\frac{g}{l\omega^2} = 1$. For low energy the mass oscillates about the minimum at $\theta = \theta_0$ whereas the motion becomes more complicated for higher energy. The bifurcation corresponds to symmetry breaking since, under spatial reflection, the equilibrium point is unchanged at low rotational frequencies but it transforms from $+\theta_0$ to $-\theta_0$ once the solution bifurcates, that is, the symmetry is broken. Also chaos can occur at the separatrix that separates the bifurcation. Note that either the Lagrange multiplier approach, or the generalized force approach, can be used to determine the applied torque required to ensure a constant ω for the cranked pendulum.

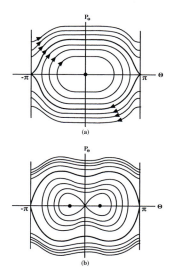

Phase-space diagrams for the plane pendulum cranked at angular velocity ω about a vertical axis. Figure (a) is for $\omega < \frac{g}{l}$ while (b) is for $\omega > \frac{g}{l}$.

12.5 Example: Nucleon orbits in deformed nuclei

Consider the rotation of axially-symmetric, prolate-deformed nucleus. Many nuclei have a prolate spheroidal shape, (the shape of a rugby ball) and they rotate perpendicular to the symmetry axis. In the non-inertial body-fixed frame, pairs of nucleons, each with angular momentum j, are bound in orbits with the projection of the angular momentum along the symmetry axis being conserved with value $\Omega = K$, which is a cyclic variable. Since the nucleus is of dimensions $10^{-14}m$, quantization is important and the quantized binding energies of the individual nucleons are separated by spacings $\leq 500keV$.

The Lagrangian and Hamiltonian are scalars and can be evaluated in any coordinate frame of reference. It is most useful to calculate the Hamiltonian for a deformed body in the non-inertial rotating body-fixed frame of reference. The body-fixed Hamiltonian corresponds to the Routhian $R_{noncyclic}$

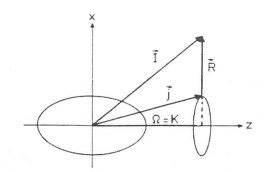

Schematic diagram for the strong coupling of a nucleon to the deformation axis. The projection of I on the symmetry axis is K, and the projection of j is Ω. For axial symmetry Noether's theorem gives that the projection of the angular momentum K on the symmetry axis is a conserved quantity.

$$R_{noncyclic} = H - \boldsymbol{\omega} \cdot \mathbf{J}$$

where it is assumed that the deformed nucleus has the symmetry axis along the z direction and rotates about the x axis. Since the Routhian is for a non-inertial rotating frame of reference it does not include the total energy but, if the shape is constant in time, then $R_{noncyclic}$ and the corresponding body-fixed Hamiltonian are conserved and the energy levels for the nucleons bound in the spheroidal potential well can be calculated using a conventional quantum mechanical model.

For a prolate spheroidal deformed potential well, the nucleon orbits that have the angular momentum nearly aligned to the symmetry axis correspond to nucleon trajectories that are restricted to the narrowest *part of the spheroid, whereas trajectories with the angular momentum vector close to perpendicular to the symmetry axis have trajectories that probe the largest radii of the spheroid. The Heisenberg Uncertainty Principle, mentioned in chapter 3.11.3, describes how orbits restricted to the smallest dimension will have the highest linear momentum, and corresponding kinetic energy, and vise versa for the larger sized orbits. Thus the binding energy of different nucleon trajectories in the spheroidal potential well depends on the angle between the angular momentum vector and the symmetry axis of the spheroid as well as the deformation of the spheroid. A quantal nuclear model Hamiltonian is solved for assumed spheroidal-shaped potential wells. The corresponding orbits each have angular momenta $\mathbf{j}_i$ for which the projection of the angular momentum along the symmetry axis Ω_i is conserved, but the projection of $\mathbf{j}_i$ in the laboratory frame j_z is not conserved since the potential well is not spherically symmetric. However, the total Hamiltonian is spherically symmetric in the laboratory frame, which is satisfied by allowing the deformed spheroidal potential well to rotate freely in the laboratory frame, and then $j_i^2, j_{i,z}$, and Ω_i all are conserved quantities. The attractive residual nucleon-nucleon pairing interaction results in pairs of nucleons being bound in time-reversed orbits $(j \times j)^0$, that is, with resultant total spin zero, in this spheroidal nuclear potential. Excitation of an even-even nucleus can break one pair and then the total projection of the angular momentum along the symmetry axis is $K = |\Omega_1 \pm \Omega_2|$, depending on whether the projections are parallel or antiparallel. More excitation energy can break several pairs and the projections continue to be additive. The binding energies calculated in the spheroidal potential well must be added to the rotational energy $E_{Rot} = \frac{\mathcal{J}}{2}\omega^2$ to get the total energy, where $\mathcal{J}$ is the moment of inertia. Nuclear structure measurements are in good agreement with the predictions of nuclear structure calculations that employ the Routhian approach.*

12.10   Effective gravitational force near the surface of the Earth

Consider that the translational acceleration of the center of the Earth can be neglected, and thus a set of non-rotating axes through the center of the Earth can be assumed to be approximately an inertial frame. The effects of the motion of the Earth around the Sun, or the motion of the Solar system in our Galaxy, are small compared with the effects due to the rotation of the Earth.

Consider a rotating frame attached to the surface of the earth as shown in figure 12.5. The vector with respect to the center of the Earth $\mathbf{r}$ can be decomposed into a vector to the origin of the reference frame fixed to the surface of the Earth $\mathbf{R}$, plus the vector with respect to this surface reference frame $\mathbf{r}'$.

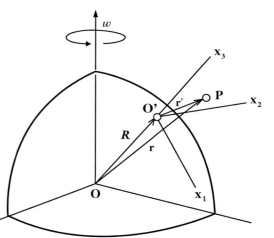

$$\mathbf{r} = \mathbf{R} + \mathbf{r}' \qquad (12.48)$$

If the external force is separated into the gravitational term $m\mathbf{g}$, plus some other physical force $\mathbf{F}$, then the acceleration in the non-inertial surface frame of reference is

Figure 12.5: Rotating frame at the surface of the Earth.

$$\mathbf{a}' = \frac{\mathbf{F}}{m} + \mathbf{g} - (\mathbf{A} + 2\boldsymbol{\omega} \times \mathbf{v}' + \boldsymbol{\omega} \times (\boldsymbol{\omega} \times \mathbf{r}') + \dot{\boldsymbol{\omega}} \times \mathbf{r}') \quad (12.49)$$

But

$$\mathbf{V} = \left(\frac{d\mathbf{R}}{dt}\right)_{fixed} = \left(\frac{d\mathbf{R}}{dt}\right)_{rotating} + \boldsymbol{\omega} \times \mathbf{R} = \boldsymbol{\omega} \times \mathbf{R} \qquad (12.50)$$

since in the rotating frame $\left(\frac{d\mathbf{R}}{dt}\right)_{rotating} = 0$. Also the acceleration

$$\mathbf{A} = \left(\frac{d\mathbf{V}}{dt}\right)_{fixed} = \left(\frac{d\mathbf{V}}{dt}\right)_{rotating} + \boldsymbol{\omega} \times \mathbf{V} = \boldsymbol{\omega} \times (\boldsymbol{\omega} \times \mathbf{R}) \qquad (12.51)$$

since $\left(\frac{d\mathbf{V}}{dt}\right)_{rotating} = 0$. Substituting this into the above equation gives

$$
\begin{aligned}
\mathbf{a}' &= \frac{\mathbf{F}}{m} + \mathbf{g} - (2\boldsymbol{\omega} \times \mathbf{v}' + \boldsymbol{\omega} \times (\boldsymbol{\omega} \times [\mathbf{r}' + \mathbf{R}]) + \dot{\boldsymbol{\omega}} \times \mathbf{r}') \\
&= \frac{\mathbf{F}}{m} + \mathbf{g} - (2\boldsymbol{\omega} \times \mathbf{v}' + \boldsymbol{\omega} \times (\boldsymbol{\omega} \times \mathbf{r}) + \dot{\boldsymbol{\omega}} \times \mathbf{r}')
\end{aligned}
$$

where $\mathbf{r}$ is with respect to the center of the Earth. This is as expected directly from equation 12.36. Since the angular frequency of the earth is a constant then $\dot{\boldsymbol{\omega}} \times \mathbf{r}' = 0$. Thus the acceleration can be written as

$$
\mathbf{a}' = \frac{\mathbf{F}}{m} + [\mathbf{g} - \boldsymbol{\omega} \times (\boldsymbol{\omega} \times \mathbf{r})] - 2\boldsymbol{\omega} \times \mathbf{v}' \tag{12.52}
$$

The term in the square brackets combines the gravitational acceleration plus the centrifugal acceleration.

A measurement of the Earth's gravitational acceleration actually measures the term in the square brackets in equation 12.52, that is, an effective gravitational acceleration where

$$
\mathbf{g}_{eff} = \mathbf{g} - \boldsymbol{\omega} \times (\boldsymbol{\omega} \times \mathbf{r}) \tag{12.53}
$$

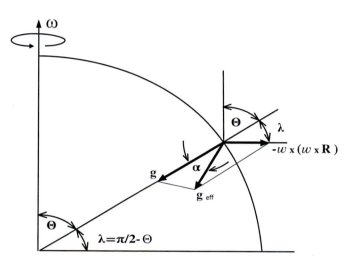

near the surface of the earth $\mathbf{r} \approx \mathbf{R}$. The effective gravitational force does not point towards the center of the Earth as shown in figure 12.6. A plumb line points, or an object falls, in the direction of $\mathbf{g}_{eff}$. The shape of the earth is such that the Earth's surface is perpendicular to $\mathbf{g}_{eff}$. This is the reason why the earth is distorted into an oblate ellipsoid, that is, it is flattened at the poles.

The angle α between $\mathbf{g}_{eff}$ and the line pointing to the center of the earth is dependent on the latitude $\lambda = \frac{\pi}{2} - \theta$. Note that the colatitude θ is taken to be zero at the North pole whereas the latitude λ is taken to be zero at the equator. The angle α can be estimated by assuming that $r' << R$, then the centrifugal term then can be approximated by

Figure 12.6: Effective gravitational acceleration.

$$
|\boldsymbol{\omega} \times (\boldsymbol{\omega} \times \mathbf{r})| \approx \omega^2 R \sin\theta = \omega^2 R \cos\lambda \tag{12.54}
$$

This is quite small for the Earth since $\omega = 0.73 \times 10^{-4}\ rads/s$ and $R = 6371km$, leading to a correction term $\omega^2 R \cos\lambda = 0.03 \cos\lambda\ m/s^2$. Since

$$
g_{eff}^{horizontal} = \omega^2 R \cos\lambda \sin\lambda \tag{12.55}
$$

and

$$
g_{eff}^{vertical} = g - \omega^2 R \cos^2\lambda \tag{12.56}
$$

Then the angle α between $\mathbf{g}_{eff}$ and $\mathbf{g}$ is given by

$$
\alpha \simeq \tan\alpha = \frac{g_{eff}^{horizontal}}{g_{eff}^{vertical}} = \frac{\omega^2 R \cos\lambda \sin\lambda}{g - \omega^2 R \cos^2\lambda} \tag{12.57}
$$

This has a maximum value at $\lambda = 45^o$ which is $\alpha = 0.0088^\circ$.

12.11   Free motion on the earth

The calculation of trajectories for objects as they move near the surface of the earth is frequently required for many applications. Such calculations require inclusion of the non-inertial Coriolis force. In the frame of reference fixed to the earth's surface, assuming that air resistance and other forces can be neglected, then the acceleration equals

$$\mathbf{a}' = \mathbf{g}_{eff} - 2\boldsymbol{\omega} \times \mathbf{v}' \qquad (12.58)$$

Neglect the centrifugal correction term since it is very small, that is, let $\mathbf{g}_{eff} = \mathbf{g}$. Using the coordinate axis shown in figure 12.7, the surface-frame vectors have components

$$\boldsymbol{\omega} = 0\widehat{\mathbf{i}'} + \omega \cos \lambda \widehat{\mathbf{j}'} + \omega \sin \lambda \widehat{\mathbf{k}'} \qquad (12.59)$$

and

$$\mathbf{g}_{eff} = -g\widehat{\mathbf{k}'} \qquad (12.60)$$

Thus the Coriolis term is

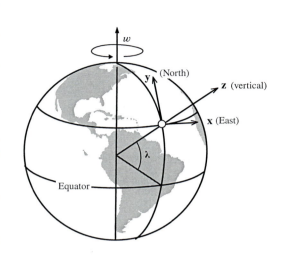

Figure 12.7: Rotating frame fixed on the surface of the Earth.

$$2\boldsymbol{\omega} \times \mathbf{v}' = 2 \begin{vmatrix} \widehat{\mathbf{i}'} & \widehat{\mathbf{j}'} & \widehat{\mathbf{k}'} \\ 0 & \omega \cos \lambda & \omega \sin \lambda \\ \dot{x}' & \dot{y}' & \dot{z}' \end{vmatrix}$$

$$= 2\left[\left(\omega \dot{z}' \cos \lambda - \omega \dot{y}' \sin \lambda\right)\widehat{\mathbf{i}'} + \left(\omega \dot{x}' \sin \lambda\right)\widehat{\mathbf{j}'} - \left(\omega \dot{x}' \cos \lambda\right)\widehat{\mathbf{k}'}\right]$$

Therefore the equations of motion are

$$m\ddot{\mathbf{r}}' = -mg\widehat{\mathbf{k}'} - 2m[\widehat{\mathbf{i}'}(\dot{z}'\omega \cos \lambda - \dot{y}'\omega \sin \lambda) + \widehat{\mathbf{j}'}\dot{x}'\omega \sin \lambda - \widehat{\mathbf{k}'}\dot{x}'\omega \cos \lambda] \qquad (12.61)$$

That is, the components of this equation of motion are

$$\ddot{x}' = -2\omega\left(\dot{z}' \cos \lambda - \dot{y}' \sin \lambda\right) \qquad (12.62)$$
$$\ddot{y}' = -2\omega\dot{x}' \sin \lambda$$
$$\ddot{z}' = -g + 2\omega\dot{x}' \cos \lambda$$

Integrating these differential equations gives

$$\dot{x}' = -2\omega\left(z' \cos \lambda - y' \sin \lambda\right) + \dot{x}'_0 \qquad (12.63)$$
$$\dot{y}' = -2\omega x' \sin \lambda + \dot{y}'_0$$
$$\dot{z}' = -gt + 2\omega x' \cos \lambda + \dot{z}'_0$$

where $\dot{x}'_0, \dot{y}'_0, \dot{z}'_0$ are the initial velocities. Substituting the above velocity relations into the equation of motion for $\ddot{x}$ gives

$$\ddot{x}' = 2\omega gt \cos \lambda - 2\omega\left(\dot{z}'_0 \cos \lambda - \dot{y}'_0 \sin \lambda\right) - 4\omega^2 x' \qquad (12.64)$$

The last term $4\omega^2 x$ is small and can be neglected leading to a simple uncoupled second-order differential equation in x. Integrating this twice assuming that $x'_0 = y'_0 = z'_0 = 0$, plus the fact that $2\omega gt \cos \lambda$ and $2\omega\left(\dot{z}'_0 \cos \lambda - \dot{y}'_0 \sin \lambda\right)$ are constant, gives

$$x' = \frac{1}{3}\omega gt^3 \cos \lambda - \omega t^2\left(\dot{z}'_0 \cos \lambda - \dot{y}'_0 \sin \lambda\right) + \dot{x}'_0 t \qquad (12.65)$$

Similarly,

$$y' = \left(\dot{y}'_0 t - \omega \dot{x}'_0 t^2 \sin \lambda\right) \qquad (12.66)$$

$$z' = -\frac{1}{2}gt^2 + \dot{z}'_0 t + \omega \dot{x}'_0 t^2 \cos \lambda \qquad (12.67)$$

Consider the following special cases;

12.6 Example: Free fall from rest

Assume that an object falls a height h starting from rest at $t = 0$, $x = 0$, $y = 0$, $z = h$. Then

$$
\begin{aligned}
x' &= \frac{1}{3}\omega g t^3 \cos\lambda \\
y' &= 0 \\
z' &= h - \frac{1}{2}g t^2
\end{aligned}
$$

Substituting for t gives

$$
x' = \frac{1}{3}\omega\cos\lambda\sqrt{\frac{8h^3}{g}}
$$

Thus the object drifts eastward as a consequence of the earth's rotation. Note that relative to the fixed frame it is obvious that the angular velocity of the body must increase as it falls to compensate for the reduced distance from the axis of rotation in order to ensure that the angular momentum is conserved.

12.7 Example: Projectile fired vertically upwards

An upward fired projectile with initial velocities $\dot{x}'_0 = \dot{y}'_0 = 0$ and $\dot{z}'_0 = v_0$ leads to the relations

$$
x' = \frac{1}{3}\omega g t^3 \cos\lambda - \omega t^2 v_0 \cos\lambda
$$

$$
y' = 0
$$

$$
z' = -\frac{1}{2}g t^2 + v_0 t
$$

Solving for t when $z' = 0$ gives $t = 0$, and $t = \frac{2v_0}{g}$. Also since the maximum height h that the projectile reaches is related by

$$
v_0 = \sqrt{2gh}
$$

then the final deflection is

$$
x' = -\frac{4}{3}\omega\cos\lambda\sqrt{\frac{8h^3}{g}}
$$

Thus the body drifts westwards.

12.8 Example: Motion parallel to Earth's surface

For motion in the horizontal $x' - y'$ plane the deflection is always to the right in the northern hemisphere of the Earth since the vertical component of ω is upwards and thus $-2\overrightarrow{\omega} \times \overrightarrow{v'}$ points to the right. In the southern hemisphere the vertical component of ω is downward and thus $-2\overrightarrow{\omega} \times \overrightarrow{v'}$ points to the left. This is also shown using the above relations for the case of a projectile fired upwards in an easterly direction with components $\dot{x}'_0, 0, \dot{z}'_0$. The resultant displacements are

$$
x' = \frac{1}{3}\omega g t^3 \cos\lambda - \omega t^2 \dot{z}'_0 \cos\lambda + \dot{x}'_0 t
$$

Similarly,

$$
y' = -\omega\dot{x}'_0 t^2 \sin\lambda
$$

$$
z' = -\frac{1}{2}g t^2 + \dot{z}'_0 t + \omega\dot{x}'_0 t^2 \cos\lambda
$$

The trajectory is non-planar and, in the northern hemisphere, the projectile drifts to the right, that is southerly.

In the battle of the River de la Plata, during World War 2, the gunners on the British light cruisers Exeter, Ajax and Achilles found that their accurately aimed salvos against the German pocket battleship Graf Spee were falling 100 yards to the left. The designers of the gun sighting mechanisms had corrected for the Coriolis effect assuming the ships would fight at latitudes near 50° north, not 50° south.

12.12   Weather systems

Weather systems on Earth provide a classic example of motion in a rotating coordinate system. In the northern hemisphere, air flowing into a low-pressure region is deflected to the right causing counterclockwise circulation, whereas air flowing out of a high-pressure region is deflected to the right causing a clockwise circulation. Trade winds on the Earth result from air rising or sinking due to thermal activity combined with the Coriolis effect. Similar behavior is observed on other planets such as the Red Spot on Jupiter.

For a fluid or gas, equation (12.36) can be written in terms of the fluid density ρ in the form

$$\rho\mathbf{a}" = -\boldsymbol{\nabla}P - \rho[2\boldsymbol{\omega} \times \mathbf{v}" - \boldsymbol{\omega} \times (\boldsymbol{\omega} \times \mathbf{r}')] \tag{12.68}$$

where the translational acceleration $\mathbf{A}$, the gravitational force, and the azimuthal acceleration ($\dot{\boldsymbol{\omega}} \times \mathbf{r}'$) terms are ignored. The external force per unit volume equals the pressure gradient $-\boldsymbol{\nabla}P$ while $\boldsymbol{\omega}$ is the rotation vector of the earth.

In fluid flow, the Rossby number Ro is defined to be

$$Ro = \frac{\text{inertial force}}{\text{Coriolis force}} \approx \frac{\mathbf{a}"}{2\boldsymbol{\omega} \times \mathbf{v}"} \tag{12.69}$$

For large dimensional pressure systems in the atmosphere, e.g. $L \simeq 1000km$, the Rossby number is $Ro \sim 0.1$ and thus the Coriolis force dominates and the radial acceleration can be neglected. This leads to a flow velocity $v \simeq 10m/s$ which is perpendicular to the pressure gradient ∇P, that is, the air flows horizontally parallel to the isobars of constant pressure which is called geostrophic flow. For much smaller dimension systems, such as at the wall of a hurricane, $L \simeq 50km$, and $v \simeq 50m/s$, the Rossby number $Ro \simeq 10$ and the Coriolis effect plays a much less significant role compared to the balance between the radial centrifugal forces and the pressure gradient. The same situation of the Coriolis forces being insignificant occurs for most small-scale vortices such as tornadoes, typical thermal vortices in the atmosphere, and for water draining a bath tub.

12.12.1   Low-pressure systems:

It is interesting to analyze the motion of air circulating around a low pressure region at large radii where the motion is tangential. As shown in figure 12.8, a parcel of air circulating anticlockwise around the low with velocity v involves a pressure difference ΔP acting on the surface area S, plus the centrifugal and Coriolis forces. Assuming that these forces are balanced such that $\mathbf{a}" \simeq 0$, then equation 12.68 simplifies to

$$\frac{v^2}{r} = \frac{1}{\rho}\nabla P - 2v\omega \sin\lambda \tag{12.70}$$

where the latitude $\lambda = \pi - \theta$. Thus the force equation can be written

$$\frac{1}{\rho}\frac{dP}{dr} = \frac{v^2}{r} + 2v\omega \sin\lambda \tag{12.71}$$

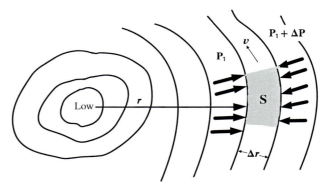

It is apparent that the combined outward Coriolis force plus outward centrifugal force, acting on the circulating air, can support a large pressure gradient.

Figure 12.8: Air flow and pressures around a low-pressure region.

The tangential velocity v can be obtained by solving this equation to give

$$v = \sqrt{(r\omega \sin\lambda)^2 + \frac{r}{\rho}\frac{dP}{dr}} - r\omega \sin\lambda \tag{12.72}$$

Note that the velocity equals zero when $r = 0$ assuming that $\frac{dP}{dr}$ is finite. That is, the velocity reaches a maximum at a radius

$$r_{peakvel} = \frac{1}{4}(1 + \frac{1}{\rho\omega \sin\lambda}\frac{dP}{dr}) \tag{12.73}$$

Figure 12.9: Hurricane Katrina over the Gulf of Mexico on 28 August 2005. [Published by the NOAA]

which occurs at the wall of the eye of the circulating low-pressure system.

Low pressure regions are produced by heating of air causing it to rise and resulting in an inflow of air to replace the rising air. Hurricanes form over warm water when the temperature exceeds $26°C$ and the moisture levels are above average. They are created at latitudes between $10° - 15°$ where the sea is warmest, but not closer to the equator where the Coriolis force drops to zero. About 90% of the heating of the air comes from the latent heat of vaporization due to the rising warm moist air condensing into water droplets in the cloud similar to what occurs in thunderstorms. For hurricanes in the northern hemisphere, the air circulates anticlockwise inwards. Near the wall of the eye of the hurricane, the air rises rapidly to high altitudes at which it then flows clockwise and outwards and subsequently back down in the outer reaches of the hurricane. Both the wind velocity and pressure are low inside the eye which can be cloud free. The strongest winds are in vortex surrounding the eye of the hurricane, while weak winds exist in the counter-rotating vortex of sinking air that occurs far outside the hurricane.

Figure 12.9 shows the satellite picture of the hurricane Katrina, recorded on 28 August 2005. The eye of the hurricane is readily apparent in this picture. The central pressure was $90200N/m^2$ ($902mb$) compared with the standard atmospheric pressure of $101300N/m^2$ ($1013mb$). This $111mb$ pressure difference produced steady winds in Katrina of $280km/hr$ ($175mph$) with gusts up to $344km/hr$ which resulted in 1833 fatalities.

Tornadoes are another example of a vortex low-pressure system that are the opposite extreme in both size and duration compared with a hurricane. Tornadoes may last only ~ 10 minutes and be quite small in radius. Pressure drops of up to $100mb$ have been recorded, but since they may only be a few 100 meters in diameter, the pressure gradient can be much higher than for hurricanes leading to localized winds thought to approach $500km/hr$. Unfortunately, the instrumentation and buildings hit by a tornado often are destroyed making study difficult. Note that the the pressure gradient in small diameter of rope tornadoes is much more destructive than for larger 1/4 mile diameter tornadoes, which results in stronger winds.

12.12.2   High-pressure systems:

In contrast to low-pressure systems, high-pressure systems are very different in that the Coriolis force points inward opposing the outward pressure gradient and centrifugal force. That is,

$$\frac{v^2}{r} = 2v\omega \sin \lambda - \frac{1}{\rho}\frac{dP}{dr} \qquad (12.74)$$

which gives that

$$v = r\omega \sin \lambda - \sqrt{(r\omega \sin \lambda)^2 - \frac{r}{\rho}\frac{dP}{dr}} \qquad (12.75)$$

This implies that the maximum pressure gradient plus centrifugal force supported by the Coriolis force is

$$\frac{r}{\rho}\frac{dP}{dr} \leq (r\omega \sin \lambda)^2 \qquad (12.76)$$

As a consequence, high pressure regions tend to have weak pressure gradients and light winds in contrast to the large pressure gradients plus concomitant damaging winds possible for low pressure systems such a hurricanes or tornados.

The circulation behavior, exhibited by weather patterns, also applies to ocean currents and other liquid flow on earth. However, the residual angular momentum of the liquid often can overcome the Coriolis terms. Thus often it will be found experimentally that water exiting the bathtub does not circulate anticlockwise in the northern hemisphere as predicted by the Coriolis force. This is because it was not stationary originally, but rotating slowly.

Reliable prediction of weather is an extremely difficult, complicated and challenging task, which is of considerable importance in modern life. As discussed in chapter 16.8, fluid flow can be much more complicated than assumed in this discussion of air flow and weather. Both turbulent and laminar flow are possible. As a consequence, computer simulations of weather phenomena are difficult because the air flow can be turbulent and the transition from order to chaotic flow is very sensitive to the initial conditions. Typically the air flow can involve both macroscopic ordered coherent structures over a wide dynamic range of dimensions, coexisting with chaotic regions. Computer simulations of fluid flow often are performed based on Lagrangian mechanics to exploit the scalar properties of the Lagrangian. Ordered coherent structures, ranging from microscopic bubbles to hurricanes, can be recognized by exploiting Lyapunov exponents to identify the ordered motion buried in the underlying chaos. Thus the techniques discussed in classical mechanics are of considerable importance outside of physics.

12.13   Foucault pendulum

A classic example of motion in non-inertial frames is the rotation of the Foucault pendulum on the surface of the earth. The Foucault pendulum is a spherical pendulum with a long suspension that oscillates in the $x - y$ plane with sufficiently small amplitude that the vertical velocity $\dot{z}$ is negligible. Assume that the pendulum is a simple pendulum of length l and mass m as shown in figure 12.10. The equation of motion is given by

$$\ddot{\mathbf{r}} = \mathbf{g} + \frac{\mathbf{T}}{m} - 2\mathbf{\Omega} \times \dot{\mathbf{r}} \qquad (12.77)$$

where $\frac{T}{m}$ is the acceleration produced by the tension in the pendulum suspension and the rotation vector of the earth is designated by $\mathbf{\Omega}$ to avoid confusion with the oscillation frequency of the pendulum ω. The effective gravitational acceleration $\mathbf{g}$ is given by

$$\mathbf{g} = \mathbf{g}_0 - \mathbf{\Omega} \times [\mathbf{\Omega} \times (\mathbf{r} + \mathbf{R})] \qquad (12.78)$$

that is, the true gravitational field $\mathbf{g}_0$ corrected for the centrifugal force.

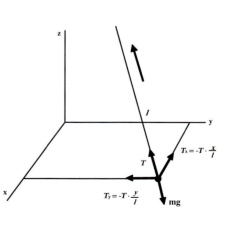

Figure 12.10: Foucault pendulum.

Assume the small angle approximation for the pendulum deflection angle β, then $T_z = T\cos\beta \simeq T$ and $T_z = mg$, thus $T \simeq mg$. Then has shown in figure 12.10, the horizontal components of the restoring force are

$$T_x = -mg\frac{x}{l} \tag{12.79}$$

$$T_y = -mg\frac{y}{l} \tag{12.80}$$

Since $\mathbf{g}$ is vertical, and neglecting terms involving $\dot{z}$, then evaluating the cross product in equation (12.78) simplifies to

$$\ddot{x} = -g\frac{x}{l} + 2\dot{y}\Omega\cos\theta \tag{12.81}$$

$$\ddot{y} = -g\frac{y}{l} - 2\dot{x}\Omega\cos\theta \tag{12.82}$$

where θ is the colatitude which is related to the latitude λ by

$$\cos\theta = \sin\lambda \tag{12.83}$$

The natural angular frequency of the simple pendulum is

$$\omega_0 = \sqrt{\frac{g}{l}} \tag{12.84}$$

while the z component of the earth's angular velocity is

$$\Omega_z = \Omega\cos\theta \tag{12.85}$$

Thus equations 12.81 and 12.82 can be written as

$$\begin{aligned}\ddot{x} - 2\Omega_z\dot{y} + \omega_0^2 x &= 0\\ \ddot{y} - 2\Omega_z\dot{x} + \omega_0^2 y &= 0\end{aligned} \tag{12.86}$$

These are two coupled equations that can be solved by making a coordinate transformation.

Define a new coordinate that is a complex number

$$\eta = x + iy \tag{12.87}$$

Multiply the second of the coupled equations 12.86 by i and add to the first equation gives

$$(\ddot{x} + i\ddot{y}) + 2i\Omega_z(\dot{x} + i\dot{y}) + \omega_0^2(x + iy) = 0$$

which can be written as a differential equation for η

$$\ddot{\eta} + 2i\Omega_z\dot{\eta} + \omega_0^2\eta = 0 \tag{12.88}$$

Note that the complex number η contains the same information regarding the position in the $x - y$ plane as equations 12.86. The plot of η in the complex plane, the Argand diagram, is a birds-eye view of the position coordinates (x, y) of the pendulum. This second-order homogeneous differential equation has two independent solutions that can be derived by guessing a solution of the form

$$\eta(t) = Ae^{-i\alpha t} \tag{12.89}$$

Substituting equation 12.89 into 12.88 gives that

$$\alpha^2 - 2\Omega_z\alpha - \omega_o^2 = 0$$

That is

$$\alpha = \Omega_z \pm \sqrt{\Omega_z^2 + \omega_0^2} \tag{12.90}$$

If the angular velocity of the pendulum $\omega_0 \gg \Omega$, then

$$\alpha \simeq \Omega_z \pm \omega_0 \tag{12.91}$$

Thus the solution is of the form

$$\eta(t) = e^{-i\Omega_z t}(A_+ e^{i\omega_0 t} + A_- e^{i\omega_0 t}) \tag{12.92}$$

This can be written as

$$\eta(t) = A e^{-i\Omega_z t} \cos(\omega_0 t + \delta) \tag{12.93}$$

where the phase δ and amplitude A depend on the initial conditions. Thus the plane of oscillation of the pendulum is defined by the ratio of the x and y coordinates, that is the phase angle $i\Omega_z t$. This phase angle rotates with angular velocity Ω_z where

$$\Omega_z = \Omega \cos\theta = \Omega \sin\lambda \tag{12.94}$$

At the north pole the earth rotates under the pendulum with angular velocity Ω and the axis of the pendulum is fixed in an inertial frame of reference. At lower latitudes, the pendulum precesses at the lower angular frequency $\Omega_z = \Omega \sin\lambda$ that goes to zero at the equator. For example, in Rochester, NY, $\lambda = 43°N$, and therefore a Foucault pendulum precesses at $\Omega_z = 0.682\Omega$. That is, the pendulum precesses $245.5°/$day.

12.14   Summary

This chapter has focussed on describing motion in non-inertial frames of reference. It has been shown that the force and acceleration in non-inertial frames can be related using either Newtonian or Lagrangian mechanics by introducing additional inertial forces in the non-inertial reference frame.

Translational acceleration of a reference frame   In a primed frame, that is undergoing translational acceleration $\mathbf{A}$, the motion in this non-inertial frame can be calculated by addition of an inertial force $-m\mathbf{A}$, that leads to an equation of motion

$$m\mathbf{a}' = \mathbf{F} - m\mathbf{A} \tag{12.6}$$

Note that the primed frame is an inertial frame if $\mathbf{A} = 0$.

Rotating reference frame   It was shown that the time derivatives of a general vector $\mathbf{G}$ in both an inertial frame and a rotating reference frame are related by

$$\left(\frac{d\mathbf{G}}{dt}\right)_{fixed} = \left(\frac{d\mathbf{G}}{dt}\right)_{rotating} + \boldsymbol{\omega} \times \mathbf{G} \tag{12.16}$$

where the $\boldsymbol{\omega} \times \mathbf{G}$ term originates from the fact that the unit vectors in the rotating reference frame are time dependent with respect to the inertial frame.

Reference frame undergoing both rotation and translation   Both Newtonian and Lagrangian mechanics were used to show that for the case of translational acceleration plus rotation, the effective force in the non-inertial (double-primed) frame can be written as

$$\mathbf{F}_{eff} = m\mathbf{a}'' = \mathbf{F} - m\left(\mathbf{A} + \boldsymbol{\omega} \times \mathbf{V} + 2\boldsymbol{\omega} \times \mathbf{v}'' + \boldsymbol{\omega} \times (\boldsymbol{\omega} \times \mathbf{r}') + \dot{\boldsymbol{\omega}} \times \mathbf{r}'\right) \tag{12.28, 12.36}$$

These inertial correction forces result from describing the system using a non-inertial frame. These inertial forces are felt when in the rotating-translating frame of reference. Thus the notion of these inertial forces can be very useful for solving problems in non-inertial frames. For the case of rotating frames, two important inertial forces are the centrifugal force, $-\boldsymbol{\omega} \times (\boldsymbol{\omega} \times \mathbf{r}')$, and the Coriolis force $-2\boldsymbol{\omega} \times \mathbf{v}''$.

Routhian reduction for rotating systems   It was shown that for non-inertial systems, identical equations of motion are derived using Newtonian, Lagrangian, Hamiltonian, and Routhian mechanics.

Terrestrial manifestations of rotation Examples of motion in rotating frames presented in the chapter included projectile motion with respect to the surface of the Earth, rotation alignment of nucleons in rotating nuclei, and weather phenomena.

Workshop exercises

1. Consider a fixed reference frame S and a rotating frame S'. The origins of the two coordinate systems always coincide. By carefully drawing a diagram, derive an expression relating the coordinates of a point P in the two systems. (This was covered in Chapter 2, but it is worth reviewing now.

2. The effective force observed in a rotating coordinate system is given by equation 12.28.

   (a) What is the significance of each term in this expression?

   (b) Suppose you wanted to measure the gravitational force, both magnitude and direction, on a body of mass m at rest on the surface of the Earth. What terms in the effective force can be neglected?

   (c) Suppose you wanted to calculate the deflection of a projectile fired horizontally along the Earth's surface. What terms in the effective force can be neglected?

   (d) Suppose you wanted to calculate the effective force on a small block of mass m placed on a frictionless turntable rotating with a time-dependent angular velocity $\omega(t)$. What terms in the effective force can be neglected?

3. A plumb line is carried along in a moving train, with m the mass of the plumb bob. Neglect any effects due to the rotation of the Earth and work in the noninertial frame of reference of the train.

   (a) Find the tension in the cord and the deflection from the local vertical if the train is moving with constant acceleration a_0.

   (b) Find the tension in the cord and the deflection from the local vertical if the train is rounding a curve of radius ρ with constant speed v_0.

4. A bead on a rotating rod is free to slide without friction. The rod has a length L and rotates about its end with angular velocity ω. The bead is initially released from rest (relative to the rod) at the midpoint of the rod.

   (a) Find the displacement of the bead along the wire as a function of time.

   (b) Find the time when the bead leaves the end of the rod.

   (c) Find the velocity (relative to the rod) of the bead when it leaves the end of the rod.

5. Here is a "thought experiment" for you to consider. Suppose you are in a small sailboat of mass M at the Earth's equator. At the equator there is very little wind (this is known as the "equatorial doldrums"), so your sailboat is, more or less, sitting still. You have a small anchor of mass m on deck and a single mast of height h in the middle of the boat. How can you use the anchor to put the boat into motion? In which direction will the boat move?

6. Does water really flow in the other direction when you flush a toilet in the southern hemisphere? What (if anything) does the Coriolis force have to do with this?

7. We are presently at a latitude λ (with respect to the equator) and Earth is rotating with constant angular velocity ω. Consider the following two scenarios: Scenario A: A particle is thrown upward with initial speed v_0. Scenario B: An identical particle is dropped (at rest) from the maximum height of the particle in Scenario A. Circle all the true statements regarding the Coriolis deflection assuming that the particles have landed for a) and b), .

   (a) The magnitude is greater in A than in B.

   (b) The direction in A and B are the same.

   (c) The direction in A does not change throughout flight.

Problems

1. If a projectile is fired due east from a point on the surface of the Earth at a northern latitude λ with a velocity of magnitude V_0 and at an inclination to the horizontal of α, show that the lateral deflection when the projectile strikes the Earth is

$$d = \frac{4V_0^3}{g^2}\omega \sin \lambda \sin^2 \alpha \cos \alpha \qquad (12.95)$$

where ω is the rotation frequency of the Earth.

2. Obtain an expression for the angular deviation of a particle projected from the North Pole in a path that lies close to the surface of the earth. Is the deviation significant for a missile that makes a 4800-km flight in 10 minutes? What is the "miss distance" if the missile is aimed directly at the target? Is the miss difference greater for a 19300-km flight at the same velocity?

3. An automobile drag racer drives a car with acceleration a and instantaneous velocity v. The tires of radius r_0 are not slipping. Derive which point on the tire has the greatest acceleration relative to the ground. What is this acceleration?

4. Shot towers were popular in the eighteenth and nineteenth centuries for dropping melted lead down tall towers to form spheres for bullets. The lead solidified while falling and often landed in water to cool the lead bullets. Many such shot towers were built in New York State. Assume a shot tower was constructed at latitude $42°N$, and that the lead fell a distance of $27m$. In what direction and by how far did the lead bullets land from the direct vertical?

Chapter 13

Rigid-body rotation

13.1 Introduction

Rigid-body rotation features prominently in science, engineering, and sports. Prior chapters have focussed primarily on motion of point particles. This chapter extends the discussion to motion of finite-sized rigid bodies. A rigid body is a collection of particles where the relative separations remain rigidly fixed. In real life, there is always some motion between individual atoms, but usually this microscopic motion can be neglected when describing macroscopic properties. Note that the concept of perfect rigidity has limitations in the theory of relativity since information cannot travel faster than the velocity of light, and thus signals cannot be transmitted instantaneously between the ends of a rigid body which is implied if the body had perfect rigidity.

The description of rigid-body rotation is most easily handled by specifying the properties of the body in the rotating body-fixed coordinate frame whereas the observables are measured in the stationary inertial laboratory coordinate frame. In the body-fixed coordinate frame, the primary observable for classical mechanics is the inertia tensor of the rigid body which is well defined and independent of the rotational motion. By contrast, in the stationary inertial frame the observables depend sensitively on the details of the rotational motion. For example, when observed in the stationary fixed frame, rapid rotation of a long thin cylindrical pencil about the longitudinal symmetry axis gives a time-averaged shape of the pencil that looks like a thin cylinder, whereas the time-averaged shape is a flat disk for rotation about an axis perpendicular to the symmetry axis of the pencil. In spite of this, the pencil always has the same unique inertia tensor in the body-fixed frame. Thus the best solution for describing rotation of a rigid body is to use a rotation matrix that transforms from the stationary fixed frame to the instantaneous body-fixed frame for which the moment of inertia tensor can be evaluated. Moreover, the problem can be greatly simplified by transforming to a body-fixed coordinate frame that is aligned with any symmetry axes of the body since then the inertia tensor can be diagonal; this is called a principal axis system.

Rigid-body rotation can be broken into the following two classifications.

1) Rotation about a fixed axis:

A body can be constrained to rotate about an axis that has a fixed location and orientation relative to the body. The hinged door is a typical example. Rotation about a fixed axis is straightforward since the axis of rotation, plus the moment of inertia about this axis, are well defined and this case was discussed in chapter 2.12.7.

2) Rotation about a point

A body can be constrained to rotate about a fixed point of the body but the orientation of this rotation axis about this point is unconstrained. One example is rotation of an object flying freely in space which can rotate about the center of mass with any orientation. Another example is a child's spinning top which has one point constrained to touch the ground but the orientation of the rotation axis is undefined.

The prior discussion in chapter 2.12.7 showed that rigid-body rotation is more complicated than assumed in introductory treatments of rigid-body rotation. It is necessary to expand the concept of moment of inertia to the concept of the inertia tensor, plus the fact that the angular momentum may not point along the rotation axis. The most general case requires consideration of rotation about a body-fixed point where the orientation of the axis of rotation is unconstrained. The concept of the inertia tensor of a rotating body is crucial for describing rigid-body motion. It will be shown that working in the body-fixed coordinate frame of a rotating body allows a description of the equations of motion in terms of the inertia tensor for a given point of the body, and that it is possible to rotate the body-fixed coordinate system into a **principal axis** system where the inertia tensor is diagonal. For any principal axis, the angular momentum is parallel to the angular velocity if it is aligned with a principal axis. The use of a principal axis system greatly simplifies treatment of rigid-body rotation and exploits the powerful and elegant matrix algebra mentioned in appendix A.

The following discussion of rigid-body rotation is broken into three topics, (1) the inertia tensor of the rigid body, (2) the transformation between the rotating body-fixed coordinate system and the laboratory frame, i.e., the Euler angles specifying the orientation of the body-fixed coordinate frame with respect to the laboratory frame, and (3) Lagrange and Euler's equations of motion for rigid-bodies. This is followed by a discussion of practical applications.

13.2   Rigid-body coordinates

Motion of a rigid body is a special case for motion of the N-body system when the relative positions of the N bodies are related. It was shown in chapter 2 that the motion of a rigid body can be broken into a combination of a linear translation of some point in the body, plus rotation of the body about an axis through that point. This is called **Chasles' Theorem.** Thus the position of every particle in the rigid body is fixed with respect to one point in the body. If the fixed point of the body is chosen to be the center of mass, then, as discussed in chapter 2, it is possible to separate the kinetic energy, linear momentum, and angular momentum into the center-of-mass motion, plus the motion about the center of mass. Thus the behavior of the body can be described completely using only six independent coordinates governed by six equations of motion, three for translation and three for rotation.

Referred to an inertial frame, the translational motion of the center of mass is governed by

$$\mathbf{F}^E = \frac{d\mathbf{P}}{dt} \tag{13.1}$$

while the rotational motion about the center of mass is determined by

$$\mathbf{N}^E = \frac{d\mathbf{L}}{dt} \tag{13.2}$$

where the external force $\mathbf{F}^E$ and external torque $\mathbf{N}^E$ are identified separately from the internal forces acting between the particles in the rigid body. It will be assumed that the internal forces are central and thus do not contribute to the angular momentum.

The location of any fixed point in the body, such as the center of mass, can be specified by three generalized cartesian coordinates with respect to a fixed frame. The rotation of the body-fixed axis system about this fixed point in the body can be described in terms of three independent angles with respect to the fixed frame. There are several possible sets of orthogonal angles that can be used to describe the rotation. This book uses the Euler angles ϕ, θ, ψ which correspond first to a rotation ϕ about the z-axis, then a rotation θ about the x axis subsequent to the first rotation, and finally a rotation ψ about the new z axis following the first two rotations. The Euler angles will be discussed in detail following introduction of the inertia tensor and angular momentum.

13.3   Rigid-body rotation about a body-fixed point

With respect to some point O fixed in the body coordinate system, the angular momentum of the body α is given by

$$\mathbf{L} = \sum_i^n \mathbf{L}_i = \sum_i^n \mathbf{r}_i \times \mathbf{p}_i \tag{13.3}$$

There are two especially convenient choices for the fixed point O. If no point in the body is fixed with respect to an inertial coordinate system, then it is best to choose O as the center of mass. If one point of the body is fixed with respect to a fixed inertial coordinate system, such as a point on the ground where a child's spinning top touches, then it is best to choose this stationary point as the body-fixed point O.

Consider a rigid body composed of N particles of mass m_α where $\alpha = 1, 2, 3, ...N$. As discussed in chapter 12.4, if the body rotates with an instantaneous angular velocity $\boldsymbol{\omega}$ about some fixed point, with respect to the body-fixed coordinate system, and this point has an instantaneous translational velocity $\mathbf{V}$ with respect to the fixed (inertial) coordinate system, see figure 13.1, then the instantaneous velocity $\mathbf{v}_\alpha$ of the αth particle in the fixed frame of reference is given by

$$\mathbf{v}_\alpha = \mathbf{V} + \mathbf{v}''_\alpha + \boldsymbol{\omega} \times \mathbf{r}'_\alpha \qquad (13.4)$$

However, for a rigid body, the velocity of a body-fixed point with respect to the body is zero, that is $\mathbf{v}''_\alpha = 0$, thus

$$\mathbf{v}_\alpha = \mathbf{V} + \boldsymbol{\omega} \times \mathbf{r}'_\alpha \qquad (13.5)$$

Consider the translational velocity of the body-fixed point O to be zero, i.e. $\mathbf{V} = 0$ and let $\mathbf{R} = 0$, then $\mathbf{r}_\alpha = \mathbf{r}'_\alpha$. These assumptions allow the linear momentum of the particle α to be written as

$$\mathbf{p}_\alpha = m_\alpha \mathbf{v}_\alpha = m_\alpha \boldsymbol{\omega} \times \mathbf{r}_\alpha \qquad (13.6)$$

Therefore

$$\mathbf{L} = \sum_\alpha^N \mathbf{r}_\alpha \times \mathbf{p}_\alpha = \sum_\alpha^N m_\alpha \mathbf{r}_\alpha \times (\boldsymbol{\omega} \times \mathbf{r}_\alpha) \qquad (13.7)$$

Using the vector identity

$$\mathbf{A} \times (\mathbf{B} \times \mathbf{A}) = A^2 \mathbf{B} - \mathbf{A} (\mathbf{A} \cdot \mathbf{B})$$

leads to

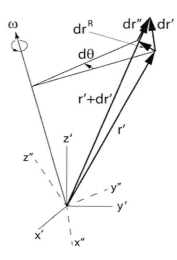

Figure 13.1: Infinitessimal displacement dr', in the primed frame, broken into a part dr^R due to rotation of the primed frame plus a part dr'' due to displacement with respect to this rotating frame.

$$\mathbf{L} = \sum_\alpha^N m_\alpha \left[r_\alpha^2 \boldsymbol{\omega} - \mathbf{r}_\alpha (\mathbf{r}_\alpha \cdot \boldsymbol{\omega}) \right] \qquad (13.8)$$

The angular momentum can be expressed in terms of components of $\boldsymbol{\omega}$ and $\mathbf{r}'_\alpha$ relative to the body-fixed frame. The following formulae can be written more compactly if $\mathbf{r}_\alpha = (x_\alpha, y_\alpha, z_\alpha)$, in the rotating body-fixed frame, is written in the form $\mathbf{r}_\alpha = (x_{\alpha,1}, x_{\alpha,2}, x_{\alpha,3})$ where the axes are defined by the numbers $1, 2, 3$ rather than x, y, z. In this notation, the angular momentum is written in component form as

$$L_i = \sum_\alpha^N m_\alpha \left[\omega_i \sum_k x_{\alpha,k}^2 - x_{\alpha,i} \left(\sum_j x_{\alpha,j} \omega_j \right) \right] \qquad (13.9)$$

Assume the Kronecker delta relation

$$\omega_i = \sum_j^3 \omega_j \delta_{ij} \qquad (13.10)$$

where

$$\begin{aligned} \delta_{ij} &= 1 & i = j \\ \delta_{ij} &= 0 & i \neq j \end{aligned}$$

Substitute (13.10) in (13.9) gives

$$\begin{aligned} L_i &= \sum_\alpha^N m_\alpha \sum_j \left[\omega_j \delta_{ij} \sum_k x_{\alpha,k}^2 - \omega_j x_{\alpha,i} x_{\alpha,j} \right] \\ &= \sum_j^3 \omega_j \left[\sum_\alpha^N m_\alpha \left(\delta_{ij} \sum_k x_{\alpha,k}^2 - x_{\alpha,i} x_{\alpha,j} \right) \right] \qquad (13.11) \end{aligned}$$

13.4   Inertia tensor

The square bracket term in (13.11) is called the **moment of inertia tensor I** which is usually referred to as the **inertia tensor**

$$I_{ij} \equiv \sum_{\alpha}^{N} m_{\alpha} \left[\delta_{ij} \left(\sum_{k}^{3} x_{\alpha,k}^2 \right) - x_{\alpha,i} x_{\alpha,j} \right] \tag{13.12}$$

In most cases it is more useful to express the components of the inertia tensor in an integral form over the mass distribution rather than a summation for N discrete bodies. That is,

$$I_{ij} = \int \rho\left(\mathbf{r}'\right) \left(\delta_{ij} \left(\sum_{k}^{3} x_k^2 \right) - x_i x_j \right) dV \tag{13.13}$$

The inertia tensor is easier to understand when written in cartesian coordinates $\mathbf{r}'_{\alpha} = (x_{\alpha}, y_{\alpha}, z_{\alpha})$ rather than in the form $\mathbf{r}'_{\alpha} = (x_{\alpha,1}, x_{\alpha,2}, x_{\alpha,3})$. Then, the diagonal **moments of inertia** of the inertia tensor are

$$I_{xx} \equiv \sum_{\alpha}^{N} m_{\alpha} \left[x_{\alpha}^2 + y_{\alpha}^2 + z_{\alpha}^2 - x_{\alpha}^2 \right] = \sum_{\alpha}^{N} m_{\alpha} \left[y_{\alpha}^2 + z_{\alpha}^2 \right] \tag{13.14}$$

$$I_{yy} \equiv \sum_{\alpha}^{N} m_{\alpha} \left[x_{\alpha}^2 + y_{\alpha}^2 + z_{\alpha}^2 - y_{\alpha}^2 \right] = \sum_{\alpha}^{N} m_{\alpha} \left[x_{\alpha}^2 + z_{\alpha}^2 \right]$$

$$I_{zz} \equiv \sum_{\alpha}^{N} m_{\alpha} \left[x_{\alpha}^2 + y_{\alpha}^2 + z_{\alpha}^2 - z_{\alpha}^2 \right] = \sum_{\alpha}^{N} m_{\alpha} \left[x_{\alpha}^2 + y_{\alpha}^2 \right]$$

while the off-diagonal **products of inertia** are

$$I_{yx} = I_{xy} \equiv - \sum_{\alpha}^{N} m_{\alpha} \left[x_{\alpha} y_{\alpha} \right] \tag{13.15}$$

$$I_{zx} = I_{xz} \equiv - \sum_{\alpha}^{N} m_{\alpha} \left[x_{\alpha} z_{\alpha} \right]$$

$$I_{zy} = I_{yz} \equiv - \sum_{\alpha}^{N} m_{\alpha} \left[y_{\alpha} z_{\alpha} \right]$$

Note that the products of inertia are symmetric in that

$$I_{ij} = I_{ji} \tag{13.16}$$

The above notation for the inertia tensor allows the angular momentum (13.12) to be written as

$$L_i = \sum_{j}^{3} I_{ij} \omega_j \tag{13.17}$$

Expanded in cartesian coordinates

$$\begin{aligned} L_x &= I_{xx}\omega_x + I_{xy}\omega_y + I_{xz}\omega_z \\ L_y &= I_{yx}\omega_x + I_{yy}\omega_y + I_{yz}\omega_z \\ L_z &= I_{zx}\omega_x + I_{zy}\omega_y + I_{zz}\omega_z \end{aligned} \tag{13.18}$$

Note that every fixed point in a body has a specific inertia tensor. The components of the inertia tensor at a specified point depend on the orientation of the coordinate frame whose origin is located at the specified fixed point. For example, the inertia tensor for a cube is very different when the fixed point is at the center of mass compared with when the fixed point is at a corner of the cube.

13.5 Matrix and tensor formulations of rigid-body rotation

The prior notation is clumsy and can be streamlined by use of matrix methods. Write the inertia tensor in a matrix form as

$$\{\mathbb{I}\} = \begin{pmatrix} I_{11} & I_{12} & I_{13} \\ I_{21} & I_{22} & I_{23} \\ I_{31} & I_{32} & I_{33} \end{pmatrix} \tag{13.19}$$

The angular velocity and angular momentum both can be written as a column vectors, that is

$$\boldsymbol{\omega} = \begin{pmatrix} \omega_1 \\ \omega_2 \\ \omega_3 \end{pmatrix} \qquad \mathbf{L} = \begin{pmatrix} L_1 \\ L_2 \\ L_3 \end{pmatrix} \tag{13.20}$$

As discussed in appendix $E2$, equation (13.18) now can be written in tensor notation as an inner product of the form

$$\mathbf{L} = \{\mathbb{I}\} \cdot \boldsymbol{\omega} \tag{13.21}$$

Note that the above notation uses boldface for the inertia tensor $\mathbb{I}$, implying a rank-2 tensor representation, while the angular velocity $\boldsymbol{\omega}$ and the angular momentum $\mathbf{L}$ are written as column vectors. The inertia tensor is a 9-component rank-2 tensor defined as the ratio of the angular momentum vector $\mathbf{L}$ and the angular velocity $\boldsymbol{\omega}$.

$$\{\mathbb{I}\} = \frac{\mathbf{L}}{\boldsymbol{\omega}} \tag{13.22}$$

Note that, as described in appendix E, the inner product of a vector $\boldsymbol{\omega}$, which is the rank 1 tensor, and a rank 2 tensor $\{\mathbb{I}\}$, leads to the vector $\mathbf{L}$. This compact notation exploits the fact that the matrix and tensor representation are completely equivalent, and are ideally suited to the description of rigid-body rotation.

13.6 Principal axis system

The *inertia tensor is a real symmetric matrix* because of the symmetry given by equation (13.16). A property of real symmetric matrices is that there exists an orientation of the coordinate frame, with its origin at the chosen body-fixed point O, such that the inertia tensor is diagonal. The coordinate system for which the inertia tensor is diagonal is called the **Principal axis system** which has three perpendicular **principal axes.** Thus, in the principal axis system, the inertia tensor has the form

$$\{\mathbf{I}\} = \begin{pmatrix} I_{11} & 0 & 0 \\ 0 & I_{22} & 0 \\ 0 & 0 & I_{33} \end{pmatrix} \tag{13.23}$$

where I_{jj} are real numbers, which are called the **principal moments of inertia** of the body, and are usually written as I_j. When the angular velocity vector $\boldsymbol{\omega}$ points along any principal axis unit vector $\hat{\jmath}$, then the angular momentum $\mathbf{L}$ is parallel to $\boldsymbol{\omega}$ and the magnitude of the principal moment of inertia about this principal axis is given by the relation

$$L_j \hat{\jmath} = I_j \omega_j \hat{\jmath} \tag{13.24}$$

The principal axes are fixed relative to the shape of the rigid body and they are invariant to the orientation of the body-fixed coordinate system used to evaluate the inertia tensor. The advantage of having the body-fixed coordinate frame aligned with the principal axis coordinate frame is that then the inertia tensor is diagonal, which greatly simplifies the matrix algebra. Even when the body-fixed coordinate system is not aligned with the principal axis frame, if the angular velocity is specified to point along a principal axis then the corresponding moment of inertia will be given by (13.24).

In principle it is possible to locate the principal axes by varying the orientation of the angular velocity vector $\boldsymbol{\omega}$ to find those orientations for which the angular momentum $\mathbf{L}$ and angular velocity $\boldsymbol{\omega}$ are parallel which characterizes the principal axes. However, the best approach is to diagonalize the inertia tensor.

13.7   Diagonalize the inertia tensor

Finding the three principal axes involves diagonalizing the inertia tensor, which is the classic eigenvalue problem discussed in appendix A. Solution of the eigenvalue problem for rigid-body motion corresponds to a rotation of the coordinate frame to the principal axes resulting in the matrix

$$\{\mathbf{I}\} \cdot \boldsymbol{\omega} = I\boldsymbol{\omega} \tag{13.25}$$

where I comprises the three-valued eigenvalues, while the corresponding vector $\boldsymbol{\omega}$ is the eigenvector. Appendix $A.4$ gives the solution of the matrix relation

$$\{\mathbf{I}\} \cdot \boldsymbol{\omega} = I\{\mathbb{I}\}\boldsymbol{\omega} \tag{13.26}$$

where I are three-valued eigen values for the principal axis moments of inertia, and $\{\mathbb{I}\}$ is the unity tensor, equation $A.2.4$.

$$\{\mathbb{I}\} \equiv \left\{ \begin{array}{ccc} 1 & 0 & 0 \\ 0 & 1 & 0 \\ 0 & 0 & 1 \end{array} \right\} \tag{13.27}$$

Rewriting (13.26) gives

$$(\{\mathbf{I}\} - I\{\mathbb{I}\}) \cdot \boldsymbol{\omega} = 0 \tag{13.28}$$

This is a matrix equation of the form $\mathbf{A} \cdot \boldsymbol{\omega} = 0$ where $\mathbf{A}$ is a 3×3 matrix and $\boldsymbol{\omega}$ is a vector with values $\omega_x, \omega_y, \omega_z$. The matrix equation $\mathbf{A} \cdot \boldsymbol{\omega} = 0$ really corresponds to three simultaneous equations for the three numbers $\omega_x, \omega_y, \omega_z$. It is a well-known property of equations like (13.28) that they have a non-zero solution if, and only if, the determinant $\det(\mathbf{A})$ is zero, that is

$$\det(\mathbf{I} - I\mathbb{I}) = 0 \tag{13.29}$$

This is called the **characteristic equation**, or **secular equation** for the matrix $\mathbf{I}$. The determinant involved is a cubic equation in the value of I that gives the three principal moments of inertia. Inserting one of the three values of I into equation (13.17) gives the corresponding eigenvector ω. Applying the above eigenvalue problem to rigid-body rotation corresponds to requiring that some arbitrary set of body-fixed axes be the principal axes of inertia. This is obtained by rotating the body-fixed axis system such that

$$
\begin{aligned}
L_1 &= I_{11}\omega_1 + I_{12}\omega_2 + I_{13}\omega_3 = I\omega_1 \\
L_2 &= I_{21}\omega_1 + I_{22}\omega_2 + I_{23}\omega_3 = I\omega_2 \\
L_3 &= I_{31}\omega_1 + I_{32}\omega_2 + I_{33}\omega_3 = I\omega_3
\end{aligned} \tag{13.30}
$$

or

$$
\begin{aligned}
(I_{11} - I)\,\omega_1 + I_{12}\omega_2 + I_{13}\omega_3 &= 0 \\
I_{21}\omega_1 + (I_{22} - I)\,\omega_2 + I_{23}\omega_3 &= 0 \\
I_{31}\omega_1 + I_{32}\omega_2 + (I_{33} - I)\,\omega_3 &= 0
\end{aligned} \tag{13.31}
$$

These equations have a non-trivial solution for the ratios $\omega_1 : \omega_2 : \omega_3$ since the determinant vanishes, that is

$$
\begin{vmatrix}
(I_{11} - I) & I_{12} & I_{13} \\
I_{21} & (I_{22} - I) & I_{23} \\
I_{31} & I_{32} & (I_{33} - I)
\end{vmatrix} = 0 \tag{13.32}
$$

The expansion of this determinant leads to a cubic equation with three roots for I. This is the **secular equation** for I whose eigenvalues are the **principal moments of inertia**.

The directions of the **principal axes**, that is the eigenvectors, can be found by substituting the corresponding solution for I into the prior equation. Thus for eigensolution I_1 the eigenvector is given by solving

$$
\begin{aligned}
(I_{11} - I_1)\,\omega_{11} + I_{12}\omega_{21} + I_{13}\omega_{31} &= 0 \\
I_{21}\omega_{11} + (I_{22} - I_1)\,\omega_{21} + I_{23}\omega_{31} &= 0 \\
I_{31}\omega_{11} + I_{32}\omega_{21} + (I_{33} - I_1)\,\omega_{31} &= 0
\end{aligned} \tag{13.33}
$$

These equations are solved for the ratios $\omega_{11} : \omega_{21} : \omega_{31}$ which are the direction numbers of the principle axis system corresponding to solution I_1. This principal axis system is defined relative to the original coordinate system. This procedure is repeated to find the orientation of the other two mutually perpendicular principal axes.

13.8 Parallel-axis theorem

The values of the components of the inertia tensor depend on both the location and the orientation about which the body rotates relative to the body-fixed coordinate system. The parallel-axis theorem is valuable for relating the inertia tensor for rotation about parallel axes passing through different points fixed with respect to the rigid body. For example, one may wish to relate the inertia tensor through the center of mass to another location that is constrained to remain stationary, like the tip of the spinning top.

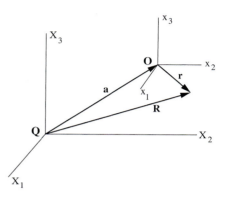

    Consider the mass α at the location $\mathbf{r} = (x_1, x_2, x_3)$ with respect to the origin of the center of mass body-fixed coordinate system O. Transform to an arbitrary but parallel body-fixed coordinate system Q, that is, the coordinate axes have the same orientation as the center of mass coordinate system. The location of the mass α with respect to this arbitrary coordinate system is $\mathbf{R} = (X_1, X_2, X_3)$. That is, the general vectors for the two coordinates systems are related by

$$\mathbf{R} = \mathbf{a} + \mathbf{r} \tag{13.34}$$

Figure 13.2: Transformation between two parallel body-coordinate systems, O and Q.

where $\mathbf{a}$ is the vector connecting the origins of the coordinate systems O and Q illustrated in figure 13.2. The elements of the inertia tensor with respect to axis system Q, are given by equation 13.12 to be

$$J_{ij} \equiv \sum_{\alpha}^{N} m_{\alpha} \left[\delta_{ij} \left(\sum_{k}^{3} X_{\alpha,k}^2 \right) - X_{\alpha,i} X_{\alpha,j} \right] \tag{13.35}$$

The components along the three axes for each of the two coordinate systems are related by

$$X_i = a_i + x_i \tag{13.36}$$

Substituting these into the above inertia tensor relation gives

$$
\begin{aligned}
J_{ij} &= \sum_{\alpha}^{N} m_{\alpha} \left[\delta_{ij} \left(\sum_{k}^{3} (x_{\alpha,k} + a_i)^2 \right) - (x_{\alpha,i} + a_i)(x_{\alpha,j} + a_i) \right] \\
&= \sum_{\alpha}^{N} m_{\alpha} \left[\delta_{ij} \left(\sum_{k}^{3} x_{\alpha,k}^2 \right) - x_{\alpha,i} x_{\alpha,j} \right] + \sum_{\alpha}^{N} m_{\alpha} \left[\delta_{ij} \left(\sum_{k}^{3} (2x_{\alpha,k} a_k + a_k^2) \right) - (a_i x_{\alpha,j} + a_j x_{\alpha,i} + a_i a_j) \right]
\end{aligned}
\tag{13.37}
$$

The first summation on the right-hand side corresponds to the elements I_{ij} of the inertia tensor in the center-of-mass frame. Thus the terms can be regrouped to give

$$J_{ij} \equiv I_{ij} + \sum_{\alpha}^{N} m_{\alpha} \left(\delta_{ij} \sum_{k}^{3} a_k^2 - a_i a_j \right) + \sum_{\alpha}^{N} m_{\alpha} \left[2\delta_{ij} \sum_{k}^{3} x_{\alpha,k} a_k - a_i x_{\alpha,j} - a_j x_{\alpha,i} \right] \tag{13.38}$$

However, each term in the last bracket involves a sum of the form $\sum_{\alpha}^{N} m_{\alpha} x_{\alpha,k}$. Take the coordinate system O to be with respect to the center of mass for which

$$\sum_{\alpha}^{N} m_{\alpha} \mathbf{r}' = 0 \tag{13.39}$$

This also applies to each component k, that is

$$\sum_{\alpha}^{N} m_{\alpha} x_{\alpha,k} = 0 \tag{13.40}$$

Therefore all of the terms in the last bracket cancel leaving

$$J_{ij} \equiv I_{ij} + \sum_{\alpha}^{N} m_{\alpha} \left(\delta_{ij} \sum_{k}^{3} a_k^2 - a_i a_j \right) \tag{13.41}$$

But $\sum_{\alpha}^{N} m_{\alpha} = M$ and $\sum_{k}^{3} a_k^2 = a^2$, thus

$$J_{ij} \equiv I_{ij} + M \left(a^2 \delta_{ij} - a_i a_j \right) \tag{13.42}$$

where I_{ij} is the center-of-mass inertia tensor. This is the general form of Steiner's **parallel-axis theorem.**
    As an example, the moment of inertia around the X_1 axis is given by

$$J_{11} \equiv I_{11} + M \left(\left(a_1^2 + a_2^2 + a_3^2 \right) \delta_{11} - a_1^2 \right) = I_{11} + M \left(a_2^2 + a_3^2 \right) \tag{13.43}$$

which corresponds to the elementary statement that the *difference* in the moments of inertia equals the mass of the body multiplied by the square of the distance between the parallel axes, x_1, X_1. Note that the minimum moment of inertia of a body is I_{ij} which is about the center of mass.

13.1 *Example: Inertia tensor of a solid cube rotating about the center of mass.*

The complicated expressions for the inertia tensor can be understood using the example of a uniform solid cube with side b, density ρ, and mass $M = \rho b^3$, rotating about different axes. Assume that the origin of the coordinate system O is at the center of mass with the axes perpendicular to the centers of the faces of the cube.

The components of the inertia tensor can be calculated using (13.13) written as an integral over the mass distribution rather than a summation.

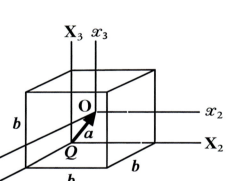

$$I_{ij} = \int \rho \left(\mathbf{r}' \right) \left(\delta_{ij} \left(\sum_{k}^{3} x_k^2 \right) - x_i x_j \right) dV$$

Thus

$$
\begin{aligned}
I_{11} &= \rho \int_{-b/2}^{b/2} \int_{-b/2}^{b/2} \int_{-b/2}^{b/2} \left(x_2^2 + x_3^2 \right) dx_3 dx_2 dx_1 \\
&= \frac{1}{6} \rho b^5 = \frac{1}{6} M b^2 = I_{22} = I_{33}
\end{aligned}
$$

Inertia tensor of a uniform solid cube of side b about the center of mass O and a corner of the cube Q. The vector a is the vector distance between O and Q.

By symmetry the diagonal moments of inertia about each face are identical. Similarly the products of inertia are given by

$$I_{12} = -\rho \int_{-b/2}^{b/2} \int_{-b/2}^{b/2} \int_{-b/2}^{b/2} \left(x_1 x_2 \right) dx_3 dx_2 dx_1 = 0$$

Thus the inertia tensor is given by

$$\mathbf{I}^{cm} = \frac{1}{6} M b^2 \begin{pmatrix} 1 & 0 & 0 \\ 0 & 1 & 0 \\ 0 & 0 & 1 \end{pmatrix}$$

Note that this inertia tensor is diagonal implying that this is the principal axis system. In this case all three principal moments of inertia are identical and perpendicular to the centers of the faces of the cube. This is as expected from the symmetry of the cubic geometry.

13.2 Example: Inertia tensor of about a corner of a solid cube.

a) Direct calculation   Let one corner of the cube be the origin of the coordinate system Q and assume that the three adjacent sides of the cube lie along the coordinate axes. The components of the inertia tensor can be calculated using (13.13). Thus

$$I_{11} = \rho \int_0^b \int_0^b \int_0^b \left(x_2^2 + x_3^2\right) dx_3 dx_2 dx_1 = \frac{2}{3}\rho b^5 = \frac{2}{3}Mb^2$$

$$I_{12} = -\rho \int_0^b \int_0^b \int_0^b \left(x_1 x_2\right) dx_3 dx_2 dx_1 = -\frac{1}{4}\rho b^5 = -\frac{1}{4}Mb^2$$

Thus, evaluating all the nine components gives

$$\mathbf{I}^{corner} = \frac{1}{12}Mb^2 \begin{pmatrix} 8 & -3 & -3 \\ -3 & 8 & -3 \\ -3 & -3 & 8 \end{pmatrix}$$

b) Parallel-axis theorem   This inertia tensor also can be calculated using the parallel-axis theorem to relate the moment of inertia about the corner, to that at the center of mass. As shown in the figure, the vector a has components

$$a_1 = a_2 = a_3 = \frac{b}{2}$$

Applying the parallel-axis theorem gives

$$J_{11} = I_{11} + M\left(a^2 - a_1^2\right) = I_{11} + M\left(a_2^2 + a_3^2\right) = \frac{1}{6}Mb^2 + \frac{1}{2}Mb^2 = \frac{2}{3}Mb^2$$

and similarly for J_{22} and J_{33}.   The off-diagonal terms are given by

$$J_{12} = I_{12} + M\left(-a_1 a_2\right) = -\frac{1}{4}Mb^2$$

Thus the inertia tensor, transposed from the center of mass, to the corner of the cube is

$$\mathbf{I}^{corner} = \begin{pmatrix} \frac{2}{3}Mb^2 & -\frac{1}{4}Mb^2 & -\frac{1}{4}Mb^2 \\ -\frac{1}{4}Mb^2 & \frac{2}{3}Mb^2 & -\frac{1}{4}Mb^2 \\ -\frac{1}{4}Mb^2 & -\frac{1}{4}Mb^2 & \frac{2}{3}Mb^2 \end{pmatrix} = \frac{1}{12}Mb^2 \begin{pmatrix} 8 & -3 & -3 \\ -3 & 8 & -3 \\ -3 & -3 & 8 \end{pmatrix}$$

This inertia tensor about the corner of the cube, is the same as that obtained by direct integration.

c) Principal moments of inertia   The coordinate axis frame used for rotation about the corner of the cube is not a principal axis frame.   Therefore let us diagonalize the inertia tensor to find the principal axis frame the principal moments of inertia about a corner.   To achieve this requires solving the secular determinant

$$\begin{vmatrix} \left(\frac{2}{3}Mb^2 - I\right) & -\frac{1}{4}Mb^2 & -\frac{1}{4}Mb^2 \\ -\frac{1}{4}Mb^2 & \left(\frac{2}{3}Mb^2 - I\right) & -\frac{1}{4}Mb^2 \\ -\frac{1}{4}Mb^2 & -\frac{1}{4}Mb^2 & \left(\frac{2}{3}Mb^2 - I\right) \end{vmatrix} = 0$$

The value of a determinant is not affected by adding or subtracting any row or column from any other row or column. Subtract row 1 from row 2 gives

$$\begin{vmatrix} \left(\frac{2}{3}Mb^2 - I\right) & -\frac{1}{4}Mb^2 & -\frac{1}{4}Mb^2 \\ -\frac{11}{12}Mb^2 + I & \left(\frac{11}{12}Mb^2 - I\right) & 0 \\ -\frac{1}{4}Mb^2 & -\frac{1}{4}Mb^2 & \left(\frac{2}{3}Mb^2 - I\right) \end{vmatrix} = 0$$

The determinant of this matrix is straightforward to evaluate and equals

$$\left(\frac{1}{6}Mb^2 - I\right)\left(\frac{11}{12}Mb^2 - I\right)\left(\frac{11}{12}Mb^2 - I\right) = 0$$

Thus the roots are

$$\mathbf{I}^{corner} = \begin{pmatrix} \frac{1}{6}Mb^2 & 0 & 0 \\ 0 & \frac{11}{12}Mb^2 & 0 \\ 0 & 0 & \frac{11}{12}Mb^2 \end{pmatrix}$$

The identical roots $I_{22} = I_{33} = \frac{11}{12}Mb^2$ imply that the principal axis associated with I_{11} must be a symmetry axis. The orientation can be found by substituting I_{11} into the above equation

$$(\{\mathbf{I}\} - I\{\mathbb{I}\}) \cdot \boldsymbol{\omega} = \frac{1}{12}Mb^2 \begin{pmatrix} 6 & -3 & -3 \\ -3 & 6 & -3 \\ -3 & -3 & 6 \end{pmatrix} \begin{pmatrix} \omega_{11} \\ \omega_{21} \\ \omega_{31} \end{pmatrix} = 0$$

where the second subscript 1 attached to ω_i signifies that this solution corresponds to I_{11}. This gives

$$\begin{aligned} 2\omega_{11} - \omega_{21} - \omega_{31} &= 0 \\ -\omega_{11} + 2\omega_{21} - \omega_{31} &= 0 \\ -\omega_{11} - \omega_{21} + 2\omega_{31} &= 0 \end{aligned}$$

Solving these three equations gives the unit vector for the first principal axis for which $I_{11} = \frac{1}{6}Mb^2$ to be $\hat{\mathbf{e}}_1 = \frac{1}{\sqrt{3}} \begin{pmatrix} 1 \\ 1 \\ 1 \end{pmatrix}$. This can be repeated to find the other two principal axes by substituting $I_{22} = \frac{11}{12}Mb^2$. This gives for the second principal moment I_{22}

$$(\{\mathbf{I}\} - I\{\mathbb{I}\}) \cdot \boldsymbol{\omega} = \frac{1}{12}Mb^2 \begin{pmatrix} -3 & -3 & -3 \\ -3 & -3 & -3 \\ -3 & -3 & -3 \end{pmatrix} \begin{pmatrix} \omega_{12} \\ \omega_{22} \\ \omega_{32} \end{pmatrix} = 0$$

This results in three identical equations for the components of ω but all three equations are the same, namely

$$\omega_{12} + \omega_{22} + \omega_{32} = 0$$

This does not uniquely determine the direction of ω. However, it does imply that ω_2 corresponding to the second principal axis has the property that

$$\hat{\boldsymbol{\omega}} \cdot \hat{\mathbf{e}}_1 = 0$$

that is, any direction of $\hat{\mathbf{e}}_2$ that is perpendicular to $\hat{\mathbf{e}}_1$ is acceptable. In other words; any two orthogonal unit vectors $\hat{\mathbf{e}}_2$ and $\hat{\mathbf{e}}_3$ that are perpendicular to $\hat{\mathbf{e}}_1$ are acceptable. This ambiguity exists whenever two eigenvalues are equal; the three principal axes are only uniquely defined if all three eigenvalues are different. The same ambiguity exist when all three eigenvalues are identical as occurs for the principal moments of inertia about the center-of-mass of a uniform solid cube. This explains why the principal moment of inertia for the diagonal of the cube, that passes through the center of mass, has the same moment as when the principal axes pass through the center of the faces of the cube.

13.9   Perpendicular-axis theorem for plane laminae

Rigid-body rotation of thin plane laminae objects is encountered frequently. Examples of such laminae bodies are a plane sheet of metal, a thin door, a bicycle wheel, a thin envelope or book. Deriving the inertia tensor for a plane lamina is relatively simple because there are limits on the possible relative magnitude of the principal moments of inertia. Consider that the principal axis are along the x, y, z, coordinate axes. Then the sum of two principal moments of inertia about the center of mass are

$$\begin{aligned} I_x + I_y &= \int \rho(y^2 + z^2)dV + \int \rho(x^2 + z^2)dV \\ &= \int \rho(x^2 + y^2)dV + 2\int \rho z^2 dV \geq \int \rho(x^2 + y^2)dV = I_z \end{aligned} \tag{13.44}$$

Note that for any body the three principal moments of inertia must satisfy the triangle rule that the sum of any pair must exceed or equal the third. Moreover, if the body is a thin lamina with thickness $z = 0$, that is, a thin plate in the $x - y$ plane, then

$$I_x + I_y = I_z \tag{13.45}$$

This perpendicular-axis theorem can be very useful for solving problems involving rotation of plane laminae.

The opposite of a plane laminae is a long thin cylindrical needle of mass m, length L, and radius r. Along the symmetry axis the principal moments are $I_z = \frac{1}{2}mr^2 \to 0$ as $r \to 0$, while perpendicular to the symmetry axis $I_x = I_y = \frac{1}{12}mL^2$. These satisfy the triangle rule.

13.3 *Example: Inertia tensor of a hula hoop*

The hula hoop is a thin plane circular ring or radius R and mass M. Assume that the symmetry axis of the circular ring is the 3 axis.

a) The principal moments of inertia about the center of mass: *The principal moment of inertia along the 3 axis is $I_{33} = MR^2$. Then equation 13.45 plus symmetry tells us that the two principal moments of inertia in the plane of the hula hoop must be $I_{11} = I_{22} = \frac{1}{2}MR^2$.*

b) The principal moments of inertia about the periphery of the ring: *Using the Parallel-axis theorem tells us that the moment perpendicular to the plane of the hula hoop $I_{33} = 2MR^2$. In the plane of the hoop the moment tangential to the hoop is $I_{11} = \frac{3}{2}MR^2$ and the moment radial to the hoop $I_{22} = \frac{1}{2}MR^2$. The hula dancer often swings the hoop about the periphery and perpendicular to the plane by swinging their hips. Another movement is jumping through the hoop by rotating the hoop tangential to the periphery. Calculation of such maneuvers requires knowledge of these principal moments of inertia.*

13.4 *Example: Inertia tensor of a thin book*

Consider a thin rectangular book of mass M, width a and length b with thickness $t \ll a$ and $t \ll b$. About the center of mass the inertia tensor perpendicular to the plane of the book is $I_{33} = \frac{M}{12}(a^2 + b^2)$. The other two moments are $I_{11} = \frac{M}{12}a^2$ and $I_{22} = \frac{M}{12}b^2$ which satisfy equation 13.45.

13.10   General properties of the inertia tensor

13.10.1   Inertial equivalence

The elements of the inertia tensor, the values of the principal moments of inertia, and the orientation of the principal axes for a rigid body, all depend on the choice of origin for the system. Recall that for the kinetic energy to be separable into translational and rotational portions, the origin of the body coordinate system must coincide with the center of mass of the body. However, for *any* choice of the origin of *any* body, there always exists an orientation of the axes that diagonalizes the inertia tensor.

The inertial properties of a body for rotation about a specific body-fixed location is defined completely by only three principal moments of inertia irrespective of the detailed shape of the body. As a result, the inertial properties of any body about a body-fixed point are equivalent to that of an ellipsoid that has the same three principal moments of inertia. The symmetry properties of this equivalent ellipsoidal body define the symmetry of the inertial properties of the body. If a body has some simple symmetry then usually it is obvious as to what will be the principal axes of the body.

Spherical top: $I_1 = I_2 = I_3$

A spherical top is a body having three degenerate principal moments of inertia. Such a body has the same symmetry as the inertia tensor about the center of a uniform sphere. For a sphere it is obvious from the symmetry that any orientation of three mutually orthogonal axes about the center of the uniform sphere are equally good principal axes. For a uniform cube the principal axes of the inertia tensor about the center of mass were shown to be aligned such that they pass through the center of each face, and the three principal moments are identical; that is, inertially it is equivalent to a spherical top. A less obvious consequence of the spherical symmetry is that any orientation of three mutually perpendicular axes about the center of mass of a uniform cube is an equally good principal axis system.

Symmetric top: $I_1 = I_2 \neq I_3$

The equivalent ellipsoid for a body with two degenerate principal moments of inertia is a spheroid which has cylindrical symmetry with the cylindrical axis aligned along the third axis. A body with $I_3 < I_1 = I_2$ is a prolate spheroid while a body with $I_3 > I_1 = I_2$ is an oblate spheroid. Examples with a prolate spheroidal equivalent inertial shape are a rugby ball, pencil, or a baseball bat. Examples of an oblate spheroid are an orange, or a frisbee. A uniform sphere, or a uniform cube, rotating about a point displaced from the center-of-mass also behave inertially like a symmetric top. The cylindrical symmetry of the equivalent spheroid makes it obvious that any mutually perpendicular axes that are normal to the axis of cylindrical symmetry are equally good principal axes even when the cross section in the $1-2$ plane is square as opposed to circular.

A **rotor** is a diatomic-molecule shaped body which is a special case of a symmetric top where $I_1 = 0$, and $I_2 = I_3$. The rotation of a rotor is perpendicular to the symmetry axis since the rotational energy and angular momentum about the symmetry axis are zero because the principal moment of inertia about the symmetry axis is zero.

Asymmetric top: $I_1 \neq I_2 \neq I_3$

A body where all three principal moments of inertia are distinct, $I_1 \neq I_2 \neq I_3$, is called an **asymmetric top.** Some molecules, and nuclei have asymmetric, triaxially-deformed, shapes.

13.10.2   Orthogonality of principal axes

The body-fixed principal axes comprise an orthogonal set, for which the vectors $\mathbf{L}$ and $\boldsymbol{\omega}$ are simply related. Components of $\mathbf{L}$ and $\boldsymbol{\omega}$ can be taken along the three body-fixed axes denoted by i. Thus for the mth principal moment I_m

$$L_{im} = I_m \omega_{im} \tag{13.46}$$

Written in terms of the inertia tensor

$$L_{im} = \sum_{k}^{3} I_{ik} \omega_{km} = I_m \omega_{im} \tag{13.47}$$

Similarly the nth principal moment can be written as

$$L_{kn} = \sum_{i}^{3} I_{ki} \omega_{in} = I_n \omega_{kn} \tag{13.48}$$

Multiply the equation 13.47 by ω_{in} and sum over i gives

$$\sum_{i,k} I_{ik} \omega_{km} \omega_{in} = \sum_{i} I_{mm} \omega_{im} \omega_{in} \tag{13.49}$$

Similarly multiplying equation 13.48 by ω_{km} and summing over k gives

$$\sum_{i,k} I_{ki} \omega_{km} \omega_{in} = \sum_{k} I_{nn} \omega_{km} \omega_{kn} \tag{13.50}$$

The left-hand sides of these equations are identical since the inertia tensor is symmetric, that is $I_{ik} = I_{ki}$. Therefore subtracting these equations gives

$$\sum_{i} I_{mm} \omega_{im} \omega_{in} - \sum_{k} I_{nn} \omega_{km} \omega_{kn} = 0 \tag{13.51}$$

That is

$$(I_{mm} - I_{nn}) \sum_{k} \omega_{km} \omega_{kn} = 0 \tag{13.52}$$

or

$$(I_{mm} - I_{nn}) \, \boldsymbol{\omega}_m \cdot \boldsymbol{\omega}_n = 0 \tag{13.53}$$

If $I_m \neq I_n$ then

$$\boldsymbol{\omega}_m \cdot \boldsymbol{\omega}_n = 0 \tag{13.54}$$

which implies that the m and n principal axes are perpendicular. However, if $I_{mm} = I_{nn}$ then equation 13.53 does not require that $\boldsymbol{\omega}_m \cdot \boldsymbol{\omega}_n = 0$, that is, these axes are not necessarily perpendicular, but, with no loss of generality, these two axes can be chosen to be perpendicular with any orientation in the plane perpendicular to the symmetry axis.

Summarizing the above discussion, the inertia tensor has the following properties.

1) Diagonalization may be accomplished by an appropriate rotation of the axes in the body.

2) The principal moments (eigenvalues) and principal axes (eigenvectors) are obtained as roots of the secular determinant and are real.

3) The principal axes (eigenvectors) are real and orthogonal.

4) For a symmetric top with two identical principal moments of inertia, any orientation of two orthogonal axes perpendicular to the symmetry axis are satisfactory eigenvectors.

5) For a spherical top with three identical principal moment of inertia, the principal axes system can have any orientation with respect to the origin.

13.11 Angular momentum L and angular velocity ω vectors

The angular momentum is a primary observable for rotation. As discussed in chapter 13.5, the angular momentum **L** is compactly and elegantly written in matrix form using the tensor algebra relation

$$\mathbf{L} = \begin{pmatrix} I_{11} & I_{12} & I_{13} \\ I_{21} & I_{22} & I_{23} \\ I_{31} & I_{32} & I_{33} \end{pmatrix} \cdot \begin{pmatrix} \omega_1 \\ \omega_2 \\ \omega_3 \end{pmatrix} = \{\mathbf{I}\} \cdot \boldsymbol{\omega} \tag{13.55}$$

where $\boldsymbol{\omega}$ is the angular velocity, $\{\mathbf{I}\}$ the inertia tensor, and **L** the corresponding angular momentum.

Two important consequences of equation 13.55 are that:

- *The angular momentum* **L** *and angular velocity* $\boldsymbol{\omega}$ *are not necessarily colinear.*

- *In general the Principal axis system of the rotating rigid body is not aligned with either the angular momentum or angular velocity vectors.*

An exception to these statements occurs when the angular velocity $\boldsymbol{\omega}$ is aligned along a principal axes for which the inertia tensor is diagonal, i.e. $I_{ij} = I_i \delta_{ij}$, and then both **L** and $\boldsymbol{\omega}$ point along this principal axis. In general the angular momentum **L** and angular velocity $\boldsymbol{\omega}$ precess around each other. An important special case is for torque-free systems where Noether's theorem implies that the angular momentum vector **L** is conserved both in magnitude and amplitude. In this case, the angular velocity $\boldsymbol{\omega}$, and the Principal axis system, both precesses around the angular momentum vector **L**. That is, the body appears to tumble with respect to the laboratory fixed frame. Understanding rigid-body rotation requires care not to confuse the body-fixed Principal axis coordinate frame, used to determine the inertia tensor, and the fixed laboratory frame where the motion is observed.

13.5 *Example: Rotation about the center of mass of a solid cube*

It is illustrative to use the inertia tensors of a uniform cube to compute the angular momentum for any applied angular velocity vector ω using equation (13.55). If the angular velocity is along the x axis, then using the inertia tensor for a solid cube, derived earlier, in equation (13.55) gives the angular momentum to be

$$\mathbf{L} = \{\mathbf{I}\} \cdot \boldsymbol{\omega} = \frac{1}{6} M b^2 \omega \begin{pmatrix} 1 & 0 & 0 \\ 0 & 1 & 0 \\ 0 & 0 & 1 \end{pmatrix} \cdot \begin{pmatrix} 1 \\ 0 \\ 0 \end{pmatrix} = \frac{1}{6} M b^2 \omega \begin{pmatrix} 1 \\ 0 \\ 0 \end{pmatrix}$$

This shows that **L** *and* $\boldsymbol{\omega}$ *are colinear and thus the x axis is a principal axis. By symmetry, the y and z body fixed axis also must be principal axes.*

Consider that the body is rotated about a diagonal of the cube for which the center of mass will be on the rotation axis. Then the angular velocity vector is written as $\boldsymbol{\omega} = \omega \frac{1}{\sqrt{3}} \begin{pmatrix} 1 \\ 1 \\ 1 \end{pmatrix}$ *where the components of* $\omega_x = \omega_y = \omega_z = \omega \frac{1}{\sqrt{3}}$ *with the angular velocity magnitude* $\sqrt{\omega_x^2 + \omega_y^2 + \omega_z^2} = \omega$.

$$\mathbf{L} = \{\mathbf{I}\} \cdot \boldsymbol{\omega} = \frac{1}{6} M b^2 \omega \frac{1}{\sqrt{3}} \begin{pmatrix} 1 & 0 & 0 \\ 0 & 1 & 0 \\ 0 & 0 & 1 \end{pmatrix} \cdot \begin{pmatrix} 1 \\ 1 \\ 1 \end{pmatrix} = \frac{1}{6} M b^2 \omega \frac{1}{\sqrt{3}} \begin{pmatrix} 1 \\ 1 \\ 1 \end{pmatrix} = \frac{1}{6} M b^2 \boldsymbol{\omega}$$

Note that $\mathbf{L}$ *and* $\boldsymbol{\omega}$ *again are colinear showing it also is a principal axis. Moreover, the magnitude of* $\mathbf{L}$ *is identical for orientations of the rotation axes* $\boldsymbol{\omega}$ *passing through the center of mass when centered on either one face, or the diagonal, of the cube implying that the principal moments of inertia about these axes are identical. This illustrates the important property that, when the three principal moments of inertia are identical, then any orientation of the coordinate system is an equally good principal axis system. That is, this corresponds to the spherical top where all orientations are principal axes, not just along the obvious symmetry axes.*

13.6 Example: Rotation about the corner of the cube

Let us repeat the above exercise for rotation about one corner of the cube. Consider that the angular velocity is along the x *axis. Then example (13.2) gives the angular momentum to be*

$$\mathbf{L} = \{\mathbf{I}\} \cdot \boldsymbol{\omega} = \frac{1}{12} M b^2 \omega \begin{pmatrix} +8 & -3 & -3 \\ -3 & +8 & -3 \\ -3 & -3 & +8 \end{pmatrix} \cdot \begin{pmatrix} 1 \\ 0 \\ 0 \end{pmatrix} = \frac{1}{12} M b^2 \omega \begin{pmatrix} +8 \\ -3 \\ -3 \end{pmatrix}$$

The angular momentum is far from being aligned with the axis $\boldsymbol{\omega}$, *that is, it is not a principal axis.*

Consider that the body is rotated with the angular velocity aligned along a diagonal of the cube through the center of mass on this axis. Then the angular velocity is written as $\boldsymbol{\omega} = \frac{1}{\sqrt{3}} \begin{pmatrix} 1 \\ 1 \\ 1 \end{pmatrix}$ *where the components of* $\omega_x = \omega_y = \omega_z = \frac{1}{\sqrt{3}}$ *ensuring that the magnitude equals* $\sqrt{\omega_x^2 + \omega_y^2 + \omega_z^2} = \omega$.

$$\mathbf{L} = \{\mathbf{I}\} \cdot \boldsymbol{\omega} = \frac{1}{12} M b^2 \omega \frac{1}{\sqrt{3}} \begin{pmatrix} +8 & -3 & -3 \\ -3 & +8 & -3 \\ -3 & -3 & +8 \end{pmatrix} \cdot \begin{pmatrix} 1 \\ 1 \\ 1 \end{pmatrix} = \frac{1}{12} M b^2 \omega \frac{1}{\sqrt{3}} \begin{pmatrix} 2 \\ 2 \\ 2 \end{pmatrix} = \frac{1}{6} M b^2 \boldsymbol{\omega}$$

This is a principal axis since $\mathbf{L}$ *and* $\boldsymbol{\omega}$ *again are colinear and the angular momentum is the same as for any axis through the center of mass of a uniform solid cube due to the high symmetry of the cube. If the angular velocity is perpendicular to the diagonal of the cube, then, for either of these perpendicular axes, the relation between L and* ω *is given by*

$$\mathbf{L} = \frac{1}{12} M b^2 \omega \frac{1}{\sqrt{2}} \begin{pmatrix} +8 & -3 & -3 \\ -3 & +8 & -3 \\ -3 & -3 & +8 \end{pmatrix} \cdot \begin{pmatrix} -1 \\ +1 \\ 0 \end{pmatrix} = \frac{1}{12} M b^2 \omega \frac{1}{\sqrt{2}} \begin{pmatrix} -11 \\ +11 \\ 0 \end{pmatrix} = \frac{11}{12} M b^2 \omega \begin{pmatrix} -1 \\ +1 \\ 0 \end{pmatrix}$$

Note that this must be a principal axis for rotation about a corner of the cube since $\mathbf{L}$ *and* $\boldsymbol{\omega}$ *are colinear. The angular momentum is the same for both possible orientations of* ω *that are perpendicular to the diagonal through the center of mass. Diagonalizing the inertia tensor in example 13.2 also gave the above result with the symmetry axis along the diagonal of the cube.*

This example illustrates that it is not necessary to diagonalize the inertia tensor matrix to obtain the principal axes. The corner of the cube has three mutually perpendicular principal axes independent of the choice of a body-fixed coordinate frame. The advantage of the principal axis coordinate frame is that the inertia tensor is diagonal making evaluation of the angular momentum trivial. That is, there is no physics associated with the orientation chosen for the body-fixed coordinate frame, this frame only determines the ratio of the components of the inertia tensor along the chosen coordinates. Note that, if a body has an obvious symmetry, then intuition is a powerful way to identify the principal axis frame.

13.12 Kinetic energy of rotating rigid body

An important observable is the kinetic energy of rotation of a rigid body. Consider a rigid body composed of N particles of mass m_α where $\alpha = 1, 2, 3, ...N$. If the body rotates with an instantaneous angular velocity $\boldsymbol{\omega}$ about some fixed point, with respect to the body coordinate system, and this point has an instantaneous translational velocity $\mathbf{V}$ with respect to the fixed (inertial) coordinate system, see figure 13.1, then the instantaneous velocity $\mathbf{v}_\alpha$ of the αth particle in the fixed frame of reference is given by

$$\mathbf{v}_\alpha = \mathbf{V} + \mathbf{v}''_\alpha + \boldsymbol{\omega} \times \mathbf{r}'_\alpha \qquad (13.56)$$

However, for a rigid body, the velocity of a body-fixed point with respect to the body is zero, that is $\mathbf{v}''_\alpha = 0$, thus

$$\mathbf{v}_\alpha = \mathbf{V} + \boldsymbol{\omega} \times \mathbf{r}'_\alpha \qquad (13.57)$$

The total kinetic energy is given by

$$
\begin{aligned}
T &= \sum_\alpha^N \frac{1}{2} m_\alpha \mathbf{v}_\alpha \cdot \mathbf{v}_\alpha = \sum_\alpha^N \frac{1}{2} m_\alpha \left(\mathbf{V} + \boldsymbol{\omega} \times \mathbf{r}'_\alpha \right) \cdot \left(\mathbf{V} + \boldsymbol{\omega} \times \mathbf{r}'_\alpha \right) \\
&= \frac{1}{2} \sum_\alpha^N m_\alpha V^2 + \sum_i^N m_\alpha \mathbf{V} \cdot \boldsymbol{\omega} \times \mathbf{r}'_\alpha + \frac{1}{2} \sum_\alpha^N m_\alpha \left(\boldsymbol{\omega} \times \mathbf{r}'_\alpha \right) \cdot \left(\boldsymbol{\omega} \times \mathbf{r}'_\alpha \right)
\end{aligned}
\qquad (13.58)
$$

This is a general expression for the kinetic energy that is valid for any choice of the origin from which the body-fixed vectors $\mathbf{r}'_\alpha$ are measured. However, if the origin is chosen to be the center of mass, then, and only then, the middle term cancels. That is, since $\mathbf{V} \cdot \boldsymbol{\omega}$ is independent of the specific particle, then

$$\sum_\alpha^N m_\alpha \mathbf{V} \cdot \boldsymbol{\omega} \times \mathbf{r}'_\alpha = \mathbf{V} \cdot \boldsymbol{\omega} \times \left(\sum_\alpha^N m_\alpha \mathbf{r}'_\alpha \right) \qquad (13.59)$$

But the definition of the center of mass is

$$\sum_\alpha m_\alpha \mathbf{r}' = M\mathbf{R} \qquad (13.60)$$

and $\mathbf{R} = 0$ in the body-fixed frame if the selected point in the body is the center of mass. Thus, *when using the center of mass frame*, the middle term of equation 13.58 is zero. Therefore, for the center of mass frame, the kinetic energy separates into two terms in the body-fixed frame

$$T = T_{trans} + T_{rot} \qquad (13.61)$$

where

$$T_{trans} = \frac{1}{2} \sum_\alpha^N m_\alpha V^2 \qquad (13.62)$$

$$T_{rot} = \frac{1}{2} \sum_\alpha^N m_i \left(\boldsymbol{\omega} \times \mathbf{r}'_\alpha \right) \cdot \left(\boldsymbol{\omega} \times \mathbf{r}'_\alpha \right)$$

The vector identity

$$(\mathbf{A} \times \mathbf{B}) \cdot (\mathbf{A} \times \mathbf{B}) = A^2 B^2 - (\mathbf{A} \cdot \mathbf{B})^2 \qquad (13.63)$$

can be used to simplify T_{rot}

$$T_{rot} = \frac{1}{2} \sum_\alpha^N m_\alpha \left[\omega^2 r'^2_\alpha - (\boldsymbol{\omega} \cdot \mathbf{r}'_\alpha)^2 \right] \qquad (13.64)$$

The rotational kinetic energy T_{rot} can be expressed in terms of components of $\boldsymbol{\omega}$ and $\mathbf{r}'_\alpha$ in the body-fixed frame. Also the following formulae are greatly simplified if $\mathbf{r}'_\alpha = (x_\alpha, y_\alpha, z_\alpha)$ in the rotating body-fixed frame is written in the form $\mathbf{r}'_\alpha = (x_{\alpha,1}, x_{\alpha,2}, x_{\alpha,3})$ where the axes are defined by the numbers $1, 2, 3$ rather than x, y, z. In this notation the rotational kinetic energy is written as

$$T_{rot} = \frac{1}{2} \sum_\alpha^N m_\alpha \left[\left(\sum_i \omega_i^2 \right) \left(\sum_k x_{\alpha,k}^2 \right) - \left(\sum_i \omega_i x_{\alpha,i} \right) \left(\sum_j \omega_j x_{\alpha,j} \right) \right] \tag{13.65}$$

Assume the Kronecker delta relation

$$\omega_i = \sum_j^3 \omega_j \delta_{ij} \tag{13.66}$$

where $\delta_{ij} = 1$ if $i = j$ and $\delta_{ij} = 0$ if $i \neq j$.

Then the kinetic energy can be written more compactly

$$
\begin{aligned}
T_{rot} &= \frac{1}{2} \sum_\alpha^N m_\alpha \left[\left(\sum_i \omega_i^2 \right) \left(\sum_k x_{\alpha,k}^2 \right) - \left(\sum_i \omega_i x_{\alpha,i} \right) \left(\sum_j \omega_j x_{\alpha,j} \right) \right] \\
&= \frac{1}{2} \sum_\alpha^N \sum_{i,j}^3 m_\alpha \left[(\omega_i \omega_j \delta_{ij}) \left(\sum_k^3 x_{\alpha,k}^2 \right) - (\omega_i x_{\alpha,i})(\omega_j x_{\alpha,j}) \right] \\
&= \frac{1}{2} \sum_{i,j}^3 \omega_i \omega_j \left[\sum_\alpha^N m_\alpha \left[\delta_{ij} \left(\sum_k^3 x_{\alpha,k}^2 \right) - x_{\alpha,i} x_{\alpha,j} \right] \right]
\end{aligned}
\tag{13.67}
$$

The term in the outer square brackets is the inertia tensor defined in equation 13.12 for a discrete body. The inertia tensor components for a continuous body are given by equation 13.13.

Thus the rotational component of the kinetic energy can be written in terms of the inertia tensor as

$$T_{rot} = \frac{1}{2} \sum_{i,j}^3 I_{ij} \omega_i \omega_j \tag{13.68}$$

Note that when the inertia tensor is diagonal ,then the evaluation of the kinetic energy simplifies to

$$T_{rot} = \frac{1}{2} \sum_i^3 I_{ii} \omega_i^2 \tag{13.69}$$

which is the familiar relation in terms of the scalar moment of inertia I discussed in elementary mechanics.

Equation 13.68 also can be factored in terms of the angular momentum $\mathbf{L}$.

$$T_{rot} = \frac{1}{2} \sum_{i,j} I_{ij} \omega_i \omega_j = \frac{1}{2} \sum_i \omega_i \sum_j I_{ij} \omega_j = \frac{1}{2} \sum_i \omega_i L_i \tag{13.70}$$

As mentioned earlier, tensor algebra is an elegant and compact way of expressing such matrix operations. Thus it is possible to express the rotational kinetic energy as

$$T_{rot} = \frac{1}{2} \begin{pmatrix} \omega_1 & \omega_2 & \omega_3 \end{pmatrix} \cdot \begin{pmatrix} I_{11} & I_{12} & I_{13} \\ I_{21} & I_{22} & I_{23} \\ I_{31} & I_{32} & I_{33} \end{pmatrix} \cdot \begin{pmatrix} \omega_1 \\ \omega_2 \\ \omega_3 \end{pmatrix} \tag{13.71}$$

$$T_{rot} \equiv \mathbf{T} = \frac{1}{2} \boldsymbol{\omega} \cdot \{\mathbf{I}\} \cdot \boldsymbol{\omega} \tag{13.72}$$

where the rotational energy $\mathbf{T}$ is a scalar. Using equation 13.55 the rotational component of the kinetic energy also can be written as

$$T_{rot} \equiv \mathbf{T} = \frac{1}{2} \boldsymbol{\omega} \cdot \mathbf{L} \tag{13.73}$$

which is the same as given by (13.70). It is interesting to realize that even though $\mathbf{L} = \{\mathbf{I}\} \cdot \boldsymbol{\omega}$ is the inner product of a tensor and a vector, it is a vector as illustrated by the fact that the inner product $T_{rot} = \frac{1}{2} \boldsymbol{\omega} \cdot \mathbf{L} = \frac{1}{2} \boldsymbol{\omega} \cdot (\{\mathbf{I}\} \cdot \boldsymbol{\omega})$ is a scalar. Note that the translational kinetic energy T_{trans} must be added to the rotational kinetic energy T_{rot} to get the total kinetic energy as given by equation 13.61.

13.13 Euler angles

The description of rigid-body rotation is greatly facilitated by transforming from the space-fixed coordinate frame $(\hat{\mathbf{x}}, \hat{\mathbf{y}}, \hat{\mathbf{z}})$ to a rotating body-fixed coordinate frame $(\hat{\mathbf{1}}, \hat{\mathbf{2}}, \hat{\mathbf{3}})$ for which the inertia tensor is diagonal. Appendix D introduced the rotation matrix $\{\boldsymbol{\lambda}\}$ which can be used to rotate between the space-fixed coordinate system, which is stationary, and the instantaneous body-fixed frame which is rotating with respect to the space-fixed frame. The transformation can be represented by a matrix equation

$$(\hat{\mathbf{1}}, \hat{\mathbf{2}}, \hat{\mathbf{3}}) = \{\boldsymbol{\lambda}\} \cdot (\hat{\mathbf{x}}, \hat{\mathbf{y}}, \hat{\mathbf{z}}) \qquad (13.74)$$

where the space-fixed system is identified by unit vectors $(\hat{\mathbf{x}}, \hat{\mathbf{y}}, \hat{\mathbf{z}})$ while $(\hat{\mathbf{1}}, \hat{\mathbf{2}}, \hat{\mathbf{3}})$ defines unit vectors in the rotated body-fixed system. The rotation matrix $\{\boldsymbol{\lambda}\}$ completely describes the instantaneous relative orientation of the two systems. Rigid-body rotation requires three independent angular parameters that specify the orientation of the rigid body such that the corresponding orthogonal transformation matrix is proper, that is, it has a determinant $|\lambda| = +1$ as given by equation $(D.33)$.

As discussed in Appendix $D.2$, the 9 component rotation matrix involves only three independent angles. There are many possible choices for these three angles. It is convenient to use the **Euler angles**, ϕ, θ, ψ, (also called Eulerian angles) shown in figure 13.3.[1] The Euler angles are generated by a series of three rotations that rotate from the space-fixed $(\hat{\mathbf{x}}, \hat{\mathbf{y}}, \hat{\mathbf{z}})$ system to the body-fixed $(\hat{\mathbf{1}}, \hat{\mathbf{2}}, \hat{\mathbf{3}})$ system. The rotation must be such that the space-fixed z axis rotates by an angle θ to align with the body-fixed 3 axis. This can be performed by rotating through an angle θ about the $\hat{\mathbf{n}} \equiv \hat{\mathbf{z}} \times \hat{\mathbf{3}}$ direction, where $\hat{\mathbf{z}}$ and $\hat{\mathbf{3}}$ designate the unit vectors along the "z" axes of the space and body fixed frames respectively. The unit vector $\hat{\mathbf{n}} \equiv \hat{\mathbf{z}} \times \hat{\mathbf{3}}$ is the vector normal to the plane

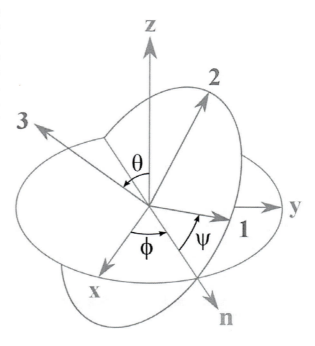

Figure 13.3: The $z - x - z$ sequence of rotations $\lambda_\phi, \lambda_\theta, \lambda_\psi$ corresponding to the Eulerian angles (ϕ, θ, ψ). The first rotation ϕ about the space-fixed $\mathbf{z}$ axis (blue) is from the x-axis (blue) to the line of nodes $\mathbf{n}$ (green). The second rotation θ about the line of nodes (green) is from the space-fixed z axis (blue) to the body-fixed 3-axis (red). The third rotation ψ about the body-fixed 3-axis (red) is from the line of nodes (green) to the body-fixed 1 axis (red).

defined by the $\hat{\mathbf{z}}$ and $\hat{\mathbf{3}}$ unit vectors and this unit vector $\hat{\mathbf{n}} = \hat{\mathbf{z}} \times \hat{\mathbf{3}}$ is called the *line of nodes.* The chosen convention is that the unit vector $\hat{\mathbf{n}} = \hat{\mathbf{z}} \times \hat{\mathbf{3}}$ is along the "x" axis of an intermediate-axis frame designated by $(\hat{\mathbf{n}}, \hat{\mathbf{y}}', \hat{\mathbf{z}})$, that is, the unit vector $\hat{\mathbf{n}} = \hat{\mathbf{z}} \times \hat{\mathbf{3}}$ plus the unit vectors $\hat{\mathbf{y}}'$ and $\hat{\mathbf{z}}$ are in the same plane as the $\hat{\mathbf{z}}$ and $\hat{\mathbf{3}}$ unit vectors. The sequence of three rotations is performed as summarized below.

1) Rotation ϕ about the space-fixed $\hat{\mathbf{z}}$ axis from the space $\hat{\mathbf{x}}$ axis to the line of nodes $\hat{\mathbf{n}}$: The first rotation $(\mathbf{x}, \mathbf{y}, \mathbf{z}) \cdot \boldsymbol{\lambda}_\phi \rightarrow (\mathbf{n}, \mathbf{y}', \mathbf{z})$ is in a right-handed direction through an angle ϕ about the *space-fixed* $\mathbf{z}$ axis. Since the rotation takes place in the $\mathbf{x} - \mathbf{y}$ plane, the transformation matrix is

$$\{\boldsymbol{\lambda}_\phi\} = \begin{pmatrix} \cos\phi & \sin\phi & 0 \\ -\sin\phi & \cos\phi & 0 \\ 0 & 0 & 1 \end{pmatrix} \qquad (13.75)$$

---

[1] The space-fixed coordinate frame and the body-fixed coordinate frames are unambiguously defined, that is, the space-fixed frame is stationary while the body-fixed frame is the principal-axis frame of the body. There are several possible intermediate frames that can be used to define the Euler angles. The $z - x - z$ sequence of rotations, used here, is used in most physics textbooks in classical mechanics. Unfortunately scientists and engineers use slightly different conventions for defining the Euler angles. As discussed in Appendix A of "Classical Mechanics" by Goldstein, nuclear and particle physicists have adopted the $z - y - z$ sequence of rotations while the US and UK aerodynamicists have adopted a $x - y - z$ sequence of rotations.

This leads to the intermediate coordinate system $(\mathbf{n}, \mathbf{y'}, \mathbf{z})$ where the rotated $\mathbf{x}$ axis now is colinear with the $\mathbf{n}$ axis of the intermediate frame, that is, the *line of nodes*.

$$(\mathbf{n}, \mathbf{y'}, \mathbf{z}) = \{\boldsymbol{\lambda}_\phi\} \cdot (\mathbf{x}, \mathbf{y}, \mathbf{z}) \tag{13.76}$$

The *precession angular velocity* $\dot{\phi}$ is the rate of change of angle of the line of nodes with respect to the space x axis about the space-fixed z axis.

2) Rotation θ about the line of nodes $\hat{\mathbf{n}}$ from the space $\hat{\mathbf{z}}$ axis to the body-fixed $\hat{\mathbf{3}}$ axis: The second rotation

$$(\mathbf{n}, \mathbf{y'}, \mathbf{z}) \cdot \lambda_\theta \rightarrow (\mathbf{n}, \mathbf{y''}, \mathbf{3}) \tag{13.77}$$

is in a right-handed direction through the angle θ about the $\hat{\mathbf{n}}$ axis (line of nodes) so that the "z" axis becomes colinear with the body-fixed $\hat{\mathbf{3}}$ axis. Because the rotation now is in the $\hat{\mathbf{z}} - \hat{\mathbf{3}}$ plane, the transformation matrix is

$$\{\boldsymbol{\lambda}_\theta\} = \begin{pmatrix} 1 & 0 & 0 \\ 0 & \cos\theta & \sin\theta \\ 0 & -\sin\theta & \cos\theta \end{pmatrix} \tag{13.78}$$

The line of nodes which is at the intersection of the space-fixed and body-fixed planes, shown in figure 13.3, points in the $\hat{\mathbf{n}} = \hat{\mathbf{z}} \times \hat{\mathbf{3}}$ direction. The new "z" axis now is the body-fixed $\hat{\mathbf{3}}$ axis. The angular velocity $\dot{\theta}$ is the rate of change of angle of the body-fixed $\hat{\mathbf{3}}$-axis relative to the space-fixed $\hat{\mathbf{z}}$-axis about the line of nodes.

3) Rotation ψ about the body-fixed $\hat{\mathbf{3}}$ axis from the line of nodes to the body-fixed $\hat{\mathbf{1}}$ axis: The third rotation

$$(\mathbf{n}, \mathbf{y''}, \mathbf{3}) \cdot \lambda_\psi \rightarrow (\hat{\mathbf{1}}, \hat{\mathbf{2}}, \hat{\mathbf{3}}) \tag{13.79}$$

is in a right-handed direction through the angle ψ about the new body-fixed $\hat{\mathbf{3}}$ axis. This third rotation transforms the rotated intermediate $(\mathbf{n}, \mathbf{y''}, \mathbf{3})$ frame to final body-fixed coordinate system $(\hat{\mathbf{1}}, \hat{\mathbf{2}}, \hat{\mathbf{3}})$. The transformation matrix is

$$\{\boldsymbol{\lambda}_\psi\} = \begin{pmatrix} \cos\psi & \sin\psi & 0 \\ -\sin\psi & \cos\psi & 0 \\ 0 & 0 & 1 \end{pmatrix} \tag{13.80}$$

The *spin angular velocity* $\dot{\psi}$ is the rate of change of the angle of the body-fixed **1**-axis with respect to the line of nodes about the body-fixed 3 axis.

The total rotation matrix $\{\boldsymbol{\lambda}\}$ is given by

$$\{\boldsymbol{\lambda}\} = \{\boldsymbol{\lambda}_\psi\} \cdot \{\boldsymbol{\lambda}_\theta\} \cdot \{\boldsymbol{\lambda}_\phi\} \tag{13.81}$$

Thus the complete rotation from the space-fixed $(\mathbf{x}, \mathbf{y}, \mathbf{z})$ axis system to the body-fixed $(\mathbf{1}, \mathbf{2}, \mathbf{3})$ axis system is given by

$$(\mathbf{1}, \mathbf{2}, \mathbf{3}) = \{\boldsymbol{\lambda}\} \cdot (\mathbf{x}, \mathbf{y}, \mathbf{z}) \tag{13.82}$$

where $\{\boldsymbol{\lambda}\}$ is given by the triple product equation (13.81) leading to the rotation matrix

$$\{\boldsymbol{\lambda}\} = \begin{pmatrix} \cos\phi\cos\psi - \sin\phi\cos\theta\sin\psi & \sin\phi\cos\psi + \cos\phi\cos\theta\sin\psi & \sin\theta\sin\psi \\ -\cos\phi\sin\psi - \sin\phi\cos\theta\cos\psi & -\sin\phi\sin\psi + \cos\phi\cos\theta\cos\psi & \sin\theta\cos\psi \\ \sin\phi\sin\theta & -\cos\phi\sin\theta & \cos\theta \end{pmatrix} \tag{13.83}$$

The inverse transformation from the body-fixed axis system to the space-fixed axis system is given by

$$(\mathbf{x}, \mathbf{y}, \mathbf{z}) = \{\boldsymbol{\lambda}\}^{-1} \cdot (\mathbf{1}, \mathbf{2}, \mathbf{3}) \tag{13.84}$$

where the inverse matrix $\{\boldsymbol{\lambda}\}^{-1}$ equals the transposed rotation matrix $\{\boldsymbol{\lambda}\}^T$, that is,

$$\{\boldsymbol{\lambda}\}^{-1} = \{\boldsymbol{\lambda}\}^T = \begin{pmatrix} \cos\phi\cos\psi - \sin\phi\cos\theta\sin\psi & -\cos\phi\sin\psi - \sin\phi\cos\theta\cos\psi & \sin\phi\sin\theta \\ \sin\phi\cos\psi + \cos\phi\cos\theta\sin\psi & -\sin\phi\sin\psi + \cos\phi\cos\theta\cos\psi & -\cos\phi\sin\theta \\ \sin\theta\sin\psi & \sin\theta\cos\psi & \cos\theta \end{pmatrix} \tag{13.85}$$

Taking the product $\{\lambda\}\{\lambda\}^{-1} = 1$ shows that the rotation matrix is a proper, orthogonal, unit matrix.

The use of three different coordinate systems, space-fixed, the intermediate line of nodes, and the body-fixed frame can be confusing at first glance. Basically the angle ϕ specifies the rotation about the *space-fixed* z axis between the *space-fixed* x axis and the *line of nodes* of the Euler angle intermediate frame. The angle ψ specifies the rotation about the *body-fixed* 3 axis between the *line of nodes* and the *body-fixed* 1 axis. Note that although the space-fixed and body-fixed axes systems each are orthogonal, the Euler angle basis in general is not orthogonal. For rigid-body rotation the rotation angle ϕ about the space-fixed z axis is time dependent, that is, the line of nodes is rotating with an angular velocity $\dot{\phi}$ with respect to the space-fixed coordinate frame. Similarly the body-fixed coordinate frame is rotating about the body-fixed 3 axis with angular velocity $\dot{\psi}$ relative to the line of nodes.

13.7 *Example: Euler angle transformation*

The definition of the Euler angles can be confusing, therefore it is useful to illustrate their use for a rotational transformation of a primed frame (x', y', z') to an unprimed frame (x, y, z). Assume the first rotation about the z' axis, is $\phi = 30°$

$$\lambda_\phi = \begin{pmatrix} \frac{\sqrt{3}}{2} & \frac{1}{2} & 0 \\ -\frac{1}{2} & \frac{\sqrt{3}}{2} & 0 \\ 0 & 0 & 1 \end{pmatrix}$$

Let the second rotation be $\theta = 45°$ about the line of nodes, that is, the intermediate x" axis. Then

$$\lambda_\theta = \begin{pmatrix} 1 & 0 & 0 \\ 0 & \frac{1}{\sqrt{2}} & \frac{1}{\sqrt{2}} \\ 0 & -\frac{1}{\sqrt{2}} & \frac{1}{\sqrt{2}} \end{pmatrix}$$

Let the third rotation be $\psi = 90°$ about the z axis.

$$\lambda_\psi = \begin{pmatrix} 0 & 1 & 0 \\ -1 & 0 & 0 \\ 0 & 0 & 1 \end{pmatrix}$$

Thus the net rotation corresponds to $\lambda = \lambda_\psi \lambda_\theta \lambda_\phi$

$$\lambda = \begin{pmatrix} \frac{\sqrt{3}}{2} & \frac{1}{2} & 0 \\ -\frac{1}{2} & \frac{\sqrt{3}}{2} & 0 \\ 0 & 0 & 1 \end{pmatrix} \begin{pmatrix} 1 & 0 & 0 \\ 0 & \frac{1}{\sqrt{2}} & \frac{1}{\sqrt{2}} \\ 0 & -\frac{1}{\sqrt{2}} & \frac{1}{\sqrt{2}} \end{pmatrix} \begin{pmatrix} 0 & 1 & 0 \\ -1 & 0 & 0 \\ 0 & 0 & 1 \end{pmatrix} = \begin{pmatrix} -\frac{1}{4}\sqrt{2} & \frac{1}{2}\sqrt{3} & \frac{1}{4}\sqrt{2} \\ -\frac{1}{4}\sqrt{6} & -\frac{1}{2} & \frac{1}{4}\sqrt{6} \\ \frac{1}{2}\sqrt{2} & 0 & \frac{1}{2}\sqrt{2} \end{pmatrix}$$

13.14   Angular velocity ω

It is useful to relate the rigid-body equations of motion in the space-fixed $(\hat{\mathbf{x}}, \hat{\mathbf{y}}, \hat{\mathbf{z}})$ coordinate system to those in the body-fixed $(\hat{\mathbf{e}}_1, \hat{\mathbf{e}}_2, \hat{\mathbf{e}}_3)$ coordinate system where the principal axis inertia tensor is defined. It was shown in appendix D that an infinitessimal rotation can be represented by a vector. Thus the time derivatives of these rotation angles can be associated with the components of the angular velocity ω, where the *precession* $\omega_\phi = \dot{\phi}$, the *nutation* $\omega_\theta = \dot{\theta}$, and the *spin* $\omega_\psi = \dot{\psi}$. Unfortunately the coordinates (ϕ, θ, ψ) are with respect to mixed coordinate frames and thus are not orthogonal axes. That is, the Euler angular velocities are expressed in different coordinate frames, where the *precession* $\dot{\phi}$ is around the space-fixed $\hat{\mathbf{z}}$ axis measured relative to the $\hat{\mathbf{x}}$-axis, the *spin* $\dot{\psi}$ is around the body-fixed $\hat{\mathbf{e}}_3$ axis relative to the rotating line-of-nodes, and the *nutation* $\dot{\theta}$ is the angular velocity between the $\hat{\mathbf{z}}$ and $\hat{\mathbf{e}}_3$ axes and points along the instantaneous line-of-nodes in the $\hat{\mathbf{e}}_3 \times \hat{\mathbf{z}}$ direction. By reference to figure 13.3 it can be seen that the components along the body-fixed axes are as given in Table 13.1.

Table 13.1; Euler angular velocity components in the body-fixed frame

Precession $\dot{\phi}$	Nutation $\dot{\theta}$	Spin $\dot{\psi}$
$\dot{\phi}_1 = \dot{\phi}\sin\theta\sin\psi$	$\dot{\theta}_1 = \dot{\theta}\cos\psi$	$\dot{\psi}_1 = 0$
$\dot{\phi}_2 = \dot{\phi}\sin\theta\cos\psi$	$\dot{\theta}_2 = -\dot{\theta}\sin\psi$	$\dot{\psi}_2 = 0$
$\dot{\phi}_3 = \dot{\phi}\cos\theta$	$\dot{\theta}_3 = 0$	$\dot{\psi}_3 = \dot{\psi}$

Note that the precession angular velocity $\dot{\phi}$ is the angular velocity that the body-fixed $\hat{\mathbf{e}}_3$ and $\hat{\mathbf{z}} \times \hat{\mathbf{3}}$ axes precess around the space-fixed $\hat{\mathbf{z}}$ axis. Table 13.1 gives the Euler angular velocities required to calculate the components of the angular velocity $\boldsymbol{\omega}$ for the body-fixed $(\mathbf{1}, \mathbf{2}, \mathbf{3})$ axis system. Collecting the individual components of $\boldsymbol{\omega}$, gives the components of the angular velocity of the body, relative to the space-fixed axes, in the *body-fixed axis system* $(1, 2, 3)$

$$\omega_1 = \dot{\phi}_1 + \dot{\theta}_1 + \dot{\psi}_1 = \dot{\phi}\sin\theta\sin\psi + \dot{\theta}\cos\psi \tag{13.86}$$

$$\omega_2 = \dot{\phi}_2 + \dot{\theta}_2 + \dot{\psi}_2 = \dot{\phi}\sin\theta\cos\psi - \dot{\theta}\sin\psi \tag{13.87}$$

$$\omega_3 = \dot{\phi}_3 + \dot{\theta}_3 + \dot{\psi}_3 = \dot{\phi}\cos\theta + \dot{\psi} \tag{13.88}$$

The angular velocity of the body about the body-fixed $\mathbf{3}$-axis, ω_3, is the sum of the projection of the precession angular velocity of the line-of-nodes $\dot{\phi}$ with respect to the space-fixed $\mathbf{x}$-axis, plus the angular velocity $\dot{\psi}$ of the body-fixed 3-axis with respect to the rotating line-of-nodes.

Similarly, the components of the body angular velocity $\boldsymbol{\omega}$ for the *space-fixed axis system* (x, y, z) can be derived to be

$$\omega_x = \dot{\theta}\cos\phi + \dot{\psi}\sin\theta\sin\phi \tag{13.89}$$

$$\omega_y = \dot{\theta}\sin\phi - \dot{\psi}\sin\theta\cos\phi \tag{13.90}$$

$$\omega_z = \dot{\phi} + \dot{\psi}\cos\theta \tag{13.91}$$

Note that when $\theta = 0$ then the Euler angles are singular in that the space-fixed z axis is parallel with the body-fixed 3 axis and there is no way of distinguishing between precession $\dot{\phi}$ and spin $\dot{\psi}$, leading to $\omega_z = \omega_3 = \dot{\phi} + \dot{\psi}$. When $\theta = \pi$ then the z axis and 3 axis are antiparallel and $\omega_z = \dot{\phi} - \dot{\psi} = -\omega_3$. The other special case is when $\cos\theta = 0$ for which the Euler angle system is orthogonal and the space-fixed $\omega_z = \dot{\phi}$, that is, it equals the precession, while the body-fixed $\omega_3 = \dot{\psi}$, that is, it equals the spin. When the Euler angle basis is not orthogonal then equations $(13.86 - 88)$ and $(13.89 - 91)$ are needed for expressing the Euler equations of motion in either the body-fixed frame or the space-fixed frame respectively.

Equations $13.86 - 88$ for the components of the angular velocity in the body-fixed frame can be expressed in terms of the Euler angle velocities in a matrix form as

$$\begin{pmatrix} \omega_1 \\ \omega_2 \\ \omega_3 \end{pmatrix} = \begin{pmatrix} \sin\theta\sin\psi & \cos\psi & 0 \\ \sin\theta\cos\psi & -\sin\psi & 0 \\ \cos\theta & 0 & 1 \end{pmatrix} \cdot \begin{pmatrix} \dot{\phi} \\ \dot{\theta} \\ \dot{\psi} \end{pmatrix} \tag{13.92}$$

again note that the transformation matrix is not orthogonal which is to be expected since the Euler angular velocities are about axes that do not form a rectangular system of coordinates. Similarly equations $13.89 - 91$ for the angular velocity in the space-fixed frame can be expressed in terms of the Euler angle velocities in matrix form as

$$\begin{pmatrix} \omega_x \\ \omega_y \\ \omega_z \end{pmatrix} = \begin{pmatrix} 0 & \cos\phi & \sin\theta\sin\phi \\ 0 & \sin\phi & \sin\theta\cos\phi \\ 1 & 0 & \cos\theta \end{pmatrix} \cdot \begin{pmatrix} \dot{\phi} \\ \dot{\theta} \\ \dot{\psi} \end{pmatrix} \tag{13.93}$$

13.15   Kinetic energy in terms of Euler angular velocities

The kinetic energy is a scalar quantity and thus is the same in both stationary and rotating frames of reference. It is much easier to evaluate the kinetic energy in the rotating Principal-axis frame since the inertia tensor is diagonal in the Principal-axis frame as given in equation 13.69

$$T_{rot} = \frac{1}{2} \sum_i^3 I_i \omega_i^2 \tag{13.94}$$

Using equation $13.86 - 88$ for the body-fixed angular velocities gives the rotational kinetic energy in terms of the Euler angular velocities and principal-frame moments of inertia to be

$$T_{rot} = \frac{1}{2} \left[I_1 \left(\dot{\phi}\sin\theta\sin\psi + \dot{\theta}\cos\psi \right)^2 + I_2 \left(\dot{\phi}\sin\theta\cos\psi - \dot{\theta}\sin\psi \right)^2 + I_3 \left(\dot{\phi}\cos\theta + \dot{\psi} \right)^2 \right] \tag{13.95}$$

13.16   Rotational invariants

The scalar properties of a rotating body, such as mass M, Lagrangian L, and Hamiltonian H, are rotationally invariant, that is, they are the same in any body-fixed or laboratory-fixed coordinate frame. This fact also applies to scalar products of all vector observables such as angular momentum. For example the scalar product

$$\mathbf{L} \cdot \mathbf{L} = l^2 \tag{13.96}$$

where l is the root mean square value of the angular momentum. An example of a scalar invariant is the scalar product of the angular velocity

$$\boldsymbol{\omega} \cdot \boldsymbol{\omega} = \omega^2 \tag{13.97}$$

where ω^2 is the mean square angular velocity. The scalar product $\omega \cdot \omega = |\omega|^2$ can be calculated using the Euler-angle velocities for the body-fixed frame, equations $13.86 - 88$, to be

$$\boldsymbol{\omega} \cdot \boldsymbol{\omega} = |\omega|^2 = \omega_1^2 + \omega_2^2 + \omega_3^2 = \dot{\phi}^2 + \dot{\theta}^2 + \dot{\psi}^2 + 2\dot{\phi}\dot{\psi}\cos\theta$$

Similarly, the scalar product can be calculated using the Euler angle velocities for the space-fixed frame using equations $13.89 - 91$.

$$\boldsymbol{\omega} \cdot \boldsymbol{\omega} = |\omega|^2 = \omega_x^2 + \omega_y^2 + \omega_z^2 = \dot{\phi}^2 + \dot{\theta}^2 + \dot{\psi}^2 + 2\dot{\phi}\dot{\psi}\cos\theta$$

This shows the obvious result that the scalar product $\omega \cdot \omega = |\omega|^2$ is invariant to rotations of the coordinate frame, that is, it is identical when evaluated in either the space-fixed, or body-fixed frames.

Note that for $\theta = 0$, the $\hat{3}$ and $\hat{z}$ axes are parallel, and perpendicular to the $\hat{\theta}$ axis, then

$$|\omega|^2 = \left(\dot{\phi} + \dot{\psi}\right)^2 + \dot{\theta}^2$$

For the case when $\theta = 180°$, the $\hat{3}$ and $\hat{z}$ axes are antiparallel, and perpendicular to the $\hat{\theta}$ axis, then

$$|\omega|^2 = \left(\dot{\phi} - \dot{\psi}\right)^2 + \dot{\theta}^2$$

For the case when $\theta = 90°$, the $\hat{3}$, $\hat{z}$, and $\hat{\theta}$ axes are mutually perpendicular, that is, orthogonal, and then

$$|\omega|^2 = \dot{\phi}^2 + \dot{\psi}^2 + \dot{\theta}^2$$

The time-averaged shape of a rapidly-rotating body, as seen in the fixed inertial frame, is very different from the actual shape of the body, and this difference depends on the rotational frequency. For example, a pencil rotating rapidly about an axis perpendicular to the body-fixed symmetry axis has an average shape that is a flat disk in the laboratory frame which bears little resemblance to a pencil. The actual shape of the pencil could be determined by taking high-speed photographs which display the instantaneous body-fixed shape of the object at given times. Unfortunately for fast rotation, such as rotation of a molecule or a nucleus, it is not possible to take photographs with sufficient speed and spatial resolution to observe the instantaneous shape of the rotating body. What is measured is the average shape of the body as seen in the fixed laboratory frame. In principle the shape observed in the fixed inertial frame can be related to the shape in the body-fixed frame, but this requires knowing the body-fixed shape which in general is not known. For example, a deformed nucleus may be both vibrating and rotating about some triaxially deformed average shape which is a function of the rotational frequency. This is not apparent from the shapes measured in the fixed frame for each of the excited states.

The fact that scalar products are rotationally invariant, provides a powerful means of transforming products of observables in the body-fixed frame, to those in the laboratory frame. In 1971 Cline developed a powerful model-independent method that utilizes rotationally-invariant products of the electromagnetic quadrupole operator $E2$ to relate the electromagnetic $E2$ properties for the observed levels of a rotating nucleus measured in the laboratory frame, to the electromagnetic $E2$ properties of the deformed rotating nucleus measured in the body-fixed frame.[Cli71, Cli72, Cli86] The method uses the fact that scalar products of the electromagnetic multipole operators are rotationally invariant. This allows transforming scalar products of a complete set of measured electromagnetic matrix elements, measured in the laboratory frame, into the electromagnetic properties in the body-fixed frame of the rotating nucleus. These rotational invariants provide a model-independent determination of the magnitude, triaxiality, and vibrational amplitudes of the average shapes in the body-fixed frame for individual observed nuclear states that may be undergoing both rotation and vibration. When the bombarding energy is below the Coulomb barrier, the scattering of a projectile nucleus by a target nucleus is due purely to the electromagnetic interaction since the distance of closest approach exceeds the range of the nuclear force. For such pure Coulomb collisions, the electromagnetic excitation of collective nuclei populates many excited states, as illustrated in figure 14.13, with cross sections that are a direct measure of the $E2$ matrix elements. These measured matrix elements are precisely those required to evaluate, in the laboratory frame, the $E2$ rotational invariants from which it is possible to deduce the intrinsic quadrupole shapes of the rotating-vibrating nuclear states in the body-fixed frame[Cli86].

13.17   Euler's equations of motion for rigid-body rotation

Rigid-body rotation can be confusing in that two coordinate frames are involved and, in general, the angular velocity and angular momentum are not aligned. The motion of the rigid body is observed in the space-fixed inertial frame whereas it is simpler to calculate the equations of motion in the body-fixed principal axis frame, for which the inertia tensor is known and is constant. The rigid body is rotating about the angular velocity vector $\boldsymbol{\omega}$, which is not aligned with the angular momentum $\mathbf{L}$. For torque-free motion, $\mathbf{L}$ is conserved and has a fixed orientation in the space-fixed axis system. Euler's equations of motion, presented below, are given in the body-fixed frame for which the inertial tensor is known since this simplifies solution of the equations of motion. However, this solution has to be rotated back into the space-fixed frame to describe the rotational motion as seen by an observer in the inertial frame.

This chapter has introduced the inertial properties of a rigid body, as well as the Euler angles for transforming between the body-fixed and inertial frames of reference. This has prepared the stage for solving the equations of motion for rigid-body motion, namely, the dynamics of rotational motion about a body-fixed point under the action of external forces. The Euler angles are used to specify the instantaneous orientation of the rigid body.

In Newtonian mechanics, the rotational motion is governed by the equivalent Newton's second law given in terms of the external torque $\mathbf{N}$ and angular momentum $\mathbf{L}$

$$\mathbf{N} = \left(\frac{d\mathbf{L}}{dt}\right)_{space} \tag{13.98}$$

Note that this relation is expressed in the inertial space-fixed frame of reference, not the non-inertial body-fixed frame. The subscript *space* is added to emphasize that this equation is written in the inertial space-fixed frame of reference. However, as already discussed, it is much more convenient to transform from the space-fixed inertial frame to the body-fixed frame for which the inertia tensor of the rigid body is known. Thus the next stage is to express the rotational motion in terms of the body-fixed frame of reference. For simplicity, translational motion will be ignored.

The rate of change of angular momentum can be written in terms of the body-fixed value, using the transformation from the space-fixed inertial frame $(\hat{\mathbf{x}}, \hat{\mathbf{y}}, \hat{\mathbf{z}})$ to the rotating frame $(\hat{\mathbf{e}}_1, \hat{\mathbf{e}}_2, \hat{\mathbf{e}}_3)$ as given in chapter 10.3,

$$\mathbf{N} = \left(\frac{d\mathbf{L}}{dt}\right)_{space} = \left(\frac{d\mathbf{L}}{dt}\right)_{body} + \boldsymbol{\omega} \times \mathbf{L} \tag{13.99}$$

However, the body axis $\hat{\mathbf{e}}_i$ is chosen to be the principal axis such that

$$L_i = I_i \omega_i \tag{13.100}$$

where the principal moments of inertia are written as I_i. Thus the equation of motion can be written using the body-fixed coordinate system as

$$\mathbf{N} = I_1 \dot{\omega}_1 \hat{\mathbf{e}}_1 + I_2 \dot{\omega}_2 \hat{\mathbf{e}}_2 + I_3 \dot{\omega}_3 \hat{\mathbf{e}}_3 + \begin{vmatrix} \hat{\mathbf{e}}_1 & \hat{\mathbf{e}}_2 & \hat{\mathbf{e}}_3 \\ \omega_1 & \omega_2 & \omega_3 \\ I_1\omega_1 & I_2\omega_2 & I_3\omega_3 \end{vmatrix} \tag{13.101}$$

$$= \left(I_1\dot{\omega}_1 - (I_2 - I_3)\,\omega_2\omega_3\right)\hat{\mathbf{e}}_1 + \left(I_2\dot{\omega}_2 - (I_3 - I_1)\,\omega_3\omega_1\right)\hat{\mathbf{e}}_2 + \left(I_3\dot{\omega}_3 - (I_1 - I_2)\,\omega_1\omega_2\right)\hat{\mathbf{e}}_3 \tag{13.102}$$

where the components in the body-fixed axes are given by

$$
\begin{aligned}
N_1 &= I_1\dot{\omega}_1 - (I_2 - I_3)\,\omega_2\omega_3 \\
N_2 &= I_2\dot{\omega}_2 - (I_3 - I_1)\,\omega_3\omega_1 \\
N_3 &= I_3\dot{\omega}_3 - (I_1 - I_2)\,\omega_1\omega_2
\end{aligned}
\tag{13.103}
$$

These are the **Euler equations for rigid body in a force field** expressed in the *body-fixed coordinate frame*. They are applicable for any applied external torque **N**.

The motion of a rigid body depends on the structure of the body only via the three principal moments of inertia $I_1, I_2,$ and I_3. Thus all bodies having the same principal moments of inertia will behave exactly the same even though the bodies may have very different shapes. As discussed earlier, the simplest geometrical shape of a body having three different principal moments is a homogeneous ellipsoid. Thus, the rigid-body motion often is described in terms of the equivalent ellipsoid that has the same principal moments.

A deficiency of Euler's equations is that the solutions yield the time variation of $\boldsymbol{\omega}$ as seen from the body-fixed reference frame axes, and not in the observers fixed inertial coordinate frame. Similarly the components of the external torques in the Euler equations are given with respect to the body-fixed axis system which implies that the orientation of the body is already known. Thus for non-zero external torques the problem cannot be solved until the the orientation is known in order to determine the components N_i^{ext}. However, these difficulties disappear when the external torques are zero, or if the motion of the body is known and it is required to compute the applied torques necessary to produce such motion.

13.18 Lagrange equations of motion for rigid-body rotation

The Euler equations of motion were derived using Newtonian concepts of torque and angular momentum. It is of interest to derive the equations of motion using Lagrangian mechanics. It is convenient to use a generalized torque N and assume that $U = 0$ in the Lagrange-Euler equations. Note that the generalized force is a torque since the corresponding generalized coordinate is an angle, and the conjugate momentum is angular momentum. If the body-fixed frame of reference is chosen to be the principal axes system, then, since the inertia tensor is diagonal in the principal axis frame, the kinetic energy is given in terms of the principal moments of inertia as

$$
T = \frac{1}{2}\sum_i I_i\omega_i^2
\tag{13.104}
$$

Using the Euler angles as generalized coordinates, then the Lagrange equation for the specific case of the ψ coordinate and including a generalized force N_ψ gives

$$
\frac{d}{dt}\frac{\partial T}{\partial \dot{\psi}} - \frac{\partial T}{\partial \psi} = N_\psi
\tag{13.105}
$$

which can be expressed as

$$
\frac{d}{dt}\sum_i^3 \frac{\partial T}{\partial \omega_i}\frac{\partial \omega_i}{\partial \dot{\psi}} - \sum_i^3 \frac{\partial T}{\partial \omega_i}\frac{\partial \omega_i}{\partial \psi} = N_\psi
\tag{13.106}
$$

Equation 13.104 gives

$$
\frac{\partial T}{\partial \omega_i} = I_i\omega_i
\tag{13.107}
$$

Differentiating the angular velocity components in the body-fixed frame, equations $(13.86 - 13.88)$, give

$\frac{\partial \omega_1}{\partial \psi} = \dot{\phi}\sin\theta\cos\psi - \dot{\theta}\sin\psi = \omega_2$	$\frac{\partial \omega_1}{\partial \dot{\psi}} = \frac{\partial \omega_2}{\partial \dot{\psi}} = 0$
$\frac{\partial \omega_2}{\partial \psi} = -\dot{\phi}\sin\theta\sin\psi - \dot{\theta}\cos\psi = -\omega_1$	$\frac{\partial \omega_1}{\partial \dot{\psi}} = \frac{\partial \omega_2}{\partial \dot{\psi}} = 0$
$\frac{\partial \omega_3}{\partial \psi} = 0$	$\frac{\partial \omega_3}{\partial \dot{\psi}} = 1$

Substituting these into the Lagrange equation (13.106) gives

$$
\frac{d}{dt}I_3\omega_3 - I_1\omega_1\omega_2 + I_2\omega_2\left(-\omega_1\right) = N_3
\tag{13.108}
$$

since the ψ and $\widehat{\mathbf{e}}_3$ axes are colinear. This can be rewritten as

$$I_3\dot{\omega}_3 - (I_1 - I_2)\,\omega_1\omega_2 = N_3 \tag{13.109}$$

Any axis could have been designated the $\widehat{\mathbf{e}}_3$ axis, thus the above equation can be generalized to all three axes to give

$$
\begin{aligned}
I_1\dot{\omega}_1 - (I_2 - I_3)\,\omega_2\omega_3 &= N_1 \\
I_2\dot{\omega}_2 - (I_3 - I_1)\,\omega_3\omega_1 &= N_2 \\
I_3\dot{\omega}_3 - (I_1 - I_2)\,\omega_1\omega_2 &= N_3
\end{aligned}
\tag{13.110}
$$

These are the **Euler's equations** given previously in (13.103). Note that although $\dot{\omega}_3$ is the equation of motion for the ψ coordinate, this is not true for the ϕ and θ rotations which are not along the body-fixed x_1 and x_2 axes as given in table 13.1.

13.8 Example: Rotation of a dumbbell

Consider the motion of the symmetric dumbbell shown in the adjacent figure. Let $|r_1| = |r_2| = b$. Let the body-fixed coordinate system have its origin at O and symmetry axis $\widehat{e}_3$ be along the weightless shaft toward m_1 and $\mathbf{v}_\alpha = v_\alpha\hat{e}_1$. The angular momentum is given by

$$\mathbf{L} = \sum_i m_i\mathbf{r}_i \times \mathbf{v}_i$$

Because $\mathbf{L}$ is perpendicular to the shaft, and $\mathbf{L}$ rotates around $\boldsymbol{\omega}$ as the shaft rotates, let $\widehat{e}_2$ be along $\mathbf{L}$.

$$\mathbf{L} = L_2\widehat{\mathbf{e}}_2$$

If α is the angle between $\boldsymbol{\omega}$ and the shaft, the components of $\boldsymbol{\omega}$ are

$$
\begin{aligned}
\omega_1 &= 0 \\
\omega_2 &= \omega\sin\alpha \\
\omega_3 &= \omega\cos\alpha
\end{aligned}
$$

Assume that the principal moments of the dumbbell are

$$
\begin{aligned}
I_1 &= (m_1 + m_2)\,b^2 \\
I_2 &= (m_1 + m_2)\,b^2 \\
I_3 &= 0
\end{aligned}
$$

Thus the angular momentum is given by

$$
\begin{aligned}
L_1 &= I_1\omega_1 = 0 \\
L_2 &= I_2\omega_2 = (m_1 + m_2)\,b^2\omega\sin\alpha \\
L_3 &= I_3\omega_3 = 0
\end{aligned}
$$

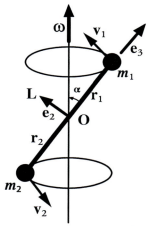

Rotation of a dumbbell.

which is consistent with the angular momentum being along the $\widehat{e}_2$ axis.

Using Euler's equations, and assuming that the angular velocity is constant, i.e. $\dot{\omega} = 0$, then the components of the torque required to satisfy this motion are

$$
\begin{aligned}
N_1 &= -(m_1 + m_2)\,b^2\omega^2\sin\alpha\cos\alpha \\
N_2 &= 0 \\
N_3 &= 0
\end{aligned}
$$

That is, this motion can only occur in the presence of the above applied torque which is in the direction $-\widehat{e}_1$, that is, mutually perpendicular to $\widehat{e}_2$ and $\widehat{e}_3$. This torque can be written as $\mathbf{N} = \boldsymbol{\omega} \times \mathbf{L}$.

13.19 Hamiltonian equations of motion for rigid-body rotation

The Hamiltonian equations of motion are expressed in terms of the Euler angles plus their corresponding canonical angular momenta $(\phi, \theta, \psi, p_\phi, p_\theta, p_\psi)$ in contrast to Lagrangian mechanics which is based on the Euler angles plus their corresponding angular velocities $(\phi, \theta, \psi, \dot\phi, \dot\theta, \dot\psi)$. The Hamiltonian approach is conveniently expressed in terms of a set of Andoyer-Deprit action-angle coordinates that include the three Euler angles, specifying the orientation of the body-fixed frame, plus the corresponding three angles specifying the orientation of the spin frame of reference. This phase space approach[Dep67] can be employed for calculations of rotational motion in celestial mechanics that can include spin-orbit coupling. This Hamiltonian approach is beyond the scope of the present textbook.

13.20 Torque-free rotation of an inertially-symmetric rigid rotor

13.20.1 Euler's equations of motion:

There are many situations where one has rigid-body motion free of external torques, that is, $\mathbf{N} = 0$. The tumbling motion of a jugglers baton, a diver, a rotating galaxy, or a frisbee, are examples of rigid-body rotation. For torque-free rotation, the body will rotate about the center of mass, and thus the inertia tensor with respect to the center of mass is required. An inertially-symmetric rigid body has two identical principal moments of inertia with $I_1 = I_2 \neq I_3$, and provides a simple example that illustrates the underlying motion. The force-free Euler equations for the symmetric body in the body-fixed principal axis system are given by

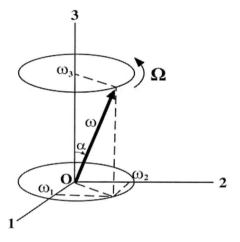

$$(I_2 - I_3)\,\omega_2\omega_3 - I_1\dot\omega_1 = 0 \qquad (13.111)$$

$$(I_3 - I_1)\,\omega_3\omega_1 - I_2\dot\omega_2 = 0 \qquad (13.112)$$

$$I_3\dot\omega_3 = 0 \qquad (13.113)$$

where $I_1 = I_2$ and $N = 0$ apply.

Note that for torque-free motion of an inertially symmetric body equation 13.113 implies that $\dot\omega_3 = 0$, i.e. ω_3 is a constant of motion and thus is a cyclic variable for the symmetric rigid body.

Equations 13.111 and 13.112 can be written as two coupled equations

$$\dot\omega_1 + \Omega\omega_2 = 0 \qquad (13.114)$$

$$\dot\omega_2 - \Omega\omega_1 = 0 \qquad (13.115)$$

Figure 13.4: The force-free symmetric top angular velocity ω precesses on a conical trajectory about the body-fixed symmetry axis $\hat{\mathbf{3}}$.

where the precession angular velocity $\boldsymbol{\Omega} = \dot\psi$ *with respect to the body-fixed frame* is defined to be

$$\boldsymbol{\Omega} \equiv \left(\frac{(I_3 - I_1)}{I_1}\omega_3 \right) \qquad (13.116)$$

Combining the time derivatives of equations 13.114 and 13.115 leads to two uncoupled equations

$$\ddot\omega_1 + \Omega^2\omega_1 = 0 \qquad (13.117)$$

$$\ddot\omega_2 + \Omega^2\omega_2 = 0 \qquad (13.118)$$

These are the differential equations for a harmonic oscillator with solutions

$$\omega_1 = A\cos\Omega t \qquad (13.119)$$

$$\omega_2 = A\sin\Omega t$$

These equations describe a vector A rotating in a circle of radius A about an axis perpendicular to $\hat{e}_3$, that is, rotating in the $\hat{e}_1 - \hat{e}_2$ plane with angular frequency $\Omega = -\dot{\psi}$. Note that

$$\omega_1^2 + \omega_2^2 = A^2 \tag{13.120}$$

which is a constant. In addition ω_3 is constant, therefore the magnitude of the total angular velocity

$$|\boldsymbol{\omega}| = \sqrt{\omega_1^2 + \omega_2^2 + \omega_3^2} = \text{ constant} \tag{13.121}$$

The motion of the torque-free symmetric body is that the angular velocity $\boldsymbol{\omega}$ precesses around the symmetry axis $\hat{e}_3$ of the body at an angle α with a constant precession frequency Ω with respect to the body-fixed frame as shown in figure 13.4. Thus, to an observer on the body, $\boldsymbol{\omega}$ traces out a cone around the body-fixed symmetry axis. Note from (13.116) that the vectors $\Omega\hat{e}_3$ and $\omega_3\hat{e}_3$ are parallel when Ω is positive, that is, $I_3 > I$ (oblate shape) and antiparallel if $I_3 < I$ (prolate shape).

For the system considered, the orientation of the angular momentum vector $\mathbf{L}$ must be stationary in the space-fixed inertial frame since the system is torque free, that is, $\mathbf{L}$ is a constant of motion. Also we have that the projection of the angular momentum on the body-fixed symmetry axis is a constant of motion, that is, it is a cyclic variable. Thus

$$L_3 = I_3\omega_3 = \frac{I_1 I_3}{(I_3 - I_1)}\Omega \tag{13.122}$$

Understanding the relation between the angular momentum and angular velocity is facilitated by considering another constant of motion for the torque-free symmetric rotor, namely the rotational kinetic energy.

$$T_{rot} = \frac{1}{2}\boldsymbol{\omega} \cdot \mathbf{L} = \text{constant} \tag{13.123}$$

Since $\mathbf{L}$ is a constant for torque-free motion, and also the magnitude of $\boldsymbol{\omega}$ was shown to be constant, therefore the angle between these two vectors must be a constant to ensure that also $T_{\mathbf{rot}} = \frac{1}{2}\boldsymbol{\omega} \cdot \mathbf{L} = \text{constant}$. That is, $\boldsymbol{\omega}$ precesses around $\mathbf{L}$ at a constant angle $(\theta - \alpha)$ such that the projection of $\boldsymbol{\omega}$ onto $\mathbf{L}$ is constant. Note that

$$\boldsymbol{\omega} \times \widehat{\mathbf{e}}_3 = \omega_2\widehat{\mathbf{e}}_1 - \omega_1\widehat{\mathbf{e}}_2 \tag{13.124}$$

and, for a symmetric rotor,

$$\mathbf{L} \cdot \boldsymbol{\omega} \times \widehat{\mathbf{e}}_3 = I_1\omega_1\omega_2 - I_2\omega_1\omega_2 = 0 \tag{13.125}$$

since $I_1 = I_2$ for the symmetric rotor. Because $\mathbf{L} \cdot \boldsymbol{\omega} \times \widehat{\mathbf{e}}_3 = 0$ for a symmetric top then $\mathbf{L}, \boldsymbol{\omega}$ and $\widehat{\mathbf{e}}_3$ are coplanar.

Figure 13.5 shows the geometry of the motion for both oblate and prolate axially-deformed bodies. To an observer in the space-fixed inertial frame, the angular velocity $\boldsymbol{\omega}$ traces out a cone that precesses with angular velocity Ω around the space fixed $\mathbf{L}$ axis called the space cone. For convenience, figure 13.5 assumes that $\mathbf{L}$ and the space-fixed inertial frame $\hat{\mathbf{z}}$ axis are colinear. The angular velocity $\boldsymbol{\omega}$ also traces out the body cone as it precesses about the body-fixed $\hat{\mathbf{e}}_3$ axis. Since $\mathbf{L}, \boldsymbol{\omega}$ and $\widehat{\mathbf{e}}_3$ are coplanar, then the $\boldsymbol{\omega}$ vector is at the intersection of the space and body cones as the body cone rolls around the space cone. That is, the space and body cones have one generatrix in common which coincides with $\boldsymbol{\omega}$. As shown in figure 13.5b, for a needle the body cone appears to roll without slipping on the outside of the space cone at the precessional velocity of $\Omega = -\omega$. By contrast, as shown in figure 13.5a for an oblate (disc-shaped) symmetric top the space cone rolls inside the body cone and the precession Ω is faster than ω.

Since no external torques are acting for torque-free motion, then the magnitude and direction of the total angular momentum are conserved. The description of the motion is simplified if $\mathbf{L}$ is taken to be along the space-fixed $\hat{\mathbf{z}}$ axis, then the Euler angle θ is the angle between the body-fixed basis vector $\hat{\mathbf{e}}_3$ and space-fixed basis vector $\hat{\mathbf{z}}$. If at some instant in the body frame, it is assumed that $\widehat{\mathbf{e}}_2$ is aligned in the plane of $\mathbf{L}, \boldsymbol{\omega}$ and $\widehat{\mathbf{e}}_3$, then

$$L_1 = 0 \qquad\qquad L_2 = L\sin\theta \qquad\qquad L_3 = L\cos\theta \tag{13.126}$$

If α is the angle between the angular velocity $\boldsymbol{\omega}$ and the body-fixed $\hat{\mathbf{e}}_3$ axis, then at the same instant

$$\omega_1 = 0 \qquad\qquad \omega_2 = \omega\sin\alpha \qquad\qquad \omega_3 = \omega\cos\alpha \tag{13.127}$$

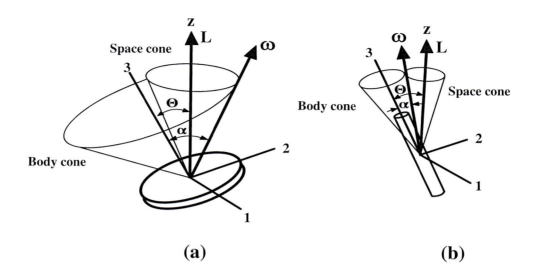

Figure 13.5: Torque-free rotation of symmetric tops; (a) circular flat disk, (b) circular rod. The space-fixed and body-fixed cones are shown by fine lines. The space-fixed axis system is designated by the unit vectors $(\hat{\mathbf{x}}, \hat{\mathbf{y}}, \hat{\mathbf{z}})$ and the body-fixed principal axis system by unit vectors $(\hat{\mathbf{1}}, \hat{\mathbf{2}}, \hat{\mathbf{3}})$.

The components of the angular momentum also can be derived from $\mathbf{L} = \mathbf{I} \cdot \boldsymbol{\omega}$ to give

$$L_1 = I_1\omega_1 = 0 \qquad\qquad L_2 = I_2\omega_2 = I_1\omega\sin\alpha \qquad\qquad L_3 = I_3\omega_3 = I_3\omega\cos\alpha \qquad (13.128)$$

Equations 13.126 and 13.128 give two relations for the ratio $\frac{L_2}{L_3}$, that is,

$$\frac{L_2}{L_3} = \tan\theta = \frac{I_1}{I_3}\tan\alpha \qquad (13.129)$$

For a *prolate spheroid* $I_1 > I_3$ therefore $\theta > \alpha$ while Ω and ω_3 have opposite signs.
For a *oblate spheroid* $I_1 < I_3$ therefore $\alpha > \theta$ while Ω and ω_3 have the same sign.

The sense of precession can be understood if the body cone rolls without slipping on the outside of the space cone with Ω in the opposite orientation to ω for the prolate case, while for the oblate case the space cone rolls inside the body cone with Ω and ω oriented in similar directions. Note from (13.129) that $\theta = 0$ if $\alpha = 0$, that is $\mathbf{L}, \boldsymbol{\omega}$ and the **3** axis are aligned corresponding to a principal axis. Similarly, $\theta = 90°$ if $\alpha = 90°$, then again $\mathbf{L}$ and $\boldsymbol{\omega}$ are aligned corresponding to them being principal axes.

Lagrangian mechanics has been used to calculate the motion with respect to the body-fixed principal axis system. However, the motion needs to be known relative to the space-fixed inertial frame where the motion is observed. This transformation can be done using the following relation

$$\left(\frac{d\hat{\mathbf{e}}_3}{dt}\right)_{space} = \left(\frac{d\hat{\mathbf{e}}_3}{dt}\right)_{body} + \boldsymbol{\omega} \times \hat{\mathbf{e}}_3 = \boldsymbol{\omega} \times \hat{\mathbf{e}}_3 \qquad (13.130)$$

since the unit vector $\hat{\mathbf{e}}_3$ is stationary in the body-fixed frame. The vector product of $\boldsymbol{\omega} \times \hat{\mathbf{e}}_3$ and $\hat{\mathbf{e}}_3$ gives

$$\hat{\mathbf{e}}_3 \times \left(\frac{d\hat{\mathbf{e}}_3}{dt}\right)_{space} = \hat{\mathbf{e}}_3 \times \boldsymbol{\omega} \times \hat{\mathbf{e}}_3 = (\hat{\mathbf{e}}_3 \cdot \hat{\mathbf{e}}_3)\boldsymbol{\omega} - (\hat{\mathbf{e}}_3 \cdot \boldsymbol{\omega})\hat{\mathbf{e}}_3 = \boldsymbol{\omega} - \omega_3\hat{\mathbf{e}}_3$$

therefore

$$\boldsymbol{\omega} = \hat{\mathbf{e}}_3 \times \left(\frac{d\hat{\mathbf{e}}_3}{dt}\right)_{space} + \omega_3\hat{\mathbf{e}}_3 \qquad (13.131)$$

The angular momentum equals $\mathbf{L} = \{\mathbf{I}\} \cdot \boldsymbol{\omega}$. Since $\hat{\mathbf{e}}_3 \times \left(\frac{d\hat{\mathbf{e}}_3}{dt}\right)_{space}$ is perpendicular to the $\hat{\mathbf{e}}_3$ axis, then for the case with $I_1 = I_2$,

$$\mathbf{L} = I_1\hat{\mathbf{e}}_3 \times \left(\frac{d\hat{\mathbf{e}}_3}{dt}\right)_{space} + I_3\omega_3\hat{\mathbf{e}}_3 \qquad (13.132)$$

Thus the angular momentum for a torque-free symmetric rigid rotor comprises two components, one being the perpendicular component that precesses around $\hat{\mathbf{e}}_3$, and the other is L_3.

In the space-fixed frame assume that the $\hat{\mathbf{z}}$ axis is colinear with $\mathbf{L}$. Then taking the scalar product of $\hat{\mathbf{e}}_3$ and $\mathbf{L}$, using equation 13.126 gives

$$L_3 = \hat{\mathbf{e}}_3 \cdot \mathbf{L} = I_1 \hat{\mathbf{e}}_3 \cdot \hat{\mathbf{e}}_3 \times \left(\frac{d\hat{\mathbf{e}}_3}{dt} \right)_{space} + I_3 \omega_3 \hat{\mathbf{e}}_3 \cdot \hat{\mathbf{e}}_3 \tag{13.133}$$

The first term on the right is zero and thus equation 13.133 and 13.126 give

$$L_3 = I_3 \omega_3 = L \cos\theta \tag{13.134}$$

The time dependence of the rotation of the body-fixed symmetry axis with respect to the space-fixed axis system can be obtained by taking the vector product $\hat{\mathbf{e}}_3 \times \mathbf{L}$ using equation 13.132 and using equation B.24 to expand the triple vector product,

$$\hat{\mathbf{e}}_3 \times \mathbf{L} = I_1 \hat{\mathbf{e}}_3 \times \left(\hat{\mathbf{e}}_3 \times \left(\frac{d\hat{\mathbf{e}}_3}{dt} \right)_{space} \right) + I_3 \omega_3 \hat{\mathbf{e}}_3 \times \hat{\mathbf{e}}_3 \tag{13.135}$$

$$= I_1 \left[\left(\hat{\mathbf{e}}_3 \cdot \left(\frac{d\hat{\mathbf{e}}_3}{dt} \right)_{space} \right) \hat{\mathbf{e}}_3 - (\hat{\mathbf{e}}_3 \cdot \hat{\mathbf{e}}_3) \left(\frac{d\hat{\mathbf{e}}_3}{dt} \right)_{space} \right] + 0$$

since $(\hat{\mathbf{e}}_3 \times \hat{\mathbf{e}}_3) = 0$. Moreover $(\hat{\mathbf{e}}_3 \cdot \hat{\mathbf{e}}_3) = 1$, and $\hat{\mathbf{e}}_3 \cdot \left(\frac{d\hat{\mathbf{e}}_3}{dt} \right)_{space} = 0$, since they are perpendicular, then

$$\left(\frac{d\hat{\mathbf{e}}_3}{dt} \right)_{space} = \frac{\mathbf{L}}{I_1} \times \hat{\mathbf{e}}_3 \tag{13.136}$$

This equation shows that the body-fixed symmetry axis $\hat{\mathbf{e}}_3$ precesses around the $\mathbf{L}$, where $\mathbf{L}$ is a constant of motion for torque-free rotation. The true rotational angular velocity $\boldsymbol{\omega}$ in the space-fixed frame, given by equations 13.131, can be evaluated using equation 13.136. Remembering that it was assumed that $\mathbf{L}$ is in the $\hat{\mathbf{z}}$ direction, that is, $\mathbf{L} = L\hat{\mathbf{z}}$, then

$$\boldsymbol{\omega} = \hat{\mathbf{e}}_3 \times \left(\frac{d\hat{\mathbf{e}}_3}{dt} \right)_{space} + \omega_3 \hat{\mathbf{e}}_3$$

$$= \frac{L}{I_1} \hat{\mathbf{e}}_3 \times (\hat{\mathbf{z}} \times \hat{\mathbf{e}}_3) + \left(\frac{L \cos\alpha}{I_3} \right) \hat{\mathbf{e}}_3$$

$$= \frac{L}{I_1} \hat{\mathbf{z}} + L \cos\alpha \left(\frac{I_1 - I_3}{I_1 I_3} \right) \hat{\mathbf{e}}_3 \tag{13.137}$$

That is, the symmetry axis of the axially-symmetric rigid rotor makes an angle θ to the angular momentum vector $L\hat{\mathbf{z}}$ and precesses around $L\hat{\mathbf{z}}$ with a constant angular velocity $\frac{L}{I_1}$ while the axial spin of the rigid body has a constant value $\frac{L}{I_3}$. Thus, in the precessing frame, the rigid body appears to rotate about its fixed symmetry axis with a constant angular velocity $\frac{L \cos\alpha}{I_3} - \frac{L \cos\alpha}{I_1} = L \cos\alpha \left(\frac{I_1 - I_3}{I_1 I_3} \right)$. The precession of the symmetry axis looks like a wobble superimposed on the spinning motion about the body-fixed symmetry axis. The angular precession rate in the space-fixed frame can be deduced by using the fact that

$$\dot{\phi} \sin\theta = \omega \sin\alpha \tag{13.138}$$

Then using equation 13.129 allows equation 13.138 to be written as

$$\dot{\phi} = \omega \sqrt{\left[1 + \left(\left(\frac{I_3}{I_1} \right)^2 - 1 \right) \cos^2\alpha \right]} \tag{13.139}$$

which gives the precession rate about the space-fixed axis in terms of the angular velocity ω. Note that the precession rate $\dot{\phi} > \omega$ if $\frac{I_3}{I_1} > 1$, that is, for oblate shapes, and $\dot{\phi} < \omega$ if $\frac{I_3}{I_1} < 1$, that is, for prolate shapes.

13.20.2 Lagrange equations of motion:

It is interesting to compare the equations of motion for torque-free rotation of an inertially-symmetric rigid rotor derived using Lagrange mechanics with that derived previously using Euler's equations based on Newtonian mechanics. Assume that the principal moments about the fixed point of the symmetric top are $I_1 = I_2 \neq I_3$ and that the kinetic energy equals the rotational kinetic energy, that is, it is assumed that the translational kinetic energy $T_{trans} = 0$. Then the kinetic energy is given by

$$T = \frac{1}{2} \sum_i I_i \omega_i^2 = \frac{1}{2} I_1 \left(\omega_1^2 + \omega_2^2 \right) + \frac{1}{2} I_3 \omega_3^2 \tag{13.140}$$

Equations $(13.86 - 88)$ for the body-fixed frame give

$$\omega_1^2 = \left(\dot{\phi} \sin\theta \sin\psi + \dot{\theta} \cos\psi \right)^2 = \dot{\phi}^2 \sin^2\theta \sin^2\psi + 2\dot{\phi}\dot{\theta} \sin\theta \sin\psi \cos\psi + \dot{\theta}^2 \cos^2\psi \tag{13.141}$$

$$\omega_2^2 = \left(\dot{\phi} \sin\theta \cos\psi - \dot{\theta} \sin\psi \right)^2 = \dot{\phi}^2 \sin^2\theta \cos^2\psi - 2\dot{\phi}\dot{\theta} \sin\theta \sin\psi \cos\psi + \dot{\theta}^2 \sin^2\psi \tag{13.142}$$

Therefore

$$\omega_1^2 + \omega_2^2 = \dot{\phi}^2 \sin^2\theta + \dot{\theta}^2 \tag{13.143}$$

and

$$\omega_3^2 = \left(\dot{\phi} \cos\theta + \dot{\psi} \right)^2 \tag{13.144}$$

Therefore the kinetic energy is

$$T = \frac{1}{2} I_1 \left(\dot{\phi}^2 \sin^2\theta + \dot{\theta}^2 \right) + \frac{1}{2} I_3 \left(\dot{\phi} \cos\theta + \dot{\psi} \right)^2 \tag{13.145}$$

Since the system is torque free, the scalar potential energy U can be assumed to be zero, and then the Lagrangian equals

$$L = \frac{1}{2} I_1 \left(\dot{\phi}^2 \sin^2\theta + \dot{\theta}^2 \right) + \frac{1}{2} I_3 \left(\dot{\phi} \cos\theta + \dot{\psi} \right)^2 \tag{13.146}$$

The angular momentum about the *space-fixed z axis* p_ϕ is conjugate to ϕ. From Lagrange's equations

$$\dot{p}_\phi = \frac{\partial L}{\partial \phi} = 0 \tag{13.147}$$

that is, the *angular momentum about the space-fixed z axis, p_ϕ is a constant of motion* given by

$$p_\phi = \frac{\partial L}{\partial \dot{\phi}} = \left(I_1 \sin^2\theta + I_3 \cos^2\theta \right) \dot{\phi} + I_3 \dot{\psi} \cos\theta = \text{constant.} \tag{13.148}$$

Similarly, the *angular momentum about the body-fixed 3 axis is conjugate to ψ.* From Lagrange's equations,

$$\dot{p}_\psi = \frac{\partial L}{\partial \psi} = 0 \tag{13.149}$$

that is, p_ψ *is a constant of motion* given by

$$p_\psi = \frac{\partial L}{\partial \dot{\psi}} = I_3 \left(\dot{\phi} \cos\theta + \dot{\psi} \right) = I_3 \omega_3 = \text{ constant} \tag{13.150}$$

The above two relations derived from the Lagrangian can be solved to give the *precession angular velocity* $\dot{\phi}$ about the space-fixed $\hat{\mathbf{z}}$ axis

$$\dot{\phi} = \frac{p_\phi - p_\psi \cos\theta}{I_1 \sin^2\theta} \tag{13.151}$$

and the *spin about the body-fixed $\hat{\mathbf{3}}$ axis* $\dot{\psi}$ which is given by

$$\dot{\psi} = \frac{p_\psi}{I_3} - \frac{(p_\phi - p_\psi \cos\theta) \cos\theta}{I_1 \sin^2\theta} \tag{13.152}$$

Since p_ϕ and p_ψ are constants of motion, then the precessional angular velocity $\dot{\phi}$ about the space-fixed $\hat{\mathbf{z}}$ axis, and the spin angular velocity $\dot{\psi}$, which is the spin frequency about the body-fixed $\hat{\mathbf{3}}$ axis, are constants that depend directly on I_1, I_3. and θ.

There is one additional constant of motion available if no dissipative forces act on the system, that is, energy conservation which implies that the total energy

$$E = \frac{1}{2}I_1 \left(\dot{\phi}^2 \sin^2\theta + \dot{\theta}^2\right) + \frac{1}{2}I_3 \left(\dot{\phi}\cos\theta + \dot{\psi}\right)^2 \tag{13.153}$$

will be a constant of motion. But the second term on the right-hand side also is a constant of motion since p_ψ and I_3 both are constants, that is

$$\frac{1}{2}I_3\omega_3^2 = \frac{1}{2}I_3 \left(\dot{\phi}\cos\theta + \dot{\psi}\right)^2 = \frac{p_\psi^2}{I_3} = \text{constant} \tag{13.154}$$

Thus energy conservation implies that the first term on the right-hand side also must be a constant given by

$$\frac{1}{2}I_1 \left(\omega_1^2 + \omega_2^2\right) = \frac{1}{2}I_1 \left(\dot{\phi}^2 \sin^2\theta + \dot{\theta}^2\right) = E - \frac{p_\psi^2}{I_3} = \text{constant} \tag{13.155}$$

These results are identical to those given in equations 13.120 and 13.121 which were derived using Euler's equations. These results illustrate that the underlying physics of the torque-free rigid rotor is more easily extracted using Lagrangian mechanics rather than using the Euler-angle approach of Newtonian mechanics.

13.9 *Example: Precession rate for torque-free rotating symmetric rigid rotor*

Table 13.2 lists the precession and spin angular velocities, in the space-fixed frame, for torque-free rotation of three extreme symmetric-top geometries spinning with constant angular momentum ω when the motion is slightly perturbed such that ω is at a small angle α to the symmetry axis. Note that this assumes the perpendicular axis theorem, equation 13.45 which states that for a thin laminae $I_1 + I_2 = I_3$ giving, for a thin circular disk, $I_1 = I_2$ and thus $I_3 = 2I_1$.

Table 13.2: Precession and spin rates for torque-free axial rotation of symmetric rigid rotors

Rigid-body symmetric shape	Principal moment ratio $\frac{I_3}{I_1}$	Precession rate $\dot{\phi}$	Spin rate $\dot{\psi}$
Symmetric needle	0	0	ω
Sphere	1	ω	0
Thin circular disk	2	2ω	$-\omega$

The precession angular velocity in the space frame ranges between 0 to 2ω depending on whether the body-fixed spin angular velocity is aligned or anti-aligned with the rotational frequency ω. For an extreme prolate spheroid $\frac{I_3}{I_1} = 0$, the body-fixed spin angular velocity $\Omega = -\omega_3$ which cancels the angular velocity ω of the rotating frame resulting in a zero precession angular velocity of the body-fixed $\hat{\mathbf{e}}_3$ axis around the space-fixed frame. The spin $\Omega = 0$ in the body-fixed frame for the rigid sphere $\frac{I_3}{I_1} = 1$, and thus the precession rate of the body-fixed $\hat{\mathbf{e}}_3$ axis of the sphere around the space-fixed frame equals ω. For oblate spheroids and thin disks, such as a frisbee, $\frac{I_3}{I_1} = 2$ making the body-fixed precession angular velocity $\Omega = +\omega$ which adds to the angular velocity ω and increases the precession rate up to 2ω as seen in the space-fixed frame. This illustrates that the spin angular velocity can add constructively or destructively with the angular velocity ω.[2]

---

[2] In his autobiography *Surely You're Joking Mr Feynman*, he wrote " I was in the [Cornell] cafeteria and some guy, fooling around, throws a plate in the air. As the plate went up in the air I saw it wobble, and noticed that the red medallion of Cornell on the plate going around. It was pretty obvious to me that the medallion went around faster than the wobbling. I started to figure out the motion of the rotating plate. I discovered that when the angle is very slight, the medallion rotates twice as fast as the wobble rate. It came out of a very complicated equation! ". The quoted ratio (2 : 1) is incorrect, it should be (1 : 2). Benjamin Chao in *Physics Today* of February 1989 speculated that Feynman's error in inverting the factor of two might be "in keeping with the spirit of the author and the book, another practical joke meant for those who do physics without experimenting". He pointed out that this story occurred on page 157 of a book of length 314 pages (1:2). Observe the dependence of the ratio of wobble to rotation angular velocities on the tilt angle θ.

13.21   Torque-free rotation of an asymmetric rigid rotor

The Euler equations of motion for the case of torque-free rotation of an asymmetric (triaxial) rigid rotor about the center of mass, with principal moments of inertia $I_1 \neq I_2 \neq I_3$, lead to more complicated motion than for the symmetric rigid rotor.[3] The general features of the motion of the asymmetric rotor can be deduced using the conservation of angular momentum and rotational kinetic energy.

Assuming that the external torques are zero then the Euler equations of motion can be written as

$$I_1\dot{\omega}_1 = (I_2 - I_3)\,\omega_2\omega_3 \qquad (13.156)$$
$$I_2\dot{\omega}_2 = (I_3 - I_1)\,\omega_3\omega_1$$
$$I_3\dot{\omega}_3 = (I_1 - I_2)\,\omega_1\omega_2$$

Since $L_i = I_i\omega_i$ for $i = 1, 2, 3$, then equation 13.156 gives

$$I_2 I_3 \dot{L}_1 = (I_2 - I_3)\,L_2 L_3 \qquad (13.157)$$
$$I_1 I_3 \dot{L}_2 = (I_3 - I_1)\,L_3 L_1$$
$$I_1 I_2 \dot{L}_3 = (I_1 - I_2)\,L_1 L_2$$

Multiply the first equation by $I_1 L_1$, the second by $I_2 L_2$ and the third by $I_3 L_3$ and sum, which gives

$$I_1 I_2 I_3 \left(L_1 \dot{L}_1 + L_2 \dot{L}_2 + L_3 \dot{L}_3 \right) = 0 \qquad (13.158)$$

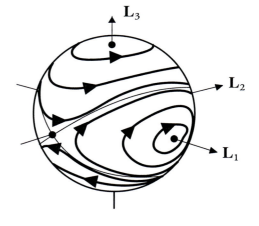

The bracket is equivalent to $\frac{d}{dt}(L_1^2 + L_2^2 + L_3^2) = 0$ which implies that *the total rotational angular momentum L is a constant of motion* as expected for this torque-free system, even though the individual components L_1, L_2, L_3 may vary. That is

$$L_1^2 + L_2^2 + L_3^2 = L^2 \qquad (13.159)$$

Figure 13.6: Rotation of an asymmetric rigid rotor. The dark lines correspond to contours of constant total rotational kinetic energy T, which has an ellipsoidal shape, projected onto the angular momentum L sphere in the body-fixed frame.

Note that *equation 13.159 is the equation of a sphere of radius L.*

Multiply the first equation of 13.157 by L_1, the second by L_2, and the third by L_3, and sum gives

$$I_2 I_3 L_1 \dot{L}_1 + I_1 I_3 L_2 \dot{L}_2 + I_1 I_2 L_3 \dot{L}_3 = 0 \qquad (13.160)$$

Divide 13.160 by $I_1 I_2 I_3$ gives $\frac{d}{dt}(\frac{L_1^2}{2I_1} + \frac{L_2^2}{2I_2} + \frac{L_3^2}{2I_3}) = 0$. This implies that the *total rotational kinetic energy T*, given by

$$\frac{L_1^2}{2I_1} + \frac{L_2^2}{2I_2} + \frac{L_3^2}{2I_3} = T \qquad (13.161)$$

is a constant of motion as expected when there are no external torques and zero energy dissipation. Note that 13.161 *is the equation of an ellipsoid.*

Equations 13.159 and 13.161 both must be satisfied by the rotational motion for any value of the total angular momentum **L** and kinetic energy T. Fig 13.6 shows a graphical representation of the intersection of the L sphere and T ellipsoid as seen *in the body-fixed frame.* The angular momentum vector **L** must follow the constant-energy contours given by where the T-ellipsoids intersect the L-sphere, shown for the case where $I_3 > I_2 > I_1$. Note that the precession of the angular momentum vector **L** follows a trajectory that has closed paths that circle around the principal axis with the smallest I, that is, $\hat{\mathbf{e}}_1$, or the principal axis with the maximum I, that is, $\hat{\mathbf{e}}_3$. However, the angular momentum vector does not have a stable minimum for precession around the intermediate principal moment of inertia axis $\hat{\mathbf{e}}_2$. In addition to the precession, the angular momentum vector **L** executes nutation, that is a nodding of the angle θ.

For any fixed value of L, the kinetic energy has upper and lower bounds given by

$$\frac{L^2}{2I_3} \leq T \leq \frac{L^2}{2I_1} \qquad (13.162)$$

---

[3]Similar discussions of the freely-rotating asymmetric top are given by Landau and Lifshitz [La60] and by Gregory [Gr06].

Thus, for a given value of L, when $T = T_{\min} = \frac{L^2}{2I_3}$, the orientation of $\mathbf{L}$ in the body-fixed frame is either $(0, 0, +L)$ or $(0, 0, -L)$, that is, aligned with the $\hat{\mathbf{e}}_3$ axis along which the principal moment of inertia is largest. For slightly higher kinetic energy the trajectory of L follows closed paths precessing around $\hat{\mathbf{e}}_3$. When the kinetic energy $T = \frac{L^2}{2I_2}$ the angular momentum vector L follows either of the two thin-line trajectories each of which are a separatrix. These do not have closed orbits around $\hat{\mathbf{e}}_2$ and they separate the closed solutions around either $\hat{\mathbf{e}}_3$ or $\hat{\mathbf{e}}_1$. For higher kinetic energy the precessing angular momentum vector follows closed trajectories around $\hat{\mathbf{e}}_1$ and becomes fully aligned with $\hat{\mathbf{e}}_1$ at the upper-bound kinetic energy.

Note that for the special case when $I_3 > I_2 = I_1$, then the asymmetric rigid rotor equals the symmetric rigid rotor for which the solutions of Euler's equations were solved exactly in chapter 13.19. For the symmetric rigid rotor the T-ellipsoid becomes a spheroid aligned with the symmetry axis and thus the intersections with the L-sphere lead to circular paths around the $\hat{\mathbf{e}}_3$ body-fixed principal axis, while the separatrix circles the equator corresponding to the $\hat{\mathbf{e}}_3$ axis separating clockwise and anticlockwise precession about $\mathbf{L}_3$. This discussion shows that energy, plus angular momentum conservation, provide the general features of the solution for the torque-free symmetric top that are in agreement with those derived using Euler's equations of motion

13.22   Stability of torque-free rotation of an asymmetric body

It is of interest to extend the prior discussion to address the stability of an asymmetric rigid rotor undergoing force-free rotation close to a principal axes, that is, when subject to small perturbations. Consider the case of a general asymmetric rigid body with $I_3 > I_2 > I_1$. Let the system start with rotation about the $\hat{\mathbf{e}}_1$ axis, that is, the principal axis associated with the moment of inertia I_1. Then

$$\boldsymbol{\omega} = \omega_1 \hat{\mathbf{e}}_1 \tag{13.163}$$

Consider that a small perturbation is applied causing the angular velocity vector to be

$$\boldsymbol{\omega} = \omega_1 \hat{\mathbf{e}}_1 + \lambda \hat{\mathbf{e}}_2 + \mu \hat{\mathbf{e}}_3 \tag{13.164}$$

where λ, μ are very small. The Euler equations (13.156) become

$$\begin{aligned}
(I_2 - I_3)\, \lambda\mu - I_1 \dot{\omega}_1 &= 0 \\
(I_3 - I_1)\, \mu\omega_1 - I_2 \dot{\lambda} &= 0 \\
(I_1 - I_2)\, \omega_1\lambda - I_3 \dot{\mu} &= 0
\end{aligned}$$

Assuming that the product $\lambda\mu$ in the first equation is negligible, then $\dot{\omega}_1 = 0$, that is, ω_1 is constant.

The other two equations can be solved to give

$$\dot{\lambda} = \left(\frac{(I_3 - I_1)}{I_2} \omega_1 \right) \mu \tag{13.165}$$

$$\dot{\mu} = \left(\frac{(I_1 - I_2)}{I_3} \omega_1 \right) \lambda \tag{13.166}$$

Take the time derivative of the first equation

$$\ddot{\lambda} = \left(\frac{(I_3 - I_1)}{I_2} \omega_1 \right) \dot{\mu} \tag{13.167}$$

and substitute for $\dot{\mu}$ gives

$$\ddot{\lambda} + \left(\frac{(I_1 - I_3)(I_1 - I_2)}{I_2 I_3} \omega_1^2 \right) \lambda = 0 \tag{13.168}$$

The solution of this equation is

$$\lambda(t) = A e^{i\Omega_{1\lambda} t} + B e^{-i\Omega_{1\lambda} t} \tag{13.169}$$

where

$$\Omega_{1\lambda} = \omega_1 \sqrt{\frac{(I_1 - I_3)(I_1 - I_2)}{I_2 I_3}} \tag{13.170}$$

Note that since it was assumed that $I_3 > I_2 > I_1$, then $\Omega_{1\lambda}$ is real. The solution for $\lambda(t)$ therefore represents a stable oscillatory motion with precession frequency $\Omega_{1\lambda}$. The identical result is obtained for $\Omega_{1\mu} = \Omega_{1\lambda} = \Omega_1$. Thus the motion corresponds to a stable minimum about the $\hat{\mathbf{e}}_1$ axis with oscillations about the $\lambda = \mu = 0$ minimum with period.

$$\Omega_1 = \omega_1 \sqrt{\frac{(I_1 - I_3)(I_1 - I_2)}{I_2 I_3}} \tag{13.171}$$

Permuting the indices gives that for perturbations applied to rotation about either the 2 or 3 axes give precession frequencies

$$\Omega_2 = \omega_2 \sqrt{\frac{(I_2 - I_1)(I_2 - I_3)}{I_1 I_3}} \tag{13.172}$$

$$\Omega_3 = \omega_3 \sqrt{\frac{(I_3 - I_2)(I_3 - I_1)}{I_1 I_2}} \tag{13.173}$$

Since $I_3 > I_2 > I_1$ then Ω_1 and Ω_3 are real while Ω_2 is imaginary. Thus, whereas rotation about either the I_3 or the I_1 axes are stable, the imaginary solution about $\hat{\mathbf{e}}_2$ corresponds to a perturbation increasing with time. Thus, only rotation about the largest or smallest moments of inertia are stable. Moreover for the symmetric rigid rotor, with $I_1 = I_2 \neq I_3$, stability exists only about the symmetry axis $\hat{\mathbf{e}}_3$ independent on whether the body is prolate or oblate. This result was implied from the discussion of energy and angular momentum conservation in chapter 13.20. Friction was not included in the above discussion. In the presence of dissipative forces, such as friction or drag, only rotation about the principal axis corresponding to the maximum moment of inertia is stable.

Stability of rigid-body rotation has broad applications to rotation of satellites, molecules and nuclei. The first U.S. satellite, Explorer 1, was launched in 1958 with the rotation axis aligned with the cylindrical axis which was the minimum principal moment of inertia. After a few hours the satellite started tumbling with increasing amplitude due to a flexible antenna dissipating and transferring energy to the perpendicular axis which had the largest moment of inertia. Torque-free motion of a deformed rigid body is a ubiquitous phenomena in many branches of science, engineering, and sports as illustrated by the following examples.

13.10 *Example: Tennis racquet dynamics*

A tennis racquet is an asymmetric body that exhibits the above rotational behavior. Assume that the head of a tennis racquet is a uniform thin circular disk of radius R and mass M which is attached to a cylindrical handle of diameter $r = \frac{R}{10}$, length $2R$, and mass M as shown in the figure. The principle moments of inertia about the three axes through the center-of-mass can be calculated by addition of the moments for the circular disk and the cylindrical handle and using both the parallel-axis and the perpendicular-axis theorems.

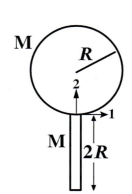

Principal rotation axes for the center of mass of a tennis racket. The 1 and 2 -axes are in the plane of the racket head and the 3 axis is perpendicular to the plane of the racket head.

Axis	Head	Handle	Racquet
1	$\frac{1}{4}MR^2 + MR^2 = \frac{5}{4}MR^2$	$\frac{4}{3}MR^2$	$\frac{31}{12}MR^2$
2	$\frac{1}{4}MR^2 + 0 = \frac{1}{4}MR^2$	$\frac{1}{200}MR^2$	$\frac{51}{200}MR^2$
3	$\frac{1}{2}MR^2 + MR^2 = \frac{3}{2}MR^2$	$\frac{4}{3}MR^2$	$\frac{17}{6}MR^2$

Note that $I_{11} : I_{22} : I_{33} = 2.5833 : 0.2550 : 2.8333$. Inserting these principle moments of inertia into equations $13.171 - 13.173$ gives the following precession frequencies

$$\Omega_1 = i0.8976\omega_1 \qquad \Omega_2 = 0.9056\omega_2 \qquad \Omega_3 = 0.9892\omega_3$$

The imaginary precession frequency Ω_1 about the 1 axis implies unstable rotation leading to tumbling whereas the minimum moment I_{22} and maximum moment I_{33} imply stable rotation about the 2 and 3 axes. This rotational behavior is easily demonstrated by throwing a tennis racquet and is called the tennis racquet theorem. The center of percussion, example 2.14, is another important inertial property of a tennis racquet.

13.11 *Example: Rotation of asymmetrically-deformed nuclei*

Some nuclei and molecules have average shapes that have significant asymmetric deformation leading to interesting quantal analogs of the rotational properties of an asymmetrically-deformed rigid body. The major difference between a quantal and a classical rotor is that the energies, and angular momentum are quantized, rather than being continuously variable quantities. Otherwise, the quantal rotors exhibit general features similar to the classical analog. Studies [Cli86] of the rotational behavior of asymmetrically-deformed nuclei exploit three aspects of classical mechanics, namely classical Coulomb trajectories, rotational invariants, and the properties of ellipsoidal rigid-bodies.

Ellipsoidal deformation can be specified by the dimensions along each of the three principle axes. Bohr and Mottelson parameterized the ellipsoidal deformation in terms of three parameters, R_0 which is the radius of the equivalent sphere, β which is a measure of the magnitude of the ellipsoidal deformation from the sphere, and γ which specifies the deviation of the shape from axial symmetry. The ellipsoidal intrinsic shape can be expressed in terms of the deviation from the equivalent sphere by the equation

$$\delta R(\theta, \phi) = R(\theta, \phi) - R_0 = R_0 \sum_{\mu=-2}^{\mu+2} \alpha_{2\mu}^* Y_{2\mu}(\theta, \phi) \tag{a}$$

where $Y_{\lambda\mu}(\theta, \phi)$ is a Laplace spherical harmonic defined as

$$Y_{\lambda\mu}(\theta, \phi) = \sqrt{\frac{(2\lambda + 1)}{4\pi} \frac{(\lambda - \mu)!}{(\lambda + \mu)!}} P_{\lambda\mu}(\cos\theta) e^{-i\mu\phi}$$

and $P_{\lambda\mu}(\cos\theta)$ is an associated Legendre function of $\cos\theta$. Spherical harmonics are the angular portion of a set of solutions to Laplace's equation. Represented in a system of spherical coordinates, Laplace's spherical harmonics $Y_{\lambda\mu}(\theta, \phi)$ are a specific set of spherical harmonics that form an orthogonal system. Spherical harmonics are important in many theoretical and practical applications.

In the principal axis frame of the body, there are three non-zero quadrupole deformation parameters which can be written in terms of the deformation parameters β, γ where $\alpha_{20} = \beta \cos\gamma$, $\alpha_{21} = \alpha_{2-1} = 0$, and $\alpha_{22} = \alpha_{2-2} = \frac{1}{\sqrt{2}} \beta \sin\gamma$. Using these in equations (a) give the three semi-axis dimensions in the principal axis frame, (primed frame),

$$\delta R_k = \sqrt{\frac{5}{4\pi}} R_0 \beta \cos(\gamma - \frac{2\pi k}{3}) \tag{b}$$

Note that for $\gamma = 0$, then $\delta R_1 = \delta R_2 = -\frac{1}{2}\sqrt{\frac{5}{4\pi}} R_0 \beta$ while $\delta R_3 = +\sqrt{\frac{5}{4\pi}} R_0 \beta$, that is the body has prolate deformation with the symmetry axis along the 3 axis. The same prolate shape is obtained for $\gamma = \frac{2\pi}{3}$ and $\gamma = \frac{4\pi}{3}$ with the prolate symmetry axes along the 1 and 2 axes respectively. For $\gamma = \frac{\pi}{3}$ then $\delta R_1 = \delta R_3 = +\frac{1}{2}\sqrt{\frac{5}{4\pi}} R_0 \beta$ while $\delta R_2 = -\sqrt{\frac{5}{4\pi}} R_0 \beta$, that is the body has oblate deformation with the symmetry axis along the 2 axis. The same oblate shape is obtained for $\gamma = \pi$ and $\gamma = \frac{5\pi}{3}$ with the oblate symmetry axes along the 3 and 1 axes respectively. For other values of γ the shape is ellipsoidal.

For the asymmetric deformed rigid body, the rotational Hamiltonian can be expressed in the form[Dav58]

$$H = \sum_{k=1}^{3} \frac{|R|^2}{4B\beta^2 \sin^2(\gamma' - \frac{2\pi k}{3})}$$

where the rotational angular momentum is $\mathbf{R}$. The principal moments of inertia are related by the triaxiality parameter γ' which they assumed is identical to the shape parameter γ. For axial symmetry the moment of inertia about the symmetry axis is taken to be zero for a quantal system since rotation of the potential well about the symmetry axis corresponds to no change in the potential well, or corresponding rotation of the bound nucleons. That is, the nucleus is not a rigid body, the nucleons only rotate to the extent that the ellipsoidal potential well is cranked around such that the nucleons must follow the rotation of the potential well. In addition, vibrational modes coexist about the average asymmetric deformation, plus octupole deformation often coexists with the above quadrupole deformed modes.

13.23   Symmetric rigid rotor subject to torque about a fixed point

The motion of a symmetric top rotating in a gravitational field, with one point at a fixed location, is encountered frequently in rotational motion. Examples are the gyroscope and a child's spinning top. Rotation of a rigid rotor subject to torque about a fixed point, is a case where it is necessary to take the inertia tensor with respect to the fixed point in the body, and not at the center of mass.

Consider the geometry, shown in figure 13.7, where the symmetric top of mass M is spinning about a fixed tip that is displaced by a distance h from the center of mass. The tip of the top is assumed to be at the origin of both the space-fixed frame (x, y, z) and the body-fixed frame $(1, 2, 3)$. Assume that the translational velocity is zero and let the principal moments about the fixed point of the symmetric top be $I_1 = I_2 \neq I_3$.

The Lagrange equations of motion can be derived assuming that the kinetic energy equals the rotational kinetic energy, that is, it is assumed that the translational kinetic energy $T_{trans} = 0$. Then the kinetic energy of an inertially-symmetric rigid rotor can be derived for the torque-free symmetric top as given in equation 13.145 to be

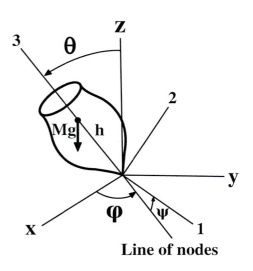

Figure 13.7: Symmetric top spinning about one fixed point.

$$T = \frac{1}{2}\sum_i I_i \omega_i^2 = \frac{1}{2}I_1\left(\omega_1^2 + \omega_2^2\right) + \frac{1}{2}I_3\omega_3^2 \qquad (13.174)$$

$$= \frac{1}{2}I_1\left(\dot{\phi}^2\sin^2\theta + \dot{\theta}^2\right) + \frac{1}{2}I_3\left(\dot{\phi}\cos\theta + \dot{\psi}\right)^2 \qquad (13.175)$$

Since the potential energy is $U = Mgh\cos\theta$ then the Lagrangian equals

$$L = \frac{1}{2}I_1\left(\dot{\phi}^2\sin^2\theta + \dot{\theta}^2\right) + \frac{1}{2}I_3\left(\dot{\phi}\cos\theta + \dot{\psi}\right)^2 - Mgh\cos\theta \qquad (13.176)$$

The angular momentum about the *space-fixed z axis p_ϕ is conjugate to* ϕ. From Lagrange's equations

$$\dot{p}_\phi = \frac{\partial L}{\partial \phi} = 0 \qquad (13.177)$$

that is, p_ϕ *is a constant of motion* given by the generalized momentum

$$p_\phi = \frac{\partial L}{\partial \dot{\phi}} = \left(I_1\sin^2\theta + I_3\cos^2\theta\right)\dot{\phi} + I_3\dot{\psi}\cos\theta = S_z = \text{ constant} \qquad (13.178)$$

where S_z is the angular momentum projection along the space-fixed z axis.

Similarly, the *angular momentum about the body-fixed 3 axis is conjugate to* ψ. From Lagrange's equations,

$$\dot{p}_\psi = \frac{\partial L}{\partial \psi} = 0 \qquad (13.179)$$

that is, p_ψ *is a constant of motion* given by the generalized momentum

$$p_\psi = \frac{\partial L}{\partial \dot{\psi}} = I_3\left(\dot{\phi}\cos\theta + \dot{\psi}\right) = B_3 = \text{ constant} \qquad (13.180)$$

where B_3 is the angular momentum projection along the body-fixed 3 axis. The above two relations can be solved to give the precessional angular velocity $\dot{\phi}$ about the space-fixed z axis

$$\dot{\phi} = \frac{p_\phi - p_\psi\cos\theta}{I_1\sin^2\theta} = \frac{S_z - B_3\cos\theta}{I_1\sin^2\theta} \qquad (13.181)$$

and the spin angular velocity $\dot{\psi}$ about the body-fixed x_3 axis

$$\dot{\psi} = \frac{p_\psi}{I_3} - \frac{(p_\phi - p_\psi\cos\theta)\cos\theta}{I_1\sin^2\theta} = \frac{B_3}{I_3} - \frac{(S_z - B_3\cos\theta)\cos\theta}{I_1\sin^2\theta} \qquad (13.182)$$

Since p_ϕ and p_ψ are constants of motion, i.e. S_3, B_3, then these rotational angular velocities depend on only I_1, I_3. and θ.

There is one further constant of motion available if no frictional forces act on the system, that is, energy conservation. This implies that the total energy

$$E = \frac{1}{2}I_1\left(\dot{\phi}^2\sin^2\theta + \dot{\theta}^2\right) + \frac{1}{2}I_3\left(\dot{\phi}\cos\theta + \dot{\psi}\right)^2 + Mgh\cos\theta \quad (13.183)$$

will be a constant of motion. But the middle term on the right-hand side also is a constant of motion

$$\frac{1}{2}I_3\left(\dot{\phi}\cos\theta + \dot{\psi}\right)^2 = \frac{p_\psi^2}{I_3} = \frac{B_3^2}{I_3} = \text{constant} \quad (13.184)$$

Thus energy conservation can be rewritten by defining an energy E' where

$$E' \equiv E - \frac{p_\psi^2}{I_3} = \frac{1}{2}I_1\left(\dot{\phi}^2\sin^2\theta + \dot{\theta}^2\right) + Mgh\cos\theta = \text{constant} \quad (13.185)$$

This can be written as

$$E' = \frac{1}{2}I_1\dot{\theta}^2 + \frac{(p_\phi - p_\psi\cos\theta)^2}{2I_1\sin^2\theta} + Mgh\cos\theta \quad (13.186)$$

which can be expressed as

$$E' = \frac{1}{2}I_1\dot{\theta}^2 + V(\theta) \quad (13.187)$$

where $V(\theta)$ is an effective potential

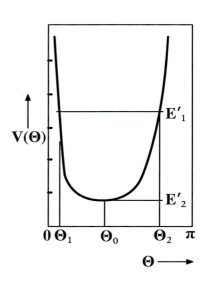

Figure 13.8: Effective potential diagram for a spinning symmetric top as a function of theta.

$$V(\theta) \equiv \frac{(p_\phi - p_\psi\cos\theta)^2}{2I_1\sin^2\theta} + Mgh\cos\theta = \frac{(S_z - B_3\cos\theta)^2}{2I_1\sin^2\theta} + Mgh\cos\theta \quad (13.188)$$

The effective potential $V(\theta)$ is shown in figure 13.8. It is clear that the motion of a symmetric top with effective energy E' is confined to angles $\theta_1 < \theta < \theta_2$.

Note that the above result also is obtained if the Routhian is used, rather than the Lagrangian, as mentioned in chapter 8.7, and defined by equation (8.65). That is, the Routhian can be written as

$$
\begin{aligned}
R(\theta, \dot{\theta}, p_\phi p_\psi)_{cyclic} &= \dot{\phi}p_\phi + \dot{\psi}p_\psi - L = H(\phi, p_\phi, \psi, p_\psi) - L(\theta, \dot{\theta})_{noncyclic} \\
&= -\frac{1}{2}I_1\dot{\theta}^2 + \frac{(p_\phi - p_\psi\cos\theta)^2}{2I_1\sin^2\theta} + \frac{p_\psi^2}{2I_3} + Mgh\cos\theta \quad (13.189)
\end{aligned}
$$

The Routhian $R(\theta, \dot{\theta}, p_\phi p_\psi)_{cyclic}$ acts like a Hamiltonian for the (ϕ, p_ϕ) and (ψ, p_ψ) variables which are constants of motion, and thus are ignorable variables. The Routhian acts as the negative Lagrangian for the remaining variable θ, with rotational kinetic energy $\frac{1}{2}I_1\dot{\theta}^2$ and effective potential energy V_{eff}

$$V_{eff} = \frac{(p_\phi - p_\psi\cos\theta)^2}{2I_1\sin^2\theta} + \frac{p_\psi^2}{I_3} + Mgh\cos\theta = V(\theta) + \frac{p_\psi^2}{I_3}$$

The equation of motion describing the system in the rotating frame is given by one Lagrange equation

$$\frac{d}{dt}\left(\frac{\partial R_{cyclic}}{\partial\dot{\theta}}\right) - \frac{\partial R_{cyclic}}{\partial\theta} = 0$$

The negative sign of the Routhian cancels out when used in the Lagrange equation. Thus, in the rotating frame of reference, the system is reduced to a single degree of freedom, the nutation angle θ, with effective energy E' given by equations $13.186 - 13.188$.

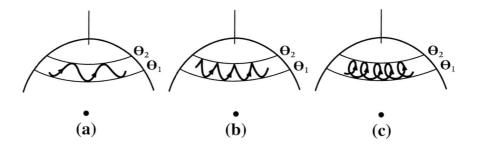

$$\textbf{(a)} \qquad\qquad \textbf{(b)} \qquad\qquad \textbf{(c)}$$

Figure 13.9: Nutational motion of the body-fixed symmetry axis projected onto the space-fixed unit sphere. The three case are (a) $\dot{\phi}$ never vanishes, (b) $\dot{\phi} = 0$ at $\theta = \theta_2$ (c) $\dot{\phi}$ changes sign between θ_1 and θ_2,

The motion of the symmetric top is simplest at the minimum value of the effective potential curve, where $E' = V_{\min}$, at which the nutation θ is restricted to a single value $\theta = \theta_0$. The motion is a steady precession at a fixed angle of inclination, that is, the "sleeping top". Solving for $\left(\frac{dV}{d\theta}\right)_{\theta=\theta_0} = 0$ gives that

$$p_\phi - p_\psi \cos\theta = \frac{p_\psi \sin^2\theta_0}{2\cos\theta_0}\left[1 \pm \sqrt{1 - \frac{4MghI_1\cos\theta_0}{p_\psi^2}}\right] \tag{13.190}$$

If $\theta_0 < \frac{\pi}{2}$, then to ensure that the solution is real requires a minimum value of the angular momentum on the body-fixed axis of $p_\psi^2 \geq 4MghI_1\cos\theta_0$. If $\theta_0 > \frac{\pi}{2}$ then there is no minimum angular momentum projection on the body-fixed axis. There are two possible solutions to the quadratic relation corresponding to either a slow or fast precessional frequency. Usually the slow precession is observed.

For the general case, where $E'_1 > V_{\min}$, the nutation angle θ between the space-fixed and body-fixed 3 axes varies in the range $\theta_1 < \theta < \theta_2$. This axis exhibits a nodding variation which is called **nutation**. Figure 13.9 shows the projection of the body-fixed symmetry axis on the unit sphere in the space-fixed frame. Note that the observed nutation behavior depends on the relative sizes of p_ϕ and $p_\psi \cos\theta$. For certain values, the precession $\dot{\phi}$ changes sign between the two limiting values of θ producing a looping motion as shown in figure 13.9c. Another condition is where the precession is zero for θ_2 producing a cusp at θ_2 as illustrated in figure 13.9b. This behavior can be demonstrated using the gyroscope or the symmetric top.

13.12 *Example: The Spinning "Jack"*

The game "Jacks" is played using metal Jacks, each of which comprises six equal masses m at the opposite ends of orthogonal axes of length l. Consider one jack spinning around the body-fixed $3-$axis with the lower mass at a fixed point on the ground, and with a steady precession around the space-fixed vertical axis z with angle θ as shown. Assume that the body-fixed axes align with the arms of the jack.

The principal moments of inertia about one mass is given by the parallel axis theorem to be $I_2 = I_1 = 4ml^2 + 6ml^2 = 10ml^2$ and $I_3 = 4ml^2$.

In the rotating body-fixed frame the torque due to gravity has components

$$\mathbf{N} = \begin{pmatrix} 6mgl\sin\theta\sin\psi \\ 6mgl\sin\theta\cos\psi \\ 0 \end{pmatrix}$$

and the components of the angular velocity are

$$\boldsymbol{\omega} = \begin{pmatrix} \dot{\phi}\sin\theta\sin\psi + \dot{\theta}\cos\psi \\ \dot{\phi}\sin\theta\cos\psi - \dot{\theta}\sin \\ \dot{\phi}\cos\theta + \dot{\psi} \end{pmatrix}$$

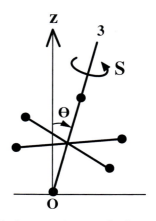

Jack comprises six bodies of mass m at each end of orthogonal arms of length l

Using Euler's equations (13.103) for the above components of N and ω in the body-fixed frame, gives

$$10\dot{\omega}_1 - 6\omega_2\omega_3 \;\; = \;\; \frac{6g}{l}\sin\theta\sin\psi \tag{a}$$

$$10\dot{\omega}_2 - 6\omega_1\omega_3 \;\; = \;\; \frac{6g}{l}\sin\theta\cos\psi \tag{b}$$

$$4\dot{\omega}_3 \;\; = \;\; 0 \tag{c}$$

Equation (c) relates the spin about the 3 axis, the precession, and the angle to the vertical θ, that is

$$\omega_3 = \dot{\phi}\cos\theta + \dot{\psi} = \; \Omega\cos\theta + s = \text{constant}$$

where $\dot{\psi} \equiv s$ is the spin and $\dot{\phi} \equiv \Omega$ is the precession angular velocity.

If the spin axis is nearly vertical, $\theta \approx 0$ and thus $\sin\theta \approx \theta$ and $\cos\theta \approx 1$. Multiply equation $(a) \times \sin\psi + (b) \times \cos\psi$ and using the equations of the components of ω gives

$$5\ddot{\theta} + \left(2\Omega s - 3\Omega^2 - \frac{3g}{l}\right)\theta = 0$$

The bracket must be positive to have stable sinusoidal oscillations. That is, the spin angular velocity s required for the jack to spin about a stable vertical axis is given by.

$$s > \frac{3\Omega}{2} + \frac{3g}{2l\Omega}$$

This example illustrates the conditions required for stable rotation of any axially-symmetric top.

13.13 *Example:* **The Tippe Top**

The Tippe Top comprises a section of a sphere, to which a short cylindrical rod is mounted on the planar section, as illustrated. When the Tippe Top is spun on a horizontal surface this top exhibits the perverse behavior of transitioning from rotation with the spherical head resting on the horizontal surface, to flipping over such that it rotates resting on its elongated cylindrical rod. The orientation of angular momentum remains roughly vertical as expected from conservation of angular momentum. This implies that the rotation with respect to the body-fixed axes must invert as the top inverts. The center of mass is raised when the top inverts; the additional potential energy is provided by a reduction in the rotational kinetic energy.

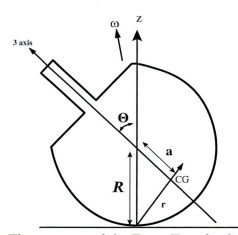

The Tippe Top behavior was first discovered in the 1890's but adequate solutions of the equations of motion have only been developed since the 1950's. Since the top precesses around the vertical axis, the point of contact is not on the symmetry axis of the top. Sliding friction between the surface of the spinning top and the horizontal surface provides a torque that causes the precession of the top to increase and eventually flip up onto the cylindrical peg. The Tippe Top is typical of many phenomena in physics where the underlying physics principle can be recognized but a detailed and rigorous solution can be complicated.

The geometry of the Tippe Top of radius R spinning on a horizontal surface with slipping friction acting between the top and the horizontal plane. The center of mass is a distance a from the center of the spherical section along the axis of symmetry of the top.

The system has five degrees of freedom, x, y which specify the location on the horizontal plane, plus the three Euler angles (φ, θ, ϕ). The paper by Cohen[Coh77] explains the motion in terms of Euler angles using the laboratory to body-fixed transformation relation. It shows that friction plays a pivotal role in the motion contrary to some earlier claims. Ciocci and Langerock[Cio07] used the Routhian R_{cyclic} to reduce the number *of degrees of freedom from 5 to 2, namely θ which is the tilt angle, and φ' which is the orientation of the tilt. This Routhian R_{cyclic} is a Lagrangian in two dimension that was used to derive the equations of motion via the Lagrange Euler equation*

$$\frac{d}{dt}\left(\frac{\partial R_{cyclic}}{\partial \dot{\theta}}\right) - \frac{\partial R_{cyclic}}{\partial \theta} = Q_\theta$$

$$\frac{d}{dt}\left(\frac{\partial R_{cyclic}}{\partial \dot{\varphi}'}\right) - \frac{\partial R_{cyclic}}{\partial \varphi'} = Q_{\varphi'}$$

where the Q_θ $Q_{\varphi'}$ are generalized torques about the 2 angles that take into account the sliding frictional forces. This sophisticated Routhian reduction approach provides an exhaustive and refined solution for the Tippe Top and confirms that sliding friction plays a key role in the unusual behavior of the Tippe Top.

13.24 The rolling wheel

As discussed in chapter 5.7, the rolling wheel is a non-holonomic system that is simple in principle, but in practice the solution can be complicated, as illustrated by the Tippe Top. Chapter 13.23 discussed the motion of a symmetric top rotating about a fixed point on the symmetry axis when subject to a torque. The rolling wheel involves rotation of a symmetric rigid body that is subject to torques. However, the point of contact of the wheel with a static plane is on the periphery of the wheel, and friction at the point of contact is assumed to ensure zero slip. Note that friction is necessary to ensure that the rotating object rolls without slipping, but the frictional force does no work for pure rolling of an undeformable rigid wheel.

The coordinate system employed is shown in Figure 13.10. For simplicity it is better to use a moving coordinate frame $(\mathbf{1}, \mathbf{2}, \mathbf{3})$ that is fixed to the orientation of the wheel with the origin at the center of mass of the wheel, but this moving reference frame *does not* include the angular velocity $\dot{\psi}$ of the disk about the **3** axis. That is, the moving $(\mathbf{1}, \mathbf{2}, \mathbf{3})$ frame has angular velocities

$$\omega_1 = \dot{\theta} \tag{13.191}$$
$$\omega_2 = \dot{\phi}\sin\theta$$
$$\omega_3 = \dot{\phi}\cos\theta$$

The frame fixed in the rotating wheel must include the additional angular velocity of the disk $\dot{\psi}$ about the $\hat{\mathbf{e}}_3$ axis, that is

$$\Omega_1 = \omega_1 = \dot{\theta} \tag{13.192}$$
$$\Omega_2 = \omega_2 = \dot{\phi}\sin\theta$$
$$\Omega_3 = \omega_3 + \dot{\psi} = \dot{\phi}\cos\theta + \dot{\psi}$$

where $\mathbf{\Omega}$ designates the angular velocity of the rotating disk, while $\boldsymbol{\omega}$ designates the rotation of the moving frame $(\mathbf{1}, \mathbf{2}, \mathbf{3})$.

The principle moments of inertia of a thin circular disk are related by the perpendicular axis theorem (chapter 13.9)

$$I_1 + I_2 = I_3$$

Since $I_1 = I_2$ for a uniform disk, therefore $I_3 = 2I_1$.

Equation 12.16 can be used to relate the vector forces $\mathbf{F}$ in the space-fixed frame to the rate of change of momenta in the moving frame $(\mathbf{1}, \mathbf{2}, \mathbf{3})$.

$$\mathbf{F} = \dot{\mathbf{p}}_{space} = \dot{\mathbf{p}}_{moving} + \boldsymbol{\omega} \times \mathbf{p} \tag{13.193}$$

This leads to the following relations for the three components in the moving frame

$$F_1 = \dot{p}_1 + \omega_2 p_3 - \omega_3 p_2 \tag{13.194}$$
$$F_2 - Mg\sin\theta = \dot{p}_2 + \omega_3 p_1 - \omega_1 p_3$$
$$F_3 - Mg\cos\theta = \dot{p}_3 + \omega_1 p_2 - \omega_2 p_1$$

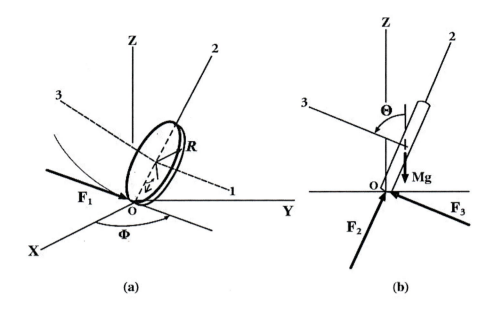

Figure 13.10: Uniform disk rolling on a horizontal plane as viewed in the (a) fixed frame, and (b) rolling disk frame. The space-fixed axis system is $(\mathbf{x}, \mathbf{y}, \mathbf{z})$, while the moving reference frame $(\mathbf{1}, \mathbf{2}, \mathbf{3})$ is centered at the center of mass of the disk with the $\mathbf{1}, \mathbf{2}$ axes in the plane of the disk. The disk is rotating with a uniform angular velocity $\dot{\psi}$ about the $\mathbf{3}$ axis and rolling in the direction that is at an angle ϕ relative to the x axis.

where F_1, F_2, F_3 are the reactive forces acting shown in figure 13.10.

Similarly, the torques $\mathbf{N}$ in the space-fixed frame can be related to the rate of change of angular momentum by

$$\mathbf{N} = \dot{\mathbf{L}}_{space} = \dot{\mathbf{L}}_{moving} + \boldsymbol{\omega} \times \mathbf{L} \tag{13.195}$$

where $L_i = \mathbf{I}_i \Omega_i$. This leads to the following relations for the three torque equations in the moving frame

$$
\begin{aligned}
N_1 &= -F_3 R = I_1 \dot{\Omega}_1 + I_3 \Omega_3 \omega_2 - I_2 \Omega_2 \omega_3 \tag{13.196}\\
N_2 &= 0 = I_1 \dot{\Omega}_2 + I_1 \Omega_1 \omega_3 - I_3 \Omega_3 \omega_1 \\
N_3 &= F_1 R = I_3 \dot{\Omega}_3 + I_2 \Omega_2 \omega_1 - I_1 \Omega_1 \omega_2
\end{aligned}
$$

The rolling constraints are

$$
\begin{aligned}
p_1 + MR\Omega_3 &= 0 \tag{13.197}\\
p_2 &= 0 \\
p_3 - MR\Omega_1 &= 0
\end{aligned}
$$

where $p_i = Mv_i$. Combining equations $13.194, 13.196, 13.197$ gives

$$
\begin{aligned}
\left(I_1 + MR^2\right) \dot{\Omega}_1 + \left(I_3 + MR^2\right) \omega_2 \Omega_3 - I_2 \omega_3 \Omega_2 &= -MgR\cos\theta \tag{13.198}\\
I_1 \dot{\Omega}_2 + I_1 \omega_3 \Omega_1 - I_3 \omega_1 \Omega_3 &= 0 \\
\left(I_3 + MR^2\right) \dot{\Omega}_3 + I_2 \omega_1 \Omega_2 - \left(I_1 + MR^2\right) \omega_2 \Omega_1 &= 0
\end{aligned}
$$

These are the torque equations about the point of contact O.

Introduction of equations 13.191 and 13.192 into equation 13.198 expresses the equations of motion in terms of the Euler angles to be

$$
\begin{aligned}
\left(I_1 + MR^2\right) \ddot{\theta} + \left(I_3 + MR^2\right) \dot{\phi}\sin\theta \left(\dot{\phi}\cos\theta + \dot{\psi}\right) - I_1 \dot{\phi}^2 \sin\theta\cos\theta &= -MgR\cos\theta \tag{13.199}\\
I_1 \ddot{\phi}\sin\theta + 2I_1 \dot{\phi}\dot{\theta}\cos\theta - I_3 \dot{\theta}\left(\dot{\phi}\cos\theta + \dot{\psi}\right) &= 0 \\
\left(I_3 + MR^2\right)\left(\ddot{\phi}\cos\theta - \dot{\phi}\dot{\theta}\sin\theta + \ddot{\psi}\right) - MR^2\dot{\theta}\dot{\phi}\sin\theta &= 0
\end{aligned}
$$

Equations 13.199 are non-linear, and a closed-form solution is possible only for limited cases such as when $\theta = 90°$.

Note that the above equations of motion also can be derived using Lagrangian mechanics knowing that

$$L = \frac{1}{2}M\left(v_1^2 + v_2^2 + v_3^2\right) + \frac{1}{2}I_1\left(\Omega_1^2 + \Omega_2^2\right) + \frac{1}{2}I_3\Omega_3^2 - MgR\cos\theta$$

The differential equations of constraint can be derived from equations 13.197 to be

$$
\begin{aligned}
dx - R\cos\phi\, d\psi &= 0 \\
dy - R\sin\phi\, d\psi &= 0
\end{aligned}
$$

Use of generalized forces plus the Lagrange-Euler equations (6.45) can be used to derive the equations of motion and solve for the components of the constraint force F_1, F_2, and F_3.

13.14 *Example: Tipping stability of a rolling wheel*

A circular wheel rolling in a vertical plane at high angular velocity initially rolls in a straight line and remains vertical. However, below a certain angular velocity, gyroscopic forces become weaker and the wheel will tip sideways and veer rapidly from the initial direction. It is interesting to estimate the minimum angular velocity of the disk such that it does not start to tip over sideways.

Note that equations 13.199 are satisfied for $\theta = \frac{\pi}{2}$, $\phi = 0$ and $\dot{\psi} = \Omega_3 = $ constant. Assume a small disturbance causes the tilt angle to be $\theta = \frac{\pi}{2} + \alpha$ where α is small and that ϕ is non-zero but small, that is $\dot{\theta} = \dot{\alpha}$ and $\dot{\phi}$ are small. Keeping only terms to first order in the third of equations 13.199, and integrating gives

$$\dot{\phi}\cos\theta + \dot{\psi} = \Omega_3 \tag{a}$$

The first two of equations 13.198 become

$$
\begin{aligned}
\left(I_1 + MR^2\right)\ddot{\alpha} + \left(I_3 + MR^2\right)\dot{\phi}\Omega_3 - MgR\alpha &= 0 \tag{b} \\
I_1\ddot{\phi} - I_3\Omega_3\dot{\alpha} &= 0 \tag{c}
\end{aligned}
$$

Integrating equation (c) gives

$$\dot{\phi} = \frac{I_3\Omega_3}{I_1}\alpha \tag{d}$$

Inserting (d) into (b) gives

$$\left(I_1 + MR^2\right)\ddot{\alpha} + \left[\left(I_3 + MR^2\right)\frac{I_3\Omega_3^2}{I_1} - MgR\right]\alpha = 0 \tag{e}$$

Equation (e) has a stable oscillatory solution when the square bracket in positive, that is,

$$\Omega_3^2 > \frac{I_1 MgR}{I_3\left(I_3 + MR^2\right)} \tag{f}$$

which gives the minimum angular velocity required for stable rolling motion. For angular velocity less than the minimum, the square bracket in equation (e) is negative leading to an exponentially decaying and divergent solution. For a uniform disk the perpendicular axis theorem gives $I_3 = 2I_1 = \frac{1}{2}MR^2$ for which equation (f) gives

$$\Omega_3^2 > \frac{g}{3R} \tag{g}$$

Therefore the critical linear velocity of the wheel is

$$v = R\Omega_3 > \sqrt{\frac{gR}{3}} \tag{h}$$

The bicycle wheel provides a common example of the tipping of a rolling wheel. For the typical 0.35m radius of a bicycle wheel, this gives a critical velocity of $v > 1.07m/s = 2.4mph$.[4]

---

[4] The stability of the bicycle is sensitive to the castor and other aspects of the steering geometry of the front wheel, in addition to the gyroscopic effects. Excellent articles on this subject have been written by D.E.H. Jones *Physics Today* **23**(4) (1970) 34, and also by J. Lowell & H.D. McKell, *American Journal of Physics* **50** (1982) 1106.

13.15 *Example: Pivoting*

A rolling and a pivoting body can lead to confusion as to whether to compute the angular momentum and kinetic energy with respect to the center of mass, or the point of contact on the circumference of the body for rolling, or of the pivot point for a fixed pivot. For pivoting or rolling of a wheel it is useful to compare the angular momentum and total energy computed with respect to (1) the center of mass of a cylinder and (2) with respect to the point of contact of the cylinder with the horizontal ground plane.

Consider a cylinder of radius R and mass m pivoting about the point of contact with the plane with angular velocity $\omega = \frac{v}{R}$ where v is the instantaneous velocity of the center of mass. The angular momentum about the pivot point is

$$\mathbf{L}_{pivot} = \mathbf{R} \times \mathbf{v} = I_{pivot}\boldsymbol{\omega}$$

The parallel-axis theorem relates the moment of inertia with respect to the pivot point and center of mass

$$I_{pivot} = mR^2 + I_{cm}$$

The angular velocities of the center of mass, and about the center of mass, are identical since the pivot point is fixed, that is

$$\omega_{pivot} = \omega_{cm} = \omega$$

Thus the angular momentum about the pivot point is given by the sum of the angular momenta

$$\mathbf{L}_{pivot} = I_{pivot}\boldsymbol{\omega} = mR^2\boldsymbol{\omega} + I_{cm}\boldsymbol{\omega}$$

That is, the angular momentum is the sum of the angular momentum of the body about the center of mass, plus the angular momentum of the center of mass about the pivot point. This is an example of Chasles theorem.

The kinetic energy is given only by the rotational energy since the pivot point is stationary

$$KE_{pivot} = \frac{1}{2}I_{pivot}\omega^2 = \frac{1}{2}mR^2\omega^2 + \frac{1}{2}I_{cm}\omega^2 = \frac{1}{2}mv^2 + \frac{1}{2}I_{cm}\omega^2$$

That is, it equals the kinetic energy of rotation about the center of mass plus the instantaneous kinetic energy for translation of the center of mass in agreement with Chasles theorem. Thus, for pivoting, the angular momentum and kinetic energy are the same if evaluated using either center of mass coordinates or using the pivot point as the reference point.

13.16 *Example: Rolling*

Consider the same system except the cylinder is rolling without slipping on a plane. The subtle difference between pivoting and rolling is that the rolling point of contact and the center of mass are moving at the same velocity in contrast to pivoting where the point of contact is stationary. Thus for rolling there is no angular momentum of the center of mass with respect to the point of contact. Therefore the angular momentum about the instantaneous point of contact is

$$\mathbf{L}_{rolling} = L_{pivot} + L_{cm} = mR^2 0 + I_{cm}\boldsymbol{\omega} = I_{cm}\boldsymbol{\omega}$$

That is, the angular momentum only includes the angular momentum about the center of mass which is smaller than the angular momentum for the same body pivoting about a point on the periphery of the cylinder.

The kinetic energy is given by

$$KE_{roll} = \frac{1}{2}mv^2 + \frac{1}{2}I_{rolling}\omega^2 = \frac{1}{2}mv^2 + \frac{1}{2}I_{cm}\omega^2$$

Thus the angular momentum is significantly smaller for rolling relative to pivoting of a given body, whereas the kinetic energy is the same for both rolling or pivoting of a given body.

13.25 Dynamic balancing of wheels

For rotating machinery It is crucial that rotors be both statically and dynamically balanced. *Static balance means that the center of mass is on the axis of rotation. Dynamic balance means that the axis of rotation is a principal axis.*

For example, consider the symmetric rotor that has its symmetry axis at an angle ϕ to the axis of rotation. In this case the system is statically balanced since the center of gravity is on the axis of rotation. However, the rotation axis is at an angle ϕ to the symmetry axis. This implies that the axle has to provide a torque to maintain rotation that is not along a principal axis. If you distort the front wheel of your car by hitting it sideways against the sidewalk curb, or if the wheel is not dynamically balanced, then you will find that the steering wheel can vibrate wildly at certain speeds due to the torques caused by dynamic imbalance shaking the steering mechanism. This can be especially bad when the rotation frequency is close to a resonant frequency of the suspension system. Insist that your automobile wheels are dynamically balanced when you change tires, static balancing will not eliminate the dynamic imbalance forces. Another example is that the ailerons, rudder, and elevator on aircraft usually are dynamically balanced to stop the build up of oscillations that can couple to flexing and flutter of the airframe which can lead to airframe failure.

13.17 *Example: Forces on the bearings of a rotating circular disk*

A homogeneous circular disk of mass M, and radius R, rotates with constant angular velocity ω about a body-fixed axis passing through the center of the circular disk as shown in the adjacent figure. The rotation axis is inclined at an angle α to the symmetry axis of the circular disk by bearings on both sides of the disk spaced a distance d apart. Determine the forces on the bearings.

Choose the body-fixed axes such that $\hat{e}_3$ is along the symmetry axis of the circular disk, and $\hat{e}_1$ points in the plane of the disk symmetry axis and the rotation axis. These axes are the principal axes for which the inertia tensor can be calculated to be

$$\mathbf{I} = \frac{MR^2}{4} \begin{pmatrix} 1 & 0 & 0 \\ 0 & 1 & 0 \\ 0 & 0 & 2 \end{pmatrix}$$

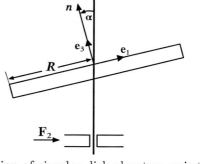

Rotation of circular disk about an axis that is at an angle α to the symmetry axis of the circular disk.

Note that for this thin plane laminae disk $I_{11} + I_{22} = I_{33}$.

The components of the angular velocity vector ω along the three body-fixed axes are given by

$$\boldsymbol{\omega} = (\omega \sin \alpha, 0, \omega \cos \alpha)$$

Since it is assumed that $\dot{\omega} = 0$ then substituting into Euler's equations (13.103) gives the torques acting to be

$$N_1 = N_3 = 0$$
$$N_2 = -\omega^2 \sin \alpha \cos \alpha \frac{1}{4} MR^2$$

That is, the torque is in the $\hat{e}_2$ direction. Thus the forces F on the bearings can be calculated since $\mathbf{N} = \mathbf{r} \times \mathbf{F}$, thus

$$|F| = \frac{|N_2|}{2d} = MR^2 \omega^2 \frac{\sin 2\alpha}{16d}$$

Estimate the size of these forces for the front wheel of your car travelling at 70 m.p.h. if the rotation axis is displaced by 2° from the symmetry axis of the wheel.

Figure 13.11: Forward two-and-a-half somersaults with two twists demonstrates unequivocally that a diver can initiate continuous twisting in midair. In the illustrated maneuver the diver does more than one full somersault before he starts to twist. To maintain the twisting the diver does not have to move his legs.[Fro80]

13.26   Rotation of deformable bodies

The discussion in this chapter has assumed that the rotating body is a rigid body. However, there is a broad and important class of problems in classical mechanics where the rotating body is deformable that leads to intriguing new phenomena. The classic example is the cat, which, if dropped upside down with zero angular momentum, is able to distort its body plus tail in order to rotate such that it lands on its feet in spite of the fact that there are no external torques acting and thus the angular momentum is conserved. Another example is the high diver doing a forward two–and-a-half somersault with two twists.[Fro80] Once the diver leaves the board then the total angular momentum must be conserved since there are no external torques acting on the system. The diver begins a somersault by rotating about a horizontal axis which is a principal axis that is perpendicular to the axis of his body passing through his hips. Initially the angular momentum, and angular velocity, are parallel and point perpendicular to the symmetry axis. Initially the diver goes into a tuck which greatly reduces his moment of inertia along the axis of his somersault which concomitantly increases his angular velocity about this axis and he performs one full somersault prior to initiating twisting. Then the diver twists its body and moves its arms to destroy the axial symmetry of his body which changes the direction of the principal axes of the inertia tensor. This causes the angular velocity to change in both direction and magnitude such that the angular momentum remains conserved. The angular velocity now is no longer parallel to the angular momentum resulting in a component along the length of the body causing it to twist while somersaulting. This twisting motion will continue until the symmetry of the diver's body is restored which is done just before entering the water. By skilled timing, and body movement, the diver restores the symmetry of his body to the optimum orientation for entering the water. Such phenomena involving deformable bodies are important to motion of ballet dancers, jugglers, astronauts in space, and satellite motion. The above rotational phenomena would be impossible if the cat or diver were rigid bodies having a fixed inertia tensor. Calculation of the dynamics of the motion of deformable bodies is complicated and beyond the scope of this book, but the concept of a time dependent transformation of the inertia tensor underlies the subsequent motion. The theory is complicated since it is difficult even to quantify what corresponds to rotation as the body morphs from one shape to another. Further information on this topic can be found in the literature. [Fro80]

13.27 Summary

This chapter has introduced the important, topic of rigid-body rotation which has many applications in physics, engineering, sports, etc.

Inertia tensor The concept of the inertia tensor was introduced where the 9 components of the inertia tensor are given by

$$I_{ij} = \int \rho \left(\mathbf{r}' \right) \left(\delta_{ij} \left(\sum_k^3 x_k^2 \right) - x_i x_j \right) dV \tag{13.14}$$

Steiner's parallel-axis theorem

$$J_{11} \equiv I_{11} + M \left(\left(a_1^2 + a_2^2 + a_3^2 \right) \delta_{11} - a_1^2 \right) = I_{11} + M \left(a_2^2 + a_3^2 \right) \tag{13.43}$$

relates the inertia tensor about the center-of-mass to that about parallel axis system not through the center of mass.

Diagonalization of the inertia tensor about any point was used to find the corresponding Principal axes of the rigid body.

Angular momentum The angular momentum $\mathbf{L}$ for rigid-body rotation is expressed in terms of the inertia tensor and angular frequency ω by

$$\mathbf{L} = \begin{pmatrix} I_{11} & I_{12} & I_{13} \\ I_{21} & I_{22} & I_{23} \\ I_{31} & I_{32} & I_{33} \end{pmatrix} \cdot \begin{pmatrix} \omega_1 \\ \omega_2 \\ \omega_3 \end{pmatrix} = \{ \mathbf{I} \} \cdot \boldsymbol{\omega} \tag{13.56}$$

Rotational kinetic energy The rotational kinetic energy is

$$T_{rot} = \frac{1}{2} \begin{pmatrix} \omega_1 & \omega_2 & \omega_3 \end{pmatrix} \cdot \begin{pmatrix} I_{11} & I_{12} & I_{13} \\ I_{21} & I_{22} & I_{23} \\ I_{31} & I_{32} & I_{33} \end{pmatrix} \cdot \begin{pmatrix} \omega_1 \\ \omega_2 \\ \omega_3 \end{pmatrix} \tag{13.72}$$

$$T_{rot} \equiv \mathbf{T} = \frac{1}{2} \boldsymbol{\omega} \cdot \{ \mathbf{I} \} \cdot \boldsymbol{\omega} = \frac{1}{2} \boldsymbol{\omega} \cdot \mathbf{L} \tag{13.73}$$

Euler angles The Euler angles relate the space-fixed and body-fixed principal axes. The angular velocity $\boldsymbol{\omega}$ expressed in terms of the Euler angles has components for the angular velocity in the *body-fixed axis system* $(1, 2, 3)$

$$\omega_1 = \dot{\phi}_1 + \dot{\theta}_1 + \dot{\psi}_1 = \dot{\phi} \sin \theta \sin \psi + \dot{\theta} \cos \psi \tag{13.86}$$

$$\omega_2 = \dot{\phi}_2 + \dot{\theta}_2 + \dot{\psi}_2 = \dot{\phi} \sin \theta \cos \psi - \dot{\theta} \sin \psi \tag{13.87}$$

$$\omega_3 = \dot{\phi}_3 + \dot{\theta}_3 + \dot{\psi}_3 = \dot{\phi} \cos \theta + \dot{\psi} \tag{13.88}$$

Similarly, the components of the angular velocity for the *space-fixed axis system* (x, y, z) are

$$\omega_x = \dot{\theta} \cos \phi + \dot{\psi} \sin \theta \sin \phi \tag{13.89}$$

$$\omega_y = \dot{\theta} \sin \phi - \dot{\psi} \sin \theta \cos \phi \tag{13.90}$$

$$\omega_z = \dot{\phi} + \dot{\psi} \cos \theta \tag{13.91}$$

Rotational invariants The powerful concept of the rotational invariance of scalar properties was introduced. Important examples of rotational invariants are the Hamiltonian, Lagrangian, and Routhian.

Euler equations of motion for rigid-body motion The dynamics of rigid-body rotational motion was explored and the Euler equations of motion were derived using both Newtonian and Lagrangian mechanics.

$$N_1^{ext} = I_1 \dot{\omega}_1 - (I_2 - I_3) \omega_2 \omega_3 \tag{13.103}$$

$$N_2^{ext} = I_2 \dot{\omega}_2 - (I_3 - I_1) \omega_3 \omega_1$$

$$N_3^{ext} = I_3 \dot{\omega}_3 - (I_1 - I_2) \omega_1 \omega_2$$

Lagrange equations of motion for rigid-body motion   The Euler equations of motion for rigid-body motion, given in equation 13.103, were derived using the Lagrange-Euler equations.

Torque-free motion of rigid bodies   The Euler equations and Lagrangian mechanics were used to study torque-free rotation of both symmetric and asymmetric bodies including discussion of the stability of torque-free rotation.

Rotating symmetric body subject to a torque   The complicated motion exhibited by a symmetric top, that is spinning about one fixed point and subject to a torque, was introduced and solved using Lagrangian mechanics.

The rolling wheel   The non-holonomic motion of rolling wheels was introduced, as well as the importance of static and dynamic balancing of rotating machinery..

Rotation of deformable bodies   The complicated non-holonomic motion involving rotation of deformable bodies was introduced.

Workshop exercises

1. Three objects are described below. Break up into three groups, one group per object, and determine the inertia tensor.

   - A very thin sheet with a mass density $\sigma = Cxy$ where C is a positive constant. The sheet lies in the xy plane and its sides are both of length a.

   - An inclined-plane shaped block of mass M is oriented with one corner at the origin as shown.

   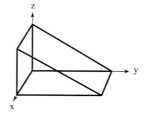

   - An equilateral triangle made up of three thin rods of length l and uniform mass density ρ.

2. Consider the objects described in problem 1.

   (a) For the first object (the thin sheet), determine the principal moments of inertia.

   (b) For the second object (the inclined plane), determine the principal axes.

   (c) For the third object (the equilateral triangle), determine the products of inertia.

3. Consider the inertia tensor.

   (a) What are the advantages of diagonalizing the inertia tensor?

   (b) How can the inertia tensor be diagonalized?

   (c) What can you say about a tensor that is real and symmetric?

4. A hollow spherical shell has a mass m and radius R.

   (a) Calculate the inertia tensor for a set of coordinates whose origin is at the center of mass of the shell.

   (b) Now suppose that the shell is rolling without slipping toward a step of height h, where $h < R$. The shell has a linear velocity v. What is the angular momentum of the shell relative to the tip of the step?

   (c) The shell now strikes the tip of the step inelastically (so that the point of contact sticks to the step, but the shell can still rotate about the tip of the step). What is the angular momentum of the shell immediately after contact?

   (d) Finally, find the minimum velocity which enables the shell to surmount the step. Express your result in terms of m, g, R, and h.

5. The vectors $\hat{x}$, $\hat{y}$, and $\hat{z}$ constitute a set of orthogonal right-handed axes. The vectors $\hat{x} + \hat{y} - 2\hat{z}$, $-\hat{x} + \hat{y}$, and $\hat{x} + \hat{y} + \hat{z}$ are also perpendicular to one another.

   (a) Write out the set of direction cosines relating the new axes to the old.

   (b) How are the Eulerian angles defined? Describe this transformation by a set of Eulerian angles.

6. A torsional pendulum consists of a vertical wire attached to a mass which can rotate about the vertical axis. Consider three torsional pendula which consist of identical wires from which identical homogeneous solid cubes are hung. One cube is hung from a corner, one from midway along an edge, and one from the middle of a face as shown. What are the ratios of the periods of the three pendula?

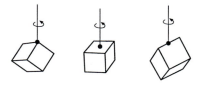

7. A dumbbell comprises two equal point masses M connected by a massless rigid rod of length $2A$ which is constrained to rotate about an axle fixed to the center of the rod at an angle θ as shown in the figure. The center of the rod is at the origin of the coordinates, the axle along the z-axis, and the dumbbell lies in the $x - y$ plane at $t = 0$. The angular velocity w is a constant in time and is directed along the z axis.

a) Calculate all elements of the inertia tensor. Be sure to specify the coordinate system used.

b) Using the calculated inertia tensor find the angular momentum of the dumbbell in the laboratory frame as a function of time.

c) Using the equation $L = r \times p$, calculate the angular momentum and show that it it is equal to the answer of part (b).

d) Calculate the torque on the axle as a function of time.

e) Calculate the kinetic energy of the dumbbell.

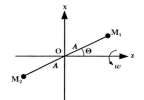

8. A heavy symmetric top has a mass m with the center of mass a distance h from the fixed point about which it spins and $I_1 = I_2 \neq I_3$. The top is precessing at a steady angular velocity Ω about the vertical space-fixed z axis. What is the minimum spin w' about the body-fixed symmetry axis, that is, the 3 axis assuming that the 3 axis is inclined at an angle $\theta = \theta$ with respect to the vertical z axis. Solve the problem at the instant when the $z, x, 3, 1$ axes all are in the same plane as shown in the figure.

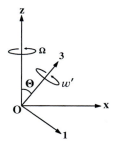

9. Consider an object with the center of mass is at the origin and inertia tensor,

$$I = I \begin{pmatrix} 1/2 & -1/2 & 0 \\ -1/2 & 1/2 & 0 \\ 0 & 0 & 1 \end{pmatrix}$$

(a) Determine the principal moments of inertia and the principal axes. Guess the object.

(b) Determine the rotation matrix R and compute $R^\dagger I R$. Do the diagonal elements match with your results from (a)? Note: columns of R are eigenvectors of I.

(c) Assume $\omega = \frac{\omega}{\sqrt{2}}(\hat{x} + \hat{z})$. Determine L in the rotating coordinate system. Are L and ω in the same direction? What does this mean?

(d) Repeat (c) for $\omega = \frac{\omega}{\sqrt{2}}(\hat{x} - \hat{y})$. What is different and why?

(e) For which case will there be a non-zero torque required?

(f) Determine the rotational kinetic energy for the case $\omega = \frac{\omega}{\sqrt{2}}(\hat{x} - \hat{y})$?

10. Consider a wheel (solid disk) of mass m and radius r. The wheel is subject to angular velocities $w_A = w_A \hat{n}$ where $\hat{n}$ is normal to the surface and $w_B = w_B \hat{z}$.

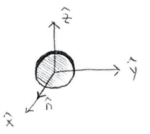

(a) Choose a set of principal axes by observation.

(b) Determine the angular velocities and angular momentum along the principal axes. Note: $I_1 = \frac{1}{2}mr^2$ and $I_2 = I_3 = \frac{1}{4}mr^2$.

(c) Determine the torque.

(d) Determine the rotation matrix that rotates the fixed coordinate system to the body coordinate system.

11. Determine the principal moments of inertia of an ellipsoid given by the equation,

$$\frac{x^2}{a^2} + \frac{y^2}{b^2} + \frac{z^2}{c^2} = 1.$$

12. Determine the principal moments of inertia of a sphere of radius R with a cavity of radius r located ϵ from the center of the sphere.

13. Three equal masses m form the vertices of an equilateral triangle of side length L. The masses are located at $\left(0, 0, \frac{L}{\sqrt{3}}\right)$, $\left(0, \frac{L}{2}, -\frac{L}{2\sqrt{3}}\right)$, and $\left(0, -\frac{L}{2}, -\frac{L}{2\sqrt{3}}\right)$, such that the center-of-mass is located at the origin.

(a) Determine the principal moments of inertia and principal axes.
Now consider the same system rotated $45°$ about the $\hat{z}$-axis. The masses are located at $\left(0, 0, \frac{L}{\sqrt{3}}\right)$, $\left(-\frac{L}{2\sqrt{2}}, \frac{L}{2\sqrt{2}}, -\frac{L}{2\sqrt{3}}\right)$, and $\left(\frac{L}{2\sqrt{2}}, -\frac{L}{2\sqrt{2}}, -\frac{L}{2\sqrt{3}}\right)$, respectively.
(b) Determine the principal moments of inertia and principal axes.
(c) Could you have answered (b) without explicitly determining the inertia tensor? How?

Problems

1. Calculate the moments of inertia I_1, I_2, I_3 for a homogeneous cone of mass M whose height is h and whose base has a radius R. Choose the x_3-axis along the symmetry axis of the cone.

   a) Choose the origin at the apex of the cone, and calculate the elements of the inertia tensor.

   b) Make a transformation such that the center of mass of the cone is the origin and find the principal moments of inertia.

2. Four masses, all of mass m, lie in the $x - y$ plane at positions $(x, y) = (a, 0), (-a, 0), (0, +2a), (0, -2a)$. These are joined by massless rods to form a rigid body

   (a) Find the inertial tensor, using the x, y, z axes as a reference system. Exhibit the tensor as a matrix.

   (b) Consider a direction given by the unit vector $\hat{n}$ that lies equally between the positive x, y, z axes; that is it makes equal angles with these three directions. Find the moment of inertia for rotation about this $\hat{n}$ axis.

   (c) Given that at a certain time t the angular velocity vector lies along the above direction $\hat{n}$, find, for that instant, the angle between the angular momentum vector and $\hat{n}$.

3. A homogeneous cube, each edge of which has a length l, initially is in a position of unstable equilibrium with one edge of the cube in contact with a horizontal plane. The cube then is given a small displacement causing it to tip over and fall. Show that the angular velocity of the cube when one face strikes the plane is given by

$$\omega^2 = A\frac{g}{l}\left(\sqrt{2} - 1\right)$$

   where $A = \frac{3}{2}$ if the edge cannot slide on the plane, and where $A = \frac{12}{5}$ if sliding can occur without friction.

4. A symmetric body moves without the influence of forces or torques. Let x_3 be the symmetry axis of the body and L be along x_3'. The angle between ω and x_3 is α. Let ω and L initially be in the $x_2 - x_3$ plane. What is the angular velocity of the symmetry axis about L in terms of I_1, I_3, ω, and α?

5. Consider a thin rectangular plate with dimensions a by b and mass M. Determine the torque necessary to rotate the thin plate with angular velocity ω about a diagonal. Explain the physical behavior for the case when $a = b$.

Chapter 14

Coupled linear oscillators

14.1 Introduction

Chapter 3 discussed the behavior of a single linearly-damped linear oscillator subject to a harmonic force. No account was taken for the influence of the single oscillator on the driver for the case of forced oscillations. Many systems in nature comprise complicated free or forced oscillations of coupled-oscillator systems. Examples of coupled oscillators are; automobile suspension systems, electronic circuits, electromagnetic fields, musical instruments, atoms bound in a crystal, neural circuits in the brain, networks of pacemaker cells in the heart, etc. Energy can be transferred back and forth between coupled oscillators as the motion evolves. It is possible to describe the motion of coupled linear oscillators in terms of a sum over independent normal coordinates, i.e. normal modes, even though the motion may be very complicated. These normal modes are constructed from the original coordinates in such a way that the normal modes are uncoupled. The topic of finding the normal modes of coupled oscillator systems is a ubiquitous problem encountered in all branches of science and engineering. As discussed in chapter 3, oscillatory motion of non-linear systems can be complicated. Fortunately most oscillatory systems are approximately linear when the amplitude of oscillation is small. This discussion assumes that the oscillation amplitudes are sufficiently small to ensure linearity.

14.2 Two coupled linear oscillators

Consider the two-coupled linear oscillator, shown in figure 14.1, which comprises two identical masses each connected to fixed locations by identical springs having a force constant κ. A spring with force constant κ' couples the two oscillators. The equilibrium lengths of the outer two springs are l while that of the coupling spring is l'. The problem is simplified by restricting the motion to be along the line connecting the masses and assuming fixed endpoints. The small displacements of m_1 and m_2 are taken to be x_1 and x_2 with respect to the equilibrium positions l and $l+l'$ respectively. The restoring force on m_1 is $-\kappa x_1 - \kappa'(x_1 - x_2)$ while the restoring force on m_2 is $-\kappa x_2 - \kappa'(x_2 - x_1)$. This coupled double-oscillator system exhibits basic features of coupled linear oscillator systems.

Assuming $m_1 = m_2 = m$, then the equations of motion are

$$
\begin{aligned}
m\ddot{x}_1 + (\kappa + \kappa')x_1 - \kappa'x_2 &= 0 \\
m\ddot{x}_2 + (\kappa + \kappa')x_2 - \kappa'x_1 &= 0
\end{aligned}
\tag{14.1}
$$

Assume that the motion for these coupled equations is oscil-

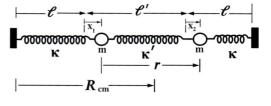

Figure 14.1: Two coupled linear oscillators. The equilibrium spring-lengths are l for the outer springs and l' for the coupling spring. The displacement from the stable locations are given by x_1 and x_2. The separation between the two masses is r and the location of the center-of-mass is R_{cm}.

latory with a solution of the form

$$x_1 = B_1 e^{i\omega t} \tag{14.2}$$
$$x_2 = B_2 e^{i\omega t}$$

where the constants B may be complex to take into account both the magnitude and phase. Substituting these possible solutions into the equations of motion gives

$$-m\omega^2 B_1 e^{i\omega t} + (\kappa + \kappa') B_1 e^{i\omega t} - \kappa' B_2 e^{i\omega t} = 0 \tag{14.3}$$
$$-m\omega^2 B_2 e^{i\omega t} + (\kappa + \kappa') B_2 e^{i\omega t} - \kappa' B_1 e^{i\omega t} = 0$$

Collecting terms, and cancelling the common exponential factor, gives

$$(\kappa + \kappa' - m\omega^2) B_1 - \kappa' B_2 = 0 \tag{14.4}$$
$$(\kappa + \kappa' - m\omega^2) B_2 - \kappa' B_1 = 0$$

The existence of a non-trivial solution of these two simultaneous equations requires that the determinant of the coefficients of B_1 and B_2 must vanish, that is

$$\begin{vmatrix} \kappa + \kappa' - m\omega^2 & -\kappa' \\ -\kappa' & \kappa + \kappa' - m\omega^2 \end{vmatrix} = 0 \tag{14.5}$$

The expansion of this secular determinant yields

$$(\kappa + \kappa' - m\omega^2)^2 - \kappa'^2 = 0 \tag{14.6}$$

Solving for ω gives

$$\omega = \sqrt{\frac{\kappa + \kappa' \pm \kappa'}{m}} \tag{14.7}$$

That is, there are two characteristic frequencies (or eigenfrequencies) for the system

$$\omega_1 = \sqrt{\frac{\kappa + 2\kappa'}{m}} \tag{14.8}$$

$$\omega_2 = \sqrt{\frac{\kappa}{m}} \tag{14.9}$$

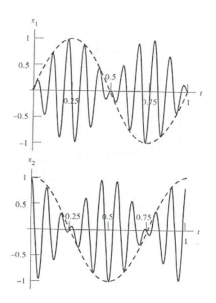

Figure 14.2: Displacement of each of two coupled linear harmonic oscillators with $\kappa = 4$ and $\kappa' = 1$ in relative units.

Since superposition applies for these linear equations, then the general solution can be written as a sum of the terms that account for the two possible values of ω.

Figure 14.2 shows the solutions for a case where $\kappa = 4$ and $\kappa' = 1$, in arbitrary units, with the initial condition that $x_2 = D$, and $x_1 = \dot{x}_1 = \dot{x}_2 = 0$. The two characteristic frequencies are $\omega_1 = \sqrt{\frac{6}{m}}$ and $\omega_2 = \sqrt{\frac{4}{m}}$. The characteristic beats phenomenon is exhibited where the envelope over one complete cycle of the low frequency encompasses several higher frequency oscillations. That is, the solution is

$$x_2(t) = \frac{D}{4} \left[e^{i\omega_1 t} + e^{-i\omega_1 t} + e^{i\omega_2 t} + e^{-i\omega_2 t} \right] = D \cos \left[\left(\frac{\omega_1 + \omega_2}{2} \right) t \right] \cos \left[\left(\frac{\omega_1 - \omega_2}{2} \right) t \right] \tag{14.10}$$

while

$$x_1(t) = \frac{D}{4} \left[e^{i\omega_1 t} + e^{-i\omega_1 t} - e^{i\omega_2 t} - e^{-i\omega_2 t} \right] = D \sin \left[\left(\frac{\omega_1 + \omega_2}{2} \right) t \right] \sin \left[\left(\frac{\omega_1 - \omega_2}{2} \right) t \right] \tag{14.11}$$

The energy in the two-coupled oscillators flows back and forth between the coupled oscillators as illustrated in figure 14.2.

A better understanding of the energy flow occurring between the two coupled oscillators is given by using a (x_1, x_2) configuration-space plot, shown in figure 14.3. The flow of energy occurring between the two coupled oscillators can be represented by choosing normal-mode coordinates η_1 and η_2 that are rotated by $45°$ with respect to the spatial coordinates (x_1, x_2). These normal-mode coordinates (η_1, η_2) correspond to the two normal modes of the coupled double-oscillator system.

14.3 Normal modes

The **normal modes** of the two-coupled oscillator system are obtained by a transformation to a pair of **normal coordinates** (η_1, η_2) that are independent and correspond to the two normal modes. The pair of normal coordinates for this case are

$$
\begin{aligned}
\eta_1 &\equiv x_1 - x_2 \\
\eta_2 &\equiv x_1 + x_2
\end{aligned}
\tag{14.12}
$$

that is

$$
\begin{aligned}
x_1 &= \frac{1}{2}(\eta_2 + \eta_1) \\
\end{aligned}
\tag{14.13}
$$

$$
x_2 = \frac{1}{2}(\eta_2 - \eta_1)
$$

Substitute these into the equations of motion (14.1), gives

$$
\begin{aligned}
m\left(\ddot{\eta}_1 + \ddot{\eta}_2\right) + \left(\kappa + 2\kappa'\right)\eta_1 + \kappa'\eta_2 &= 0 \\
m\left(\ddot{\eta}_1 - \ddot{\eta}_2\right) + \left(\kappa + 2\kappa'\right)\eta_1 - \kappa'\eta_2 &= 0
\end{aligned}
\tag{14.14}
$$

Adding and subtracting these two equations gives

$$
\begin{aligned}
m\ddot{\eta}_1 + \left(\kappa + 2\kappa'\right)\eta_1 &= 0 \\
m\ddot{\eta}_2 + \kappa\eta_2 &= 0
\end{aligned}
\tag{14.15}
$$

Note that the two coordinates η_1 and η_2 are uncoupled and therefore are independent. The solutions of these equations are

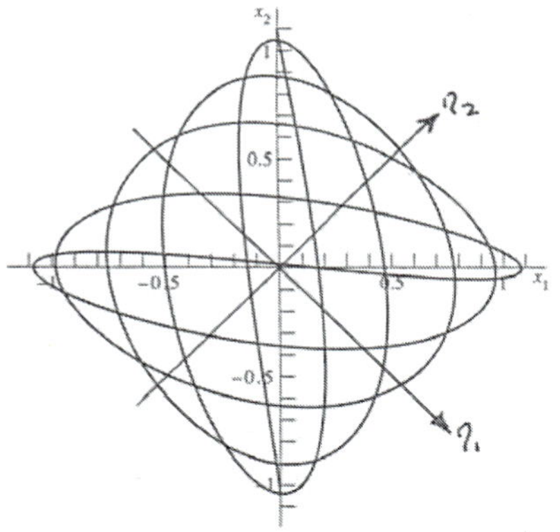

$$
\begin{aligned}
\eta_1(t) &= C_1^+ e^{i\omega_1 t} + C_1^- e^{-i\omega_1 t} \\
\eta_2(t) &= C_2^+ e^{i\omega_2 t} + C_2^- e^{-i\omega_2 t}
\end{aligned}
\tag{14.16}
$$

where η_1 corresponds to angular frequencies ω_1, and η_2 corresponds to ω_2. The two coordinates η_1 and η_2 are called the **normal coordinates and the two solutions are the normal modes with corresponding angular frequencies, ω_1 and ω_2.**

The (η_1, η_2) axes of the two normal modes correspond to a rotation of $45°$ in configuration space, figure 14.3. The initial conditions chosen correspond to $\eta_1 = -\eta_2$ and thus both modes are excited with equal intensity. Note that there are 5 lobes along the η_2 axis versus 4 lobes along the η_1 axis reflecting the ratio of the eigenfrequencies ω_1 and ω_2. Also note that the diamond shape of the motion in the (x_1, x_2) configuration space illustrates that the extrema amplitudes for x_2 are a maximum when x_1 is zero, and vise versa. This is equivalent to the statement that the energies in the two modes are coupled with the energy for the first oscillator being a maximum when the energy is a minimum for the second oscillator, and vise versa. By contrast, in the (η_1, η_2) configuration space, the motion is bounded by a rectangle parallel to the (η_1, η_2) axes reflecting the fact that the extrema amplitudes, and corresponding energies, for the η_1 normal mode are constant and independent of the motion for the η_2 normal mode, and vise versa. The decoupling of the two normal modes is best illustrated by considering the case when only one of these two normal modes is excited. For the initial conditions $x_1(0) = -x_2(0)$, and $\dot{x}_1(0) = -\dot{x}_2(0)$, then $\eta_2(t) = 0$. That is, only the $\eta_1(t)$ normal mode is excited with frequency ω_1 which corresponds to motion confined to the η_1 axis of figure 14.3.

Figure 14.3: Motion of two coupled harmonic oscillators in the (x_1, x_2) spatial configuration space and in terms of the normal modes (η_1, η_2). Initial conditions are $x_2 = D, x_1 = \dot{x}_1 = \dot{x}_2 = 0$.

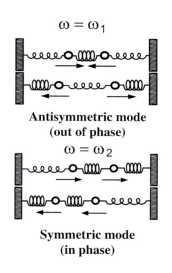

Figure 14.4: Normal modes for two coupled oscillators.

As shown in figure 14.4, $\eta_1(t)$ is the *antisymmetric mode* in which the two masses oscillate out of phase such as to keep the center of mass of the two masses stationary. For the initial conditions $x_1(0) = x_2(0)$, and $\dot{x}_1(0) = \dot{x}_2(0)$, then $\eta_1(t) = 0$, that is, only the $\eta_2(t)$ normal mode is excited. The $\eta_2(t)$ normal mode is the *symmetric mode* where the two masses oscillate in phase with frequency ω_2; it corresponds to motion along the η_2 axis. For the symmetric phase, both masses move together leading to a constant extension of the coupling spring. As a result the frequency ω_2 of the symmetric mode $\eta_2(t)$ is lower than the frequency ω_1 of the asymmetric mode $\eta_1(t)$. That is, the asymmetric mode is stiffer since all three springs provide active restoring forces, compared to the symmetric mode where the coupling spring is uncompressed. In general, for attractive forces the lowest frequency always occurs for the mode with the highest symmetry.

14.4   Center of mass oscillations

Transforming the coordinates into the center of mass of the two oscillating masses elucidates an interesting feature of the normal modes for the two-coupled linear oscillator. As illustrated in figure 14.1, the center-of-mass coordinate for the two mass system is

$$2R_{cm} = l + x_1 + l + l' + x_2 = 2l + l' + \eta_2$$

while the relative separation distance is

$$r = (l + l' + x_2) - (l + x_1) = l' - \eta_1$$

That is, the two normal modes are

$$\begin{aligned}\eta_1 &= l' - r \\ \eta_2 &= 2R_{cm} - 2l - l'\end{aligned} \qquad (14.17)$$

The η_1 mode, which has angular frequency $\omega_1 = \sqrt{\frac{\kappa + 2\kappa'}{M}}$ corresponds to an oscillations of the relative separation r, while the center-of-mass location R_{cm} is stationary. By contrast, the η_2 mode, with angular frequency $\omega_2 = \sqrt{\frac{\kappa}{M}}$, corresponds to an oscillation of the center of mass R_{cm} with the relative separation r being a constant.

Figure 14.5 illustrates the decoupled center-of-mass R_{cm}, and relative motions r for both normal modes of the coupled double-oscillator system. The difference in angular frequencies and amplitudes is readily apparent.

It is of interest to consider the special case where the spring constant $\kappa = 0$ for the two outside springs. Then the angular frequencies are $\omega_1 = \sqrt{\frac{2\kappa'}{M}}$ and $\omega_2 = 0$ for the two normal modes. When $\kappa = 0$ the η_2 mode is a spurious center-of-mass mode since it corresponds to an oscillation with $\omega_2 = 0$ in spite of the fact that there are no forces acting on the center of mass. That is, the center-of-mass momentum must be a constant of motion. This spurious center-of-mass oscillation is a consequence of measuring the displacements (x_1, x_2) with respect to an arbitrary external reference that is not related to the center of mass of the coupled system. Spurious center-of-mass modes are encountered frequently in many-body coupled oscillator systems such as molecules and nuclei. In such cases it is necessary to project out the center-of-mass motion to eliminate such spurious solutions as will be discussed later.

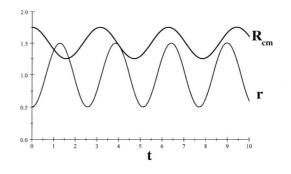

Figure 14.5: Time dependence of the center-of-mass R_{cm} and relative separation r for two coupled linear oscillators assuming spring constants of $\kappa = 4M$ and $\kappa' = M$.

14.5 Weak coupling

If one of the two coupled linear oscillator masses is held fixed, then the other free mass will oscillate with a frequency.

$$\omega_0 = \sqrt{\frac{\kappa + \kappa'}{M}} \tag{14.18}$$

The effect of coupling of the two oscillators is to split the degeneracy of the frequency for each mass to

$$\omega_1 = \sqrt{\frac{\kappa + 2\kappa'}{M}} > \omega_0 = \sqrt{\frac{\kappa + \kappa'}{M}} > \omega_2 = \sqrt{\frac{\kappa}{M}}. \tag{14.19}$$

Thus the degeneracy is broken, and the two normal modes have frequencies straddling the single-oscillator frequency.

It is interesting to consider the case where the coupling is weak because this situation occurs frequently in nature. The coupling is weak if the coupling constant $\kappa' << \kappa$. Then

$$\omega_1 = \sqrt{\frac{\kappa + 2\kappa'}{M}} = \sqrt{\frac{\kappa}{M}}\sqrt{1 + 4\varepsilon} \tag{14.20}$$

where

$$\varepsilon \equiv \frac{\kappa'}{2\kappa} << 1 \tag{14.21}$$

Thus

$$\omega_1 \approx \sqrt{\frac{\kappa}{M}}\left(1 + 2\varepsilon\right) \tag{14.22}$$

The natural frequency of a single oscillator was shown to be

$$\omega_0 = \sqrt{\frac{\kappa + \kappa'}{M}} \approx \sqrt{\frac{\kappa}{M}}\left(1 + \varepsilon\right) \tag{14.23}$$

that is

$$\sqrt{\frac{\kappa}{M}} = \omega_0\left(1 - \varepsilon\right) \tag{14.24}$$

Thus the frequencies for the normal modes for weak coupling can be written as

$$\omega_1 = \sqrt{\frac{\kappa}{M}}\left(1 + 2\varepsilon\right)$$
$$\approx \omega_0\left(1 - \varepsilon\right)\left(1 + 2\varepsilon\right) \approx \omega_0\left(1 + \varepsilon\right) \tag{14.25}$$

while

$$\omega_2 = \sqrt{\frac{\kappa}{M}} \approx \omega_0\left(1 - \varepsilon\right) \tag{14.26}$$

That is the two solutions are split equally spaced about the single uncoupled oscillator value given by $\omega_0 = \sqrt{\frac{\kappa + \kappa'}{M}} \approx \sqrt{\frac{\kappa}{M}}\left(1 + \varepsilon\right)$. Note that the single uncoupled oscillator frequency ω_0 depends on the coupling strength κ'.

This splitting of the characteristic frequencies is a feature exhibited by many systems of n identical oscillators where half of the frequencies are shifted upwards and half downward. If n is odd, then the central frequency is unshifted as illustrated for the case of $n = 3$. An example of this behavior is the Zeeman effect where the magnetic field couples the atomic motion resulting in a hyperfine splitting of the energy levels as illustrated.

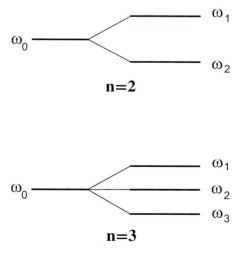

Figure 14.6: Normal-mode frequencies for n=2 and n=3 weakly-coupled oscillators.

There are myriad examples involving weakly-coupled oscillators in many aspects of the natural world. The example of collective modes in nuclear physics, illustrated in example 14.13, is typical of applications to physics, while there are many examples applied to musical instruments, acoustics, and engineering. Weakly-coupled oscillators are a dominant theme throughout biology as illustrated by congregations of synchronously flashing fireflies, crickets that chirp in unison, an audience clapping at the end of a performance, networks of pacemaker cells in the heart, insulin-secreting cells in the pancreas, and neural networks in the brain and spinal cord that control rhythmic behaviors such as breathing, walking, and eating. Synchronous motion of a large number of weakly-coupled oscillators often leads to large collective motion of weakly-coupled systems as discussed in chapter 14.12.

14.1 *Example: The Grand Piano*

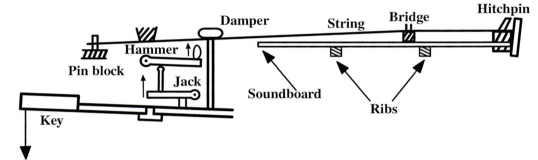

Schematic diagram of the action for a grand piano, including the strings, bridge and sounding board. Note that there are either two or three parallel strings per note that are hit by a single hammer.

The grand piano provides an excellent example of a weakly-coupled harmonic oscillator system that has normal modes. There are either two or three parallel strings per note that are stretched tightly parallel to the top of the horizontal sounding board. The strings press downwards on the bridge that is attached to the top of the sounding board. The strings for each note are excited when struck vertically upwards by a single hammer. In the base section of the piano each note comprises two strings tuned to nearly the same frequency. The coupling of the motion of the strings is via the bridge plus sounding board. Normally, the hammer strikes both strings simultaneously exciting the vertical symmetric mode, not the vertical antisymmetric mode. The bridge is connected to the sounding board which moves the largest amount for the symmetric mode where both strings move the bridge in phase. This strong coupling produces a loud sound. The antisymmetric mode does not move the sounding board much since the strings at the bridge move out of phase. Consequently, the symmetric mode, that is strongly coupled to the sounding board, damps out more rapidly than the antisymmetric mode which is weakly coupled to the sound board and thus has a longer time constant for decay since the radiated sound energy is lower than the symmetric mode.

The una-corda pedal (soft pedal) for a grand piano moves the action sideways such that the hammer strikes only one of the two strings, or two of the three strings, resulting in both the symmetric and antisymmetric modes being excited equally. The una-corda pedal produces a characteristically different tone than when the hammer simultaneously hits the coupled strings; that is, it produces a smaller transient component. The symmetric mode rapidly damps due to energy propagation by the sounding board. Thus the longer lasting antisymmetric mode becomes more prominent when both modes are equally excited using the una-corda pedal. The symmetric and antisymmetric modes have slightly different frequencies and produce beats which also contributes to the different timbre produced using the una-corda pedal. For the mid and upper frequency range, the piano has three strings per note which have one symmetric mode and two separate antisymmetric modes. To further complicate matters, the strings also can oscillate horizontally which couples weakly to the bridge plus sounding board. The strengths that these different modes are excited depend on subtle differences in the shape and roughness of the hammer head striking the strings. Primarily the hammer excites the two vertical modes rather than the horizontal modes.

14.6 General analytic theory for coupled linear oscillators

The above discussion of a coupled double-oscillator system has shown that it is possible to select symmetric and antisymmetric normal modes that are independent and each have characteristic frequencies. The normal coordinates for these two normal modes correspond to linear superpositions of the spatial amplitudes of the two oscillators and can be obtained by a rotation into the appropriate normal coordinate system. Extension of this to systems comprising n coupled linear oscillators, requires development of a general analytic theory, that is capable of finding the normal modes plus their eigenvalues and eigenvectors. As illustrated for the double oscillator, the solution of many coupled linear oscillators is a classic eigenvalue problem where one has to rotate to the principal axis system to project out the normal modes. The following discussion presents a general approach to the problem of finding the normal coordinates for a system of n coupled linear oscillators.

Consider a conservative system of n coupled oscillators, described in terms of generalized coordinates q_k and t, with subscript $k = 1, 2, 3 ... n$ for a system with n degrees of freedom. The coupled oscillators are assumed to have a stable equilibrium with generalized coordinates q_{k0} at equilibrium. In addition, *it is assumed that the oscillation amplitudes are sufficiently small to ensure that the system is linear.*

For the equilibrium position $q_k = q_{k0}$ the Lagrange equations must satisfy

$$\dot{q}_k = 0 \qquad (14.27)$$
$$\ddot{q}_k = 0$$

Every non-zero term of the form $\frac{d}{dt}\frac{\partial L}{\partial \dot{q}_k}$ in Lagrange's equations must contain at least either $\dot{q}_k$ or $\ddot{q}_k$ which are zero at equilibrium; thus all such terms vanish at equilibrium. That is at equilibrium

$$\left(\frac{\partial L}{\partial q_k}\right)_0 = \left(\frac{\partial T}{\partial q_k}\right)_0 - \left(\frac{\partial U}{\partial q_k}\right)_0 = 0 \qquad (14.28)$$

where the subscript 0 designates at equilibrium.

14.6.1 Kinetic energy tensor T

In chapter 7.6 it was shown that, in terms of fixed rectangular coordinates, the kinetic energy for N bodies, with n generalized coordinates, is expressed as

$$T = \frac{1}{2}\sum_{\alpha=1}^{N}\sum_{i=1}^{3} m_\alpha \dot{x}_{\alpha,i}^2 \qquad (14.29)$$

Expressing these in terms of generalized coordinates $x_{\alpha,i} = x_{\alpha,i}(q_j, t)$ where $j = 1, 2, ... n$, then the generalized velocities are given by

$$\dot{x}_{\alpha,i} = \sum_{j=1}^{n} \frac{\partial x_{\alpha,i}}{\partial q_j}\dot{q}_j + \frac{\partial x_{\alpha,i}}{\partial t} \qquad (14.30)$$

As discussed in chapter 7.6, if the system is **scleronomic** then the partial time derivative

$$\frac{\partial x_{\alpha,i}}{\partial t} = 0 \qquad (14.31)$$

Thus the kinetic energy, equation 14.29, of a scleronomic system can be written as a homogeneous quadratic function of the generalized velocities

$$T = \frac{1}{2}\sum_{j,k}^{n} T_{jk}\dot{q}_j\dot{q}_k \qquad (14.32)$$

where the components of the kinetic energy tensor **T** are

$$T_{jk} \equiv \sum_{\alpha}^{N} m_\alpha \sum_{i}^{3} \frac{\partial x_{\alpha,i}}{\partial q_j}\frac{\partial x_{\alpha,i}}{\partial q_k} \qquad (14.33)$$

Note that if the velocities $\dot{q}$ correspond to translational velocity, then the kinetic energy tensor **T** corresponds to an effective mass tensor, whereas if the velocities correspond to angular rotational velocities, then the kinetic energy tensor **T** corresponds to the inertia tensor.

It is possible to make an expansion of the T_{jk} about the equilibrium values of the form

$$T_{jk}(q_1, q_2, ..q_n) = T_{jk}(q_{i0}) + \sum_l \left(\frac{\partial T_{jk}}{\partial q_l} \right)_0 q_l + ... \tag{14.34}$$

Only the first-order term will be kept since the second and higher terms are of the same order as the higher-order terms ignored in the Taylor expansion of the potential. Thus, at the equilibrium point, assume that $\left(\frac{\partial T}{\partial q_k} \right)_0 = 0$ where $k = 1, 2, 3, ...n$.

14.6.2   Potential energy tensor V

Equations 14.28 plus 14.34 imply that

$$\left(\frac{\partial U}{\partial q_k} \right)_0 = 0 \tag{14.35}$$

where $k = 1, 2, 3, ...n$.

Make a Taylor expansion about equilibrium for the potential energy, assuming for simplicity that the coordinates have been translated to ensure that $q_k = 0$ at equilibrium. This gives

$$U(q_1, q_2, ..q_n) = U_0 + \sum_k \left(\frac{\partial U}{\partial q_k} \right)_0 q_k + \frac{1}{2} \sum_{j,k} \left(\frac{\partial^2 U}{\partial q_j \partial q_k} \right)_0 q_j q_k + .. \tag{14.36}$$

The linear term is zero since $\left(\frac{\partial U}{\partial q_k} \right)_0 = 0$ at the equilibrium point, and without loss of generality, the potential can be measured with respect to U_0. Assume that the amplitudes are small, then the expansion can be restricted to the quadratic term, corresponding to the simple linear oscillator potential

$$U(q_1, q_2, ..q_n) - U_0 = U'(q_1, q_2, ..q_n) = \frac{1}{2} \sum_{j,k} \left(\frac{\partial^2 U}{\partial q_j \partial q_k} \right)_0 q_j q_k = \frac{1}{2} \sum_{j,k} V_{jk} q_j q_k \tag{14.37}$$

That is

$$U'(q_1, q_2, ..q_n) = \frac{1}{2} \sum_{j,k} V_{jk} q_j q_k \tag{14.38}$$

where the components of the potential energy tensor $\mathbf{V}$ are defined as

$$V_{jk} \equiv \left(\frac{\partial^2 U'}{\partial q_j \partial q_k} \right)_0 \tag{14.39}$$

Note that the order of differentiation is unimportant and thus the quantity V_{jk} is symmetric

$$V_{jk} = V_{kj} \tag{14.40}$$

The motion of the system has been specified for small oscillations around the equilibrium position and it has been shown that $U'(q_1, q_2, ...q_n)$ has a minimum value at equilibrium which is taken to be zero for convenience.

In conclusion, equations (14.32) and (14.38) give

$$T = \frac{1}{2} \sum_{j,k}^n T_{jk} \dot{q}_j \dot{q}_k \tag{14.41}$$

$$U' = \frac{1}{2} \sum_{j,k}^n V_{jk} q_j q_k \tag{14.42}$$

where the components of the kinetic energy tensor $\mathbf{T}$ and potential energy tensor $\mathbf{V}$ are

$$T_{jk} \equiv \left(\sum_\alpha^N m_\alpha \sum_i^3 \frac{\partial x_{\alpha,i}}{\partial q_j} \frac{\partial x_{\alpha,i}}{\partial q_k} \right)_0 \tag{14.43}$$

$$V_{jk} \equiv \left(\frac{\partial^2 U'}{\partial q_j \partial q_k} \right)_0 \tag{14.44}$$

Note that q_j and q_k may have different units, but all the terms in the summations for both T and U', have units of energy. The V_{jk} and T_{jk} values are evaluated at the equilibrium point, and thus both V_{jk} and T_{jk} are $n \times n$ arrays of values evaluated at the equilibrium location.

14.6.3 Equations of motion

Both the kinetic energy and potential energy terms are products of the coordinates leading to a set of coupled equations that are complicated to solve. The problem is greatly simplified by selecting a set of normal coordinates for which both T and U are diagonal, then the coupling terms disappear. Thus a coordinate transformation must be found that simultaneously diagonalizes T_{jk} and V_{jk} in order to obtain a set of normal coordinates.

The kinetic energy T *is only a function of generalized velocities* $\dot{q}_k$ while the conservative potential energy *is only a function of the generalized coordinates* q_k. Thus the Lagrange equations

$$\frac{\partial L}{\partial q_k} - \frac{d}{dt}\frac{\partial L}{\partial \dot{q}_k} = 0 \tag{14.45}$$

reduce to

$$\frac{\partial U}{\partial q_k} + \frac{d}{dt}\frac{\partial T}{\partial \dot{q}_k} = 0 \tag{14.46}$$

But

$$\frac{\partial U}{\partial q_k} = \sum_j^n V_{jk} q_j \tag{14.47}$$

and

$$\frac{\partial T}{\partial \dot{q}_k} = \sum_j^n T_{jk} \dot{q}_j \tag{14.48}$$

Thus the Lagrange equations reduce to the following set of equations of motion,

$$\sum_j^n \left(V_{jk} q_j + T_{jk} \ddot{q}_j \right) = 0 \tag{14.49}$$

For each k, where $1 \le k \le n$, there exists a set of n second-order linear homogeneous differential equations with constant coefficients. Since the system is oscillatory, it is natural to try a solution of the form

$$q_j(t) = a_j e^{i(\omega t - \delta)} \tag{14.50}$$

Assuming that the system is conservative, then this implies that ω is real, since an imaginary term for ω would lead to an exponential damping term. The arbitrary constants are the real amplitude a_j and the phase δ. Substitution of this trial solution for each k leads to a set of equations

$$\sum_j \left(V_{jk} - \omega^2 T_{jk} \right) a_j = 0 \tag{14.51}$$

where the common factor $e^{i(\omega t - \delta)}$ has been removed. Equation 14.51 corresponds to a set of n linear homogeneous algebraic equations that the a_j amplitudes must satisfy for each k. For a non-trivial solution to exist, the determinant of the coefficients must vanish, that is

$$\begin{vmatrix} V_{11} - \omega^2 T_{11} & V_{12} - \omega^2 T_{12} & V_{13} - \omega^2 T_{13} & \dots \\ V_{12} - \omega^2 T_{12} & V_{22} - \omega^2 T_{22} & V_{23} - \omega^2 T_{23} & \dots \\ V_{13} - \omega^2 T_{13} & V_{23} - \omega^2 T_{23} & V_{33} - \omega^2 T_{33} & \dots \\ \dots & \dots & \dots & \dots \end{vmatrix} = 0 \tag{14.52}$$

where the symmetry $V_{jk} = V_{kj}$ has been included. This is the standard eigenvalue problem for which the above determinant gives the **secular equation** or the **characteristic equation.** It is an equation of degree n in ω^2. The n roots of this equation are ω_r^2 where ω_r are the **characteristic frequencies** or **eigenfrequencies** of the normal modes.

Substitution of ω_r^2 into equation 14.52 determines the ratio $a_{1,r} : a_{2,r} : a_{3,r} : \dots : a_{n,r}$ for this solution which defines the components of the n-dimensional **eigenvector** $\mathbf{a}_r$. That is, solution of the secular equations have determined the eigenvalues and eigenvectors of the n solutions of the coupled-channel system.

14.6.4 Superposition

The equations of motion $\sum_j (V_{jk}q_j + T_{jk}\ddot{q}_j) = 0$ are linear equations that satisfy superposition. Thus the most general solution $q_j(t)$ can be a superposition of the n eigenvectors $\mathbf{a}_{jr}$, that is

$$q_j(t) = \sum_r^n a_{jr}e^{i(\omega_r t - \delta_r)} \tag{14.53}$$

Only the real part of $q_j(t)$ is meaningful, that is,

$$q_j(t) = \text{Re}\sum_r^n a_{jr}e^{i(\omega_r t - \delta_r)} = \sum_r^n a_{jr}\cos(\omega_r t - \delta_r) \tag{14.54}$$

Thus the most general solution of these linear equations involves a sum over the eigenvectors of the system which are cosine functions of the corresponding eigenfrequencies.

14.6.5 Eigenfunction orthonormality

It can be shown that the eigenvectors are orthogonal. In addition, the above procedure only determines ratios of amplitudes, thus there is an indeterminacy that can be used to normalize the a_{jr}. Thus the eigenvectors form an orthonormal set. Orthonormality of the eigenfunctions for the rank 3 inertia tensor was illustrated in chapter 13.10.2. Similar arguments apply that allow extending orthonormality to higher rank cases such that for n-body coupled oscillators.

The **eigenfunction orthogonality** for n coupled oscillators can be proved by writing equation 14.51 for both the sth root and the rth root. That is,

$$\sum_j V_{jk}a_{ks} = \omega_s^2 \sum_j T_{jk}a_{ks} \tag{14.55}$$

$$\sum_j V_{jk}a_{jr} = \omega_r^2 \sum_j T_{jk}a_{jr} \tag{14.56}$$

Multiply equation 14.55 by a_{jr} and sum over k. Similarly multiply equation 14.56 by a_{ks} and sum over k. These summations lead to

$$\sum_{jk} V_{jk}a_{jr}a_{ks} = \omega_s^2 \sum_{jk} T_{jk}a_{jr}a_{ks} \tag{14.57}$$

$$\sum_{jk} V_{jk}a_{jr}a_{ks} = \omega_r^2 \sum_{jk} T_{jk}a_{jr}a_{ks} \tag{14.58}$$

Note that the left-hand sides of these two equations are identical. Thus taking the difference between these equations gives

$$(\omega_r^2 - \omega_s^2)\sum_{jk} T_{jk}a_{jr}a_{ks} = 0 \tag{14.59}$$

Note that if $(\omega_r^2 - \omega_s^2) \neq 0$, that is, assuming that the eigenfrequencies are not degenerate, then to ensure that equation 14.59 is zero requires that

$$\sum_{jk} T_{jk}a_{jr}a_{ks} = 0 \qquad\qquad r \neq s \tag{14.60}$$

This shows that the eigenfunctions are orthogonal. If the eigenfrequencies are degenerate, i.e. $\omega_r^2 = \omega_s^2$, then, with no loss of generality, the axes r and s can be chosen to be orthogonal.

The **eigenfunction normalization** can be chosen freely since only ratios of the eigenfunction components a_{jr} are determined when ω_r is used in equation 14.51. The kinetic energy, given by equation 14.32 must be positive, or zero for the case of a static system. That is

$$T = \frac{1}{2}\sum_{j,k}^n T_{jk}\dot{q}_j\dot{q}_k \geq 0 \tag{14.61}$$

Use the time derivative of equation 14.54 to determine $\dot{q}_r$ and insert into equation 14.61 gives that the kinetic energy is

$$T = \frac{1}{2} \sum_{j,k}^{n} T_{jk} \dot{q}_j \dot{q}_k = \frac{1}{2} \sum_{j,k}^{n} T_{jk} \sum_{r,s} \omega_r \omega_s a_{jr} \cos\left(\omega_r t - \delta_r\right) a_{ks} \cos\left(\omega_s t - \delta_s\right) \tag{14.62}$$

For the diagonal term $r = s$

$$T = \frac{1}{2} \sum_{j,k}^{n} T_{jk} \dot{q}_j \dot{q}_k = \left[\frac{1}{2} \sum_{r}^{n} \omega_r^2 \cos^2\left(\omega_r t - \delta_r\right) \right] \sum_{j,k} T_{jk} a_{jr} a_{kr} \geq 0 \tag{14.63}$$

Since the term in the square brackets must be positive, then

$$\sum_{j,k} T_{jk} a_{jr} a_{kr} \geq 0 \tag{14.64}$$

Since this sum must be a positive number, and the magnitude of the amplitudes can be chosen freely, then it is possible to **normalize** the eigenfunction amplitudes to unity. That is, choose that

$$\sum_{j,k} T_{jk} a_{jr} a_{ks} = 1 \tag{14.65}$$

The orthogonality equation, 14.60 and the normalization equation 14.65 can be combined into a single orthonormalization equation

$$\sum_{j,k} T_{jk} a_{jr} a_{ks} = \delta_{rs} \tag{14.66}$$

This has shown that the eigenvectors form an orthonormal set.

Since the jth component of the rth eigenvector is a_{jr}, then the rth eigenvector can be written in the form

$$\mathbf{a}_r = \sum_{j} a_{jr} \widehat{\mathbf{e}}_j \tag{14.67}$$

where $\widehat{\mathbf{e}}_j$ are the unit vectors for the generalized coordinates.

14.6.6 Normal coordinates

The above general solution of the coupled-oscillator problem is best expressed in terms of the normal coordinates which are independent. It is more transparent if the superposition of the normal modes are written in the form

$$q_j\left(t\right) = \sum_{r}^{n} \beta_r a_{jr} e^{i\omega_r t} \tag{14.68}$$

where the complex factor β_r includes the arbitrary scale factor to allow for arbitrary amplitudes q_j as well as the fact that the amplitudes a_{jr} have been normalized and the phase factor δ_r has been chosen.

Define

$$\eta_r\left(t\right) \equiv \beta_r e^{i\omega_r t} \tag{14.69}$$

then equation 14.68 can be written as

$$q_j\left(t\right) = \sum_{r}^{n} a_{jr} \eta_r\left(t\right) \tag{14.70}$$

Equation 14.70 can be expressed schematically as the matrix multiplication

$$\mathbf{q} = \{\mathbf{a}\} \cdot \boldsymbol{\eta} \tag{14.71}$$

The $\eta_r\left(t\right)$ are the **normal coordinates** which can be expressed in the form

$$\boldsymbol{\eta} = \{\mathbf{a}\}^{-1} \mathbf{q} \tag{14.72}$$

Each normal mode η_r corresponds to a single eigenfrequency, ω_r which satisfies the linear oscillator equation

$$\ddot{\eta}_r + \omega_r^2 \eta_r = 0 \tag{14.73}$$

14.7   Two-body coupled oscillator systems

The two-body coupled oscillator is the simplest coupled-oscillator system that illustrates the general features of coupled oscillators. The following four examples involve parallel and series couplings of two linear oscillators or two plane pendula.

14.2 *Example: Two coupled linear oscillators*

The coupled double-oscillator problem, figure 14.1 discussed in chapter 14.2, can be used to demonstrate that the general analytic theory gives the same solution as obtained by direct solution of the equations of motion in chapter 14.2.

1)   The first stage is to determine the potential and kinetic energies using an appropriate set of generalized coordinates, which here are x_1 and x_2. The potential energy is

$$U = \frac{1}{2}\kappa x_1^2 + \frac{1}{2}\kappa x_2^2 + \frac{1}{2}\kappa' \left(x_2 - x_1 \right)^2 = \frac{1}{2}\left(\kappa + \kappa' \right) x_1^2 + \frac{1}{2}\left(\kappa + \kappa' \right) x_2^2 - \kappa' x_1 x_2$$

while the kinetic energy is given by

$$T = \frac{1}{2}m\dot{x}_1^2 + \frac{1}{2}m\dot{x}_2^2$$

2)   The second stage is to evaluate the potential energy V and kinetic energy T tensors. The potential energy tensor V is nondiagonal since V_{jk} gives

$$V_{11} \equiv \left(\frac{\partial^2 U}{\partial q_1 \partial q_1} \right)_0 = \kappa + \kappa' = V_{22}$$

$$V_{12} = \left(\frac{\partial^2 U}{\partial q_1 \partial q_2} \right)_0 = -\kappa' = V_{21}$$

That is, the potential energy tensor V is

$$\mathbf{V} = \left\{ \begin{array}{cc} \kappa + \kappa' & -\kappa' \\ -\kappa' & \kappa + \kappa' \end{array} \right\}$$

Similarly, the kinetic energy is given by

$$T = \frac{1}{2}m\dot{x}_1^2 + \frac{1}{2}m\dot{x}_2^2 = \frac{1}{2}\sum_{j,k} T_{jk}\dot{q}_j\dot{q}_k$$

Since $T_{11} = T_{22} = m$ and $T_{12} = T_{21} = 0$ then the kinetic energy tensor T is

$$\mathbf{T} = \left\{ \begin{array}{cc} m & 0 \\ 0 & m \end{array} \right\}$$

Note that for this case, the kinetic energy tensor T equals the mass tensor, which is diagonal, whereas the potential energy tensor equals the spring constant tensor, which is nondiagonal.

3)   The third stage is to use the potential energy V and kinetic energy T tensors to evaluate the secular determinant using equations 14.52

$$\left| \begin{array}{cc} \kappa + \kappa' - m\omega^2 & -\kappa' \\ -\kappa' & \kappa + \kappa' - m\omega^2 \end{array} \right| = 0$$

The expansion of this secular determinant yields

$$\left(\kappa + \kappa' - m\omega^2 \right)^2 - \kappa'^2 = 0$$

That is

$$\left(\kappa + \kappa' - m\omega^2 \right) = \pm\kappa'$$

Solving for ω_r gives

$$\omega_r = \sqrt{\frac{\kappa + \kappa' \pm \kappa'}{m}}$$

The solutions are

$$\omega_1 = \sqrt{\frac{\kappa + 2\kappa'}{m}}$$

$$\omega_2 = \sqrt{\frac{\kappa}{m}}$$

which is the same as derived previously, (equations $14.7 - 9$).

4)   The fourth step is to insert either one of these eigenfrequencies into the secular equation

$$\sum_j \left(V_{jk} - \omega_r^2 T_{jk}\right) a_{jr} = 0 \qquad (a)$$

Consider the secular equation a for $k = 1$

$$\left(\kappa + \kappa' - \omega_r^2 M\right) a_{1r} - \kappa' a_{2r} = 0$$

Then for the first eigenfrequency ω_1, that is, $k = 1, r = 1$

$$\left(\kappa + \kappa' - \kappa - 2\kappa'\right) a_{11} - \kappa' a_{21} = 0$$

which simplifies to

$$a_{jr} = a_{11} = -a_{21}$$

Similarly, for the other eigenfrequency ω_2, that is, $k = 1, r = 2$

$$\left(\kappa + \kappa' - \kappa\right) a_{12} - \kappa' a_{22} = 0$$

which simplifies to

$$a_{jr} = a_{12} = a_{22}$$

5)   The final stage is to write the general coordinates in terms of the normal coordinates $\eta_r(t) \equiv \beta_r e^{i\omega_r t}$. Thus

$$x_1 = a_{11}\eta_1 + a_{12}\eta_2 = a_{11}\eta_1 + a_{22}\eta_2$$

and

$$x_2 = a_{21}\eta_1 + a_{22}\eta_2 = -a_{11}\eta_1 + a_{22}\eta_2$$

Adding or subtracting gives that the normal modes are

$$\eta_1 = \frac{1}{2a_{11}}(x_1 - x_2)$$

$$\eta_2 = \frac{1}{2a_{22}}(x_2 + x_1)$$

Thus the symmetric normal mode η_2 corresponds to an oscillation of the center-of-mass with the lower frequency $\omega_2 = \sqrt{\frac{\kappa}{m}}$. This frequency is the same as for one single mass on a spring of spring constant κ which is as expected since they vibrate in unison and thus the coupling spring force does not act. The antisymmetric mode η_1 has the higher frequency $\omega_1 = \sqrt{\frac{\kappa + 2\kappa'}{m}}$ since the restoring force includes both the main spring plus the coupling spring.

The above example illustrates that the general analytic theory for coupled linear oscillators gives the same answer as obtained in chapter 14.2 using Newton's equations of motion. However, the general analytic theory is a more powerful technique for solving complicated coupled oscillator systems. Thus the general analytic theory will be used for solving all the following coupled oscillator problems.

14.3 Example: Two equal masses series-coupled by two equal springs

Consider the series-coupled system shown in the figure.

1) The first stage is to determine the potential and kinetic energies using an appropriate set of generalized coordinates, which here are x_1 and x_2. The potential energy is

$$U = \frac{1}{2}\kappa x_1^2 + \frac{1}{2}\kappa (x_2 - x_1)^2 = \kappa x_1^2 + \frac{1}{2}\kappa x_2^2 - \kappa x_1 x_2$$

while the kinetic energy is given by

$$T = \frac{1}{2}m\dot{x}_1^2 + \frac{1}{2}m\dot{x}_2^2$$

Two equal masses series-coupled by two equal springs.

2) The second stage is to evaluate the potential energy V and mass T tensors. The potential energy tensor V is nondiagonal since V_{jk} gives

$$V_{11} \equiv \left(\frac{\partial^2 U}{\partial q_1 \partial q_1}\right)_0 = 2\kappa$$

$$V_{12} = \left(\frac{\partial^2 U}{\partial q_1 \partial q_2}\right)_0 = -\kappa = V_{21}$$

$$V_{22} = \left(\frac{\partial^2 U}{\partial q_2 \partial q_2}\right)_0 = \kappa$$

That is, the potential energy tensor V is

$$\mathbf{V} = \left\{ \begin{array}{cc} 2\kappa & -\kappa \\ -\kappa & \kappa \end{array} \right\}$$

Similarly, since the kinetic energy is given by

$$T = \frac{1}{2}m\dot{x}_1^2 + \frac{1}{2}m\dot{x}_2^2 = \frac{1}{2}\sum_{j,k} m_{jk}\dot{q}_j \dot{q}_k$$

then $T_{11} = T_{22} = m$ and $T_{12} = T_{21} = 0$. Thus the kinetic energy tensor T is

$$\mathbf{T} = \left\{ \begin{array}{cc} m & 0 \\ 0 & m \end{array} \right\}$$

Note that for this case the kinetic energy tensor is diagonal whereas the potential energy tensor is nondiagonal.

3)   The third stage is to use the potential energy V and kinetic energy T tensors to evaluate the secular determinant using equation 14.52

$$\left| \begin{array}{cc} 2\kappa - m\omega^2 & -\kappa \\ -\kappa & \kappa - m\omega^2 \end{array} \right| = 0$$

The expansion of this secular determinant yields

$$\left(2\kappa - m\omega^2\right)\left(\kappa - m\omega^2\right) - \kappa^2 = 0$$

That is

$$\omega^4 - 3\frac{\kappa}{m}\omega^2 + \frac{\kappa^2}{m^2} = 0$$

The solutions are

$$\omega_1 = \frac{\sqrt{5}+1}{2}\sqrt{\frac{\kappa}{m}} \qquad \omega_2 = \frac{\sqrt{5}-1}{2}\sqrt{\frac{\kappa}{m}}$$

4)   The fourth step is to insert these eigenfrequencies into the secular equation 14.51

$$\sum_j \left(V_{jk} - \omega_r^2 T_{jk}\right) a_{jr} = 0$$

Consider $k = 1$ in the above equation

$$\left(2\kappa - \omega_r^2 M\right) a_{1r} - \kappa a_{2r} = 0$$

Then for eigenfrequency ω_1, that is, $k = 1$, $r = 1$

$$\frac{\sqrt{5} - 1}{2} a_{11} = -a_{21}$$

Similarly, for $k = 1, r = 2$

$$\frac{\sqrt{5} + 1}{2} a_{12} = a_{22}$$

5)   *The final stage is to write the general coordinates in terms of the normal coordinates $\eta_r(t) \equiv \beta_r e^{i\omega_r t}$.*
    Thus

$$x_1 = a_{11}\eta_1 + a_{12}\eta_2 = a_{11}\eta_1 + \frac{2a_{22}}{\sqrt{5} + 1}\eta_2$$

and

$$x_2 = a_{21}\eta_1 + a_{22}\eta_2 = -\left(\frac{\sqrt{5} - 1}{2}\right) a_{11}\eta_1 + a_{22}\eta_2$$

Adding or subtracting gives that the normal modes are

$$\eta_1 = \frac{1}{a_{11}\sqrt{5}} \left(x_1 - \left(\frac{\sqrt{5} - 1}{2}\right) x_2 \right)$$

$$\eta_2 = \frac{1}{a_{22}\sqrt{5}} \left(x_1 + \left(\frac{\sqrt{5} + 1}{2}\right) x_2 \right)$$

Thus the symmetric normal mode has the lower frequency $\omega_2 = \frac{\sqrt{5}-1}{2}\sqrt{\frac{\kappa}{m}}$. The antisymmetric mode has the frequency $\omega_1 = \frac{\sqrt{5}+1}{2}\sqrt{\frac{\kappa}{m}}$ since both springs provide the restoring force. This case is interesting in that for both normal modes, the amplitudes for the motion of the two masses are different.

14.4 Example: Two parallel-coupled plane pendula

Consider the coupled double pendulum system shown in the adjacent figure, which comprises two parallel plane pendula weakly coupled by a spring. The angles θ_1 and θ_2 are chosen to be the generalized coordinates and the potential energy is chosen to be zero at equilibrium. Then the kinetic energy is

$$T = \frac{1}{2}m\left(b\dot{\theta}_1\right)^2 + \frac{1}{2}m\left(b\dot{\theta}_2\right)^2$$

As discussed in chapter 3, it is necessary to make the small-angle approximation in order to make the equations of motion for the simple pendulum linear and solvable analytically. That is,

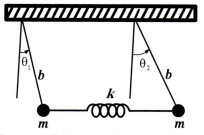

Two parallel-coupled plane pendula.

$$U = mgb\left(1 - \cos\theta_1\right) + mgb\left(1 - \cos\theta_2\right) + \frac{1}{2}\kappa\left(b\sin\theta_1 - b\sin\theta_2\right)^2$$

$$\simeq \frac{mgb}{2}\left(\theta_1^2 + \theta_2^2\right) + \frac{\kappa b^2}{2}\left(\theta_1 - \theta_2\right)^2$$

assuming the small angle approximation $\sin\theta \approx \theta$ and $(1 - \cos\theta_1) = \frac{\theta^2}{2}$.
    The second stage is to evaluate the kinetic energy T and potential energy V tensors

$$\mathbf{T} = \left\{ \begin{matrix} mb^2 & 0 \\ 0 & mb^2 \end{matrix} \right\} \qquad\qquad \mathbf{V} = \left\{ \begin{matrix} mgb + \kappa b^2 & -\kappa b^2 \\ -\kappa b^2 & mgb + \kappa b^2 \end{matrix} \right\}$$

Note that for this case the kinetic energy tensor is diagonal whereas the potential energy tensor is nondiagonal. The third stage is to evaluate the secular determinant

$$\begin{vmatrix} mgb + \kappa b^2 - \omega^2 mb^2 & -\kappa b^2 \\ -\kappa b^2 & mgb + \kappa b^2 - \omega^2 mb^2 \end{vmatrix} = 0$$

which gives the characteristic equation

$$\left(mgb + \kappa b^2 - \omega^2 mb^2\right)^2 = \left(\kappa b^2\right)^2$$

or

$$mg + \kappa b - \omega^2 mb = \pm \kappa b$$

The two solutions are

$$\omega_1^2 = \frac{g}{b} \qquad\qquad \omega_2^2 = \frac{g}{b} + \frac{2\kappa}{m}$$

The fourth step is to insert these eigenfrequencies into equation 14.51

$$\sum_j^n \left(V_{jk} - \omega_r^2 T_{jk}\right) a_{jr} = 0$$

Consider $k = 1$

$$\left(mgb + \kappa b^2 - \omega_r^2 mb^2\right) a_{1r} - \kappa b^2 a_{2r} = 0$$

Then for the first eigenfrequency, ω_1, *the subscripts are* $k = 1, r = 1$

$$\left(mgb + \kappa b^2 - \frac{g}{b} mb^2\right) a_{11} - \kappa b^2 a_{21} = 0$$

which simplifies to

$$a_{11} = a_{21}$$

Similarly, for $k = 1, r = 2$

$$\left(mgb + \kappa b^2 - \left(\frac{g}{b} + \frac{2\kappa}{m}\right) mb^2\right) a_{12} - \kappa b^2 a_{22} = 0$$

which simplifies to

$$a_{12} = -a_{22}$$

The final stage is to write the general coordinates in terms of the normal coordinates

$$\theta_1 = a_{11}\eta_1 + a_{12}\eta_2 = a_{11}\eta_1 - a_{22}\eta_2$$

and

$$\theta_2 = a_{21}\eta_1 + a_{22}\eta_2 = a_{11}\eta_1 + a_{22}\eta_2$$

Adding or subtracting these equations gives that the normal modes are

$$\eta_1 = \frac{1}{2a_{11}}\left(\theta_1 + \theta_2\right) \qquad\qquad \eta_2 = \frac{1}{2a_{22}}\left(\theta_2 - \theta_1\right)$$

As for the case of the double oscillator discussed in example 14.2, the symmetric normal mode corresponds to an oscillation of the center-of-mass, with zero relative motion of the two pendula, which has the lower frequency $\omega_1 = \sqrt{\frac{g}{b}}$. *This frequency is the same as for one independent pendulum as expected since they vibrate in unison and thus the only restoring force is gravity. The antisymmetric mode corresponds to relative motion of the two pendula with stationary center-of-mass and has the frequency* $\omega_2 = \sqrt{\left(\frac{g}{b} + \frac{2\kappa}{m}\right)}$ *since the restoring force includes both the coupling spring and gravity.*

This example introduces the role of degeneracy which occurs in this system if the coupling of the pendula is zero, that is, $\kappa = 0$, *leading to both frequencies being equal, i.e.* $\omega_1 = \omega_2 = \sqrt{\frac{g}{b}}$. *When* $\kappa = 0$, *then both* $\{\mathbf{T}\}$ *and* $\{\mathbf{V}\}$ *are diagonal and thus in the* (θ_1, θ_2) *space the two pendula are independent normal modes. However, the symmetric and asymmetric normal modes, as derived above, are equally good normal modes. In fact, since the modes are degenerate, any linear combination of the motion of the independent pendula are equally good normal modes and thus one can use any set of orthogonal normal modes to describe the motion.*

14.5 *Example: The series-coupled double plane pendula*

The double-pendula system comprises one plane pendulum attached to the end of another plane pendulum both oscillating in the same plane. The kinetic and potential energies for this system are given in example 6.21 to be

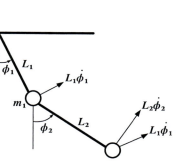

$$T = \frac{1}{2}(m_1 + m_2)L_1^2\dot{\phi}_1^2 + m_2L_1L_2\dot{\phi}_1\dot{\phi}_2\cos(\phi_1 - \phi_2) + \frac{1}{2}m_2L_2^2\dot{\phi}_2^2$$
$$U = (m_1 + m_2)gL_1(1 - \cos\phi_1) + m_2gL_2(1 - \cos\phi_2)$$

a) Small-amplitude linear regime

Use of the small-angle approximation makes this system linear and solvable analytically. That is, T and U become

$$U = \frac{1}{2}(m_1 + m_2)gL_1\phi_1^2 + \frac{1}{2}m_2gL_2\phi_2^2$$
$$T = \frac{1}{2}(m_1 + m_2)L_1^2\dot{\phi}_1^2 + m_2L_1L_2\dot{\phi}_1\dot{\phi}_2 + \frac{1}{2}m_2L_2^2\dot{\phi}_2^2$$

Two series-coupled plane pendula.

Thus the kinetic energy and potential energy tensors are

$$\mathbf{T} = \left\{ \begin{array}{cc} (m_1 + m_2)L_1^2 & m_2L_1L_2 \\ m_2L_1L_2 & m_2L_2^2 \end{array} \right\} \qquad \mathbf{V} = \left\{ \begin{array}{cc} (m_1 + m_2)gL_1 & 0 \\ 0 & m_2gL_2 \end{array} \right\}$$

Note that **T** *is nondiagonal, whereas* **V** *is diagonal which is opposite to the case of the two parallel-coupled plane pendula.*

The solution of this case is simpler if it is assumed that $L_1 = L_2 = L$ and $m_1 = m_2 = m$. Then

$$\mathbf{T} = mL^2 \left\{ \begin{array}{cc} 2 & 1 \\ 1 & 1 \end{array} \right\} \qquad \mathbf{V} = mL^2 \left\{ \begin{array}{cc} 2\omega_0^2 & 0 \\ 0 & \omega_0^2 \end{array} \right\}$$

where $\omega_0 = \sqrt{\frac{g}{L}}$ which is the frequency of a single pendulum. The next stage is to evaluate the secular determinant

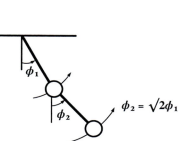

$$mL^2 \left| \begin{array}{cc} 2(\omega_0^2 - \omega^2) & -\omega^2 \\ -\omega^2 & (\omega_0^2 - \omega^2) \end{array} \right| = 0$$

The eigenvalues are

$$\omega_1^2 = (2 - \sqrt{2})\omega_0^2 \qquad\qquad \omega_2^2 = (2 + \sqrt{2})\omega_0^2$$

As shown in the adjacent figure, the normal modes for this system are

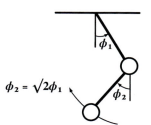

$$\eta_1 = \frac{1}{2a_{11}}(\phi_1 + \frac{\phi_2}{\sqrt{2}}) \qquad\qquad \eta_2 = \frac{1}{2a_{22}}(\phi_1 - \frac{\phi_2}{\sqrt{2}})$$

Normal modes for two series-coupled plane pendula.

The second mass has a $\sqrt{2}$ larger amplitude that is in phase for solution 1 and out of phase for solution 2.

b) Large amplitude chaotic regime

Stachowiak and Okada [Sta05] used computer simulations to numerically analyze the behavior of this system with increase in the oscillation amplitudes. Poincaré sections, bifurcation diagrams, and Lyapunov exponents all confirm that this system evolves from regular normal-mode oscillatory behavior in the linear regime at low energy, to chaotic behavior at high excitation energies where non-linearity dominates. This behavior is analogous to that of the driven, linearly-damped, harmonic pendulum described in chapter 3.5

14.8   Three-body coupled linear oscillator systems

Chapter 14.7 discussed parallel and series arrangements of two coupled oscillators. Extending from two to three coupled linear oscillators introduces interesting new characteristics of coupled oscillator systems. For more than two coupled oscillators, coupled oscillator systems separate into two classifications depending on whether each oscillator is coupled to the remaining $n - 1$ oscillators, or when the coupling is only to the nearest neighbors as illustrated below.

14.6 *Example: Three plane pendula; mean-field linear coupling*

Consider three identical pendula with mass m and length b, suspended from a common support that yields slightly to pendulum motion leading to a coupling between all three pendula as illustrated in the adjacent figure. Assume that the motion of the three pendula all are in the same plane. This case is analogous to the piano where three strings in the treble section are coupled by the slightly-yielding common bridge plus sounding board leading to coupling between each of the three coupled oscillators. This case illustrates the important concept of degeneracy.

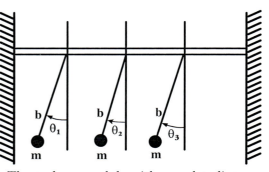

Three plane pendula with complete linear coupling.

The generalized coordinates are the angles $\theta_1, \theta_2,$ and θ_3. Assume that the support yields such that the actual deflection angle for pendulum 1 is

$$\theta_1' = \theta_1 - \frac{\varepsilon}{2} (\theta_2 + \theta_3)$$

where the coupling coefficient ε is small and involves all the pendula, not just the nearest neighbors. Assume that the same coupling relation exists for the other angle coordinates. The gravitational potential energy of each pendulum is given by

$$U_1 = mgb(1 - \cos\theta_1) \approx \frac{1}{2}mgb\theta_1^2$$

assuming the small angle approximation. Ignoring terms of order ε^2 gives that the potential energy

$$U = \frac{mgb}{2} \left(\theta_1'^2 + \theta_2'^2 + \theta_3'^2\right) = \frac{mgb}{2} \left(\theta_1^2 + \theta_2^2 + \theta_3^2 - 2\varepsilon\theta_1\theta_2 - 2\varepsilon\theta_1\theta_3 - 2\varepsilon\theta_2\theta_3\right)$$

The kinetic energy evaluated at the equilibrium location is

$$T = \frac{1}{2}m \left(b\dot\theta_1\right)^2 + \frac{1}{2}m \left(b\dot\theta_2\right)^2 + \frac{1}{2}m \left(b\dot\theta_3\right)^2$$

The next stage is to evaluate the $\{\mathbf{T}\}$ and $\{\mathbf{V}\}$ tensors

$$\mathbf{T} = mb^2 \left\{\begin{array}{ccc} 1 & 0 & 0 \\ 0 & 1 & 0 \\ 0 & 0 & 1 \end{array}\right\} \qquad \mathbf{V} = mgb \left\{\begin{array}{ccc} 1 & -\varepsilon & -\varepsilon \\ -\varepsilon & 1 & -\varepsilon \\ -\varepsilon & -\varepsilon & 1 \end{array}\right\}$$

The third stage is to evaluate the secular determinant which can be written as

$$mgb \left| \begin{array}{ccc} 1 - \frac{b}{g}\omega^2 & -\varepsilon & -\varepsilon \\ -\varepsilon & 1 - \frac{b}{g}\omega^2 & -\varepsilon \\ -\varepsilon & -\varepsilon & 1 - \frac{b}{g}\omega^2 \end{array} \right| = 0$$

Expanding and factoring gives

$$\left(\frac{b}{g}\omega^2 - 1 - \varepsilon\right) \left(\frac{b}{g}\omega^2 - 1 - \varepsilon\right) \left(\frac{b}{g}\omega^2 - 1 + 2\varepsilon\right) = 0$$

The roots are

$$\omega_1 = \sqrt{\frac{g}{b}}\sqrt{1+\varepsilon} \qquad \omega_2 = \sqrt{\frac{g}{b}}\sqrt{1+\varepsilon} \qquad \omega_3 = \sqrt{\frac{g}{b}}\sqrt{1-2\varepsilon}$$

This case results in two degenerate eigenfrequencies, $\omega_1 = \omega_2$ while ω_3 is the lowest eigenfrequency. The eigenvectors can be determined by substitution of the eigenfrequencies into

$$\sum_j^n \left(V_{jk} - \omega_r^2 T_{jk}\right) a_{jr} = 0$$

Consider the lowest eigenfrequency ω_3, i.e. $r = 3$, for $k = 1$, and substitute for $\omega_3 = \sqrt{\frac{g}{b}}\sqrt{1-2\varepsilon}$ gives

$$2\varepsilon a_{13} - \varepsilon a_{23} - \varepsilon a_{33} = 0$$

while for $r = 3, k = 2$

$$-\varepsilon a_{13} + 2\varepsilon a_{23} - \varepsilon a_{33} = 0$$

Solving these gives

$$a_{13} = a_{23} = a_{33}$$

Assuming that the eigenfunction is normalized to unity

$$a_{13}^2 + a_{23}^2 + a_{33}^2 = 1$$

then for the third eigenvector a_3

$$a_{13} = a_{23} = a_{33} = \frac{1}{\sqrt{3}}$$

This solution corresponds to all three pendula oscillating in phase with the same amplitude, that is, a coherent oscillation.

*Derivation of the eigenfunctions for the other two eigenfrequencies is complicated because of the degeneracy $\omega_1 = \omega_2$, there are only five independent equations to specify the six unknowns for the eigenvectors a_1 and a_2. That is, the eigenvectors can be chosen freely as long as the orthogonality and normalization are satisfied. For example, setting $a_{31} = 0$, to remove the indeterminacy, results in the **a** matrix*

$$\{\mathbf{a}\} = \left\{ \begin{matrix} \frac{1}{2}\sqrt{2} & \frac{1}{6}\sqrt{6} & \frac{1}{3}\sqrt{3} \\ -\frac{1}{2}\sqrt{2} & \frac{1}{6}\sqrt{6} & \frac{1}{3}\sqrt{3} \\ 0 & -\frac{1}{3}\sqrt{6} & \frac{1}{3}\sqrt{3} \end{matrix} \right\}$$

and thus the solution is given by

$$\left\{ \begin{matrix} \theta_1 \\ \theta_2 \\ \theta_3 \end{matrix} \right\} = \left\{ \begin{matrix} \frac{1}{2}\sqrt{2} & \frac{1}{6}\sqrt{6} & \frac{1}{3}\sqrt{3} \\ -\frac{1}{2}\sqrt{2} & \frac{1}{6}\sqrt{6} & \frac{1}{3}\sqrt{3} \\ 0 & -\frac{1}{3}\sqrt{6} & \frac{1}{3}\sqrt{3} \end{matrix} \right\} \left\{ \begin{matrix} \eta_1 \\ \eta_2 \\ \eta_3 \end{matrix} \right\}$$

The normal modes are obtained by taking the inverse matrix $\{\mathbf{a}\}^{-1}$ and using $\{\boldsymbol{\eta}\} = \{\mathbf{a}\}^{-1}\{\boldsymbol{\theta}\}$. Note that since $\{\mathbf{a}\}$ is real and orthogonal, then $\{\mathbf{a}\}^{-1}$ equals the transpose of $\{\mathbf{a}\}$. That is;

$$\left\{ \begin{matrix} \eta_1 \\ \eta_2 \\ \eta_3 \end{matrix} \right\} = \left\{ \begin{matrix} \frac{1}{2}\sqrt{2} & -\frac{1}{2}\sqrt{2} & 0 \\ \frac{1}{6}\sqrt{6} & \frac{1}{6}\sqrt{6} & -\frac{1}{3}\sqrt{6} \\ \frac{1}{3}\sqrt{3} & \frac{1}{3}\sqrt{3} & \frac{1}{3}\sqrt{3} \end{matrix} \right\} \left\{ \begin{matrix} \theta_1 \\ \theta_2 \\ \theta_3 \end{matrix} \right\}$$

The normal mode η_3 has eigenfrequency

$$\omega_3 = \sqrt{\frac{g}{b}}\sqrt{1-2\varepsilon}$$

and eigenvector

$$\boldsymbol{\eta}_3 = \frac{1}{\sqrt{3}}(\theta_1, \theta_2, \theta_3)$$

This corresponds to the in-phase oscillation of all three pendula.

The other two degenerate solutions are

$$\boldsymbol{\eta}_1 = \frac{1}{\sqrt{2}}\,(\theta_1, -\theta_2, 0) \qquad\qquad \boldsymbol{\eta}_2 = \frac{1}{\sqrt{6}}\,(\theta_1, \theta_2, -2\theta_3)$$

with eigenvalues

$$\omega_1 = \omega_2 = \sqrt{\frac{g}{b}}\sqrt{1+\varepsilon}$$

These two degenerate normal modes correspond to two pendula oscillating out of phase with the same amplitude, or two oscillating in phase with the same amplitude and the third out of phase with twice the amplitude. An important result of this toy model is that the most symmetric mode η_3 is pushed far from all the other modes. Note that for this example, the coherent mode a_3 corresponds to the center-of-mass oscillation with no relative motion between the three pendula. This is in contrast to the eigenvectors a_1 and a_2 which both correspond to relative motion of the pendula such that there is zero center-of-mass motion. This mean-field coupling behavior is exhibited by collective motion in nuclei as discussed in example 14.12.

14.7 *Example: Three plane pendula; nearest-neighbor coupling*

There is a large and important class of coupled oscillators where the coupling is only between nearest neighbors; a crystalline lattice is a classic example. A toy model for such a system is the case of three identical pendula coupled by two identical springs, where only the nearest neighbors are coupled as shown in the adjacent figure. Assume the identical pendula are of length b and mass m. As in the last example, the kinetic energy evaluated at the equilibrium location is

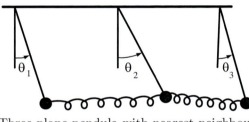

Three plane pendula with nearest-neighbour coupling.

$$T = \frac{1}{2}mb^2\dot{\theta}_1^2 + \frac{1}{2}mb^2\dot{\theta}_2^2 + \frac{1}{2}mb^2\dot{\theta}_3^2$$

The gravitational potential energy of each pendulum equals $mgb(1-\cos\theta) \approx \frac{1}{2}mgb\theta^2$ thus

$$U_{grav} = \frac{1}{2}mgb(\theta_1^2 + \theta_2^2 + \theta_3^2)$$

while the potential energy in the springs is given by

$$U_{spring} = \frac{1}{2}\kappa b^2\left[(\theta_2 - \theta_1)^2 + (\theta_3 - \theta_2)^2\right] = \frac{1}{2}\kappa b^2\left[\theta_1^2 + 2\theta_2^2 + \theta_3^2 - 2\theta_1\theta_2 - 2\theta_2\theta_3\right]$$

Thus the total potential energy is given by

$$U = \frac{1}{2}mgb(\theta_1^2 + \theta_2^2 + \theta_3^2) + \frac{1}{2}\kappa b^2\left[\theta_1^2 + 2\theta_2^2 + \theta_3^2 - 2\theta_1\theta_2 - 2\theta_2\theta_3\right]$$

The Lagrangian then becomes

$$L = \frac{1}{2}mb^2\left(\dot{\theta}_1^2 + \dot{\theta}_2^2 + \dot{\theta}_3^2\right) - \frac{1}{2}\left(mgb + \kappa b^2\right)\theta_1^2 + \frac{1}{2}\left(mgb + 2\kappa b^2\right)\theta_2^2 + \frac{1}{2}\left(mgb + \kappa b^2\right)\theta_3^2 - \kappa b^2\left(\theta_1\theta_2 + \theta_2\theta_3\right)$$

Using this in the Euler-Lagrange equations gives the equations of motion

$$\begin{aligned}
mb^2\ddot{\theta}_1 - (mgb + \kappa b^2)\theta_1 + \kappa b^2\theta_2 &= 0 \\
mb^2\ddot{\theta}_2 - (mgb + 2\kappa b^2)\theta_2 + \kappa b^2\left(\theta_1 + \theta_3\right) &= 0 \\
mb^2\ddot{\theta}_3 - (mgb + \kappa b^2)\theta_3 + \kappa b^2\theta_2 &= 0
\end{aligned}$$

The general analytic approach requires the T and V energy tensors given by

$$\mathbf{T} = mb^2\left\{\begin{array}{ccc} 1 & 0 & 0 \\ 0 & 1 & 0 \\ 0 & 0 & 1 \end{array}\right\} \qquad\qquad \mathbf{V} = \left\{\begin{array}{ccc} mgb + \kappa b^2 & -\kappa b^2 & 0 \\ -\kappa b^2 & mgb + 2\kappa b^2 & -\kappa b^2 \\ 0 & -\kappa b^2 & mgb + \kappa b^2 \end{array}\right\}$$

Note that in contrast to the prior case of three fully-coupled pendula, for the nearest neighbor case the potential energy tensor $\{\mathbf{V}\}$ *is non-zero only on the diagonal and* ± 1 *components parallel to the diagonal.*

The third stage is to evaluate the secular determinant of the $\left(\mathbf{V} - \omega^2 \mathbf{T}\right)$ *matrix, that is*

$$\begin{vmatrix} mgb + \kappa b^2 - \omega^2 mb^2 & -\kappa b^2 & 0 \\ -\kappa b^2 & mgb + 2\kappa b^2 - \omega^2 mb^2 & -\kappa b^2 \\ 0 & -\kappa b^2 & mgb + \kappa b^2 - \omega^2 mb^2 \end{vmatrix} = 0$$

This results in the characteristic equation

$$\left(mgb - \omega^2 mb^2\right)\left(mgb + \kappa b^2 - \omega^2 mb^2\right)\left(mgb + 3\kappa b^2 - \omega^2 mb^2\right) = 0$$

which results in the three non-degenerate eigenfrequencies for the normal modes.

The normal modes are similar to the prior case of complete linear coupling, as shown in the adjacent figure.

$\omega_1 = \sqrt{\frac{g}{b}}$ *This lowest mode* η_1 *involves the three pendula oscillating in phase such that the springs are not stretched or compressed thus the period of this coherent oscillation is the same as an independent pendulum of mass* m *and length* b*. That is*

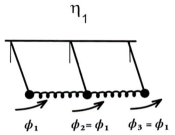

$$\boldsymbol{\eta}_1 = \frac{1}{\sqrt{3}}\left(\theta_1, \theta_2, \theta_3\right)$$

$\omega_2 = \sqrt{\frac{g}{b} + \frac{\kappa}{m}}$*. This second mode* η_2 *has the central mass stationary with the outer pendula oscillating with the same amplitude and out of phase. That is*

$$\boldsymbol{\eta}_2 = \frac{1}{\sqrt{2}}\left(\theta_1, 0, -\theta_3\right)$$

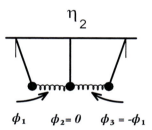

$\omega_3 = \sqrt{\frac{g}{b} + \frac{3\kappa}{m}}$ *. This third mode* η_3 *involves the outer pendula in phase with the same amplitude while the central pendulum oscillating with angle* $\theta_3 = -2\theta_1$*. That is*

$$\boldsymbol{\eta}_3 = \frac{1}{\sqrt{6}}\left(\theta_1, -2\theta_2, \theta_3\right)$$

Similar to the prior case of three completely-coupled pendula, the coherent normal mode $\boldsymbol{\eta}_1$ *corresponds to an oscillation of the center-of-mass with no relative motion, while* $\boldsymbol{\eta}_2$ *and* $\boldsymbol{\eta}_3$ *correspond to relative motion of the pendula with stationary center of mass motion. In contrast to the prior example of complete coupling, for nearest neighbor coupling the two higher lying solutions are not degenerate. That is, the nearest neighbor coupling solutions differ from when all masses are linearly coupled.*

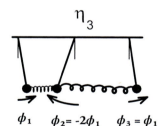

Normal modes of three plane pendula with nearest-neighbour coupling.

It is interesting to note that this example combines two coupling mechanisms that can be used to predict the solutions for two extreme cases by switching off one of these coupling mechanisms. Switching off the coupling springs, by setting $\kappa = 0$*, makes all three normal frequencies degenerate with* $\omega_1 = \omega_2 = \omega_3 = \sqrt{\frac{g}{b}}$*. This corresponds to three independent identical pendula each with frequency* $\omega = \sqrt{\frac{g}{b}}$*. Also the three linear combinations* η_1, η_2, η_3 *also have this same frequency, in particular* η_1 *corresponds to an in-phase oscillation of the three pendula. The three uncoupled pendula are independent and any combination the three modes is allowed since the three frequencies are degenerate.*

The other extreme is to let $\frac{g}{b} = 0$*, that is switch off the gravitational field or let* $b \to \infty$*, then the only coupling is due to the two springs. This results in* $\omega_1 = 0$ *because there is no restoring force acting on the coherent motion of the three in-phase coupled oscillators; as a result, oscillatory motion cannot be sustained since it corresponds to the center of mass oscillation with no external forces acting which is spurious. That is, this spurious solution corresponds to constant linear translation.*

14.8 *Example: System of three bodies coupled by six springs*

Consider the completely-coupled mechanical system shown in the adjacent figure.

1)   The first stage is to determine the potential and kinetic energies using an appropriate set of generalized coordinates, which here are x_1 and x_2. The potential energy is the sum of the potential energies for each of the six springs

$$U = \frac{3}{2}\kappa x_1^2 + \frac{3}{2}\kappa x_2^2 + \frac{3}{2}\kappa x_3^2 - \kappa x_1 x_2 - \kappa x_1 x_3 - \kappa x_2 x_3$$

while the kinetic energy is given by

$$T = \frac{1}{2}m\dot{x}_1^2 + \frac{1}{2}m\dot{x}_2^2 + \frac{1}{2}m\dot{x}_3^2$$

2)   The second stage is to evaluate the potential energy V and kinetic energy T tensors.

$$\mathbf{V} = \left\{ \begin{array}{ccc} 3\kappa & -\kappa & -\kappa \\ -\kappa & 3\kappa & -\kappa \\ -\kappa & -\kappa & 3\kappa \end{array} \right\} \qquad \mathbf{T} = \left\{ \begin{array}{ccc} M & 0 & 0 \\ 0 & M & 0 \\ 0 & 0 & M \end{array} \right\}$$

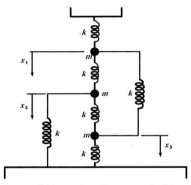

Note that for this case the kinetic energy tensor is diagonal whereas the potential energy tensor is nondiagonal and corresponds to complete coupling of the three coordinates.

3)   The third stage is to use the potential V and kinetic T energy tensors to evaluate the secular determinant giving

$$\begin{vmatrix} (3\kappa - m\omega^2) & -\kappa & -\kappa \\ -\kappa & (3\kappa - m\omega^2) & -\kappa \\ -\kappa & -\kappa & (3\kappa - m\omega^2) \end{vmatrix} = 0$$

System of three bodies coupled by six springs.

The expansion of this secular determinant yields

$$\left(\kappa - m\omega^2\right)\left(4\kappa - m\omega^2\right)\left(4\kappa - mM\omega^2\right) = 0$$

The solution for this complete-coupled system has two degenerate eigenvalues.

$$\omega_1 = \omega_2 = 2\sqrt{\frac{\kappa}{m}} \qquad\qquad \omega_3 = \sqrt{\frac{\kappa}{m}}$$

4)   The fourth step is to insert these eigenfrequencies into the secular equation

$$\sum_j \left(V_{jk} - \omega_r^2 T_{jk}\right) a_{jr} = 0$$

to determine the coefficients a_{jr}.

5)   The final stage is to write the general coordinates in terms of the normal coordinates.

The result is that the angular frequency $\omega_3 = \sqrt{\frac{\kappa}{m}}$ corresponds to a normal mode for which the three masses oscillate in phase corresponding to a center-of-mass oscillation with no relative motion of the masses.

$$\eta_3 = \frac{1}{\sqrt{3}}\left(x_1 + x_2 + x_3\right)$$

For this coherent motion only one spring per mass is stretched resulting in the same frequency as one mass on a spring. The other two solutions correspond to the three masses oscillating out of phase which implies all three springs are stretched and thus the angular frequency is higher. Since the two eigenvalues $\omega_1 = \omega_2 = 2\sqrt{\frac{\kappa}{m}}$ are degenerate then there are only five independent equations to specify the six unknowns for the degenerate eigenvalues. Thus it is possible to select a combination of the eigenvectors η_1 and η_2 such that the combination is orthogonal to η_3. Choose $a_{31} = 0$ to removes the indeterminacy. Then adding or subtracting gives that the normal modes are

$$\eta_1 = \frac{1}{\sqrt{2}}\left(x_1 - x_2 + 0\right) \qquad\qquad \eta_2 = \frac{1}{\sqrt{2}}\left(x_1 + x_2 - 2x_3\right)$$

These two degenerate normal modes correspond to relative motion of the masses with stationary center-of-mass.

14.9 Molecular coupled oscillator systems

There are many examples of coupled oscillations in atomic and molecular physics most of which involve nearest-neighbor coupling. The following two examples are for molecular coupled oscillators. The triatomic molecule is a typical linearly-coupled molecular oscillator. The benzene molecule is an elementary example of a ring structure coupled oscillator.

14.9 Example: Linear triatomic molecular CO_2

Molecules provide excellent examples of vibrational modes involving nearest neighbor coupling. Depending on the atomic structure, triatomic molecules can be either linear, like CO_2, or bent like water, H_2O which has a bend angle of $\theta = 109°$. A molecule with n atoms has $3n$ degrees of freedom. There are three degrees of freedom for translation and three degrees of freedom for rotation leaving $3n - 6$ degrees of freedom for vibrations. A triatomic molecule has three vibrational modes, two longitudinal and one transverse. Consider the normal modes for vibration of the linear molecule CO_2

Longitudinal modes

The coordinate system used is illustrated in the adjacent figure.
The Lagrangian for this system is

$$L = \left(\frac{m}{2}\dot{x}_1^2 + \frac{M}{2}\dot{x}_2^2 + \frac{m}{2}\dot{x}_3^2\right) - \frac{\kappa}{2}[(x_2 - x_1)^2 + (x_3 - x_2)^2]$$

Evaluating the kinetic energy tensor gives

$$\mathbf{T} = \left\{ \begin{array}{ccc} m & 0 & 0 \\ 0 & M & 0 \\ 0 & 0 & m \end{array} \right\}$$

while the potential energy tensor gives

$$\mathbf{V} = \kappa \left\{ \begin{array}{ccc} 1 & -1 & 0 \\ -1 & 2 & -1 \\ 0 & -1 & 1 \end{array} \right\}$$

The secular equation becomes

$$\left| \begin{array}{ccc} \left(-m\omega^2 + \kappa\right) & -\kappa & 0 \\ -\kappa & \left(-M\omega^2 + 2\kappa\right) & -\kappa \\ 0 & -\kappa & \left(-m\omega^2 + \kappa\right) \end{array} \right| = 0$$

Note that the same answer is obtained using Newtonian mechanics. That is, the force equation gives

$$\begin{aligned} m\ddot{x}_1 - \kappa(x_2 - x_1) &= 0 \\ M\ddot{x}_2 + \kappa(x_2 - x_1) - \kappa(x_3 - x_2) &= 0 \\ m\ddot{x}_3 - \kappa(x_3 - x_2) &= 0 \end{aligned}$$

Let the solution be of the form

$$x_j = a_j e^{i\omega t} \qquad j = 1, 2, 3$$

Substitute this solution gives

$$\begin{aligned} \left(-m\omega^2 + \kappa\right)a_1 - \kappa a_2 &= 0 \\ -\kappa a_1 + \left(-M\omega^2 + 2\kappa\right)a_2 - \kappa a_3 &= 0 \\ -\kappa a_2 + \left(m\omega^2 + \kappa\right)a_3 &= 0 \end{aligned}$$

This leads to the same secular determinant as given above with the matrix elements clustered along the diagonal for nearest-neighbor problems.

Expanding the determinant and collecting terms yields

$$\omega^2 \left(-m\omega^2 + \kappa\right) \left(-mM\omega^2 + \kappa M + 2\kappa m\right) = 0$$

Equating either of the three factors to zero gives

$$\omega_1 = 0$$

$$\omega_2 = \sqrt{\frac{\kappa}{m}}$$

$$\omega_3 = \sqrt{\left(\frac{\kappa}{m} + \frac{2\kappa}{M}\right)}$$

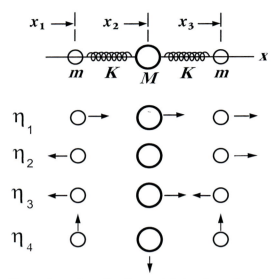

Normal modes of a linear triatomic molecule

The solutions are:

1) $\omega_1 = 0$; This solution gives $\eta_1 = a\{1,1,1\}$. This mode is not an oscillation at all, but is a pure translation of the system as a whole as shown in the adjacent figure. There is no change in the restoring forces since the system moves such as not to change the length of the springs, that is, they stay in their equilibrium positions. This motion corresponds to a spurious oscillation of the center of mass that results from referencing the three atom locations with respect to some fixed reference point. This reference point should have been chosen as the center of mass since the motion of the center-of-mass already has been taken into account separately. Spurious center of mass oscillations occur any time that the reference point is not at the center of mass for an isolated system with no external forces acting.

2) $\omega_2 = \sqrt{\frac{\kappa}{m}}$: This solution corresponds to $\eta_2 = a\{1,0,-1\}$ and is shown in the adjacent figure. The central mass M remains stationary while the two end masses vibrate longitudinally in opposite directions with the same amplitude. This mode has a stationary center of mass. For CO_2 the electrical geometry is $O^-C^{++}O^-$. Mode 2 for CO_2 does not radiate electromagnetically because the center of charge is stationary with respect to the center of mass, that is, the electric dipole moment is constant.

3) $\omega_3 = \sqrt{\left(\frac{\kappa}{m} + \frac{2\kappa}{M}\right)}$: This solution corresponds to $\eta_3 = a\left\{1, -2\left(\frac{m}{M}\right), 1\right\}$. As shown in the adjacent figure, this motion corresponds to the two end masses vibrating in unison while the central mass vibrates oppositely with a different amplitude such that the center-of-mass is stationary. This CO_2 mode does radiate electromagnetically since it corresponds to an oscillating electric dipole.

It is interesting to note that the ratio $\frac{\omega_3}{\omega_2} = 1.915$ for CO_2 and the ratio of the two modes is independent of the potential energy tensor V. That is

$$\frac{\omega_3}{\omega_2} = \sqrt{\left(1 + 2\frac{m}{M}\right)}$$

Transverse modes

The solutions are:

4) $\omega_4 = \sqrt{2\left(\frac{2m+M}{M}\right)\frac{\kappa}{m}}$. This is the only non-spurious transverse mode η_4 which corresponds to the two outside masses vibrating in unison transverse to the symmetry axis while the central mass vibrates oppositely. This mode radiates electric dipole radiation since the electric dipole is oscillating.

5) $\omega_5 = 0$. This transverse solution η_5 has all three nuclei vibrating in unison transverse to the symmetry axis and corresponds to a spurious center of mass oscillation.

6) $\omega_6 = 0$. This transverse solution η_6 corresponds to a stationary central mass with the two outside masses vibrating oppositely. This corresponds to a rotational oscillation of the molecule which is spurious since there are no torques acting on the molecule for a central force. Rotational motion usually is taken into account separately.

The normal modes for the bent triatomic molecule are similar except that the oscillator coupling strength is reduced by the factor $\cos\theta$ where θ is the bend angle.

14.10 *Example: Benzene ring*

The benzene ring comprises six carbon atoms bound in a plane hexagonal ring. A classical analog of the benzene ring comprises 6 identical masses m on a frictionless ring bound by 6 identical springs with linear spring constant K, as illustrated in the adjacent figure. Consider only the in-plane motion, then the kinetic energy is given by

$$T = \frac{1}{2}mr^2 \sum_{i=1}^{6} \dot{\theta}_i^2$$

The potential energy equals

$$U = \frac{1}{2}Kr^2 \sum_{i=1}^{6} (\theta_{i+1} - \theta_i)^2 = Kr^2 \left[\sum_{i=1}^{6} \theta_i^2 - \theta_1\theta_2 - \theta_2\theta_3 - \theta_3\theta_4 - \theta_4\theta_5 - \theta_5\theta_6 - \theta_6\theta_1 \right]$$

where $i = 7 \equiv 1$. Thus the kinetic energy and potential energy tensors are given by

$$T = mr^2 \begin{pmatrix} 1 & 0 & 0 & 0 & 0 & 0 \\ 0 & 1 & 0 & 0 & 0 & 0 \\ 0 & 0 & 1 & 0 & 0 & 0 \\ 0 & 0 & 0 & 1 & 0 & 0 \\ 0 & 0 & 0 & 0 & 1 & 0 \\ 0 & 0 & 0 & 0 & 0 & 1 \end{pmatrix} \qquad U = Kr^2 \begin{pmatrix} 2 & -1 & 0 & 0 & 0 & -1 \\ -1 & 2 & -1 & 0 & 0 & 0 \\ 0 & -1 & 2 & -1 & 0 & 0 \\ 0 & 0 & -1 & 2 & -1 & 0 \\ 0 & 0 & 0 & -1 & 2 & -1 \\ -1 & 0 & 0 & 0 & -1 & 2 \end{pmatrix}$$

This nearest-neighbor system includes non-zero $(n,1)$ and $(1,n)$ elements due to the ring structure. Define $x = \frac{m\omega^2}{K} - 2$ then the solution of the set of linear homogeneous equations requires that

$$\begin{vmatrix} x & 1 & 0 & 0 & 0 & 1 \\ 1 & x & 1 & 0 & 0 & 0 \\ 0 & 1 & x & 1 & 0 & 0 \\ 0 & 0 & 1 & x & 1 & 0 \\ 0 & 0 & 0 & 1 & x & 1 \\ 1 & 0 & 0 & 0 & 1 & x \end{vmatrix} = 0$$

that is

$$(x-2)(x-1)^2(x+1)^2(x+2) = 0$$

The eigenvalues and eigenfunctions are given in the table

Classical analog of a benzene molecular ring.

n	x_n	ω_n^2	Normal modes
1	2	$\frac{4K}{m}$	$\theta_1 - \theta_2 + \theta_3 - \theta_4 + \theta_5 - \theta_6$
2	1	$\frac{3K}{m}$	$-\theta_1 + \theta_3 - \theta_4 + \theta_6$
3	1	$\frac{3K}{m}$	$-\theta_1 + \theta_2 - \theta_4 + \theta_5$
4	-1	$\frac{K}{m}$	$\theta_1 - \theta_3 - \theta_4 + \theta_6$
5	-1	$\frac{K}{m}$	$-\theta_1 - \theta_2 + \theta_4 + \theta_5$
6	-2	0	$\theta_1 + \theta_2 + \theta_3 + \theta_4 + \theta_5 + \theta_6$

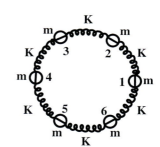

Note the following properties of the normal modes and their frequencies.

$n = 1$: Adjacent masses vibrate $180°$ out of phase, thus each spring has maximal compression or extension, leading to the energy of this normal mode being the highest.

$n = 2, 3$: These two solutions are degenerate and correspond to two pairs of masses vibrating out of phase while the third pair of masses are stationary. Thus the energy of this normal mode is slightly lower than the $n = 1$ normal mode. Any combination of these degenerate normal modes are equally good solutions.

$n = 4, 5$: From the figure it can be seen that both of these solutions correspond to a center of mass oscillation and thus these modes are spurious.

$n = 6$: This vibrational mode has zero energy corresponding to zero restoring force and all six masses moving uniformly in the same direction. This mode corresponds to the rotation of the benzene molecule about the symmetry axis of the ring which usually is taken into account assuming a separate rotational component.

This classical analog of the benzene molecule is interesting because it simultaneously exhibits degenerate normal modes, spurious center of mass oscillation, and a rotational mode.

14.10   Discrete Lattice Chain

Crystalline lattices and linear molecules are important classes of coupled oscillator systems where nearest neighbor interactions dominate. A crystalline lattice comprises thousands of coupled oscillators in a three-dimensional matrix with atomic spacing of a few $10^{-10}m$. Even though a full description of the dynamics of crystalline lattices demands a quantal treatment, a classical treatment is of interest since classical mechanics underlies many features of the motion of atoms in a crystalline lattice. The linear discrete lattice chain is the simplest example of many-body coupled oscillator systems that can illuminate the physics underlying a range of interesting phenomena in solid-state physics. As illustrated in example 2.7, the linear approximation usually is applicable for small-amplitude displacements of nearest-neighbor interacting systems which greatly simplifies treatment of the lattice chain. The linear discrete lattice chain involves three independent polarization modes, one longitudinal mode, plus two perpendicular transverse modes. The $3n$ degrees of freedom for the n atoms, on a discrete linear lattice chain, are partitioned with n degrees of freedom for each of the three polarization modes. These three polarization modes each have n normal modes, or n travelling waves, and exhibit quantization, dispersion, and can have a complex wave number.

14.10.1   Longitudinal motion

The equations of motion for longitudinal modes of the lattice chain can be derived by considering a linear chain of n identical masses, of mass m, separated by a uniform spacing d as shown in Fig 14.7. Assume that the n masses are coupled by $n+1$ springs, with spring constant κ, where both ends of the chain are fixed, that is, the displacements $q_0 = q_{n+1} = 0$ and velocities $\dot{q}_0 = \dot{q}_{n+1} = 0$. The force required to stretch a length d of the chain a longitudinal displacements, q_j for mass j, is $F_j = \kappa q_j$. Thus the potential energy for stretching the spring for segment $(q_{j-1} - q_j)$ is $U_j = \frac{\kappa}{2}(q_{j-1} - q_j)$. The total potential and kinetic energies are

$$U = \frac{\kappa}{2} \sum_{j=1}^{n+1} (q_{j-1} - q_j)^2 \tag{14.74}$$

$$T = \frac{1}{2}m \sum_{j=1}^{n} \dot{q}_j^2 \tag{14.75}$$

Since $\dot{q}_{n+1} = 0$ the kinetic energy and Lagrangian can be extended to $j = n+1$, that is, the Lagrangian can be written as

$$L = \frac{1}{2} \sum_{j=1}^{n+1} \left(m\dot{q}_j^2 - \kappa (q_{j-1} - q_j)^2 \right) \tag{14.76}$$

Using this Lagrangian in the Lagrange-Euler equations gives the following second-order equation of motion for longitudinal oscillations

$$\ddot{q}_j = \omega_o^2 (q_{j-1} - 2q_j + q_{j+1}) \tag{14.77}$$

where $j = 1, 2, ...n$ and where

$$\omega_o \equiv \sqrt{\frac{\kappa}{m}} \tag{14.78}$$

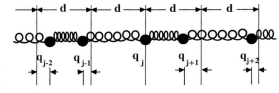

Figure 14.7: Portion of a lattice chain of identical masses m connected by identical springs of spring constant κ. The displacement of the jth mass from the equilibrium position is q_j assumed to be positive to the right.

14.10.2   Transverse motion

The equations of motion for transverse motion on a linear discrete lattice chain, illustrated in figure 14.8, can be derived by considering the displacements q_j of the ith mass for n identical masses, with mass m, separated by equal spacings d and assuming that the tension in the string is $\tau = \left(\frac{\partial U}{\partial x}\right)$. Assuming that the transverse deflections q_j are small, then the $j-1$ to j spring is stretched to a length

$$d' = \sqrt{d^2 + (q_j - q_{j-1})^2} \tag{14.79}$$

Thus the incremental stretching is

$$\delta d \sim \frac{(q_j - q_{j-1})^2}{2d} \tag{14.80}$$

The work done against the tension τ is $\tau \cdot \delta d$ per segment. Thus the total potential energy is

$$U = \frac{\tau}{2d} \sum_{j=1}^{n+1} (q_{j-1} - q_j)^2 \tag{14.81}$$

where q_0 and q_{n+1} are identically zero.

The kinetic energy is

$$T = \frac{1}{2} m \sum_{j=1}^{n} \dot{q}_j^2 \tag{14.82}$$

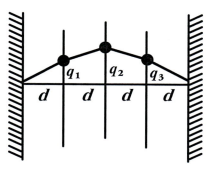

Since $\dot{q}_{n+1} = 0$, the kinetic energy and Lagrangian summations can be extended to $j = n + 1$, that is

$$L = \frac{1}{2} \sum_{j=1}^{n+1} \left(m \dot{q}_j^2 - \frac{\tau}{d} (q_{j-1} - q_j)^2 \right) \tag{14.83}$$

Figure 14.8: Transverse motion of a linear discrete lattice chain

Using this Lagrangian in the Lagrange Euler equations gives the following second-order equation of motion for transverse oscillations

$$\ddot{q}_j = \omega_o^2 (q_{j-1} - 2q_j + q_{j+1}) \tag{14.84}$$

where $j = 1, 2, \ldots n$ and

$$\omega_o \equiv \sqrt{\frac{\tau}{dm}} \tag{14.85}$$

The normal modes for the transverse modes comprise standing waves that satisfy the same boundary conditions as for the longitudinal modes. The n equations of motion for longitudinal motion, equation 14.77, or transverse motion, equation 14.84, are identical in form. The major difference is that ω_0 for the transverse normal modes $\omega_o \equiv \sqrt{\frac{\tau}{dm}}$ differs from that for the longitudinal modes which is $\omega_o \equiv \sqrt{\frac{\kappa}{m}}$. Thus the following discussion of the normal modes on a discrete lattice chain is identical in form for both transverse and longitudinal waves.

14.10.3 Normal modes

The normal modes of the n equations of motion on the discrete lattice chain, are either longitudinal or transverse standing waves that satisfy the boundary conditions at the extreme ends of the lattice chain. The solutions can be given by assuming that the n identical masses on the chain oscillate with a common frequency ω. Then the displacement amplitude for the jth mass can be written in the form

$$q_j(t) = a_j e^{i\omega t} \tag{14.86}$$

where the amplitude a_j can be complex. Substitution into the preceding n equations of motion, $14.77, 14.84$, yields the following recursion relation

$$\left(-\omega^2 + 2\omega_o^2\right) a_j - \omega_0^2 (a_{j-1} + a_{j+1}) = 0 \tag{14.87}$$

where $j = 1, 2, \ldots n$. Note that the boundary conditions, $q_0 = 0$ and $q_{n+1} = 0$ require that $a_o = a_{n+1} = 0$.

The above recursion relation corresponds to a system of n homogeneous algebraic equations with n unknowns $a_1, a_2, \ldots a_n$. A non-trivial solution is given by setting the determinant of its coefficients equal to zero

$$\begin{vmatrix} -\omega^2 + 2\omega_o^2 & -\omega_o^2 & 0 & 0 \\ -\omega_o^2 & -\omega^2 + 2\omega_o^2 & -\omega_o^2 & 0 \\ 0 & -\omega_o^2 & -\omega^2 + 2\omega_o^2 & -\omega_o^2 \\ \ldots\ldots & \ldots\ldots & \ldots\ldots & \ldots\ldots \\ 0 & 0 & -\omega_o^2 & -\omega^2 + 2\omega_o^2 \end{vmatrix} = 0 \tag{14.88}$$

This secular determinant corresponds to the special case of nearest neighbor interactions with the kinetic energy tensor $\mathbf{T}$ being diagonal and the potential energy tensor $\mathbf{V}$ involving coupling only to adjacent masses. The secular determinant is of order n and thus determines exactly n eigen frequencies ω_r for each polarization mode.

For large n, the solution of this problem is more efficiently obtained by using a recursion relation approach, rather than solving the above secular determinant. The trick is to assume that the phase differences ϕ_r between the motion of adjacent masses all are identical for a given polarization. Then the amplitude for the jth mass for the rth frequency mode ω_r is of the form

$$a_{jr} = a_r e^{i(j\phi_r - \delta_r)} \tag{14.89}$$

Insert the above into the recursion relation (14.87) gives

$$\left(-\omega_r^2 + 2\omega_o^2\right) - \omega_0^2 \left[e^{-i\phi_r} + e^{i\phi_r}\right] = 0 \tag{14.90}$$

which reduces to

$$\omega_r^2 = 2\omega_o^2 - 2\omega_o^2 \cos\phi_r = 4\omega_o^2 \sin^2 \frac{\phi_r}{2}$$

that is

$$\omega_r = 2\omega_o \sin \frac{\phi_r}{2} \tag{14.91}$$

where $r = 1, 2, 3, ...n$.

Now it is necessary to determine the phase angle ϕ_r which can be done by applying the boundary conditions for standing waves on the lattice chain. These boundary conditions for stationary modes require that the ends of the lattice chain are nodes, that is $a_{o,r} = a_{(n+1),r} = 0$. Using the fact that only the real part of a_{jr} has physical meaning, leads to the amplitude for the jth mass for the rth mode to be

$$a_{j,r} = a_r \cos\left(j\phi_r - \delta_r\right) \tag{14.92}$$

The boundary condition $a_{0r} = 0$ requires that the phase $\delta_r = \frac{\pi}{2}$. That is

$$a_{jr} = a_r \cos\left(j\phi_r - \frac{\pi}{2}\right) = a_r \sin j\phi_r \tag{14.93}$$

where $r = 1, 2, ..., n$.

The boundary condition for $j = n + 1$, gives

$$a_{(n+1)r} = 0 = a_r \sin\left(n + 1\right)\phi_r \tag{14.94}$$

Therefore

$$\left(n + 1\right)\phi_r = r\pi \tag{14.95}$$

where $r = 1, 2, 3, ..., n$. That is

$$\phi_r = \frac{r\pi}{n+1} = \frac{r\pi d}{(n+1)d} = \frac{r\pi d}{D} = \frac{k_r d}{2} \tag{14.96}$$

where $D = (n+1)d$ is the total length of the discrete lattice chain.

The n eigen frequencies for a given polarization are given by

$$\omega_r = 2\omega_o \sin \frac{r\pi}{2(n+1)} = 2\omega_o \sin \frac{r\pi d}{2(n+1)d} = 2\omega_o \sin \frac{r\pi d}{2D} = 2\omega_o \sin \frac{k_r d}{2} \tag{14.97}$$

where the corresponding wavenumber k_r is given by

$$k_r = \frac{r\pi}{(n+1)d} = \frac{r\pi}{D} = \frac{2\pi}{\lambda_r} \tag{14.98}$$

This implies that the normal modes are quantized with half-wavelengths $\frac{\lambda_r}{2} = \frac{D}{r}$.

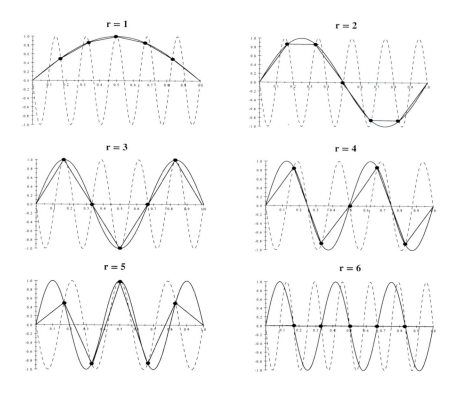

Figure 14.9: Plots of the maximal vibrational amplitudes a_r for the rth frequency sinusoidal mode, versus distance along the chain, for transverse normal modes of a vibrating discrete lattice with $n = 5$. Only $r = 1, 2, 3, 4, 5$, are distinct modes because $r = 6$ is a null mode. Note that the modes with $r = 7, 8, 9, 10, 11, 12$, shown dashed, duplicate the locations of the mass displacement given by the lower-order modes.

Combining equations 14.96 and 14.93 gives the maximum amplitudes for the eigenvectors to be

$$a_{jr} = a_r \sin j \frac{k_r d}{2} \tag{14.99}$$

For n independent linear oscillators there are only n independent normal modes, that is, for $r = n + 1$ the sine function in equation 14.97 must be zero. Beyond $r = n$ the equations do not describe physically new situations. This is illustrated by figure 14.9 which shows the transverse modes of a lattice chain with $n = 5$. There are only $n = 5$ independent normal modes of this system since $r = n + 1 = 6$ corresponds to a null mode with all $q_j(t) = 0$. Also note that the solutions for $r > n + 1$, shown dashed, replicate the mass locations of modes with $r < n + 1$, that is, the modes with $r > 6$ are replicas of the lower-order modes.

Note that ω_r has a maximum value $\omega_r \leq 2\omega_0$ since the sine function cannot exceed unity. This leads to a maximum frequency $\omega_c = 2\omega_0$, called the cut-off frequency, which occurs when $k_r d = \pi$. That is, the null-mode occurs when $r = n + 1$ for which equation 14.99 equals zero. The range of n quantized normal modes that can occur is intuitive. That is, the longest half-wavelength $\frac{\lambda_{\max}}{2} = D = (n+1)d$ equals the total length of the discrete lattice chain. The shortest half-wavelength $\frac{\lambda_{cut-off}}{2} = d$ is set by the lattice spacing. Thus the discrete wavenumbers of the normal modes, for each polarization, range from k_1 to nk_1 where n is an integer.

Assuming real k_r, the normal coordinate η_r and corresponding frequency ω_r are,

$$\eta_r = a_r e^{i\omega_r t} \tag{14.100}$$

Equations 14.97 and 14.99 give the angular frequency and displacement. Note that superposition applies since this system is linear. Therefore the most general solution for each polarization can be any superposition of the form

$$q_j(t) = \sum_{r=1}^{n} \eta_r \sin \left[\frac{r\pi j}{(n+1)} \right] \tag{14.101}$$

14.10.4   Travelling waves

Travelling waves are equally good solutions of the equations of motion 14.77, 14.84 as are the normal modes. Travelling waves on the one-dimensional lattice chain will be of the form

$$q(x,t) = Ce^{i(\omega t \pm kx)} \tag{14.102}$$

where the distance along the chain $x = \nu d$, that is, it is quantized in units of the cell spacing d, with ν being an integer. The positive sign in the exponent corresponds to a wave travelling in the $-x$ direction while the negative sign corresponds to a wave travelling in the $+x$ direction. The velocity of a fixed phase of the travelling wave must satisfy that $\omega t \pm kx$ is a constant. This will occur if the *phase velocity* of the wave is given by

$$v^{phase} = \frac{dx}{dt} = \frac{\omega}{k} \tag{14.103}$$

The wave has a frequency $f = \frac{\omega}{2\pi}$ and wavelength $\lambda = \frac{2\pi}{k}$, thus the phase velocity $v_{phase} = \frac{\omega}{k} = \lambda f$.

   Inserting the travelling wave 14.102 into the transverse equation of motion 14.84 for the discrete lattice chain gives

$$-\omega^2 q_r = \omega_0^2(e^{-\phi_r} - 2 + e^{\phi_r})q_r \tag{14.104}$$

where $j = 1, 2, ...n$. That is

$$\omega_r = \pm 2\omega_0 \sin \frac{\phi_r}{2} \tag{14.105}$$

   The phase ϕ_r is determined by the Born-von Karman periodic boundary condition that assumes that the chain is duplicated indefinitely on either side of $k = \pm\frac{\pi}{d}$. Thus, for n discrete masses, k must satisfy the condition that $q_r = q_{r+n}$. That is

$$e^{ik_r nd} = 1 \tag{14.106}$$

That is

$$k_r = \frac{2\pi r}{nd} \tag{14.107}$$

   Note that the periodic boundary condition gives n discrete modes for wavenumbers between

$$-\frac{\pi}{d} \le k_r \le +\frac{\pi}{d} \tag{14.108}$$

where the index

$$r = -\frac{n}{2}, -\frac{n}{2}+1, ..., \frac{n}{2}-1, \frac{n}{2}$$

Thus equation 14.105 becomes

$$\omega_r = \pm 2\omega_0 \sin \frac{k_r d}{2} \tag{14.109}$$

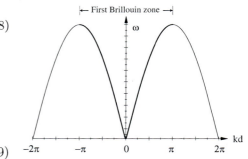

Equation 14.109 is a dispersion relation that is identical to equation 14.97 derived during the discussion of the normal modes of the lattice chain. This confirms that the travelling waves on the lattice chain are equally good solutions as the normal standing-wave modes. Clearly, superposition of the standing-wave normal modes can lead to travelling waves and vice versa.

Figure 14.10: Plot of the dispersion curve (ω versus k) for a monoatomic linear lattice chain subject to only nearest neighbor interactions. The first Brillouin zone is the segment between $-\frac{\pi}{d} \le k \le \frac{\pi}{d}$ which covers all independent solutions.

14.10.5   Dispersion

The lattice chain is an interesting example of a dispersive system in that ω_r is a function of k_r. Figure 14.10 shows a plot of the dispersion curve (ω versus k) for a monoatomic linear lattice chain subject to only nearest neighbor interactions. Note that ω depends linearly on k for small k and that $\frac{d\omega}{dk} = 0$ at the boundaries of the first Brillouin zone.

The lattice chain has a phase velocity for the rth wave given by

$$v_r^{phase} = \frac{\omega_r}{k_r} = \omega_0 d \frac{\left|\sin \frac{k_r d}{2}\right|}{\frac{k_r d}{2}} \qquad (14.110)$$

while the group velocity is

$$v_r^{group} = \left(\frac{d\omega}{dk}\right)_r = \omega_0 d \cos \frac{k_r d}{2} \qquad (14.111)$$

Note that in the limit when $\frac{k_r d}{2} \to 0$, the phase velocity and group velocity are identical, that is, $v_r^{phase} = v_r^{group} = \omega_0 d$.

14.10.6 Complex wavenumber

The maximum allowed frequency, which is called the cut-off frequency, $\omega_c = 2\omega_0$, occurs when $k_r d = \pi$, that is, $\frac{\lambda}{2} = d$. That is, the minimum half-wavelength equals the spacing d between the discrete masses. At the cut-off frequency, the phase velocity is $v_r^{phase} = \frac{2}{\pi}\omega_0 d$ and the group velocity $v_r^{group} = 0$.

It is interesting to note that ω_r can exceed the cut-off frequency $\omega_c = 2\omega_0$ if k_r is assumed to be complex, that is, if

$$k_r = \kappa_r - i\Gamma_r \qquad (14.112)$$

Then

$$\omega_r = 2\omega_0 \sin \frac{k_r d}{2} = 2\omega_0 \sin \frac{d}{2}(\kappa_r - i\Gamma_r) = 2\omega_0 \left(\sin \frac{\kappa_r d}{2} \cosh \frac{\Gamma_r d}{2} - i \cos \frac{\kappa_r d}{2} \sinh \frac{\Gamma_r d}{2}\right) \qquad (14.113)$$

To ensure that ω_r is real, the imaginary term must be zero, that is

$$\cos \frac{\kappa_r d}{2} = 0 \qquad (14.114)$$

Therefore

$$\sin \frac{\kappa_r d}{2} = 1 \qquad (14.115)$$

that is, $k_r = \frac{\pi}{d}$, and the dispersion relation between ω and k for $\omega > 2\omega_0$ becomes

$$\omega_r = 2\omega_0 \cosh \frac{\Gamma_r d}{2} \qquad (14.116)$$

which increases with Γ. Thus, when $\omega > \omega_c = 2\omega_0$ then the amplitude of the wave is of the form

$$q_r(t) = a_r e^{-\Gamma_r x} e^{i(\omega_r t - \kappa_r x)} \qquad (14.117)$$

which corresponds to a spatially damped oscillatory wave with phase velocity

$$v_r^{phase} = \frac{\omega_r}{\kappa_r} \qquad (14.118)$$

and damping factor Γ_r.

There are many examples in physics where the wavenumber is complex as exhibited by the discrete lattice chain for $\frac{\lambda}{2} \le d$. Other examples are electromagnetic waves in conductors or plasma (example 3.5), matter waves tunnelling through a potential barrier, or standing waves on musical instruments which have a complex wavenumber k due to damping.

This simple toy model of the discrete linear lattice chain has illustrated that classical mechanics explains many features of the many-body nearest-neighbor coupled linear oscillator system, including normal modes, standing and travelling waves, cut-off frequency dispersion, and complex wavenumber. These phenomena feature prominently in applications of the quantal discrete coupled-oscillator system to solid-state physics.

14.11   Damped coupled linear oscillators

The discussion of coupled linear oscillators has neglected non-conservative damping forces which always exist to some extent in physical systems. In general, dissipative forces are non linear which greatly complicates solving the equations of motion for such coupled oscillator systems. However, for some systems the dissipative forces depend linearly on velocity which allows use of the Rayleigh dissipation function, described in chapter 10.4. The most general definition of the Rayleigh dissipation function, 10.4, was given to be

$$\mathcal{R} = \frac{1}{2} \sum_{i=1}^{n} \sum_{j=1}^{n} c_{ij} \dot{q}_i \dot{q}_j \tag{14.119}$$

For this special case, it was shown in chapter 10 that the Lagrange equations can be written in terms of the Rayleigh dissipation function as

$$\left\{ \frac{d}{dt} \left(\frac{\partial L}{\partial \dot{q}_j} \right) - \frac{\partial L}{\partial q_j} \right\} + \frac{\partial \mathcal{R}}{\partial \dot{q}_j} = Q_j \tag{14.120}$$

where Q_j are generalized forces acting on the system that are not absorbed into the potential U. Using equations $14.43, 14.44$, and 14.120, allows the equations of motion for damped coupled linear oscillators to be written in a matrix form as

$$\{\mathbf{T}\}\,\ddot{\mathbf{q}} + \{\mathbf{C}\}\,\dot{\mathbf{q}} + \{\mathbf{V}\}\,\mathbf{q} = \{\mathbf{Q}\} \tag{14.121}$$

where the symmetric matrices $\{\mathbf{T}\}, \{\mathbf{C}\}$, and $\{\mathbf{V}\}$ are positive definite for positive definite systems. Rayleigh pointed out that in the special case where the damping matrix $\{\mathbf{C}\}$ is a linear combination of the $\{\mathbf{T}\}$ and $\{\mathbf{V}\}$ matrices, then the matrix $\{\mathbf{C}\}$ is diagonal leading to a separation of the damped system into normal modes. As discussed in chapter 4 many systems in nature are linear for small amplitude oscillations allowing use of the Rayleigh dissipation function which provides an analytic solution. However, in general, except for when $\{\mathbf{C}\}$ is small, this separation into normal modes is not possible for damped systems and the solutions must be obtained numerically.

The following example illustrates approaches used to handle linearly-damped coupled-oscillator systems.

14.11 *Example: Two linearly-damped coupled linear oscillators*

Consider the two coupled oscillator system shown where the two carts have spring constants k_1, k_2 and linear damping constants $c_1 c_2$. As discussed in example 14.3, the kinetic energy tensor is given by

$$T = \frac{1}{2} m_1 \dot{q}_1^2 + \frac{1}{2} m_2 \dot{q}_2^2 \qquad (a)$$

and the potential energy is given by

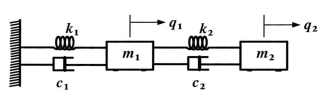

Two linearly-damped coupled linear oscillators.

$$
\begin{aligned}
U & = \frac{1}{2} \left[k_1 q_1^2 + k_2 (q_2 - q_1)^2 \right] \\
& = \frac{1}{2} \left[(k_1 + k_2) q_1^2 - 2 k_2 q_1 q_2 + k_2 q_2^2 \right] \qquad (b)
\end{aligned}
$$

Similarly the Rayleigh dissipation function has the form

$$\mathcal{R} = \frac{1}{2} \left[c_1 \dot{q}_1^2 + c_2 (\dot{q}_2^2 - \dot{q}_1^2) \right] = \frac{1}{2} \left[(c_1 + c_2) \dot{q}_1^2 - 2 c_2 \dot{q}_1 \dot{q}_2 + c_2 \dot{q}_2^2 \right] \qquad (c)$$

Inserting a, b, and c into equation 14.120 gives the two equations of motion to be

$$
\begin{aligned}
m_1 \ddot{q}_1 + (c_1 + c_2) \dot{q}_1 - c_2 \dot{q}_2 + (k_1 + k_2) q_1 - k_2 q_2 & = 0 \\
m_2 \ddot{q}_2 - c_2 \dot{q}_1 + c_2 \dot{q}_2 - k_2 q_1 + k_2 q_2 & = 0
\end{aligned}
$$

When the drag is zero the solution of these two coupled equations can be separated into two independent normal modes of the system as described earlier. Usually it is not possible to separate the motion into decoupled normal modes except for certain cases where the dissipative forces can be described by Rayleigh's dissipation function.

14.12   Collective synchronization of coupled oscillators

Collective synchronization of coupled oscillators is a multifaceted phenomenon where large ensembles of coupled oscillators, with comparable natural frequencies, self synchronize leading to coherent collective modes of motion. Biological examples include congregations of synchronously flashing fireflies, crickets that chirp in unison, an audience clapping at the end of a performance, networks of pacemaker cells in the heart, insulin-secreting cells in the pancreas, as well as neural networks in the brain and spinal cord that control rhythmic behaviors such as breathing, walking, and eating. Example 14.13 illustrates an application to nuclei.

An ensemble of coupled oscillators will have a frequency distribution with a finite width. It is interesting to elucidate how an ensemble of coupled oscillators, that have a finite width frequency distribution, can self synchronize their motion to a unique common frequency, and how that synchronization is maintained over long time periods. The answers to these issues provide insight into the dynamics of coupled oscillators.

The discussion of coupled oscillators has implicitly assumed n identical undamped linear oscillators that have identical, infinitely-sharp, natural frequencies ω_i. In nature typical coupled oscillators can have a finite-width frequency distribution $g(\omega)$ about some average value, due to the natural variability of the oscillator parameters for biological systems, the manufacturing tolerances for mechanical oscillators, or the natural Lorentzian frequency distribution associated with the uncertainty principle that occurs even for atomic clocks where the oscillator frequencies are defined directly by the physical constants. Assume that the ensemble of coupled oscillators has a frequency distribution $g(\omega)$ about some average value.

Undamped linear oscillators have elliptical closed-path trajectories in phase space whereas dissipation leads to a spiral attractor unless the system is driven such as to preserve the total energy. As described in chapter 4.4 many systems in nature, especially biological systems, have closed limit cycles in phase space where the energy lost to dissipation is replenished by a driving mechanism. The simplest systems for understanding collective synchronization of coupled oscillators are those that involve closed limit cycles in phase space.

N. Wiener first recognized the ubiquity of collective synchronization in the natural world, but his mathematical approach, based on Fourier integrals, was not suited to this problem. A more fruitful approach was pioneered in 1975 by an undergraduate student A.T. Winfree[Win67] who recognized that the long-time behavior of a large ensemble of limit-cycle oscillators can be characterized in the simplest terms by considering only the phase of closed phase-space trajectories. He assumed that the instantaneous state of an ensemble of oscillators can be represented by points distributed around the circular phase-space diagram shown in figure 14.11. For uncoupled oscillators these points will be distributed randomly around the circle, whereas coupling of the oscillators will result in a spatial correlation of the points. That is, the dynamics of the phases can be visualized as a swarm of points running around the unit circle in the complex plane of the phase space diagram. The complex order parameter of this swarm can be defined to be the magnitude and phase of the centroid of this swarm

$$re^{i\psi} = \frac{1}{N}\sum_{j=1}^{N} e^{i\theta_j} \tag{14.122}$$

The centroid of the ensemble of points on the phase diagram has a magnitude r, designating the offset of the centroid from the center of the circular phase diagram, and ψ which is the phase of this centroid. A uniform distribution of points around the unit circle will lead to a centroid $r = 0$. Correlated motion leads to a bunching of the points around some phase value leading to a non-zero centroid r and angle ψ. If the swarm acts like a fully-coupled single oscillator then $r \approx 1$ with an appropriate phase ψ.

The **Kuramoto model**[Kur75, Str00] incorporates Winfree's intuition by mapping the limit cycles onto a simple circular phase diagram and incorporating the long-term dynamics of coupled oscillators in terms of the relative phases for a mean-field system. That is, the angular velocity of the phase $\dot{\phi}_i$ for the ith oscillator is

$$\dot{\phi}_i = \omega_i + \sum_{j=1}^{N} \Gamma_{ij}(\phi_j - \phi_i) \tag{14.123}$$

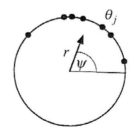

Figure 14.11: Order parameter for weakly-coupled oscillators.

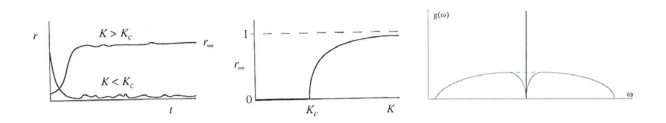

Figure 14.12: Kuramoto model of collective synchronization of coupled oscillators. The left and center plots show the time and coupling strength dependence of the order parameter r. The right plot shows the frequency dependence including coupling (solid line) and without coupling (dashed line).

where $i = 1, 2, , , N$. Kuramoto recognized that mean-field coupling was the most tractable system to solve, that is, a system where the coupling is applicable equally to all the oscillators. Moreover, he assumed an equally-weighted, pure sinusoidal coupling for the coupling term $\Gamma_{ij}(\theta_j - \theta_i)$ between the coupled oscillators. That is, he assumed

$$\Gamma_{ij}(\phi_j - \phi_i) = \frac{K}{N} \sin(\phi_j - \phi_i) \tag{14.124}$$

where $K \geq 0$ is the coupling strength, and the factor $\frac{1}{N}$ ensures that the model is well behaved as $N \to \infty$. Kuramoto assumed that the frequency distribution $g(\omega)$ was unimodular and symmetric about the mean frequency Ω, that is $g(\Omega + \omega) = g(\Omega - \omega)$.

This problem can be simplified by exploiting the rotational symmetry and transforming to a frame of reference that is rotating at an angular frequency Ω. That is, use the transformation $\theta_i = \phi_i - \Omega t$ where θ_i is measured in the rotating frame. This makes $g(\omega)$ unimodular with a symmetric frequency distribution about $\omega = 0$. The phase velocity in this rotating frame is

$$\dot{\theta}_i = \omega_i + \sum_{j=1}^{N} \frac{K}{N} \sin(\theta_j - \theta_i) \tag{14.125}$$

Kuramoto observed that the phase-space distribution can be expressed in terms of the order parameters r, ψ in that equation 14.122 can be multiplied on both sides by $e^{-i\theta_i}$ to give

$$re^{i(\psi - \theta_i)} = \frac{1}{N} \sum_{j=1}^{N} e^{i(\theta_j - \theta_i)} \tag{14.126}$$

Equating the imaginary parts yields

$$r \sin(\psi - \theta_i) = \frac{1}{N} \sum_{j=1}^{N} \sin(\theta_j - \theta_i) \tag{14.127}$$

This allows equation 14.125 to be written as

$$\dot{\theta}_i = \omega_i + Kr \sin(\psi - \theta_i) \tag{14.128}$$

for $i = 1, 2, , N$. Equation 14.128 reflects the mean-field aspect of the model in that *each oscillator θ_i is attracted to the phase of the mean field ψ rather than to the phase of another individual oscillator.*

Simulations showed that the evolution of the order parameter with coupling strength K is as illustrated in figure 14.12. This simulation shows (1) for all K, when below a certain threshold K_c, the order parameter decays to an incoherent jitter as expected for random scatter of N points. (2) When $K > K_c$ this incoherent state becomes unstable and the order parameter r grows exponentially reflecting the nucleation of small clusters of oscillators that are mutually synchronized. (3) The population of individual oscillators splits into two groups. The oscillators near the center of the distribution lock together in phase at the mean angular frequency Ω and co-rotate with average phase $\psi(t)$, whereas those frequencies lying further from the center continue to rotate independently at their natural frequencies and drift relative to the coherent cluster frequency Ω. As a consequence this mixed state is only partially synchronized as illustrated on the right side of figure 14.12. The synchronized fraction has a δ-function behavior for the frequency distribution which grows in intensity with further increase in K. The unsynchronized component has nearly the original frequency distribution $g(\omega)$ except that it is depleted in the region of the locked frequency due to strength absorbed by the δ-function component.

Kuramoto's toy model nicely illustrates the essential features of the evolution of collective synchronization with coupling strength. It has been applied to the study neuronal synchronization in the brain[Cum07]. The model illustrates that the collective synchronization of coupled oscillators leads to a component that has a single frequency for correlated motion which can be much narrower than the inherent frequency distribution of the ensemble of coupled oscillators.

14.12 *Example: Collective motion in nuclei*

The nucleus is an unusual quantal system that involves the coupled motion of the many nucleons. It exhibits features characteristic of the many-body classical coupled oscillator with coupling between all the valence nucleons. Nuclear structure can be described by a shell model of individual nucleons bound in weakly interacting orbits in a central average mean field that is produced by the summed attraction of all the nucleons in the nucleus. However, nuclei also exhibit features characteristic of collective rotation and vibration of a quantal fluid. For example, beautiful rotational bands up to spin over $60\hbar$ are observed in heavy nuclei. These rotational bands are similar to those observed in the rotational structure of diatomic molecules. Actinide nuclei also can fission into two large fragments which is another manifestation of collective motion.

Figure 14.13 shows the case of collective bands in ^{238}U populated by Coulomb exciting a $1355MeV$ ^{238}U beam by a ^{208}Pb target. This case exhibits both quadrupole and octupole collective rotational bands up to spin 40. The inset shows the moment of inertia plotted versus the angular rotational energy $\hbar\omega$. The electromagnetic E2 transition rates correspond to collective motion of ≈ 32 nucleons. Collective motion of many nucleons is the antithesis of shell model motion where the nucleons are assumed to follow independent orbiting motion like planets around the Sun. Although the nucleus is a quantal system, this strange dichotomy can be understood in terms of a classical rotating system having weak linear coupling between each of many similar harmonic oscillators; which in this case, are nucleons bound in a spheroidally-deformed shell-model potential well.

The essential general feature of weakly-coupled identical oscillators is illustrated by the solutions of the three linearly-coupled identical oscillators where the most symmetric state is displaced in frequency from the remaining states. For n identical oscillators, one state is displaced significantly in energy from the remaining $n-1$ degenerate states. This most symmetric state is pushed downwards in energy if the residual coupling force is attractive, and it is pushed upwards if the coupling force is repulsive. This symmetric state corresponds to the coherent oscillation of all the coupled oscillators, and carries all of the strength for the corresponding dominant multipole for the coupling force. In the nucleus this state corresponds to coherent shape oscillations of many nucleons.

The weak residual electric quadrupole and octupole nucleon-nucleon correlations in the nucleon-nucleon interactions generate collective quadrupole and octupole motion in nuclei. The collective synchronization of such coherent quadrupole and octupole excitation leads to collective bands of states, that correspond to synchronized in-phase motion of the protons and neutrons in the valence oscillator shell. These modes correspond to rotations and vibrations about the center of mass. The attractive residual nucleon-nucleon interaction couples the many individual particle excitations in a given shell producing one coherent state that is pushed downwards in energy far from the remaining $n-1$ degenerate states. This coherent state involves correlated motion of the nucleons that corresponds to a macroscopic oscillation of a charged fluid. For non-closed shell nuclei like ^{238}U, the dominant quadrupole multipole in the residual nucleon-nucleon interaction leads to the ground state being a coherent state corresponding to ≈ 16 protons plus ≈ 20 neutrons oscillating in phase. The collective motion of the charged protons leads to electromagnetic E2 radiation with a transition decay amplitude being about 16 times larger than for a single proton. This corresponds to radiative decay probability being enhanced by a factor of ≈ 256 relative to radiation by a single proton. This collective state corresponds to a macroscopic quadrupole deformation at low excitation energies that exhibits both collective rotational and vibrational degrees of freedom as shown in the figure. This coherent state is

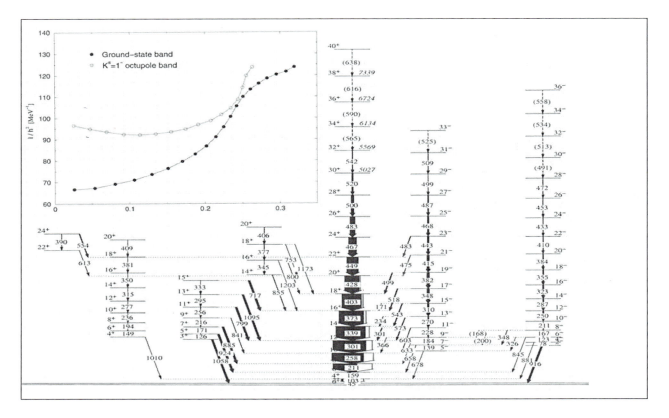

Figure 14.13: Collective rotational bands in the nucleus ^{238}U excited by Coulomb excitation. [Sim98]

analogous to the correlated flow of individual water molecules in a tidal wave. The weaker octupole term in the residual interaction leads to an octupole [pear-shaped] coupled oscillator coherent state lying slightly above the quadrupole coherent state. In contrast to the rotational motion of strongly-deformed quadrupole-deformed nuclei, the octupole deformation exhibits more vibrational-like properties than rotational motion of a charged tidal wave. The observed large increase in moment of inertia at higher rotational frequencies, shown in the insert, is due to the Coriolis force aligning the individual valence nucleons along the rotational axis. Thus, although the nucleus ^{238}U is the epitome of a complicated many-body quantal system, it is apparent that basic classical mechanics of coupled oscillators, and rotation, underlie the physics phenomena exhibited by synchronized collective motion in the nuclear many-body system.

The close correspondence between classical mechanics predictions, and the observed excitation phenomena observed for the ^{238}U nucleus, is surprising for a system that is the epitome of a many-body quantal fluid. The following list identifies other manifestations of classical mechanics discussed in this book, that were exploited for study of such correlated motion of many-body nuclear systems.

1. *Coincident detection of the excited nuclei recoiling in vacuum was used to identify the exact scattering angles, plus recoil velocities, of the scattered nuclei. This specifies the hyperbolic Rutherford trajectory for each scattered nucleus, the nuclear masses, and their recoil velocities. The deexcitation $\gamma-$rays, emitted in flight by each recoiling nucleus, were detected in coincidence with the scattered nuclei. Knowledge of the recoil velocities and scattering angles enabled correction for the Doppler shift in energy of each detected coincident γ-ray to enhance the experimental energy resolution achieved by the γ-ray detectors.*

2. *The transition energies and angular distribution of the deexcitation γ-rays determined the energies, spins, and parities of the excited states in ^{235}U.*

3. *The measured yields of the coincident deexcitation γ-rays determined the excitation cross section as a function of the nuclear scattering angle.*

4. *A full quantal calculation for this system is beyond the capabilities of modern computers since the experiment involves excitation of ~ 100 excited levels, coupled by about ~ 1000 electromagnetic matrix elements, and the scattering involves inclusion of thousands of partial wave due to the long range of the Coulomb potential for the heavy mass of the scattered nuclei. Therefore a semi-classical approximation is used for the quantal calculation of the electromagnetic excitation cross sections as a function of time as the scattered nuclei traverse Rutherford's hyperbolic Coulomb scattering trajectory for each scattered nucleus.*

5. *The measured cross section for the deexcitation γ-rays are compared with the predicted cross sections to determine the ~ 1000 electromagnetic matrix elements connecting the states in ^{235}U.*

6. *The measured electromagnetic matrix elements have been measured in the laboratory frame of reference. Much more insight into the collective motion in ^{235}U is obtained by transforming the electromagnetic matrix elements into the body-fixed frame of reference for this rotating deformed body. Rotational invariants, described in chapter 13.16, are used to derive the electromagnetic properties in the rotating body-fixed frame of reference which unambiguously determines the electromagnetic shape for each excited nuclear state observed in ^{235}U.*

7. *Hamiltonian mechanics, based on the Routhian $R_{noncyclic}$, is used to make theoretical model calculations of the nuclear structure of ^{235}U in the rotating body-fixed frame for comparison with the experimental data derived from this experiment.*

This experiment illustrates that classical mechanics plays a key role in all aspects of the study of the nuclear structure of the many-body nuclear quantal system.

14.13   Summary

This chapter has focussed on many–body coupled linear oscillator systems which are a ubiquitous feature in nature. A summary of the main conclusions are the following.

Normal modes: It was shown that coupled linear oscillators exhibit normal modes and normal coordinates that correspond to independent modes of oscillation with characteristic eigenfrequencies ω_i.

General analytic theory for coupled linear oscillators   Lagrangian mechanics was used to derive the general analytic procedure for solution of the many-body coupled oscillator problem which reduces to the conventional eigenvalue problem. A summary of the procedure for solving coupled oscillator problems is as follows:.

1) Choose generalized coordinates q_j and evaluate T and U.

$$T = \frac{1}{2} \sum_{j,k}^{n} T_{jk} \dot{q}_j \dot{q}_k \tag{14.41}$$

and

$$U = \frac{1}{2} \sum_{j,k}^{n} V_{jk} q_j q_k \tag{14.42}$$

where the components of the **T** and **V** tensors are

$$T_{jk} \equiv \sum_{\alpha}^{N} m_\alpha \sum_{i}^{3} \frac{\partial x_{\alpha,i}}{\partial q_j} \frac{\partial x_{\alpha,i}}{\partial q_k} \tag{14.43}$$

and

$$V_{jk} \equiv \left(\frac{\partial^2 U}{\partial q_j \partial q_k} \right)_0 \tag{14.44}$$

2) Determine the eigenvalues ω_r using the secular determinant.

$$
\begin{vmatrix}
V_{11} - \omega^2 T_{11} & V_{12} - \omega^2 T_{12} & V_{13} - \omega^2 T_{13} & \dots \\
V_{12} - \omega^2 T_{12} & V_{22} - \omega^2 T_{22} & V_{23} - \omega^2 T_{23} & \dots \\
V_{13} - \omega^2 T_{13} & V_{23} - \omega^2 T_{23} & V_{33} - \omega^2 T_{33} & \dots \\
\dots & \dots & \dots & \dots
\end{vmatrix} = 0
\tag{14.52}
$$

3) The eigenvectors are obtained by inserting the eigenvalues ω_r into

$$
\sum_j^n \left(V_{jk} - \omega_r^2 T_{jk} \right) a_{jr} = 0
\tag{14.51}
$$

4) From the initial conditions determine the complex scale factors β_r where

$$
\eta_r (t) \equiv \beta_r e^{i\omega_r t}
\tag{14.58}
$$

5) Determine the normal coordinates where each η_r is a normal mode. The normal coordinates can be expressed as

$$
\eta = \{\mathbf{a}\}^{-1} \mathbf{q}
\tag{14.61}
$$

Few-body coupled oscillator systems   The general analytic theory was used to determine the solutions for parallel and series couplings of two and three linear oscillators. The phenomena observed include degenerate and non-degenerate eigenvalues and spurious center-of-mass oscillatory modes. There are two broad classifications for three or more coupled oscillators, that is, either complete coupling of all oscillators, or coupling of the nearest-neighbor oscillators. It is observed that the eigenvalue corresponding to the most coherent motion of the coupled oscillators corresponds to the most collective motion and its eigenvalue is displaced the most in energy from the remaining eigenvalues. For some systems this coherent collective mode corresponded to a center-of-mass motion with no internal excitation of the other modes, while the other eigenvalues corresponded to modes with internal excitation of the oscillators such that the center of mass is stationary. The above procedure has been applied to two classification of coupling, complete coupling of many oscillators, and nearest neighbor coupling. Both degenerate and spurious center-of-mass modes were observed. Strong collective shape degrees of freedom in nuclei are examples of complete coupling due to the weak residual interactions between nucleons in the nucleus. It was seen that, for many coupled oscillators, one coherent state separates from the other states and this coherent state carries the bulk of the collective strength.

Discrete lattice chain   Transverse and longitudinal modes of motion on the discrete lattice chain were discussed because of the important role it plays in nature, such as in crystalline lattice structures. Both normal modes and travelling waves were discussed including the phenomena of dispersion and cut-off frequencies. Molecules and the crystalline lattice chains are examples where nearest neighbor coupling is manifest. It was shown that, for the $n-$oscillator discrete lattice chain, there are only n independent longitudinal modes plus n modes for the two transverse polarizations, and that the angular frequency $\omega_r \leq 2\omega_0$ that is, a cut-off frequency exists.

Damped coupled linear oscillators   It was shown that linearly-damped coupled oscillator systems can be solved analytically using the concept of the Rayleigh dissipation function.

Collective synchronization of coupled oscillators   The Kuramoto schematic phase model was used to illustrate how weak residual forces can cause collective synchronization of the motion of many coupled oscillators. This is applicable to biological systems as well as mechanical systems.

Workshop exercises

1. Consider two masses (each of mass M) connected by a spring to each other and by springs to fixed positions. Motion is only allowed along one dimension. (This is exactly the same system that is discussed in chapter 14.2 on coupled oscillations.) Let each of the two oscillator springs have a force constant κ and let the force constant of the coupling spring be κ_{12}. Let x_1 and x_2 be the coordinates as described in the textbook.

   (a) Draw a picture of the two masses displaced by a small amount. Using the picture, try to make sense of the equations of motion as given in the text:

   $$M\ddot{x}_1 + (\kappa + \kappa')x_1 - \kappa'x_2 = 0, ::::: M\ddot{x}_2 + (\kappa + \kappa')x_2 - \kappa'x_1 = 0$$

   (b) Each of the trial solutions is written in the form $Be^{i\omega t}$. Why are the trial solutions written this way? Are there any other ways to write the trial solution?

   (c) For a nontrivial solution to exist for the pair of simultaneous equations resulting from the substitution of the trial solution, the determinant of the coefficients of B_1 and B_2 must vanish. Why must this be the case? Is a similar statement true when considering three masses? What about n masses?

   (d) Suppose you had the actual two-mass system sitting in front of you. How could you create antisymmetric motion? How could you create symmetric motion? Can you describe each of these motions using a set of suitable initial conditions?

2. Two particles, each with mass m, move in one dimension in a region near a local minimum of the potential energy where the potential energy is approximately given by

   $$U = \frac{1}{2}k(7x_1^2 + 4x_2^2 + 4x_1x_2)$$

   where k is a constant.

   (a) Determine the frequencies of oscillation.

   (b) Determine the normal coordinates.

3. What is degeneracy? When does it arise?

4. The Lagrangian of three coupled oscillators is given by:

   $$\sum_{n=1}^{3}\left[\frac{m\dot{x}_n^2}{2} - \frac{kx_n^2}{2}\right] + k'(x_1x_2 + x_2x_3).$$

   Find $x_2(t)$ for the following initial conditions (at $t = 0$):

   $$(x_1, x_2, x_3) = (x_0, 0, 0), :::::: (\dot{x}_1, \dot{x}_2, \dot{x}_3) = (0, 0, v_0).$$

5. A mechanical analog of the benzene molecule comprises a discrete lattice chain of 6 point masses M connected in a plane hexagonal ring by 6 identical springs each with spring constant κ and length d.

   a) List the wave numbers of the allowed undamped longitudinal standing waves.

   b) Calculate the phase velocity and group velocity for longitudinal travelling waves on the ring.

   c) Determine the time dependence of a longitudinal standing wave for a angular frequency $\omega = 2\omega_{cutoff}$, that is, twice the cut-off frequency.

6. Consider a one dimensional, two-mass, three-spring system governed by the matrix A,

   $$A = \begin{pmatrix} 4 & -2 \\ -2 & 7 \end{pmatrix}$$

   such that $Ax = \omega^2 x$,

   (a) Determine the eigenfrequencies and normal coordinates.

   (b) Choose a set of initial conditions such that the system oscillates at its highest eigenfrequency.

   (c) Determine the solutions $x_1(t)$ and $x_2(t)$.

Problems

1. Four identical masses m are connected by four identical springs, spring constant κ, and constrained to move on a frictionless circle of radius b as shown on the left in the figure.

   a) How many normal modes of small oscillation are there?

   b) What are the eigenfrequencies of the small oscillations?

   c) Describe the motion of the four masses for each eigenfrequency.

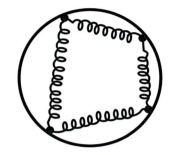

2. Consider the two identical coupled oscillators given on the right in the figure assuming $\kappa_1 = \kappa_2 = \kappa$. Let both oscillators be linearly damped with a damping constant β. A force $F = F_0 \cos(\omega t)$ is applied to mass m_1. Write down the pair of coupled differential equations that describe the motion. Obtain a solution by expressing the differential equations in terms of the normal coordinates. Show that the normal coordinates η_1 and η_2 exhibit resonance peaks at the characteristic frequencies ω_1 and ω_2 respectively.

$$
\begin{array}{ccc}
m_1 = M & & m_2 = M \\
k_1 = K & k_{12} & k_2 = K
\end{array}
$$

3. As shown on the left below the mass M moves horizontally along a frictionless rail. A pendulum is hung from M with a weightless rod of length b with a mass m at its end.

   a) Prove that the eigenfrequencies are

   $$\omega_1 = 0 \qquad\qquad \omega_2 = \sqrt{\frac{g}{Mb}(M+m)}$$

   b) Describe the normal modes.

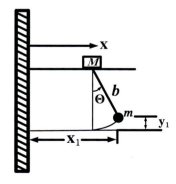

Chapter 15

Advanced Hamiltonian mechanics

15.1 Introduction

This study of classical mechanics has involved climbing a vast mountain of knowledge, while the pathway to the top has led us to elegant and beautiful theories that underlie much of modern physics. Being so close to the summit provides the opportunity to take a few extra steps in order to provide a glimpse of applications to physics at the summit. These are described in chapters $15-18$.

Hamilton's development of Hamiltonian mechanics in 1834 is the crowning achievement for applying variational principles to classical mechanics. A fundamental advantage of Hamiltonian mechanics is that it uses the conjugate coordinates $\mathbf{q}, \mathbf{p}$, plus time t, which is a considerable advantage in most branches of physics and engineering. Compared to Lagrangian mechanics, Hamiltonian mechanics has a significantly broader arsenal of powerful techniques that can be exploited to obtain an analytical solution of the integrals of the motion for complicated systems. In addition, Hamiltonian dynamics provides a means of determining the unknown variables for which the solution assumes a soluble form, and is ideal for study of the fundamental underlying physics in applications to fields such as quantum or statistical physics. As a consequence, Hamiltonian mechanics has become the preeminent variational approach used in modern physics. This chapter introduces the following four techniques in Hamiltonian mechanics: (1) the elegant Poisson bracket representation of Hamiltonian mechanics, which played a pivotal role in the development of quantum theory; (2) the powerful Hamilton-Jacobi theory coupled with Jacobi's development of canonical transformation theory; (3) action-angle variable theory; and (4) canonical perturbation theory.

Prior to further development of the theory of Hamiltonian mechanics, it is useful to summarize the major formula relevant to Hamiltonian mechanics that have been presented in chapters $7, 8$, and 9.

Action functional S:

As discussed in chapter 9.2, Hamiltonian mechanics is built upon Hamilton's action functional

$$S(\mathbf{q}, \mathbf{p}, t) = \int_{t_1}^{t_2} L(\mathbf{q}, \dot{\mathbf{q}}, t) dt \tag{15.1}$$

Hamilton's Principle of least action states that

$$\delta S(\mathbf{q}, \mathbf{p}, t) = \delta \int_{t_1}^{t_2} L(\mathbf{q}, \dot{\mathbf{q}}, t) dt = 0 \tag{15.2}$$

Generalized momentum p:

In chapter 7.2, the generalized (canonical) momentum was defined in terms of the Lagrangian L to be

$$p_i \equiv \frac{\partial L(\mathbf{q}, \dot{\mathbf{q}}, t)}{\partial \dot{q}_i} \tag{15.3}$$

Chapter 9.2 defined the generalized momentum in terms of the action functional S to be

$$p_j = \frac{\partial S(\mathbf{q}, \mathbf{p}, t)}{\partial q_j} \tag{15.4}$$

Generalized energy $h(\mathbf{q}, \dot{q}, t)$:
Jacobi's **Generalized Energy** $h(\mathbf{q}, \dot{q}, t)$ was defined in equation 7.37 as

$$h(\mathbf{q}, \dot{\mathbf{q}}, t) \equiv \sum_j \left(\dot{q}_j \frac{\partial L(\mathbf{q}, \dot{\mathbf{q}}, t)}{\partial \dot{q}_j} \right) - L(\mathbf{q}, \dot{\mathbf{q}}, t) \tag{15.5}$$

Hamiltonian function:
The Hamiltonian $H(\mathbf{q}, \mathbf{p}, t)$ was defined in terms of the generalized energy $h(\mathbf{q}, \dot{\mathbf{q}}, t)$ plus the generalized momentum. That is

$$H(\mathbf{q}, \mathbf{p}, t) \equiv h(\mathbf{q}, \dot{\mathbf{q}}, t) = \sum_j p_j \dot{q}_j - L(\mathbf{q}, \dot{\mathbf{q}}, t) = \mathbf{p} \cdot \dot{\mathbf{q}} - L(\mathbf{q}, \dot{\mathbf{q}}, t) \tag{15.6}$$

where $\mathbf{p}, \mathbf{q}$ correspond to n-dimensional vectors, e.g. $\mathbf{q} \equiv (q_1, q_2, ..., q_n)$ and the scalar product $\mathbf{p} \cdot \dot{\mathbf{q}} = \sum_i p_i \dot{q}_i$. Chapter 8.2 used a Legendre transformation to derive this relation between the Hamiltonian and Lagrangian functions. Note that whereas the Lagrangian $L(\mathbf{q}, \dot{\mathbf{q}}, t)$ is expressed in terms of the coordinates $\mathbf{q}$, plus conjugate velocities $\dot{\mathbf{q}}$, the Hamiltonian $H(\mathbf{q}, \mathbf{p}, t)$ is expressed in terms of the coordinates $\mathbf{q}$ plus their conjugate momenta $\mathbf{p}$. For scleronomic systems, plus assuming the standard Lagrangian, then equations 7.44 and 7.29 give that the Hamiltonian simplifies to equal the total mechanical energy, that is, $H = T + U$.

Generalized energy theorem:
The equations of motion lead to the generalized energy theorem which states that the time dependence of the Hamiltonian is related to the time dependence of the Lagrangian.

$$\frac{dH(\mathbf{q}, \mathbf{p}, t)}{dt} = \sum_j \dot{q}_j \left[Q_j^{EXC} + \sum_{k=1}^m \lambda_k \frac{\partial g_k}{\partial q_j}(\mathbf{q}, t) \right] - \frac{\partial L(\mathbf{q}, \dot{\mathbf{q}}, t)}{\partial t} \tag{15.7}$$

Note that if all the generalized non-potential forces and Lagrange multiplier terms are zero, and if the Lagrangian is not an explicit function of time, then the Hamiltonian is a constant of motion.

Hamilton's equations of motion:
Chapter 8.3 showed that a Legendre transform plus the Lagrange-Euler equations led to Hamilton's equations of motion. Hamilton derived these equations of motion directly from the action functional, as shown in chapter 9.2.

$$\dot{q}_j = \frac{\partial H(\mathbf{q}, \mathbf{p}, t)}{\partial p_j} \tag{15.8}$$

$$\dot{p}_j = -\frac{\partial H}{\partial q_j}(\mathbf{q}, \mathbf{p}, t) + \left[\sum_{k=1}^m \lambda_k \frac{\partial g_k}{\partial q_j} + Q_j^{EXC} \right] \tag{15.9}$$

$$\frac{\partial H(\mathbf{q}, \mathbf{p}, t)}{\partial t} = -\frac{\partial L(\mathbf{q}, \dot{\mathbf{q}}, t)}{\partial t} \tag{15.10}$$

Note the symmetry of Hamilton's two canonical equations. The canonical variables p_k, q_k are treated as independent canonical variables. Lagrange was the first to derive the canonical equations but he did not recognize them as a basic set of equations of motion. Hamilton derived the canonical equations of motion from his fundamental variational principle and made them the basis for a far-reaching theory of dynamics. Hamilton's equations give $2s$ first-order differential equations for p_k, q_k for each of the s degrees of freedom. Lagrange's equations give s second-order differential equations for the variables $q_k, \dot{q}_k$.

Hamilton-Jacobi equation:
Hamilton used Hamilton's Principle to derive the Hamilton-Jacobi equation.

$$\frac{\partial S}{\partial t} + H(\mathbf{q}, \mathbf{p}, t) = 0 \tag{15.11}$$

The solution of Hamilton's equations is trivial if the Hamiltonian is a constant of motion, or when a set of generalized coordinate can be identified for which all the coordinates q_i are constant, or are cyclic (also called *ignorable* coordinates). Jacobi developed the mathematical framework of canonical transformations required to exploit the Hamilton-Jacobi equation.

15.2 Poisson bracket representation of Hamiltonian mechanics

15.2.1 Poisson Brackets

Poisson brackets were developed by Poisson, who was a student of Lagrange. Hamilton's canonical equations of motion describe the time evolution of the canonical variables (q, p) in phase space. Jacobi showed that the framework of Hamiltonian mechanics can be restated in terms of the elegant and powerful Poisson bracket formalism. The Poisson bracket representation of Hamiltonian mechanics provides a direct link between classical mechanics and quantum mechanics.

The Poisson bracket of any two continuous functions of generalized coordinates $F(p, q)$ and $G(p, q)$, is defined to be

$$[F, G]_{qp} \equiv \sum_i \left(\frac{\partial F}{\partial q_i} \frac{\partial G}{\partial p_i} - \frac{\partial F}{\partial p_i} \frac{\partial G}{\partial q_i} \right) \tag{15.12}$$

Note that the above definition of the Poisson bracket leads to the following identity, antisymmetry, linearity, Leibniz rules, and Jacobi Identity.

$$[F, F] = 0 \tag{15.13}$$

$$[F, G] = -[G, F] \tag{15.14}$$

$$[G, F + Y] = [G, F] + [G, Y] \tag{15.15}$$

$$[G, FY] = [G, F]Y + F[G, Y] \tag{15.16}$$

$$0 = [F, [G, Y]] + [G, [Y, F]] + [Y[F, G]] \tag{15.17}$$

where G, H, and Y are functions of the canonical variables plus time. Jacobi's identity; (15.17) states that the sum of the cyclic permutation of the double Poisson brackets of three functions is zero. Jacobi's identity plays a useful role in Hamiltonian mechanics as will be shown.

15.2.2 Fundamental Poisson brackets:

The Poisson brackets of the canonical variables themselves are called the *fundamental Poisson brackets*. They are

$$[q_k, q_l]_{qp} = \sum_i \left(\frac{\partial q_k}{\partial q_i} \frac{\partial q_l}{\partial p_i} - \frac{\partial q_k}{\partial p_i} \frac{\partial q_l}{\partial q_i} \right) = \sum_i (\delta_{ki} \cdot 0 - 0 \cdot \delta_{li}) = 0 \tag{15.18}$$

$$[p_k, p_l]_{qp} = \sum_i \left(\frac{\partial p_k}{\partial q_i} \frac{\partial p_l}{\partial p_i} - \frac{\partial p_k}{\partial p_i} \frac{\partial p_l}{\partial q_i} \right) = \sum_i (0 \cdot \delta_{li} - \delta_{ki} \cdot 0) = 0 \tag{15.19}$$

$$[q_k, p_l]_{qp} = \sum_i \left(\frac{\partial q_k}{\partial q_i} \frac{\partial p_l}{\partial p_i} - \frac{\partial q_k}{\partial p_i} \frac{\partial p_l}{\partial q_i} \right) = \sum_i (\delta_{ki} \cdot \delta_{li} - 0 \cdot 0) = \delta_{kl} \tag{15.20}$$

In summary, the fundamental Poisson brackets equal

$$[q_k, q_l]_{qp} = 0 \tag{15.21}$$

$$[p_k, p_l]_{qp} = 0 \tag{15.22}$$

$$[q_k, p_l]_{qp} = -[p_l, q_k]_{qp} = \delta_{kl} \tag{15.23}$$

Note that the Poisson bracket is antisymmetric under interchange in p and q. It is interesting that the only non-zero fundamental Poisson bracket is for conjugate variables where $k = l$, that is

$$[q_k, p_k]_{pq} = 1 \tag{15.24}$$

15.2.3   Poisson bracket invariance to canonical transformations

The Poisson brackets are invariant under a canonical transformation from one set of canonical variables (q_k, p_k) to a new set of canonical variables (Q_k, P_k) where $Q_k \to Q_k(\mathbf{q}, \mathbf{p})$ and $P_k \to P_k(\mathbf{q}, \mathbf{p})$. This is shown by transforming equation 15.12 to the new variables by the following derivation

$$[F, G]_{qp} = \sum_j \left(\frac{\partial F}{\partial q_j} \frac{\partial G}{\partial p_j} - \frac{\partial F}{\partial p_j} \frac{\partial G}{\partial q_j} \right) \tag{15.25}$$

$$= \sum_{jk} \left(\frac{\partial F}{\partial q_j} \left(\frac{\partial G}{\partial Q_k} \frac{\partial Q_k}{\partial p_j} + \frac{\partial G}{\partial P_k} \frac{\partial P_k}{\partial p_j} \right) - \frac{\partial F}{\partial p_j} \left(\frac{\partial G}{\partial Q_k} \frac{\partial Q_k}{\partial q_j} + \frac{\partial G}{\partial P_k} \frac{\partial P_k}{\partial q_j} \right) \right) \tag{15.26}$$

The terms can be rearranged to give

$$[F, G]_{qp} = \sum_k \left(\frac{\partial G}{\partial Q_k} [F, Q_k]_{qp} + \frac{\partial G}{\partial P_k} [F, P_k]_{qp} \right) \tag{15.27}$$

Let $F = Q_k$ and replace G by F, and use the fact that the fundamental Poisson brackets $[Q_k, Q_j]_{qp} = 0$ and $[Q_k, P_j]_{qp} = \delta_{jk}$, then equation 15.25 reduces to

$$[Q_k, F]_{qp} = \sum_j \left(\frac{\partial F}{\partial Q_j} [Q_k, Q_j] + \frac{\partial F}{\partial P_j} [Q_k, P_j] \right) = \sum_j \frac{\partial F}{\partial P_j} \delta_{jk} \tag{15.28}$$

That is

$$[F, Q_k] = -\frac{\partial F}{\partial P_k} \tag{15.29}$$

Similarly

$$[P_k, F]_{qp} = \sum_j \left(\frac{\partial F}{\partial Q_j} [P_k, Q_j]_{qp} + \frac{\partial F}{\partial P_j} [P_k, P_j]_{qp} \right) \tag{15.30}$$

leading to

$$[F, P_k]_{qp} = \frac{\partial F}{\partial Q_k} \tag{15.31}$$

Substituting equations (15.29) and (15.31) into equation (15.27) gives

$$[F, G]_{qp} = \sum_k \left(\frac{\partial F}{\partial Q_k} \frac{\partial G}{\partial P_k} - \frac{\partial F}{\partial P_k} \frac{\partial G}{\partial Q_k} \right) = [F, G]_{QP} \tag{15.32}$$

Thus the canonical variable subscripts (q, p) and (Q, P) can be ignored since the Poisson bracket is invariant to any canonical transformation of canonical variables. The counter argument is that if the Poisson bracket is independent of the transformation, then the transformation is canonical.

15.1 *Example: Check that a transformation is canonical*

The independence of Poisson brackets to canonical transformations can be used to test if a transformation is canonical. Assume that the transformation equations between two sets of coordinates are given by

$$Q = \ln \left(1 + q^{\frac{1}{2}} \cos p \right) \qquad\qquad P = 2 \left(1 + q^{\frac{1}{2}} \cos p \right) q^{\frac{1}{2}} \sin p$$

Evaluating the Poisson brackets gives $[Q, Q] = 0$, $[P, P] = 0$ while

$$
\begin{aligned}
[Q, P] &= \frac{\partial Q}{\partial q} \frac{\partial P}{\partial p} - \frac{\partial P}{\partial q} \frac{\partial Q}{\partial p} \\
&= \frac{q^{-\frac{1}{2}} \cos p}{1 + q^{\frac{1}{2}} \cos p} [-q \sin^2 p + (1 + q^{\frac{1}{2}} \cos p) q^{\frac{1}{2}} \cos p] + \frac{q^{\frac{1}{2}} \sin^2 p}{1 + q^{\frac{1}{2}} \cos p} [\cos p + (1 + q^{\frac{1}{2}} \cos p) q^{-\frac{1}{2}}] = 1
\end{aligned}
$$

Therefore if q, p are canonical with a Poisson bracket $[q, p] = 1$, then so are Q, P since $[Q, P] = 1 = [q, p]$.

Since it has been shown that this transformation is canonical, it is possible to go further and determine the function that generates this transformation. Solving the transformation equations for q and p give

$$q = \left(e^Q - 1\right)^2 \sec^2 p \qquad\qquad P = 2e^Q \left(e^Q - 1\right) \tan p$$

Since the transformation is canonical, there exists a generating function $F_3(Q, p)$ such that

$$q = -\frac{\partial F_3}{\partial p} \qquad\qquad P = -\frac{\partial F_3}{\partial Q}$$

The transformation function $F_3(Q, p)$ can be obtained using

$$
\begin{aligned}
dF_3(Q, p) &= \frac{\partial F_3}{\partial Q} dQ + \frac{\partial F_3}{\partial p} dp = -P dQ - q dp \\
&= -d\left[\left(e^Q - 1\right)^2\right] \tan p - \left(e^Q - 1\right)^2 d \tan p = -d\left[\left(e^Q - 1\right)^2 \tan p\right]
\end{aligned}
$$

This then gives that the required generating function is

$$F_3(Q, p) = \left(e^Q - 1\right)^2 \tan p$$

This example illustrates how to determine a useful generating function and prove that the transformation is canonical.

15.2.4   Correspondence of the commutator and the Poisson Bracket

In classical mechanics there is a formal correspondence between the Poisson bracket and the commutator. This can be shown by deriving the Poisson Bracket of four functions taken in two pairs. The derivation requires deriving the two possible Poisson Brackets involving three functions.

$$
\begin{aligned}
[F_1 F_2, G] &= \sum_j \left\{ \left(\frac{\partial F_1}{\partial q_j} F_2 + F_1 \frac{\partial F_2}{\partial q_j} \right) \frac{\partial G}{\partial p_j} - \left(\frac{\partial F_1}{\partial p_j} F_2 + F_1 \frac{\partial F_2}{\partial p_j} \right) \frac{\partial G}{\partial q_j} \right\} \\
&= [F_1, G] F_2 + F_1 [F_2, G] \qquad\qquad (15.33) \\
[F, G_1 G_2] &= [F, G_1] G_2 + G_1 [F, G_2] \qquad\qquad (15.34)
\end{aligned}
$$

These two Poisson Brackets for three functions can be used to derive the Poisson Bracket of four functions, taken in pairs. This can be accomplished two ways using either equation 15.33 or 15.34.

$$
\begin{aligned}
[F_1 F_2, G_1 G_2] &= [F_1, G_1 G_2] F_2 + F_1 [F_2, G_1 G_2] \\
&= \{[F_1, G_1] G_2 + G_1 [F_1, G_2]\} F_2 + F_1 \{[F_2, G_1] G_2 + G_1 [F_2, G_2]\} \\
&= [F_1, G_1] G_2 F_2 + G_1 [F_1, G_2] F_2 + F_1 [F_2, G_1] G_2 + F_1 G_1 [F_2, G_2] \qquad (15.35)
\end{aligned}
$$

The alternative approach gives

$$
\begin{aligned}
[F_1 F_2, G_1 G_2] &= [F_1 F_2, G_1] G_2 + G_1 [F_1 F_2, G_2] \\
&= [F_1, G_1] F_2 G_2 + F_1 [F_2, G_1] G_2 + G_1 [F_1, G_2] F_2 + G_1 F_1 [F_2, G_2] \qquad (15.36)
\end{aligned}
$$

These two alternate derivations give different relations for the same Poisson Bracket. Equating the alternative equations 15.35 and 15.36 gives that

$$[F_1, G_1] (F_2 G_2 - G_2 F_2) = (F_1 G_1 - G_1 F_1) [F_2, G_2]$$

This can be factored into separate relations, the left-hand side for body 1, and the right-hand side for body 2.

$$\frac{(F_1 G_1 - G_1 F_1)}{[F_1, G_1]} = \frac{(F_2 G_2 - G_2 F_2)}{[F_2, G_2]} = \lambda \qquad\qquad (15.37)$$

Since the left-hand ratio holds for F_1, G_1 independent of F_2, G_2, and vise versa, then they must equal a constant λ that does not depend on F_1, G_1, does not depend on F_2, G_2, and λ must commute with $(F_1 G_1 - G_1 F_1)$. That is, λ must be a constant number independent of these variables.

$$(F_1 G_1 - G_1 F_1) = \lambda [F_1, G_1] \equiv \lambda \sum_i \left(\frac{\partial F_1}{\partial q_i} \frac{\partial G_1}{\partial p_i} - \frac{\partial F_1}{\partial p_i} \frac{\partial G_1}{\partial q_i} \right) \tag{15.38}$$

Equation 15.38 is an especially important result which states that to *within a multiplicative constant number* λ, *there is a one-to-one correspondence between the Poisson Bracket and the commutator of two independent functions.* An important implication is that *if two functions, $F_i G_k$ have a Poisson Bracket that is zero, then the commutator of the two functions also must be zero, that is, F_i and G_k commute.*

Consider the special case where the variables F_1 and G_1 correspond to the fundamental canonical variables, (q_k, p_l). Then the commutators of the fundamental canonical variables are given by

$$
\begin{aligned}
q_k p_l - p_l q_k &= \lambda [q_k, p_l] = \lambda \delta_{kl} & (15.39) \\
q_k q_l - q_l q_k &= \lambda [q_k, q_l] = 0 & (15.40) \\
p_k p_l - p_l p_k &= \lambda [p_k, p_l] = 0 & (15.41)
\end{aligned}
$$

In 1925, Paul Dirac, a 23-year old graduate student at Bristol, recognized that the formal correspondence between the Poisson bracket in classical mechanics, and the corresponding commutator, provides a logical and consistent way to bridge the chasm between the Hamiltonian formulation of classical mechanics, and quantum mechanics. He realized that making the assumption that the constant $\lambda \equiv i\hbar$, leads to Heisenberg's fundamental commutation relations in quantum mechanics, as is discussed in chapter 18.3.1. Assuming that $\lambda \equiv i\hbar$ provides a logical and consistent way that builds quantization directly into classical mechanics, rather than using ad-hoc, case-dependent, hypotheses as was used by the older quantum theory of Bohr.

15.2.5 Observables in Hamiltonian mechanics

Poisson brackets, and the corresponding commutation relations, are especially useful for elucidating which observables are constants of motion, and whether any two observables can be measured simultaneously and exactly. The properties of any observable are determined by the following two criteria.

Time dependence:

The total time differential of a function $G(q_i, p_i, t)$ is defined by

$$\frac{dG}{dt} = \frac{\partial G}{\partial t} + \sum_i \left(\frac{\partial G}{\partial q_i} \dot{q}_i + \frac{\partial G}{\partial p_i} \dot{p}_i \right) \tag{15.42}$$

Hamilton's canonical equations give that

$$\dot{q}_i = \frac{\partial H}{\partial p_i} \tag{15.43}$$

$$\dot{p}_i = -\frac{\partial H}{\partial q_i} \tag{15.44}$$

Substituting these in the above relation gives

$$\frac{dG}{dt} = \frac{\partial G}{\partial t} + \sum_i \left(\frac{\partial G}{\partial q_i} \frac{\partial H}{\partial p_i} - \frac{\partial G}{\partial p_i} \frac{\partial H}{\partial q_i} \right)$$

that is

$$\frac{dG}{dt} = \frac{\partial G}{\partial t} + [G, H] \tag{15.45}$$

This important equation states that the total time derivative of any function $G(q, p, t)$ can be expressed in terms of the partial time derivative plus the Poisson bracket of $G(q, p, t)$ with the Hamiltonian.

Any observable $G(p, q, t)$ will be a constant of motion if $\frac{dG}{dt} = 0$, and thus equation (15.45) gives

$$\frac{\partial G}{\partial t} + [G, H] = 0 \qquad \text{(If } G \text{ is a constant of motion)}$$

That is, it is a constant of motion when

$$\frac{\partial G}{\partial t} = [H, G] \tag{15.46}$$

Moreover, this can be extended further to the statement that *if the constant of motion G is not explicitly time dependent* then

$$[G, H] = 0 \tag{15.47}$$

The Poisson bracket with the Hamiltonian is zero for a constant of motion G that is not explicitly time dependent. Often it is more useful to turn this statement around with the statement that *if $[G, H] = 0$, and $\frac{\partial G}{\partial t} = 0$, then $\frac{dG}{dt} = 0$, implying that G is a constant of motion*.

Independence

Consider two observables $F(p, q, t)$ and $G(p, q, t)$. The independence of these two observables is determined by the Poisson bracket

$$[F, G] = -[G, F] \tag{15.48}$$

If this Poisson bracket is zero, that is, if the two observables $F(p, q, t)$ and $G(p, q, t)$ commute, then their values are independent and can be measured independently. However, if the Poisson bracket $[F, G] \neq 0$, that is $F(p, q, t)$ and $G(p, q, t)$ do not commute, then F and G are correlated since interchanging the order of the Poisson bracket changes the sign which implies that the measured value for F depends on whether G is simultaneously measured.

A useful property of Poisson brackets is that if F and G both are constants of motion, then the double Poisson bracket $[H, [F, G]] = 0$. This can be proved using Jacobi's identity

$$[F, [G, H]] + [G, [H, F]] + [H, [F, G]] = 0 \tag{15.49}$$

If $[G, H] = 0$ and $[F, H] = 0$, then $[H, [F, G]] = 0$, that is, the Poisson bracket $[F, G]$ commutes with H. Note that if F and G do not depend explicitly on time, that is $\frac{\partial F}{\partial t} = \frac{\partial G}{\partial t} = 0$, then combining equations (15.45) and (15.49) leads to Poisson's Theorem that relates the total time derivatives.

$$\frac{d}{dt}[F, G] = \left[\frac{dF}{dt}, G\right] + \left[F, \frac{dG}{dt}\right] \tag{15.50}$$

This implies that if F and G are invariants, that is $\frac{dF}{dt} = \frac{dG}{dt} = 0$, then the Poisson bracket $[F, G]$ is an invariant if F and G are not explicitly time dependent.

15.2 Example: Angular momentum:

Angular momentum, L, provides an example of the use of Poisson brackets to elucidate which observables can be determined simultaneously. Consider that the Hamiltonian is time independent with a spherically symmetric potential $U(r)$. Then it is best to treat such a spherically symmetric potential using spherical coordinates since the Hamiltonian is independent of both θ and ϕ.

The Poisson Brackets in classical mechanics can be used to tell us if two observables will commute. Since $U(r)$ is time independent, then the Hamiltonian in spherical coordinates is

$$H = T + U = \frac{1}{2m}\left(p_r^2 + \frac{p_\theta^2}{r^2} + \frac{p_\phi^2}{r^2 \sin^2 \theta}\right) + U(r)$$

Evaluate the Poisson bracket using the above Hamiltonian gives

$$[p_\phi, H] = 0$$

Since p_ϕ is not an explicit function of time, $\frac{\partial p_\phi}{\partial t} = 0$, then $\frac{dp_\phi}{dt} = 0$, that is, the angular momentum about the z axis $L_z = p_\phi$ is a constant of motion.

The Poisson bracket of the total angular momentum L^2 commutes with the Hamiltonian, that is

$$[L^2, H] = \left[p_\theta^2 + \frac{p_\phi^2}{\sin^2 \theta}, H \right] = 0$$

Since the total angular momentum $L^2 = p_\theta^2 + \frac{p_\phi^2}{\sin^2 \theta}$ is not explicitly time dependent, then it also must be a constant of motion. Note that Noether's theorem gives that both the angular momenta L^2 and L_z are constants of motion. Also since the Poisson brackets are

$$[L_z, H] = 0$$
$$[L^2, H] = 0$$

then Jacobi's identity, equation 15.17, can be used to imply that

$$[H, [L^2, L_z]] = 0$$

That is, the Poisson bracket $[L^2, L_z]$ is a constant of motion. Note that if L^2 and L_z commute, that is, $[L^2, L_z] = 0$, then they can be measured simultaneously with unlimited accuracy, and this also satisfies that $[L^2, L_z]$ commutes with H.

The (x, y, z) components of the angular momentum L are given by

$$L_x = \sum_{i=1}^{n} (\mathbf{r} \times \mathbf{p})_x = \sum_{i=1}^{n} (y_i p_{z,i} - z_i p_{y,i})$$

$$L_y = \sum_{i=1}^{n} (\mathbf{r} \times \mathbf{p})_y = \sum_{i=1}^{n} (z_i p_{x,i} - x_i p_{z,i})$$

$$L_z = \sum_{i=1}^{n} (\mathbf{r} \times \mathbf{p})_z = \sum_{i=1}^{n} (x_i p_{y,i} - y_i p_{x,i})$$

Evaluate the Poisson bracket

$$[L_x, L_y] = \sum_{i=1}^{n} \left[\left(\frac{\partial L_x}{\partial x_i} \frac{\partial L_y}{\partial p_{x,i}} - \frac{\partial L_x}{\partial p_{x,i}} \frac{\partial L_y}{\partial x_i} \right) + \left(\frac{\partial L_x}{\partial y_i} \frac{\partial L_y}{\partial p_{y,i}} - \frac{\partial L_x}{\partial p_{y,i}} \frac{\partial L_y}{\partial y_i} \right) + \left(\frac{\partial L_x}{\partial z_i} \frac{\partial L_y}{\partial p_{z,i}} - \frac{\partial L_x}{\partial p_{z,i}} \frac{\partial L_y}{\partial z_i} \right) \right]$$

$$= \sum_{i=1}^{n} [(0) + (0) + (x_i p_{y,i} - y_i p_{x,i})] = L_z$$

Similarly, Poisson brackets for L_x, L_y, L_z are

$$[L_x, L_y] = L_z$$
$$[L_y, L_z] = L_x$$
$$[L_z, L_x] = L_y$$

where x, y, and z are taken in a right-handed cyclic order. This usually is written in the form

$$[L_i, L_j] = \epsilon_{ijk} L_k$$

where the Levi-Civita density ϵ_{ijk} equals zero if two of the ijk indices are identical, otherwise it is $+1$ for a cyclic permutation of i, j, k, and -1 for a non-cyclic permutation.

Note that since these Poisson brackets are nonzero, the components of the angular momentum L_x, L_y, L_z do not commute and thus simultaneously they cannot be measured precisely. Thus we see that although L^2 and L_i are simultaneous constants of motion, where the subscript i can be either x, y, or z, only one component L_i can be measured simultaneously with L^2. This behavior is exhibited by rigid-body rotation where the body precesses around one component of the total angular momentum, L_z, such that the total angular momentum, L^2, plus the component along one axis, L_z are constants of motion. Then $L_x^2 + L_y^2 = L^2 - L_z^2$ is constant but not the individual L_x or L_y.

15.2.6 Hamilton's equations of motion

An especially important application of Poisson brackets is that Hamilton's canonical equations of motion can be expressed directly in the Poisson bracket form. The Poisson bracket representation of Hamiltonian mechanics has important implications to quantum mechanics as will be described in chapter 18.

In equation (15.45) assume that G is a fundamental coordinate, that is, $G \equiv q_k$. Since q_k is not explicitly time dependent, then

$$
\begin{aligned}
\frac{dq_k}{dt} &= \frac{\partial q_k}{\partial t} + [q_k, H] && (15.51) \\
&= 0 + \sum_i \left(\frac{\partial q_k}{\partial q_i} \frac{\partial H}{\partial p_i} - \frac{\partial q_k}{\partial p_i} \frac{\partial H}{\partial q_i} \right) \\
&= \sum_i \left(\delta_{ik} \frac{\partial H}{\partial p_i} - 0 \cdot \frac{\partial H}{\partial q_i} \right) \\
&= \frac{\partial H}{\partial p_k} && (15.52)
\end{aligned}
$$

That is

$$
\dot{q}_k = [q_k, H] = \frac{\partial H}{\partial p_k} \tag{15.53}
$$

Similarly consider the fundamental canonical momentum $G \equiv p_k$. Since it is not explicitly time dependent, then

$$
\begin{aligned}
\frac{dp_k}{dt} &= \frac{\partial p_k}{\partial t} + [p_k, H] && (15.54) \\
&= 0 + \sum_i \left(\frac{\partial p_k}{\partial q_i} \frac{\partial H}{\partial p_i} - \frac{\partial p_k}{\partial p_i} \frac{\partial H}{\partial q_i} \right) \\
&= \sum_i \left(0 \cdot \frac{\partial H}{\partial p_i} - \delta_{ik} \cdot \frac{\partial H}{\partial q_i} \right) \\
&= -\frac{\partial H}{\partial q_k} && (15.55)
\end{aligned}
$$

That is

$$
\dot{p}_k = [p_k, H] = -\frac{\partial H}{\partial q_k} \tag{15.56}
$$

Thus, it is seen that the Poisson bracket form of the equations of motion includes the Hamilton equations of motion. That is,

$$
\dot{q}_k = [q_k, H] = \frac{\partial H}{\partial p_k} \tag{15.57}
$$

$$
\dot{p}_k = [p_k, H] = -\frac{\partial H}{\partial q_k} \tag{15.58}
$$

The above shows that the full structure of Hamilton's equations of motion can be expressed directly in terms of Poisson brackets.

The elegant formulation of Poisson brackets has the same form in all canonical coordinates as the Hamiltonian formulation. However, the normal Hamilton canonical equations in classical mechanics assume implicitly that one can specify the exact position and momentum of a particle simultaneously at any point in time which is applicable only to classical mechanics variables that are continuous functions of the coordinates, and not to quantized systems. The important feature of the Poisson Bracket representation of Hamilton's equations is that it generalizes Hamilton's equations into a form (15.57, 15.58) where the Poisson bracket is equally consistent with both classical and quantum mechanics in that it allows for non-commuting canonical variables and Heisenberg's Uncertainty Principle. Thus the generalization of Hamilton's equations, via use of the Poisson brackets, provides one of the most powerful analytic tools applicable to both classical and quantal dynamics. It played a pivotal role in derivation of quantum theory as described in chapter 18.

15.3 Example: Lorentz force in electromagnetism

Consider a charge q, and mass m, in a constant electromagnetic fields with scalar potential Φ and vector potential A. Chapter 6.10 showed that the Lagrangian for electromagnetism can be written as

$$L = \frac{1}{2}m\dot{\mathbf{x}} \cdot \dot{\mathbf{x}} - q(\mathbf{\Phi} - \mathbf{A} \cdot \dot{\mathbf{x}})$$

The generalized momentum then is given by

$$\mathbf{p} = \frac{\partial L}{\partial \dot{\mathbf{x}}} = m\dot{\mathbf{x}} + q\mathbf{A}$$

Thus the Hamiltonian can be written as

$$H = (\mathbf{p} \cdot \dot{\mathbf{x}}) - L = \frac{(\mathbf{p}-q\mathbf{A})^2}{2m} + q\Phi$$

The Hamilton equations of motion give

$$\dot{\mathbf{x}} = [\mathbf{x}, H] = \frac{(\mathbf{p}-q\mathbf{A})}{m}$$

and

$$\dot{\mathbf{p}} = [\mathbf{p}, H] = -q\mathbf{\nabla}\Phi + \frac{q}{m}\left\{(\mathbf{p}-q\mathbf{A}) \times (\mathbf{\nabla} \times \mathbf{A})\right\}$$

Define the magnetic field to be

$$\mathbf{B} \equiv \mathbf{\nabla} \times \mathbf{A}$$

and the electric field to be

$$\mathbf{E} = -\mathbf{\nabla}\Phi - \frac{\partial \mathbf{A}}{\partial t}$$

then the Lorentz force can be written as

$$\mathbf{F} = \dot{\mathbf{p}} = q\left(\mathbf{E} + \dot{\mathbf{x}} \times \mathbf{B}\right)$$

15.4 Example: Wavemotion:

Assume that one is dealing with traveling waves of the form $\Psi = Ae^{i\left(\frac{1}{m}xp_x - \omega t\right)}$ for a one-dimensional conservative system of many identical coupled linear oscillators. Then evaluating the following Poisson brackets gives

$$
\begin{aligned}
{[p_x, H]} &= 0 \\
{[x, H]} &= 0 \\
{[\omega, H]} &= 0 \\
{[t, H]} &= 0
\end{aligned}
$$

Thus p_x, x, ω, and t are constants of motion. However,

$$
\begin{aligned}
{[p_x, x]} &\neq 0 \\
{[\omega, t]} &\neq 0
\end{aligned}
$$

Thus one cannot simultaneously measure the conjugate variables $(p_x x)$ or (ω, t). This is the Uncertainty Principle that is manifest by all forms of wave motion in classical and quantal mechanics as discussed in chapter 3.11.3.

15.5 Example: Two-dimensional, anisotropic, linear oscillator

Consider a mass m bound by an anisotropic, two-dimensional, linear oscillator potential. As discussed in chapter 11, the motion can be described as lying entirely in the $x - y$ plane that is perpendicular to the angular momentum J. It is interesting to derive the equations of motion for this system using the Poisson bracket representation of Hamiltonian mechanics.

The kinetic energy is given by

$$T(\dot{x}, \dot{y}) = \frac{1}{2}m\left(\dot{x}^2 + \dot{y}^2\right)$$

The linear binding is reproduced assuming a quadratic scalar potential energy of the form

$$U(x, y) = \frac{1}{2}k\left(x^2 + y^2\right) + \eta xy$$

where η is the anharmonic strength that coupled the modes of the isotropic linear oscillator.

a) NORMAL MODES: As discussed in chapter 14, a transformation to the normal modes of the system is given by using variables (α, β) where $\alpha \equiv \frac{1}{\sqrt{2}}(x + y)$ and $\beta \equiv \frac{1}{\sqrt{2}}(x - y)$, that is

$$x \equiv \frac{1}{\sqrt{2}}(\alpha + \beta) \qquad\qquad y \equiv \frac{1}{\sqrt{2}}(\alpha - \beta)$$

Express the kinetic and potential energies in terms of the new coordinates gives

$$T(\dot{x}, \dot{y}) = \frac{1}{4}m\left[\left(\dot{\alpha} + \dot{\beta}\right)^2 + \left(\dot{\alpha} - \dot{\beta}\right)^2\right] = \frac{1}{2}m\left(\dot{\alpha}^2 + \dot{\beta}^2\right)$$

$$U = \frac{1}{4}k\left[(\alpha + \beta)^2 + (\alpha - \beta)^2\right] + \frac{1}{2}\eta\left(\alpha^2 - \beta^2\right) = \frac{1}{2}(k + \eta)\alpha^2 + \frac{1}{2}(k - \eta)\beta^2$$

Note that the coordinate transformation makes the Lagrangian separable, that is

$$L = \frac{1}{2}m\left(\dot{\alpha}^2 + \dot{\beta}^2\right) - \frac{1}{2}(k + \eta)\alpha^2 + \frac{1}{2}(k - \eta)\beta^2 = L_\alpha + L_\beta$$

where

$$L_\alpha = \frac{1}{2}m\dot{\alpha}^2 - \frac{1}{2}(k + \eta)\alpha^2 \qquad\qquad L_\beta = \frac{1}{2}m\dot{\beta}^2 - \frac{1}{2}(k - \eta)\beta^2$$

This shows that that the transformation has separated the system into two normal modes that are harmonic oscillators with angular frequencies

$$\omega_1 = \sqrt{\frac{k + \eta}{m}} \qquad\qquad \omega_2 = \sqrt{\frac{k - \eta}{m}}$$

Note that the non-isotropic harmonic oscillator reduces to the isotropic linear oscillator when $\eta = 0$.

b) HAMILTONIAN: The canonical momenta are given by

$$p_\alpha = \frac{\partial L}{\partial \dot{\alpha}} = m\dot{\alpha}$$

$$p_\beta = \frac{\partial L}{\partial \dot{\beta}} = m\dot{\beta}$$

The definition of the Hamiltonian gives

$$H = p_\alpha \dot{\alpha} + p_\beta \dot{\beta} - L = \frac{1}{2m}\left(p_\alpha^2 + p_\beta^2\right) + \frac{1}{2}(k + \eta)\alpha^2 + \frac{1}{2}(k - \eta)\beta^2$$

Note that this can be factored as

$$H = H_\alpha + H_\beta$$

where

$$H_\alpha = \frac{1}{2m}p_\alpha^2 + \frac{1}{2}(k + \eta)\alpha^2 \qquad\qquad H_\beta = \frac{1}{2m}p_\beta^2 + \frac{1}{2}(k - \eta)\beta^2$$

Using the Poisson Bracket expression for the time dependence, equation 15.45, and using the fact that the Hamiltonian is not explicitly time dependent, that is, $\frac{\partial H}{\partial t} = 0$, gives

$$
\begin{aligned}
\frac{dH_\alpha}{dt} &= \frac{\partial H_\alpha}{\partial t} + [H_\alpha, H] = 0 + [H_\alpha, H_\alpha + H_\beta] = [H_\alpha, H_\beta] \\
&= \frac{\partial H_\alpha}{\partial \alpha}\frac{\partial H_\beta}{\partial p_\alpha} + \frac{\partial H_\alpha}{\partial \beta}\frac{\partial H_\beta}{\partial p_\beta} - \frac{\partial H_\alpha}{\partial p_\alpha}\frac{\partial H_\beta}{\partial \alpha} - \frac{\partial H_\alpha}{\partial p_\beta}\frac{\partial H_\beta}{\partial \beta} = 0
\end{aligned}
$$

Similarly $\frac{dH_\beta}{dt} = 0$. This implies that the Hamiltonians for both normal modes, H_α and H_β, are time-independent constants of motion which are equal to the total energy for each mode.

c) ANGULAR MOMENTUM: The angular momentum for motion in the $\alpha\beta$ plane is perpendicular to the $\alpha\beta$ plane with a magnitude of

$$ J = m\left(\alpha p_\beta - \beta p_\alpha\right) $$

The time dependence of the angular momentum is given by

$$
\begin{aligned}
\frac{dJ}{dt} &= \frac{\partial J}{\partial t} + [J, H] = 0 + \frac{\partial J}{\partial \alpha}\frac{\partial H}{\partial p_\alpha} - \frac{\partial J}{\partial p_\alpha}\frac{\partial H}{\partial \alpha} + \frac{\partial J}{\partial \beta}\frac{\partial H}{\partial p_\beta} - \frac{\partial J}{\partial p_\beta}\frac{\partial H}{\partial \beta} \\
&= p_\beta p_\alpha + mk\beta\alpha + m\eta\beta\alpha - p_\alpha p_\beta - mk\alpha\beta + m\eta\beta\alpha = 2m\eta\beta\alpha
\end{aligned}
$$

Note that if $\eta = 0$, then the two eigenfrequencies, are degenerate, $\omega_\alpha = \omega_\beta$, that is, the system reduces to the isotropic harmonic oscillator in the $\alpha\beta$ plane that was discussed in chapter 11.9. In addition, $\frac{dJ}{dt} = 0$ for $\eta = 0$, that is, the angular momentum J in the $\alpha\beta$ plane is a constant of motion when $\eta = 0$.

d) SYMMETRY TENSOR: The symmetry tensor was defined in chapter 11.9.3 to be

$$ A'_{ij} = \frac{p_i p_j}{2m} + \frac{1}{2}kx_i x_j $$

where i and j can correspond to either α or β. The symmetry tensor defines the orientation of the major axis of the elliptical orbit for the two-dimensional, isotropic, linear oscillator as described in chapter 11.9.3.

The isotropic oscillator has been shown to have two normal modes that are degenerate, therefore α and β are equally good normal modes. The Hamiltonian showed that, for $\eta = 0$, the Hamiltonian gives that the total energy is conserved, as well as the energies for each of the two normal modes which are.

$$ E_\alpha = \frac{p_\alpha^2}{2m} + \frac{1}{2}k\alpha^2 \qquad\qquad E_\beta = \frac{p_\beta^2}{2m} + \frac{1}{2}k\beta^2 $$

Consider the matrix element

$$ A'_{ij} = \frac{p_i p_j}{2m} + \frac{1}{2}kx_i x_j $$

where i, j each can represent α or β. Then for each matrix element

$$ \frac{dA'_{ij}}{dt} = \frac{\partial A'_{ij}}{\partial t} + [A_{ij}, H] = 0 + \frac{\partial A'_{ij}}{\partial \alpha}\frac{\partial H}{\partial p_\alpha} - \frac{\partial A'_{ij}}{\partial p_\alpha}\frac{\partial H}{\partial \alpha} + \frac{\partial A'_{ij}}{\partial \beta}\frac{\partial H}{\partial p_\beta} - \frac{\partial A'_{ij}}{\partial p_\beta}\frac{\partial H}{\partial \beta} = 0 $$

That is, each matrix element A'_{12}, commutes with the Hamiltonian

$$ \left[A'_{ij}, H\right] = 0 $$

Thus the Poisson Brackets representation of Hamiltonian mechanics has been used to prove that the symmetry tensor $A'_{ij} = \frac{p_i p_j}{2m} + \frac{1}{2}kx_i x_j$ is a constant of motion for the isotropic harmonic oscillator. That is, all the elements $A'_{\alpha\alpha}$, $A'_{\beta\beta}$, and $A'_{\alpha\beta}$ of the symmetric tensor $\mathbf{A}'$ commute with the Hamiltonian.

Note that the three constants of motion, L, A' and H for the isotropic, two-dimensional, linear oscillator form a closed algebra under the Poisson Bracket formalism.

15.6 *Example: The eccentricity vector*

Chapter 11.8.4 showed that Hamilton's eccentricity vector for the inverse square-law attractive force,

$$ \mathbf{A} \equiv (\mathbf{p} \times \mathbf{L}) + (\mu k \hat{\mathbf{r}}) $$

is a constant of motion that specifies the major axis of the elliptical orbit. The eccentricity vector for the inverse-square-law force can be investigated using Poisson Brackets as was done for the symmetry tensor above. It can be shown that

$$
\begin{aligned}
[L_i, A_j] &= \epsilon_{ijk} A_k \\
[A_i, A_j] &= -2\left(\frac{\mathbf{p}^2}{2\mu} + \frac{k}{r}\right)\epsilon_{ijk} L_k
\end{aligned}
\tag{a}
$$

Note that the bracket on the right-hand side of equation (a) equals the Hamiltonian H for the inverse square-law attractive force, and thus the Poisson bracket equals

$$
[A_i, A_j] = -2\left(\frac{\mathbf{p}^2}{2\mu} + \frac{k}{r}\right)\epsilon_{ijk} L_k = -2H\epsilon_{ijk} L_k
$$

For the Hamiltonian H it can be shown that the Poisson bracket

$$
[H, \mathbf{A}] = 0
$$

That is, the eccentricity vector commutes with the Hamiltonian and thus it is a constant of motion. Previously this result was obtained directly using the equations of motion as given in equation 11.87. Note that the three constants of motion, L, A and H form a closed algebra under the Poisson Bracket formalism similar to the triad of constants of motion, L, A' and H that occur for the two-dimensional, isotropic linear oscillator described above. Examples 15.5 and 15.6 illustrate that the Poisson Brackets representation of Hamiltonian mechanics is a powerful probe of the underlying physics, as well as confirming the results obtained directly from the equations of motion as described in chapter 11.8.4 and 11.9.3.

15.2.7 Liouville's Theorem

Liouvilles Theorem illustrates an application of Poisson Brackets to Hamiltonian phase space that has important implications for statistical physics. The trajectory of a single particle in phase space is completely determined by the equations of motion if the initial conditions are known. However, many-body systems have so many degrees of freedom it becomes impractical to solve all the equations of motion of the many bodies. An example is a statistical ensemble in a gas, a plasma, or a beam of particles. Usually it is not possible to specify the exact point in phase space for such complicated systems. However, it is possible to define an ensemble of points in phase space that encompasses all possible trajectories for the complicated system. That is, the statistical distribution of particles in phase space can be specified.

Consider a density ρ of representative points in $(\mathbf{q}, \mathbf{p})$ phase space. The number N of systems in the volume element dv is

$$
N = \rho dv
\tag{15.59}
$$

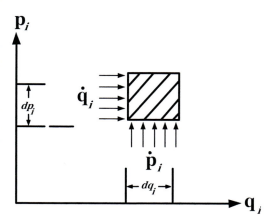

Figure 15.1: Infinitessimal element of area in phase space

where it is assumed that the infinitessimal volume element $dv = dq_1, dq_2....dq_s, dp_1, dp_2....dp_s$ contains many possible systems so that ρ can be considered a continuous distribution. For the conjugate variables (q_i, p_i) shown in figure 15.1, the number of representative points moving across the left-hand edge into the area per unit time is

$$
\rho \dot{q}_i dp_i
\tag{15.60}
$$

The number of representative points flowing out of the area along the right-hand edge is

$$
\left[\rho \dot{q}_i + \frac{\partial}{\partial q_i}(\rho \dot{q}_i)\, dq_i\right] dp_i
\tag{15.61}
$$

Hence the net increase in ρ in the infinitessimal rectangular element $dq_i dp_i$ due to flow in the horizontal direction is

$$-\frac{\partial}{\partial q_i}\left(\rho \dot{q}_i\right) dq_i dp_i \tag{15.62}$$

Similarly, the net gain due to flow in the vertical direction is

$$-\frac{\partial}{\partial p_i}\left(\rho \dot{p}_i\right) dp_i dq_i \tag{15.63}$$

Thus the total increase in the element $dq_i dp_i$ per unit time is therefore

$$-\left[\frac{\partial}{\partial q_i}\left(\rho \dot{q}_i\right) + \frac{\partial}{\partial p_i}\left(\rho \dot{p}_i\right)\right] dp_i dq_i \tag{15.64}$$

Assume that the total number of points must be conserved, then the total increase in the number of points inside the element $dq_i dp_i$ must equal the net changes in ρ on the infinitessimal surface element per unit time. That is

$$\left(\frac{\partial \rho}{\partial t}\right) dq_i dp_i \tag{15.65}$$

Thus summing over all possible values of i gives

$$\frac{\partial \rho}{\partial t} + \sum_i \left[\frac{\partial}{\partial q_i}\left(\rho \dot{q}_i\right) + \frac{\partial}{\partial p_i}\left(\rho \dot{p}_i\right)\right] = 0 \tag{15.66}$$

or

$$\frac{\partial \rho}{\partial t} + \sum_i \left[\dot{q}_i \frac{\partial \rho}{\partial q_i} + \dot{p}_i \frac{\partial \rho}{\partial p_i}\right] + \rho \sum_i \left[\frac{\partial \dot{p}_i}{\partial p_i} + \frac{\partial \dot{q}_i}{\partial q_i}\right] = 0 \tag{15.67}$$

Inserting Hamilton's canonical equations into both brackets and differentiating the last bracket results in

$$\frac{\partial \rho}{\partial t} + \sum_i \left[\frac{\partial H}{\partial p_i}\frac{\partial \rho}{\partial q_i} - \frac{\partial H}{\partial q_i}\frac{\partial \rho}{\partial p_i}\right] + \rho \sum_i \left[\frac{\partial^2 H}{\partial p_i \partial q_i} - \frac{\partial^2 H}{\partial p_i \partial q_i}\right] = 0 \tag{15.68}$$

The two terms in the last bracket cancel and thus

$$\frac{\partial \rho}{\partial t} + \sum_i \left[\frac{\partial H}{\partial p_i}\frac{\partial \rho}{\partial q_i} - \frac{\partial H}{\partial q_i}\frac{\partial \rho}{\partial p_i}\right] = \frac{\partial \rho}{\partial t} + [\rho, H] = 0 \tag{15.69}$$

However, this just equals $\frac{d\rho}{dt}$, therefore

$$\frac{d\rho}{dt} = \frac{\partial \rho}{\partial t} + [\rho, H] = 0 \tag{15.70}$$

This is called **Liouville's theorem** which states that the rate of change of density of representative points vanishes, that is, the density of points is a constant in the Hamiltonian phase space along a specific trajectory. Liouville's theorem means that the system acts like an incompressible fluid that moves such as to occupy an equal volume in phase space at every instant, even though the shape of the phase-space volume may change, that is, the phase-space density of the fluid remains constant. Equation (15.70) is another illustration of the basic Poisson bracket relation (15.45) and the usefulness of Poisson brackets in physics.

Liouville's theorem is crucially important to statistical mechanics of ensembles where the exact knowledge of the system is unknown, only statistical averages are known. An example is in focussing of beams of charged particles by beam handling systems. At a focus of the beam, the transverse width in x is minimized, while the width in p_x is largest since the beam is converging to the focus, whereas a parallel beam has maximum width x and minimum spreading width p_x. However, the product $x p_x$ remains constant throughout the focussing system. For a two dimensional beam, this applies equally for the y and p_y coordinates, etc. It is obvious that the final beam quality for any beam transport system is ultimately limited by the emittance of the source of the beam, that is, the initial area of the phase space distribution. Note that Liouville's theorem only applies to Hamiltonian $q_i - p_i$ phase space, not to $x - \dot{x}$ Lagrangian state space. As a consequence, Hamiltonian dynamics, rather than Lagrange dynamics, is used to discuss ensembles in statistical physics.

Note that Liouville's theorem is applicable only for conservative systems, that is, where Hamilton's equations of motion apply. For dissipative systems the phase space volume shrinks with time rather than being a constant of the motion.

15.3 Canonical transformations in Hamiltonian mechanics

Hamiltonian mechanics is an especially elegant and powerful way to derive the equations of motion for complicated systems. Unfortunately, integrating the equations of motion to derive a solution can be a challenge. Hamilton recognized this difficulty, so he proposed using generating functions to make canonical transformations which transform the equations into a known soluble form. Jacobi, a contemporary mathematician, recognized the importance of Hamilton's pioneering developments in Hamiltonian mechanics, and therefore he developed a sophisticated mathematical framework for exploiting the generating function formalism in order to make the canonical transformations required to solve Hamilton's equations of motion.

In the Lagrange formulation, transforming coordinates $(q_i, \dot{q}_i)$ to cyclic generalized coordinates $(Q_i, \dot{Q}_i)$, simplifies finding the Euler-Lagrange equations of motion. For the Hamiltonian formulation, the concept of coordinate transformations is extended to include simultaneous canonical transformation of both the spatial coordinates q_i and the conjugate momenta p_i from (q_i, p_i) to (Q_i, P_i), where both of the canonical variables are treated equally in the transformation. Compared to Lagrangian mechanics, Hamiltonian mechanics has twice as many variables which is an asset, rather than a liability, since it widens the realm of possible canonical transformations.

Hamiltonian mechanics has the advantage that generating functions can be exploited to make canonical transformations to find solutions, which avoids having to use direct integration. Canonical transformations are the foundation of Hamiltonian mechanics; they underlie Hamilton-Jacobi theory and action-angle variable theory, both of which are powerful means for exploiting Hamiltonian mechanics to solve problems in physics and engineering. The concept underlying canonical transformations is that, if the equations of motion are simplified by using a new set of generalized variables $(\mathbf{Q}, \mathbf{P})$, compared to using the original set of variables $(\mathbf{q}, \mathbf{p})$, then an advantage has been gained. The solution, expressed in terms of the generalized variables $(\mathbf{Q}, \mathbf{P})$, can be transformed back to express the solution in terms of the original coordinates, $(\mathbf{q}, \mathbf{p})$.

Only a specialized subset of transformations will be considered, namely *canonical transformations* that preserve the canonical form of Hamilton's equations of motion. That is, given that the original set of variables (q_i, p_i) satisfy Hamilton's equations

$$\dot{\mathbf{q}} = \frac{\partial H(\mathbf{q}, \mathbf{p}, t)}{\partial \mathbf{p}} \qquad -\dot{\mathbf{p}} = \frac{\partial H(\mathbf{q}, \mathbf{p}, t)}{\partial \mathbf{q}} \tag{15.71}$$

for some Hamiltonian $H(\mathbf{q}, \mathbf{p}, t)$, then the transformation to coordinates $Q_i(q_k, p_k, t), P_i(q_k, p_k, t)$ is canonical if, and only if, there exists a function $\mathcal{H}(\mathbf{Q}, \mathbf{P}, t)$ such that the $\mathbf{P}$ and $\mathbf{Q}$ are still governed by Hamilton's equations. That is,

$$\dot{\mathbf{Q}} = \frac{\partial \mathcal{H}(\mathbf{Q}, \mathbf{P}, t)}{\partial \mathbf{P}} \qquad -\dot{\mathbf{P}} = \frac{\partial \mathcal{H}(\mathbf{Q}, \mathbf{P}, t)}{\partial \mathbf{Q}} \tag{15.72}$$

where $\mathcal{H}(\mathbf{Q}, \mathbf{P}, t)$ plays the role of the Hamiltonian for the new variables. Note that $\mathcal{H}(\mathbf{Q}, \mathbf{P}, t)$ may be very different from the old Hamiltonian $H(\mathbf{q}, \mathbf{p}, t)$. The invariance of the Poisson bracket to canonical transformations, chapter 15.2.3, provides a powerful test that the transformation is canonical.

Hamilton's Principle of least action, discussed in chapter 9, states that

$$\delta S = \delta \int_{t_1}^{t_2} L(\mathbf{q}, \dot{\mathbf{q}}, t) dt = \delta \int_{t_1}^{t_2} [\mathbf{p} \cdot \dot{\mathbf{q}} - H(\mathbf{q}, \mathbf{p}, t)] \, dt = 0 \tag{15.73}$$

Similarly, applying Hamilton's Principle of least action to the new Lagrangian $\mathcal{L}(\mathbf{Q}, \dot{\mathbf{Q}}, t)$ gives

$$\delta S = \delta \int_{t_1}^{t_2} \mathcal{L}(\mathbf{Q}, \dot{\mathbf{Q}}, t) dt = \delta \int_{t_1}^{t_2} \left[\mathbf{P} \cdot \dot{\mathbf{Q}} - \mathcal{H}(\mathbf{Q}, \mathbf{P}, t) \right] dt = 0 \tag{15.74}$$

The discussion of gauge-invariant Lagrangians, chapter 9.3, showed that L and $\mathcal{L}$ can be related by the total time derivative of a generating function F where

$$\frac{dF}{dt} = \mathcal{L} - L \tag{15.75}$$

The generating function F can be any well-behaved function with continuous second derivatives of both the old and new canonical variables $\mathbf{p}, \mathbf{q}, \mathbf{P}, \mathbf{Q}$ and t. Thus the integrands of (15.73) and (15.74) are related by

$$\mathbf{p} \cdot \dot{\mathbf{q}} - H(\mathbf{q}, \mathbf{p}, t) = \lambda \left[\mathbf{P} \cdot \dot{\mathbf{Q}} - \mathcal{H}(\mathbf{Q}, \mathbf{P}, t) \right] + \frac{dF}{dt} \tag{15.76}$$

where λ is a possible scale transformation. A scale transformation, such as changing units, is trivial, and will be assumed to be absorbed into the coordinates, making $\lambda = 1$. Assuming that $\lambda \neq 1$ is called an extended canonical transformation.

15.3.1   Generating functions

The generating function F has to be chosen such that the transformation from the initial variables $(\mathbf{q}, \mathbf{p})$ to the final variables $(\mathbf{Q}, \mathbf{P})$ is a canonical transformation. The chosen generating function contributes to (15.76) only if it is a function of the old plus new variables. The four possible types of generating functions of the first kind, are $F_1(\mathbf{q}, \mathbf{Q}, t)$, $F_2(\mathbf{q}, \mathbf{P}, t)$, $F_3(\mathbf{p}, \mathbf{Q}, t)$, and $F_4(\mathbf{p}, \mathbf{P}, t)$. These four generating functions lead to relatively simple canonical transformations, are shown below.

Type 1: $F = F_1(\mathbf{q}, \mathbf{Q}, t)$:

The total time derivative of the generating function $F = F_1(\mathbf{q}, \mathbf{Q}, t)$ is given by

$$\frac{dF(\mathbf{q}, \mathbf{Q}, t)}{dt} = \left[\frac{\partial F_1(\mathbf{q}, \mathbf{Q}, t)}{\partial \mathbf{q}} \cdot \dot{\mathbf{q}} + \frac{\partial F_1(\mathbf{q}, \mathbf{Q}, t)}{\partial \mathbf{Q}} \cdot \dot{\mathbf{Q}} \right] + \frac{\partial F_1(\mathbf{q}, \mathbf{Q}, t)}{\partial t} \tag{15.77}$$

Insert equation (15.77) into equation (15.76), and assume that the trivial scale factor $\lambda = 1$, then

$$\left[\mathbf{p} - \frac{\partial F_1(\mathbf{q}, \mathbf{Q}, t)}{\partial \mathbf{q}} \right] \cdot \dot{\mathbf{q}} - H(\mathbf{q}, \mathbf{p}, t) = \left[\mathbf{P} + \frac{\partial F_1(\mathbf{q}, \mathbf{Q}, t)}{\partial \mathbf{Q}} \right] \cdot \dot{\mathbf{Q}} - \mathcal{H}(\mathbf{Q}, \mathbf{P}, t) + \frac{\partial F_1(\mathbf{q}, \mathbf{Q}, t)}{\partial t}$$

Assume that the generating function F_1 determines the canonical variables $\mathbf{p}$ and $\mathbf{P}$ to be

$$\mathbf{p} = \frac{\partial F_1(\mathbf{q}, \mathbf{Q}, t)}{\partial \mathbf{q}} \qquad\qquad \mathbf{P} = -\frac{\partial F_1(\mathbf{q}, \mathbf{Q}, t)}{\partial \mathbf{Q}} \tag{15.78}$$

then the terms in each square bracket cancel, leading to the required canonical transformation

$$\mathcal{H}(\mathbf{Q}, \mathbf{P}, t) = H(\mathbf{q}, \mathbf{p}, t) + \frac{\partial F_1(\mathbf{q}, \mathbf{Q}, t)}{\partial t} \tag{15.79}$$

Type 2: $F = F_2(\mathbf{q}, \mathbf{P}, t) - \mathbf{Q} \cdot \mathbf{P}$:

The total time derivative of the generating function $F = F_2(\mathbf{q}, \mathbf{P}, t) - \mathbf{Q} \cdot \mathbf{P}$ is given by

$$\frac{dF}{dt} = \left[\frac{\partial F_2(\mathbf{q}, \mathbf{P}, t)}{\partial \mathbf{q}} \cdot \dot{\mathbf{q}} + \frac{\partial F_2(\mathbf{q}, \mathbf{P}, t)}{\partial \mathbf{P}} \cdot \dot{\mathbf{P}} - \mathbf{P} \cdot \dot{\mathbf{Q}} - \dot{\mathbf{P}} \cdot \mathbf{Q} \right] + \frac{\partial F_2(\mathbf{q}, \mathbf{P}, t)}{\partial t} \tag{15.80}$$

Insert this into equation (15.76), and assume that the trivial scale factor $\lambda = 1$, then

$$\left(\mathbf{p} - \frac{\partial F_2(\mathbf{q}, \mathbf{P}, t)}{\partial \mathbf{q}} \right) \cdot \dot{\mathbf{q}} - H(\mathbf{q}, \mathbf{p}, t) = \mathbf{P} \cdot \dot{\mathbf{Q}} - \mathbf{P} \cdot \dot{\mathbf{Q}} + \left[\frac{\partial F_2(\mathbf{q}, \mathbf{P}, t)}{\partial \mathbf{P}} - \mathbf{Q} \right] \cdot \dot{\mathbf{P}} - \mathcal{H}(\mathbf{Q}, \mathbf{P}, t) + \frac{\partial F_2(\mathbf{q}, \mathbf{P}, t)}{\partial t}$$

Assume that the generating function F_2 determines the canonical variables $\mathbf{p}$ and $\mathbf{Q}$ to be

$$\mathbf{p} = \frac{\partial F_2(\mathbf{q}, \mathbf{P}, t)}{\partial \mathbf{q}} \qquad\qquad \mathbf{Q} = \frac{\partial F_2(\mathbf{q}, \mathbf{P}, t)}{\partial \mathbf{P}} \tag{15.81}$$

then the terms in brackets cancel, leading to the required transformation

$$\mathcal{H}(\mathbf{Q}, \mathbf{P}, t) = H(\mathbf{q}, \mathbf{p}, t) + \frac{\partial F_2(\mathbf{q}, \mathbf{P}, t)}{\partial t} \tag{15.82}$$

Type 3: $F = F_3(\mathbf{p}, \mathbf{Q}, t) + \mathbf{q} \cdot \mathbf{p}$:

The total time derivative of the generating function $F = F_3(\mathbf{p}, \mathbf{Q}, t) + \mathbf{q} \cdot \mathbf{p}$ is given by

$$\frac{dF}{dt} = \left[\frac{\partial F_3(\mathbf{p}, \mathbf{Q}, t)}{\partial \mathbf{p}} \cdot \dot{\mathbf{p}} + \frac{\partial F_3(\mathbf{p}, \mathbf{Q}, t)}{\partial \mathbf{Q}} \cdot \dot{\mathbf{Q}} + \dot{\mathbf{q}} \cdot \mathbf{p} + \mathbf{q} \cdot \dot{\mathbf{p}} \right] + \frac{\partial F_3(\mathbf{p}, \mathbf{Q}, t)}{\partial t} \tag{15.83}$$

Insert this into equation (15.76), and assume that the trivial scale factor $\lambda = 1$, then

$$-\left[\mathbf{q} + \frac{\partial F_3(\mathbf{p}, \mathbf{Q}, t)}{\partial \mathbf{p}} \right] \cdot \dot{\mathbf{p}} - H(\mathbf{q}, \mathbf{p}, t) = \left[\mathbf{P} + \frac{\partial F_3(\mathbf{p}, \mathbf{Q}, t)}{\partial \mathbf{Q}} \right] \cdot \dot{\mathbf{Q}} - \mathcal{H}(\mathbf{Q}, \mathbf{P}, t) + \frac{\partial F_3(\mathbf{p}, \mathbf{Q}, t)}{\partial t}$$

Assume that the generating function F_3 determines the canonical variables $\mathbf{q}$ and $\mathbf{P}$ to be

$$\mathbf{q} = -\frac{\partial F_3(\mathbf{p}, \mathbf{Q}, t)}{\partial \mathbf{p}} \qquad\qquad \mathbf{P} = -\frac{\partial F_3(\mathbf{p}, \mathbf{Q}, t)}{\partial \mathbf{Q}} \tag{15.84}$$

then the terms in brackets cancel, leading to the required transformation

$$\mathcal{H}(\mathbf{Q}, \mathbf{P}, t) = H(\mathbf{q}, \mathbf{p}, t) + \frac{\partial F_3(\mathbf{p}, \mathbf{Q}, t)}{\partial t} \tag{15.85}$$

Type 4: $F = F_4(\mathbf{p}, \mathbf{P}, t) + \mathbf{q} \cdot \mathbf{p} - \mathbf{Q} \cdot \mathbf{P}$:

The total time derivative of the generating function $F = F_4(\mathbf{p}, \mathbf{P}, t) + \mathbf{q} \cdot \mathbf{p} - \mathbf{Q} \cdot \mathbf{P}$ is given by

$$\frac{dF}{dt} = \left[\frac{\partial F_4(\mathbf{p}, \mathbf{P}, t)}{\partial \mathbf{p}} \cdot \dot{\mathbf{p}} + \frac{\partial F_4(\mathbf{p}, \mathbf{P}, t)}{\partial \mathbf{P}} \cdot \dot{\mathbf{P}} + \dot{\mathbf{q}} \cdot \mathbf{p} + \mathbf{q} \cdot \dot{\mathbf{p}} - \dot{\mathbf{Q}} \cdot \mathbf{P} - \mathbf{Q} \cdot \dot{\mathbf{P}} \right] + \frac{\partial F_4(\mathbf{p}, \mathbf{P}, t)}{\partial t} \tag{15.86}$$

Insert this into equation (15.76), and assume that the trivial scale factor $\lambda = 1$, then

$$-\left[\mathbf{q} + \frac{\partial F_4(\mathbf{p}, \mathbf{P}, t)}{\partial \mathbf{p}} \right] \cdot \dot{\mathbf{p}} - H(\mathbf{q}, \mathbf{p}, t) = \left[\frac{\partial F_4(\mathbf{p}, \mathbf{P}, t)}{\partial \mathbf{P}} - \mathbf{Q} \right] \cdot \dot{\mathbf{P}} - \mathcal{H}(\mathbf{Q}, \mathbf{P}, t) + \frac{\partial F_4(\mathbf{p}, \mathbf{P}, t)}{\partial t}$$

Assume that the generating function F_4 determines the canonical variables $\mathbf{q}$ and $\mathbf{Q}$ to be

$$\mathbf{q} = -\frac{\partial F_4(\mathbf{p}, \mathbf{P}, t)}{\partial \mathbf{p}} \qquad\qquad \mathbf{Q} = \frac{\partial F_4(\mathbf{p}, \mathbf{P}, t)}{\partial \mathbf{P}} \tag{15.87}$$

then the terms in brackets cancel, leading to the required transformation

$$\mathcal{H}(\mathbf{Q}, \mathbf{P}, t) = H(\mathbf{q}, \mathbf{p}, t) + \frac{\partial F_4(\mathbf{p}, \mathbf{P}, t)}{\partial t} \tag{15.88}$$

Note that the last three generating functions require the inclusion of additional bilinear products of q, p, Q, P in order for the terms to cancel to give the required result. The addition of the bilinear terms, ensures that the resultant generating function F is the same using any of the four generating functions F_1, F_2, F_3, F_4. Frequently the $F_2(\mathbf{q}, \mathbf{P}, t)$ generating function is the most convenient. The four possible generating functions of the first kind, given above, are related by Legendre transformations. A canonical transformation does not have to conform to only one of the four generating functions F_k for all the degrees of freedom, they can be a mixture of different flavors for the different degrees of freedom. The properties of the generating functions are summarized in table 15.1.

Table 15.1 Canonical transformation generating functions

Generating function	Generating function derivatives		Trivial special examples		
$F = F_1(\mathbf{q}, \mathbf{Q}, t)$	$p_i = \frac{\partial F_1}{\partial q_i}$	$P_i = -\frac{\partial F_1}{\partial Q_i}$	$F_1 = q_i Q_i$	$Q_i = p_i$	$P_i = -q_i$
$F = F_2(\mathbf{q}, \mathbf{P}, t) - \mathbf{Q} \cdot \mathbf{P}$	$p_i = \frac{\partial F_2}{\partial q_i}$	$Q_i = \frac{\partial F_2}{\partial P_i}$	$F_2 = q_i P_i$	$Q_i = q_i$	$P_i = p_i$
$F = F_3(\mathbf{p}, \mathbf{Q}, t) + \mathbf{q} \cdot \mathbf{p}$	$q_i = -\frac{\partial F_3}{\partial p_i}$	$P_i = -\frac{\partial F_3}{\partial Q_i}$	$F_3 = p_i Q_i$	$Q_i = -q_i$	$P_i = -p_i$
$F = F_4(\mathbf{p}, \mathbf{P}, t) + \mathbf{q} \cdot \mathbf{p} - \mathbf{Q} \cdot \mathbf{P}$	$q_i = -\frac{\partial F_4}{\partial p_i}$	$Q_i = \frac{\partial F_4}{\partial P_i}$	$F_4 = p_i P_i$	$Q_i = p_i$	$P_i = -q_i$

The partial derivatives of the generating functions F_i determine the corresponding conjugate variables not explicitly included in the generating function F_i. Note that, for the first trivial example $F_1 = q_iQ_i$, the old momenta become the new coordinates, $Q_i = p_i$, and vice versa, $P_i = -q_i$. This illustrates that it is better to name them "conjugate variables" rather than "momenta" and "coordinates".

In summary, Jacobi has developed a mathematical framework for finding the generating function F required to make a canonical transformation to a new Hamiltonian $\mathcal{H}(\mathbf{Q}, \mathbf{P}, t)$, that has a known solution. That is,

$$\mathcal{H}(\mathbf{Q}, \mathbf{P}, t) = H(\mathbf{q}, \mathbf{p}, t) + \frac{\partial F}{\partial t} \tag{15.89}$$

When $\mathcal{H}(\mathbf{Q}, \mathbf{P}, t)$ is a constant, then a solution has been obtained. The inverse transformation for this solution $\mathbf{Q}(t), \mathbf{P}(t) \rightarrow \mathbf{q}(t), \mathbf{p}(t)$ now can be used to express the final solution in terms of the original variables of the system.

Note the special case when $\mathcal{H}(\mathbf{Q}, \mathbf{P}, t) = 0$, then equation 15.89 has been reduced to the Hamilton-Jacobi relation (15.11)

$$H(\mathbf{q}, \mathbf{p}, t) + \frac{\partial S}{\partial t} = 0 \tag{15.11}$$

In this case, the generating function F determines the action functional S required to solve the Hamilton-Jacobi equation (15.110). Since equation (15.89) has transformed the Hamiltonian $H(\mathbf{q}, \mathbf{p}, t) \rightarrow \mathcal{H}(\mathbf{Q}, \mathbf{P}, t)$, for which $\mathcal{H}(\mathbf{Q}, \mathbf{P}, t) = 0$, then the solution $\mathbf{Q}(t), \mathbf{P}(t)$ for the Hamiltonian $\mathcal{H}(\mathbf{Q}, \mathbf{P}, t) = 0$ is obtained easily. This approach underlies Hamilton-Jacobi theory presented in chapter 15.4.

15.3.2 Applications of canonical transformations

The canonical transformation procedure may appear unnecessarily complicated for solving the examples given in this book, but it is essential for solving the complicated systems that occur in nature. For example, canonical transformations can be used to transform time-dependent, (non-autonomous) Hamiltonians to time-independent, (autonomous) Hamiltonians for which the solutions are known. Example 15.19 describes such a system. Canonical transformations provide a remarkably powerful approach for solving the equations of motion in Hamiltonian mechanics, especially when using the Hamilton-Jacobi approach discussed in chapter 15.4.

15.7 Example: The identity canonical transformation

The identity transformation $F_2(\mathbf{q}, \mathbf{P}) = \mathbf{q} \cdot \mathbf{P}$ satisfies (15.89) if the following relations are satisfied $p_i = \frac{\partial F_2}{\partial q_i} = P_i$, $Q_i = \frac{\partial F_2}{\partial P_i} = q_i$, $\mathcal{H}=H$. Note that the new and old coordinates are identical, hence $F_2 = q_iP_i$ generates the identity transformation $q_i = Q_i$, $p_i = P_i$.

15.8 Example: The point canonical transformation

Consider the point transformation $F_2(\mathbf{q} \cdot \mathbf{P}) = f(\mathbf{q},t) \cdot \mathbf{P}$ where $f(\mathbf{q},t)$ is some function of $\mathbf{q}$. This transformation satisfies (15.89) if the following relations are satisfied $Q_i = \frac{\partial F_2}{\partial P_i} = f_i(q_i)$, $p_i = \frac{\partial F_2}{\partial q_i} = \frac{\partial f_i(q_i,t)}{\partial q_i}$, $\mathcal{H}=H$. Point transformations correspond to point-to-point transformations of coordinates.

15.9 Example: The exchange canonical transformation

The identity transformation $F_1(\mathbf{q}, \mathbf{Q}) = \mathbf{q} \cdot \mathbf{Q}$ satisfies (15.89) if the following relations are satisfied $p_i = \frac{\partial F_1}{\partial q_i} = Q_i$, $P_i = -\frac{\partial F_1}{\partial Q_i} = -q_i$, $\mathcal{H}=H$ That is, the coordinates and momenta have been interchanged.

15.10 Example: Infinitessimal point canonical transformation

Consider an infinitessimal point canonical transformation, that is infinitesimally close to a point identity.

$$F_2(\mathbf{q} \cdot \mathbf{P}, t) = \mathbf{q} \cdot \mathbf{P} + \epsilon G(\mathbf{q}, \mathbf{P}, t)$$

satisfies (15.89) if the following relations are satisfied

$$Q_i = \frac{\partial F_2}{\partial P_i} = q_i + \epsilon \frac{\partial G(\mathbf{q}, \mathbf{P}, t)}{\partial P_i}$$

$$p_i = \frac{\partial F_2}{\partial q_i} = P_i + \epsilon \frac{\partial G(\mathbf{q}, \mathbf{P}, t)}{\partial q_i}$$

Thus the infinitessimal changes in q_i and p_i are given by

$$\delta q_i(\mathbf{q}, \mathbf{p}, t) = Q_i - q_i = \epsilon \frac{\partial G(\mathbf{q}, \mathbf{P}, t)}{\partial P_i} = \epsilon \frac{\partial G(\mathbf{q}, \mathbf{P}, t)}{\partial p_i} + O(\epsilon^2)$$

$$\delta p_i(\mathbf{q}, \mathbf{p}, t) = P_i - p_i = -\epsilon \frac{\partial G(\mathbf{q}, \mathbf{P}, t)}{\partial q_i} = -\epsilon \frac{\partial G(\mathbf{q}, \mathbf{P}, t)}{\partial p_i} + O(\epsilon^2)$$

Thus $G(\mathbf{q}, \mathbf{P}, t)$ is the generator of the infinitessimal canonical transformation.

15.11 Example: 1-D harmonic oscillator via a canonical transformation

The classic one-dimensional harmonic oscillator provides an example of the use of canonical transformations. Consider the Hamiltonian where $\omega^2 = \frac{k}{m}$ then

$$H = \frac{p^2}{2m} + \frac{kq^2}{2} = \frac{1}{2m}\left(p^2 + m^2\omega^2 q^2\right)$$

This form of the Hamiltonian is a sum of two squares suggesting a canonical transformation for which H is cyclic in a new coordinate. A guess for a canonical transformation is of the form $p = m\omega q \cot Q$ which is of the $F_1(\mathbf{q}, \mathbf{Q})$ type where F_1 equals $F_1(q, Q) = \frac{m\omega q^2}{2} \cot Q$. Using (15.78) gives

$$p = \frac{\partial F_1(q, Q)}{\partial q_i} = m\omega q \cot Q$$

$$P = -\frac{\partial F_1(q, Q)}{\partial Q} = \frac{m}{2} \frac{\omega q^2}{\sin^2 Q}$$

Solving for the coordinates (p, q) yields

$$q = \sqrt{\frac{2P}{m\omega}} \sin Q \tag{a}$$

$$p = \sqrt{2m\omega P} \cos Q \tag{b}$$

Inserting these into H gives

$$\mathcal{H} = \omega P(\cos^2 Q + \sin^2 Q) = \omega P$$

which implies that Q is a cyclic coordinate.

The Hamiltonian is conservative, since it does not explicitly depend on time, and it equals the total energy since the transformation to generalized coordinates is time independent. Thus

$$\mathcal{H} = E = \omega P$$

Since

$$\dot{Q} = \frac{\partial \mathcal{H}}{\partial P} = \omega$$

then

$$Q = \omega t + \phi$$

Substituting Q into (a) gives the well known solution of the one-dimensional harmonic oscillator

$$q = \sqrt{\frac{2E}{m\omega^2}} \sin(\omega t + \phi)$$

15.4 Hamilton-Jacobi theory

Hamilton used the Principle of Least Action to derive the Hamilton-Jacobi relation (chapter 15.3)

$$H(\mathbf{q}, \mathbf{p}, t) + \frac{\partial S}{\partial t} = 0 \tag{15.11}$$

where $\mathbf{q}, \mathbf{p}$ refer to the $1 \leq i \leq n$ variables q_i, p_i and $S(q_j(t_1), t_1, q_j(t_2), t_2)$ is the action functional. Integration of this first-order partial differential equation is non trivial which is a major handicap for practical exploitation of the Hamilton-Jacobi equation. This stimulated Jacobi to develop the mathematical framework for canonical transformation that are required to solve the Hamilton-Jacobi equation. Jacobi's approach is to exploit generating functions for making a canonical transformation to a new Hamiltonian $\mathcal{H}(\mathbf{Q}, \mathbf{P}, t)$ that equals zero.

$$\mathcal{H}(\mathbf{Q}, \mathbf{P}, t) = H(\mathbf{q}, \mathbf{p}, t) + \frac{\partial S}{\partial t} = 0 \tag{15.90}$$

The generating function for solving the Hamilton-Jacobi equation then equals the action functional S.

The Hamilton-Jacobi theory is based on selecting a canonical transformation to new coordinates (Q, P, t) all of which are either constant, or the Q_i are cyclic, which implies that the corresponding momenta P_i are constants. In either case, a solution to the equations of motion is obtained. A remarkable feature of Hamilton-Jacobi theory is that the canonical transformation is completely characterized by a single generating function, S. The canonical equations likewise are characterized by a single Hamiltonian function, H. Moreover, the generating function S, and Hamiltonian function H, are linked together by equation 15.11. The underlying goal of Hamilton-Jacobi theory is to transform the Hamiltonian to a known form such that the canonical equations become directly integrable. Since this transformation depends on a single scalar function, the problem is reduced to solving a single partial differential equation.

15.4.1 Time-dependent Hamiltonian

Jacobi's complete integral $S(q_i, P_i, t)$

The principle underlying Jacobi's approach to Hamilton-Jacobi theory is to provide a recipe for finding the generating function $F = S$ needed to transform the Hamiltonian $H(\mathbf{q}, \mathbf{p}, t)$ to the new Hamiltonian $\mathcal{H}(\mathbf{Q}, \mathbf{P}, t)$ using equation 15.90. When the derivatives of the transformed Hamiltonian $\mathcal{H}(\mathbf{Q}, \mathbf{P}, t)$ are zero, then the equations of motion become

$$\dot{Q}_i = \frac{\partial \mathcal{H}}{\partial P_i} = 0 \tag{15.91}$$

$$\dot{P}_i = -\frac{\partial \mathcal{H}}{\partial Q_i} = 0 \tag{15.92}$$

and thus Q_i and P_i are constants of motion. The new Hamiltonian $\mathcal{H}$ must be related to the original Hamiltonian H by a canonical transformation for which

$$\mathcal{H}(\mathbf{Q}, \mathbf{P}, t) = H(\mathbf{q}, \mathbf{p}, t) + \frac{\partial S}{\partial t} \tag{15.93}$$

Equations 15.91 and 15.92 are automatically satisfied if the new Hamiltonian $\mathcal{H} = 0$ since then equation 15.93 gives that the generating function S satisfies equation 15.90.

Any of the four types of generating function can be used. Jacobi chose the type 2 generating function as being the most useful for many practical cases, that is, $S(q_i, P_i, t)$ which is called **Jacobi's complete integral**.

For generating functions F_1 and F_2 the generalized momenta are derived from the action by the derivative

$$p_i = \frac{\partial S}{\partial q_i} \tag{15.4}$$

Use this generalized momentum to replace p_i in the Hamiltonian H, given in equation (15.93), leads to the **Hamilton-Jacobi equation** expressed in terms of the action S.

$$H(q_1, ... q_n; \frac{\partial S}{\partial q_1}, ..., \frac{\partial S}{\partial q_n}; t) + \frac{\partial S}{\partial t} = 0 \tag{15.94}$$

The Hamilton-Jacobi equation, (15.94), can be written more compactly using tensors $\mathbf{q}$ and $\boldsymbol{\nabla}S$ to designate $(q_1, ..q_n)$ and $\frac{\partial S}{\partial q_1}, ..., \frac{\partial S}{\partial q_n}$ respectively. That is

$$H(\mathbf{q}, \boldsymbol{\nabla}S, t) + \frac{\partial S}{\partial t} = 0 \tag{15.95}$$

Equation (15.95) is a first-order partial differential equation in $n+1$ variables which are the old spatial coordinates q_i plus time t. The new momenta P_i have not been specified except that they are constants since $\mathcal{H} = 0$.

Assume the existence of a solution of (15.95) of the form $S(q_i, P_i, t) = S(q_1, ..q_n; \alpha_1, ..\alpha_{n+1}; t)$ where the generalized momenta $P_i = \alpha_1, \alpha_2, ...\alpha$ plus t are the $n+1$ *independent constants of integration* in the transformed frame. One constant of integration is irrelevant to the solution since only partial derivatives of $S(q_i, P_i, t)$ with respect to q_i and t are involved. Thus, if S is a solution of the first-order partial differential equation, then so is $S + \alpha$ where α is a constant. Thus it can be assumed that one of the $n+1$ constants of integration is just an additive constant which can be ignored leading effectively to a solution

$$S(q_i, P_i, t) = S(q_1, ...q_n; \alpha_1, ...\alpha_n; t) \tag{15.96}$$

where none of the n independent constants are solely additive. Such generating function solutions are called *complete solutions* of the first-order partial differential equations since all constants of integration are known.

It is possible to assume that the n generalized momenta, P_i are constants α_i, where the α_i are the constants. This allows the generalized momentum to be written as

$$p_i = \frac{\partial S(\mathbf{q}, \boldsymbol{\alpha}, t)}{\partial q_i} \tag{15.97}$$

Similarly, Hamilton's equations of motion give the conjugate coordinate $\mathbf{Q} = \boldsymbol{\beta}$, where β_i are constants. That is

$$Q_i = \beta_i = \frac{\partial S(\mathbf{q}, \boldsymbol{\alpha}, t)}{\partial \alpha_i} \tag{15.98}$$

The above procedure has determined the complete set of $2n$ constants $(\mathbf{Q} = \boldsymbol{\beta}, \mathbf{P} = \boldsymbol{\alpha})$. It is possible to invert the canonical transformation to express the above solution, which is expressed in terms of $Q_i = \beta_i$ and $P_i = \alpha_i$, back to the original coordinates, that is, $q_j = q_j(\alpha, \beta, t)$ and momenta $p_j = p_j(\alpha, \beta, t)$ which is the required solution.

Hamilton's principle function $S_H(\mathbf{q}_i, t; \mathbf{q}_o t_o)$

Hamilton's approach to solving the Hamilton-Jacobi equation (15.95) is to seek a canonical transformation from variables $(\mathbf{p}, \mathbf{q})$ at time t, to a new set of constant quantities, which may be the initial values $(\mathbf{q}_0, \mathbf{p}_0)$ at time $t = 0$. Hamilton's principle function $S_H(q_i, t; q_o t_o)$ is the generating function for this canonical transformation from the variables $(\mathbf{q}, \mathbf{p})$ at time t to the initial variables $(\mathbf{q}_0, \mathbf{p}_0)$ at time t_0. Hamilton's principle function $S_H(q_i, t; q_o t_o)$ is directly related to Jacobi's complete integral $S(q_i, P_i, t)$.

Note that S_H is the generating function of a canonical transformation from the present time $(\mathbf{q}, \mathbf{p}, t)$ variables to the initial $(\mathbf{q}_0, \mathbf{p}_0, t_0)$, whereas Jacobi's S is the generating function of a canonical transformation from the present $(\mathbf{q}, \mathbf{p}, t)$ variables to the constant variables $(\mathbf{Q} = \boldsymbol{\beta}, \mathbf{P} = \boldsymbol{\alpha})$. For the Hamilton approach, the canonical transformation can be accomplished in two steps using S by first transforming from $(\mathbf{q}, \mathbf{p}, t)$ at time t, to $(\boldsymbol{\beta}, \boldsymbol{\alpha})$, then transforming from $(\boldsymbol{\beta}, \boldsymbol{\alpha})$ to $(\mathbf{q}_0, \mathbf{p}_0, t_0)$. That is, this two-step process corresponds to

$$S_H(\mathbf{q}, t; \mathbf{q}_o t_o) = S(\mathbf{q}, \boldsymbol{\alpha}, t) - S(\mathbf{q}_0, \boldsymbol{\alpha}, t_0) \tag{15.99}$$

Hamilton's principle function $S_H(\mathbf{q}, t; \mathbf{q}_o t_o)$ is related to Jacobi's complete integral $S(\mathbf{q}, \boldsymbol{\alpha}, t)$, and it will not be discussed further in this book.

15.4.2   Time-independent Hamiltonian

Frequently the Hamiltonian does not explicitly depend on time. For the standard Lagrangian with time-independent constraints and transformation, then $H(\mathbf{q}, \mathbf{p}, t) = E$ which is the total energy. For this case, the Hamilton-Jacobi equation simplifies to give

$$\frac{\partial S}{\partial t} = -H(\mathbf{q}, \mathbf{p}, t) = -E(\boldsymbol{\alpha}) \tag{15.100}$$

The integration of the time dependence is trivial, and thus the action integral for a time-independent Hamiltonian equals

$$S(\mathbf{q}, \boldsymbol{\alpha}, t) = W(\mathbf{q}, \boldsymbol{\alpha}) - E(\boldsymbol{\alpha}) t \tag{15.101}$$

That is, the action integral has separated into a time independent term $W(\mathbf{q}, \boldsymbol{\alpha})$ which is called **Hamilton's characteristic function** plus a time-dependent term $-E(\boldsymbol{\alpha}) t$. Thus using equations 15.97, 15.101 gives that the generalized momentum is

$$p_i = \frac{\partial W(\mathbf{q}, \boldsymbol{\alpha})}{\partial q_i} \tag{15.102}$$

The physical significance of Hamilton's characteristic function $W(\mathbf{q}, \boldsymbol{\alpha})$ can be understood by taking the total time derivative

$$\frac{dW}{dt} = \sum_i \frac{\partial W(\mathbf{q}, \boldsymbol{\alpha})}{\partial q_i} \dot{q}_i = \sum_i p_i \dot{q}_i$$

Taking the time integral then gives

$$W(\mathbf{q}, \boldsymbol{\alpha}) = \int \sum p_i \dot{q}_i dt = \int \sum p_i dq_i \tag{15.103}$$

Note that this equals the abbreviated action described in chapter 9.2.3, that is $W(\mathbf{q}, \boldsymbol{\alpha}) = S_0(\mathbf{q}, \boldsymbol{\alpha})$.

Inserting the action $S(\mathbf{q}, \boldsymbol{\alpha})$ into the Hamilton-Jacobi equation (15.12) gives

$$H(\mathbf{q}; \frac{\partial W(\mathbf{q}, \boldsymbol{\alpha})}{\partial q}) = E(\boldsymbol{\alpha}) \tag{15.104}$$

This is called the **time-independent Hamilton-Jacobi equation.** Usually it is convenient to have E equal the total energy. However, sometimes it is more convenient to exclude the kth energy $E(\alpha_k)$ in the set, in which case $E = E(\alpha_1, \alpha_2, ...\alpha_{k-1})$; the Routhian exploits this feature.

The equations of the canonical transformation expressed in terms of $W(\mathbf{q}, \boldsymbol{\alpha})$ are

$$p_i = \frac{\partial W(\mathbf{q}, \boldsymbol{\alpha})}{\partial q_i} \qquad\qquad \beta_i + \frac{\partial E(\boldsymbol{\alpha})}{\partial \alpha_i} t = \frac{\partial W(\mathbf{q}, \boldsymbol{\alpha})}{\partial \alpha_i} \tag{15.105}$$

These equations show that Hamilton's characteristic function $W(\mathbf{q}, \boldsymbol{\alpha})$ is itself the generating function of a time-independent canonical transformation from the old variables (q, p) to a set of new variables

$$Q_i = \beta_i + \frac{\partial E(\boldsymbol{\alpha})}{\partial \alpha_i} t \qquad\qquad P_i = \alpha_i \tag{15.106}$$

Table 15.2 summarizes the time-dependent and time-independent forms of the Hamilton-Jacobi equation.

Table 15.2; Hamilton-Jacobi formulations

Hamiltonian	Time dependent $H(q, p, t)$	Time independent $H(q, p)$
Transformed Hamiltonian	$\mathcal{H} = 0$	$\mathcal{H}$ is cyclic
Canonical transformed variables	All $Q_i P_i$ are constants of motion	All P_i are constants of motion
Transformed equations of motion	$\dot{Q}_i = \frac{\partial \mathcal{H}}{\partial P_i} = 0$, therefore $Q_i = \beta_i$ $\dot{P}_i = -\frac{\partial \mathcal{H}}{\partial Q_i} = 0$, therefore $P_i = \alpha_i$	$\dot{Q}_i = \frac{\partial \mathcal{H}}{\partial P_i} = v_i$, therefore $Q_i = v_i t + \beta_i$ $\dot{P}_i = -\frac{\partial \mathcal{H}}{\partial Q_i} = 0$, therefore $P_i = \alpha_i$
Generating function	Jacobi's complete integral $S(\mathbf{q}, \mathbf{P}, t)$	Characteristic Function $W(\mathbf{q}, \mathbf{P})$
Hamilton-Jacobi equation	$H(q_1, ...q_n; \frac{\partial S}{\partial q_1}, ..., \frac{\partial S}{\partial q_n}; t) + \frac{\partial S}{\partial t} = 0$	$H(q_1, ...q_n; \frac{\partial W}{\partial q_1}, ..., \frac{\partial W}{\partial q_n}) = E$
Transformation equations	$p_i = \frac{\partial S}{\partial q_i}$ $Q_i = \frac{\partial S}{\partial \alpha_i} = \beta_i$	$p_i = \frac{\partial W}{\partial q_i}$ $Q_i = \frac{\partial W}{\partial \alpha_i} = v_i t + \beta_i$

15.4.3 Separation of variables

Exploitation of the Hamilton-Jacobi theory requires finding a suitable action function S. When the Hamiltonian is time independent, then equation 15.101 shows that the time dependence of the action integral separates out from the dependence on the spatial variables. For many systems, the Hamilton's characteristic function $W(\mathbf{q}, \mathbf{P})$ separates into a simple sum of terms each of which is a function of a single variable. That is,

$$W(\mathbf{q}, \boldsymbol{\alpha}) = W_1(q_1) + W_2(q_2) + \cdots \cdot W_n(q_n) \tag{15.107}$$

where each function in the summation on the right depends only on a single variable. Then equation (15.100) reduces to

$$H(q_1, ...q_n; \frac{\partial W}{\partial q_1}, ..., \frac{\partial W}{\partial q_n}) = E \tag{15.108}$$

where E is the constant denoting the total energy.

Hamilton's characteristic function $W(\mathbf{q}, \mathbf{P})$ can be used with equations (15.101), (15.102), (15.91), (15.92), and (15.93) to derive

$$p_i = \frac{\partial W(\mathbf{q}, \boldsymbol{\alpha})}{\partial q_i} \qquad\qquad Q_i = \frac{\partial W(\mathbf{q}, \boldsymbol{\alpha})}{\partial P_i} \tag{15.109}$$

$$\dot{Q}_i = \frac{\partial \mathcal{H}}{\partial P_i} = 0 \qquad\qquad \dot{P}_i = \frac{\partial \mathcal{H}}{\partial Q_i} = 0 \tag{15.110}$$

$$\mathcal{H} = H + \frac{\partial S}{\partial t} = H - E = 0 \tag{15.111}$$

which has reduced the problem to a simple sum of one-dimensional first-order differential equations.

If the ith variable is cyclic, then the Hamiltonian is not a function of q_i and the ith term in Hamilton's characteristic function equals $W_i = \alpha_i q_i$ which separates out from the summation in equation 15.107. That is, all cyclic variables can be factored out of $W(\mathbf{q}, \boldsymbol{\alpha})$ which greatly simplifies solution of the Hamilton-Jacobi equation. As a consequence, the ability of the Hamilton-Jacobi method to make a canonical transformation to separate the system into many cyclic or independent variables, which can be solved trivially, is a remarkably powerful way for solving the equations of motion in Hamiltonian mechanics.

15.12 *Example: Free particle*

Consider the motion of a free particle of mass m in a force-free region. Then equation 15.93 reduces to

$$H(q_1, ...q_n; \frac{\partial S}{\partial q_1}, ..., \frac{\partial S}{\partial q_n}; t) + \frac{\partial S}{\partial t} = 0$$

Since no forces act, and the momentum $\mathbf{p} = \boldsymbol{\nabla} S$, thus the Hamilton-Jacobi equation reduces to

$$\frac{1}{2m}\nabla^2 S + \frac{\partial S}{\partial t} = 0 \tag{A}$$

The Hamiltonian is time independent, thus equation 15.101 applies

$$S(\mathbf{q}, t) = W(\mathbf{q}, \boldsymbol{\alpha}) - E(\boldsymbol{\alpha})t$$

Since the Hamiltonian does not explicitly depend on the coordinates (x, y, z), then the coordinates are cyclic and separation of the variables, 15.107, gives that the action

$$S = \boldsymbol{\alpha} \cdot \mathbf{r} - Et \tag{B}$$

For equation B to be a solution of equation A requires that

$$E = \frac{1}{2m}\boldsymbol{\alpha}^2 \tag{C}$$

Therefore

$$S = \boldsymbol{\alpha} \cdot \mathbf{r} - \frac{1}{2m}\boldsymbol{\alpha}^2 t \tag{D}$$

Since

$$\dot{\mathbf{Q}} = \frac{\partial S}{\partial \boldsymbol{\alpha}} = \mathbf{r} - \frac{\boldsymbol{\alpha}}{m} t$$

the equation of motion and the conjugate momentum are given by

$$\mathbf{r} = \dot{\mathbf{Q}} + \frac{\boldsymbol{\alpha}}{m} t \qquad\qquad \mathbf{p} = \boldsymbol{\nabla} S = \boldsymbol{\alpha}$$

Thus the Hamilton-Jacobi relation has given both the equation of motion and the linear momentum **p**.

15.13 *Example: Point particle in a uniform gravitational field*

The Hamiltonian is

$$H = \frac{1}{2m}(p_x^2 + p_y^2 + p_z^2) + mgz$$

Since the system is conservative, then the Hamilton-Jacobi equation can be written in terms of Hamilton's characteristic function W

$$E = \frac{1}{2m}\left[\left(\frac{\partial W}{\partial x}\right)^2 + \left(\frac{\partial W}{\partial y}\right)^2 + \left(\frac{\partial W}{\partial z}\right)^2\right] + mgz$$

Assuming that the variables can be separated $W = X(x) + Y(y) + Z(z)$ *leads to*

$$
\begin{aligned}
p_x &= \frac{\partial X(x)}{\partial x} = \alpha_x \\
p_y &= \frac{\partial Y(y)}{\partial y} = \alpha_y \\
p_z &= \frac{\partial Z(z)}{\partial z} = \sqrt{2m(E - mgz) - \alpha_x^2 - \alpha_y^2}
\end{aligned}
$$

Thus by integration the total W *equals*

$$W = \int_{x_0}^{x} \alpha_x dx + \int_{y_0}^{y} \alpha_y dy + \int_{z_0}^{z} \left(\sqrt{2m(E - mgz) - \alpha_x^2 - \alpha_y^2}\right) dz$$

Therefore using (15.106) *gives*

$$
\begin{aligned}
\beta_z &= t - t_0 = \int_{z_0}^{z} \frac{m\, dz}{\sqrt{2m(E - mgz) - \alpha_x^2 - \alpha_y^2}} \\
\beta_x &= \text{constant} = (x - x_0) - \int_{z_0}^{z} \frac{\alpha_x dz}{\sqrt{2m(E - mgz) - \alpha_x^2 - \alpha_y^2}} \\
\beta_y &= \text{constant} = (y - y_0) - \int_{z_0}^{z} \frac{\alpha_y dz}{\sqrt{2m(E - mgz) - \alpha_x^2 - \alpha_y^2}}
\end{aligned}
$$

If x_0, y_0, z_0 *is the position of the particle at time* $t = t_0$ *then* $\beta_x = \beta_y = 0$, *and from* (15.106)

$$
\begin{aligned}
x - x_0 &= \left(\frac{\alpha_x}{m}\right)(t - t_0) \\
y - y_0 &= \left(\frac{\alpha_y}{m}\right)(t - t_0) \\
z - z_0 &= \left(\frac{\sqrt{2m(E - mgz) - \alpha_x^2 - \alpha_y^2}}{m}\right)(t - t_0) - \frac{1}{2}g(t - t_0)^2
\end{aligned}
$$

This corresponds to a parabola as should be expected for this trivial example.

15.14 *Example: One-dimensional harmonic oscillator*

As discussed in example 15.11 the Hamiltonian for the one-dimensional harmonic oscillator can be written as

$$H = \frac{1}{2m}\left(p^2 + m^2\omega^2 q^2\right) = E$$

assuming it is conservative and where $\omega = \sqrt{\frac{k}{m}}$.

Hamilton's characteristic function W *can be used where*

$$S(q, E, t) = W(q, E) - Et$$

$$p_i = \frac{\partial W}{\partial q_i}$$

Inserting the generalized momentum p_i *into the Hamiltonian gives*

$$\frac{1}{2m}\left(\left[\frac{\partial W}{\partial q}\right]^2 + m^2\omega^2 q^2\right) = E$$

Integration of this equation gives

$$W = \sqrt{2mE}\int dq\sqrt{1 - \frac{m\omega^2 q^2}{2E}}$$

That is

$$S = \sqrt{2mE}\int dq\sqrt{1 - \frac{m\omega^2 q^2}{2E}} - Et$$

Note that

$$\frac{\partial S(q, E, t)}{\partial E} = \sqrt{\frac{2m}{E}}\int\frac{dq}{\sqrt{1 - \frac{m\omega^2 q^2}{2E}}} - t$$

This can be integrated to give

$$t = \frac{1}{\omega}\arcsin\left(q\sqrt{\frac{m\omega^2}{2E}}\right) + t_0$$

That is

$$q = \sqrt{\frac{2E}{m\omega^2}}\sin\omega(t - t_0)$$

This is the familiar solution of the undamped harmonic oscillator.

15.15 *Example: The central force problem*

The problem of a particle acted upon by a central force occurs frequently in physics. Consider the mass m *acted upon by a time-independent central potential energy* $U(r)$. *The Hamiltonian is time independent and can be written in spherical coordinates as*

$$H = \frac{1}{2m}\left(p_r^2 + \frac{1}{r^2}p_\theta^2 + \frac{1}{r^2\sin^2\theta}p_\phi^2\right) + U(r) = E$$

The time-independent Hamilton-Jacobi equation is conservative, thus

$$\frac{1}{2m}\left[\left(\frac{\partial W}{\partial r}\right)^2 + \frac{1}{r^2}\left(\frac{\partial W}{\partial \theta}\right)^2 + \frac{1}{r^2\sin^2\theta}\left(\frac{\partial W}{\partial \phi}\right)^2\right] + U(r) = E$$

Try a separable solution for Hamilton's characteristic function W *of the form*

$$W = R(r) + \Theta(\theta) + \Phi(\phi)$$

The Hamilton-Jacobi equation then becomes

$$\frac{1}{2m}\left[\left(\frac{\partial R}{\partial r}\right)^2 + \frac{1}{r^2}\left(\frac{\partial \Theta}{\partial \theta}\right)^2 + \frac{1}{r^2 \sin^2 \theta}\left(\frac{\partial \Phi}{\partial \phi}\right)^2\right] + U(r) = E$$

This can be rearranged into the form

$$2mr^2 \sin^2 \theta \left\{\frac{1}{2m}\left[\left(\frac{\partial R}{\partial r}\right)^2 + \frac{1}{r^2}\left(\frac{\partial \Theta}{\partial \theta}\right)^2\right] + U(r) + E\right\} = -\left(\frac{\partial \Phi}{\partial \phi}\right)^2$$

The left-hand side is independent of ϕ whereas the right-hand side is independent of r and θ. Both sides must equal a constant which is set to equal $-L_z^2$, that is

$$\frac{1}{2m}\left[\left(\frac{\partial R}{\partial r}\right)^2 + \frac{1}{r^2}\left(\frac{\partial \Theta}{\partial \theta}\right)^2\right] + U(r) + \frac{L_z^2}{2mr^2 \sin^2 \theta} = E$$

$$\left(\frac{\partial \Phi}{\partial \phi}\right)^2 = L_z^2$$

The equation in r and θ can be rearranged in the form

$$2mr^2\left[\frac{1}{2m}\left(\frac{\partial R}{\partial r}\right)^2 + U(r) - E\right] = -\left[\left(\frac{\partial \Theta}{\partial \theta}\right)^2 + \frac{L_z^2}{\sin^2 \theta}\right]$$

The left-hand side is independent of θ and the right-hand side is independent of r so both must equal a constant which is set to be $-L^2$

$$\frac{1}{2m}\left(\frac{\partial R}{\partial r}\right)^2 + U(r) + \frac{L^2}{2mr^2} = E$$

$$\left(\frac{\partial \Theta}{\partial \theta}\right)^2 + \frac{L_z^2}{\sin^2 \theta} = L^2$$

The variables now are completely separated and, by rearrangement plus integration, one obtains

$$R(r) = \sqrt{2m}\int \sqrt{E - U(r) - \frac{L^2}{2mr^2}}\, dr$$

$$\Theta(\theta) = \int \sqrt{L^2 - \frac{L_z^2}{\sin^2 \theta}}\, d\theta$$

$$\Phi(\phi) = L_z \phi$$

Substituting these into $W = R(r) + \Theta(\theta) + \Phi(\phi)$ gives

$$W = \sqrt{2m}\int \sqrt{E - U(r) - \frac{L^2}{2mr^2}}\, dr + \int \sqrt{L^2 - \frac{L_z^2}{\sin^2 \theta}}\, d\theta + L_z \phi$$

Hamilton's characteristic function W is the generating function from coordinates $(r, \theta, \phi, p_r, p_\theta, p_\phi)$ to new coordinates, which are cyclic, and new momenta that are constant and taken to be the separation constants E, L, L_z.

$$p_r = \frac{\partial W}{\partial r} = \sqrt{2m}\sqrt{E - U(r) - \frac{L^2}{2mr^2}}$$

$$p_\theta = \frac{\partial W}{\partial \theta} = \sqrt{L^2 - \frac{L_z^2}{\sin^2 \theta}}$$

$$p_\phi = \frac{\partial W}{\partial \phi} = L_z$$

Similarly, using (15.109) gives the new coordinates E, L, L_z

$$\beta_E + t = \frac{\partial W}{\partial E} = \sqrt{\frac{m}{2}} \int \frac{dr}{\sqrt{E - U(r) - \frac{L^2}{2mr^2}}}$$

$$\beta_L = \frac{\partial W}{\partial L} = \sqrt{2m} \int \frac{dr}{\sqrt{E - U(r) - \frac{L^2}{2mr^2}}} \left(\frac{-L}{2mr^2}\right) + \int \frac{L d\theta}{\sqrt{L^2 - \frac{L_z^2}{\sin^2 \theta}}}$$

$$\beta_{L_z} = \frac{\partial W}{\partial L_z} = \int \frac{d\theta}{\sqrt{L^2 - \frac{L_z^2}{\sin^2 \theta}}} \left(\frac{-L}{2mr^2}\right) + \phi$$

These equations lead to the elliptical, parabolic, or hyperbolic orbits discussed in chapter 11.

15.16 *Example: Linearly-damped, one-dimensional, harmonic oscillator*

A canonical treatment of the linearly-damped harmonic oscillator provides an example that combines use of non-standard Lagrangian and Hamiltonians, a canonical transformation to an autonomous system, and use of Hamilton-Jacobi theory to solve this transformed system. It shows that Hamilton-Jacobi theory can be used to determine directly the solutions for the linearly-damped harmonic oscillator.

Non-standard Hamiltonian:

In chapter 3.5, the equation of motion for the linearly-damped, one-dimensional, harmonic oscillator was given to be

$$\frac{m}{2} \left[\ddot{q} + \Gamma \dot{q} + \omega_0^2 q\right] = 0 \tag{a}$$

Example 10.3 showed that three non-standard Lagrangians give equation of motion α when used with the standard Euler-Lagrange variational equations. One of these was the Bateman[Bat31] time-dependent Lagrangian

$$L_2(q, \dot{q}, t) = \frac{m}{2} e^{\Gamma t} \left[\dot{q}^2 - \omega_0^2 q^2\right] \tag{b}$$

This Lagrangian gave the generalized momentum to be

$$p = \frac{\partial L_2}{\partial \dot{q}} = m\dot{q}e^{\Gamma t} \tag{c}$$

which was used with equation 15.3 to derive the Hamiltonian

$$H_2(q, p, t) = p\dot{q} - L_2(q, \dot{q}, t) = e^{-\Gamma t} \frac{p^2}{2m} + \frac{1}{2} m\omega_0^2 q^2 e^{\Gamma t} \tag{d}$$

Note that both the Lagrangian and Hamiltonian are explicitly time dependent and thus they are not conserved quantities. This is as expected for this dissipative system.

Hamilton-Jacobi theory:

The form of the non-autonomous Hamiltonian (d) suggests use of the generating function for a canonical transformation to an autonomous Hamiltonian, for which H is a constant of motion.

$$S(q, P, t) = F_2(q, P, t) = qPe^{\frac{\Gamma t}{2}} = QP \tag{d}$$

Then the canonical transformation gives

$$p = \frac{\partial S}{\partial q} = Pe^{\frac{\Gamma t}{2}} \tag{e}$$

$$Q = \frac{\partial S}{\partial P} = qe^{\frac{\Gamma t}{2}}$$

Insert this canonical transformation into the above Hamiltonian leads to the transformed Hamiltonian that is autonomous.

$$\mathcal{H}(Q, P, t) = H_2(q, p, t) + \frac{\partial F_2}{\partial t} = \frac{P^2}{2m} + \frac{\Gamma}{2} QP + \frac{m\omega_0^2}{2} Q^2 \tag{f}$$

That is, the transformed Hamiltonian $\mathcal{H}(Q, P, t)$ is not explicitly time dependent, and thus is conserved. Expressed in the original canonical variables (q, p), the transformed Hamiltonian $\mathcal{H}(Q, P, t)$

$$\mathcal{H}(Q, P, t) = \frac{p^2}{2m} e^{-\Gamma t} + \frac{\Gamma}{2} qp + \frac{m\omega_0^2}{2} q^2 e^{\Gamma t}$$

is a constant of motion which was not readily apparent when using the original Hamiltonian. This unexpected result illustrates the usefulness of canonical transformations for solving dissipative systems. The Hamilton-Jacobi theory now can be used to solve the equations of motion for the transformed variables (Q, P) plus the transformed Hamiltonian $\mathcal{H}(Q, P, t)$. The derivative of the generating function

$$\frac{\partial S}{\partial Q} = P \tag{g}$$

Use equation (g) to substitute for P in the Hamiltonian $\mathcal{H}(Q, P, t)$ (equation (f)), then the Hamilton-Jacobi method gives

$$\frac{1}{2m} \left(\frac{\partial S}{\partial Q} \right)^2 + \frac{\Gamma}{2} Q \frac{\partial S}{\partial Q} + \frac{m\omega_0^2}{2} Q^2 + \frac{\partial S}{\partial t} = 0$$

This equation is separable as described in 15.107 and thus let

$$S(Q, \alpha, t) = W(Q, \alpha) - \alpha t$$

where α is a separation constant. Then

$$\left[\frac{1}{2m} \left(\frac{\partial W}{\partial Q} \right)^2 + \Gamma Q \frac{\partial W}{\partial Q} + \frac{m\omega_0^2}{2} Q^2 \right] = \alpha \tag{h}$$

To simplify the equations define the variable x as

$$x \equiv \sqrt{m\omega_0} Q \tag{i}$$

then equation (h) can be written as

$$\left(\frac{\partial W}{\partial x} \right)^2 + Ax \frac{\partial W}{\partial x} + \left(x^2 - B \right) = 0 \tag{j}$$

where $A = \frac{\Gamma}{\omega_0}$ and $B = \frac{2\alpha}{\omega_0}$. Assume initial conditions $q(0) = q_0$ and $\dot{q}(0) = 0$
For this case the separation constant $\alpha > 0$, therefore $B > 0$. Note that equation (j) is a simple second-order algebraic relation, the solution of which is

$$\frac{\partial W}{\partial x} = -\frac{\alpha x}{2} \pm \sqrt{B - \left[1 - \left(\frac{A}{2} \right)^2 \right] x^2} \tag{k}$$

The choice of the sign is irrelevant for this case and thus the positive sign is chosen. There are three possible cases for the solution depending on whether the square-root term is real, zero, or imaginary.
   Case 1: $\frac{A}{2} < 1$, that is, $\frac{\lambda}{2m\omega_0} < 1$
   Define $C = \sqrt{\left[1 - \left(\frac{A}{2} \right)^2 \right]}$ Then equation (k) can be integrated to give

$$S = -\alpha t - \frac{Ax^2}{4} + \int \sqrt{(B - C^2 x^2)} dx \tag{l}$$

and

$$\beta = \frac{\partial S}{\partial \alpha} = -t + \frac{1}{\omega_0} \int \frac{dx}{\sqrt{(B - C^2 x^2)}}$$

This integral gives

$$\sin^{-1} \left(\frac{Cx}{\sqrt{B}} \right) = C\omega_0 (t + \beta) \equiv \omega t + \delta$$

where

$$\omega = \omega_0 C = \omega_0 \sqrt{1 - \left(\frac{\Gamma}{2\omega_0}\right)^2} = \sqrt{\omega_0^2 - \left(\frac{\Gamma}{2}\right)^2} \qquad (m)$$

Transforming back to the original variable q gives

$$q(t) = G e^{-\frac{\Gamma t}{2}} \sin(\omega t + \delta) \qquad (n)$$

where G and δ are given by the initial conditions. Equation m is identical to the solution for the underdamped linearly-damped linear oscillator given previously in equation 3.35.

Case 2: $\frac{A}{2} = 1$, *that is,* $\frac{\Gamma}{2\omega_0} = 1$

In this case $C = \sqrt{\left[1 - \left(\frac{A}{2}\right)^2\right]} = 0$ *and thus equation k simplifies to*

$$S = -\alpha t - \frac{Ax^2}{4} + x\sqrt{B}$$

and

$$\beta = \frac{\partial S}{\partial \alpha} = -t + \frac{x}{\omega_0 \sqrt{B}}$$

Therefore the solution is

$$q(t) = e^{-\frac{\Gamma t}{2}} (F + Gt) \qquad (o)$$

where F and G are constants given by the initial conditions. This is the solution for the critically-damped linearly-damped, linear oscillator given previously in equation 3.38.

Case 3: $\frac{A}{2} > 1$, *that is,* $\frac{\Gamma}{2\omega_0} > 1$

Define a real constant D where $D = \sqrt{\left[\left(\frac{A}{2}\right)^2 - 1\right]} = iC$, *then*

$$S = -\alpha t - \frac{Ax^2}{4} + \int \sqrt{(B + D^2 x^2)} dx$$

Then

$$\beta = \frac{\partial S}{\partial \alpha} = -t + \frac{1}{\omega_0} \int \frac{dx}{\sqrt{(B + D^2 x^2)}}$$

This last integral gives

$$\sinh^{-1}\left(\frac{Dx}{\sqrt{B}}\right) = D\omega_0 (t + \beta) \equiv \omega t + \delta$$

where

$$\omega = \omega_0 C = \omega_0 \sqrt{\left(\frac{\lambda}{2m\omega_0}\right)^2 - 1}$$

Then the original variable gives

$$q(t) = G e^{-\frac{\Gamma t}{2}} \sinh(\omega t + \delta) \qquad (l)$$

This is the classic solution of the overdamped linearly-damped, linear harmonic oscillator given previously in equation 3.37. The canonical transformation from a non-autonomous to an autonomous system allowed use of Hamiltonian mechanics to solve the damped oscillator problem.

Note that this example used Bateman's non-standard Lagrangian, and corresponding Hamiltonian, for handling a dissipative linear oscillator system where the dissipation depends linearly on velocity. This non-standard Lagrangian led to the correct equations of motion and solutions when applied using either the time-dependent Lagrangian, or time-dependent Hamiltonian, and these solutions agree with those given in chapter 3.5 which were derived using Newtonian mechanics.

15.4.4   Visual representation of the action function S.

The important role of the action integral S can be illuminated by considering the case of a single point mass m moving in a time independent potential $U(r)$. Then the action reduces to

$$S(q, \alpha, t) = W(q, \alpha) - Et \qquad (15.112)$$

Let $q_1 = x$, $q_2 = y$, $q_3 = z$, $p_1 = p_x$, $p_2 = p_y$, $p_3 = p_z$. The momentum components are given by

$$p_i = \frac{\partial W(q, \alpha)}{\partial q_i} \qquad (15.113)$$

which corresponds to

$$\mathbf{p} = \boldsymbol{\nabla} W = \boldsymbol{\nabla} S \qquad (15.114)$$

That is, the time-independent Hamilton-Jacobi equation is

$$\frac{1}{2m} |\boldsymbol{\nabla} W|^2 + U(r) = E \qquad (15.115)$$

This implies that the particle momentum is given by the gradient of Hamilton's characteristic function and is perpendicular to surfaces of constant W as illustrated in

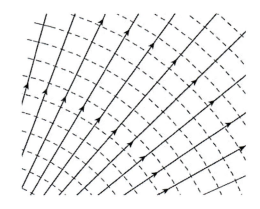

Figure 15.2: Surfaces of constant action integral S (dashed lines) and the corresponding particle momenta (solid lines) with arrows showing the direction.

figure 15.2. The constant W surfaces are time dependent as given by equation (15.101). Thus, if at time $t = 0$ the equi-action surface $S_0(q, t) = W_0(q, P_i) = 0$, then at $t = 1$ the same surface $S_0(q, t) = 0$ now coincides with the $S_0(q, t) = E$ surface etc. That is, the equi-action surfaces move through space separately from the motion of the single point mass.

   The above pictorial representation is analogous to the situation for motion of a wavefront for electromagnetic waves in optics, or matter waves in quantum physics where the wave equation separates into the form $\phi = \phi_0 e^{\frac{iS}{\hbar}} = \phi_0 e^{i(\mathbf{k} \cdot \mathbf{r} - \omega t)}$. Hamilton's goal was to create a unified theory for optics that was equally applicable to particle motion in classical mechanics. Thus the optical-mechanical analogy of the Hamilton-Jacobi theory has culminated in a universal theory that describes wave-particle duality; this was a Holy Grail of classical mechanics since Newton's time. It played an important role in development of the Schrödinger representation of quantum mechanics.

15.4.5   Advantages of Hamilton-Jacobi theory

Initially, only a few scientists, like Jacobi, recognized the advantages of Hamiltonian mechanics. In 1843 Jacobi made some brilliant mathematical developments in Hamilton-Jacobi theory that greatly enhanced exploitation of Hamiltonian mechanics. Hamilton-Jacobi theory now serves as a foundation for contemporary physics, such as quantum and statistical mechanics. A major advantage of Hamilton-Jacobi theory, compared to other formulations of analytic mechanics, is that it provides a *single, first-order* partial differential equation for the action S, which is a function of the n generalized coordinates $\mathbf{q}$ and time t. The generalized momenta no longer appear explicitly in the Hamiltonian in equations 15.94, 15.95. Note that the generalized momentum do not explicitly appear in the equivalent Euler-Lagrange equations of Lagrangian mechanics, but these comprise a system of n *second-order*, partial differential equations for the time evolution of the generalized coordinate $\mathbf{q}$. Hamilton's equations of motion are a system of $2n$ *first-order equations* for the time evolution of the generalized coordinates and their conjugate momenta.

   An important advantage of the Hamilton-Jacobi theory is that it provides a formulation of classical mechanics in which motion of a particle can be represented by a wave. In this sense, the Hamilton-Jacobi equation fulfilled a long-held goal of theoretical physics, that dates back to Johann Bernoulli, of finding an analogy between the propagation of light and the motion of a particle. This goal motivated Hamilton to develop Hamiltonian mechanics. A consequence of this wave-particle analogy is that the Hamilton-Jacobi formalism featured prominently in the derivation of the Schrödinger equation during the development of quantum-wave mechanics.

15.5 Action-angle variables

15.5.1 Canonical transformation

Systems possessing periodic solutions are a ubiquitous feature in physics. The periodic motion can be either an oscillation, for which the trajectory in phase space is a closed loop (libration), or rolling (rotational) motion as discussed in chapter 3.4.4. For many problems involving periodic motion, the interest often lies in the frequencies of motion rather than the detailed shape of the trajectories in phase space. The action-angle variable approach uses a canonical transformation to action and angle variables which provide a powerful, and elegant method to exploit Hamiltonian mechanics. In particular, it can determine the frequencies of periodic motion without having to calculate the exact trajectories for the motion. This method was introduced by the French astronomer Ch. E. Delaunay$(1816 - 1872)$ for applications to orbits in celestial mechanics, but it has equally important applications beyond celestial mechanics such as to bound solutions of the atom in quantum mechanics.

The action-angle method replaces the momenta in the Hamilton-Jacobi procedure by the **action phase integral** for the closed loop (libration) trajectory in phase space defined by

$$J_i \equiv \oint p_i dq_i \tag{15.116}$$

where for each cyclic variable the integral is taken over one complete period of oscillation. The cyclic variable I_i is called the **action variable** where

$$I_i \equiv \frac{1}{2\pi} J_i = \frac{1}{2\pi} \oint p_i dq_i \tag{15.117}$$

The canonical variable to the action variable **I** is the angle variable ϕ. Note that the name "action variable" is used to differentiate **I** from the action functional $S = \int L dt$ which has the same units; i.e. angular momentum.

The general principle underlying the use of action-angle variables is illustrated by considering one body, of mass m, subject to a one-dimensional bound conservative potential energy $U(q)$. The Hamiltonian is given by

$$H(p,q) = \frac{p^2}{2m} + U(q) \tag{15.118}$$

This bound system has a (q,p) phase space contour for each energy $H = E$.

$$p(q,E) = \pm\sqrt{2m(E - U(q))} \tag{15.119}$$

For an oscillatory system the two-valued momentum of equation 15.119 is non-trivial to handle. By contrast, the area $J \equiv \oint p dq$ of the closed loop in phase space is a single-valued scalar quantity that depends on E and $U(q)$. Moreover, Liouville's theorem states that the area of the closed contour in phase space $J \equiv \oint p dq$ is invariant to canonical transformations. These facts suggest the use of a new pair of conjugate variables, (ϕ, I), where $I(E)$ uniquely labels the trajectory, and corresponding area, of a closed loop in phase space for each value of E, and the single-valued function ϕ is a corresponding angle that specifies the exact point along the phase-space contour as illustrated in Fig 15.3.

For simplicity consider the linear harmonic oscillator where

$$U(q) = \frac{1}{2}m\omega^2 q^2 \tag{15.120}$$

Then the Hamiltonian, 15.118 equals

$$H(p,q) = \frac{p^2}{2m} + \frac{1}{2}m\omega^2 q^2 \tag{15.121}$$

Hamilton's equations of motion give that

$$\dot{p} = -\frac{\partial H}{\partial q} = -m\omega^2 q \tag{15.122}$$

$$\dot{q} = \frac{\partial H}{\partial p} = \frac{p}{m} \tag{15.123}$$

The solution of equations 15.122 and 15.123 is of the form

$$q \;=\; C\cos(\omega(t - t_0)) \tag{15.124}$$
$$p \;=\; -m\omega C\sin\omega(t - t_0) \tag{15.125}$$

where C, and t_0 are integration constants. For the harmonic oscillator, equations 15.124 and 15.125 correspond to the usual elliptical contours in phase space, as illustrated in figure 15.3.

The action-angle canonical transformation involves making the transform

$$(q, p) \to (\phi, I) \tag{15.126}$$

where I is defined by equation 15.117 and the angle ϕ being the corresponding canonical angle. The logical approach to this canonical transformation for the harmonic oscillator is to define q and p in terms of ϕ and I

$$q \;=\; \sqrt{\frac{2I}{m\omega}}\cos\phi \tag{15.127}$$
$$p \;=\; \sqrt{2mI\omega}\sin\phi \tag{15.128}$$

Note that the Poisson bracket is unity

$$[q, p]_{(\phi, I)} = 1$$

which implies that the above transformation is canonical, and thus the phase space area $I(E) \equiv \frac{1}{2\pi}\oint p\,dq$ is conserved.

For this canonical transformation the transformed Hamiltonian $\mathcal{H}(\phi, I)$ is

$$\mathcal{H}(\phi, I) = \frac{1}{2m}(2m\omega I)\sin^2\phi + \frac{1}{2}m\omega^2\frac{2I}{m\omega}\cos^2\phi = \omega I \tag{15.129}$$

Note that this Hamiltonian is a constant that is independent of the angle ϕ, and thus Hamilton's equations of motion give

$$\dot{I} \;=\; -\frac{\partial \mathcal{H}(\phi, I)}{\partial \phi} = 0 \tag{15.130}$$
$$\dot{\phi} \;=\; \frac{\partial \mathcal{H}(\phi, I)}{\partial I} = \omega \tag{15.131}$$

Thus we have mapped the harmonic oscillator to new coordinates (ϕ, I) where

$$I \;=\; \frac{\mathcal{H}(\phi, I)}{\omega} = \frac{E}{\omega} \tag{15.132}$$
$$\phi \;=\; \omega(t - t_0) \tag{15.133}$$

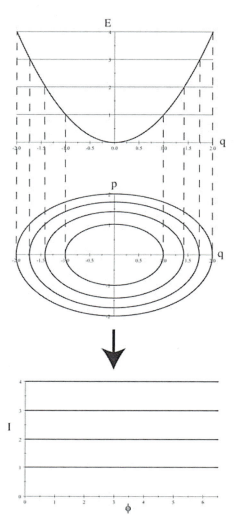

Figure 15.3: The potential energy $V(q)$, (upper) and corresponding phase space (p, q) (middle) for the harmonic oscillator at four equally spaced total energies E. The corresponding action-angles $(I\ \phi)$ resulting from a canonical transformation of this system are shown in the lower plot.

That is, the phase space has been mapped from ellipses, with area proportional to E in the (q, p) phase space, to a cylindrical (ϕ, I) phase space where $I = \frac{E}{\omega}$ are constant values that are independent of the angle, while ϕ increases linearly with time. Thus the variables (q, p) are periodic with modulus $\Delta\phi = 2\pi$.

$$q(\phi + 2\pi, I) \;=\; q(\phi, I) \tag{15.134}$$
$$p(\phi + 2\pi, I) \;=\; p(\phi, I) \tag{15.135}$$

The period τ of the periodic oscillatory motion is given simply by $\Delta\phi = 2\pi = \omega\tau$ which is the well known result for the harmonic oscillator. Note that the action-angle variable canonical transformation has determined the frequency of the periodic motion without solving the detailed trajectory of the motion.

The above example of the harmonic oscillator has shown that, for integrable periodic systems, it is possible to identify a canonical transformation to (ϕ, I) such that the Hamiltonian is independent of the angle ϕ which specifies the instantaneous location on the constant energy contour I. If the phase space contour is a separatrix, then it divides phase space into invariant regions containing phase-space contours with differing behavior. The action-angle variables are not useful for separatrix contours. For rolling motion, the system rotates with continuously increasing, or decreasing angle, and there is no natural boundary for the action angle variable since the phase space trajectory is continuous and not closed. However, the action-angle approach still is valid if the motion involves periodic as well as rolling motion.

The example of the one-dimensional, one-body, harmonic oscillator can be expanded to the more general case for many bodies in three dimensions. This is illustrated by considering multiple periodic systems for which the Hamiltonian is conservative and where the equations of the canonical transformation are separable. The generalized momenta then can be written as

$$p_i = \frac{\partial W_i(q_i; \alpha_1, \alpha_2, ..\alpha_n)}{\partial q_i} \tag{15.136}$$

for which each p_i is a function of q_i and the n integration constants α_j

$$p_i = p_i(q_i, \alpha_1, \alpha_2, ..\alpha_n) \tag{15.137}$$

The momentum $p_i(q_i, \alpha_1, \alpha_2, ..\alpha_n)$ represents the trajectory of the system in the (q_i, p_i) phase space that is characterized by Hamilton's characteristic function $W(q, J)$. Combining equations $15.116, 15.136$ gives

$$J_i \equiv \oint \frac{\partial W_i(q_i; \alpha_1, \alpha_2, ..\alpha_n)}{\partial q_i} dq_i \tag{15.138}$$

Since q_i is merely a variable of integration, each active action variable J_i is a function of the n constants of integration in the Hamilton-Jacobi equation. Because of the independence of the separable-variable pairs (q_i, p_i), the J_i form n independent functions of the α_i, and hence are suitable for use as a new set of constant momenta. Thus the characteristic function W can be written as

$$W(q_1, ...q_n; J_1, ...J_n) = \sum_j W_j(q_j; J_1, ...J_n) \tag{15.139}$$

while the Hamiltonian is only a function of the momenta $H(J_1, ...J_n)$

The generalized coordinate, conjugate to J, is known as the **angle variable** ϕ_i which is defined by the transformation equation

$$\phi_i = \frac{\partial W}{\partial J_i} = \sum_{j=1}^{n} \frac{\partial W_j(q_j; J_1, ...J_n)}{\partial J_i} \tag{15.140}$$

The corresponding equation of motion for ϕ is given by

$$\dot{\phi}_i = \frac{\partial H(J)}{\partial J_i} = 2\pi\omega_i(J_1, ...J_n) \tag{15.141}$$

where $\omega_i(J)$ are constant functions of the action variables J_j with a solution

$$\phi_i = 2\pi\omega_i t + \beta_i \tag{15.142}$$

that is, they are linear functions of time. The constants ω_i can be identified with the frequencies of the multiple periodic motions.

The action-angle variables appear to be no different than a particular set of transformed coordinates. Their merit appears when the physical interpretation is assigned to ω_i. Consider the change $\delta\phi_i$ as the q_j are changed infinitesimally

$$\delta\phi_i = \sum_j \frac{\partial \phi_i}{\partial q_j} \partial q_j = \sum_j \frac{\partial^2 W}{\partial J_i \partial q_j} \partial q_j \tag{15.143}$$

The derivative with respect to q_i vanishes except for the W_j component of W. Thus equation 15.143 reduces to

$$\delta\phi_i = \frac{\partial}{\partial J_i} \sum_j p_j(q_j, J) \, dq_j \tag{15.144}$$

Therefore, the total change in ϕ, as the system goes through one complete cycle is

$$\Delta\phi_i = \sum_j \frac{\partial}{\partial J_i} \oint p_j(q_j, J) \, dq_j = 2\pi\delta_{ij} \tag{15.145}$$

where $\frac{\partial}{\partial J_i}$ is outside the integral since the J_i are constants for cyclic motion. Thus $\Delta\phi_i = 2\pi = \omega_i\tau_i$ where τ_i is the period for one cycle of oscillation, where the angular frequency ω_i is given by

$$\frac{\omega_i}{2\pi} = \nu_i = \frac{1}{\tau_i} \tag{15.146}$$

Thus the frequency ν associated with the periodic motion is the reciprocal of the period τ. The secret here is that the derivative of H with respect to the action variable J given by equation (15.141) directly determines the frequency of the periodic motion without the need to solve the complete equations of motion. Note that multiple periodic motion can be represented by a Fourier expansion of the form

$$q_k = \sum_{j_1=-\infty}^{\infty} \sum_{j_2=-\infty}^{\infty} \cdots \sum_{j_n=-\infty}^{\infty} a_{j_1,..,j_n}^k e^{2\pi i(j_1\omega_1 + j_2\omega_2 + j_3\omega_3 + .. + j_n\omega_n)} \tag{15.147}$$

Although the action-angle approach to Hamilton-Jacobi theory does not produce complete equations of motion, it does provide the frequency decomposition that often is the physics of interest. The reason that the powerful action-angle variable approach has been introduced here is that it is used extensively in celestial mechanics. The action-angle concept also played a key role in the development of quantum mechanics, in that Sommerfeld recognized that Bohr's ad hoc assumption that angular momentum is quantized, could be expressed in terms of quantization of the angle variable as is mentioned in chapter 18.

15.5.2   Adiabatic invariance of the action variables

When the Hamiltonian depends on time it can be quite difficult to solve for the motion because it is difficult to find constants of motion for time-dependent systems.   However, if the time dependence is sufficiently slow, that is, if the motion is adiabatic, then there exist dynamical variables that are almost constant which can be used to solve for the motion. In particular, such approximate constants are the familiar action-angle integrals. The adiabatic invariance of the action variables played an important role in the development of quantum mechanics during the 1911 Solvay Conference. This was a time when physicists were grappling with the concepts of quantum mechanics. Einstein used the following classical mechanics example of adiabatic invariance, applied to the simple pendulum, in order to illustrate the concept of adiabatic invariance of the action. This example demonstrates the power of using action-angle variables.

15.17 *Example: Adiabatic invariance for the simple pendulum*

Consider that the pendulum is made up of a point mass M suspended from a pivot by a light string of length L that is swinging freely in a vertical plane. Derive the dependence of the amplitude of the oscillations θ, assuming θ is small, if the string is very slowly shortened by a factor of 2, that is, assume that the change in length during one period of the oscillation is very small.

The tension in the string T is given by

$$T = Mg\langle\cos\theta\rangle + \left\langle\frac{ML^2\dot{\theta}^2}{L}\right\rangle$$

Let the pendulum angle be oscillatory

$$\theta = \theta_0\cos(\omega t + \varphi_0)$$

Then the average mean square amplitude and velocity over one period are

$$\langle \theta^2 \rangle = \langle [\theta_0 \cos(\omega t + \varphi_0)]^2 \rangle = \frac{\theta_0^2}{2}$$

$$\langle \dot{\theta}^2 \rangle = \langle [-\theta_0 \omega \sin(\omega t + \varphi_0)]^2 \rangle = \frac{\omega^2 \theta_0^2}{2}$$

Since, for the simple pendulum, $\omega^2 = \frac{g}{L}$, then the tension in the string

$$T = Mg(1 - \frac{\langle \theta^2 \rangle}{2}) + ML\langle \dot{\theta}^2 \rangle = Mg(1 + \frac{\theta_0^2}{4})$$

Assuming that θ_0 is a small angle, and that the change in length $-\Delta L$ is very small during one period τ, then the work done is

$$\Delta W = T\Delta L = -Mg\Delta L - Mg\frac{\theta_0^2}{4}\Delta L \tag{a}$$

while the change in internal oscillator energy is

$$\Delta(-MgL\cos\theta_0) = \Delta\left[-MgL(1 - \frac{\theta_0^2}{2})\right] = -Mg\Delta L + \frac{1}{2}Mg\Delta(L\theta_0^2) = -Mg\Delta L + \frac{1}{2}Mg\theta_0^2\Delta L + MgL\theta_0\Delta\theta_0 \tag{b}$$

The work done must balance the increment in internal energy therefore

$$L\theta_0\Delta\theta_0 + \frac{3\theta_0^2\Delta L}{4} = 0$$

or

$$L\theta_0^2\Delta\ln(\theta_0 L^{\frac{3}{4}}) = 0$$

Therefore it follows that

$$(\theta_0 L^{\frac{3}{4}}) = \text{constant} \tag{c}$$

or

$$\theta_0 \propto L^{-\frac{3}{4}}$$

Thus shortening the length of the pendulum string from L to $\frac{L}{2}$ adiabatically corresponds to the amplitude increasing by a factor 1.68.

Consider the action-angle integral for one closed period $\tau = \frac{2\pi}{\omega}$ for this problem

$$\begin{aligned} J &= \oint P_\theta d\theta \\ &= \oint ML^2\dot{\theta} \cdot \dot{\theta}dt \\ &= ML^2\langle \dot{\theta}^2 \rangle\frac{2\pi}{\omega} \\ &= \pi ML^2\theta_0^2\omega \\ &= \pi Mg^{\frac{1}{2}}\theta_0^2 L^{\frac{3}{2}} = \text{constant} \end{aligned}$$

where that last step is due to equation (c).

The above example shows that the action integral $J = \text{constant}$, that is, it is invariant to an adiabatic change. In retrospect this result is as expected in that the action integral should be minimized.

15.6   Canonical perturbation theory

Most examples in classical mechanics discussed so far have been capable of exact solutions. In real life, the majority of problems cannot be solved exactly. For example, in celestial mechanics the two-body Kepler problem can be solved exactly, but solution of the three-body problem is intractable. Typical systems in celestial mechanics are never as simple as the two-body Kepler system because of the influence of additional bodies. Fortunately in most cases the influence of additional bodies is sufficiently small to allow use of perturbation theory. That is, the restricted three-body approximation can be employed for which the system is reduced to considering it as an exactly solvable two-body problem, subject to a small perturbation to this solvable two-body system. Note that even though the change in the Hamiltonian due to the perturbing term may be small, the impact on the motion can be especially large near a resonance.

Consider the Hamiltonian, subject to a time-dependent perturbation, is written as

$$H(q, p, t) = H_0(q, p, t) + \Delta H(q, p, t)$$

where $H_0(q, p, t)$ designates the unperturbed Hamiltonian and $\Delta H(q, p, t)$ designates the perturbing term. For the unperturbed system the Hamilton-Jacobi equation is given by

$$\mathcal{H}(Q_i, P_i, t) = H_0(q_1, ... q_n; \frac{\partial S}{\partial q_1} ..., \frac{\partial S}{\partial q_n}; t) + \frac{\partial S}{\partial t} = 0 \qquad (15.90)$$

where $S(q_i, P_i, t)$ is the generating function for the canonical transformation $(q, p) \rightarrow (Q, P)$. The perturbed $S(q_i, P_i, t)$ remains a canonical transformation, but the transformed Hamiltonian $\mathcal{H}(Q_i, P_i, t) \neq 0$. That is,

$$\mathcal{H}(Q_i, P_i, t) = H_0 + \Delta H(q, p, t) + \frac{\partial S}{\partial t} = \Delta H(q, p, t) \qquad (15.148)$$

The equations of motion satisfied by the transformed variables now are

$$\dot{Q}_i = \frac{\partial \Delta H}{\partial P_i} \qquad (15.149)$$
$$\dot{P}_i = \frac{\partial \Delta H}{\partial Q_i}$$

These equations remain as difficult to solve as the full Hamiltonian. However, the perturbation technique assumes that ΔH is small, and that one can neglect the change of (Q_i, P_i) over the perturbing interval. Therefore, to a first approximation, the unperturbed values of $\frac{\partial \Delta H}{\partial P_i}$ and $\frac{\partial \Delta H}{\partial Q_i}$ can be used in equations 15.149. A detailed explanation of canonical perturbation theory is presented in chapter 12 of Goldstein[Go50].

15.18 *Example: Harmonic oscillator perturbation*

(a) Consider first the Hamilton-Jacobi equation for the generating function $S(q, \alpha, t)$ for the case of a single free particle subject to the Hamiltonian $H = \frac{1}{2}p^2$. Find the canonical transformation $q = q(\beta, \alpha)$ and $p = p(\beta, \alpha)$ where β and α are the transformed coordinate and momentum respectively.

The Hamilton-Jacobi equation

$$\frac{\partial S}{\partial t} + H(q, p, t) = 0$$

Using $p = \frac{\partial S}{\partial q}$ in the Hamiltonian $H = \frac{1}{2}p^2$ gives

$$\frac{\partial S}{\partial t} + \frac{1}{2}\left(\frac{\partial S}{\partial q}\right)^2 = 0$$

Since H does not depend on q, t explicitly, then the two terms on the left hand side of the equation can be set equal to $-\gamma, \gamma$ respectively, where γ is at most a function of p. Then the generating function is

$$S = \sqrt{2\gamma}q - \gamma t$$

Set $\alpha = \sqrt{2\gamma}$ then the generating function can be written as

$$S = \alpha q - \frac{1}{2}\alpha^2 t$$

The constant α can be identified with the new momentum P. Then the transformation equations become

$$p = \frac{\partial S}{\partial q} = \alpha \qquad\qquad Q = \frac{\partial S}{\partial P} = \frac{\partial S}{\partial \alpha} = q - \alpha t = \beta$$

That is

$$q = \beta + \alpha t$$

which corresponds to motion with a uniform velocity α in the q, p system.

(b) Consider that the Hamiltonian is perturbed by addition of potential $U = \frac{q^2}{2}$ which corresponds to the harmonic oscillator. Then

$$H = \frac{1}{2}p^2 + \frac{q^2}{2}$$

Consider the transformed Hamiltonian

$$\mathcal{H} = H + \frac{\partial S}{\partial t} = \frac{1}{2}p^2 + \frac{q^2}{2} - \frac{\alpha^2}{2} = \frac{q^2}{2} = \frac{1}{2}(\beta + \alpha t)^2$$

Hamilton's equations of motion

$$\dot{Q} = \frac{\partial \mathcal{H}}{\partial P} \qquad\qquad \dot{P} = -\frac{\partial \mathcal{H}}{\partial Q}$$

give that

$$\begin{aligned} \dot{\beta} &= (\beta + \alpha t)\, t \\ \dot{\alpha} &= -(\beta + \alpha t) \end{aligned}$$

These two equations can be solved to give

$$\ddot{\alpha} + \alpha = 0$$

which is the equation of a harmonic oscillator showing that α is harmonic of the form $\alpha = \alpha_0 \sin(t + \delta)$ where α_0, δ are constants of motion. Thus

$$\beta = -\dot{\alpha} - t = -\alpha_0 [\cos(t + \delta) + t \sin(t + \delta)]$$

The transformation equations then give

$$\begin{aligned} p &= \alpha = \alpha_0 \sin(t + \delta) \\ q &= \beta + \alpha t = -\dot{\alpha} = -\alpha_0 \cos(t + \delta) \end{aligned}$$

Hence the solution for the perturbed system is harmonic, which is to be expected since the potential has a quadratic dependence of position.

15.19 *Example: Lindblad resonance in planetary and galactic motion*

Use of canonical perturbation theory in celestial mechanics has been exploited by Professor Alice Quillen and her group. They combine use of action-angle variables and Hamilton-Jacobi theory to investigate the role of Lindblad resonance to planetary motion, and also for stellar motion in galaxies. A Lindblad resonance is an orbital resonance in which the orbital period of a celestial body is a simple multiple of some forcing frequency. Even for very weak perturbing forces, such resonance behavior can lead to orbit capture and chaotic motion.

For planetary motion the planet masses are about $1/1000$ that of the central star, so the perturbations to Kepler orbits are small. However, Lindblad resonance for planetary motion led to Saturn's rings which result from perturbations produced by the moons of Saturn that skulpt and clear dust rings. Stellar orbits in disk galaxies are perturbed a few percent by non axially-symmetric galactic features such as spiral arms or bars. Lindblad resonances perturb stellar motion and drive spiral density waves at distances from the center of a galactic disk where the natural frequency of the radial component of a star's orbital velocity is close to the frequency of the fluctuations in the gravitational field due to passage through spiral arms or bars. If a stars orbital speed around a galactic center is greater than that of the part of a spiral arm through which it is traversing, then an inner Lindblad resonance occurs which speeds up the star's orbital speed moving the orbit outwards. If the orbital speed is less than that of a spiral arm, an inner Lindblad resonance occurs causing inward movement of the orbit.

15.7   Symplectic representation

The Hamilton's first-order equations of motion are symmetric if the generalized and constraint force terms, in equation 15.9, are excluded.

$$\dot{\mathbf{q}} = \frac{\partial H}{\partial \mathbf{p}} \qquad\qquad -\dot{\mathbf{p}} = \frac{\partial H}{\partial \mathbf{q}}$$

This stimulated attempts to treat the canonical variables $(\mathbf{q}, \mathbf{p})$ in a symmetric form using group theory. Some graduate textbooks in classical mechanics have adopted use of symplectic symmetry in order to unify the presentation of Hamiltonian mechanics. For a system of n degrees of freedom, a column matrix $\boldsymbol{\eta}$ is constructed that has $2n$ elements where

$$\eta_j = q_j \qquad\qquad \eta_{n+j} = p_j \qquad\qquad j \leq n \tag{15.150}$$

Therefore the column matrix

$$\left(\frac{\partial H}{\partial \boldsymbol{\eta}}\right)_j = \frac{\partial H}{\partial q_j} \qquad \left(\frac{\partial H}{\partial \boldsymbol{\eta}}\right)_{n+j} = \frac{\partial H}{\partial p_j} \qquad\qquad j \leq n \tag{15.151}$$

The symplectic matrix $\mathbf{J}$ is defined as being a $2n$ by $2n$ skew-symmetric, orthogonal matrix that is broken into four $n \times n$ null or unit matrices according to the scheme

$$\mathbf{J} = \begin{pmatrix} [\mathbf{0}] & +[\mathbf{1}] \\ -[\mathbf{1}] & [\mathbf{0}] \end{pmatrix} \tag{15.152}$$

where $[\mathbf{0}]$ is the n-dimension null matrix, for which all elements are zero. Also $[\mathbf{1}]$ is the n-dimensional unit matrix, for which the diagonal matrix elements are unity and all off-diagonal matrix elements are zero. The $\mathbf{J}$ matrix accounts for the opposite signs used in the equations for $\dot{\mathbf{q}}$ and $\dot{\mathbf{p}}$. The symplectic representation allows the Hamilton's equations of motion to be written in the compact form

$$\dot{\boldsymbol{\eta}} = \mathbf{J}\frac{\partial H}{\partial \boldsymbol{\eta}} \tag{15.153}$$

This textbook does not use the elegant symplectic representation since it ignores the important generalized forces and Lagrange multiplier forces.

15.8   Comparison of the Lagrangian and Hamiltonian formulations

Common features

The discussion of Lagrangian and Hamiltonian dynamics has illustrated the power of such algebraic formulations. Both approaches are based on application of variational principles to scalar energy which gives the freedom to concentrate solely on active forces and to ignore internal forces. Both methods can handle many-body systems and exploit canonical transformations, which are impractical or impossible using the vectorial Newtonian mechanics. These algebraic approaches simplify the calculation of the motion for constrained systems by representing the vector force fields, as well as the corresponding equations of motion, in terms of either the Lagrangian function $L(\mathbf{q}, \dot{\mathbf{q}}, t)$ or the action functional $S(\mathbf{q}, \mathbf{p}, t)$ which are related by the definite integral

$$S(\mathbf{q}, \mathbf{p}, t) = \int_{t_1}^{t_2} L(\mathbf{q}, \dot{\mathbf{q}}, t) dt \tag{15.1}$$

The Lagrangian function $L(\mathbf{q}, \dot{\mathbf{q}}, t)$, and the action functional $S(\mathbf{q}, \mathbf{p}, t)$, are scalar functions under rotation, but they determine the vector force fields and the corresponding equations of motion. Thus the use of rotationally-invariant functions $L(\mathbf{q}, \dot{\mathbf{q}}, t)$ and $S(\mathbf{q}, \mathbf{p}, t)$ provide a simple representation of the vector force fields. This is analogous to the use of scalar potential fields $\phi(\mathbf{q}, t)$ to represent the electrostatic and gravitational vector force fields. Like scalar potential fields, Lagrangian and Hamiltonian mechanics represents the observables as derivatives of $L(\mathbf{q}, \dot{\mathbf{q}}, t)$ and $S(\mathbf{q}, \mathbf{p}, t)$, and the absolute values of $L(\mathbf{q}, \dot{\mathbf{q}}, t)$ and $S(\mathbf{q}, \mathbf{p}, t)$ are undefined; only differences in $L(\mathbf{q}, \dot{\mathbf{q}}, t)$ and $S(\mathbf{q}, \mathbf{p}, t)$ are observable. For example, the generalized momenta are given by the derivatives $p_i \equiv \frac{\partial L}{\partial \dot{q}_i}$ and $p_j = \frac{\partial S}{\partial q_j}$. The physical significance of the least action $S(\mathbf{q}, \boldsymbol{\alpha}, t)$ is illustrated when the canonically transformed momenta $\mathbf{P} = \boldsymbol{\alpha}$ is a constant. Then the generalized momenta and the Hamilton-Jacobi equation, imply that the total time derivative of the action equals

$$\frac{dS}{dt} = \frac{\partial S}{\partial q_i}\dot{q}_i + \frac{\partial S}{\partial t} = p_i q_i - H = L \tag{15.154}$$

The indefinite integral of this equation reproduces the definite integral (15.1) to within an arbitrary constant, i.e.

$$S(\mathbf{q}, \mathbf{p}) = \int L(\mathbf{q}, \dot{\mathbf{q}}, t)dt + \text{constant} \tag{15.155}$$

Lagrangian formulation:

Consider a system with n independent generalized coordinates, plus m constraint forces that are not required to be known. The Lagrangian approach can reduce the system to a minimal system of $s = n - m$ independent generalized coordinates leading to $s = n - m$ *second-order* differential equations. By comparison, the Newtonian approach uses $n + m$ unknowns. Alternatively, the Lagrange multipliers approach allows determination of the holonomic constraint forces resulting in $s = n + m$ second order equations to determine $s = n + m$ unknowns. The Lagrangian potential function is limited to conservative forces, but generalized forces can be used to handle non-conservative and non-holonomic forces. The advantage of the Lagrange equations of motion is that they can deal with any type of force, conservative or non-conservative, and they directly determine $q, \dot{q}$ rather than q, p which then requires relating p to $\dot{q}$. The Lagrange approach is superior to the Hamiltonian approach if a numerical solution is required for typical undergraduate problems in classical mechanics. However, Hamiltonian mechanics has a clear advantage for addressing more profound and philosophical questions in physics.

Hamiltonian formulation:

For a system with n independent generalized coordinates, and m constraint forces, the Hamiltonian approach determines $2n$ *first-order* differential equations. In contrast to Lagrangian mechanics, where the Lagrangian is a function of the coordinates and their velocities, the Hamiltonian uses the variables $\mathbf{q}$ and $\mathbf{p}$, rather than velocity. The Hamiltonian has twice as many independent variables as the Lagrangian which is a great advantage, not a disadvantage, since it broadens the realm of possible transformations that can be used to simplify the solutions. Hamiltonian mechanics uses the conjugate coordinates $\mathbf{q}, \mathbf{p}$, corresponding to phase space. This is an advantage in most branches of physics and engineering. Compared to Lagrangian mechanics, Hamiltonian mechanics has a significantly broader arsenal of powerful techniques that can be exploited to obtain an analytical solution of the integrals of the motion for complicated systems. These techniques include, the Poisson bracket formulation, canonical transformations, the Hamilton-Jacobi approach, the action-angle variables, and canonical perturbation theory. In addition, Hamiltonian dynamics also provides a means of determining the unknown variables for which the solution assumes a soluble form, and it is ideal for study of the fundamental underlying physics in applications to other fields such as quantum or statistical physics. However, the Hamiltonian approach endemically assumes that the system is conservative putting it at a disadvantage with respect to the Lagrangian approach. The appealing symmetry of the Hamiltonian equations, plus their ability to utilize canonical transformations, makes it the formalism of choice for examination of system dynamics. For example, Hamilton-Jacobi theory, action-angle variables and canonical perturbation theory are used extensively to solve complicated multibody orbit perturbations in celestial mechanics by finding a canonical transformation that transforms the perturbed Hamiltonian to a solved unperturbed Hamiltonian.

The Hamiltonian formalism features prominently in quantum mechanics since there are well established rules for transforming the classical coordinates and momenta into linear operators used in quantum mechanics. The variables $\mathbf{q}, \dot{\mathbf{q}}$ used in Lagrangian mechanics do not have simple analogs in quantum physics. As a consequence, the Poisson bracket formulation, and action-angle variables of Hamiltonian mechanics played a key role in development of matrix mechanics by Heisenberg, Born, and Dirac, while the Hamilton-Jacobi formulation played a key role in development of Schrödinger's wave mechanics. Similarly, Hamiltonian mechanics is the preeminent variational approached used in statistical mechanics.

15.9   Summary

This chapter has gone beyond what is normally covered in an undergraduate course in classical mechanics, in order to illustrate the power of the remarkable arsenal of methods available for solution of the equations of motion using Hamiltonian mechanics. This has included the Poisson bracket representation of Hamiltonian formulation of mechanics, canonical transformations, Hamilton-Jacobi theory, action-angle variables, and canonical perturbation theory. The purpose was to illustrate the power of variational principles in Hamiltonian mechanics and how they relate to fields such as quantum mechanics. The following are the key points made in this chapter.

Poisson brackets: The elegant and powerful Poisson bracket formalism of Hamiltonian mechanics was introduced. The Poisson bracket of any two continuous functions of generalized coordinates $F(p, q)$ and $G(p, q)$, is defined to be

$$[F, G]_{pq} \equiv \sum_i \left(\frac{\partial F}{\partial q_i} \frac{\partial G}{\partial p_i} - \frac{\partial F}{\partial p_i} \frac{\partial G}{\partial q_i} \right) \tag{15.13}$$

The fundamental Poisson brackets equal

$$[q_k, q_l] = 0 \tag{15.21}$$

$$[p_k, p_l] = 0 \tag{15.22}$$

$$[q_k, p_l] = -[p_l, q_k] = \delta_{kl} \tag{15.23}$$

The Poisson bracket is invariant to a canonical transformation from (q, p) to (Q, P). That is

$$[F, G]_{qp} = \sum_k \left(\frac{\partial F}{\partial Q_k} \frac{\partial G}{\partial P_k} - \frac{\partial F}{\partial P_k} \frac{\partial G}{\partial Q_k} \right) = [F, G]_{QP} \tag{15.32}$$

There is a one-to-one correspondence between the commutator and Poisson Bracket of two independent functions,

$$(F_1 G_1 - G_1 F_1) = \lambda [F_1, G_1] \tag{15.38}$$

where λ is an independent constant. In particular $F_1 G_1$ commute of the Poisson Bracket $[F_1, G_1] = 0$.

Poisson Bracket representation of Hamiltonian mechanics: It has been shown that the Poisson bracket formalism contains the Hamiltonian equations of motion and is invariant to canonical transformations. Also this formalism extends Hamilton's canonical equations to non-commuting canonical variables.

Hamilton's equations of motion can be expressed directly in terms of the Poisson brackets

$$\dot{q}_k = [q_k, H] = \frac{\partial H}{\partial p_k} \tag{15.57}$$

$$\dot{p}_k = [p_k, H] = -\frac{\partial H}{\partial q_k} \tag{15.58}$$

An important result is that the total time derivative of any operator is given by

$$\frac{dG}{dt} = \frac{\partial G}{\partial t} + [G, H] \tag{15.45}$$

Poisson brackets provide a powerful means of determining which observables are time independent and whether different observables can be measured simultaneously with unlimited precision. It was shown that the Poisson bracket is invariant to canonical transformations, which is a valuable feature for Hamiltonian mechanics. Poisson brackets were used to prove Liouville's theorem which plays an important role in the use of Hamiltonian phase space in statistical mechanics. The Poisson bracket is equally applicable to continuous solutions in classical mechanics as well as discrete solutions in quantized systems.

Canonical transformations: A transformation between a canonical set of variables (q, p) with Hamiltonian $H(q, p, t)$ to another set of canonical variable (Q, P) with Hamiltonian $\mathcal{H}(Q, P, t)$ can be achieved using a generating functions F such that

$$\mathcal{H}(Q, P, t) = H(q, p, t) + \frac{\partial F}{\partial t} \tag{15.89}$$

Possible generating functions are summarized in the following table.

Generating function	Generating function derivatives		Trivial special case		
$F = F_1(\mathbf{q}, \mathbf{Q}, t)$	$p_i = \frac{\partial F_1}{\partial q_i}$	$P_i = -\frac{\partial F_1}{\partial Q_i}$	$F_1 = q_i Q_i$	$Q_i = p_i$	$P_i = -q_i$
$F = F_2(\mathbf{q}, \mathbf{P}, t) - \mathbf{Q} \cdot \mathbf{P}$	$p_i = \frac{\partial F_2}{\partial q_i}$	$Q_i = \frac{\partial F_2}{\partial P_i}$	$F_2 = q_i P_i$	$Q_i = q_i$	$P_i = p_i$
$F = F_3(\mathbf{p}, \mathbf{Q}, t) + \mathbf{q} \cdot \mathbf{p}$	$q_i = -\frac{\partial F_3}{\partial p_i}$	$P_i = -\frac{\partial F_3}{\partial Q_i}$	$F_3 = p_i Q_i$	$Q_i = -q_i$	$P_i = -p_i$
$F = F_4(\mathbf{p}, \mathbf{P}, t) + \mathbf{q} \cdot \mathbf{p} - \mathbf{Q} \cdot \mathbf{P}$	$q_i = -\frac{\partial F_4}{\partial p_i}$	$Q_i = \frac{\partial F_4}{\partial P_i}$	$F_1 = p_i P_i$	$Q_i = p_i$	$P_i = -q_i$

If the canonical transformation makes $\mathcal{H}(Q, P, t) = 0$ then the conjugate variables (Q, P) are constants of motion. Similarly if $\mathcal{H}(Q, P, t)$ is a cyclic function then the corresponding P are constants of motion.

Hamilton-Jacobi theory: Hamilton-Jacobi theory determines the generating function required to perform canonical transformations that leads to a powerful method for obtaining the equations of motion for a system. The Hamilton-Jacobi theory uses the action function $S \equiv F_2$ as a generating function, and the canonical momentum is given by

$$p_i = \frac{\partial S}{\partial q_i} \tag{15.4}$$

This can be used to replace p_i in the Hamiltonian H leading to the **Hamilton-Jacobi equation**

$$H(q; \frac{\partial S}{\partial q}; t) + \frac{\partial S}{\partial t} = 0 \tag{15.94}$$

Solutions of the Hamilton-Jacobi equation were obtained by separation of variables. The close optical-mechanical analogy of the Hamilton-Jacobi theory is an important advantage of this formalism that led to it playing a pivotal role in the development of wave mechanics by Schrödinger.

Action-angle variables: The action-angle variables exploits a canonical transformation from $(q, p) \rightarrow (\phi, I)$ where

$$I_i \equiv \frac{1}{2\pi} J_i = \frac{1}{2\pi} \oint p_i dq_i \tag{15.117}$$

For periodic motion the phase-space trajectory is closed with area given by J and this area is conserved for the above canonical transformation. For a conserved Hamiltonian the action variable I is independent of the angle variable ϕ. The time dependence of the angle variable ϕ directly determines the frequency of the periodic motion without recourse to calculation of the detailed trajectory of the periodic motion.

Canonical perturbation theory: Canonical perturbation theory is a valuable method of handling multi-body interactions. The adiabatic invariance of the action-angle variables provides a powerful approach for exploiting canonical perturbation theory.

Comparison of Lagrangian and Hamiltonian formulations: The remarkable power, and intellectual beauty, provided by use of variational principles to exploit the underlying principles of natural economy in nature, has had a long and rich history. It has led to profound developments in many branches of theoretical physics. However, it is noted that although the above algebraic formulations of classical mechanics have been used for over two centuries, the important limitations of these algebraic formulations to non-linear systems remain a challenge that still is being addressed.

It has been shown that the Lagrangian and Hamiltonian formulations represent the vector force fields, and the corresponding equations of motion, in terms of the Lagrangian function $L(\mathbf{q}, \dot{\mathbf{q}}, t)$, or the action functional $S(\mathbf{q}, \mathbf{p}, t)$, which are scalars under rotation. The Lagrangian function $L(\mathbf{q}, \dot{\mathbf{q}}, t)$ is related to the action functional $S(\mathbf{q}, \mathbf{p}, t)$ by

$$S(\mathbf{q}, \mathbf{p}, t) = \int_{t_1}^{t_2} L(\mathbf{q}, \dot{\mathbf{q}}, t) dt \qquad (15.1)$$

These functions are analogous to electric potential, in that the observables are derived by taking derivatives of the Lagrangian function $L(\mathbf{q}, \dot{\mathbf{q}}, t)$ or the action functional $S(\mathbf{q}, \mathbf{p}, t)$. The Lagrangian formulation is more convenient for deriving the equations of motion for simple mechanical systems. The Hamiltonian formulation has a greater arsenal of techniques for solving complicated problems plus it uses the canonical variables (q_i, p_i) which are the variables of choice for applications to quantum mechanics and statistical mechanics.

Workshop exercises

1. Poisson brackets are a powerful means of elucidating when observables are constant of motion and whether two observables can be simultaneously measured with unlimited precision. Consider a spherically symmetric Hamiltonian

$$H = \frac{1}{2m} \left(p_r^2 + \frac{p_\theta^2}{r^2} + \frac{p_\phi^2}{r^2 \sin^2 \theta} \right) + U(r)$$

for a mass m where $U(r$ is a central potential. Use the Poisson bracket plus the time dependence to determine the following:

   (a) Does p_ϕ commute with H and is it a constant of motion?

   (b) Does $p_\theta^2 + \frac{p_\phi^2}{\sin^2 \theta}$ commute with H and is it a constant of motion?

   (c) Does p_r commute with H and is it a constant of motion?

   (d) Does p_ϕ commute with p_θ and what does the result imply?

2. Consider the Poisson brackets for angular momentum L

   (a) Show $\{L_i, r_j\} = \epsilon_{ijk} r_k$, where the Levi-Cevita tensor is,

$$\epsilon_{ijk} = \begin{cases} +1 & \text{if } ijk \text{ are cyclically permuted} \\ -1 & \text{if } ijk \text{ are anti-cyclically permuted} \\ 0 & \text{if } i = j \text{ or } i = k \text{ or } j = k \end{cases}$$

   (b) Show $\{L_i, p_j\} = \epsilon_{ijk} p_k$.

   (c) Show $\{L_i, L_j\} = \epsilon_{ijk} L_k$. The following identity may be useful: $\epsilon_{ijk}\epsilon_{ilm} = \delta_{jl}\delta_{km} - \delta_{jm}\delta_{kl}$.

   (d) Show $\{L_i, L^2\} = 0$.

3. Consider the Hamiltonian of a two-dimensional harmonic oscillator,

$$H = \frac{\mathbf{p}^2}{2m} + \frac{1}{2}m \left(\omega_1^2 r_1^2 + \omega_2^2 r_2^2 \right)$$

What condition is satisfied if L^2 a conserved quantity?

Problems

1. Consider the motion of a particle of mass m in an isotropic harmonic oscillator potential $U = \frac{1}{2}kr^2$ and take the orbital plane to be the $x - y$ plane. The Hamiltonian is then

$$H \equiv S_0 = \frac{1}{2m}(p_x^2 + p_y^2) + \frac{1}{2}k(x^2 + y^2)$$

Introduce the three quantities

$$
\begin{aligned}
S_1 &= \frac{1}{2m}(p_x^2 - p_y^2) + \frac{1}{2}k(x^2 - y^2) \\
S_2 &= \frac{1}{m}p_x p_y + kxy \\
S_3 &= \omega(xp_y - yp_x)
\end{aligned}
$$

with $\omega = \sqrt{\frac{k}{m}}$. Use Poisson brackets to solve the following:

   a) Show that $[S_0, S_i] = 0$ for $i = 1, 2, 3$ proving that (S_1, S_2, S_3) are constants of motion.

   b) Show that

$$
\begin{aligned}
\left[S_1, S_2\right] &= 2\omega S_3 \\
\left[S_2, S_3\right] &= 2\omega S_1 \\
\left[S_3, S_1\right] &= 2\omega S_2
\end{aligned}
$$

so that $(2\omega)^{-1}(S_1, S_2, S_3)$ have the same Poisson bracket relations as the components of a 3-dimensional angular momentum.

   c) Show that

$$S_0^2 = S_1^2 + S_2^2 + S_3^2$$

2. Assume that the transformation equations between the two sets of coordinates (q, p) and (Q, P) are

$$
\begin{aligned}
Q &= \ln(1 + q^{\frac{1}{2}} \cos p) \\
P &= 2(1 + q^{\frac{1}{2}} \cos p)q^{\frac{1}{2}} \sin p)
\end{aligned}
$$

   a) Assuming that q, p are canonical variables, i.e. $[q, p] = 1$, show directly from the above transformation equations that Q, P are canonical variables.

   b) Show that the generating function that generates this transformation between the two sets of canonical variables is

$$F_3 = -[e^Q - 1]^2 \tan p$$

3. Consider a bound two-body system comprising a mass m in an orbit at a distance r from a mass M. The attractive central force binding the two-body system is

$$\mathbf{F} = \frac{k}{r^2}\hat{\mathbf{r}}$$

where k is negative. Use Poisson brackets to prove that the eccentricity vector $A = p \times L + \mu k \hat{r}$ is a conserved quantity.

4. (a) Consider the case of a single mass m where the Hamiltonian $H = \frac{1}{2}p^2$. Use the generating function $S(q, P, t)$ to solve the Hamilton-Jacobi equation with the canonical transformation $q = q(Q, P)$ and $p = p(Q, P)$ and determine the equations relating the (q, p) variables to the transformed coordinate and momentum (Q, P).

   (b) If there is a perturbing Hamiltonian $\Delta H = \frac{1}{2}q^2$, then P will not be constant. Express the transformed Hamiltonian H (using the transformation given above in terms of P, Q, and t). Solve for $Q(t)$ and $P(t)$ and show that the perturbed solution $q[Q(t), P(t)], p[Q(t), P(t)]$ is simple harmonic.

Chapter 16

Analytical formulations for continuous systems

16.1 Introduction

Lagrangian and Hamiltonian mechanics have been used to determine the equations of motion for discrete systems having a finite number of discrete variables q_i where $1 \leq i \leq n$. There are important classes of systems where it is more convenient to treat the system as being continuous. For example, the interatomic spacing in solids is a few $10^{-10}m$ which is negligible compared with the size of typical macroscopic, three-dimensional solid objects. As a consequence, for wavelengths much greater than the atomic spacing in solids, it is useful to treat macroscopic crystalline lattice systems as continuous three-dimensional uniform solids, rather than as three-dimensional discrete lattice chains. Fluid and gas dynamics are other examples of continuous mechanical systems. Another important class of continuous systems involves the theory of fields, such as electromagnetic fields. Lagrangian and Hamiltonian mechanics of the continua extend classical mechanics into the advanced topic of field theory. This chapter goes beyond the scope of a typical undergraduate classical mechanics course in order to provide a brief glimpse of how Lagrangian and Hamiltonian mechanics can underlie advanced and important aspects of the mechanics of the continua, including field theory.

16.2 The continuous uniform linear chain

The Lagrangian for the discrete lattice chain, for longitudinal modes, is given by equation 14.76 to be

$$L = \frac{1}{2} \sum_{j=1}^{n+1} \left(m \dot{q}_j^2 - \kappa \left(q_{j-1} - q_j \right)^2 \right) \tag{16.1}$$

where the n masses are attached in series to $n+1$ identical springs of length d and spring constant κ. Assume that the spring has a uniform cross-section area A and length d. Then each spring volume element $\Delta\tau = Ad$ has a mass m, that is, the volume mass density $\rho = \frac{m}{\Delta\tau}$ or $m = \rho\Delta\tau$. Chapter 16.5.3 will show that the spring constant $\kappa = \frac{EA}{d}$ where E is Young's modulus, A is the cross sectional area of the chain element, and d is the length of the element. Then the spring constant can be written as $\kappa = \frac{E\Delta\tau}{d^2}$. Therefore equation 16.1 can be expressed as a sum over volume elements $\Delta\tau = Ad$

$$L = \frac{1}{2} \sum_{j=1}^{n+1} \left(\rho \dot{q}_j^2 - E \left(\frac{q_{j-1} - q_j}{d} \right)^2 \right) \Delta\tau \tag{16.2}$$

In the limit that $n \to \infty$ and the spacing $d = dx \to 0$, then the summation in equation 16.2 can be written as a volume integral where $x = jd$ is the distance along the linear chain and the volume element $\boldsymbol{\Delta\tau \to 0}$. Then the Lagrangian can be written as the integral over the volume element $d\tau$ rather than a summation over $\Delta\tau$. That is,

$$L = \frac{1}{2} \int \left(\rho\dot{q}^2 - E\left(\frac{dq(x,t)}{dx}\right)^2 \right) d\tau \qquad (16.3)$$

The discrete-chain coordinate $q(t)$ is assumed to be a continuous function $q(x,t)$ for the uniform chain. Thus the integral form of the Lagrangian can be expressed as

$$L = \frac{1}{2} \int \left(\rho\dot{q}^2 - E\left(\frac{dq(x,t)}{dx}\right)^2 \right) d\tau = \int \mathfrak{L}d\tau \qquad (16.4)$$

where the function $\mathfrak{L}$ is called the **Lagrangian density** defined by

$$\mathfrak{L} \equiv \frac{1}{2} \left(\rho\dot{q}^2 - E\left(\frac{dq(x,t)}{dx}\right)^2 \right) \qquad (16.5)$$

The variable x in the Lagrangian density is not a generalized coordinate; it only serves the role of a continuous index played previously by the index j. For the discrete case, each value of j defined a different generalized coordinate q_i. Now for each value of x there is a continuous function $q(x,t)$ which is a function of both position and time.

Lagrange's equations of motion applied to the continuous Lagrangian in equation 16.4 gives

$$\rho\frac{d^2q}{dt^2} - E\frac{d^2q}{dx^2} = 0 \qquad (16.6)$$

This is the familiar wave equation in one dimension for a longitudinal wave on the continuous chain with a phase velocity

$$v_{phase} = \sqrt{\frac{E}{\rho}} \qquad (16.7)$$

The continuous linear chain also can exhibit transverse modes which have a Lagrangian density were the Young's modulus E is replaced by the tension τ in the chain, and ρ is replaced by the linear mass density μ of the chain, leading to a phase velocity for a transverse wave $v_{phase} = \sqrt{\frac{\tau}{\mu}}$.

16.3   The Lagrangian density formulation for continuous systems

16.3.1   One spatial dimension

In general the Lagrangian density can be a function of q, ∇q, $\frac{dq}{dt}$, x, y, z, and t. It is of interest that Hamilton's principle leads to a set of partial differential equations of motion, based on the Lagrangian density, that are analogous to the Lagrange equations of motion for discrete systems. When deriving the Lagrangian equations of motion in terms of the Lagrangian density using Hamilton's principle, the notation is simplified if the system is limited to one spatial coordinate x. In addition, it is convenient to use the compact notation where the spatial derivative is written $q' \equiv \frac{dq}{dx}$ and the time derivative is $\dot{q} \equiv \frac{dq}{dt}$, and the one-dimensional Lagrangian density is assumed to be a function $\mathfrak{L}(q, q', \dot{q}, x, t)$. The appearance of the derivative $q' \equiv \frac{dq}{dx}$ as an argument of the Lagrange density is a consequence of the continuous dependence of q on x. In principle, higher-order derivatives could occur but they do not arise in most problems of physical interest.

Assuming that the one spatial dimension is x, then Hamilton's principle of least action can be expressed in terms of the Lagrangian density as

$$\delta S = \delta \int_{t_1}^{t_2} L(q, \dot{q}, t)dt = \delta \int_{t_1}^{t_2} \int_{x_1}^{x_2} \mathfrak{L}(q, q', \dot{q}, x, t)dxdt \qquad (16.8)$$

Following the same approach used in chapter 5.2, it is assumed that the stationary path for the action integral is described by the function $q(x,t)$. Define a neighboring function using a parametric representation $q(x,t;\epsilon)$ such that when $\epsilon = 0$, the extremum function $q = q(x,t)$ yields the stationary action integral S.

Assume that an infinitessimal fraction ϵ of a neighboring function $\eta(x,t)$ is added to the extremum path $q(x,t)$. That is, assume

$$q(x,t;\epsilon) = q(x,t) + \epsilon\eta(x,t) \tag{16.9}$$

$$q'(x,t;\epsilon) \equiv \frac{dq(x,t;\epsilon)}{dx} = \frac{dq(x,t)}{dx} + \epsilon\frac{d\eta(x,t)}{dx} = q'(x,t) + \epsilon\eta'(x,t) \tag{16.10}$$

$$\dot{q}(x,t;\epsilon) \equiv \frac{dq(x,t;\epsilon)}{dt} = \frac{dq(x,t)}{dt} + \epsilon\frac{d\eta(x,t)}{dt} = \dot{q}(x,t) + \epsilon\dot{\eta}(x,t) \tag{16.11}$$

where it is assumed that both the extremum function $q(x,t)$ and the auxiliary function $\eta(x,t)$ are well behaved functions of x and t, with continuous first derivatives, and that $\eta(x,t) = 0$ at (x_1,t_1) and (x_2,t_2) because, for all possible paths, the function $q(x,t;\epsilon)$ must be identical with $q(x,t)$ at the end points of the path, i.e. $\eta(x_1,t_1) = \eta(x_2,t_2) = 0$.

A parametric family of curves $S(\epsilon)$, as a function of the admixture coefficient ϵ, is described by the function

$$S(\epsilon) = \int_{t_1}^{t_2}\int_{x_1}^{x_2}\mathfrak{L}(q(x,t;\epsilon),q'(x,t;\epsilon),\dot{q}(x,t;\epsilon),x,t)dxdt \tag{16.12}$$

Then Hamilton's principle requires that the action integral be a stationary function value for $\epsilon = 0$, that is, $S(\epsilon)$ is independent of ϵ which is satisfied if

$$\frac{\partial S(\epsilon)}{\partial\epsilon} = \int_{t_1}^{t_2}\int_{x_1}^{x_2}\left(\frac{\partial\mathfrak{L}}{\partial q}\frac{\partial q}{\partial\epsilon} + \frac{\partial\mathfrak{L}}{\partial\dot{q}}\frac{\partial\dot{q}}{\partial\epsilon} + \frac{\partial\mathfrak{L}}{\partial q'}\frac{\partial q'}{\partial\epsilon}\right)dxdt = 0 \tag{16.13}$$

Equations $16.9, 16.10,$ and 16.11 give the partial differentials

$$\frac{\partial q}{\partial\epsilon} = \eta(x,t) \tag{16.14}$$

$$\frac{\partial q'}{\partial\epsilon} = \eta'(x,t) \tag{16.15}$$

$$\frac{\partial\dot{q}}{\partial\epsilon} = \dot{\eta}(x,t) \tag{16.16}$$

Integration by parts in both the x and t terms in equation 16.13, plus using the fact that $\eta(x_1,t_1) = \eta(x_2,t_2) = 0$ at both end points, yields

$$\int_{t_1}^{t_2}\frac{\partial\mathfrak{L}}{\partial\dot{q}}\frac{\partial\dot{q}}{\partial\epsilon}dt = -\int_{t_1}^{t_2}\frac{\partial}{\partial t}\left(\frac{\partial\mathfrak{L}}{\partial\dot{q}}\right)\frac{\partial q}{\partial\epsilon}dt \tag{16.17}$$

$$\int_{x_1}^{x_2}\frac{\partial\mathfrak{L}}{\partial q'}\frac{\partial q'}{\partial\epsilon}dx = -\int_{x_1}^{x_2}\frac{\partial}{\partial x}\left(\frac{\partial\mathfrak{L}}{\partial q'}\right)\frac{\partial q}{\partial\epsilon}dx \tag{16.18}$$

Therefore Hamilton's principle, equation 16.13 becomes

$$\frac{\partial S(\epsilon)}{\partial\epsilon} = \int_{t_1}^{t_2}\int_{x_1}^{x_2}\left[\frac{\partial\mathfrak{L}}{\partial q} - \frac{\partial}{\partial t}\left(\frac{\partial\mathfrak{L}}{\partial\dot{q}}\right) - \frac{\partial}{\partial x}\left(\frac{\partial\mathfrak{L}}{\partial q'}\right)\right]\eta(x,t)dxdt = 0 \tag{16.19}$$

Since the auxiliary function $\eta(x,t)$ is arbitrary, then the integrand term in the square brackets of equation 16.19 must equal zero. That is,

$$\frac{\partial}{\partial t}\left(\frac{\partial\mathfrak{L}}{\partial\dot{q}}\right) + \frac{\partial}{\partial x}\left(\frac{\partial\mathfrak{L}}{\partial q'}\right) - \frac{\partial\mathfrak{L}}{\partial q} = 0 \tag{16.20}$$

Equation 16.20 gives the equations of motion in terms of the Lagrangian density that has been derived based on Hamilton's principle.

16.3.2   Three spatial dimensions

Equation 16.4 expresses the Lagrangian as an integral of the Lagrangian density over a single continuous index $q(x,t)$ where the Lagrangian density is a function $\mathfrak{L}(q,\frac{dq}{dt},\frac{dq}{dx},x,t)$. The derivation of the Lagrangian equations of motion in terms of the Lagrangian density for three spatial dimensions involves the straightforward addition of the y, and z coordinates. That is, in three dimensions the vector displacement is expressed by the vector $\mathbf{q}(x, y, z, t)$ and the Lagrangian density is related to the Lagrangian by integration over three dimensions. That is, they are related by the equation

$$L = \int \mathfrak{L}(\mathbf{q}, \frac{d\mathbf{q}}{dt}, \nabla \cdot \mathbf{q}, x, y, z, t) d\tau \tag{16.21}$$

where, in cartesian coordinates, the volume element $d\tau = dx dy dz$. The Lagrangian density is a function $\mathfrak{L}(\mathbf{q}, \frac{d\mathbf{q}}{dt}, \nabla \cdot \mathbf{q}, x, y, z, t)$ where the one field quantity $q(x, t)$ has been extended to a spatial vector $\mathbf{q}(x, y, z, t)$ and the spatial derivatives q' have been transformed into $\nabla \cdot \mathbf{q}$. Applying the method used for the one-dimensional spatial system, to the three-dimensional system, leads to the following set of equations of motion

$$\frac{\partial}{\partial t}\left(\frac{\partial \mathfrak{L}}{\partial \frac{\partial \mathbf{q}}{\partial t}}\right) + \frac{\partial}{\partial x}\left(\frac{\partial \mathfrak{L}}{\partial \frac{\partial \mathbf{q}}{\partial x}}\right) + \frac{\partial}{\partial y}\left(\frac{\partial \mathfrak{L}}{\partial \frac{\partial \mathbf{q}}{\partial y}}\right) + \frac{\partial}{\partial z}\left(\frac{\partial \mathfrak{L}}{\partial \frac{\partial \mathbf{q}}{\partial z}}\right) - \frac{\partial \mathfrak{L}}{\partial \mathbf{q}} = 0 \tag{16.22}$$

where the x, y, z spatial derivatives have been written explicitly for clarity.

Note that the equations of motion, equation 16.22, treat the spatial and time coordinates symmetrically. This symmetry between space and time is unchanged by multiplying the spatial and time coordinate by arbitrary numerical factors. This suggests the possibility of introducing a four-dimensional coordinate system

$$\phi_\mu \equiv \{x, y, z, \alpha t\}$$

where the parameter α is freely chosen. Using this 4-dimensional formalism allows equation 16.22 to be written more compactly as

$$\sum_\mu^4 \frac{\partial}{\partial \phi_\mu}\left(\frac{\partial \mathfrak{L}}{\partial \frac{\partial \mathbf{q}}{\partial \phi_\mu}}\right) - \frac{\partial \mathfrak{L}}{\partial \mathbf{q}} = 0 \tag{16.23}$$

As discussed in chapter 17, relativistic mechanics treats time and space symmetrically, that is, a four-dimensional vector $\mathbf{q}(x, y, z, t)$ can be used that treats time and the three spatial dimensions symmetrically and equally. This four-dimensional space-time formulation allows the first four terms in equation 16.22 to be condensed into a single term which illustrates the symmetry underlying equation 16.23. If the Lagrangian density is Lorentz invariant, and if $\alpha = ic$, then equation 16.23 is covariant. Thus the Lagrangian density formulation is ideally suited to the development of relativistically covariant descriptions of fields.

16.4 The Hamiltonian density formulation for continuous systems

Chapter 16.3 illustrates, in general terms, how field theory can be expressed in a Lagrangian formulation via use of the Lagrange density. It is equally possible to obtain a Hamiltonian formulation for continuous systems analogous to that obtained for discrete systems. As summarized in chapter 8, the Hamiltonian and Hamilton's canonical equations of motion are related directly to the Lagrangian by use of a Legendre transformation. The Hamiltonian is defined as being

$$H \equiv \sum_i \left(\dot{q}_i \frac{\partial L}{\partial \dot{q}_i}\right) - L \tag{16.24}$$

The generalized momentum is defined to be

$$p_i \equiv \frac{\partial L}{\partial \dot{q}_i} \tag{16.25}$$

Equation (16.25) allows the Hamiltonian (16.24) to be written in terms of the conjugate momenta as

$$H(q_i, p_i, t) = \sum_i p_i \dot{q}_i - L(q_i, \dot{q}_i, t) = \sum_i (p_i \dot{q}_i - L_i(q_i, \dot{q}_i, t)) \tag{16.26}$$

where the Lagrangian has been partitioned into the terms for each of the individual coordinates, that is, $L(q_i, \dot{q}_i, t) = \sum_i L_i(q_i, \dot{q}_i, t)$.

In the limit that the coordinates q, p are continuous, then the summation in equation 16.26 can be transformed into a volume integral over the Lagrangian density $\mathcal{L}$. In addition, a momentum density can be represented by the vector field $\boldsymbol{\pi}$ where

$$\boldsymbol{\pi} \equiv \frac{\partial \mathcal{L}}{\partial \dot{\mathbf{q}}} \tag{16.27}$$

Then the obvious definition of the Hamiltonian density $\mathfrak{H}$ is

$$H = \int \mathfrak{H} d\tau = \int (\boldsymbol{\pi} \cdot \dot{\mathbf{q}} - \mathcal{L}) \, d\tau \tag{16.28}$$

where the Hamiltonian density is defined to be

$$\mathfrak{H} = \boldsymbol{\pi} \cdot \dot{\mathbf{q}} - \mathcal{L} \tag{16.29}$$

Unfortunately the Hamiltonian density formulation does not treat space and time symmetrically making it more difficult to develop relativistically covariant descriptions of fields. Hamilton's principle can be used to derive the Hamilton equations of motion in terms of the Hamiltonian density analogous to the approach used to derive the Lagrangian density equations of motion. As described in Classical Mechanics 2nd edition by Goldstein, the resultant Hamilton equations of motion for one dimension are

$$\frac{\partial \mathfrak{H}}{\partial \pi} = \dot{q} \tag{16.30}$$

$$\frac{\partial \mathfrak{H}}{\partial q} - \frac{d}{dx} \frac{\partial \mathfrak{H}}{\partial q'} = -\dot{\pi} \tag{16.31}$$

$$\frac{\partial \mathfrak{H}}{\partial t} = -\frac{\partial \mathcal{L}}{\partial t} \tag{16.32}$$

Note that equation 16.31 differs from that for discontinuous systems.

16.5 Linear elastic solids

Elasticity is a property of matter where the atomic forces in matter act to restore the shape of a solid when distorted due to the application of external forces. A perfectly elastic material returns to its original shape if the external force producing the deformation is removed. Materials are elastic when the external forces do not exceed the elastic limit. Above the elastic limit, solids can exhibit plastic flow and concomitant heat dissipation. Such non-elastic behavior in solids occurs when they are subject to strong external forces.

The discussion of linear systems, in chapters 3 and 14, focussed on one dimensional systems, such as the linear chain, where the transverse rigidity of the chain was ignored. An extension of the one-dimensional linear chain to two-dimensional membranes, such as a drum skin, is straightforward if the membrane is thin enough so that the rigidity of the membrane can be ignored. Elasticity for three-dimensional solids requires accounting for the strong elastic forces exerted against any change in shape in addition to elastic forces opposing change in volume. The stiffness of solids to changes in shape, or volume, is best represented using the concepts of stress and strain.

Forces in matter can be divided into two classes; (1) body forces, such as gravity, which act on each volume element, and (2) surface forces which are the forces that act on both sides of any infinitessimal surface element inside the solid. Surface forces can have components along the normal to the infinitessimal surface, as well as shear components in the plane of the surface element. Typically solids are elastic to both normal and shear components of the surface forces whereas shear forces in liquids and gases lead to fluid flow plus viscous forces due to energy dissipation. As described below, the forces acting on an infinitessimal surface element are best expressed in terms of the stress tensor, while the relative distortion of the shape, or volume, of the body are best expressed in terms of the strain tensor. The moduli of elasticity relate the ratio of the corresponding stress and strain tensors. The moduli of elasticity are constant in linear elastic solids and thus the stress is proportional to the strain providing that the strains do not exceed the elastic limit.

16.5.1   Stress tensor

Consider an infinitessimal surface area dA of an arbitrary closed volume element dV inside the medium. The surface area element is defined as a vector $d\mathbf{A} = \hat{\mathbf{n}}dA$ where $\hat{\mathbf{n}}$ is the outward normal to the closed surface that encloses the volume element. Assume that $d\mathbf{F}$ is the force element exerted by the outside on the material inside the volume element. The stress tensor $\mathbf{T}$ is defined as the ratio of $d\mathbf{F}$ and $d\mathbf{A}$ where the force vector $d\mathbf{F}$ is given by the inner product of the stress tensor $\mathbf{T}$ and the surface element vector $d\mathbf{A}$. That is,

$$d\mathbf{F} = \mathbf{T}{\cdot}d\mathbf{A} \tag{16.33}$$

Since both $d\mathbf{F}$ and $d\mathbf{A}$ are vectors, then equation 16.33 implies that the stress tensor must be a second-rank tensor as described in appendix E, that is, the stress tensor is analogous to the rotation matrix or the inertia tensor. Note that if $d\mathbf{F}$ and $\hat{\mathbf{n}}dA$ are colinear, then the stress tensor $\mathbf{T}$ reduces to the conventional pressure P. The general stress tensor equals the momentum flux density and has the dimensions of pressure.

16.5.2   Strain tensor

Forces applied to a solid body can lead to translational, or rotational acceleration, in addition to changing the shape or volume of the body. Elastic forces do not act when an overall displacement $\boldsymbol{\xi}$ of an infinitesimal volume occurs, such as is involved in translational or rotational motion. Elastic forces act to oppose position-dependent differences in the displacement vector $\boldsymbol{\xi}$, that is, the strain depends on the tensor product $\boldsymbol{\nabla} \otimes \boldsymbol{\xi}$. For an elastic medium, the strain depends only on the applied stress and not on the prior loading history.

Consider that the matter at the location $\mathbf{r}$ is subject to an elastic displacement $\boldsymbol{\xi}$, and similarly at a displaced location $\mathbf{r}' = \mathbf{r} + \sum_i \frac{\partial \boldsymbol{\xi}}{\partial x_i}dx_i$ where x_i are cartesian coordinates. The net relative displacement between $\mathbf{r}$ and $\mathbf{r}'$ is given by

$$d\xi^2 = \sum_i (dx_i + d\xi_i)^2 - \sum_i (dx_i)^2 = \sum_{ik}\left[2\left(\frac{d\xi_i}{dx_k} + \frac{d\xi_k}{dx_i}\right) + \frac{d\xi_m}{dx_i}\frac{d\xi_m}{dx_k}\right]dx_idx_k \tag{16.34}$$

Ignoring the second order term $\frac{d\xi_m}{dx_i}\frac{d\xi_m}{dx_k}$ equation gives that the ith component of $d\xi_i$ is

$$d\xi_i = \sum_k \frac{1}{2}\left(\frac{d\xi_i}{dx_k} + \frac{d\xi_k}{dx_i}\right)dx_idx_k \tag{16.35}$$

Define the elements of the strain tensor to be given by

$$\sigma_{ik} = \frac{1}{2}\left(\frac{d\xi_i}{dx_k} + \frac{d\xi_k}{dx_i}\right) \tag{16.36}$$

then

$$d\xi_i = \sum_k \sigma_{ik}dx_idx_k \tag{16.37}$$

Thus the strain tensor $\boldsymbol{\sigma}$ is a rank-2 tensor defined as the ratio of the strain vector $\boldsymbol{\xi}$ and the infinitessimal area vector $d\mathbf{A}$.

$$d\boldsymbol{\xi} = \boldsymbol{\sigma}{\cdot}d\mathbf{A} \tag{16.38}$$

where the component form of the rank -2 strain tensor is

$$\boldsymbol{\sigma} = \frac{1}{2}\begin{vmatrix} \frac{d\xi_1}{dx_1} & \frac{d\xi_1}{dx_2} & \frac{d\xi_1}{dx_3} \\ \frac{d\xi_2}{dx_1} & \frac{d\xi_2}{dx_2} & \frac{d\xi_2}{dx_3} \\ \frac{d\xi_3}{dx_1} & \frac{d\xi_3}{dx_2} & \frac{d\xi_3}{dx_3} \end{vmatrix} \tag{16.39}$$

The potential-energy density for linear elastic forces is quadratic in the strain components. That is, it is of the form

$$U = \sum_{ijkl} \frac{1}{2}C_{ijkl}\sigma_{ij}\sigma_{kl} \tag{16.40}$$

where C_{ijkl} is a rank-4 tensor. No preferential directions remain for a homogeneous isotropic elastic body which allows for two contractions, thereby reducing the potential energy density to the inner product

$$U = \sum_{ik} \frac{1}{2}D_{ik}\left(\sigma_{ik}\right)^2 \tag{16.41}$$

16.5.3 Moduli of elasticity

The **modulus of elasticity** of a body is defined to be the slope of the stress-strain curve and thus, in principle, it is a complicated rank-4 tensor that characterizes the elastic properties of a material. Thus the general theory of elasticity is complicated because the elastic properties depend on the orientation of the microscopic composition of the elastic matter. The theory simplifies considerably for homogeneous, isotropic linear materials below the elastic limit, where the strain is proportional to the applied stress. That is, the modulus of elasticity then reduces by contractions to a constant scalar value that depends on the properties of the matter involved.

The potential energy density for homogeneous, isotropic, linear material, equation 16.41, can be separated into diagonal and off-diagonal components of the strain tensor. That is,

$$U = \frac{1}{2}\left[\lambda\sum_i (\sigma_{ii})^2 + 2\mu\sum_{ik}(\sigma_{ik})^2\right] \tag{16.42}$$

The diagonal first term is the dilation term which corresponds to changes in the volume with no changes in shape. The off-diagonal second term involves the shear terms that correspond to changes of the shape of the body that also changes the volume. The constants λ and μ are Lamé's moduli of elasticity which are positive. The various moduli of elasticity, corresponding to different distortions in the shape and volume of any solid body, can be derived from Lamé's moduli for the material.

The components of the elastic forces can be derived from the gradient of the elastic potential energy, equation 16.42 by use of Gauss' law plus vector differential calculus. The components of the elastic force, derived from the strain tensor $\boldsymbol{\sigma}$, can be associated with the corresponding components of the stress tensor **T**. Thus, for homogeneous isotropic linear materials, the components of the stress tensor are related to the strain tensor by the relation

$$T_{ij} = \lambda\delta_{ij}\sum_k \frac{\partial\xi_k}{\partial x_k} + \mu\left(\frac{d\xi_i}{dx_j} + \frac{d\xi_j}{dx_i}\right) = \lambda\delta_{ij}\sum_k \sigma_{kk} + 2\mu\sigma_{ij} \tag{16.43}$$

where it has been assumed that $\sigma_{ij} = \sigma_{ji}$. The two moduli of elasticity λ and μ are material-dependent constants. Equation 16.43 can be written in tensor notation as

$$\mathbf{T} = \lambda tr(\boldsymbol{\sigma})\mathbf{I} + 2\mu\boldsymbol{\sigma} \tag{16.44}$$

where $tr(\sigma)$ is the trace of the strain tensor and I is the identity matrix.

Equation 16.44 can be inverted to give the strain tensor components in terms of the stress tensor components.

$$\sigma_{ij} = \frac{1}{2\mu}\left[T_{ij} - \frac{\lambda}{(3\lambda + 2\mu)}\sum_k T_{kk}\delta_{ij}\right] \tag{16.45}$$

The various moduli of elasticity relate combinations of different stress and strain tensor components. The following five elastic moduli are used frequently to describe elasticity in homogeneous isotropic media, and all are related to Lamé's two moduli of elasticity.

1) *Young's modulus* E describes tensile elasticity which is axial stiffness of the length of a body to deformation along the axis of the applied tensile force.

$$E \equiv \frac{T_{11}}{\sigma_{11}} = \frac{\mu(3\lambda + 2\mu)}{(\lambda + \mu)} \tag{16.46}$$

2) *Bulk modulus* $B = \frac{\Delta V}{V}$ defines the relative dilation or compression of a bodies volume to pressure applied uniformly in all directions.

$$B = \lambda + \frac{2}{3}\mu \tag{16.47}$$

The bulk modulus is an extension of Young's modulus to three dimensions and typically is larger than E. The inverse of the bulk modulus is called the compressibility of the material.

3) *Shear modulus* G describes the shear stiffness of a body to volume-preserving shear deformations. The shear strain σ becomes a deformation angle given by the ratio of the displacement along the axis of the shear force and the perpendicular moment arm. The shear modulus G equals Lamé's constant μ. That is,

$$G = \mu \tag{16.48}$$

4) *Poisson's ratio* ν is the negative ratio of the transverse to axial strain. It is a measure of the volume conserving tendency of a body to contract in the directions perpendicular to the axis along which it is stretched. In terms of Lamé's constants, Poisson's ratio equals

$$\nu = \frac{\lambda}{2\,(\lambda + \mu)} \tag{16.49}$$

Note that for a stable, isotropic elastic material, Poisson's ratio is bounded between $-1.0 \leq \nu \leq 0.5$ to ensure that the B, μ and λ moduli have positive values. At the incompressible limit, $\nu = 0.5$, and the bulk modulus and Lame parameter λ are infinite, that is, the compressibility is zero. Typical solids have Poisson's ratios of $\nu \approx 0.05$ if hard and $\nu = 0.25$ if soft.

The stiffness of elastic solids in terms of the elastic moduli of solids can be complicated due to the geometry and composition of solid bodies. Often it is more convenient to express the stiffness in terms of the **spring constant** κ where

$$\kappa = \frac{dF}{dx} \tag{16.50}$$

The spring constant is inversely proportional to the length of the spring because the strain of the material is defined to be the *fractional* deformation, not the *absolute* deformation.

16.5.4   Equations of motion in a uniform elastic media

The divergence theorem (H.8) relates the volume integral of the divergence of $\mathbf{T}$ to the vector force density $\mathbf{F}$ acting on the closed surface.

$$\mathbf{F} = \oint \mathbf{T}{\cdot}d\mathbf{A} = \int \boldsymbol{\nabla} \cdot \mathbf{T} d\tau = \int \mathbf{f} d\tau \tag{16.51}$$

That is, the inner product of the del operator, $\boldsymbol{\nabla}$, and the rank-2 stress tensor $\mathbf{T}$, give the vector force density $\mathbf{f}$. This force acting on the enclosed mass $\oint \rho d\tau$, for the closed volume, leads to an acceleration $\frac{\partial^2 \boldsymbol{\xi}}{\partial t^2}$. Thus

$$\mathbf{F} = \oint \mathbf{T}{\cdot}d\mathbf{A} = \int \boldsymbol{\nabla} \cdot \mathbf{T} d\tau = \oint \rho \frac{\partial^2 \boldsymbol{\xi}}{\partial t^2} d\tau \tag{16.52}$$

Use equation 16.44 to relate the stress tensor $\mathbf{T}$ to the moduli of elasticity gives

$$\rho \frac{\partial^2 \boldsymbol{\xi}_i}{\partial t^2} = \sum_j \left[(\lambda + \mu) \frac{\partial^2 \boldsymbol{\xi}_j}{\partial x_i \partial x_j} + \mu \frac{\partial^2 \boldsymbol{\xi}_i}{\partial x_j^2} \right] \tag{16.53}$$

where $i = 1, 2, 3$. In general this equation is difficult to solve. However, for the simple case of a plane wave in the $i = 1$ direction, the problem reduces to the following three equations

$$\rho \frac{\partial^2 \boldsymbol{\xi}_1}{\partial t^2} = (\lambda + 2\mu) \frac{\partial^2 \boldsymbol{\xi}_1}{\partial x_1^2} \tag{16.54}$$

$$\rho \frac{\partial^2 \boldsymbol{\xi}_2}{\partial t^2} = \mu \frac{\partial^2 \boldsymbol{\xi}_2}{\partial x_1^2} \tag{16.55}$$

$$\rho \frac{\partial^2 \boldsymbol{\xi}_3}{\partial t^2} = \mu \frac{\partial^2 \boldsymbol{\xi}_3}{\partial x_1^2} \tag{16.56}$$

Equation 16.54 corresponds to a longitudinal wave travelling with velocity $v = \sqrt{\frac{(\lambda + 2\mu)}{\rho}}$. Equations 16.55, 16.56 correspond to two perpendicular transverse waves travelling with velocity $v = \sqrt{\frac{\mu}{\rho}}$. This illustrates the important fact that longitudinal waves travel faster than transverse waves in an elastic solid. Seismic waves in the Earth, generated by earthquakes, exhibit this property. Note that shearing stresses do not exist in ideal liquids and gases since they cannot maintain shear forces and thus $\mu = 0$.

16.6 Electromagnetic field theory

16.6.1 Maxwell stress tensor

Analytical formulations for continuous systems, developed for describing elasticity, are generally applicable when applied to other fields, such as the electromagnetic field. The use of the Maxwell's stress tensor $\mathbf{T}$, to describe momentum in the electromagnetic field, is an important example of the application of continuum mechanics in field theory.

The Lorentz force can be written as

$$\mathbf{F} = \int \rho \left(\mathbf{E} + \mathbf{v} \times \mathbf{B} \right) d\tau = \int \left(\rho \mathbf{E} + \mathbf{J} \times \mathbf{B} \right) d\tau = \int \mathbf{f} d\tau \tag{16.57}$$

where the force density $\mathbf{f}$ is defined to be

$$\mathbf{f} = \left(\rho \mathbf{E} + \mathbf{J} \times \mathbf{B} \right) \tag{16.58}$$

Maxwell's equations

$$\rho = \epsilon_0 \boldsymbol{\nabla} \cdot \mathbf{E} \qquad\qquad \mathbf{J} = \frac{1}{\mu_0} \boldsymbol{\nabla} \times \mathbf{B} - \epsilon_0 \frac{\partial \mathbf{E}}{\partial t} \tag{16.59}$$

can be used to eliminate the charge and current densities in equation 16.57

$$\mathbf{f} = \epsilon_0 \left(\boldsymbol{\nabla} \cdot \mathbf{E} \right) \mathbf{E} + \left(\frac{1}{\mu_0} \boldsymbol{\nabla} \times \mathbf{B} - \epsilon_0 \frac{\partial \mathbf{E}}{\partial t} \right) \times \mathbf{B} \tag{16.60}$$

Vector calculus gives that

$$\frac{\partial}{\partial t} \left(\mathbf{E} \times \mathbf{B} \right) = \frac{\partial \mathbf{E}}{\partial t} \times \mathbf{B} + \mathbf{E} \times \frac{\partial \mathbf{B}}{\partial t} \tag{16.61}$$

while Faraday's law gives

$$\frac{\partial \mathbf{B}}{\partial t} = -\boldsymbol{\nabla} \times \mathbf{E} \tag{16.62}$$

Equation 16.62 allows equation 16.61 to be rewritten as

$$\frac{\partial \mathbf{E}}{\partial t} \times \mathbf{B} = +\frac{\partial}{\partial t} \left(\mathbf{E} \times \mathbf{B} \right) - \mathbf{E} \times \frac{\partial \mathbf{B}}{\partial t} = +\frac{\partial}{\partial t} \left(\mathbf{E} \times \mathbf{B} \right) + \mathbf{E} \times \left(\boldsymbol{\nabla} \times \mathbf{E} \right) \tag{16.63}$$

Equation 16.63 can be inserted into equation 16.60. In addition, a term $\frac{1}{\mu_0} \left(\boldsymbol{\nabla} \cdot \mathbf{B} \right) \mathbf{B}$ can be added since $\boldsymbol{\nabla} \cdot \mathbf{B} = 0$ which allows equation 16.60 to be written in the symmetric form

$$\mathbf{f} = \epsilon_0 \left(\boldsymbol{\nabla} \cdot \mathbf{E} \right) \mathbf{E} + \frac{1}{\mu_0} \left(\boldsymbol{\nabla} \cdot \mathbf{B} \right) \mathbf{B} + \frac{1}{\mu_0} \left(\boldsymbol{\nabla} \times \mathbf{B} \right) \times \mathbf{B} - \epsilon_0 \frac{\partial \mathbf{E}}{\partial t} \times \mathbf{B} \tag{16.64}$$

$$= \epsilon_0 \left(\boldsymbol{\nabla} \cdot \mathbf{E} \right) \mathbf{E} + \frac{1}{\mu_0} \left(\boldsymbol{\nabla} \cdot \mathbf{B} \right) \mathbf{B} + \frac{1}{\mu_0} \left(\boldsymbol{\nabla} \times \mathbf{B} \right) \times \mathbf{B} - \epsilon_0 \frac{\partial}{\partial t} \left(\mathbf{E} \times \mathbf{B} \right) - \epsilon_0 \mathbf{E} \times \left(\boldsymbol{\nabla} \times \mathbf{E} \right) \tag{16.65}$$

Using the vector identity

$$\boldsymbol{\nabla} \left(\mathbf{A} \cdot \mathbf{B} \right) = \mathbf{A} \times \left(\boldsymbol{\nabla} \times \mathbf{B} \right) + \mathbf{B} \times \left(\boldsymbol{\nabla} \times \mathbf{A} \right) + \left(\mathbf{A} \cdot \boldsymbol{\nabla} \right) \mathbf{B} + \left(\mathbf{B} \cdot \boldsymbol{\nabla} \right) \mathbf{A} \tag{16.66}$$

Let $\mathbf{A} = \mathbf{B} = \mathbf{E}$, then

$$\boldsymbol{\nabla} \left(E^2 \right) = 2 \mathbf{E} \times \left(\boldsymbol{\nabla} \times \mathbf{E} \right) + 2 \left(\mathbf{E} \cdot \boldsymbol{\nabla} \right) \mathbf{E} \tag{16.67}$$

That is

$$\mathbf{E} \times \left(\boldsymbol{\nabla} \times \mathbf{E} \right) = \frac{1}{2} \boldsymbol{\nabla} \left(E^2 \right) - \left(\mathbf{E} \cdot \boldsymbol{\nabla} \right) \mathbf{E} \tag{16.68}$$

Similarly

$$\mathbf{B} \times \left(\boldsymbol{\nabla} \times \mathbf{B} \right) = \frac{1}{2} \boldsymbol{\nabla} \left(B^2 \right) - \left(\mathbf{B} \cdot \boldsymbol{\nabla} \right) \mathbf{B} \tag{16.69}$$

Inserting equations 16.68 and 16.69 into equation 16.65 gives

$$\mathbf{f} = \epsilon_0 \left[\left(\boldsymbol{\nabla} \cdot \mathbf{E} \right) \mathbf{E} + \left(\mathbf{E} \cdot \boldsymbol{\nabla} \right) \mathbf{E} - \frac{1}{2} \boldsymbol{\nabla} E^2 \right] + \frac{1}{\mu_0} \left[\left(\boldsymbol{\nabla} \cdot \mathbf{B} \right) \mathbf{B} + \left(\mathbf{B} \cdot \boldsymbol{\nabla} \right) \mathbf{B} - \frac{1}{2} \boldsymbol{\nabla} B^2 \right] - \epsilon_0 \frac{\partial}{\partial t} \left(\mathbf{E} \times \mathbf{B} \right) \tag{16.70}$$

This complicated formula can be simplified by defining the rank-2 **Maxwell stress tensor T** which has components

$$T_{ij} \equiv \epsilon_0 \left(E_i E_j - \frac{1}{2}\delta_{ij} E^2 \right) + \frac{1}{\mu_0} \left(B_i B_j - \frac{1}{2}\delta_{ij} B^2 \right) \tag{16.71}$$

The inner product of the del operator and the Maxwell stress tensor is a vector with j components of

$$(\boldsymbol{\nabla} \cdot \mathbf{T})_j = \epsilon_0 \left[(\boldsymbol{\nabla} \cdot \mathbf{E}) E_j + (\mathbf{E} \cdot \boldsymbol{\nabla}) E_j - \frac{1}{2}\nabla_j^2 E^2 \right] + \frac{1}{\mu_0} \left[(\boldsymbol{\nabla} \cdot \mathbf{B}) B_j + (\mathbf{B} \cdot \boldsymbol{\nabla}) B_j - \frac{1}{2}\nabla_j^2 B^2 \right] \tag{16.72}$$

The above definition of the Maxwell stress tensor, plus the Poynting vector $\mathbf{S} = \frac{1}{\mu_0}(\mathbf{E} \times \mathbf{B})$, allows the force density equation 16.58 to be written in the form

$$\mathbf{f} = \boldsymbol{\nabla} \cdot \mathbf{T} - \epsilon_0 \mu_0 \frac{\partial \mathbf{S}}{\partial t} \tag{16.73}$$

The divergence theorem allows the total force, acting of the volume τ, to be written in the form

$$\mathbf{F} = \int \left(\boldsymbol{\nabla} \cdot \mathbf{T} - \epsilon_0 \mu_0 \frac{\partial \mathbf{S}}{\partial t} \right) d\tau \tag{16.74}$$

$$= \oint \mathbf{T} \cdot d\mathbf{a} - \epsilon_0 \mu_0 \frac{d}{dt} \int \mathbf{S} d\tau \tag{16.75}$$

Note that, if the Poynting vector is time independent, then the second term in equation 16.75 is zero and the Maxwell stress tensor **T** is the force per unit area, (stress) acting on the surface. The fact that **T** is a rank-2 tensor is apparent since the stress represents the ratio of the force-density vector $d\mathbf{f}$ and the infinitessimal area vector $d\mathbf{a}$, which do not necessarily point in the same directions.

16.6.2   Momentum in the electromagnetic field

Chapter 7.2 showed that the electromagnetic field carries a linear momentum $q\mathbf{A}$ where q is the charge on a body and $\mathbf{A}$ is the electromagnetic vector potential. It is useful to use the Maxwell stress tensor to express the momentum density directly in terms of the electric and magnetic fields.

Newton's law of motion can be used to write equation equation 16.75 as

$$\mathbf{F} = \frac{d\mathbf{p}_{mech}}{dt} = \oint \mathbf{T} \cdot d\mathbf{a} - \epsilon_0 \mu_0 \frac{d}{dt} \int \mathbf{S} d\tau \tag{16.76}$$

where **p** is the total mechanical linear momentum of the volume τ. Equation 16.76 implies that the electromagnetic field carries a linear momentum

$$\mathbf{p}_{field} = \epsilon_0 \mu_0 \int \mathbf{S} d\tau \tag{16.77}$$

The $\oint \mathbf{T} \cdot d\mathbf{a}$ term in equation 16.76 is the momentum per unit time flowing into the closed surface.

In field theory it can be useful to describe the behavior in terms of the momentum flux density $\boldsymbol{\pi}$. Thus the momentum flux density $\boldsymbol{\pi}_{field}$ in the electromagnetic field is

$$\boldsymbol{\pi}_{field} = \epsilon_0 \mu_0 \mathbf{S} \tag{16.78}$$

Then equation 16.76 implies that the total momentum flux density $\boldsymbol{\pi} = \boldsymbol{\pi}_{mech} + \boldsymbol{\pi}_{field}$ is related to Maxwell's stress tensor by

$$\frac{\partial}{\partial t} (\boldsymbol{\pi}_{mech} + \boldsymbol{\pi}_{field}) = \boldsymbol{\nabla} \cdot \mathbf{T} \tag{16.79}$$

That is, like the elasticity stress tensor, the divergence of Maxwell's stress tensor **T** equals the rate of change of the total momentum density, that is, $-\mathbf{T}$ is the momentum flux density.

This discussion of the Maxwell stress tensor and its relation to momentum in the electromagnetic field illustrates the role that analytical formulations of classical mechanics can play in field theory.

16.7 Ideal fluid dynamics

The distinction between a solid and a fluid is that a fluid flows under shear stress whereas the elasticity of solids oppose distortion and flow. Shear stress in a fluid is opposed by dissipative viscous forces, which depend on velocity, as opposed to elastic solids where the shear stress is opposed by the elastic forces which depend on the displacement. An ideal fluid is one where the viscous forces are negligible, and thus the shear stress Lamé parameter $\mu = 0$.

16.7.1 Continuity equation

Fluid dynamics requires a different philosophical approach than that used to describe the motion of an ensemble of known solid bodies. The prior discussions of classical mechanics used, as variables, the coordinates of each member of an ensemble of particles with known masses. This approach is not viable for fluids which involve an enormous number of individual atoms as the fundamental bodies of the fluid. The best philosophical approach for describing fluid dynamics is to employ continuum mechanics using definite fixed volume elements $d\tau$ and describe the fluid in terms of macroscopic variables of the fluid such as mass density ρ, pressure P, and fluid velocity $\mathbf{v}$.

Conservation of fluid mass requires that the rate of change of mass in a fixed volume must equal the net inflow of mass.

$$\frac{d}{dt}\int_\tau \rho d\tau + \oint \rho \mathbf{v}\cdot d\mathbf{a} = \mathbf{0} \tag{16.80}$$

Using the divergence theorem ($H2$) allows this to be written as

$$\int_\tau \left(\frac{\partial \rho}{\partial t} + \boldsymbol{\nabla}\cdot(\rho\mathbf{v})\right)d\tau = 0 \tag{16.81}$$

Mass conservation must hold for any arbitrary volume, therefore the *continuity equation* can be written in the differential form

$$\frac{\partial \rho}{\partial t} + \boldsymbol{\nabla}\cdot(\rho\mathbf{v}) = 0 \tag{16.82}$$

16.7.2 Euler's hydrodynamic equation

The fluid surrounding a volume τ exerts a net force $\mathbf{F}$ that equals the surface integral of the pressure $\mathbf{P}$. This force can be transformed to a volume integral of $\boldsymbol{\nabla}P$. The net force then will lead to an acceleration of the volume element. That is

$$\mathbf{F} = -\oint P d\mathbf{a} = -\int \boldsymbol{\nabla}P d\tau = \int \rho \frac{d\mathbf{v}}{dt}d\tau \tag{16.83}$$

Thus the force density $\mathbf{f}$ is given by

$$\mathbf{f} = -\boldsymbol{\nabla}\mathbf{P} = \rho\frac{d\mathbf{v}}{dt} \tag{16.84}$$

Note that the acceleration $\frac{d\mathbf{v}}{dt}$ in equation 16.83 refers to the rate of change of velocity for *individual atoms in the fluid*, not the rate of change of fluid velocity at a *fixed point in space*. These two accelerations are related by noting that, during the time dt, the change in velocity $d\mathbf{v}$ of a given fluid particle is composed of two parts, namely (1) the change during dt in the velocity at a fixed point in space, and (2) the difference between the velocities at that same instant in time at two points displaced a distance $d\mathbf{r}$ apart, where $d\mathbf{r}$ is the distance moved by a given fluid particle during the time dt. The first part is given by $\frac{\partial \mathbf{v}}{\partial t}dt$ at a given point (x, y, z) in space. The second part equals

$$dx\frac{\partial \mathbf{v}}{\partial x} + dy\frac{\partial \mathbf{v}}{\partial y} + dz\frac{\partial \mathbf{v}}{\partial z} = (d\mathbf{r}\cdot\boldsymbol{\nabla})\mathbf{v} \tag{16.85}$$

Thus

$$d\mathbf{v} = \frac{\partial \mathbf{v}}{\partial t}dt + (d\mathbf{r}\cdot\boldsymbol{\nabla})\mathbf{v} \tag{16.86}$$

Divide both sides by dt gives that the acceleration of the atoms in the fluid equals

$$\frac{d\mathbf{v}}{dt} = \frac{\partial \mathbf{v}}{\partial t} + (\mathbf{v} \cdot \boldsymbol{\nabla}) \mathbf{v} \tag{16.87}$$

Substitute equation 16.87 into 16.84 gives

$$\frac{\partial \mathbf{v}}{\partial t} + (\mathbf{v} \cdot \boldsymbol{\nabla}) \mathbf{v} = -\frac{1}{\rho} \boldsymbol{\nabla} P \tag{16.88}$$

This is Euler's equation for hydrodynamics. The two terms on the left represent the acceleration in the individual fluid components while the right-hand side lists the force density producing the acceleration.

Additional forces can be added to the right-hand side. For example, the gravitational force density $\rho \mathbf{g}$ can be expressed in terms of the gravitational scalar potential V to be

$$\rho \mathbf{g} = -\rho \boldsymbol{\nabla} V \tag{16.89}$$

Inclusion of the gravitational field force density in Euler's equation gives

$$\frac{\partial \mathbf{v}}{\partial t} + (\mathbf{v} \cdot \boldsymbol{\nabla}) \mathbf{v} = -\frac{1}{\rho} \boldsymbol{\nabla} (P + \rho V) \tag{16.90}$$

16.7.3   Irrotational flow and Bernoulli's equation

Streamlined flow corresponds to irrotational flow, that is, $\boldsymbol{\nabla} \times \mathbf{v} = \mathbf{0}$. Since irrotational flow is curl free, the velocity streamlines can be represented by a scalar potential field ϕ. That is

$$\mathbf{v} = -\boldsymbol{\nabla} \phi \tag{16.91}$$

This scalar potential field ϕ can be used to derive the vector velocity field for irrotational flow.

Note that the $(\mathbf{v} \cdot \boldsymbol{\nabla}) \mathbf{v}$ term in Euler's equation (16.90) can be rewritten using the vector identity

$$(\mathbf{v} \cdot \boldsymbol{\nabla}) \mathbf{v} = \frac{1}{2} \boldsymbol{\nabla} (v^2) - \mathbf{v} \times \boldsymbol{\nabla} \times \mathbf{v} \tag{16.92}$$

Inserting equation 16.92 into Euler's equation 16.90 then gives.

$$\frac{\partial \mathbf{v}}{\partial t} = \mathbf{v} \times \boldsymbol{\nabla} \times \mathbf{v} - \frac{1}{\rho} \boldsymbol{\nabla} \left(\frac{1}{2} \rho v^2 + P + \rho V \right) \tag{16.93}$$

Potential flow corresponds to time independent irrotational flow, that is, both $\frac{\partial \mathbf{v}}{\partial t} = 0$ and $\boldsymbol{\nabla} \times \mathbf{v} = 0$. For potential flow equation 16.93 reduces to

$$\boldsymbol{\nabla} \left(\frac{1}{2} \rho v^2 + P + \rho V \right) = 0$$

which implies that

$$\left(\frac{1}{2} \rho v^2 + P + \rho V \right) = \text{constant} \tag{16.94}$$

This is the famous Bernoulli's equation that relates the interplay of the fluid velocity, pressure and gravitational energy. Bernoulli's equation plays important roles in both hydrodynamics and aerodynamics.

16.7.4   Gas flow

Fluid dynamics applied to gases is a straightforward extension of fluid dynamics that employs standard thermodynamical concepts. The following example illustrates the application of fluid mechanics for calculating the velocity of sound in a gas.

16.1 Example: Acoustic waves in a gas

Propagation of acoustic waves in a gas provides an example of using the three-dimensional Lagrangian density. Only longitudinal waves occur in a gas and the velocity is given by thermodynamics of the gas. Let the displacement of each gas molecule be designated by the general coordinate $\mathbf{q}$ *with corresponding velocity* $\dot{\mathbf{q}}$. *Let the gas density be* ρ, *then the kinetic energy density* (KED) *of an infinitessimal volume of gas* $\Delta\tau$ *is given by*

$$\Delta\left(KED\right) = \frac{1}{2}\rho_0\dot{\mathbf{q}}^2$$

The rapid contractions and expansions of the gas in an acoustic wave occur adiabatically such that the product PV^γ *is a constant, where* $\gamma = \frac{\text{specific heat at constant pressure}}{\text{specific heat at constant volume}}$. *Therefore the change in potential energy density* $\Delta(PED)$ *is given to second order by*

$$\Delta\left(PED\right) = \frac{1}{\tau_0}\int_{V_0}^{V_0+\Delta V} Pd\tau = \frac{P_0}{\tau_0}\Delta\tau + \frac{1}{2\tau_0}\left(\frac{\partial P}{\partial\tau}\right)_0(\Delta\tau)^2 = \frac{P_0}{\tau_0}\Delta\tau - \frac{1}{2\tau_0}\left(\gamma\frac{P_0}{\tau_0}\right)(\Delta\tau)^2$$

Since the volume and density are related by

$$\tau_o = \frac{M}{\rho_0}$$

then the fractional change in the density σ *is related to the density by*

$$\rho = \rho_0(1+\sigma)$$

This implies that the potential energy density (PED) *is given by*

$$\Delta\left(PED\right) = \left[P_0\sigma + \gamma\frac{P_0}{2}\sigma^2\right]$$

The mass flowing out of the volume V_0 *must equal the fractional change in density of the volume, that is*

$$\rho_0\int\mathbf{q}\cdot\mathbf{dS} = -\rho_0\int\sigma d\tau$$

The divergence theorem gives that

$$\int\mathbf{q}\cdot\mathbf{dS} = \int\nabla\cdot\mathbf{q}d\tau = -\int\sigma d\tau$$

Thus the density σ *is given by minus the divergence of* $\mathbf{q}$

$$\sigma = -\nabla\cdot\mathbf{q}$$

This allows the potential energy density to be written as

$$\Delta(PED) = -P_0\nabla\cdot\mathbf{q} + \frac{\gamma P_0}{2}(\nabla\cdot\mathbf{q})^2$$

Combining the kinetic energy density and the potential energy density gives the complete Lagrangian density for an acoustic wave in a gas to be

$$\mathcal{L} = \frac{1}{2}\rho_0\dot{\mathbf{q}}^2 + P_0\nabla\cdot\mathbf{q} - \frac{\gamma P_0}{2}(\nabla\cdot\mathbf{q})^2$$

Inserting this Lagrangian density in the corresponding equations of motion, equation 16.23, gives that

$$\nabla^2\mathbf{q} - \frac{\rho_0}{\gamma P_0}\frac{d^2\mathbf{q}}{dt^2} = 0$$

where P_0 *and* ρ_0 *are the ambient pressure and density of the gas. This is the wave equation where the phase velocity of sound is given by*

$$v_{phase} = \sqrt{\frac{\gamma P_0}{\rho_0}}$$

16.8   Viscous fluid dynamics

Viscous fluid dynamics is a branch of classical mechanics that plays a pivotal role in a wide range of aspects of life, such as blood flow in human anatomy, weather, hydraulic engineering, and transportation by land, sea, and air. Viscous fluid flow provides natures most common manifestation of nonlinearity and turbulence in classical mechanics, and provides an excellent illustration of possible solutions of non-linear equations of motion introduced in chapter 4. A detailed description of turbulence remains a challenging problem and this subject has the reputation of being the last great unsolved problem in classical mechanics. There is an apocryphal story that Werner Heisenberg was asked, if given the opportunity, what would he like to ask God. His reply was "When I meet God, I am going to ask him two questions: Why relativity? and why turbulence?, I really believe he will only have an answer to the first".

In contrast to solids, fluids do not have elastic restoring forces to support shear stress because the fluid flows. Shear stresses in fluids are balance by viscous forces which are velocity dependent. There are two mechanisms that lead to shear stress acting between adjacent fluid layers in relative motion. The first mechanism involves laminar flow where the viscous forces produce shear stress between adjacent layers of the fluid which are moving parallel along adjacent streamlines at differing velocities. Viscous forces typically dominate laminar flow. High viscosity fluids like honey exhibit laminar flow and are more difficult to stir or pour compared with low-viscosity fluids like water. The second mechanism involves turbulent flow where shear stress is due to momentum transfer between adjacent layers when the flow breaks up into large-scale coherent vortex structures which carry most of the kinetic energy. These eddies lead to transverse motion that transfers momentum plus heat between adjacent layers and leads to higher drag. The wing-tip vortex produced by the wing tip of an aircraft is an example of a dynamically-distinct, large-scale, coherent vortex structure which has considerable angular momentum and decays by fragmentation into a cascade of smaller scale structures.

16.8.1   Navier-Stokes equation

Viscous forces acting on the small-scale coherent structures eventually dissipate the energy in turbulent motion. The viscous drag can be handled in terms of a stress tensor $\mathbf{T}$ analogous to its use when accounting for the elastic restoring forces in elasticity as discussed in chapter 16.5.3. That is, the viscous force density is related to the deceleration of the volume element by

$$\frac{\partial}{\partial t}\left(\rho\mathbf{v}\right) = -\boldsymbol{\nabla}\cdot\mathbf{T} \tag{16.95}$$

where the components of the stress tensor are

$$T_{ki} = T_{ik} = P\delta_{ik} + \rho v_i v_k \tag{16.96}$$

Note that the stress tensor gives the momentum flux density tensor, which involves a diagonal term proportional to pressure P, plus a viscous drag term that is is proportional to the product of two velocities.

The Navier-Stokes equations are the fundamental equations characterizing fluid flow. They are based on application of Newton's second law of motion to fluids together with the assumption that the fluid stress is the sum of a diffusing viscous term plus a pressure term. Combining Euler's equation, 16.90, with 16.95 gives the Navier-Stokes equation

$$\rho\left[\frac{\partial\mathbf{v}}{\partial t} + \mathbf{v}\cdot\boldsymbol{\nabla}\mathbf{v}\right] = -\boldsymbol{\nabla}P + \boldsymbol{\nabla}\cdot\mathbf{T} + \mathbf{f} \tag{16.97}$$

where ρ is the fluid density, $\mathbf{v}$ is the flow velocity vector, P the pressure, $\mathbf{T}$ is the shear stress tensor viscous drag term, and $\mathbf{f}$ represents external body forces per unit volume such as gravity acting on the fluid.

For incompressible flow the stress tensor term simplifies to $\boldsymbol{\nabla}\cdot\mathbf{T} = \mu\nabla^2\mathbf{v}$. Then the Navier-Stokes equation simplifies to

$$\rho\left[\frac{\partial\mathbf{v}}{\partial t} + \mathbf{v}\cdot\boldsymbol{\nabla}\mathbf{v}\right] = -\boldsymbol{\nabla}P + \mu\nabla^2\mathbf{v} + \mathbf{f} \tag{16.98}$$

where $\mu\nabla^2\mathbf{v}$ is the viscosity drag term. The left-hand side of equation 16.98 represents the rate of change of momentum per unit volume while the right-hand side represents the summation of the forces per unit volume that are acting.

The Navier-Stokes equations are nonlinear due to the $(\mathbf{v} \cdot \boldsymbol{\nabla}) \mathbf{v}$ term as well as being a function of velocity. This non-linearity leads to a wide spectrum of dynamic behavior ranging from ordered laminar flow to chaotic turbulence. Numerical solution of the Navier-Stokes equations is extremely difficult because of the wide dynamic range of the dimensions of the coherent structures involved in turbulent motion. For example, simulation calculations require use of a high resolution mesh which is a challenge to the capabilities of current generation computers.

The microscopic boundary condition at the interface of the solid and fluid is that the fluid molecules have zero average tangential velocity relative to the normal to the solid-fluid interface. This implies that there is a boundary layer for which there is a gradient in the tangential velocity of the fluid between the solid-fluid interface and the free-steam velocity. This velocity gradient produces vorticity in the fluid. When the viscous forces are negligible then the angular momentum in any coherent vortex structure is conserved leading to the vortex motion being preserved as it propagates.

16.8.2   Reynolds number

Fluid flow can be characterized by the Reynolds number Re which is a dimensionless number that is a measure of the ratio of the inertial forces $\rho v^2/L$ to viscous forces $\mu v/L^2$. That is,

$$\mathrm{Re} \equiv \frac{\text{Inertial forces}}{\text{Viscous forces}} = \frac{\rho v L}{\mu} = \frac{v L}{\eta} \qquad (16.99)$$

where v is the relative velocity between the free fluid flow and the solid surface, L is a characteristic linear dimension, μ is the dynamic viscosity of the fluid, η is the kinematic viscosity ($\eta = \frac{\mu}{\rho}$), and ρ is the density of the fluid. The Law of Similarity implies that at a given Reynolds number, for a specific shaped solid body, the fluid flow behaves identically independent of the size of the body. Thus one can use small models in wind tunnels, or water-flow tanks, to accurately model fluid flow that can be scaled up to a full-sized aircraft or boats by scaling v and L to give the same Reynolds number.

16.8.3   Laminar and turbulent fluid flow

Fluid flow over a cylinder illustrates the general features of fluid flow. The drag force F_D acting on a cylinder of diameter D and length l, with the cylindrical axis perpendicular to the fluid flow, is given by

$$F_D = \frac{1}{2}\rho v^2 C_D D l \qquad (16.100)$$

where C_D is the coefficient of drag. Figure 16.1*upper* shows the dependence of the drag coefficient C_D as a function of the Reynolds number, for fluid flow that is transverse to a smooth circular cylinder. The lower part of figure 16.1 shows the streamlines for flow around the cylinder at various Reynolds numbers for the points identified by the letters A, B, C, D, and E on the plot of the drag coefficient versus Reynolds number for a smooth cylinder.

A) At low velocities, where Re ≤ 1, the flow is laminar around the cylinder in that the low vorticity is damped by the viscous forces and the $\frac{\partial \mathbf{v}}{\partial t}$ term in equation 16.98 can be ignored. The coefficient of drag C_D

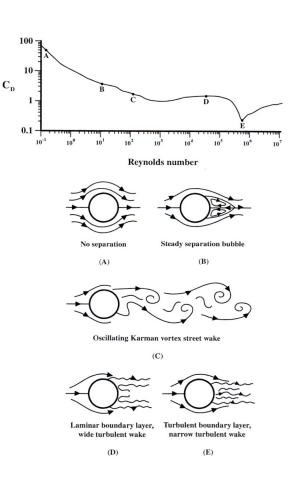

Figure 16.1: Upper: The dependence of the coefficient of drag C_D on Reynolds number Re for fluid flow perpendicular to a smooth circular cylinder of diameter D and length l. Lower: Typical flow patterns for flow past a circular cylinder at various Reynolds numbers as indicated in the upper figure.

varies inversely with Re leading to the drag forces that are roughly linear with velocity as described in chapter 2.10.5. The size and velocities of raindrops in a light rain shower correspond to such Reynolds numbers.

B) For $10 < \mathrm{Re} < 30$ the flow has two turbulent vortices immediately behind the body in the wake of the cylinder, but the flow still is primarily laminar as illustrated.

C) For $40 < \mathrm{Re} < 250$ the pair of vortices peel off alternately producing a regular periodic sequence of vortices although the flow still is laminar. This vortex sheet is called a von Kármán vortex sheet for which the velocity at a given position, relative to the cylinder, is time dependent in contrast to the situation at lower Reynolds numbers.

D) For $10^3 < \mathrm{Re} < 10^5$ viscous forces are negligible relative to the inertial effects of the vortices and boundary-layer vortices have less time to diffuse into the larger region of the fluid, thus the boundary layer is thinner. The boundary-layer flow exhibits a small scale chaotic turbulence in three dimensions superimposed on regular alternating vortex structures. In this range C_D is roughly constant and thus the drag forces are proportional to the square of the velocity. This regime of Reynold numbers corresponds to typical velocities of moving automobiles.

E) For $\mathrm{Re} \approx 10^6$, which is typical of a flying aircraft, the inertial effects dominate except in the narrow boundary layer close to the solid-fluid interface. The chaotic region works its way further forward on the cylinder reducing the volume of the chaotic turbulent boundary layer which results in a significant decreases in C_D. For a sailplane wing flying at about $50 knots$, the boundary layer at the leading edge of the cylinder reduces to the order of a millimeter in thickness at the leading edge and a centimeter at the trailing edge. At these Reynold's numbers the airflow comprises a thin boundary layer, where viscous effects are important, plus fluid flow in the bulk of the fluid where the vortex inertial terms dominate and viscous forces can be ignored. That is, the viscous stress tensor term $\nabla \cdot \mathbf{T}$, on the right-hand side of equation 16.97, can be ignored, and the Navier-Stokes equation reduces to the simpler Euler equation for such inviscid fluid flow.

The importance of the inertia of the vortices is illustrated by the persistence of the vortex structure and turbulence over a wide range of length scales characteristic of turbulent flow. The dynamic range of the dimension of coherent vortex structures is enormous. For example, in the atmosphere the vortex size ranges from $10^5 m$ in diameter for hurricanes down to $10^{-3} m$ in thin boundary layers adjacent to an aircraft wing. The transition from laminar to turbulent flow is illustrated by water flow over the hull of a ship which involves laminar flow at the bow followed by turbulent flow behind the bow wave and at the stern of the ship. The broad extent of the white foam of seawater along the side and the stern of a ship illustrates the considerable energy dissipation produced by the turbulence. The boundary layer of a stalled aircraft wing is another example. At a high angle of attack, the airflow on the lower surface of the wing remains laminar, that is, the stream velocity profile, relative to the wing, increases smoothly from zero at the wing surface outwards until it meets the ambient air velocity on the outer surface of the boundary layer which is the order of a millimeter thick. The flow on the top surface of the wing initially is laminar before becoming turbulent at which point the boundary layer rapidly increases in thickness. Further back the airflow detaches from the wing surface and large-scale vortex structures lead to a wide boundary layer comparable in thickness to the chord of the wing with vortex motion that leads to the airflow reversing its direction adjacent to the upper surface of the wing which greatly increases drag. When the vortices begin to shed off the bounded surface they do so at a certain frequency which can cause vibrations that can lead to structural failure if the frequency of the shedding vortices is close to the resonance frequency of the structure.

Considerable time and effort are expended by aerodynamicists and hydrodynamicists designing aircraft wings and ship hulls to maximize the length of laminar region of the boundary layer to minimize drag. When the Reynolds number is large the slightest imperfections in the shape of wing, such as a speck of dust, can trigger the transition from laminar to turbulent flow. The boundaries between adjacent large-scale coherent structures are sensitively identified in computer simulations by large divergence of the streamlines at any separatrix. A large positive, finite-time, Lyapunov exponent identifies divergence of the streamlines which occurs at a separatrix between adjacent large-scale coherent vortex structures, whereas the Lyapunov exponents are negative for converging streamlines within any coherent structure. Computations of turbulent flow often combine the use of finite-time Lyapunov exponents to identify coherent structures, plus Lagrangian mechanics for the equations of motion since the Lagrangian is a scalar function, it is frame independent, and it gives far better results for fluid motion than using Newtonian mechanics. Thus the Lagrangian approach in the continua is used extensively for calculations in aerodynamics, hydrodynamics, and studies of atmospheric phenomena such as convection, hurricanes, tornadoes, etc.

16.9 Summary and implications

The goal of this chapter is to provide a glimpse into the classical mechanics of the continua which introduces the Lagrangian density and Hamiltonian density formulations of classical mechanics.

Lagrangian density formulation: In three dimensional Lagrangian density $\mathfrak{L}(\mathbf{q}, \frac{d\mathbf{q}}{dt}, \nabla \cdot \mathbf{q}, x, y, z, t)$ is related to the Lagrangian L by taking the volume integral of the Lagrangian density.

$$L = \int \mathfrak{L}(\mathbf{q}, \frac{d\mathbf{q}}{dt}, \nabla \cdot \mathbf{q}, x, y, z, t) d\tau \tag{16.21}$$

Applying Hamilton's Principle to the three-dimensional Lagrangian density leads to the following set of differential equations of motion

$$\frac{\partial}{\partial t}\left(\frac{\partial \mathfrak{L}}{\frac{\partial \mathbf{q}}{\partial t}}\right) + \frac{\partial}{\partial x}\left(\frac{\partial \mathfrak{L}}{\frac{\partial \mathbf{q}}{\partial x}}\right) + \frac{\partial}{\partial y}\left(\frac{\partial \mathfrak{L}}{\frac{\partial \mathbf{q}}{\partial y}}\right) + \frac{\partial}{\partial z}\left(\frac{\partial \mathfrak{L}}{\frac{\partial \mathbf{q}}{\partial z}}\right) - \frac{\partial \mathfrak{L}}{\partial \mathbf{q}} = 0 \tag{16.22}$$

Hamiltonian density formulation: In the limit that the coordinates q, p are continuous, then the Hamiltonian density can be expressed in terms of a volume integral over the momentum density π and the Lagrangian density $\mathfrak{L}$ where

$$\pi \equiv \frac{\partial \mathfrak{L}}{\partial \dot{\mathbf{q}}} \tag{16.27}$$

Then the obvious definition of the Hamiltonian density $\mathfrak{H}$ is

$$H = \int \mathfrak{H} dV = \int (\pi \cdot \dot{\mathbf{q}} - \mathfrak{L}) d\tau \tag{16.28}$$

where the Hamiltonian density is given by

$$\mathfrak{H} = \pi \cdot \dot{\mathbf{q}} - \mathfrak{L} \tag{16.29}$$

These Lagrangian and Hamiltonian density formulations are of considerable importance to field theory and fluid mechanics.

Linear elastic solids: The theory of continuous systems was applied to the case of linear elastic solids. The **stress tensor T** is a rank 2 tensor defined as the ratio of the force vector $d\mathbf{F}$ and the surface element vector $d\mathbf{A}$. That is, the force vector is given by the inner product of the stress tensor **T** and the surface element vector $d\mathbf{A}$.

$$d\mathbf{F} = \mathbf{T} \cdot d\mathbf{A} \tag{16.33}$$

The **strain tensor σ** also is a rank 2 tensor defined as the ratio of the strain vector ξ and infinitessimal area $d\mathbf{A}$.

$$d\xi = \sigma \cdot d\mathbf{A} \tag{16.38}$$

where the component form of the rank 2 strain tensor is

$$\sigma = \frac{1}{2}\begin{vmatrix} \frac{d\xi_1}{dx_1} & \frac{d\xi_1}{dx_2} & \frac{d\xi_1}{dx_3} \\ \frac{d\xi_2}{dx_1} & \frac{d\xi_2}{dx_2} & \frac{d\xi_2}{dx_3} \\ \frac{d\xi_3}{dx_1} & \frac{d\xi_3}{dx_2} & \frac{d\xi_3}{dx_3} \end{vmatrix} \tag{16.39}$$

The modulus of elasticity is defined as the slope of the stress-strain curve. For linear, homogeneous, elastic matter, the potential energy density U separates into diagonal and off-diagonal components of the strain tensor

$$U = \frac{1}{2}\left[\lambda \sum_i (\sigma_{ii})^2 + 2\mu \sum_{ik} (\sigma_{ik})^2\right] \tag{16.42}$$

where the constants λ and μ are Lamé's moduli of elasticity which are positive. The stress tensor is related to the strain tensor by

$$T_{ij} = \lambda \delta_{ij} \sum_k \frac{\partial \xi_k}{\partial x_k} + \mu \left(\frac{d\xi_i}{dx_j} + \frac{d\xi_j}{dx_i}\right) = \lambda \delta_{ij} \sum_k \sigma_{kk} + 2\mu \sigma_{ij} \tag{16.43}$$

Electromagnetic field theory: The rank 2 Maxwell stress tensor **T** has components

$$T_{ij} \equiv \epsilon_0 \left(E_i E_j - \frac{1}{2} \delta_{ij} E^2 \right) + \frac{1}{\mu_0} \left(B_i B_j - \frac{1}{2} \delta_{ij} B^2 \right) \tag{16.71}$$

The divergence theorem allows the total electromagnetic force, acting of the volume τ, to be written as

$$\mathbf{F} = \int \left(\boldsymbol{\nabla} \cdot \mathbf{T} - \epsilon_0 \mu_0 \frac{\partial \mathbf{S}}{\partial t} \right) d\tau = \oint \mathbf{T} \cdot d\mathbf{a} - \epsilon_0 \mu_0 \frac{d}{dt} \int \mathbf{S} d\tau \tag{16.74}$$

The total momentum flux density is given by

$$\frac{\partial}{\partial t} \left(\boldsymbol{\pi}_{mech} + \boldsymbol{\pi}_{field} \right) = \boldsymbol{\nabla} \cdot \mathbf{T} \tag{16.79}$$

where the electromagnetic field momentum density is given by the Poynting vector **S** as $\boldsymbol{\pi}_{field} = \epsilon_0 \mu_0 \mathbf{S}$.

Ideal fluid dynamics: Mass conservation leads to the continuity equation

$$\frac{\partial \rho}{\partial t} + \boldsymbol{\nabla} \cdot (\rho \mathbf{v}) = 0 \tag{16.82}$$

Euler's hydrodynamic equation gives

$$\frac{\partial \mathbf{v}}{\partial t} + (\mathbf{v} \cdot \boldsymbol{\nabla}) \mathbf{v} = -\frac{1}{\rho} \boldsymbol{\nabla} (P + \rho V) \tag{16.90}$$

where V is the scalar gravitational potential. If the flow is irrotational and time independent then

$$\left(\frac{1}{2} \rho v^2 + P + \rho V \right) = \text{constant} \tag{16.94}$$

Viscous fluid dynamics: For incompressible flow the stress tensor term simplifies to $\boldsymbol{\nabla} \cdot \mathbf{T} = \mu \boldsymbol{\nabla}^2 \mathbf{v}$. Then the Navier-Stokes equation becomes

$$\rho \left[\frac{\partial \mathbf{v}}{\partial t} + \mathbf{v} \cdot \boldsymbol{\nabla} \mathbf{v} \right] = -\boldsymbol{\nabla} P + \mu \boldsymbol{\nabla}^2 \mathbf{v} + \mathbf{f} \tag{16.98}$$

where $\mu \boldsymbol{\nabla}^2 \mathbf{v}$ is the viscosity drag term. The left-hand side of equation 16.98 represents the rate of change of momentum per unit volume while the right-hand side represents the summation of the forces per unit volume that are acting.

   The Reynolds number is a dimensionless number that characterizes the ratio of inertial forces to viscous forces in a viscous medium. The evolution of flow from laminar flow to turbulent flow, with increase of Reynolds number, was discussed.

   The classical mechanics of continuous fields encompasses a remarkably broad range of phenomena with important applications to laminar and turbulent fluid flow, gravitation, electromagnetism, relativity, and quantum fields.

Chapter 17

Relativistic mechanics

17.1   Introduction

Newtonian mechanics incorporates the Newtonian concept of the complete separation of space and time. This theory reigned supreme from inception, in 1687, until November 1905 when Einstein pioneered the Special Theory of Relativity. Relativistic mechanics undermines the Newtonian concepts of absoluteness of time that is inherent to Newton's formulation, as well as when recast in the Lagrangian and Hamiltonian formulations of classical mechanics. Relativistic mechanics has had a profound impact on twentieth-century physics and the philosophy of science. Classical mechanics is an approximation of relativistic mechanics that is valid for velocities much less than the velocity of light in vacuum. The term "relativity" refers to the fact that physical measurements are always made relative to some chosen reference frame. Naively one may think that the transformation between different reference frames is trivial and contains little underlying physics. However, Einstein showed that the results of measurements depend on the choice of coordinate system, which revolutionized our concept of space and time.

Einstein's work on relativistic mechanics comprised two major advances. The first advance is the 1905 Special Theory of Relativity which refers to nonaccelerating frames of reference. The second major advance was the 1916 General Theory of Relativity which considers accelerating frames of reference and their relation to gravity. The Special Theory is a limiting case of the General Theory of Relativity. The mathematically complex General Theory of Relativity is required for describing accelerating frames, gravity, plus related topics like Black Holes, or extremely accurate time measurements inherent to the Global Positioning System. The present discussion will focus primarily on the mathematically simple Special Theory of Relativity since it encompasses most of the physics encountered in atomic, nuclear and high energy physics. This chapter uses the basic concepts of the Special Theory of Relativity to investigate the implications of extending Newtonian, Lagrangian and Hamiltonian formulations of classical mechanics into the relativistic domain. The Lorentz-invariant extended Hamiltonian and Lagrangian formalisms are introduced since they are applicable to the Special Theory of Relativity. The General Theory of Relativity incorporates the gravitational force as a geodesic phenomena in a four-dimensional Reimannian structure based on space, time, and matter. A superficial outline is given to the fundamental concepts and evidence that underlie the General Theory of Relativity.

17.2   Galilean Invariance

As discussed in chapter 2.3, an inertial frame is one in which Newton's Laws of motion apply. Inertial frames are non-accelerating frames so that pseudo forces are not induced. All reference frames moving at constant velocity relative to an inertial reference, are inertial frames. Newton's Laws of nature are the same in all inertial frames of reference and therefore there is no way of determining absolute motion because no inertial frame is preferred over any other. This is called Galilean-Newtonian invariance. Galilean invariance assumes that the concepts of space and time are completely separable. Time is assumed to be an absolute quantity that is invariant to transformations between coordinate systems in relative motion. Also the element of length is the same in different Galilean frames of reference.

Consider two coordinate systems shown in figure 17.1, where the primed frame is moving along the x axis of the fixed unprimed frame. A **Galilean transformation** implies that the following relations apply;

$$
\begin{aligned}
x_1' &= x_1 - vt \\
x_2' &= x_2 \\
x_3' &= x_3 \\
t' &= t
\end{aligned}
\tag{17.1}
$$

Note that at any instant t, the infinitessimal units of length in the two systems are identical since

$$
ds^2 = \sum_{i=1}^{3} dx_i^2 = \sum_{i=1}^{3} dx_i'^2 = ds'^2
\tag{17.2}
$$

These are the mathematical expression of the Newtonian idea of space and time. An immediate consequence of the Galilean transformation is that the velocity of light must differ in different inertial reference frames.

At the end of the 19th century physicists thought they had discovered a way of identifying an absolute inertial frame of reference, that is, it must be the frame of the medium that transmits light in vacuum. Maxwell's laws of electromagnetism predict that electromagnetic radiation in vacuum travels at $c = \frac{1}{\sqrt{\mu_o \varepsilon_o}} = 2.998 \times 10^8 m/s$. Maxwell did not address in what frame of reference that this speed applied. In the nineteenth

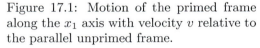

Figure 17.1: Motion of the primed frame along the x_1 axis with velocity v relative to the parallel unprimed frame.

century all wave phenomena were transmitted by some medium, such as waves on a string, water waves, sound waves in air. Physicists thus envisioned that light was transmitted by some unobserved medium which they called the *ether*. This ether had mystical properties, it existed everywhere, even in outer space, and yet had no other observed consequences. The ether obviously should be the absolute frame of reference.

In the $1880's$, Michelson and Morley performed an experiment in Cleveland to try to detect this ether. They transmitted light back and forth along two perpendicular paths in an interferometer, shown in figure 17.2, and assumed that the earth's motion about the sun led to movement through the ether.

The time taken to travel a return trip takes longer in a moving medium, if the medium moves in the direction of the motion, compared to travel in a stationary medium. For example, you lose more time moving against a headwind than you gain travelling back with the wind. The time difference Δt, for a round trip to a distance L, between travelling in the direction of motion in the ether, versus travelling the same distance perpendicular to the movement in the ether, is given by $\Delta t \approx \frac{L}{c} \left(\frac{v}{c} \right)^2$ where v is the relative velocity of the ether and c is the velocity of light.

Interference fringes between perpendicular light beams in an optical interferometer provides an extremely sensitive measure of this time difference. Michelson and Morley observed no measurable time difference at any time during the year, that

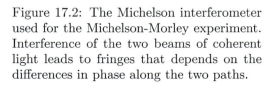

Figure 17.2: The Michelson interferometer used for the Michelson-Morley experiment. Interference of the two beams of coherent light leads to fringes that depends on the differences in phase along the two paths.

is, the relative motion of the earth within the ether is less than $1/6$ the velocity of the earth around the sun. Their conclusion was either, that the ether was dragged along with the earth, or the velocity of light was dependent on the velocity of the source, but these did not jibe with other observations. Their disappointment at the failure of this experiment to detect evidence for an absolute inertial frame is important and confounded physicists for two decades until Einstein's Special Theory of Relativity explained the result.

17.3 Special Theory of Relativity

17.3.1 Einstein Postulates

In November 1905, at the age of 26, Einstein published a seminal paper entitled "On the electrodynamics of moving bodies". He considered the relation between space and time in inertial frames of reference that are in relative motion. In this paper he made the following postulates.

1) *The laws of nature are the same in all inertial frames of reference.*
2) *The velocity of light in vacuum is the same in all inertial frames of reference.*

Note that Einstein's first postulate, coupled with Maxwell's equations, leads to the statement that the velocity of light in vacuum is a universal constant. Thus the second postulate is unnecessary since it is an obvious consequence of the first postulate plus Maxwell's equations which are basic laws of physics. This second postulate explained the null result of the Michelson-Morley experiment. However, it was not this experimental result that led Einstein to the theory of special relativity; he deduced the Special Theory of Relativity from consideration of Maxwell's equations of electromagnetism. Although Einstein's postulates appear reasonable, they lead to the following surprising implications.

17.3.2 Lorentz transformation

Galilean invariance leads to violation of the Einstein postulate that the velocity of light is a universal constant in all frames of reference. It is necessary to assume a new transformation law that renders physical laws relativistically invariant. Maxwell's equations are relativistically invariant, which led to some electromagnetic phenomena that could not be explained using Galilean invariance. In 1904 Lorentz proposed a new transformation to replace the Galilean transformation in order to explain such electromagnetic phenomena. Einstein's genius was that he derived the transformation, that had been proposed by Lorentz, directly from the postulates of the Special Theory of Relativity. The Lorentz transformation satisfies Einstein's theory of relativity, and has been confirmed to be correct by many experiments.

For the geometry shown in figure 17.1, the **Lorentz transformations** are:

$$
\begin{aligned}
x' &= \gamma(x - vt) \\
y' &= y \\
z' &= z \\
t' &= \gamma\left(t - \frac{vx}{c^2}\right)
\end{aligned}
\tag{17.3}
$$

where the Lorentz γ factor

$$
\gamma \equiv \frac{1}{\sqrt{1 - \left(\frac{v}{c}\right)^2}}
\tag{17.4}
$$

The inverse transformations are

$$
\begin{aligned}
x &= \gamma(x' + vt') \\
y &= y' \\
z &= z' \\
t &= \gamma\left(t' + \frac{vx'}{c^2}\right)
\end{aligned}
\tag{17.5}
$$

The Lorentz γ factor, defined above, is the key feature differentiating the Lorentz transformations from the Galilean transformation. Note that $\gamma \geq 1$; also $\gamma \to 1.0$ as $v \to 0$, and increases to infinity as $\frac{v}{c} \to 1$ as illustrated in figure 17.3. A useful fact that will be used later is that for $\frac{v}{c} \ll 1$;

$$
\gamma \to 1 + \frac{1}{2}\left(\frac{v}{c}\right)^2 \qquad \text{Limit for } v \ll c
$$

Note that for $v \ll c$ then $\gamma = 1$ and the Lorentz transformation is identical to the Galilean transformation.

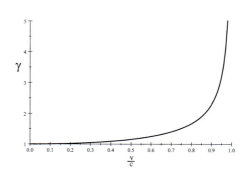

Figure 17.3: The dependence of the Lorentz γ factor on $\frac{v}{c}$.

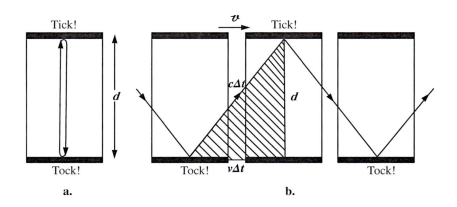

Figure 17.4: The observer and mirror are at rest in the left-hand frame (a). The light beam takes a time $\Delta t = \frac{d}{c}$ to travel to the mirror. In the right-hand frame (b) the source and mirror are travelling at a velocity v relative to the observer. The light travels further in the right-hand frame of reference (b) than is the stationary frame (a). Since Einstein states that the velocity of light is the same in both frames of reference then the time interval must by larger in frame (b) since the light travels further than in (a).

17.3.3 Time Dilation:

Consider that a clock is *fixed at x_o' in a moving frame* and measures the time interval between two events in the moving frame, i.e. $\Delta t_p' = t_1' - t_2'$. According to the Lorentz transformation, the times in the fixed frame are given by:

$$t_1 = \gamma \left(t_1' + \frac{vx_0'}{c^2} \right) \tag{17.6}$$

$$t_2 = \gamma \left(t_2' + \frac{vx_0'}{c^2} \right)$$

Thus the time interval is given by:

$$t_2 - t_1 = \gamma \left(t_2' - t_1' \right) \tag{17.7}$$

The time between events in the rest frame of the clock, $\Delta \tau \equiv \Delta t_p'$ is called the *proper time* which always is the shortest time measured for a given event and is represented by the symbol τ. That is

$$\Delta t = \gamma \Delta t_p' = \gamma \Delta \tau \tag{17.8}$$

Note that the time interval for any other frame of reference, moving with respect to the clock frame, will show larger time intervals because $\gamma \geq 1.0$ which implies that the fixed frame perceives that the moving clock is slow by the factor γ.

The plausibility of this *time dilation* can be understood by looking at the simple geometry of the space ship example shown in Figure 17.4. Pretend that the clock in the proper frame of the space ship is based on the time for the light to travel to and from the mirror in the space ship. In this proper frame the light has the shortest distance to travel, and the proper transit time is

$$\Delta \tau = \frac{2d}{c} \tag{17.9}$$

In the fixed frame, b, the component of velocity in the direction of the mirror is $\sqrt{c^2 - v^2}$ using the Pythagorus theorem, assuming that the light cannot travel faster the c. Thus the transit time towards and back from the mirror must be

$$\Delta t = \frac{2d}{c\sqrt{1 - \left(\frac{v}{c} \right)^2}} = \gamma \Delta \tau \tag{17.10}$$

which is the predicted time dilation.

There are many experimental verifications of time dilation in physics. For example, a stationary muon has a mean lifetime of $\tau_p = 2\mu\sec$, whereas the lifetime of a fast moving muon, produced in the upper atmosphere by high-energy cosmic rays, was observed in 1941 to be longer and given by $\gamma\tau_p$ as described in example 17.1. In 1972 Hafely and Keating used four accurate cesium atomic clocks to confirm time dilation. Two clocks were flown on regularly scheduled airlines travelling around the World, one westward and the other eastward. The other two clocks were used for reference. The westward moving clock was slow by $(273 \pm 7)nsec$ compared to the predicted value of $(275 \pm 10)n\sec$. The Global Positioning System of 24 geosynchronous satellites is used for locating positions to within a few meters. It has an accuracy of a few nanoseconds which requires allowance for time dilation and is a daily tribute to the correctness of Einstein's Theory of Relativity.

17.3.4 Length Contraction

The Lorentz transformation leads to a contraction of the apparent length of an object in a moving frame as seen from a fixed frame. The length of a ruler in its own frame of reference is called the *proper length*. Consider an accurately measured rod of known proper length $L_p = x_2' - x_1'$ that is, at rest in the moving primed frame. The locations of both ends of this rod are measured at a *given time in the stationary frame*, $t_1 = t_2$, by taking a photograph of the moving rod. The corresponding locations in the moving frame are:

$$
\begin{aligned}
x_2' &= \gamma(x_2 - vt_2) \\
x_1' &= \gamma(x_1 - vt_1)
\end{aligned}
\tag{17.11}
$$

Since $t_2 = t_1$, the measured lengths in the two frames are related by:

$$
x_2' - x_1' = \gamma(x_2 - x_1)
\tag{17.12}
$$

That is, the lengths are related by:

$$
L = \frac{1}{\gamma}L_p
\tag{17.13}
$$

Note that the moving rod appears shorter in the direction of motion. As $v \to c$ the apparent length shrinks to zero in the direction of motion while the dimensions perpendicular to the direction of motion are unchanged. This is called the *Lorentz contraction*. If you could ride your bicycle at close to the speed of light, you would observe that stationary cars, buildings, people, all would appear to be squeezed thin along the direction that you are travelling. Also objects that are further away down any side street would be distorted in the direction of travel. A photograph taken by a stationary observer would show the moving bicycle to be Lorentz contracted along the direction of travel and the stationary objects would be normal.

17.3.5 Simultaneity

The Lorentz transformations imply a new philosophy of space and time. A surprising consequence is that the concept of simultaneity is frame dependent in contrast to the prediction of Newtonian mechanics.

Consider that two events occur in frame S at (x_1, t_1) and (x_2, t_2). In frame S' these two events occur at (x_1', t_1') and (x_2', t_2'). From the Lorentz transformation the time difference is

$$
t_2' - t_1' = \gamma\left[(t_2 - t_1) - \frac{v(x_2 - x_1)}{c^2}\right]
\tag{17.14}
$$

If an event is simultaneous in frame S, that is $(t_2 - t_1) = 0$ then

$$
t_2' - t_1' = \gamma\left[\frac{v(x_1 - x_2)}{c^2}\right]
\tag{17.15}
$$

Thus the event is not simultaneous in frame S' if $(x_2 - x_1) = L_p \neq 0$. That is, an event that is simultaneous in one frame is not simultaneous in the other frame if the events are spatially separated. The equivalent statement is that for two clocks, spatially separated by a distance L_p, which are synchronized in their rest frame, then in a moving frame they are not simultaneous.

Figure 17.5: If lightning strikes the front and rear of the carriage simultaneously, according to the man in the fixed frame, then the woman in the moving frame sees the flash from the front first since she is moving towards that approaching wavefront during the transit time of the light. Thus if the length of the carriage in the stationary frame is $(x_2 - x_1) = L_P$ then the time difference is $\Delta t' = \gamma L_p \frac{v}{c^2}$.

Einstein discussed the example shown in figure 17.5, where lightning strikes both ends of a train simultaneously in the stationary earth frame of reference. A woman on the train will see that the strikes are not simultaneous since the wavefront from the front of the carriage will be seen first because she is moving forward during the time the light from the two lightning flashes is travelling towards her. As a consequence she observes that the two lightning flashes are not simultaneous. This explains why measurement of the length of a moving rod, performed by simultaneously locating both ends in the fixed frame, implies that the measurement occurs at different times for both ends in the moving frame resulting in a shorter apparent length. The lack of simultaneity explains why one can get the apparent inconsistency that the moving bicyclist sees that the stationary street block to be length contracted, while in contrast, a pedestrian sees that the bicycle is length contracted.

The concept of causality breaks down since $(x'_2 - x'_1)$ can be either positive or negative, therefore the corresponding Δt can be positive of negative. A consequence of the lack of simultaneity is that the image shown by a photograph of a rapidly moving object is not a true representation of the moving object. Not only is the body contracted in the direction of travel, but also it appears distorted because light arriving from the far side of the body had to be emitted earlier, that is, when the body was at an earlier location, in order to reach the observer simultaneously with light from the near side. The relativistic snake paradox, addressed in Chapter 17 workshop exercise 1, is an example of the role of simultaneity in relativistic mechanics.

17.1 *Example: Muon lifetime*

Many people had trouble comprehending time dilation and Lorentz contraction predicted by the Special Theory of Relativity. The predictions appear to be crazy, but there are many examples where time dilation and Lorentz contraction are observed experimentally such as the decay in flight of the muon. At rest, the muon decays with a mean lifetime of 2 μ sec. Muons are created high in the atmosphere due to cosmic ray bombardment. A typical muon travels at $v = 0.998c$ which corresponds to $\gamma = 15$. Time dilation implies that the lifetime of the moving muon in the earth's frame of reference is 30 μsec. The speed of the muon is essentially c in both frames of reference, and it would travel 600m in 2 μs and 9000m in 30 μs. In fact, it is observed that the muon does travel, on average, 9000m in the earth frame of reference before decaying. Is this inconsistent with the view of someone travelling with the muon? In the muon's moving frame, the lifetime is only 2 μs, but the Lorentz contraction of distance means that 9000m in the earth frame appears to be only 600m in the muon moving frame; a distance it travels is 2 μ sec. Thus in both frames of reference we have consistent explanations, that is, the muon travels the height of the mountain in one lifetime.

17.2 *Example: Relativistic Doppler Effect*

The relativistic Doppler effect is encountered frequently in physics and astronomy. Consider monochromatic electromagnetic radiation from a source, such as a star, that is moving towards the detector at a velocity v. During the time Δt in the frame of the receiver, the source emits n cycles of the sinusoidal waveform. Thus the length of this waveform, as seen by the receiver, is $n\lambda$ which equals

$$n\lambda = (c - v)\Delta t$$

The frequency as measured by the receiver is

$$\nu = \frac{c}{\lambda} = \frac{cn}{(c - v)\Delta t}$$

According to the source, it emits n waves of frequency ν_0 during the proper time interval $\Delta t'$, that is

$$n = \nu_0 \Delta t'$$

This proper time interval $\Delta t'$, in the source frame, corresponds to a time interval Δt in the receiver frame where

$$\Delta t = \gamma \Delta t'$$

Thus the frequency measured by the receiver is

$$\nu = \frac{1}{(1 - \frac{v}{c})} \frac{\nu_0}{\gamma} = \frac{\sqrt{1 - (\frac{v}{c})^2}}{(1 - \frac{v}{c})} \nu_0 = \sqrt{\frac{1 + \beta}{1 - \beta}} \nu_0$$

where $\beta \equiv \frac{v}{c}$. This formula for source and receiver approaching each other also gives the correct answer for source and receiver receding if the sign of β is changed.

This relativistic Doppler Effect accounts for the red shift observed for light emitted by receding stars and galaxies, as well as many examples in atomic and nuclear physics involving moving sources of electromagnetic radiation.

17.3 *Example: Twin paradox*

A problem that troubled physicists for many years is called the twin paradox. Consider two identical twins, Jack and Jill. Assume that Jill travels in a space ship at a speed of $\gamma = 4$ for 20 years, as measured by Jack's clock, and then returns taking another 20 years, according to Jack. Thus, Jack has aged 40 years by the time his twin sister returns home. However, Jill's clock measures $20/4 = 5$ years for each half of the trip so that she thinks she travelled for 10 years total time according to her clock. Thus she has aged only 10 years on the trip, that is, now she is 30 years younger that her twin brother. Note that, according to Jill, the distance she travelled out and back was 1/4 the distance according to Jack, so she perceives no inconsistency in her clock, and the speed of the space ship. This was called a paradox because some people claimed that Jill will perceive that the earth and Jack moved away at the same relative speed in the opposite direction and thus according to Jill, Jack should be 30 years younger, not her. Moreover, some claimed that this problem is symmetric and therefore both twins must still be the same age since there is no way of telling who was moving away from whom. This argument is incorrect because Jill was able to sense that she accelerated to $\gamma = 4$ which destroys the symmetry argument. The effect is observed with accelerated beams of unstable nuclei such as the muon and was confirmed by the results of the experiment where cesium atomic clocks were flown around the Earth. Thus the Twin paradox is not a paradox; the fact is that Jill will be younger than her twin brother.

17.4   Relativistic kinematics

17.4.1   Velocity transformations

Consider the two parallel coordinate frames with the primed frame moving at a velocity v along the x_1' axis as shown in figure 17.1. Velocities of an object measured in both frames are defined to be

$$u_i = \frac{dx_i}{dt}$$

$$u_i' = \frac{dx_i'}{dt'}$$

(17.16)

Using the Lorentz transformations 17.3, 17.5 between the two frames moving with relative velocity v along the x_1 axis, gives that the velocity along the x_1' axis is

$$u_1' = \frac{dx_1'}{dt'} = \frac{dx_1 - v dt}{dt - \frac{v}{c^2} dx_1} = \frac{u_1 - v}{1 - \frac{u_1 v}{c^2}}$$

(17.17)

Similarly we get the velocities along the perpendicular x_2' and x_3' axes to be

$$u_2' = \frac{dx_2'}{dt'} = \frac{u_2}{1 - \frac{u_1 v}{c^2}}$$

(17.18)

$$u_3' = \frac{dx_3'}{dt'} = \frac{u_3}{1 - \frac{u_1 v}{c^2}}$$

When $\frac{u_1 v}{c^2} \to 0$ these velocity transformations become the usual Galilean relations for velocity addition. Do not confuse $\mathbf{u}$ and $\mathbf{u}'$ with $\mathbf{v}$; that is, $\mathbf{u}$ and $\mathbf{u}'$ are the velocities of some object measured in the unprimed and primed frames of reference respectively, whereas $\mathbf{v}$ is the relative velocity of the origin of one frame with respect to the origin of the other frame.

17.4.2   Momentum

Using the classical definition of momentum, that is $\mathbf{p} = m\mathbf{u}$, the linear momentum is not conserved using the above relativistic velocity transformations if the mass m is a scalar quantity. This problem originates from the fact that both $\mathbf{x}$ and t have non-trivial transformations and thus $\mathbf{u} = \frac{d\mathbf{x}}{dt}$ is frame dependent.

Linear momentum conservation can be retained by redefining momentum in a form that is identical in all frames of reference, that is by referring to the *proper time* τ as measured in the rest frame of the moving object. Therefore we define relativistic linear momentum as

$$\mathbf{p} \equiv m\frac{d\mathbf{x}}{d\tau} = m\frac{d\mathbf{x}}{dt}\frac{dt}{d\tau}$$

(17.19)

But we know the time dilation relation

$$dt = \frac{d\tau}{\sqrt{\left(1 - \frac{u^2}{c^2}\right)}} = \gamma_u d\tau$$

(17.20)

Note that the γ_u in this relation refers to the velocity u between the moving object and the frame; *this is quite different* from the $\gamma = \frac{1}{\sqrt{\left(1 - \frac{v^2}{c^2}\right)}}$ which refers to the transformation between the two frames of reference. Thus the new relativistic definition of momentum is

$$\mathbf{p} \equiv m\frac{d\mathbf{x}}{d\tau} = m\gamma_u\frac{d\mathbf{x}}{dt} = \gamma_u m\mathbf{u}$$

(17.21)

The relativistic definition of linear momentum is the same as the classical definition with the rest mass m replaced by the relativistic mass γm.[1]

---

[1]Note that, until recently, the rest mass was denoted by m_0 and the relativistic mass was referred to as m. Modern texts denote the rest mass by m and the relativistic mass by γm. This book follows the modern nomenclature for rest mass to avoid confusion.

17.4.3 Center of momentum coordinate system

The classical relations for handling the kinematics of colliding objects, carry over to special relativity when the relativistic definition of linear momentum, equation 17.21, is assumed. That is, one can continue to apply conservation of linear momentum. However, there is one important conceptual difference for relativistic dynamics in that the center of mass no longer is a meaningful concept due to the interrelation of mass and energy. However, this problem is eliminated by considering the center of momentum coordinate system which, as in the non-relativistic case, is the frame where the total linear momentum of the system is zero. Using the concept of center of momentum incorporates the formalism of classical non-relativistic kinematics.

17.4.4 Force

Newton's second law $\mathbf{F} = \frac{d\mathbf{p}}{dt}$ is covariant under a Galilean transformation. In special relativity this definition also applies using the relativistic definition of momentum $\mathbf{p}$. The fact that the **relativistic momentum p** is conserved in the force-free situation, leads naturally to using the definition of force to be

$$\mathbf{F} = \frac{d\mathbf{p}}{dt} \tag{17.22}$$

Then the relativistic momentum is conserved if $\mathbf{F} = 0$.

17.4.5 Energy

The classical definition of work done is defined by

$$W_{12} = \int_1^2 \mathbf{F} \cdot d\mathbf{r} = T_2 - T_1 \tag{17.23}$$

Assume $T_1 = 0$, let $d\mathbf{r} = \mathbf{u}dt$ and insert the relativistic force relation in equation 17.23, gives

$$W = T = \int_0^t \frac{d}{dt}\left(\gamma_u m\mathbf{u}\right) \cdot \mathbf{u}dt = m\int_0^u u\, d\left(\gamma_u u\right) \tag{17.24}$$

Integrate by parts, followed by algebraic manipulation, gives

$$T = \gamma_u m u^2 - m\int_0^u \frac{u\,du}{\sqrt{1 - \frac{u^2}{c^2}}} = \gamma_u m u^2 + mc^2\sqrt{1 - \frac{u^2}{c^2}} - mc^2$$

$$= \frac{mu^2}{\sqrt{1 - \frac{u^2}{c^2}}} + \frac{mc^2}{\sqrt{1 - \frac{u^2}{c^2}}}\left(1 - \frac{u^2}{c^2}\right) - mc^2 = mc^2\left(\gamma_u - 1\right) \tag{17.25}$$

Define the **rest energy** E_0

$$E_0 \equiv mc^2 \tag{17.26}$$

and **total relativistic energy** E

$$E \equiv \gamma_u mc^2 \tag{17.27}$$

then equation 17.25 can be written as

$$E = T + E_0 = \gamma_u mc^2 \tag{17.28}$$

This is the famous Einstein relativistic energy that relates the equivalence of mass and energy. The total relativistic energy E is a conserved quantity in nature. It is an extension of the conservation of energy and manifestations of the equivalence of energy and mass occur extensively in the real world.

In nuclear physics we often convert mass to energy and back again to mass. For example, gamma rays with energies greater than $1.022 MeV$, which are pure electromagnetic energy, can be converted to an electron plus positron both of which have rest mass. The positron can then annihilate a different electron in another atom resulting in emission of two $511 keV$ gamma rays in back to back directions to conserve linear momentum. A dramatic example of Einstein's equation is a nuclear reactor. One gram of material, the mass of a paper clip, provides $E = 9 \times 10^{13}$ joules. This is the daily output of a $1GWatt$ nuclear power station or the explosive power of the Nagasaki or Hiroshima bombs.

As the velocity of a particle v approaches c, then γ and the relativistic mass γm both approach infinity. This means that the force needed to accelerate the mass also approaches infinity, and thus no particle can exceed the velocity of light. The energy continues to increase not by increasing the velocity but by increase of the relativistic mass. Although the relativistic relation for kinetic energy is quite different from the Newtonian relation, the Newtonian form is obtained for the case of $u << c$ in that

$$T = mc^2(1 - \frac{u^2}{c^2})^{-\frac{1}{2}} - mc^2 = mc^2(1 + \frac{1}{2}\frac{u^2}{c^2} + \cdots) - mc^2 = \frac{1}{2}mu^2 \qquad (17.29)$$

An especially useful relativistic relation that can be derived from the above is

$$E^2 = p^2c^2 + E_0^2 \qquad (17.30)$$

This is useful because it provides a simple relation between total energy of a particle and its relativistic linear momentum plus rest energy.

17.4 Example: Rocket propulsion

Consider a rocket, having initial mass M, is accelerated in a straight line in free space by exhausting propellant at a constant speed v_p relative to the rocket. Let u be the speed of the rocket relative to it's initial rest frame S, when its rest mass has decreased to m. At this instant the rocket is at rest in the inertial frame S'. At a proper time $\tau + d\tau$ the rest mass is $m - dm$ and it has acquired a velocity increment du relative to S' and propellant of rest mass dm_p has been expelled with velocity v_p relative to S'. At proper time τ in S' the rest mass is mc^2. At the time $\tau + d\tau$, energy conservation requires that

$$\gamma_{u'}(m - dm)c^2 + \gamma_{v_p}m_p c^2 = mc^2$$

At the same instant, conservation of linear momentum requires

$$\gamma_{u'}(m - dm)du' - \gamma_p v_p dm_p = 0$$

To first order these two equations simplify to

$$dm_p = \sqrt{1 - \left(\frac{v_p}{c}\right)^2}dm$$
$$mdu' = dm_p\gamma_{v_p}v_p$$

Therefore

$$mdu' = v_p dm \qquad (a)$$

The velocity increment du' in frame S' can be transformed back to frame S using equation 17.5, that is

$$d + du = \frac{u + du'}{1 + \frac{udu'}{c^2}} \approx u + \left(1 - \left(\frac{u}{c}\right)^2\right)du' \qquad (b)$$

Equations a and b yield a differential equation for $u(m)$ of

$$\frac{du}{1 - \left(\frac{u}{c}\right)^2} = v_p\frac{dm}{m}$$

Integrate the left-hand side between 0 and u and the right-hand side between M and m gives

$$\frac{1}{2}c\ln\left(\frac{1 + \frac{u}{c}}{1 - \frac{u}{c}}\right) = -v_p\ln\left(\frac{m}{M}\right)$$

This reduces to

$$\frac{u}{c} = \frac{1 - \left(\frac{m}{M}\right)^{2v_p/c}}{1 + \left(\frac{m}{M}\right)^{2v_p/c}}$$

When $\frac{u}{c} \to 0$ this equation reduces to the non-relativistic answer given in equation 2.123.

17.5 Geometry of space-time

17.5.1 Four-dimensional space-time

In 1906 Poincaré showed that the Lorentz transformation can be regarded as a rotation in a 4-dimensional Euclidean space-time introduced by adding an imaginary fourth space-time coordinate ict to the three real spatial coordinates. In 1908 Minkowski reformulated Einstein's Special Theory of Relativity in this 4-dimensional Euclidean space-time vector space and concluded that the spatial variables q_i, where $(i = 1, 2, 3)$, plus the time $q_0 = ict$ are equivalent variables and should be treated equally using a covariant representation of both space and time. The idea of using an imaginary time axis ict to make space-time Euclidean was elegant, but it obscured the non-Euclidean nature of space-time as well as causing difficulties when generalized to non-inertial accelerating frames in the General Theory of Relativity. As a consequence, the use of the imaginary ict has been abandoned in modern work. Minkowski developed an alternative non-Euclidean metric that treats all four coordinates (ct, x, y, z) as a four-dimensional Minkowski metric with all coordinates being real, and introduces the required minus sign explicitly.

Analogous to the usual 3-dimensional cartesian coordinates, the displacement four vector $d\mathbf{s}$ is defined using the four components along the **four unit vectors** in either the unprimed or primed coordinate frames.

$$d\mathbf{s} = dx^0\hat{\mathbf{e}}_0 + dx^1\hat{\mathbf{e}}_1 + dx^2\hat{\mathbf{e}}_2 + dx^3\hat{\mathbf{e}}_3 = dx'^0\hat{\mathbf{e}}'_0 + dx'^1\hat{\mathbf{e}}'_1 + dx'^2\hat{\mathbf{e}}'_2 + dx'^3\hat{\mathbf{e}}'_3 \tag{17.31}$$

The convention used is that greek subscripts (covariant) or superscripts (contravariant) designate a four vector with $0 \le \mu \le 3$. The covariant unit vectors $\hat{\mathbf{e}}_\mu$ are written with the subscript μ which has 4 values $0 \le \mu \le 3$. As described in appendix $E3$, using the Einstein convention the components are written with the contravariant superscript dx^μ where the time axis $x^0 = ct$, while the spatial coordinates, expressed in cartesian coordinates, are $x^1 = x$, $x^2 = y$, and $x^3 = z$. With respect to a different (primed) unit vector basis $\hat{\mathbf{e}}'_\mu$, the displacement must be unchanged as given by equation 17.31. In addition, equation 17.43 shows that the magnitude $|ds|^2$ of the displacement four vector is invariant to a Lorentz transformation.

The most general Lorentz transformation between inertial coordinate systems S and S', in relative motion with velocity $\mathbf{v}$, assuming that the two sets of axes are aligned, and that their origins overlap when $t = t' = 0$, is given by the symmetric matrix λ where

$$x'^\mu = \sum_\nu \lambda_{\mu\nu} x^\nu \tag{17.32}$$

This Lorentz transformation of the *four vector* $\mathbb{X}$ components can be written in matrix form as

$$\mathbb{X}' = \boldsymbol{\lambda}\mathbb{X} \tag{17.33}$$

Assuming that the two sets of axes are aligned, then the elements of the Lorentz transformation $\lambda_{\mu\nu}$ are given by

$$\mathbb{X}' = \begin{pmatrix} ct' \\ x'^1 \\ x'^2 \\ x'^3 \end{pmatrix} = \begin{pmatrix} \gamma & -\gamma\beta_1 & -\gamma\beta_2 & -\gamma\beta_3 \\ -\gamma\beta_1 & 1+(\gamma-1)\frac{\beta_1^2}{\beta^2} & (\gamma-1)\frac{\beta_1\beta_2}{\beta^2} & (\gamma-1)\frac{\beta_1\beta_3}{\beta^2} \\ -\gamma\beta_2 & (\gamma-1)\frac{\beta_1\beta_2}{\beta^2} & 1+(\gamma-1)\frac{\beta_2^2}{\beta^2} & (\gamma-1)\frac{\beta_2\beta_3}{\beta^2} \\ -\gamma\beta_3 & (\gamma-1)\frac{\beta_1\beta_3}{\beta^2} & (\gamma-1)\frac{\beta_2\beta_3}{\beta^2} & 1+(\gamma-1)\frac{\beta_3^2}{\beta^2} \end{pmatrix} \cdot \begin{pmatrix} ct \\ x^1 \\ x^2 \\ x^3 \end{pmatrix} \tag{17.34}$$

where $\beta = \frac{v}{c}$ and $\gamma = \frac{1}{\sqrt{1-\beta^2}}$ and assuming that the origin of S transforms to the origin of S' at $(0,0,0,0)$.

For the case illustrated in figure 17.1, where the corresponding axes of the two frames are parallel and in relative motion with velocity v in the x_1 direction, then the Lorentz transformation matrix 17.34 reduces to

$$\begin{pmatrix} ct' \\ x'^1 \\ x'^2 \\ x'^3 \end{pmatrix} = \begin{pmatrix} \gamma & -\beta\gamma & 0 & 0 \\ -\beta\gamma & \gamma & 0 & 0 \\ 0 & 0 & 1 & 0 \\ 0 & 0 & 0 & 1 \end{pmatrix} \cdot \begin{pmatrix} ct \\ x^1 \\ x^2 \\ x^3 \end{pmatrix} \tag{17.35}$$

This Lorentz transformation matrix is called a *standard boost* since it only boosts from one frame to another parallel frame. In general a rotation matrix also is incorporated into the transformation matrix λ for the spatial variables.

17.5.2 Four-vector scalar products

Scalar products of vectors and tensors usually are invariant to rotations in three-dimensional space providing an easy way to solve problems. The scalar, or inner, product of two four vectors is defined by

$$\mathbb{X} \cdot \mathbb{Y} = g_{\mu\nu}X^{\mu}Y^{\nu} = \begin{pmatrix} X^0 & X^1 & X^2 & X^3 \end{pmatrix} \cdot \begin{pmatrix} 1 & 0 & 0 & 0 \\ 0 & -1 & 0 & 0 \\ 0 & 0 & -1 & 0 \\ 0 & 0 & 0 & -1 \end{pmatrix} \cdot \begin{pmatrix} Y^0 \\ Y^1 \\ Y^2 \\ Y^3 \end{pmatrix} \quad (17.36)$$

$$= X^0Y^0 - X^1Y^1 - X^2Y^2 - X^3Y^3$$

The correct sign of the inner product is obtained by inclusion of the *Minkowski metric g* defined by

$$g_{\mu\nu} \equiv \hat{\mathbf{e}}_{\mu} \cdot \hat{\mathbf{e}}_{\nu} \quad (17.37)$$

that is, it can be represented by the matrix

$$g \equiv \begin{pmatrix} 1 & 0 & 0 & 0 \\ 0 & -1 & 0 & 0 \\ 0 & 0 & -1 & 0 \\ 0 & 0 & 0 & -1 \end{pmatrix} \quad (17.38)$$

The sign convention used in the Minkowski metric, equation 17.38, has been chosen with the time coordinate $(ct)^2$ positive which makes $(ds)^2 > 0$ for objects moving at less than the speed of light and corresponds to ds being real.[2]

The presence of the Minkowski metric matrix, in the inner product of four vectors, complicates General Relativity and thus the Einstein convention has been adopted where the components of the *contravariant four-vector* $\mathbb{X}$ are written with superscripts X^{μ}. See also appendix E. The corresponding *covariant four-vector* components are written with the subscript X_{μ} which is related to the contravariant four-vector components X^{ν} using the $\mu\nu$ component of the covariant Minkowski metric matrix $\mathbf{g}$. That is

$$X_{\mu} = \sum_{\nu=0}^{3} g_{\mu\nu}X^{\nu} \quad (17.39)$$

The contravariant metric component $g^{\mu\nu}$ is defined as the $\mu\nu$ component of the inverse metric matrix $\mathbf{g}^{-1}$ where

$$\mathbf{g}\mathbf{g}^{-1} = \mathbf{I} = \mathbf{g}^{-1}\mathbf{g} \quad (17.40)$$

where $\mathbf{I}$ is the four-vector identity matrix. The contravariant components of the four vector can be expressed in terms of the covariant components as

$$X^{\mu} = \sum_{\nu=0}^{3} g^{\mu\nu}X_{\nu} \quad (17.41)$$

Thus equations 17.39 and 17.41 can be used to transform between covariant and contravariant four vectors, that is, to raise or lower the index μ.

The scalar inner product of two four vectors can be written compactly as the scalar product of a covariant four vector and a contravariant four vector. The Minkowski metric matrix can be absorbed into either $\mathbb{X}$ or $\mathbb{Y}$ thus

$$\mathbb{X} \cdot \mathbb{Y} = \sum_{\mu=0}^{3}\sum_{\nu=0}^{3} g_{\mu\nu}X^{\mu}Y^{\nu} = \sum_{\nu=0}^{3} X_{\nu}Y^{\nu} = \sum_{\mu=0}^{3} X^{\mu}Y_{\mu} \quad (17.42)$$

If this covariant expression is Lorentz invariant in one coordinate system, then it is Lorentz invariant in all coordinate systems obtained by proper Lorentz transformations.

---

[2] Older textbooks, such as all editions of Marion, and the first two editions of Goldstein, use the Euclidean Poincaré 4-dimensional space-time with the imaginary time axis *ict*. About half the scientific community, and modern physics textbooks including this textbook and the 3rd edition of Goldstein, use the Bjorken - Drell $+, -, -, -$, sign convention given in equation 17.38 where $x_0 \equiv ct$, and x_1, x_2, x_3 are the spatial coordinates. The other half of the community, including mathematicians and gravitation physicists, use the opposite $-, +, +, +$, sign convention. Further confusion is caused by a few books that assign the time axis ct to be x_4 rather than x_0.

The scalar inner product of the invariant space-time interval is an especially important example.

$$(ds)^2 \equiv \mathbb{X} \cdot \mathbb{X} = c^2 (dt)^2 - (d\mathbf{r})^2 = (cdt)^2 - \sum_{i=1}^{3} dx_i^2 = (cd\tau)^2 \qquad (17.43)$$

This is invariant to a Lorentz transformation as can be shown by applying the Lorentz standard boost transformation given above. In particular, if S' is the rest frame of the clock, then the invariant space-time interval ds is simply given by the proper time interval $d\tau$.

17.5.3 Minkowski space-time

Figure 17.6 illustrates a three-dimensional (ct, x^1, x^2) representation of the 4−dimensional space-time diagram where it is assumed that $x^3 = 0$. The fact that the velocity of light has a fixed velocity leads to the concept of the light cone defined by the locus of $|x| = ct$.

Inside the light cone

The vertex of the cones represent the present. Locations inside the upper cone represent the future while the past is represented by locations inside the lower cone. Note that $(ds)^2 = c^2 (dt)^2 - (d\mathbf{r})^2 > 0$ inside both the future and past light cones. Thus the space-time interval $c\Delta t$ is real and positive for the future, whereas it is real and negative for the past relative to the vertex of the light cone. A **world line** is the trajectory a particle follows is a function of time in Minkowski space. In the interior of the future light cone $\Delta t > 0$ and, since it is real, it can be asserted unambiguously that any point inside this forward cone must occur later than at the vertex of the cone, that is, it is the absolute future. A Lorentz transformation can rotate Minkowski space such that the axis x_0 goes through any point within this light cone and then the "world line" is pure **time like**. Similarly, any point inside the backward light cone unambiguously occurred before the vertex, i.e. it is absolute past.

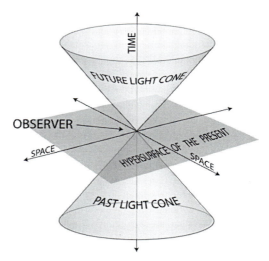

Figure 17.6: The light cone in the ct, x_1, x_2 space is defined by the condition $\mathbb{X} \cdot \mathbb{X} = c^2 t^2 - r^2 = 0$ and divides space-time into the forward and backward light cones, with $t > 0$ and $t < 0$ respectively; the interiors of the forward and backward light cones are called absolute future and absolute past.

Outside the light cone

Outside of the light cone, has $(ds)^2 = c^2 (dt)^2 - (d\mathbf{r})^2 < 0$ and thus Δs is imaginary and is called **space like**. A space-like plane hypersurface in spatial coordinates is shown for the present time in the unprimed frame. A rotation in Minkowski space can be made to s' such that the space-like hypersurface now is tilted relative to the hypersurface shown and thus any point P outside the light cone can be made to occur later, simultaneous, or earlier than at the vertex depending on the orientation of the space-like hypersurface. This startling situation implies that the time ordering of two points, each outside the others light cone, can be reversed which has profound implications related to the concept of simultaneity and the notion of causality.

For the special case of two events lying on the light cone $\sum_{\mu}^{4} x_\mu^2 = c^2 t^2 - (x_1^2 + x_2^2 + x_3^2) = 0$ and thus these events are separated by a light ray travelling at velocity c. Only events separated by time-like intervals can be connected causally. The world line of a particle must lie within its light cone. The division of intervals into space-like and time-like, because of their invariance, is an absolute concept. That is, it is independent of the frame of reference.

The concept of proper time can be expanded by considering a clock at rest in frame S' which is moving with uniform velocity v with respect to a rest frame S. The clock at rest in the S' frame measures the *proper time* τ, then the time observed in the fixed frame can be obtained by looking at the interval ds. Because of the invariance of the interval, ds^2 then

$$ds^2 = c^2 d\tau^2 = c^2 dt^2 - \left[dx_1^2 + dx_2^2 + dx_3^2\right] \tag{17.44}$$

That is,

$$d\tau = dt \left[1 - \frac{\left(dx_1^2 + dx_2^2 + dx_3^2\right)}{c^2 dt^2}\right]^{\frac{1}{2}} = dt \left[1 - \frac{v^2}{c^2}\right]^{\frac{1}{2}} = \frac{dt}{\gamma} \tag{17.45}$$

that is $dt = \gamma d\tau$ which satisfies the normal expression for time dilation, 17.8.

17.5.4 Momentum-energy four vector

The previous four-vector discussion can be elegantly exploited using the covariant Minkowski space-time representation. Separating the spatial and time of the differential four vector gives

$$d\mathbb{X} = (cdt, d\mathbf{x}) \tag{17.46}$$

Remember that the square of the four-dimensional space-time element of length $(ds)^2$ is invariant (17.43), and is simply related to the proper time element $d\tau$. Thus the scalar product

$$d\mathbb{X} \cdot d\mathbb{X} = ds^2 = c^2 d\tau^2 = c^2 dt^2 - \left[dx_1^2 + dx_2^2 + dx_3^2\right] \tag{17.47}$$

Thus the proper time is an invariant.

The ratio of the four-vector element $d\mathbb{X}$ and the invariant proper time interval $d\tau$, is a four-vector called the *four-vector velocity* $\mathbb{U}$ where

$$\mathbb{U} = \frac{d\mathbb{X}}{d\tau} = \left(c\frac{dt}{d\tau}, \frac{d\mathbf{x}}{d\tau}\right) = \gamma_u \left(c, \frac{d\mathbf{x}}{dt}\right) = \gamma_u \left(c, \mathbf{u}\right) \tag{17.48}$$

where $\mathbf{u}$ is the particle velocity, and $\gamma_u = \frac{1}{\sqrt{\left(1 - \frac{u^2}{c^2}\right)}}$.

The four-vector momentum $\mathbb{P}$ can be obtained from the four-vector velocity by multiplying it by the scalar rest mass m

$$\mathbb{P} = m\mathbb{U} = (\gamma_u mc, \gamma_u m\mathbf{u}) \tag{17.49}$$

However,

$$\gamma_u mc = \frac{E}{c} \tag{17.50}$$

thus the momentum four vector can be written as

$$\mathbb{P} = \left(\frac{E}{c}, \mathbf{p}\right) \tag{17.51}$$

where the vector $\mathbf{p}$ represents the three spatial components of the relativistic momentum. It is interesting to realize that the Theory of Relativity couples not only the spatial and time coordinates, but also, it couples their conjugate variables linear momentum $\mathbf{p}$ and total energy, $\frac{E}{c}$.

An additional feature of this momentum-energy four vector $\mathbb{P}$, is that the scalar inner product $\mathbb{P} \cdot \mathbb{P}$ is invariant to Lorentz transformations and equals $(mc)^2$ in the rest frame

$$\mathbb{P} \cdot \mathbb{P} = \sum_{\mu=0}^{3}\sum_{\nu=0}^{3} g_{\mu\nu} P^\mu P^\nu = \sum_{\mu=0}^{3}\sum_{\nu=0}^{3} P_\mu P^\nu = (\frac{E}{c})^2 - |\mathbf{p}|^2 = m^2 c^2 \tag{17.52}$$

which leads to the well-known equation

$$E^2 = p^2 c^2 + E_0^2 \tag{17.53}$$

The Lorentz transformation matrix λ can be applied to $\mathbb{P}$

$$\mathbb{P}' = \boldsymbol{\lambda}\mathbb{P} \tag{17.54}$$

The Lorentz invariant four-vector representation is illustrated by applying the Lorentz transformation shown in figure 17.1, which gives, $p_1' = \gamma\left(p_1 - \left(\frac{v}{c}\right)^2 E\right)$, $p_2' = p_2$, $p_3' = p_3$, and $E' = \gamma\left(E - vp_1\right)$.

17.6 Lorentz-invariant formulation of Lagrangian mechanics

17.6.1 Parametric formulation

The Lagrangian and Hamiltonian formalisms in classical mechanics are based on the Newtonian concept of absolute time t which serves as the system evolution parameter in Hamilton's Principle. This approach violates the Special Theory of Relativity. The extended Lagrangian and Hamiltonian formalism is a parametric approach, pioneered by Lanczos[La49], that introduces a system evolution parameter s that serves as the independent variable in the action integral, and all the space-time variables $q_i(s), t(s)$ are dependent on the evolution parameter s. This extended Lagrangian and Hamiltonian formalism renders it to a form that is compatible with the Special Theory of Relativity. The importance of the Lorentz-invariant extended formulation of Lagrangian and Hamiltonian mechanics has been recognized for decades.[La49, Go50, Sy60] Recently there has been a resurgence of interest in the extended Lagrangian and Hamiltonian formalism stimulated by the papers of Struckmeier[Str05, Str08] and this formalism has featured prominently in recent textbooks by Johns[Jo05] and Greiner[Gr10]. This parametric approach develops manifestly-covariant Lagrangian and Hamiltonian formalisms that treat equally all $2n+1$ space-time canonical variables. It provides a plausible manifestly-covariant Lagrangian for the one-body system, but serious problems exist extending this to the N-body system when $N > 1$. Generalizing the Lagrangian and Hamiltonian formalisms into the domain of the Special Theory of Relativity is of fundamental importance to physics, while the parametric approach gives insight into the philosophy underlying use of variational methods in classical mechanics.[3]

In conventional Lagrangian mechanics, the equations of motion for the n generalized coordinates are derived by minimizing the action integral, that is, Hamilton's Principle.

$$\delta S(\mathbf{q}, \dot{\mathbf{q}}, t) = \delta \int_a^b L(\mathbf{q}(t), \dot{\mathbf{q}}(t), t) dt = 0 \qquad (17.55)$$

where $L(\mathbf{q}(t), \dot{\mathbf{q}}(t), t)$ denotes the conventional Lagrangian. This approach implicitly assumes the Newtonian concept of absolute time t which is chosen to be the independent variable that characterizes the evolution parameter of the system. The actual path $[\mathbf{q}(t), \dot{\mathbf{q}}(t)]$ the system follows is defined by the extremum of the action integral $S(\mathbf{q}, \dot{\mathbf{q}}, t)$ which leads to the corresponding Euler-Lagrange equations. This assumption is contrary to the Theory of Relativity which requires that the space and time variables be treated equally, that is, the Lagrangian formalism must be covariant.

17.6.2 Extended Lagrangian

Lanczos[La49] proposed making the Lagrangian covariant by introducing a general evolution parameter s, and treating the time as a dependent variable $t(s)$ on an equal footing with the configuration space variables $q^i(s)$. That is, the time becomes a dependent variable $q_0(s) = ct(s)$ similar to the spatial variables $q_\mu(s)$ where $1 \le \mu \le n$. The dynamical system then is described as motion confined to a hypersurface within an extended space where the value of the extended Hamiltonian and the evolution parameter s constitute an additional pair of canonically conjugate variables in the extended space. That is, the canonical momentum p_0, corresponding to $q_0 = ct$, is $p_0 = \frac{E}{c}$ similar to the momentum-energy four vector, equation 17.51.

An *extended Lagrangian* $\mathbb{L}(\mathbf{q}(s), \frac{d\mathbf{q}(s)}{ds}, t(s), \frac{dt(s)}{ds})$ can be defined which can be written compactly as $\mathbb{L}(q^\mu(s), \frac{dq^\mu(s)}{ds})$ where the index $0 \le \mu \le n$ denotes the entire range of space-time variables.

This extended Lagrangian can be used in an extended action functional $\mathbb{S}(\mathbf{q}, \frac{d\mathbf{q}}{ds}, t, \frac{dt}{ds})$ to give an extended version of Hamilton's Principle[4]

$$\delta \mathbb{S}(\mathbf{q}, \frac{d\mathbf{q}}{ds}, t, \frac{dt}{ds}) = \delta \int_a^b \mathbb{L}(q^\mu(s), \frac{dq^\mu(s)}{ds}) ds = 0 \qquad (17.56)$$

---

[3]Chapters 17.6 and 17.7 reproduce the Struckmeier presentation.[Str08]

[4]These formula involve total and partial derivatives with respect to both time, t and parameter s. For clarity, the derivatives are written out in full because Lanczos[La49] and Johns[Jo05] use the opposite convention for the dot and prime superscripts as abbreviations for the differentials with respect to t and s. The blackboard bold format is used to designate the extended versions of the action $\mathbb{S}$, Lagrangian $\mathbb{L}$ and Hamiltonian $\mathbb{H}$.

The conventional action S, and extended action $\mathbb{S}$, address alternate characterizations of the same underlying physical system, and thus the action principle implies that $\delta S = \delta \mathbb{S} = 0$ must hold simultaneously. That is,

$$\delta \int_a^b L(\mathbf{q}, \frac{d\mathbf{q}}{dt}, t) \frac{dt}{ds} ds = \delta \int_a^b \mathbb{L}(\mathbf{q}, \frac{d\mathbf{q}}{ds}, t, \frac{dt}{ds}) ds \tag{17.57}$$

As discussed in chapter 9.3, there is a continuous spectrum of equivalent gauge-invariant Lagrangians for which the Euler-Lagrange equations lead to identical equations of motion. Equation 17.57 is satisfied if the conventional and extended Lagrangians are related by

$$\mathbb{L}(\mathbf{q}, \frac{d\mathbf{q}}{ds}, t, \frac{dt}{ds}) = L(\mathbf{q}, \frac{d\mathbf{q}}{dt}, t) \frac{dt}{ds} + \frac{d\Lambda(\mathbf{q}, t)}{ds} \tag{17.58}$$

where $\Lambda(\mathbf{q}, t)$ is a continuous function of $\mathbf{q}$ and t that has continuous second derivatives. It is acceptable to assume that $\frac{d\Lambda(\mathbf{q}, t)}{ds} = 0$, then the extended and conventional Lagrangians have a unique relation requiring no simultaneous transformation of the dynamical variables. That is, assume

$$\mathbb{L}(\mathbf{q}, \frac{d\mathbf{q}}{ds}, t, \frac{dt}{ds}) = L(\mathbf{q}, \frac{d\mathbf{q}}{dt}, t) \frac{dt}{ds} \tag{17.59}$$

Note that the time derivative of $\mathbf{q}$ can be expressed in terms of the s derivatives by

$$\frac{d\mathbf{q}}{dt} = \frac{d\mathbf{q}/ds}{dt/ds} \tag{17.60}$$

Thus, for a conventional Lagrangian with n variables, the corresponding extended Lagrangian is a function of $n+1$ variables while the conventional and extended Lagrangians are related using equations 17.59, and 17.60.

The derivatives of the relation between the extended and conventional Lagrangians lead to

$$\frac{\partial \mathbb{L}}{\partial q^\mu} = \frac{\partial L}{\partial q^\mu} \frac{dt}{ds} \tag{17.61}$$

$$\frac{\partial \mathbb{L}}{\partial t} = \frac{\partial L}{\partial t} \frac{dt}{ds} \tag{17.62}$$

$$\frac{\partial \mathbb{L}}{\partial \left(\frac{dq^\mu}{ds}\right)} = \frac{\partial L}{\partial \left(\frac{dq^\mu}{dt}\right)} \tag{17.63}$$

$$\frac{\partial \mathbb{L}}{\partial \left(\frac{dt}{ds}\right)} = L - \sum_{\mu=1}^n \frac{\partial L}{\partial \left(\frac{dq^\mu}{dt}\right)} \frac{dq^\mu}{dt} \tag{17.64}$$

where $1 \leq \mu \leq n$ since the $\mu = 0$ time derivatives are written explicitly in equations 17.62, 17.64.

Equations 17.63 – 17.64, summed over the extended range $0 \leq \mu \leq n$ of time and spatial dynamical variables, imply

$$\sum_{\mu=0}^n \frac{\partial \mathbb{L}}{\partial \left(\frac{dq^\mu}{ds}\right)} \left(\frac{dq^\mu}{ds}\right) = L \frac{dt}{ds} - \sum_{\mu=1}^n \frac{\partial L}{\partial \left(\frac{dq^\mu}{dt}\right)} \frac{dq^\mu}{dt} \frac{dt}{ds} + \sum_{i=1}^n \frac{\partial L}{\partial \left(\frac{dq^\mu}{dt}\right)} \frac{dq^\mu}{ds} = \mathbb{L} \tag{17.65}$$

Equation 17.65 can be written in the form

$$\mathbb{L} - \sum_{\mu=0}^n \frac{\partial \mathbb{L}}{\partial \left(\frac{dq^\mu}{ds}\right)} \frac{dq^\mu}{ds} = \begin{cases} \not\equiv 0 \text{ if } \mathbb{L} \text{ is not homogeneous in } \frac{dq^\mu}{ds} \\ \equiv 0 \text{ if } \mathbb{L} \text{ is homogeneous in } \frac{dq^\mu}{ds} \end{cases} \tag{17.66}$$

If the extended Lagrangian $\mathbb{L}(\mathbf{q}, \frac{d\mathbf{q}}{ds}, t, \frac{dt}{ds})$ is homogeneous to first order in the $n+1$ variables $\frac{dq^\mu}{ds}$, then Euler's theorem on homogeneous functions trivially implies the relation given in equation 17.66. Struckmeier[Str08] identified a subtle but important point that if $\mathbb{L}$ is not homogeneous in $\frac{dq^\mu}{ds}$, then equation 17.66 is not an identity but is an implicit equation that is always satisfied as the system evolves according to the solution of the extended Euler-Lagrange equations. Then equation 17.59 is satisfied without it being a homogeneous form in the $n+1$ velocities $\frac{dq^\mu}{ds}$. This introduces a new class of non-homogeneous Lagrangians. The relativistic free particle, discussed in example 17.5, is a case of a non-homogeneous extended Lagrangian.

17.6.3 Extended generalized momenta

The generalized momentum is defined by

$$p_\mu = \frac{\partial L}{\partial \left(\frac{\partial q^\mu}{\partial t} \right)} \tag{17.67}$$

Assume that the definitions of the extended Lagrangian $\mathbb{L}$, and the extended Hamiltonian $\mathbb{H}$, are related by a Legendre transformation, and are based on variational principles, analogous to the relation that exists between the conventional Lagrangian L and Hamiltonian H. The Legendre transformation requires defining the extended generalized (canonical) momentum-energy four vector $\mathbb{P}(s) = (\frac{\mathrm{E}(s)}{c}, \mathbf{p}(s))$. The momentum components of the momentum-energy four vector $\mathbb{P}(s) = (\frac{\mathrm{E}(s)}{c}, \mathbf{p}(s))$ are given by the $1 \leq \mu \leq n$ components using equation 17.63.

$$p_\mu(s) = \frac{\partial \mathbb{L}}{\partial \left(\frac{dq^\mu}{ds} \right)} = \frac{\partial L}{\partial \left(\frac{dq^\mu}{dt} \right)} \tag{17.68}$$

The $\mu = 0$ component of the momentum-energy four vector can be derived by recognizing that the right-hand side of equation 17.64 is equal to $-H(p_\mu, q^\mu, t)$. That is, the corresponding generalized momentum p_0, that is conjugate to $q_0 = ct$, is given by

$$p_0 = \frac{\partial \mathbb{L}}{\partial \left(\frac{dq^0}{ds} \right)} = \frac{1}{c} \left(\frac{\partial \mathbb{L}}{\partial \left(\frac{dt}{ds} \right)} \right) = \frac{1}{c} \left(L - \sum_{\mu=1}^{n} \frac{\partial L}{\partial \left(\frac{dq^\mu}{dt} \right)} \frac{dq^\mu}{dt} \right) = -\frac{H(p_\mu, q^\mu, t)}{c} \tag{17.69}$$

17.6.4 Extended Lagrange equations of motion

By direct analogy with the non-relativistic action integral 17.55, the extremum for the relativistic action integral $\mathbb{S}(\mathbf{q}, \frac{d\mathbf{q}}{ds}, t, \frac{dt}{ds})$ is obtained using the Euler-Lagrange equations derived from equation 17.56 where the independent variable is s. This implies that for $0 \leq \mu \leq n$

$$\frac{d}{ds} \left(\frac{\partial \mathbb{L}}{\partial \left(\frac{dq^\mu}{ds} \right)} \right) - \frac{\partial \mathbb{L}}{\partial q^\mu} = \mathbb{Q}_\mu^{EX} = \sum_{k=1}^{m} \frac{dt}{ds} \lambda_k \frac{\partial g_k}{\partial q^\mu} + Q_\mu^{EXC} \frac{dt}{ds} \tag{17.70}$$

where the extended generalized force $\mathbb{Q}_\mu^{EX}$ shown on the right-hand side of equation 17.70, accounts for all forces not included in the potential energy term in the Lagrangian. The extended generalized force $\mathbb{Q}_\mu^{EX}$ can be factored into two terms as discussed in chapter 6, equation 6.60. The Lagrange multiplier term includes $1 \leq k \leq m$ holonomic constraint forces where the m holonomic constraints, which do no work, are expressed in terms of the m algebraic equations of holonomic constraint g_k. The Q_μ^{EXC} term includes the remaining constraint forces and generalized forces that are not included in the Lagrange multiplier term or the potential energy term of the Lagrangian.

For the case where $\mu = 0$, since $q_0 = ct$, then equation 17.70 reduces to

$$\frac{d}{ds} \left(\frac{\partial \mathbb{L}}{\partial \left(\frac{dt}{ds} \right)} \right) - \frac{\partial \mathbb{L}}{\partial t} = \sum_{k=1}^{m} \frac{dt}{ds} \lambda_k \frac{\partial g_k}{\partial t} - \sum_{\nu=1}^{n} Q_\nu^{EXC} \frac{dq^\nu}{ds} \tag{17.71}$$

These Euler-Lagrange equations of motion 17.70, 17.71 determine the $1 \leq \mu \leq n$ generalized coordinates $q^\mu(s)$, plus $q^0 = ct(s)$ in terms of the independent variable s.

If the holonomic equations of constraint are time independent, that is $\frac{\partial g_k}{\partial t} = 0$ and if $\mathbb{Q}_0^{EXC} = 0$, then the $\mu = 0$ term of the Euler-Lagrange equations simplifies to

$$\frac{d}{ds} \left(\frac{\partial \mathbb{L}}{\partial \left(\frac{dt}{ds} \right)} \right) - \frac{\partial \mathbb{L}}{\partial t} = 0 \tag{17.72}$$

One interpretation is to select L to be primary. Then $\mathbb{L}$ is derived from L using equation 17.59 and $\mathbb{L}$ must satisfy the identity given by equation 17.66 while the Euler-Lagrange equations containing $\frac{dt}{ds}$ yield an identity which implies that L does not provide an equation of motion in terms of $t(s)$. Conversely, if $\mathbb{L}$ is chosen to be primary, then $\mathbb{L}$ is no longer a homogeneous function and equation 17.66 serves as a constraint on the motion that can be used to deduce L, while $\frac{dt}{ds}$ yields a non-trivial equation of motion in terms of $t(s)$. In both cases the occurrence of a constraint surface results from the fact that the extended space has $2n+2$ variables to describe $2n+1$ degrees of freedom, that is, one more degree of freedom than required for the actual system.

17.5 *Example: Lagrangian for a relativistic free particle*

The standard Lagrangian $L = T - U$ is not Lorentz invariant. The extended Lagrangian $L(q, \frac{d\mathbf{q}}{ds}, t, \frac{dt}{ds})$ introduces the independent variable s which treats both the space variables $q(s)$ and time variable $q_0 = ct(s)$ equally. This can be achieved by defining the non-standard Lagrangian

$$\mathbb{L}(\mathbf{q}, \frac{d\mathbf{q}}{ds}, t, \frac{dt}{ds}) = \frac{1}{2}mc^2 \left[\frac{1}{c^2} \left(\frac{d\mathbf{q}}{ds} \right)^2 - (\frac{dt}{ds})^2 - 1 \right] \tag{α}$$

The constant third term in the bracket is included to ensure that the extended Lagrangian converges to the standard Lagrangian in the limit $\frac{dt}{ds} \to 1$.

Note that the extended Lagrangian (α) is not homogeneous to first order in the velocities $\frac{d\mathbf{q}}{ds}$ as is required. Equation 17.66 must be used to ensure that equation (α) is homogeneous. That is, it must satisfy the constraint relation

$$\left(\frac{dt}{ds} \right)^2 - \frac{1}{c^2} \left(\frac{d\mathbf{q}}{ds} \right)^2 - 1 = 0 \tag{β}$$

Inserting (β) into the extended Lagrangian (α) yields that the square bracket in equation α must equal 2. Thus

$$|\mathbb{L}| = \frac{1}{2}mc^2 \left[-2 \right] = -mc^2 \tag{γ}$$

The constraint equation (β) implies that

$$\frac{ds}{dt} = \sqrt{1 - \frac{1}{c^2} \left(\frac{d\mathbf{q}}{dt} \right)^2} = \frac{1}{\gamma} \tag{δ}$$

Using equation (δ) gives that the relativistic Lagrangian is

$$L = \frac{\mathbb{L}}{\gamma} = -\frac{mc^2}{\gamma} = -mc^2\sqrt{1 - \beta^2} \tag{ϵ}$$

Equation (ϵ) is the conventional relativistic Lagrangian derived by assuming that the system evolution parameter s is transformed to be along the world line ds, where the invariant length ds replaces the proper time interval

$$ds = cd\tau = \frac{cdt}{\gamma} \tag{ε}$$

The definition of the generalized (canonical) momentum

$$p_i = \frac{\partial L}{d\dot{q}_i} = \gamma m \dot{q}_i \tag{ζ}$$

leads to the relativistic expression for momentum given in equation 17.21.

The relativistic Lagrangian is an important example of a non-standard Lagrangian. Equation (α) does not equal the difference between the kinetic and potential energies, that is, the relativistic expression for kinetic energy is given by 17.28 to be

$$T = (\gamma - 1) mc^2 \tag{η}$$

The non-standard relativistic Lagrangian (ϵ) can be used with the Euler-Lagrange equations to derive the second-order equations of motion for both relativistic and non-relativistic problems within the Special Theory of Relativity.

17.6 Example: Relativistic particle in an external electromagnetic field

A charged particle moving at relativistic speed in an external electromagnetic field provides an example of the use of the relativistic Lagrangian.

In the discussion of classical mechanics it was shown that the velocity-dependent Lorentz force can be absorbed into the scalar electric potential Φ plus the vector magnetic potential $\mathbf{A}$. That is, the potential energy is given by equation 7.6 to be $U = q(\Phi - \mathbf{A} \cdot \mathbf{v})$. Including this in the Lagrangian, 17.71, gives

$$L = -\frac{mc^2}{\gamma} - U = -mc^2\sqrt{1 - \beta^2} - q\Phi + q\mathbf{A} \cdot \mathbf{v}$$

The three spatial partial derivatives can be written in vector notation as

$$\frac{\partial L}{\partial \mathbf{r}} = -q\boldsymbol{\nabla}\Phi + \frac{q}{c}\boldsymbol{\nabla}(\mathbf{v} \cdot \mathbf{A}) \qquad (a)$$

and the generalized momentum is given by

$$\mathbf{p} = \frac{\partial L}{d\mathbf{v}} = \gamma m\mathbf{v} + q\mathbf{A}$$

which is identical to the non-relativistic answer given by equation 7.6. That is, it includes the momentum of the electromagnetic field plus the classical linear momentum of the moving particle.

The total time derivative of the generalized momentum is

$$\frac{d\mathbf{p}}{dt} = \frac{d}{dt}\left(\frac{\partial L}{d\mathbf{v}}\right) = \frac{d}{dt}(\gamma m\mathbf{v}) + q\frac{d\mathbf{A}}{dt} \qquad (b)$$

where the last term is given by the chain rule

$$\frac{d\mathbf{A}}{dt} = \frac{\partial \mathbf{A}}{\partial t} + (\mathbf{v} \cdot \boldsymbol{\nabla})\mathbf{A} \qquad (c)$$

Using equations a, b, c in the Euler-Lagrange equation gives

$$\frac{d}{dt}\left(\frac{\partial L}{d\mathbf{v}}\right) = \frac{\partial L}{\partial \mathbf{r}}$$

$$\frac{d}{dt}(\gamma m\mathbf{v}) + q\frac{d\mathbf{A}}{dt} = -q\boldsymbol{\nabla}\Phi + q\boldsymbol{\nabla}(\mathbf{v} \cdot \mathbf{A})$$

Collecting terms and using the well-known vector-product identity, plus the definition $\mathbf{B} = \boldsymbol{\nabla} \times \mathbf{A}$, gives

$$\frac{d}{dt}(\gamma m\mathbf{v}) = -\left[q\boldsymbol{\nabla}\Phi - q\frac{\partial \mathbf{A}}{\partial t}\right] + q\left[\boldsymbol{\nabla}(\mathbf{v} \cdot \mathbf{A}) - (\mathbf{v} \cdot \boldsymbol{\nabla})\mathbf{A}\right]$$

$$= -q\left[\boldsymbol{\nabla}\Phi - \frac{\partial \mathbf{A}}{\partial t}\right] + q\left[\mathbf{v} \times \boldsymbol{\nabla} \times \mathbf{A}\right]$$

$$\mathbf{F} = q\left[\mathbf{E} + \mathbf{v} \times \mathbf{B}\right]$$

If we adopt the definition that the relativistic canonical momentum is $p = \gamma mv$ then the left hand side is the relativistic force while the right-hand side is the well-known Lorentz force of electromagnetism. Thus the extended Lagrangian formulation correctly reproduces the well-known Lorentz force for a charged particle moving in an electromagnetic field.

17.7 Lorentz-invariant formulations of Hamiltonian mechanics

17.7.1 Extended canonical formalism

A Lorentz-invariant formulation of Hamiltonian mechanics can be developed that is built upon the extended Lagrangian formalism assuming that the Hamiltonian and Lagrangian are related by a Legendre transformation. That is,

$$H(\mathbf{q}, \mathbf{p}, t) = \sum_{\mu=1}^{n} p_\mu \frac{\partial q^\mu}{\partial t} - L(\mathbf{q}, \frac{\partial \mathbf{q}}{\partial t}, t) \tag{17.73}$$

where the generalized momentum is defined by

$$p_\mu = \frac{\partial L}{\partial \left(\frac{\partial q^\mu}{\partial t} \right)} \tag{17.74}$$

Struckmeier[Str08] assumes that the definitions of the extended Lagrangian $\mathbb{L}$, and the extended Hamiltonian $\mathbb{H}$, are related by a Legendre transformation, and are based on variational principles, analogous to the relation that exists between the conventional Lagrangian L and Hamiltonian H. The Legendre transformation requires defining the extended generalized (canonical) momentum-energy four vector $\mathbb{P}(s) = (\frac{\mathbb{E}(s)}{c}, \mathbf{p}(s))$. The momentum components of the momentum-energy four vector $\mathbb{P}(s) = (\frac{\mathbb{E}(s)}{c}, \mathbf{p}(s))$ are given by the $1 \leq \mu \leq n$ components using either the conventional or the extended Lagrangians as given in equation 17.68

$$p_\mu(s) = \frac{\partial \mathbb{L}}{\partial \left(\frac{dq^\mu}{ds} \right)} = \frac{\partial L}{\partial \left(\frac{dq^\mu}{dt} \right)} \tag{17.68}$$

The $\mu = 0$ component of the momentum-energy four vector is given by equation 17.69

$$p_0 = \frac{1}{c} \left(\frac{\partial \mathbb{L}}{\partial \left(\frac{dt}{ds} \right)} \right) = -\frac{H(p_\mu, q^\mu, t)}{c} = -\frac{\mathcal{E}(s)}{c} \tag{17.75}$$

where $\mathcal{E}(s)$ represents the instantaneous generalized energy of the conventional Hamiltonian at the point s, but not the functional form of $H(\mathbf{q}(s), \mathbf{p}(s), t(s))$. That is

$$\mathcal{E}(s) \stackrel{\neq}{=} H(\mathbf{q}(s), \mathbf{p}(s), t(s)) \tag{17.76}$$

Note that $\mathcal{E}(s)$ does not give the function $H(\mathbf{q}, \mathbf{p}, t)$. Equations 17.68 and 17.69 give that

$$p_0(s) = -\frac{\mathcal{E}(s)}{c} \tag{17.77}$$

The extended Hamiltonian $\mathbb{H}(\mathbf{q}, \mathbf{p}, t, \mathcal{E}(s))$, in an extended phase space, can be defined by the Legendre transformation and the four-vector $\mathbb{P}$ to be

$$\begin{aligned} \mathbb{H}(\mathbf{q}, \mathbf{p}, t, \mathcal{E}(s)) &= (\mathbb{P} \cdot \mathbf{q}) - \mathbb{L}(\mathbf{q}, \frac{d\mathbf{q}}{ds}, t, \frac{dt}{ds}) \tag{17.78} \\ &= \sum_{\mu=0}^{n} p_\mu \left(\frac{dq^\mu}{ds} \right) - \mathbb{L}(\mathbf{q}, \frac{d\mathbf{q}}{ds}, t, \frac{dt}{ds}) \\ &= \sum_{\mu=1}^{n} p_\mu \left(\frac{dq^\mu}{ds} \right) - \mathcal{E} \frac{dt}{ds} - \mathbb{L}(\mathbf{q}, \frac{d\mathbf{q}}{ds}, t, \frac{dt}{ds}) \tag{17.79} \end{aligned}$$

where the p_0 term has been written explicitly as $-\mathcal{E}\frac{dt}{ds}$ in equation 17.79. The extended Hamiltonian $\mathbb{H}((\mathbf{q}, \mathbf{p}, t, \mathcal{E}(s))$ can carry all the information on the dynamical system that is carried by the extended Lagrangian $\mathbb{L}(\mathbf{q}, \frac{d\mathbf{q}}{ds}, t, \frac{dt}{ds})$, if the Hesse matrix is non-singular. That is, if

$$\det \left(\frac{\partial^2 \mathbb{L}}{\partial \left(\frac{dq^\mu}{ds} \right) \partial \left(\frac{dq_\nu}{ds} \right)} \right) \neq 0 \tag{17.80}$$

If the extended Lagrangian $\mathbb{L}(\mathbf{q}, \frac{d\mathbf{q}}{ds}, t, \frac{dt}{ds})$ is not homogeneous in the $n+1$ velocities $\frac{dq^\mu}{ds}$, then the extended set of Euler-Lagrange equations 17.72 is not redundant. Thus equation 17.66 is not an identity but it can be regarded as an implicit equation that is always satisfied by the extended set of Euler-Lagrange equations. As a result, the Legendre transformation to an extended Hamiltonian exists. That is, equation 17.66 is identical to the Legendre transform for $\mathbb{H}((\mathbf{q}, \mathbf{p}, t, \mathcal{E}(s))$ which was shown to equal zero. Therefore

$$\mathbb{H}(\mathbf{q}(s), \mathbf{p}(s), t(s), \mathcal{E}(s)) = 0 \tag{17.81}$$

which means that the extended Hamiltonian $\mathbb{H}((\mathbf{q}, \mathbf{p}, t, \mathcal{E}(s))$ directly defines the restricted hypersurface on which the particle motion is confined.

The extended canonical equations of motion, derived using the extended Hamiltonian $\mathbb{H}(\mathbf{q}(s), \mathbf{p}(s), t(s), \mathcal{E}(s))$ with the usual Hamiltonian mechanics relations, are:

$$\frac{\partial \mathbb{H}}{\partial p_\mu} = \frac{dq^\mu}{ds} \tag{17.82}$$

$$\frac{\partial \mathbb{H}}{\partial q^\mu} = -\frac{dp_\mu}{ds} \tag{17.83}$$

$$\frac{\partial \mathbb{H}}{\partial t} = \frac{d\mathcal{E}}{ds} \tag{17.84}$$

$$\frac{\partial \mathbb{H}}{\partial \mathcal{E}} = -\frac{dt}{ds} \tag{17.85}$$

These canonical equations give that the total derivative of $\mathbb{H}((\mathbf{q}(s), \mathbf{p}(s), t(s), \mathcal{E}(s))$ with respect to s, is

$$\begin{aligned}
\frac{d\mathbb{H}}{ds} &= \frac{\partial \mathbb{H}}{\partial p_\mu}\frac{dp_\mu}{ds} + \frac{\partial \mathbb{H}}{\partial q^\mu}\frac{dq^\mu}{ds} + \frac{\partial \mathbb{H}}{\partial t}\frac{dt}{ds} + \frac{\partial \mathbb{H}}{\partial \mathcal{E}}\frac{d\mathcal{E}}{ds} \\
&= \frac{dq^\mu}{ds}\frac{dp_\mu}{ds} - \frac{dp_\mu}{ds}\frac{dq^\mu}{ds} + \frac{d\mathcal{E}}{ds}\frac{dt}{ds} - \frac{dt}{ds}\frac{d\mathcal{E}}{ds} = 0
\end{aligned} \tag{17.86}$$

That is, in contrast to the total time derivative of $H(\mathbf{q}, \mathbf{p}, t)$, the total s derivative of the extended Hamiltonian $\mathbb{H}((\mathbf{q}(s), \mathbf{p}(s), t(s), \mathcal{E}(s))$ always vanishes, that is, $\mathbb{H}(\mathbf{q}(s), \mathbf{p}(s), t(s), \mathcal{E}(s))$ is autonomous which is ideal for use with Hamilton's equations of motion. The constraints give that $\mathbb{H}(\mathbf{q}(s), \mathbf{p}(s), t(s), \mathcal{E}(s)) = 0$, (equation 17.81) and $\frac{d\mathbb{H}}{ds} = 0$, (equation 17.86) implying that the correlation between the extended and conventional Hamiltonians is given by

$$\begin{aligned}
\mathbb{H}((\mathbf{q}(s), \mathbf{p}(s), t(s), \mathcal{E}(s)) &= \sum_{\mu=1}^{n} p_\mu \left(\frac{dq^\mu}{ds}\right) - \mathcal{E}\frac{dt}{ds} - \mathbb{L}(\mathbf{q}, \frac{d\mathbf{q}}{ds}, t, \frac{dt}{ds}) \tag{17.87} \\
&= \sum_{\mu=1}^{n} p_\mu \left(\frac{dq^\mu}{ds}\right) - \mathcal{E}\frac{dt}{ds} - L(\mathbf{q}, \frac{d\mathbf{q}}{ds}, t,)\frac{dt}{ds} \tag{17.88} \\
&= \sum_{\mu=1}^{n} p_\mu \left(\frac{dq^\mu}{ds}\right) - \mathcal{E}\frac{dt}{ds} + \left[H(\mathbf{q}, \mathbf{p}, t) - \sum_{\mu=1}^{n} p_\mu \left(\frac{dq^\mu}{dt}\right)\right]\frac{dt}{ds} \tag{17.89} \\
&= (H(\mathbf{q}, \mathbf{p}, t) - \mathcal{E})\frac{dt}{ds} = 0 \tag{17.90}
\end{aligned}$$

since only the term with $\mu = 0$ does not cancel in equation 17.79. Equations 17.81 and 17.90 give that both the left and right-hand sides of equation 17.90 are zero while equation 17.86 implies that $\mathbb{H}((\mathbf{q}(s), \mathbf{p}(s), t(s), \mathcal{E}(s))$ is a constant of motion, that is, s is a cyclic variable for $\mathbb{H}((\mathbf{q}(s), \mathbf{p}(s), t(s), \mathcal{E}(s))$. Formally one can consider the extended Hamiltonian is a constant which equals zero

$$\mathbb{H}(\mathbf{q}, \mathbf{p}, t, \mathcal{E}(s)) = \mathbb{E}(s) = 0 \tag{17.91}$$

Equations 17.84, 17.85 imply that $(\mathcal{E}, t)$ form a pair of canonically conjugate variables in addition to the newly-introduced canonically-conjugate variables $(\mathbb{E}(s), s)$. Equation 17.90 shows that the motion in the $2n + 2$ extended phase space is constrained to the surface reflecting the fact that the observed system has one less degree of freedom than used by the extended Hamiltonian.

In summary, the Lorentz-invariant extended canonical formalism leads to Hamilton's first-order equations of motion in terms of derivatives with respect to s, where s is related to the proper time τ for a relativistic system.

17.7.2   Extended Poisson Bracket representation

Struckmeier[Str08] investigated the usefulness of the extended formalism when applied to the Poisson bracket representation of Hamiltonian mechanics. The extended Poisson bracket for two differentiable functions F and G is defined as

$$[[F, G]] = \sum_{j=1}^{n} \left(\frac{\partial F}{\partial q^j} \frac{\partial G}{\partial p_j} - \frac{\partial F}{\partial p_j} \frac{\partial G}{\partial q^j} \right) - \frac{\partial F}{\partial t} \frac{\partial G}{\partial H} + \frac{\partial F}{\partial H} \frac{\partial G}{\partial t} \tag{17.92}$$

As for the conventional Poisson bracket discussed in chapter 15, the extended Poisson also leads to the fundamental Poisson bracket relations

$$[[q^i, q^j]] = 0 \qquad\qquad [[p_i, p_j]] = 0 \qquad\qquad [[q^i, p_j]] = \delta^i_j \tag{17.93}$$

where $i, j = 0, 1, ..., n$. These are identical to the non-extended fundamental Poisson brackets.

The discussion of observables in Hamiltonian mechanics in chapter 15.2.5 can be trivially expanded to the extended Poisson bracket representation. In particular, the total s derivative of the function G is given by

$$\frac{dG}{ds} = \frac{\partial G}{\partial s} + [[G, \mathbb{H}]] \tag{17.94}$$

If G commutes with the extended Hamiltonian, that is, the Poisson bracket equals zero, and if $\frac{\partial G}{\partial s} = 0$, then $\frac{dG}{ds} = 0$. That is, the observable G is a constant of motion.

Substitute the fundamental variables for G gives

$$\frac{dp_\mu}{ds} = [[p_\mu, \mathbb{H}]] = -\frac{\partial \mathbb{H}}{\partial q^\mu} \qquad\qquad \frac{dq^\mu}{ds} = [[q^\mu, \mathbb{H}]] = \frac{\partial \mathbb{H}}{\partial p_\mu} \tag{17.95}$$

where $i, j = 0, 1, ..., n$.   These are Hamilton's extended canonical equations of motion expressed in terms of the system evolution parameter s. The extended Poisson bracket representation is a trivial extension of the conventional canonical equations presented in chapter 15.3.

17.7.3   Extended canonical transformation and Hamilton-Jacobi theory

Struckmeier[Str08] presented plausible extended versions of canonical transformation and Hamilton-Jacobi theories that can be used to provide a Lorentz-invariant formulation of Hamiltonian mechanics for relativistic one-body systems. A detailed description can be found in Struckmeier[Str08].[5]

17.7.4   Validity of the extended Hamilton-Lagrange formalism

It has been shown that the extended Lagrangian and Hamiltonian formalism, based on the parametric model of Lanczos[La49], leads to a plausible manifestly-covariant approach for the one-body system. The general features developed for handling Lagrangian and Hamiltonian mechanics carry over to the Special Theory of Relativity assuming the use of a non-standard, extended Lagrangian or Hamiltonian. This expansion of the range of validity of the well-known Hamiltonian and Lagrangian mechanics into the relativistic domain is important, and reduces any Lorentz transformation to a canonical transformation. The validity of this extended Hamilton-Lagrange formalism has been criticized, and problems exist extending this approach to the N-body system for $N > 1$. For example, as discussed by Goldstein[Go50] and Johns[Jo05], each of the N moving bodies have their own world lines and momenta. Defining the total momentum $\mathbf{P}$ requires knowing simultaneously the momenta of the individual bodies, but simultaneity is body dependent and thus even the total momentum is not a simple four vector. A general method is required that will allow using a manifestly-covariant Lagrangian or Hamiltonian for the N-body system. For the one-body system, the extended Hamilton-Lagrange formalism provides a powerful and logical approach to exploit analytical mechanics in the relativistic domain that retains the form of the conventional Lagrangian/Hamiltonian formalisms. Note that Noether's theorem relating energy and time is readily apparent using the extended formalism.

---

[5]Note that Greiner[Gr10] includes a reproduction of the Struckmeier paper[Str08].

17.7 Example: The Bohr-Sommerfeld hydrogen atom

The classical relativistic hydrogen atom was first solved by Sommerfeld in 1916. Sommerfeld used Bohr's "old quantum theory" plus Hamiltonian mechanics to make an important step in the development of quantum mechanics by obtaining the first-order expressions for the fine structure of the hydrogen atom. As in the non-relativistic case, the motion is confined to a plane allowing use of planar polar coordinates. Thus the relativistic Lagrangian is given by

$$L = -\frac{mc^2}{\gamma} - U = -mc^2\sqrt{1 - \frac{\dot{r}^2 + r^2\dot{\theta}^2}{c^2}} + \frac{ke^2}{r}$$

The canonical momenta are given by

$$p_\theta = \frac{\partial L}{\partial \dot{\theta}} = m\gamma r^2 \dot{\theta}$$

$$p_r = \frac{\partial L}{\partial \dot{r}} = m\gamma \dot{r}$$

$$\dot{p}_\theta = \frac{\partial L}{\partial \theta} = 0$$

$$\dot{p}_r = \frac{\partial L}{\partial r} = m\gamma r\dot{\theta}^2 + k\frac{e^2}{r^2}$$

As for the non-relativistic case, θ is a cyclic variable and thus the angular momentum $p_\theta = m\gamma r^2\dot{\theta}$ is conserved.

The relativistic Hamiltonian for the Coulomb potential between an electron and the proton, assuming that the motion is confined to a plane, which allows use of planar polar coordinates, leads to

$$H = \sqrt{p_r^2 c^2 + \frac{p_\theta^2 c^2}{r^2} + m^2 c^4} - \frac{ke^2}{r}$$

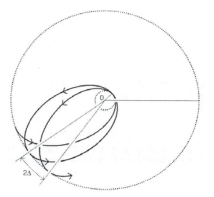

The advance of the perihelion of bound orbits due to the dependence of the relativistic mass on velocity.

The same equations of motion are obtained using Hamiltonian mechanics, that is:

$$\dot{\theta} = \frac{\partial H}{\partial p_\theta} = \frac{p_\theta}{m\gamma r^2}$$

$$\dot{r} = \frac{\partial H}{\partial p_r} = \frac{p_r}{m\gamma}$$

$$\dot{p}_\theta = -\frac{\partial H}{\partial \theta} = 0$$

$$\dot{p}_r = -\frac{\partial H}{\partial r} = m\gamma r\dot{\theta}^2 + k\frac{e^2}{r^2}$$

The radial dependence can be solved using either Lagrangian or Hamiltonian mechanics, but the solution is non-trivial. Using the same techniques applied to solve Kepler's problem, leads to the radial solution

$$r = \frac{q}{1 + \epsilon\cos[\Gamma(\theta - \theta_0)]} \qquad \Gamma = \sqrt{1 - \frac{e^4}{c^2 p_\theta^2}} \qquad q = \frac{c^2\Gamma^2 p_\theta^2}{e^2 E} \qquad \epsilon = \sqrt{1 + \frac{\Gamma^2(1 - \frac{m^2 c^4}{E^2})}{1 - \Gamma^2}}$$

The apses are $r_{\min} = \frac{q}{(1+\epsilon)}$ for $\Gamma(\theta - \theta_0) = 0, 2\pi, 4\pi$, and $r_{\max} = \frac{q}{(1-\epsilon)}$ for $\Gamma(\theta - \theta_0) = \pi, 3\pi,$. The perihelion advances between cycles due to the change in relativistic mass during the trajectory as shown in the adjacent figure. This precession leads to the fine structure observed in the optical spectra of the hydrogen atom. The same precession of the perihelion occurs for planetary motion, however, there is a comparable size effect due to gravity that requires use of general relativity to compute the trajectories.

17.8   The General Theory of Relativity

Einstein's General Theory of Relativity expands the scope of relativistic mechanics to include non-inertial accelerating frames plus a unified theory of gravitation. That is, the General Theory of Relativity incorporates both the Special Theory of Relativity as well as Newton's Law of Universal Gravitation. It provides a unified theory of gravitation that is a geometric property of space and time. In particular, the curvature of space-time is directly related to the four-momentum of matter and radiation. Unfortunately, Einstein's equations of general relativity are nonlinear partial differential equations that are difficult to solve exactly, and the theory requires knowledge of Riemannian geometry that goes beyond the scope of this book. The following summarizes the fundamental variational concepts underlying the theory, and the experimental evidence in support of the General Theory of Relativity.

17.8.1   The fundamental concepts

Einstein incorporated the following concepts in the General Theory of Relativity.

Mach's principle:

The 1883 work "The Science of Mechanics" by the philosopher/physicist, Ernst Mach, criticized Newton's concept of an absolute frame of reference, and suggested that local physical laws are determined by the large-scale structure of the universe. Mach's Principle assumes that local motion of a rotating frame is determined by the large-scale distribution of matter, that is, relative to the fixed stars. Einstein's interpretation of Mach's statement was that the inertial properties of a body is determined by the presence of other bodies in the universe, and he named this concept "Mach's Principle".

Equivalence principle:

The equivalence principle comprises closely-related concepts dealing with the equivalence of gravitational and inertial mass. The **weak equivalence principle** states that the inertial mass and gravitational mass of a body are identical, leading to acceleration that is independent of the nature of the body. Galileo demonstated this at the Leaning Tower of Pisa. Recent measurements have shown that this weak equivalence principle is obeyed to a sensitivity of 5×10^{-13}.   **Einstein's equivalence principle** states that the outcome of any local non-gravitational experiment, in a freely falling laboratory, is independent of the velocity of the laboratory and its location in space-time. This principle implies that the result of local experiments must be independent of the velocity of the apparatus. Einstein's equivalence principle has been tested by searching for variations of dimensionless fundamental constants such as the fine structure constant. The **strong equivalence principle** combines the weak equivalence and Einstein equivalence principles, and implies that the gravitational constant is constant everywhere in the universe. The strong equivalence principle suggests that gravity is geometrical in nature and does not involve any fifth force in nature. Tests of the strong equivalence principle have involved searches for variations in the gravitational constant G and masses of fundamental particles throughout the life of the universe.

Principle of covariance

A physical law that is expressed in a covariant formulation has the same mathematical form in all coordinate systems, and is usually expressed in terms of tensor fields. In the Special Theory of Relativity, the Lorentz, rotational, translational and reflection transformations between inertial coordinate frames are covariant. The covariant quantities are the 4-scalars, and 4-vectors in Minkowski space-time. Einstein recognized that the principle of covariance, that is built into the Special Theory of Relativity, should apply equally to accelerated relative motion in the General Theory of Relativity. He exploited tensor calculus to extend the Lorentz covariance to the more general local covariance in the General Theory of Relativity. The reduction locally of the general metric tensor to the Minkowski metric corresponds to free-falling motion, that is geodesic motion, and thus encompasses gravitation.

Principle of minimal gravitational coupling

The principle of minimal gravitational coupling requires that the total Lagrangian for the field equations of general relativity consist of two additive parts, one part corresponding to the free gravitational Lagrangian, and the other part to external source fields in curved space-time.

Correspondence principle

The Correspondence Principle states that the predictions of any new scientific theory must reduce to the predictions of well established earlier theories under circumstances for which the preceding theory was known to be valid. The Correspondence Principle is an important concept used both in quantum mechanics and relativistic mechanics. Einstein's Special Theory of Relativity satisfies the Correspondence Principle because it reduces to classical mechanics in the limit of velocities small compared to the speed of light. The Correspondence Principle requires that the General Theory of Relativity reduce to the Special Theory of Relativity for inertial frames, and should approximate Newton's Theory of Gravitation in weak fields and at low velocities.

17.8.2 Einstein's postulates for the General Theory of Relativity

Einstein realized that the Equivalence Principle relating the gravitational and inertial masses implies that the constancy of the velocity of light in vacuum cannot hold in the presence of a gravitational field. That is, the Minkowskian line element must be replaced by a more general line element that takes gravity into account. Einstein proposed that the Minkowskian line element in four-dimensional space-time, be replaced by introducing a four-dimensional Riemannian geometrical structure where space, time, and matter are combined. As described by Lancos[La49], [Har03], [Mu08] this astonishingly bold proposal implies that planetary motion is described as purely a geodesic phenomenon in a certain four-space of Riemannian structure, where the geodesic is the equation of a curve on a manifold for any possible set of coordinates. This implies that the concept of "gravitational force" is discarded, and planetary motion is a manifestation of a pure geodesic phenomenon for forceless motion in a four-dimensional Riemannian structure.

Chapters $6 - 9$ showed that the Lagrangian and Hamiltonian representations of variational principles are powerful approaches for determining the equation governing geodesic constrained motion that are independent of the chosen frame of reference as is also required by the General Theory of Relativity. Thus variational principles provide a theoretical representation for the General Theory of Relativity. The Einstein-Hilbert action is defined as

$$\mathbb{S} = \int \left[\frac{1}{16\pi G c^{-4}} \mathcal{R} + \mathcal{L}_M \right] \sqrt{-g} d^4 x \tag{17.96}$$

where G is Einstein's gravitational constant, $\mathcal{R}$ is the Ricci scalar, $\mathcal{L}_M$ accounts for matter fields, and g is the determinant of the metric tensor.matrix. Variational principles applied to the Einstein-Hilbert action lead to Einstein's sophisticated and advanced relativistic field equations of the General Theory of Relativity. Thus the variational approach unifies relativistic mechanics and classical field theories, such as mechanics and electromagnatism, which also were formulated in terms of least action. In relativistic mechanics, the use of action identifies the gravitational coupling of the metric to matter as well as identifying conserved quantities and symmetries using Noether's theorem. The Einstein-Hilbert action expands the scope of variational principles to include general relativity illustrating the crucial role played by variational principles in physics.

To summarize, the Special Theory of Relativity implies that the Newtonian concepts of absolute frame of reference and separation of space and time are invalid. The General Theory of Relativity goes beyond the Special Theory by implying that the gravitational force, and the resultant planetary motion, can be described as pure geodesic phenomena for forceless motion in a four-dimensional Riemannian structure.

17.8.3 Experimental evidence in support of the General Theory of Relativity

The following experimental evidence in support of Einstein's Theory of General Relativity is compelling.

Kepler problem In 1915 Einstein showed that relativistic mechanics explained the anomalous precession of the perihelion of the planet mercury, that is, the axes of the elliptical Kepler orbit are observed to precess.

Deflection of light   Einstein's prediction of the deflection of light in a gravitational field was confirmed by Eddington during the solar eclipse of 29 May 1919. Pictures of stars in the region around the Sun showed that their apparent locations were slightly shifted because the light from the stars had been curved by passing close to the sun's gravitational field.

Gravitational lensing   The deflection of light by the gravitational attraction of a massive object situated between a distant star and the observer has resulted in the observation of multiple images of a distant quasar.

Gravitational time dilation and frequency shift   Processes occurring in a high gravitation field are slower than in a weak gravitational field; this is called gravitational time dilation. In addition, light climbing out of a gravitational well is red shifted. The gravitational time dilation has been measured many times and the successful operation of the Global Position System provides an ongoing validation. The gravitational red shift has been confirmed in the laboratory using the precise Mössbauer effect in nuclear physics. Tests in stronger gravitational fields are provided by studies of binary pulsars.

Black holes   When the mass to radius ratio of a massive object becomes sufficiently large, general relativity predicts formation of a black hole, which is a region of space from which neither light nor matter can escape. Supermassive black holes, with a mass that can be $10^6 - 10^9$ solar masses, are thought to have played an important role in formation of the galaxies.

Gravitational waves detection   In 1916 Einstein predicted the existence of gravitational waves on the basis of the theory of general relativity. The first implied detection of gravitational waves were made in 1976 by Hulse and Taylor who detected a decrease in the orbital period due to significant energy loss which presumably was associated with emission of gravity waves by the compact neutron star in the binary pulsar $PSR1913 + 16$. The most compelling direct evidence for observation of a gravitational wave was made on 15 September 2015 by the LIGO Laser Interferometer Gravitational-Wave Observatories. The waveform detected by the two LIGO observatories matched the predictions of General Relativity for gravitational waves emanating from the inward spiral plus merger of a pair of black holes of around 36 and 29 solar masses, followed by the resultant binary black hole. The gravitational wave emitted by this cataclysmic merger reached Earth as a ripple in space-time that changed the length of the $4km$ LIGO arm by a thousandth of the width of the proton. The gravitational energy emitted was $3.0^{+0.5}_{-0.5}c^2$ solar masses. A second observation of gravitational waves was made on 26 December 2015, and four similar observations were made during 2017. The detection of such miniscule changes in space-time is a truly remarkable achievement. This direct detection of gravitational waves resulted in the award of the 2017 Nobel Prize to Rainer Weiss, Barry Barish, and Kip Thorn. Gravitational wave detection has opened an exciting and powerful new frontier in astrophysics that could lead to exciting new physics.

17.9   Implications of relativistic theory to classical mechanics

Einstein's theories of relativity have had an enormous impact on twentieth century physics and the philosophy of science. Relativistic mechanics is crucial to an understanding of the physics of the atom, nucleus and the substructure of the nucleons, but the impacts are minimal in everyday experience. The Special Theory of Relativity replaces Newton's Laws of motion; i.e. Newton's law is only an approximation applicable for low velocities. The General Theory of Relativity replaces Newton's Law of Gravitation and provides a natural explanation of the equivalence principle. Einstein's theories of relativity imply a profound and fundamental change in the view of the separation of space, time, and mass, that contradicts the basic tenets that are the foundation of Newtonian mechanics. The Newtonian concepts of absolute frame of reference, plus the separation of space, time, and mass, are invalid at high velocities. Lagrangian and Hamiltonian variational approaches to classical mechanics provide the formalism necessary for handling relativistic mechanics. The present chapter has shown that logical extensions of Lagrangian and Hamiltonian mechanics lead to the relativistically-invariant extended Lagrangian and Hamiltonian formulations of mechanics which are adequate for handling one-body systems. However, major unsolved problems remain applying these formulations to systems that have more than one body.

17.10 Summary

Special theory of relativity: The Special Theory of Relativity is based on Einstein's postulates;
  1) *The laws of nature are the same in all inertial frames of reference.*
  2) *The velocity of light in vacuum is the same in all inertial frames of reference.*
For a primed frame moving along the x_1 axis with velocity v Einstein's postulates imply the following Lorentz transformations between the moving (primed) and stationary (unprimed) frames

$$
\begin{aligned}
x' &= \gamma\,(x - vt) & x &= \gamma\,(x' + vt') \\
y' &= y & y &= y' \\
z' &= z & z &= z' \\
t' &= \gamma\left(t - \frac{vx}{c^2}\right) & t &= \gamma\left(t' + \frac{vx'}{c^2}\right)
\end{aligned}
$$

where the Lorentz γ factor $\gamma \equiv \dfrac{1}{\sqrt{1-\left(\frac{v}{c}\right)^2}}$

Lorentz transformations were used to illustrate Lorentz contraction, time dilation, and simultaneity. An elementary review was given of relativistic kinematics including discussion of velocity transformation, linear momentum, center-of-momentum frame, forces and energy.

Geometry of space-time: The concepts of four-dimensional space-time were introduced. A discussion of four-vector scalar products introduced the use of contravariant and covariant tensors plus the Minkowski metric g where the scalar product was defined. The Minkowski representation of space time and the momentum-energy four vector also were introduced.

Lorentz-invariant formulation of Lagrangian mechanics: The Lorentz-invariant extended Lagrangian formalism, developed by Struckmeier[Str08], based on the parametric approach pioneered by Lanczos[La49], provides a viable Lorentz-invariant extension of conventional Lagrangian mechanics that is applicable for one-body motion in the realm of the Special Theory of Relativity.

Lorentz-invariant formulation of Hamiltonian mechanics: The Lorentz-invariant extended Hamiltonian formalism, developed by Struckmeier based on the parametric approach pioneered by Lanczos, was introduced. It provides a viable Lorentz-invariant extension of conventional Hamiltonian mechanics that is applicable for one-body motion in the realm of the Special Theory of Relativity. In particular, it was shown that the Lorentz-invariant extended Hamiltonian is conserved making it ideally suited for solving complicated systems using Hamiltonian mechanics via use of the Poisson-bracket representation of Hamiltonian mechanics, canonical transformations, and the Hamilton-Jacobi techniques.

The General Theory of Relativity: An elementary summary was given of the fundamental concepts of the General Theory of Relativity and the resultant unified description of the gravitational force plus planetary motion as geodesic motion in a four-dimensional Riemannian structure. Variational mechanics were shown to be ideally suited to applications of the General Theory of Relativity.

Philosophical implications: Newton's equations of motion, and his Law of Gravitation, that reigned supreme from 1687 to 1905, have been toppled from the throne by Einstein's theories of relativistic mechanics. By contrast, the complete independence to coordinate frames in Lagrangian, and Hamiltonian formulations of classical mechanics, plus the underlying Principle of Least Action, are equally valid in both the relativistic and non-relativistic regimes. As a consequence, relativistic Lagrangian and Hamiltonian formulations underlie much of modern physics, especially quantum physics, which explains why relativistic mechanics plays such an important role in classical dynamics.

Workshop exercises

1. A relativistic snake of proper length $100cm$ is travelling to the right across a butcher's table at $v = 0.6c$. You hold two meat cleavers, one in each hand which are $100cm$ apart. You strike the table simultaneously with both cleavers at the moment when the left cleaver lands just behind the tail of the snake. You rationalize that since the snake is moving with $\beta = 0.6$, then the length of the snake is Lorentz contracted by the factor $\gamma = \frac{5}{4}$ and thus the Lorentz-contracted length of the snake is $80cm$ and thus will not be harmed. However, the snake reasons that relative to it the cleavers are moving at $\beta = 0.6$ and thus are only $80cm$ apart when they strike the $100cm$ long snake and thus it will be severed. Use the Lorentz transformation to resolve this paradox.

2. Explain what is meant by the following statement: "Lorentz transformations are orthogonal transformations in Minkowski space."

3. Which of the following are invariant quantities in space-time?

   (a) Energy
   (b) Momentum
   (c) Mass
   (d) Force
   (e) Charge
   (f) The length of a vector
   (g) The length of a four-vector

4. What does it mean for two events to have a spacelike interval? What does it mean for them to have a timelike interval? Draw a picture to support your answer. In which case can events be causally connected?

Problems

1. A supply rocket flies past two markers on the Space Station that are $50m$ apart in a time of $0.2\mu s$ as measured by an observer on the Space station.

   (a) What is the separation of the two markers as seen by the pilot riding in the supply rocket?
   (b) What is the elapsed time as measured by the pilot in the supply rocket?
   (c) What are the speeds calculated by the observer in the Space Station and the pilot of the supply rocket?

2. The Compton effect involves a photon of incident energy E_i being scattered by an electron of mass m_e which initially is stationary. The photon scattered at an angle θ with respect to the incident photon has a final energy E_f. Using the special theory of relativity derive a formula that related E_f and E_i to θ.

3. Pair creation involves production of an electron-positron pair by a photon. Show that such a process is impossible unless some other body, such as a nucleus, is involved. Suppose that the nucleus has a mass M and the electron mass m_e. What is the minimum energy that the photon must have in order to produce an electron-positron pair?

4. A K meson of rest energy $494MeV$ decays into a μ meson of rest energy $106MeV$ and a neutrino of zero rest energy. Find the kinetic energies of the μ meson and the neutrino into which the K meson decays while at rest.

Chapter 18

The transition to quantum physics

18.1   Introduction

Classical mechanics, including extensions to relativistic velocities, embrace an unusually broad range of topics ranging from astrophysics to nuclear and particle physics, from one-body to many-body statistical mechanics. It is interesting to discuss the role of classical mechanics in the development of quantum mechanics which plays a crucial role in physics. A valid question is "why discuss quantum mechanics in a classical mechanics course?". The answer is that quantum mechanics supersedes classical mechanics as the fundamental theory of mechanics. Classical mechanics is an approximation applicable for situations where quantization is unimportant. Thus there must be a correspondence principle that relates quantum mechanics to classical mechanics, analogous to the relation between relativistic and non-relativistic mechanics. It is illuminating to study the role played by the Hamiltonian formulation of classical mechanics in the development of quantal theory and statistical mechanics. The Hamiltonian formulation is expressed in terms of the phase-space variables $\mathbf{q}, \mathbf{p}$ for which there are well-established rules for transforming to quantal linear operators.

18.2   Brief summary of the origins of quantum theory

The last decade of the 19th century saw the culmination of classical physics. By 1900 scientists thought that the basic laws of mechanics, electromagnetism, and statistical mechanics were understood and worried that future physics would be reduced to confirming theories to the fifth decimal place, with few major new discoveries to be made. However, technical developments such as photography, vacuum pumps, induction coil, etc., led to important discoveries that revolutionized physics and toppled classical mechanics from its throne at the beginning of the 20th century. Table 18.1 summarizes some of the major milestones leading up to the development of quantum mechanics.

   Max Planck searched for an explanation of the spectral shape of the black-body electromagnetic radiation. He found an interpolation between two conflicting theories, one that reproduced the short wavelength behavior, and the other the long wavelength behavior. Planck's interpolation required assuming that electromagnetic radiation was not emitted with a continuous range of energies, but that electromagnetic radiation is emitted in discrete bundles of energy called quanta. In December 1900 he presented his theory which reproduced precisely the measured black body spectral distribution by assuming that the energy carried by a single quantum must be an integer multiple of $h\nu$:

$$E = h\nu = \frac{hc}{\lambda} \tag{18.1}$$

where ν is the frequency of the electromagnetic radiation and Planck's constant, $h = 6.626 10^{-34} J \cdot sec$ was the best fit parameter of the interpolation. That is, Planck assumed that energy comes in discrete bundles of energy equal to $h\nu$ which are called quanta. By making this extreme assumption, in an act of desperation, Planck was able to reproduce the experimental black body radiation spectrum. The assumption that energy was exchanged in bundles hinted that the classical laws of physics were inadequate in the microscopic domain. The older generation physicists initially refused to believe Planck's hypothesis which underlies quantum theory. It was the new generation physicists, like Einstein, Bohr, Heisenberg, Born, Schrödinger, and Dirac, who developed Planck's hypothesis leading to the revolutionary quantum theory.

In 1905, Einstein predicted the existence of the photon, derived the theory of specific heat, as well as deriving the Theory of Special Relativity. It is remarkable to realize that he developed these three revolutionary theories in one year, when he was only 26 years old. Einstein uncovered an inconsistency in Planck's derivation of the black body spectral distribution in that it assumed the statistical part of the energy is quantized, whereas the electromagnetic radiation assumed Maxwell's equations with oscillator energies being continuous. Planck demanded that light of frequency ν be packaged in quanta whose energies were multiples of $h\nu$, but Planck never thought that light would have particle-like behavior. Newton believed that light involved corpuscles, and Hamilton developed the Hamilton-Jacobi theory seeking to describe light in terms of the corpuscle theory. However, Maxwell had convinced physicists that light was a wave phenomena; interference plus diffraction effects were convincing manifestations of the wave-like properties of light. In order to reproduce Planck's prediction, Einstein had to treat black-body radiation as if it consisted of a gas of photons, each photon having energy $E = h\nu$. This was a revolutionary concept that returned to Newton's corpuscle theory of light. Einstein realized that there were direct tests of his photon hypothesis, one of which is the photo-electric effect. According to Einstein, each photon has an energy $E = h\nu$, in contrast to the classical case where the energy of the photoelectron depends on the intensity of the light. Einstein predicted that the ejected electron will have a kinetic energy

$$KE = h\nu - W \tag{18.2}$$

where W is the work function which is the energy needed to remove an electron from a solid.

Many older scientists, including Planck, accepted Einstein's theory of relativity but were skeptical of the photon concept, even after Einstein's photon concept was vindicated in 1915 by Millikan who showed that, as predicted, the energy of the ejected photoelectron depended on the frequency, and not intensity, of the light. In 1923 Compton's demonstrated that electromagnetic radiation scattered by free electrons obeyed simple two-body scattering laws which finally convinced the many skeptics of the existence of the photon.

Table 18.1: Chronology of the development of quantum mechanics

Date	Author	Development
1887	Hertz	Discovered the photo-electric effect
1895	Röntgen	Discovered x-rays
1896	Becquerel	Discovered radioactivity
1897	J.J. Thomson	Discovered the first fundamental particle, the electron
1898	Pierre & Marie Curie	Showed that thorium is radioactive which founded nuclear physics
1900	Planck	Quantization $E = h\nu$ explained the black-body spectrum
1905	Einstein	Theory of special relativity
1905	Einstein	Predicted the existence of the photon
1906	Einstein	Used Planck's constant to explain specific heats of solids
1909	Millikan	The oil drop experiment measured the charge on the electron
1911	Rutherford	Discovered the atomic nucleus with radius $10^{-15}m$
1912	Bohr	Bohr model of the atom explained the quantized states of hydrogen
1914	Moseley	X-ray spectra determined the atomic number of the elements.
1915	Millikan	Used the photo-electric effect to confirm the photon hypothesis.
1915	Wilson-Sommerfeld	Proposed quantization of the action-angle integral
1921	Stern-Gerlach	Observed space quantization in non-uniform magnetic field
1923	Compton	Compton scattering of x-rays confirmed the photon hypothesis
1924	de Broglie	Postulated wave-particle duality for matter and EM waves
1924	Bohr	Explicit statement of the correspondence principle
1925	Pauli	Postulated the exclusion principle
1925	Goudsmit-Uhlenbeck	Postulated the spin of the electron of $s = \frac{1}{2}\hbar$
1925	Heisenberg	Matrix mechanics representation of quantum theory
1925	Dirac	Related Poisson brackets and commutation relations
1926	Schrödinger	Wave mechanics
1927	G.P. Thomson/Davisson	Electron diffraction proved wave nature of electron
1928	Dirac	Developed the Dirac relativistic wave equation

18.2.1 Bohr model of the atom

The Rutherford scattering experiment, performed at Manchester in 1911, discovered that the Au atom comprised a positively charge nucleus of radius $\approx 10^{-14}m$ which is much smaller than the $1.35 \times 10^{-10}m$ radius of the Au atom. Stimulated by this discovery, Niels Bohr joined Rutherford at Manchester in 1912 where he developed the Bohr model of the atom. This theory was remarkably successful in spite of having serious inconsistencies and deficiencies. Bohr's model assumptions were:

1) Electromagnetic radiation is quantized with $E = h\nu$.

2) Electromagnetic radiation exhibits behavior characteristic of the emission of photons with energy $E = h\nu$ and momentum $p = \frac{h\nu}{c}$. That is, it exhibits both wave-like and particle-like behavior.

3) Electrons are in stationary orbits that do not radiate, which contradicts the predictions of classical electromagnetism.

4) The orbits are quantized such that the electron angular momentum is an integer multiple of $\frac{h}{2\pi} = \hbar$.

5) Atomic electromagnetic radiation is emitted with photon energy equal to the difference in binding energy between the two atomic levels involved. $h\nu = E_1 - E_2$

The first two assumptions are due to Planck and Einstein, while the last three were made by Niels Bohr.

The deficiencies of the Bohr model were the philosophical problems of violating the tenets of classical physics in explaining hydrogen-like atoms, that is, the theory was prescriptive, not deductive. The Bohr model was based implicitly on the assumption that quantum theory contains classical mechanics as a limiting case. Bohr explicitly stated this assumption which he called the **correspondence principle,** and which played a pivotal role in the development of the older quantum theory. In 1924 Bohr justified the inconsistencies of the old quantum theory by writing "As frequently emphasized, these principles, although they are formulated by the help of classical conceptions, are to be regarded purely as laws of quantum theory, which give us, not withstanding the formal nature of quantum theory, a hope in the future of a consistent theory, which at the same time reproduces the characteristic features of quantum theory, important for its applicability, and, nevertheless, can be regarded as a rational generalization of classical electrodynamics."

The old quantum theory was remarkably successful in reproducing the black-body spectrum, specific heats of solids, the hydrogen atom, and the periodic table of the elements. Unfortunately, from a methodological point of view, the theory was a hodgepodge of hypotheses, principles, theorems, and computational recipes, rather than a logical consistent theory. Every problem was first solved in terms of classical mechanics, and then would pass through a mysterious quantization procedure involving the correspondence principle. Although built on the foundation of classical mechanics, it required Bohr's hypotheses which violated the laws of classical mechanics and predictions of Maxwell's equations.

18.2.2 Quantization

By 1912 Planck, and others, had abandoned the concept that quantum theory was a branch of classical mechanics, and were searching to see if classical mechanics was a special case of a more general quantum physics, or quantum physics was a science altogether outside of classical mechanics. Also they were trying to find a consistent and rational reason for quantization to replace the ad hoc assumption of Bohr.

In 1912 Sommerfeld proposed that, in every elementary process, the atom gains or loses a definite amount of action between times t_0 and t of

$$S = \int_{t_0}^{t} L(t')dt' \tag{18.3}$$

where S is the quantal analogue of the classical action function. It has been shown that the classical principle of least action states that the action function is stationary for small variations of the trajectory. In 1915 Wilson and Sommerfeld recognized that the quantization of angular momentum could be expressed in terms of the action-angle integral, that is equation 15.116. They postulated that, for every coordinate, the action-angle variable is quantized

$$\oint p_k dq_k = nh \tag{18.4}$$

where the action-angle variable integral is over one complete period of the motion. That is, they postulated that Hamilton's phase space is quantized, but the microscopic granularity is such that the quantization is only manifest for atomic-sized domains. That is, n is a small integer for atomic systems in contrast to $n \approx 10^{64}$ for the Earth-Sun two-body system.

Sommerfeld recognized that quantization of more than one degree of freedom is needed to obtain more accurate description of the hydrogen atom. Sommerfeld reproduced the experimental data by assuming quantization of the three degrees of freedom,

$$\oint p_r dr = n_1 h \qquad\qquad \oint p_\theta d\theta = n_2 h \qquad\qquad \oint p_\phi d\phi = n_3 h \qquad (18.5)$$

and solving Hamilton-Jacobi theory by separation of variables. In 1916 the Bohr-Sommerfeld model solved the classical orbits for the hydrogen atom, including relativistic corrections as described in example 17.7. This reproduced fine structure observed in the optical spectra of hydrogen. The use of the canonical transformation to action-angle variables proved to be the ideal approach for solving many such problems in quantum mechanics. In 1921, Stern and Gerlach demonstrated space quantization by observing the splitting of atomic beams deflected by non-uniform magnetic fields. This result was a major triumph for quantum theory. Sommerfeld declared that "With their bold experimental method, Stern and Gerlach demonstrated not only the existence of space quantization, they also proved the atomic nature of the magnetic moment, its quantum-theoretic origin, and its relation to the atomic structure of electricity."

In 1925, Pauli's Exclusion Principle proposed that no more than one electron can have identical quantum numbers and that the atomic electronic state is specified by four quantum numbers. Two students, Goudsmit and Uhlenbeck suggested that a fourth two-valued quantum number was the electron spin of $\pm\frac{\hbar}{2}$. This provided a plausible explanation for the structure of multi-electron atoms.

18.2.3 Wave-particle duality

In his 1924 doctoral thesis, Prince Louis de Broglie proposed the hypothesis of wave-particle duality which was a pivotal development in quantum theory. de Broglie used the classical concept of a matter wavepacket, analogous to classical wave packets discussed in chapter 3.11. He assumed that both the group and signal velocities of a matter wave packet must equal the velocity of the corresponding particle. By analogy with Einstein's relation for the photon, and using the Theory of Special Relativity, de Broglie assumed that

$$\hbar\omega = E = \frac{mc^2}{\sqrt{\left(1 - \frac{v^2}{c^2}\right)}} \qquad (18.6)$$

The group velocity is required to equal the velocity of the mass m

$$v_{group} = \left(\frac{d\omega}{dk}\right) = \left(\frac{d\omega}{dv}\right)\left(\frac{dv}{dk}\right) = v \qquad (18.7)$$

This gives

$$\frac{dk}{dv} = \frac{1}{v}\left(\frac{d\omega}{dv}\right) = \left(\frac{m}{\hbar}\right)\left(1 - \frac{v^2}{c^2}\right)^{-\frac{3}{2}} \qquad (18.8)$$

Integration of this equation assuming that $k = 0$ when $v = 0$, then gives

$$\hbar\mathbf{k} = \frac{m\mathbf{v}}{\sqrt{\left(1 - \frac{\mathbf{v}\cdot\mathbf{v}}{c^2}\right)}} = \mathbf{p} \qquad (18.9)$$

This relation, derived by de Broglie, is required to ensure that the particle travels at the group velocity of the wave packet characterizing the particle. Note that although the relations used to characterize the matter waves are purely classical, the physical content of such waves is beyond classical physics. In 1927 C. Davisson and G.P. Thomson independently observed electron diffraction confirming wave/particle duality for the electron. Ironically, J.J. Thomson discovered that the electron was a particle, whereas his son attributed it to an electron wave.

Heisenberg developed the modern matrix formulation of quantum theory in 1925; he was 24 years old at the time. A few months later Schrödinger's developed wave mechanics based on de Broglie's concept of wave-particle duality. The matrix mechanics, and wave mechanics, quantum theories are radically different. Heisenberg's algebraic approach employs non-commuting quantities and unfamiliar mathematical techniques that emphasized the discreteness characteristic of the corpuscle aspect. In contrast, Schrödinger used the familiar analytical approach that is an extension of classical laws of motion and waves which stressed the element of continuity.

18.3 Hamiltonian in quantum theory

18.3.1 Heisenberg's matrix-mechanics representation

The algebraic Heisenberg representation of quantum theory is analogous to the algebraic Hamiltonian representation of classical mechanics, and shows best how quantum theory evolved from, and is related to, classical mechanics. Heisenberg decided to ignore the prevailing conceptual theories, such as classical mechanics, and based his quantum theory on observables. This approach was influenced by the success of Bohr's older quantum theory and Einstein's theory of relativity. He abandoned the classical notions that the canonical variables p_k, q_k can be measured directly and simultaneously. Secondly he wished to absorb the correspondence principle directly into the theory instead of it being an ad hoc procedure tailored to each application. Heisenberg considered the Fourier decomposition of transition amplitudes between discrete states and found that the product of the conjugate variables do not commute. Heisenberg derived, for the first time, the correct energy levels of the one-dimensional harmonic oscillator as $E_n = \hbar\omega(n + \frac{1}{2})$ which was a significant achievement. Born recognized that Heisenberg's strange multiplication and commutation rules for two variables, corresponded to matrix algebra. Prior to 1925, matrix algebra was an obscure branch of pure mathematics not known or used by the physics community. Heisenberg, Born, and the young mathematician Jordan, developed the commutation rules of matrix mechanics. Heisenberg's approach represents the classical position and momentum coordinates q, p by matrices $\mathbf{q}$ and $\mathbf{p}$, with corresponding matrix elements $q_{mn}e^{i\omega_{mn}t}$ and $p_{mn}e^{i\omega_{mn}t}$. Born showed that the trace of the matrix

$$H(\mathbf{pq}) = \mathbf{p\dot{q}} - L \tag{18.10}$$

gives the Hamiltonian function $H(\mathbf{p}, \mathbf{q})$ of the matrices $\mathbf{q}$ and $\mathbf{p}$ which leads to Hamilton's canonical equations

$$\mathbf{\dot{q}} = \frac{\partial H}{\partial \mathbf{p}} \qquad \mathbf{\dot{p}} = -\frac{\partial H}{\partial \mathbf{q}} \tag{18.11}$$

Heisenberg and Born also showed that the commutator of $\mathbf{q}, \mathbf{p}$ equals

$$\begin{aligned} q_k p_l - p_l q_k &= i\hbar\delta_{kl} \\ q_k q_l - q_l q_k &= 0 \\ p_k p_l - p_l p_k &= 0 \end{aligned} \tag{18.12}$$

Born realized that equation (18.12) is the only fundamental equation for introducing $\hbar$ into the theory in a logical and consistent way.

Chapter 15.2.4 discussed the formal correspondence between the Poisson bracket, defined in chapter 15.3, and the commutator in classical mechanics. It was shown that the commutator of two functions equals a constant multiplicative factor λ times the corresponding Poisson Bracket. That is

$$(F_j G_k - G_k F_j) = \lambda [F_j, G_k] \tag{18.13}$$

where the multiplicative factor λ is a number independent of F_j, G_k, and the commutator.

In 1925, Paul Dirac, a 23-year old graduate student at Bristol, recognized the crucial importance of the above correspondence between the commutator and the Poisson Bracket of two functions, to relating classical mechanics and quantum mechanics. Dirac noted that if the constant λ is assigned the value $\lambda = i\hbar$, then equation 18.13 directly relates Heisenberg's commutation relations between the fundamental canonical variables (q_j, p_k) to the corresponding classical Poisson Bracket $[q_j, p_k]$. That is,

$$\begin{aligned} q_k p_l - p_l q_k &= i\hbar [q_k, p_l] = i\hbar\delta_{kl} \tag{18.14} \\ q_k q_l - q_l q_k &= i\hbar [q_k, q_l] = 0 \tag{18.15} \\ p_k p_l - p_l p_k &= i\hbar [p_k, p_l] = 0 \tag{18.16} \end{aligned}$$

Dirac recognized that the correspondence between the classical Poisson bracket, and quantum commutator, given by equation (18.13), provides a logical and consistent way that builds quantization directly into the theory, rather than using an ad-hoc, case-dependent, hypothesis as used by the older quantum theory of

Bohr. The basis of Dirac's quantization principle, involves replacing the classical Poisson Bracket, $[F_j, G_k]$ by the commutator, $\frac{1}{i\hbar}(F_j, G_k - G_k F_j)$. That is,

$$[F_j, G_k] \Longrightarrow \frac{1}{i\hbar}(F_j G_k - G_k F_j) \tag{18.17}$$

Hamilton's canonical equations, as introduced in chapter 15, are only applicable to classical mechanics since they assume that the exact position and conjugate momentum can be specified both exactly and simultaneously which contradicts the Heisenberg's Uncertainty Principle. In contrast, the Poisson bracket generalization of Hamilton's equations allows for non-commuting variables plus the corresponding uncertainty principle. That is, the transformation from classical mechanics to quantum mechanics can be accomplished simply by replacing the classical Poisson Bracket by the quantum commutator, as proposed by Dirac. The formal analogy between classical Hamiltonian mechanics, and the Heisenberg representation of quantum mechanics is strikingly apparent using the correspondence between the Poisson Bracket representation of Hamiltonian mechanics and Heisenberg's matrix mechanics.

The direct relation between the quantum commutator, and the corresponding classical Poisson Bracket, applies to many observables. For example, the quantum analogs of Hamilton's equations of motion are given by use of Hamilton's equations of motion, $15.53, 15.56$, and replacing each Poisson Bracket by the corresponding commutator. That is

$$\frac{dq_k}{dt} = \frac{\partial H}{\partial p_k} = [q_k, H] = \frac{1}{i\hbar}(q_k H - H q_k) \tag{18.18}$$

$$\frac{dp_k}{dt} = -\frac{\partial H}{\partial q_k} = [p_k, H] = \frac{1}{i\hbar}(p_k H - H p_k) \tag{18.19}$$

Chapter 15.2.5 discussed the time dependence of observables in Hamiltonian mechanics. Equation 15.45 gave the total time derivative of any observable G to be

$$\frac{dG}{dt} = \frac{\partial G}{\partial t} + [G, H] \tag{18.20}$$

Equation 18.17 can be used to replace the Poisson Bracket by the quantum commutator, which gives the corresponding time dependence of observables in quantum physics.

$$\frac{dG}{dt} = \frac{\partial G}{\partial t} + \frac{1}{i\hbar}(GH - HG) \tag{18.21}$$

In quantum mechanics, equation 18.21 is called the *Heisenberg equation*. Note that if the observable G is chosen to be a fundamental canonical variable, then $\frac{\partial q_k}{\partial t} = 0 = \frac{\partial p_k}{\partial t}$ and equation 15.20 reduces to Hamilton's equations 18.18 and 18.19.

The analogies between classical mechanics and quantum mechanics extend further. For example, if G is a constant of motion, that is $\frac{dG}{dt} = 0$, then Heisenberg's equation of motion gives

$$\frac{\partial G}{\partial t} + \frac{1}{i\hbar}(GH - HG) = 0 \tag{18.22}$$

Moreover, if G is not an explicit function of time, then

$$0 = \frac{1}{i\hbar}(GH - HG) \tag{18.23}$$

That is, the transition to quantum physics shows that, if G is a constant of motion, and is not explicitly time dependent, then G commutes with the Hamiltonian H.

The above discussion has illustrated the close and beautiful correspondence between the Poisson Bracket representation of classical Hamiltonian mechanics, and the Heisenberg representation of quantum mechanics. Dirac provided the elegant and simple correspondence principle connecting the Poisson bracket representation of classical Hamiltonian mechanics, to the Heisenberg representation of quantum mechanics.

18.3.2 Schrödinger's wave-mechanics representation

The wave mechanics formulation of quantum mechanics, by the Austrian theorist Schrödinger, was built on the wave-particle duality concept that was proposed in 1924 by Louis de Broglie. Schrödinger developed his wave mechanics representation of quantum physics a year after the development of matrix mechanics by Heisenberg and Born. The Schrödinger wave equation is based on the non-relativistic Hamilton-Jacobi representation of a wave equation, melded with the operator formalism of Born and Wiener. The 39-year old Schrödinger was an expert in classical mechanics and wave theory, which was invaluable when he developed the important Schrödinger equation. As mentioned in chapter 15.4.4, the Hamilton-Jacobi theory is a formalism of classical mechanics that allows the motion of a particle to be represented by a wave. That is, the wavefronts are surfaces of constant action S, and the particle momenta are normal to these constant-action surfaces, that is, $\mathbf{p} = \boldsymbol{\nabla} S$. The wave-particle duality of Hamilton-Jacobi theory is a natural way to handle the wave-particle duality proposed by de Broglie.

Consider the classical Hamilton-Jacobi equation for one body, given by 18.20.

$$\frac{\partial S}{\partial t} + H(\mathbf{q}, \boldsymbol{\nabla} S, t) = 0 \tag{18.24}$$

If the Hamiltonian is time independent, then equation 15.90 gives that

$$\frac{\partial S}{\partial t} = -H(\mathbf{q}, \mathbf{p}, t) = -E(\boldsymbol{\alpha}) \tag{18.25}$$

The integration of the time dependence is trivial, and thus the action integral for a time-independent Hamiltonian is

$$S(\mathbf{q}, \boldsymbol{\alpha}, t) = W(\mathbf{q}, \boldsymbol{\alpha}) - E(\boldsymbol{\alpha}) t \tag{18.26}$$

A formal transformation gives

$$E = -\frac{\partial S}{\partial t} \qquad\qquad \mathbf{p} = \boldsymbol{\nabla} S \tag{18.27}$$

Consider that the classical time-independent Hamiltonian, for motion of a single particle, is represented by the Hamilton-Jacobi equation.

$$H = \frac{\mathbf{p}^2}{2\mu} + U(q) = -\frac{\partial S}{\partial t} \tag{18.28}$$

Substitute for $\mathbf{p}$ leads to the classical Hamilton-Jacobi relation in terms of the action S

$$\frac{1}{2\mu}(\boldsymbol{\nabla} S \cdot \boldsymbol{\nabla} S) + U(q) = -\frac{\partial S}{\partial t} \tag{18.29}$$

By analogy with the Hamilton-Jacobi equation, Schrödinger proposed the quantum operator equation

$$i\hbar\frac{\partial \psi}{\partial t} = \hat{H}\psi \tag{18.30}$$

where $\hat{H}$ is an operator given by

$$\hat{H} = -\frac{\hbar^2}{2\mu}\nabla^2 + U(r) \tag{18.31}$$

In 1926, Max Born and Norbert Wiener introduced the operator formalism into matrix mechanics for prediction of observables and this has become an integral part of quantum theory. In the operator formalism, the observables are represented by operators that project the corresponding observable from the wavefunction. That is, the quantum operator formalism for the assumed momentum and energy operators, that operate on the wavefunction ψ, are

$$p_j = \frac{\hbar}{i}\frac{\partial}{\partial q_j} \qquad\qquad E = -\frac{\hbar}{i}\frac{\partial}{\partial t} \tag{18.32}$$

Formal transformations of $\mathbf{p}$ and E in the Hamiltonian (18.26) leads to the time-independent Schrödinger equation

$$-\frac{\hbar^2}{2\mu}\frac{\partial^2 \psi}{\partial q^2} + U(q)\psi = E\psi \tag{18.33}$$

Assume that the wavefunction is of the form

$$\psi = Ae^{\frac{iS}{\hbar}} \tag{18.34}$$

where the action S gives the phase of the wavefront, and A the amplitude of the wave, as described in chapter 15.4.4. The time dependence, that characterizes the motion of the wavefront, is contained in the time dependence of S. This form for the wavefunction has the advantage that the wavefunction frequently factors into a product of terms, e.g. $\psi = R(r)\Theta(\theta)\Phi(\phi)$ which corresponds to a summation of the exponents $S = W_r + W_\theta + W_\phi - Et$. This summation form is exploited by separation of the variables, as discussed in chapter 15.4.3.

Insert ψ (18.33) into equation (18.28), plus using the fact that

$$\frac{\partial^2 \psi}{\partial q^2} = \frac{\partial}{\partial q}\left(\frac{\partial \psi}{\partial S}\frac{\partial S}{\partial q}\right) = \frac{\partial}{\partial q}\left(\frac{i}{\hbar}\psi\frac{\partial S}{\partial q}\right) = -\frac{1}{\hbar^2}\psi\left(\frac{\partial S}{\partial q}\right)^2 + \frac{i}{\hbar}\psi\frac{\partial^2 S}{\partial q^2} \tag{18.35}$$

leads to

$$-\frac{\partial S}{\partial t} = \frac{1}{2\mu}\left(\boldsymbol{\nabla}S \cdot \boldsymbol{\nabla}S\right) + U(q) - \frac{i\hbar}{2\mu}\nabla^2 S = E \tag{18.36}$$

Note that if Planck's constant $\hbar = 0$, then the imaginary term in equation (18.35) is zero, leading to 18.35 being real, and identical to the Hamilton-Jacobi result, equation 18.23. The fact that equation 18.35 equals the Hamilton-Jacobi equation in the limit $\hbar \to 0$, illustrates the close analogy between the wave-particle duality of the classical Hamilton-Jacobi theory, and de Broglie's wave-particle duality in Schrödinger's quantum wave-mechanics representation.

The Schrödinger approach was accepted in 1925 and exploited extensively with tremendous success, since it is much easier to grasp conceptually than is the algebraic approach of Heisenberg. Initially there was much conflict between the proponents of these two contradictory approaches, but this was resolved by Schrödinger who showed in 1926 that there is a formal mathematical identity between wave mechanics and matrix mechanics. That is, these quantal two representations of Hamiltonian mechanics are equivalent, even though they are built on either the Poisson bracket representation, or the Hamilton-Jacobi representation. Wave mechanics is based intimately on the quantization rule of the action variable. Heisenberg's Uncertainty Principle is automatically satisfied by Schrödinger's wave mechanics since the uncertainty principle is a feature of all wave motion, as described in chapter 3.

In 1928 Dirac developed a relativistic wave equation which includes spin as an integral part. This Dirac equation remains the fundamental wave equation of quantum mechanics. Unfortunately it is difficult to apply.

Today the powerful and efficient Heisenberg representation is the dominant approach used in the field of physics, whereas chemists tend to prefer the more intuitive Schrödinger wave mechanics approach. In either case, the important role of Hamiltonian mechanics in quantum theory is undeniable.

18.4   Lagrangian representation in quantum theory

The classical notion of canonical coordinates and momenta, has a simple quantum analog which has allowed the Hamiltonian theory of classical mechanics, that is based on canonical coordinates, to serve as the foundation for the development of quantum mechanics. The alternative Lagrangian formulation for classical dynamics is described in terms of coordinates and velocities, instead of coordinates and momenta. The Lagrangian and Hamiltonian formulations are closely related, and it may appear that the Lagrangian approach is more fundamental. The Lagrangian method allows collecting together all the equations of motion and expressing them as stationary properties of the action integral, and thus it may appear desirable to base quantum mechanics on the Lagrangian theory of classical mechanics. Unfortunately, the Lagrangian equations of motion involve partial derivatives with respect to coordinates, and their velocities, and the meaning ascribed to such derivatives is difficult in quantum mechanics. The close correspondence between Poisson brackets and the commutation rules leads naturally to Hamiltonian mechanics. However, Dirac showed that Lagrangian mechanics can be carried over to quantum mechanics using canonical transformations such that the classical Lagrangian is considered to be a function of coordinates at time t and $t + dt$ rather than of coordinates and velocities.

The motivation for Feynman's 1942 Ph.D thesis, entitled *"The Principle of Least Action in Quantum Mechanics"*, was to quantize the classical action at a distance in electrodynamics. This theory adopted an overall space-time viewpoint for which the classical Hamiltonian approach, as used in conventional formulations of quantum mechanics, is inapplicable. Feynman used the Lagrangian, plus the principle of least action, to underlie his development of quantum field theory. To paraphrase Feynman's Nobel Lecture, he used a physical approach that is quite different from the customary Hamiltonian point of view for which the system is discussed in great detail as a function of time. That is, you have the field at this moment, then a differential equation gives you the field at a later moment and so on; that is, the Hamiltonian approach is a time differential method. In Feynman's least-action approach the action describes the character of the path throughout all of space and time. The behavior of nature is determined by saying that the whole space-time path has a certain character. The use of action involves both advanced and retarded terms that make it difficult to transform back to the Hamiltonian form. The Feynman space-time approach is far beyond the scope of this course. This topic will be developed in advanced graduate courses on quantum field theory.

18.5   Correspondence Principle

The Correspondence Principle implies that any new theory in physics must reduce to preceding theories that have been proven to be valid. For example, Einstein's Special Theory of Relativity satisfies the Correspondence Principle since it reduces to classical mechanics for velocities small compared with the velocity of light. Similarly, the General Theory of Relativity reduces to Newton's Law of Gravitation in the limit of weak gravitational fields. Bohr's Correspondence Principle requires that the predictions of quantum mechanics must reproduce the predictions of classical physics in the limit of large quantum numbers. Bohr's Correspondence Principle played a pivotal role in the development of the old quantum theory, from it's inception in 1912, until 1925 when the old quantum theory was superseded by the current matrix and wave mechanics representations of quantum mechanics.

Quantum theory now is a well-established field of physics that is equally as fundamental as is classical mechanics. The Correspondence Principle now is used to project out the analogous classical-mechanics phenomena that underlie the observed properties of quantal systems. For example, this book has studied the classical-mechanics analogs of the observed behavior for typical quantal systems, such as the vibrational and rotational modes of the molecule, and the vibrational modes of the crystalline lattice. The nucleus is the epitome of a many-body, strongly-interacting, quantal system. Example 14.12 showed that there is a close correspondence between classical-mechanics predictions, and quantal predictions, for both the rotational and vibrational collective modes of the nucleus, as well as for the single-particle motion of the nucleons in the nuclear mean field, such as the onset of Coriolis-induced alignment. This use of the Correspondence Principle can provide considerable insight into the underlying classical physics embedded in quantal systems.

18.6 Summary

The important point of this discussion is that variational formulations of classical mechanics provide a rational, and direct basis, for the development of quantum mechanics. It has been shown that the final form of quantum mechanics is closely related to the Hamiltonian formulation of classical mechanics. Quantum mechanics supersedes classical mechanics as the fundamental theory of mechanics in that classical mechanics only applies for situations where quantization is unimportant, and is the limiting case of quantum mechanics when $\hbar \to 0$, which is in agreement with the Bohr's Correspondence Principle. The Dirac relativistic theory of quantum mechanics is the ultimate quantal theory for the relativistic regime.

This discussion has barely scratched the surface of the correspondence between classical and quantal mechanics, which goes far beyond the scope of this course. The goal of this chapter is to illustrate that classical mechanics, in particular, Hamiltonian mechanics, underlies much of what you will learn in your quantum physics courses. An interesting similarity between quantum mechanics and classical mechanics is that physicists usually use the more visual Schrödinger wave representation in order to describe quantum physics to the non-expert, which is analogous to the similar use of Newtonian physics in classical mechanics. However, practicing physicists invariably use the more abstract Heisenberg matrix mechanics to solve problems in quantum mechanics, analogous to widespread use of the variational approach in classical mechanics, because the analytical approaches are more powerful and have fundamental advantages. Quantal problems in molecular, atomic, nuclear, and subnuclear systems, usually involve finding the normal modes of a quantal system, that is, finding the eigen-energies, eigen-functions, spin, parity, and other observables for the discrete quantized levels. Solving the equations of motion for the modes of quantal systems is similar to solving the many-body coupled-oscillator problem in classical mechanics, where it was shown that use of matrix mechanics is the most powerful representation. It is ironic that the introduction of matrix methods to classical mechanics is a by-product of the development of matrix mechanics by Heisenberg, Born and Jordan. This illustrates that classical mechanics not only played a pivotal role in the development of quantum mechanics, but it also has benefitted considerably from the development of quantum mechanics; that is, the synergistic relation between these two complementary branches of physics has been beneficial to both classical and quantum mechanics.

Recommended reading
"Quantum Mechanics" by P.A.M. Dirac, Oxford Press, 1947,
"Conceptual Development of Quantum Mechanics" by Max Jammer, Mc Graw Hill 1966.

Chapter 19

Epilogue

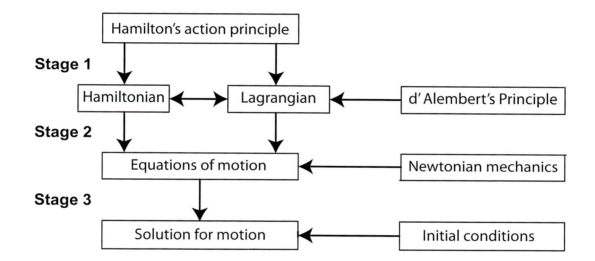

Figure 19.1: Philosophical road map of the hierarchy of stages involved in analytical mechanics. Hamilton's Action Principle is the foundation of analytical mechanics. Stage 1 uses Hamilton's Principle to derive the Lagranian and Hamiltonian. Stage 2 uses either the Lagrangian or Hamiltonian to derive the equations of motion for the system. Stage 3 uses these equations of motion to solve for the actual motion using the assumed initial conditions. The Lagrangian approach can be derived directly based on d'Alembert's Principle. Newtonian mechanics can be derived directly based on Newton's Laws of Motion.

This book has introduced powerful analytical methods in physics that are based on applications of variational principles to Hamilton's Action Principle. These methods were pioneered in classical mechanics by Leibniz, Lagrange, Euler, Hamilton, and Jacobi, during the remarkable Age of Enlightenment, and reached full fruition at the start of the 20th century.

The philosophical roadmap, shown above, illustrates the hierarchy of philosophical approaches available when using Hamilton's Action Principle to derive the equations of motion of a system. The primary **Stage1** uses Hamilton's Action functional, $S = \int_{t_i}^{t_f} L(\mathbf{q}, \dot{\mathbf{q}}, t) dt$ to derive the Lagrangian, and Hamiltonian functionals. **Stage1** provides the most fundamental and sophisticated level of understanding and involves specifying all the active degrees of freedom, as well as the interactions involved. **Stage2** uses the Lagrangian or Hamiltonian functionals, derived at **Stage1**, in order to derive the equations of motion for the system of interest. **Stage3** then uses the derived equations of motion to solve for the motion of the system, subject to a given set of initial boundary conditions.

Newton postulated equations of motion for nonrelativistic classical mechanics that are identical to those derived by applying variational principles to Hamilton's Principle. However, Newton's Laws of Motion are applicable only to nonrelativistic classical mechanics, and cannot exploit the advantages of using the more fundamental Hamilton's Action Principle, Lagrangian, and Hamiltonian. Newtonian mechanics requires that all the active forces be included in the equations of motion, and involves dealing with vector quantities which is more difficult than using the scalar functionals, action, Lagrangian, or Hamiltonian. Lagrangian mechanics based on d'Alembert's Principle does not exploit the advantages provided by Hamilton's Action Principle.

Considerable advantages result from deriving the equations of motion based on Hamilton's Principle, rather than basing them on the Newton's postulated Laws of Motion. It is significantly easier to use variational principles to handle the scalar functionals, action, Lagrangian, and Hamiltonian, rather than starting with Newton's vector differential equations-of-motion. The three hierarchical stages of analytical mechanics facilitate accommodating extra degrees of freedom, symmetries, constraints, and other interactions. For example, the symmetries identified by Noether's theorem are more easily recognized during the primary "action" and secondary "Hamiltonian/Lagrangian" stages, rather than at the subsequent "equations-of-motion" stage. Constraint forces, and approximations, introduced at the **Stage1** or **Stage2**, are easier to implement than at the subsequent **Stage3**. The correspondence of Hamilton's Action in classical and quantal mechanics, as well as relativistic invariance, are crucial advantages for using the analytical approach in relativistic mechanics, fluid motion, quantum, and field theory.

Philosophically, Newtonian mechanics is straightforward to understand since it uses vector differential equations of motion that relate the instantaneous forces to the instantaneous accelerations. Moreover, the concepts of momentum plus force are intuitive to visualize, and both cause and effect are embedded in Newtonian mechanics. Unfortunately, Newtonian mechanics is incompatible with quantum physics, it violates the relativistic concepts of space-time, and fails to provide the unified description of the gravitational force plus planetary motion as geodesic motion in a four-dimensional Riemannian structure.

The remarkable philosophical implications embedded in applying variational principles to Hamilton's Principle, are based on the astonishing assumption that motion of a constrained system in nature follows a path that minimizes the action integral. As a consequence, solving the equations of motion is reduced to finding the optimum path that minimizes the action integral. The fact that nature follows optimization principles is nonintuitive, and was considered to be metaphysical by many scientists and philosophers during the 19th century, which delayed full acceptance of analytical mechanics until the development of the Theory of Relativity and quantum mechanics. Variational formulations now have become the preeminent approach in modern physics and they have toppled Newtonian mechanics from the throne of classical mechanics that it occupied for two centuries.

The scope of this book extends beyond the typical classical mechanics textbook in order to illustrate how Lagrangian and Hamiltonian dynamics provides the foundation upon which modern physics is built. Knowledge of analytical mechanics is essential for the study of modern physics. The techniques and physics discussed in this book reappear in different guises in many fields, but the basic physics is unchanged illustrating the intellectual beauty, the philosophical implications, and the unity of the field of physics. The breadth of physics addressed by variational principles in classical mechanics, and the underlying unity of the field, are epitomized by the wide range of dimensions, energies, and complexity involved. The dimensions range from as large as $10^{27}m$, to quantal analogues of classical mechanics of systems spanning in size down to the Planck length of $1.62 \times 10^{-35}m$. Individual particles have been detected with kinetic energies ranging from zero to greater than 10^{15} eV. The complexity of classical mechanics spans from one body to the statistical mechanics of many-body systems. As a consequence, analytical variational methods have become the premier approach to describe systems from the very largest to the smallest, and from one-body to many-body dynamical systems.

The goal of this book has been to illustrate the astonishing power of analytical variational methods for understanding the physics underlying classical mechanics, as well as extensions to modern physics. However, the present narrative remains unfinished in that fundamental philosophical and technical questions have not been addressed. For example, analytical mechanics is based on the validity of the assumed principle of economy. This book has not addressed the philosophical question, *"is the principle of economy a fundamental law of nature, or is it a fortuitous consequence of the fundamental laws of nature?"*

In summary, Hamilton's action principle, which is built into Lagrangian and Hamiltonian mechanics, coupled with the availability of a wide arsenal of variational principles and mathematical techniques, provides a remarkably powerful approach for deriving the equations of motions required to determine the response of systems in a broad and diverse range of applications in science and engineering.

Appendix A

Matrix algebra

A.1   Mathematical methods for mechanics

Development of classical mechanics has involved a close and synergistic interweaving of physics and mathematics, that continues to play a key role in these fields. The concepts of scalar and vector fields play a pivotal role in describing the force fields and particle motion in both the Newtonian formulation of classical mechanics and electromagnetism. Thus it is imperative that you be familiar with the sophisticated mathematical formalism used to treat multivariate scalar and vector fields in classical mechanics. Ordinary and partial differential equations up to second order, as well as integration of algebraic and trigonometric functions play a major role in classical mechanics. It is assumed that you already have a working knowledge of differential and integral calculus in sufficient depth to handle this material. Computer codes, such as Mathematica, MatLab, and Maple, or symbolic calculators, can be used to obtain mathematical solutions for complicated cases.

The following 9 appendices provide brief summaries of matrix algebra, vector algebra, orthogonal coordinate systems, coordinate transformations, tensor algebra, multivariate calculus, vector differential plus integral calculus, Fourier analysis and time-sampled waveform analysis. The manipulation of scalar and vector fields is greatly facilitated by transforming to orthogonal curvilinear coordinate systems that match the symmetries of the problem. These appendices discuss the necessity to account for the time dependence of the orthogonal unit vectors for curvilinear coordinate systems. It is assumed that, except for coordinate transformations and tensor algebra, you have been introduced to these topics in linear algebra and other physics courses, and thus the purpose of these appendices is to serve as a reference and brief review.

A.2   Matrices

Matrix algebra provides an elegant and powerful representation of multivariate operators, and coordinate transformations that feature prominently in classical mechanics. For example they play a pivotal role in finding the eigenvalues and eigenfunctions for coupled equations that occur in rigid-body rotation, and coupled oscillator systems. An understanding of the role of matrix mechanics in classical mechanics facilitates understanding of the equally important role played by matrix mechanics in quantal physics.

It is interesting that although determinants were used by physicists in the late 19th century, the concept of matrix algebra was developed by Arthur Cayley in England in 1855, but many of these ideas were the work of Hamilton, and the discussion of matrix algebra was buried in a more general discussion of determinants. Matrix algebra was an esoteric branch of mathematics, little known by the physics community, until 1925 when Heisenberg proposed his innovative new quantum theory. The striking feature of this new theory was its representation of physical quantities by sets of time-dependent complex numbers and a peculiar multiplication rule. Max Born recognized that Heisenberg's multiplication rule is just the standard "row times column" multiplication rule of matrix algebra; a topic that he had encountered as a young student in a mathematics course. In 1924 Richard Courant had just completed the first volume of the new text *Methods of Mathematical Physics* during which Pascual Jordan had served as his young assistant working on matrix manipulation. Fortuitously, Jordan and Born happened to share a carriage on a train to Hanover during which Jordan overheard Born talk about his problems trying to work with matrices. Jordan introduced himself to Born and offered to help. This led to publication, in September 1925, of the famous Born-Jordan paper[Bor25a] that gave the first rigorous formulation of matrix mechanics in physics. This was followed in November by the Born-Heisenberg-Jordan sequel[Bor25b] that established a logical consistent general method for solving matrix mechanics problems plus a connection between the mathematics of matrix mechanics and linear algebra. Matrix algebra developed into an important tool in mathematics and physics during World War 2 and now it is an integral part of undergraduate linear algebra courses.

Most applications of matrix algebra in this book are restricted to real, symmetric, square matrices. The size of a matrix is defined by the rank, which equals the row rank and column rank, i.e. the number of independent row vectors or column vectors in the square matrix. It is presumed that you have studied matrices in a linear algebra course. Thus the goal of this review is to list simple manipulation of symmetric matrices and matrix diagonalization that will be used in this course. You are referred to a linear algebra textbook if you need further details.

Matrix definition

A matrix is a rectangular array of numbers with M rows and N columns. The notation used for an element of a matrix is A_{ij} where i designates the row and j designates the column of this matrix element in the matrix $\mathbf{A}$. Convention denotes a matrix $\mathbf{A}$ as

$$\mathbf{A} \equiv \begin{pmatrix} A_{11} & A_{12} & \ldots & A_{1(N-1)} & A_{1N} \\ A_{21} & A_{22} & .. & A_{2(N-1)} & A_{2N} \\ : & : & A_{ij} & : & : \\ A_{(M-1)1} & A_{(M-1)2} & .. & A_{(M-1)(N-1)} & A_{(M-1)N} \\ A_{M1} & A_{M2} & \ldots & A_{M(N-1)} & A_{MN} \end{pmatrix} \tag{A.1}$$

Matrices can be square, $M = N$, or rectangular $M \neq N$. Matrices having only one row or column are called row or column vectors respectively, and need only a single subscript label. For example,

$$\mathbf{A} = \begin{pmatrix} A_1 \\ A_2 \\ : \\ A_{M-1} \\ A_M \end{pmatrix} \tag{A.2}$$

Matrix manipulation

Matrices are defined to obey certain rules for matrix manipulation as given below.

1) Multiplication of a matrix by a scalar λ simply multiplies each matrix element by λ.

$$C_{ij} = \lambda A_{ij} \tag{A.3}$$

2) Addition of two matrices $\mathbf{A}$ and $\mathbf{B}$ having the same rank, i.e. the number of columns, is given by

$$C_{ij} = A_{ij} + B_{ij} \tag{A.4}$$

3) Multiplication of a matrix $\mathbf{A}$ by a matrix $\mathbf{B}$ is defined only if the number of columns in $\mathbf{A}$ equals the number of rows in $\mathbf{B}$. The product matrix $\mathbf{C}$ is given by the *matrix product*

$$\mathbf{C} = \mathbf{A} \cdot \mathbf{B} \tag{A.5}$$

$$C_{ij} = [AB]_{ij} = \sum_k A_{ik} B_{kj} \tag{A.6}$$

For example, if both $\mathbf{A}$ and $\mathbf{B}$ are rank three symmetric matrices then

$$\begin{aligned} \mathbf{C} &= \mathbf{A} \cdot \mathbf{B} = \begin{pmatrix} A_{11} & A_{12} & A_{13} \\ A_{21} & A_{22} & A_{23} \\ A_{31} & A_{32} & A_{33} \end{pmatrix} \cdot \begin{pmatrix} B_{11} & B_{12} & B_{13} \\ B_{21} & B_{22} & B_{23} \\ B_{31} & B_{32} & B_{33} \end{pmatrix} \\ &= \begin{pmatrix} A_{11}B_{11} + A_{12}B_{21} + A_{13}B_{31} & A_{11}B_{12} + A_{12}B_{22} + A_{13}B_{32} & A_{11}B_{13} + A_{12}B_{23} + A_{13}B_{33} \\ A_{21}B_{11} + A_{22}B_{21} + A_{23}B_{31} & A_{21}B_{12} + A_{22}B_{22} + A_{23}B_{32} & A_{21}B_{13} + A_{22}B_{23} + A_{23}B_{33} \\ A_{31}B_{11} + A_{32}B_{21} + A_{33}B_{31} & A_{31}B_{12} + A_{32}B_{22} + A_{33}B_{32} & A_{31}B_{13} + A_{32}B_{23} + A_{33}B_{33} \end{pmatrix} \end{aligned}$$

In general, multiplication of matrices $\mathbf{A}$ and $\mathbf{B}$ is noncommutative, i.e.

$$\mathbf{A} \cdot \mathbf{B} \neq \mathbf{B} \cdot \mathbf{A} \tag{A.7}$$

In the special case when $\mathbf{A} \cdot \mathbf{B} = \mathbf{B} \cdot \mathbf{A}$ then the matrices are said to commute.

Transposed matrix $\mathbf{A}^T$

The *transpose* of a matrix $\mathbf{A}$ will be denoted by $\mathbf{A}^T$ and is given by interchanging rows and columns, that is

$$\left(A^T\right)_{ij} = A_{ji} \tag{A.8}$$

The transpose of a column vector is a row vector. Note that older texts use the symbol $\tilde{\mathbf{A}}$ for the transpose.

Identity (unity) matrix $\mathbb{I}$

The *identity (unity) matrix* $\mathbb{I}$ is diagonal with diagonal elements equal to 1, that is

$$\mathbb{I}_{ij} = \delta_{ij} \tag{A.9}$$

where the Kronecker delta symbol is defined by

$$\begin{aligned} \delta_{ik} &= 0 & \text{if } i \neq k \\ &= 1 & \text{if } i = k \end{aligned} \tag{A.10}$$

Inverse matrix $\mathbf{A}^{-1}$

If a matrix is non-singular, that is, its determinant is non-zero, then it is possible to define an *inverse* matrix $\mathbf{A}^{-1}$. A square matrix has an inverse matrix for which the product

$$\mathbf{A} \cdot \mathbf{A}^{-1} = \mathbb{I} \tag{A.11}$$

Orthogonal matrix

A matrix with *real elements* is *orthogonal* if

$$\mathbf{A}^T = \mathbf{A}^{-1} \tag{A.12}$$

That is

$$\sum_k \left(A^T\right)_{ik} A_{kj} = \sum_k A_{ki} A_{kj} = \delta_{ij} \tag{A.13}$$

Adjoint matrix $\mathbf{A}^\dagger$

For a matrix with *complex elements*, the *adjoint* matrix, denoted by $\mathbf{A}^\dagger$ is defined as the transpose of the complex conjugate

$$\left(\mathbf{A}^\dagger\right)_{ij} = \mathbf{A}_{ji}^* \tag{A.14}$$

Hermitian matrix

The *Hermitian conjugate* of a complex matrix $\mathbf{H}$ is denoted as $\mathbf{H}^\dagger$ and is defined as

$$\mathbf{H}^\dagger = \left(\mathbf{H}^T\right)^* = \left(\mathbf{H}^*\right)^T \tag{A.15}$$

Therefore

$$H_{ij}^\dagger = H_{ji}^* \tag{A.16}$$

A matrix is *Hermitian* if it is equal to its adjoint

$$\mathbf{H}^\dagger = \mathbf{H} \tag{A.17}$$

that is

$$H_{ij}^\dagger = H_{ji}^* = H_{ij} \tag{A.18}$$

A matrix that is both Hermitian and has real elements is a symmetric matrix since complex conjugation has no effect.

Unitary matrix

A matrix with *complex* elements is *unitary* if its inverse is equal to the adjoint matrix

$$\mathbf{U}^\dagger = \mathbf{U}^{-1} \tag{A.19}$$

which is equivalent to

$$\mathbf{U}^\dagger \mathbf{U} = \mathbb{I} \tag{A.20}$$

A unitary matrix with real elements is an orthogonal matrix as given in equation *A.12*.

Trace of a square matrix $Tr\mathbf{A}$

The *trace* of a square matrix, denoted by $Tr\mathbf{A}$, is defined as the sum of the diagonal matrix elements.

$$Tr\mathbf{A} = \sum_{i=1}^{N} A_{ii} \tag{A.21}$$

Inner product of column vectors

Real vectors   The generalization of the scalar (dot) product in Euclidean space is called the **inner product**. Exploiting the rules of matrix multiplication requires taking the transpose of the first column vector to form a row vector which then is multiplied by the second column vector using the conventional rules for matrix multiplication. That is, for rank N vectors

$$[\mathbf{X}] \cdot [\mathbf{Y}] = \begin{pmatrix} X_1 \\ X_2 \\ \vdots \\ X_N \end{pmatrix} \cdot \begin{pmatrix} Y_1 \\ Y_2 \\ \vdots \\ Y_N \end{pmatrix} = [\mathbf{X}]^T [\mathbf{Y}] = \begin{pmatrix} X_1 & X_2 & .. & X_N \end{pmatrix} \begin{pmatrix} Y_1 \\ Y_2 \\ \vdots \\ Y_N \end{pmatrix} = \sum_{i=1}^{N} X_i Y_i \tag{A.22}$$

For rank $N = 3$ this inner product agrees with the conventional definition of the scalar product and gives a result that is a scalar. For the special case when $[\mathbf{A}] \cdot [\mathbf{B}] = 0$ then the two matrices are called *orthogonal*. The magnitude squared of a column vector is given by the inner product

$$[\mathbf{X}] \cdot [\mathbf{X}] = \sum_{i=1}^{N} (X_i)^2 \geq 0 \tag{A.23}$$

Note that this is only positive.

Complex vectors   For vectors having complex matrix elements the inner product is generalized to a form that is consistent with equation *A.22* when the column vector matrix elements are real.

$$[\mathbf{X}]^* \cdot [\mathbf{Y}] = [\mathbf{X}]^\dagger [\mathbf{Y}] = \begin{pmatrix} X_1^* & X_2^* & .. & X_{N-1}^* & X_N^* \end{pmatrix} \begin{pmatrix} Y_1 \\ Y_2 \\ \vdots \\ Y_{N-1} \\ Y_N \end{pmatrix} = \sum_{i=1}^{N} X_i^* Y_i \tag{A.24}$$

For the special case

$$[\mathbf{X}]^* \cdot [\mathbf{X}] = [\mathbf{X}]^\dagger [\mathbf{X}] = \sum_{i=1}^{N} X_i^* X_i \geq 0 \tag{A.25}$$

A.3 Determinants

Definition

The determinant of a square matrix with N rows equals a single number derived using the matrix elements of the matrix. The determinant is denoted as det $\mathbf{A}$ or $|\mathbf{A}|$ where

$$|\mathbf{A}| = \sum_{j=1}^{N} \varepsilon(j_1, j_2, ...j_N) A_{1j_1} A_{2j_2} ... A_{Nj_N} \tag{A.26}$$

where $\varepsilon(j_1, j_2, ...j_N)$ is the permutation index which is either even or odd depending on the number of permutations required to go from the normal order $(1, 2, 3, ...N)$ to the sequence $(j_1 j_2 j_3 ... j_N)$.

For example for $N = 3$ the determinant is

$$|\mathbf{A}| = A_{11}A_{22}A_{33} + A_{12}A_{23}A_{31} + A_{13}A_{21}A_{32} - A_{13}A_{22}A_{31} - A_{11}A_{23}A_{32} - A_{12}A_{21}A_{33} \tag{A.27}$$

Properties

1. The value of a determinant $|A| = 0$, if

   (a) all elements of a row (column) are zero.

   (b) all elements of a row (column) are identical with, or multiples of, the corresponding elements of another row (column).

2. The value of a determinant is unchanged if

   (a) rows and columns are interchanged.

   (b) a linear combination of any number of rows is added to any one row.

3. The value of a determinant changes sign if two rows, or any two columns, are interchanged.

4. Transposing a square matrix does not change its determinant. $\left|\mathbf{A}^T\right| = |\mathbf{A}|$

5. If any row (column) is multiplied by a constant factor then the value of the determinant is multiplied by the same factor.

6. The determinant of a diagonal matrix equals the product of the diagonal matrix elements. That is, when $A_{ij} = \lambda_i \delta_{ij}$ then $|\mathbf{A}| = \lambda_1 \lambda_2 \lambda_3 ... \lambda_N$

7. The determinant of the identity (unity) matrix $|\mathbb{I}| = 1$.

8. The determinant of the null matrix, for which all matrix elements are zero, $|\mathbf{0}| = 0$

9. A *singular* matrix has a determinant equal to zero.

10. If each element of any row (column) appears as the sum (difference) of two or more quantities, then the determinant can be written as a sum (difference) of two or more determinants of the same order. For example for order $N = 2$,

$$\begin{vmatrix} A_{11} \pm B_{11} & A_{12} \pm B_{12} \\ A_{21} & A_{22} \end{vmatrix} = \begin{vmatrix} A_{11} & A_{12} \\ A_{21} & A_{22} \end{vmatrix} \pm \begin{vmatrix} B_{11} & B_{12} \\ A_{21} & A_{22} \end{vmatrix}$$

11 A determinant of a matrix product equals the product of the determinants. That is, if $\mathbf{C} = \mathbf{AB}$ then $|\mathbf{C}| = |\mathbf{A}|\,|\mathbf{B}|$

Cofactor of a square matrix

For a square matrix having N rows the cofactor is obtained by removing the ith row and the jth column and then collapsing the remaining matrix elements into a square matrix with $N-1$ rows while preserving the order of the matrix elements. This is called the complementary minor which is denoted as $A^{(ij)}$. The matrix elements of the cofactor square matrix $\mathbf{a}$ are obtained by multiplying the determinant of the (ij) complementary minor by the phase factor $(-1)^{i+j}$. That is

$$a_{ij} = (-1)^{i+j} \left| A^{(ij)} \right| \tag{A.28}$$

The cofactor matrix has the property that

$$\sum_{k=1}^{N} A_{ik} a_{jk} = \delta_{ij} |\mathbf{A}| = \sum_{k=1}^{N} A_{ki} a_{kj} \tag{A.29}$$

Cofactors are used to expand the determinant of a square matrix in order to evaluate the determinant.

Inverse of a non-singular matrix

The (i,j) matrix elements of the inverse matrix $\mathbf{A}^{-1}$ of a non-singular matrix $\mathbf{A}$ are given by the ratio of the cofactor a_{ji} and the determinant $|\mathbf{A}|$, that is

$$A_{ij}^{-1} = \frac{1}{|\mathbf{A}|} a_{ji} \tag{A.30}$$

Equations $A.28$ and $A.29$ can be used to evaluate the i, j element of the matrix product $\left(\mathbf{A}^{-1}\mathbf{A}\right)$

$$\left(\mathbf{A}^{-1}\mathbf{A}\right)_{ij} = \sum_{k=1}^{N} A_{ik}^{-1} A_{kj} = \frac{1}{|\mathbf{A}|} \sum_{k=1}^{N} a_{ji} A_{kj} = \frac{1}{|\mathbf{A}|} \delta_{ji} |\mathbf{A}| = \delta_{ij} = \mathbb{I}_{ij} \tag{A.31}$$

This agrees with equation $A11$ that $\mathbf{A} \cdot \mathbf{A}^{-1} = \mathbb{I}$.

The inverse of rank 2 or 3 matrices is required frequently when determining the eigen-solutions for rigid-body rotation, or coupled oscillator, problems in classical mechanics as described in chapters 11 and 12. Therefore it is convenient to list explicitly the inverse matrices for both rank 2 and rank 3 matrices.

Inverse for rank 2 matrices:

$$\mathbf{A}^{-1} = \begin{bmatrix} a & b \\ c & d \end{bmatrix}^{-1} = \frac{1}{|\mathbf{A}|} \begin{bmatrix} d & -b \\ -c & a \end{bmatrix} = \frac{1}{(ad-bc)} \begin{bmatrix} d & -b \\ -c & a \end{bmatrix} \tag{A.32}$$

where the determinant of $\mathbf{A}$ is written explicitly in equation $A32$.

Inverse for rank 3 matrices:

$$\begin{aligned}
\mathbf{A}^{-1} &= \begin{bmatrix} a & b & c \\ d & e & f \\ g & h & i \end{bmatrix}^{-1} = \frac{1}{|\mathbf{A}|} \begin{bmatrix} A & B & C \\ D & E & F \\ G & H & I \end{bmatrix}^{T} = \frac{1}{|\mathbf{A}|} \begin{bmatrix} A & D & G \\ B & E & H \\ C & F & I \end{bmatrix} \\
&= \frac{1}{aA+bB+cC} \begin{bmatrix} A=(ei-fh) & D=-(bi-ch) & G=(bf-ce) \\ B=-(di-fg) & E=(ai-cg) & H=-(af-cd) \\ C=(dh-eg) & F=-(ah-bg) & I=(ae-bd) \end{bmatrix}
\end{aligned} \tag{A.33}$$

where the functions $A, B, C, D, E, F, G, H, I$, are equal to rank 2 determinants listed in equation $A33$.

A.4 Reduction of a matrix to diagonal form

Solving coupled linear equations can be reduced to diagonalization of a matrix. Consider the matrix $\mathbf{A}$ operating on the vector $\mathbf{X}$ to produce a vector $\mathbf{Y}$, that are expressed as components with respect to the unprimed coordinate frame, i.e.

$$\mathbf{A} \cdot \mathbf{X} = \mathbf{Y} \tag{A.34}$$

Consider that the unitary real matrix $\mathbf{R}$ with rank n, rotates the n-dimensional un-primed coordinate frame into the primed coordinate frame such that $\mathbf{A}$, $\mathbf{X}$ and $\mathbf{Y}$ are transformed to $\mathbf{A}'$, $\mathbf{X}'$ and $\mathbf{Y}'$ in the rotated primed coordinate frame. Then

$$\begin{aligned} \mathbf{X}' &= \mathbf{R} \cdot \mathbf{X} \\ \mathbf{Y}' &= \mathbf{R} \cdot \mathbf{Y} \end{aligned} \tag{A.35}$$

With respect to the primed coordinate frame equation $(A.34)$ becomes

$$\begin{aligned} \mathbf{R} \cdot (\mathbf{A} \cdot \mathbf{X}) &= \mathbf{R} \cdot \mathbf{Y} \tag{A.36} \\ \mathbf{R} \cdot \mathbf{A} \cdot \mathbf{R}^{-1} \cdot \mathbf{R} \cdot \mathbf{X} &= \mathbf{R} \cdot \mathbf{Y} \tag{A.37} \\ \mathbf{R} \cdot \mathbf{A} \cdot \mathbf{R}^{-1} \cdot \mathbf{X}' &= \mathbf{A}' \cdot \mathbf{X}' = \mathbf{Y}' \tag{A.38} \end{aligned}$$

using the fact that the identity matrix $\mathbf{I} = \mathbf{R} \cdot \mathbf{R}^{-1} = \mathbf{R} \cdot \mathbf{R}^T$ since the rotation matrix in n dimensions is orthogonal.

Thus we have that the rotated matrix

$$\mathbf{A}' = \mathbf{R} \cdot \mathbf{A} \cdot \mathbf{R}^T \tag{A.39}$$

Let us assume that this transformed matrix is diagonal, then it can be written as the product of the unit matrix $\mathbb{I}$ and a vector of scalar numbers called the characteristic roots λ as

$$\mathbf{A}' = \mathbf{R} \cdot \mathbf{A} \cdot \mathbf{R}^T = \lambda \mathbb{I} \tag{A.40}$$

using the fact that $\mathbf{R}^T = \mathbf{R}^{-1}$ then gives

$$\mathbf{R}^T \cdot (\lambda \mathbb{I}) = \mathbf{A}' \cdot \mathbf{R}^T \tag{A.41}$$

Let both sides of equation $A.41$ act on $\mathbf{X}'$ which gives

$$\lambda \mathbb{I} \cdot \mathbf{X}' = \mathbf{A}' \cdot \mathbf{X}' \tag{A.42}$$

or

$$\left[\lambda \mathbb{I} - \mathbf{A}' \right] \mathbf{X}' = \mathbf{0} \tag{A.43}$$

This represents a set of n homogeneous linear algebraic equations in n unknowns $\mathbf{X}'$ where λ is a set of characteristic roots, (eigenvalues) with corresponding eigenfunctions $\mathbf{X}'$. Ignoring the trivial case of $\mathbf{X}'$ being zero, then $(A.43)$ requires that the *secular determinant* of the bracket be zero, that is

$$\left| \lambda \mathbb{I} - \mathbf{A}' \right| = \mathbf{0} \tag{A.44}$$

The determinant can be expanded and factored into the form

$$(\lambda - \lambda_1)(\lambda - \lambda_2)(\lambda - \lambda_3) \ldots (\lambda - \lambda_n) = 0 \tag{A.45}$$

where the n eigenvalues are $\lambda = \lambda_1, \lambda_2, \ldots \lambda_n$ of the matrix $\mathbf{A}'$.

The eigenvectors $\mathbf{X}'$ corresponding to each eigenvalue are determined by substituting a given eigenvalue λ_i into the relation

$$\mathbf{X}'^T \cdot \mathbf{A}' \cdot \mathbf{X}' = [\lambda_i \delta_{ij}] \tag{A.46}$$

If all the eigenvalues are distinct, i.e. different, then this set of n equations completely determines the ratio of the components of each eigenvector along the axes of the coordinate frame. However, when two or more eigenvalues are identical, then the reduction to a true diagonal form is not possible and one has the freedom to select an appropriate eigenvector that is orthogonal to the remaining axes.

In summary, the matrix can only be fully diagonalized if (a) all the eigenvalues are distinct, (b) the real matrix is symmetric, (c) it is unitary.

A frequent application of matrices in classical mechanics is for solving a system of homogeneous linear equations of the form

$$
\begin{array}{ccccc}
A_{11}x_1 & +A_{12}x_2 & \ldots\ldots & +A_{1n}x_n & = & 0 \\
A_{11}x_1 & +A_{12}x_2 & \ldots\ldots & +A_{1n}x_n & = & 0 \\
\ldots\ldots & \ldots\ldots & \ldots\ldots & \ldots\ldots & = & \ldots\ldots \\
A_{n1}x_1 & +A_{n2}x_2 & \ldots\ldots & +A_{nn}x_n & = & 0
\end{array}
\tag{A.47}
$$

Making the following definitions

$$
\mathbf{A} = \begin{pmatrix}
A_{11} & A_{12} & \ldots & A_{1n} \\
A_{21} & A_{22} & \ldots & A_{2n} \\
\ldots & \ldots & \ldots & \ldots \\
A_{n1} & A_{n2} & \ldots & A_{nn}
\end{pmatrix}
\tag{A.48}
$$

$$
\mathbf{X} = \begin{pmatrix}
x_1 \\
x_2 \\
\ldots \\
x_n
\end{pmatrix}
\tag{A.49}
$$

Then the set of linear equations can be written in a compact form using the matrices

$$
\mathbf{A} \cdot \mathbf{X} = 0
\tag{A.50}
$$

which can be solved using equation $(A.43)$. Ensure that you are able to diagonalize a matrices with rank 2 and 3. You can use Mathematica, Maple, MatLab, or other such mathematical computer programs to diagonalize larger matrices.

A.1 *Example: Eigenvalues and eigenvectors of a real symmetric matrix*

Consider the matrix

$$
\mathbf{A} = \begin{pmatrix}
0 & 1 & 0 \\
1 & 0 & 0 \\
0 & 0 & 0
\end{pmatrix}
$$

The secular determinant is given by $(A.42)$

$$
\begin{vmatrix}
-\lambda & 1 & 0 \\
1 & -\lambda & 0 \\
0 & 0 & -\lambda
\end{vmatrix} = 0
$$

This expands to

$$
-\lambda(\lambda + 1)(\lambda - 1) = 0
$$

Thus the three eigen values are $\lambda = -1, 0, 1$.

To find each eigenvectors we substitute the corresponding eigenvalue into equation $(A.48)$.

$$
\begin{pmatrix}
-\lambda & 1 & 0 \\
1 & -\lambda & 0 \\
0 & 0 & -\lambda
\end{pmatrix}
\begin{pmatrix}
x \\
y \\
z
\end{pmatrix} =
\begin{pmatrix}
0 \\
0 \\
0
\end{pmatrix}
$$

The eigenvalue $\lambda = -1$ *yields* $x + y = 0$ *and* $z = 0$. *Thus the eigen vector is* $r_1 = (\frac{1}{\sqrt{2}}, \frac{-1}{\sqrt{2}}, 0)$. *The eigenvalue* $\lambda = 0$ *yields* $x = 0$ *and* $y = 0$. *Thus the eigen vector is* $r_2 = (0, 0, 1)$. *The eigenvalue* $\lambda = 1$ *yields* $-x + y = 0$ *and* $z = 0$. *Thus the eigen vector is* $r_3 = (\frac{1}{\sqrt{2}}, \frac{1}{\sqrt{2}}, 0)$. *The orthogonality of these three eigen vectors, which correspond to three distinct eigenvalues, can be verified.*

A.2 Example: Degenerate eigenvalues of real symmetric matrix

This example illustrates how to generate eigenvectors corresponding to degenerate eigenvalues. Consider the matrix

$$\mathbf{A} = \begin{pmatrix} 1 & 0 & 0 \\ 0 & 0 & 1 \\ 0 & 1 & 0 \end{pmatrix}$$

The secular determinant is given by (A.42)

$$\begin{vmatrix} 1-\lambda & 0 & 0 \\ 0 & -\lambda & 1 \\ 0 & 1 & -\lambda \end{vmatrix} = 0$$

This expands to

$$(1-\lambda)\,(\lambda+1)(\lambda-1) = 0$$

Thus the three eigen values are $\lambda = -1, 1, 1$.

The eigenvectors are determined by substituting the corresponding eigenvalue into equation (A.42).

$$\begin{pmatrix} 1-\lambda & 0 & 0 \\ 0 & -\lambda & 1 \\ 0 & 0 & -\lambda \end{pmatrix} \cdot \begin{pmatrix} x \\ y \\ z \end{pmatrix} = \begin{pmatrix} 0 \\ 0 \\ 0 \end{pmatrix}$$

The eigenvalue $\lambda = -1$ *yields* $2x = 0$ *and* $y + z = 0$. *Thus the eigen vector is* $r_1 = (0, \frac{1}{\sqrt{2}}, \frac{-1}{\sqrt{2}})$. *The eigenvalue* $\lambda = 1$ *yields* $-y + z = 0$. *The eigenvector* r_2 *must be perpendicular to* r_1 *and there are an infinite number of choices. Let us assume that* $r_2 = (0, \frac{1}{\sqrt{2}}, \frac{1}{\sqrt{2}})$ *which satisfies equation* (A.50) *then the eigenvector* r_3 *must be perpendicular to both* r_1 *and* r_2. *For rank three this is found using*

$$\mathbf{r}_3 = \mathbf{r}_1 \times \mathbf{r}_2 = (1, 0, 0)$$

Appendix B

Vector algebra

B.1 Linear operations

The important force fields in classical mechanics, namely, gravitation, electric, and magnetic, are vector fields that have a position-dependent magnitude and direction. Thus, it is useful to summarize the algebra of vector fields.

A vector $\mathbf{a}$ has both a magnitude $|a|$ and a direction defined by the *unit vector* $\hat{\mathbf{e}}_a$, that is, the vector can be written as a bold character $\mathbf{a}$ where

$$\mathbf{a} = a \cdot \hat{\mathbf{e}}_a \qquad (B.1)$$

where by convention the implied modulus sign is omitted. The hat symbol on the vector $\hat{\mathbf{e}}_a$ designates that this is a unit vector with modulus $|\hat{\mathbf{e}}_a| = 1$.

Vector force fields are assumed to be linear, and consequently they obey the principle of superposition, are commutative, associative, and distributive as illustrated below for three vectors $\mathbf{a}, \mathbf{b}, \mathbf{c}$ plus a scalar multiplier γ.

$$
\begin{aligned}
\mathbf{a} \pm \mathbf{b} &= \pm \mathbf{b} + \mathbf{a} \\
\mathbf{a} + (\mathbf{b} + \mathbf{c}) &= (\mathbf{a} + \mathbf{b}) + \mathbf{c} \\
\gamma(\mathbf{a} + \mathbf{b}) &= \gamma \mathbf{a} + \gamma \mathbf{b}
\end{aligned}
\qquad (B.2)
$$

The manipulation of vectors is greatly facilitated by use of components along an orthogonal coordinate system defined by three orthogonal unit vectors $(\hat{\mathbf{e}}_1, \hat{\mathbf{e}}_2, \hat{\mathbf{e}}_3)$. For example the cartesian coordinate system is defined by three unit vectors which, by convention, are called $(\hat{\mathbf{i}}, \hat{\mathbf{j}}, \hat{\mathbf{k}})$.

B.2 Scalar product

Multiplication of two vectors can produce a 9−component tensor that can be represented by a 3×3 matrix as discussed in appendix E. There are two special cases for vector multiplication that are important for vector algebra; the first is the scalar product, and the second is the vector product.

The *scalar product* of two vectors is defined to be

$$\mathbf{a} \cdot \mathbf{b} = |a|\,|b| \cos\theta \qquad (B.3)$$

where θ is the angle between the two vectors. It is a scalar and thus is independent of the orientation of the coordinate axis system. Note that the scalar product commutes, is distributive, and associative with a scalar multiplier, that is

$$
\begin{aligned}
\mathbf{a} \cdot \mathbf{b} &= \mathbf{b} \cdot \mathbf{a} \\
\mathbf{a} \cdot (\mathbf{b} + \mathbf{c}) &= \mathbf{a} \cdot \mathbf{b} + \mathbf{a} \cdot \mathbf{c} \\
(\lambda \mathbf{a}) \cdot \mathbf{b} &= \lambda(\mathbf{b} \cdot \mathbf{a})
\end{aligned}
\qquad (B.4)
$$

Note that $\mathbf{a} \cdot \mathbf{a} = |a|^2$ and if $\mathbf{a}$ and $\mathbf{b}$ are perpendicular then $\cos\theta = 0$ and thus $\mathbf{a} \cdot \mathbf{b} = 0$

If the three unit vectors $(\hat{\mathbf{e}}_1, \hat{\mathbf{e}}_2, \hat{\mathbf{e}}_3)$ form an orthonormal basis, that is, they are orthogonal unit vectors, then from equations $B.3$ and $B.4$

$$\hat{\mathbf{e}}_i \cdot \hat{\mathbf{e}}_k = \delta_{ik} \tag{B.5}$$

If $\hat{\mathbf{a}}$ is the unit vector for the vector $\mathbf{a}$ then the scalar product of a vector $\mathbf{a}$ with one of these unit vectors $\hat{\mathbf{e}}_n$ gives the cosine of the angle between the vector $\mathbf{a}$ and $\hat{\mathbf{e}}_n$, that is

$$
\begin{aligned}
\mathbf{a} \cdot \hat{\mathbf{e}}_1 &= |a|\,(\hat{\mathbf{a}} \cdot \hat{\mathbf{e}}_1) = |a| \cos \alpha \\
\mathbf{a} \cdot \hat{\mathbf{e}}_2 &= |a|\,(\hat{\mathbf{a}} \cdot \hat{\mathbf{e}}_2) = |a| \cos \beta \\
\mathbf{a} \cdot \hat{\mathbf{e}}_3 &= |a|\,(\hat{\mathbf{a}} \cdot \hat{\mathbf{e}}_3) = |a| \cos \gamma
\end{aligned}
\tag{B.6}
$$

where the cosines are called the direction cosines since they define the direction of the vector $\mathbf{a}$ with respect to each orthogonal basis unit vector. Moreover, $\mathbf{a} \cdot \hat{\mathbf{e}}_1 = |a|\,\hat{\mathbf{a}} \cdot \hat{\mathbf{e}}_1 = |a| \cos \alpha$ is the component of $\mathbf{a}$ along the $\hat{\mathbf{e}}_1$ axis. Thus the three components of the vector $\mathbf{a}$ is fully defined by the magnitude $|a|$ and the direction cosines, corresponding to the angles α, β, γ. That is,

$$
\begin{aligned}
a_1 &= |a|\,(\hat{\mathbf{a}} \cdot \hat{\mathbf{e}}_1) = |a| \cos \alpha \\
a_2 &= |a|\,(\hat{\mathbf{a}} \cdot \hat{\mathbf{e}}_2) = |a| \cos \beta \\
a_3 &= |a|\,(\hat{\mathbf{a}} \cdot \hat{\mathbf{e}}_3) = |a| \cos \gamma
\end{aligned}
\tag{B.7}
$$

If the three unit vectors $(\hat{\mathbf{e}}_1, \hat{\mathbf{e}}_2, \hat{\mathbf{e}}_3)$ form an orthonormal basis then the vector is fully defined by

$$\mathbf{a} = a_1 \hat{\mathbf{e}}_1 + a_2 \hat{\mathbf{e}}_2 + a_3 \hat{\mathbf{e}}_3 \tag{B.8}$$

Consider two vectors

$$
\begin{aligned}
\mathbf{a} &= a_1 \hat{\mathbf{e}}_1 + a_2 \hat{\mathbf{e}}_2 + a_3 \hat{\mathbf{e}}_3 \\
\mathbf{b} &= b_1 \hat{\mathbf{e}}_1 + b_2 \hat{\mathbf{e}}_2 + b_3 \hat{\mathbf{e}}_3
\end{aligned}
$$

Then using $B.5$

$$\mathbf{a} \cdot \mathbf{b} = a_1 b_1 + a_2 b_2 + a_3 b_3 = |a|\,|b| \cos \theta \tag{B.9}$$

where θ is the angle between the two vectors. In particular, since the direction cosine $\cos \alpha_a = \frac{a_1}{|a|}$, then equation $B.9$ gives

$$\cos \theta = \cos \alpha_a \cos \alpha_b + \cos \beta_a \cos \beta_b + \cos \gamma_a \cos \gamma_b \tag{B.10}$$

Note that when $\theta = 0$ then $B.10$ gives

$$\cos^2 \alpha + \cos^2 \beta + \cos^2 \gamma = 1 \tag{B.11}$$

B.3   Vector product

The vector product of two vectors is defined to be

$$\mathbf{c} = \mathbf{a} \times \mathbf{b} = |a|\,|b| \sin \theta \hat{\mathbf{n}} \tag{B.12}$$

where θ is the angle between the vectors and $\hat{\mathbf{n}}$ is a unit vector perpendicular to the plane defined by $\mathbf{a}$ and $\mathbf{b}$ such that the unit vectors $\left(\hat{\mathbf{a}}, \hat{\mathbf{b}}, \hat{\mathbf{n}}\right)$ obey a right-handed screw rule. The vector product acts like a pseudovector which comprises a normal vector multiplied by a sign factor that depends on the handedness of the system as described in appendix $D.3$.

The components of $\mathbf{c}$ are defined by the relation

$$c_i \equiv \sum_{jk} \varepsilon_{ijk} a_j b_k \tag{B.13}$$

where the (Levi-Civita) permutation symbol ε_{ijk} has the following properties

$$
\begin{aligned}
\varepsilon_{ijk} &= 0 && \text{if an index is equal to any another index} \\
\varepsilon_{ijk} &= +1 && \text{if } i, j, k, \text{ form an even permutation of } 1, 2, 3 \\
\varepsilon_{ijk} &= -1 && \text{if } i, j, k, \text{ form an odd permutation of } 1, 2, 3
\end{aligned}
\tag{B.14}
$$

For example, if the three unit vectors $(\hat{\mathbf{e}}_1, \hat{\mathbf{e}}_2, \hat{\mathbf{e}}_3)$ form an orthonormal basis, then $\hat{\mathbf{e}}_i \equiv \sum_{jk} \varepsilon_{ijk} \hat{\mathbf{e}}_j \hat{\mathbf{e}}_k$, i.e.

$$\hat{\mathbf{e}}_1 \times \hat{\mathbf{e}}_2 = \hat{\mathbf{e}}_3 \qquad \hat{\mathbf{e}}_2 \times \hat{\mathbf{e}}_3 = \hat{\mathbf{e}}_1 \qquad \hat{\mathbf{e}}_3 \times \hat{\mathbf{e}}_1 = \hat{\mathbf{e}}_2 \tag{B.15}$$

$$\hat{\mathbf{e}}_2 \times \hat{\mathbf{e}}_1 = -\hat{\mathbf{e}}_3 \qquad \hat{\mathbf{e}}_3 \times \hat{\mathbf{e}}_2 = -\hat{\mathbf{e}}_1 \qquad \hat{\mathbf{e}}_1 \times \hat{\mathbf{e}}_3 = -\hat{\mathbf{e}}_2 \tag{B.16}$$

$$\hat{\mathbf{e}}_1 \times \hat{\mathbf{e}}_1 = 0 \qquad \hat{\mathbf{e}}_2 \times \hat{\mathbf{e}}_2 = 0 \qquad \hat{\mathbf{e}}_3 \times \hat{\mathbf{e}}_0 = 0 \tag{B.17}$$

The vector product anticommutes in that

$$\mathbf{a} \times \mathbf{b} = -\mathbf{b} \times \mathbf{a} \tag{B.18}$$

However, it is distributive and associative with a scalar multiplier

$$\mathbf{a} \times (\mathbf{b} + \mathbf{c}) = \mathbf{a} \times \mathbf{b} + \mathbf{a} \times \mathbf{c} \tag{B.19}$$

$$(\lambda \mathbf{a}) \times \mathbf{b} = \lambda (\mathbf{a} \times \mathbf{b}) \tag{B.20}$$

Note that when $\sin\theta = 0$ then $\mathbf{a} \times \mathbf{b} = 0$ and in particular, $\mathbf{a} \times \mathbf{a} = 0$.

Consider two vectors

$$\mathbf{a} = a_1 \hat{\mathbf{e}}_1 + a_2 \hat{\mathbf{e}}_2 + a_3 \hat{\mathbf{e}}_3$$

$$\mathbf{b} = b_1 \hat{\mathbf{e}}_1 + b_2 \hat{\mathbf{e}}_2 + b_3 \hat{\mathbf{e}}_3$$

Then using equations $B.12$ and $B.15 - B.17$

$$\mathbf{a} \times \mathbf{b} = |a| \, |b| \sin\theta = \begin{vmatrix} \hat{\mathbf{e}}_1 & \hat{\mathbf{e}}_2 & \hat{\mathbf{e}}_3 \\ a_1 & a_2 & a_3 \\ b_1 & b_2 & b_3 \end{vmatrix} = \hat{\mathbf{e}}_1 (a_2 b_3 - a_3 b_2) + \hat{\mathbf{e}}_2 (a_3 b_1 - a_1 b_3) + \hat{\mathbf{e}}_3 (a_1 b_2 - a_2 b_1)$$

where θ is the angle between the two vectors and the determinant is evaluated for the top row. Examples of vector products are torque $\mathbf{N} = \mathbf{r} \times \mathbf{F}$, angular momentum $\mathbf{L} = \mathbf{r} \times \mathbf{p}$, and the magnetic force $\mathbf{F}_B = q\mathbf{v} \times \mathbf{B}$.

B.4 Triple products

The following scalar and vector triple products can be formed from the product of three vectors and are used frequently.

Scalar triple products

There are several permutations of scalar triple products of three vectors $[\mathbf{a}, \mathbf{b}, \mathbf{c}]$ that are identical.

$$\mathbf{a} \cdot (\mathbf{b} \times \mathbf{c}) = \mathbf{c} \cdot (\mathbf{a} \times \mathbf{b}) = \mathbf{b} \cdot (\mathbf{c} \times \mathbf{a}) = (\mathbf{a} \times \mathbf{b}) \cdot \mathbf{c} = -\mathbf{a} \cdot (\mathbf{c} \times \mathbf{b}) \tag{B.21}$$

That is, the scalar product is invariant to cyclic permutations of the three vectors but changes sign for interchange of two vectors. The scalar product is unchanged by swapping the scalar (*dot*) and vector (*cross*).

Because of the symmetry the scalar triple product can be denoted as $[\mathbf{a}, \mathbf{b}, \mathbf{c}]$ and

$$\begin{aligned} [\mathbf{a}, \mathbf{b}, \mathbf{c}] &> 0 & &\text{if } [\mathbf{a}, \mathbf{b}, \mathbf{c}] \text{ is right-handed} \\ [\mathbf{a}, \mathbf{b}, \mathbf{c}] &= 0 & &\text{if } [\mathbf{a}, \mathbf{b}, \mathbf{c}] \text{ is coplanar} \\ [\mathbf{a}, \mathbf{b}, \mathbf{c}] &< 0 & &\text{if } [\mathbf{a}, \mathbf{b}, \mathbf{c}] \text{ is left-handed} \end{aligned} \tag{B.22}$$

The scalar triple product can be written in terms of the components using a determinant

$$[\mathbf{a}, \mathbf{b}, \mathbf{c}] = \begin{vmatrix} a_1 & a_2 & a_3 \\ b_1 & b_2 & b_3 \\ c_1 & c_2 & c_3 \end{vmatrix} \tag{B.23}$$

Vector triple product

The vector triple product $\mathbf{a} \times (\mathbf{b} \times \mathbf{c})$ is a vector. Since $(\mathbf{b} \times \mathbf{c})$ is perpendicular to the plane of $\mathbf{b}, \mathbf{c}$, then $\mathbf{a} \times (\mathbf{b} \times \mathbf{c})$ must lie in the plane containing $\mathbf{b}, \mathbf{c}$. Therefore the triple product can be expanded in terms of $\mathbf{b}, \mathbf{c}$, as given by the following identity

$$\mathbf{a} \times (\mathbf{b} \times \mathbf{c}) = (\mathbf{a} \cdot \mathbf{c}) \, \mathbf{b} - (\mathbf{a} \cdot \mathbf{b}) \, \mathbf{c} \qquad \text{(B.24)}$$

Workshop exercises

1. Partition the following exercises among the group. Once you have completed your problem, check with a classmate before writing it on the board. After you have verified that you have found the correct solution, write your answer in the space provided on the board, taking care to include the steps that you used to arrive at your solution. The following information is needed.

$$\mathbf{a} = 3\mathbf{i} + 2\mathbf{j} - 9\mathbf{k} \qquad \mathbf{b} = -2\mathbf{i} + 3\mathbf{k} \qquad \mathbf{c} = -2\mathbf{i} + \mathbf{j} - 6\mathbf{k} \qquad \mathbf{d} = \mathbf{i} + 9\mathbf{j} + 4\mathbf{k}$$

$$\mathbf{E} = \begin{pmatrix} 2 & 7 & -4 \\ 3 & 1 & -2 \\ -2 & 0 & 5 \end{pmatrix} \quad \mathbf{F} = \begin{pmatrix} 3 & 4 \\ 5 & 6 \end{pmatrix} \quad \mathbf{G} = \begin{pmatrix} 2 & -4 \\ 7 & 1 \\ -1 & 1 \end{pmatrix} \quad \mathbf{H} = \begin{pmatrix} -8 & -1 & -3 \\ -4 & 2 & -2 \\ -1 & 0 & 0 \end{pmatrix}$$

Calculate each of the following

1	$\lvert \mathbf{a} - (\mathbf{b} + 3\mathbf{c}) \rvert$	7	$(\mathbf{EH})^T$
2	Component of $\mathbf{c}$ along $\mathbf{a}$	8	$\lvert \mathbf{HE} \rvert$
3	Angle between $\mathbf{c}$ and $\mathbf{d}$	9	$\mathbf{EHG}$
4	$(\mathbf{b} \times \mathbf{d}) \cdot \mathbf{a}$	10	$\mathbf{EG} - \mathbf{HG}$
5	$(\mathbf{b} \times \mathbf{d}) \times \mathbf{a}$	11	$\mathbf{EH} - \mathbf{H}^T \mathbf{E}^T$
6	$\mathbf{b} \times (\mathbf{d} \times \mathbf{a})$	12	$\mathbf{F}^{-1}$

Problems

[1] For what values of a are the vectors $\mathbf{A} = 2a\hat{\imath} - 2\hat{\jmath} + a\hat{k}$ and $\mathbf{B} = a\hat{\imath} + 2a\hat{\jmath} + 2\hat{k}$ perpendicular?

[2] Show that the triple scalar product $(A \times B) \cdot C$ can be written as

$$(\mathbf{A} \times \mathbf{B}) \cdot \mathbf{C} = \begin{vmatrix} A_1 & A_2 & A_3 \\ B_1 & B_2 & B_3 \\ C_1 & C_2 & C_3 \end{vmatrix}$$

Show also that the product is unaffected by interchange of the scalar and vector product operations or by change in the order of A, B, C as long as they are in cyclic order, that is

$$(\mathbf{A} \times \mathbf{B}) \cdot \mathbf{C} = \mathbf{A} \cdot (\mathbf{B} \times \mathbf{C}) = \mathbf{B} \cdot (\mathbf{C} \times \mathbf{A}) = (\mathbf{C} \times \mathbf{A}) \cdot \mathbf{B}$$

Therefore we may use the notation ABC to denote the triple scalar product. Finally give a geometric interpretation of ABC by computing the volume of the parallelepiped defined by the three vectors $\mathbf{A}, \mathbf{B}, \mathbf{C}$.

Appendix C

Orthogonal coordinate systems

The methods of vector analysis provide a convenient representation of physical laws. However, the manipulation of scalar and vector fields is greatly facilitated by use of components with respect to an orthogonal coordinate system.

C.1 Cartesian coordinates (x, y, z)

Cartesian coordinates (rectangular) provide the simplest orthogonal rectangular coordinate system. The unit vectors specifying the direction along the three orthogonal axes are taken to be $(\hat{\mathbf{i}}, \hat{\mathbf{j}}, \hat{\mathbf{k}})$. In cartesian coordinates scalar and vector functions are written as

$$\phi = \phi(x, y, z) \tag{C.1}$$
$$\mathbf{r} = x\hat{\mathbf{i}} + y\hat{\mathbf{j}} + z\hat{\mathbf{k}} \tag{C.2}$$

Calculation of the time derivatives of the position vector is especially simple using cartesian coordinates because the unit vectors $(\hat{\mathbf{i}}, \hat{\mathbf{j}}, \hat{\mathbf{k}})$ are constant and independent in time. That is;

$$\frac{d\hat{\mathbf{i}}}{dt} = \frac{d\hat{\mathbf{j}}}{dt} = \frac{d\hat{\mathbf{k}}}{dt} = 0$$

Since the time derivatives of the unit vectors are all zero then the velocity $\dot{\mathbf{r}} = \frac{d\mathbf{r}}{dt}$ reduces to the partial time derivatives of x, y, and z. That is,

$$\dot{\mathbf{r}} = \dot{x}\hat{\mathbf{i}} + \dot{y}\hat{\mathbf{j}} + \dot{z}\hat{\mathbf{k}} \tag{C.3}$$

Similarly the acceleration is given by

$$\ddot{\mathbf{r}} = \ddot{x}\hat{\mathbf{i}} + \ddot{y}\hat{\mathbf{j}} + \ddot{z}\hat{\mathbf{k}} \tag{C.4}$$

C.2 Curvilinear coordinate systems

There are many examples in physics where the symmetry of the problem makes it more convenient to solve motion at a point $P(x, y, z)$ using non-cartesian curvilinear coordinate systems. For example, problems having spherical symmetry are most conveniently handled using a **spherical coordinate system** (r, θ, ϕ) with the origin at the center of spherical symmetry. Such problems occur frequently in electrostatics and gravitation; e.g. solutions of the atom, or planetary systems. Note that a cartesian coordinate system still is required to define the origin plus the polar and azimuthal angles θ, ϕ. Using spherical coordinates for a spherically symmetry system allows the problem to be factored into a cyclic angular part, the solution which involves spherical harmonics that are common to all such spherically-symmetric problems, plus a one-dimensional radial part that contains the specifics of the particular spherically-symmetric potential. Similarly, for problems involving cylindrical symmetry, it is much more convenient to use a **cylindrical coordinate system** (ρ, ϕ, z). Again it is necessary to use a cartesian coordinate system to define the origin and angle ϕ. Motion in a plane can be handled using two dimensional **polar coordinates**.

Curvilinear coordinate systems introduce a complication in that the *unit vectors are time dependent* in contrast to cartesian coordinate system where the unit vectors $(\hat{\mathbf{i}}, \hat{\mathbf{j}}, \hat{\mathbf{k}})$ are independent and constant in time. The introduction of this time dependence warrants further discussion.

Each of the three axes q_i in curvilinear coordinate systems can be expressed in cartesian coordinates (x, y, z) as surfaces of constant q_i given by the function

$$q_i = f_i(x, y, z) \tag{C.5}$$

where $i = 1, 2$, or 3. An element of length ds_i perpendicular to the surface q_i is the distance between the surfaces q_i and $q_i + dq_i$ which can be expressed as

$$ds_i = h_i dq_i \tag{C.6}$$

where h_i is a function of (q_1, q_2, q_3). In cartesian coordinates h_1, h_2, and h_3 are all unity. The unit-length vectors $\hat{q}_1$, $\hat{q}_2$, $\hat{q}_3$, are perpendicular to the respective q_1, q_2, q_3 surfaces, and are oriented to have increasing indices such that $\hat{q}_1 \times \hat{q}_2 = \hat{q}_3$. The correspondence of the curvilinear coordinates, unit vectors, and transform coefficients to cartesian, polar, cylindrical and spherical coordinates is given in table $C.1$.

Curvilinear	q_1	q_2	q_3	$\hat{\mathbf{q}}_1$	$\hat{\mathbf{q}}_2$	$\hat{\mathbf{q}}_3$	h_1	h_2	h_3
Cartesian	x	y	z	$\hat{\imath}$	$\hat{\jmath}$	$\hat{\mathbf{k}}$	1	1	1
Polar	r	θ		$\hat{\mathbf{r}}$	$\hat{\boldsymbol{\theta}}$		1	r	
Cylindrical	ρ	φ	z	$\hat{\boldsymbol{\rho}}$	$\hat{\boldsymbol{\varphi}}$	$\hat{\mathbf{z}}$	1	ρ	1
Spherical	r	θ	φ	$\hat{\mathbf{r}}$	$\hat{\boldsymbol{\theta}}$	$\hat{\boldsymbol{\varphi}}$	1	r	$r\sin\theta$

Table $C.1$: Curvilinear coordinates

The differential distance and volume elements are given by

$$\begin{aligned} d\mathbf{s} &= ds_1\hat{\mathbf{q}}_1 + ds_2\hat{\mathbf{q}}_2 + ds_3\hat{\mathbf{q}}_3 = h_1 dq_1\hat{\mathbf{q}}_1 + h_2 dq_2\hat{\mathbf{q}}_2 + h_3 dq_3\hat{\mathbf{q}}_3 \tag{C.7} \\ d\tau &= ds_1 ds_2 ds_3 = h_1 h_2 h_3 (dq_1 dq_2 dq_3) \tag{C.8} \end{aligned}$$

These are evaluated below for polar, cylindrical, and spherical coordinates.

C.2.1 Two-dimensional polar coordinates (r, θ)

The complication and implications of time-dependent unit vectors are best illustrated by considering two-dimensional polar coordinates which is the simplest curvilinear coordinate system. Polar coordinates are a special case of cylindrical coordinates, when z is held fixed, or a special case of spherical coordinate system, when ϕ is held fixed.

Consider the motion of a point P as it moves along a curve $\mathbf{s}(t)$ such that in the time interval dt it moves from $P^{(1)}$ to $P^{(2)}$ as shown in figure $C.2$. The two-dimensional polar coordinates have *unit vectors* $\hat{\mathbf{r}}$, $\hat{\boldsymbol{\theta}}$, which are orthogonal and change from $\hat{\mathbf{r}}_1$, $\hat{\boldsymbol{\theta}}_1$, to $\hat{\mathbf{r}}_2$, $\hat{\boldsymbol{\theta}}_2$, in the time dt. Note that for these polar coordinates the angle unit vector $\hat{\boldsymbol{\theta}}$ is taken to be *tangential* to the rotation since this is the direction of motion of a point on the circumference at radius r.

The net changes shown in figure of table $C.2$ are

$$d\hat{\mathbf{r}} = \hat{\mathbf{r}}_2 - \hat{\mathbf{r}}_1 = d\hat{\mathbf{r}} = |\hat{\mathbf{r}}|\, d\theta\hat{\boldsymbol{\theta}} = d\theta\hat{\boldsymbol{\theta}} \tag{C.9}$$

since the unit vector $\hat{\mathbf{r}}$ is a constant with $|\hat{\mathbf{r}}| = 1$. Note that the infinitessimal $d\hat{\mathbf{r}}$ is perpendicular to the unit vector $\hat{\mathbf{r}}$, that is, $d\hat{\mathbf{r}}$ points in the tangential direction $\hat{\boldsymbol{\theta}}$.

Similarly, the infinitessimal

$$d\hat{\boldsymbol{\theta}} = \hat{\boldsymbol{\theta}}_2 - \hat{\boldsymbol{\theta}}_1 = d\hat{\boldsymbol{\theta}} = -d\theta\hat{\mathbf{r}} \tag{C.10}$$

which is perpendicular to the tangential $\hat{\boldsymbol{\theta}}$ unit vector and therefore points in the direction $-\hat{\mathbf{r}}$. The minus sign causes $-d\theta\hat{\mathbf{r}}$ to be directed in the opposite direction to $\hat{\mathbf{r}}$.

The net distance element $d\mathbf{s}$ is given by

$$d\mathbf{s} = dr\hat{\mathbf{r}} + r d\hat{\mathbf{r}} = dr\hat{\mathbf{r}} + r d\theta \hat{\boldsymbol{\theta}} \tag{C.11}$$

This agrees with the prediction obtained using table $C.1$.

The time derivatives of the unit vectors are given by equations $(C.9)$ and $(C.10)$ to be,

$$\frac{d\hat{\mathbf{r}}}{dt} = \frac{d\theta}{dt}\hat{\boldsymbol{\theta}} \tag{C.12}$$

$$\frac{d\hat{\boldsymbol{\theta}}}{dt} = -\frac{d\theta}{dt}\hat{\mathbf{r}} \tag{C.13}$$

Note that *the time derivatives of unit vectors are perpendicular to the corresponding unit vector, and the unit vectors are coupled.*

Consider that the velocity $\mathbf{v}$ is expressed as

$$\mathbf{v} = \frac{d\mathbf{r}}{dt} = \frac{d}{dt}(r\hat{\mathbf{r}}) = \frac{dr}{dt}\hat{\mathbf{r}} + r\frac{d\hat{\mathbf{r}}}{dt} = \dot{r}\hat{\mathbf{r}} + r\dot{\theta}\hat{\boldsymbol{\theta}} \tag{C.14}$$

The velocity is resolved into a radial component $\dot{r}$ and an angular, transverse, component $r\dot{\theta}$.

Similarly the acceleration is given by

$$
\begin{aligned}
\mathbf{a} &= \frac{d\mathbf{v}}{dt} = \frac{d\dot{r}}{dt}\hat{\mathbf{r}} + \dot{r}\frac{d\hat{\mathbf{r}}}{dt} + \frac{dr}{dt}\dot{\theta}\hat{\boldsymbol{\theta}} + r\frac{d\dot{\theta}}{dt}\hat{\boldsymbol{\theta}} + r\dot{\theta}\frac{d\hat{\boldsymbol{\theta}}}{dt} \\
&= \left(\ddot{r} - r\dot{\theta}^2\right)\hat{\mathbf{r}} + \left(r\ddot{\theta} + 2\dot{r}\dot{\theta}\right)\hat{\boldsymbol{\theta}}
\end{aligned}
\tag{C.15}
$$

where the $r\dot{\theta}^2\hat{\mathbf{r}}$ term is the effective centripetal acceleration while the $2\dot{r}\dot{\theta}\hat{\boldsymbol{\theta}}$ term is called the Coriolis term. For the case when $\dot{r} = \ddot{r} = 0$, then the first bracket in $C.15$ is the centripetal acceleration while the second bracket is the tangential acceleration.

This discussion has shown that in contrast to the time independence of the cartesian unit basis vectors, *the unit basis vectors for curvilinear coordinates are time dependent which leads to components of the velocity and acceleration involving coupled coordinates.*

Coordinates	r, θ
Distance element	$d\mathbf{s} = dr\hat{\mathbf{r}} + r d\theta \hat{\boldsymbol{\theta}}$
Area element	$da = r\,dr\,d\theta$
Unit vectors	$\hat{\mathbf{r}} = \hat{\imath}\cos\theta + \hat{\jmath}\sin\theta$ $\hat{\boldsymbol{\theta}} = -\hat{\imath}\sin\theta + \hat{\jmath}\cos\theta$
Time derivatives of unit vectors	$\frac{d\hat{\mathbf{r}}}{dt} = \dot{\theta}\hat{\boldsymbol{\theta}}$ $\frac{d\hat{\boldsymbol{\theta}}}{dt} = -\dot{\theta}\hat{\mathbf{r}}$
Velocity	$\mathbf{v} = \dot{r}\hat{\mathbf{r}} + r\dot{\theta}\hat{\boldsymbol{\theta}}$
Kinetic energy	$\frac{m}{2}\left(\dot{r}^2 + r^2\dot{\theta}^2\right)$
Acceleration	$\mathbf{a} = \left(\ddot{r} - r\dot{\theta}^2\right)\hat{\mathbf{r}}$ $+ \left(r\ddot{\theta} + 2\dot{r}\dot{\theta}\right)\hat{\boldsymbol{\theta}}$

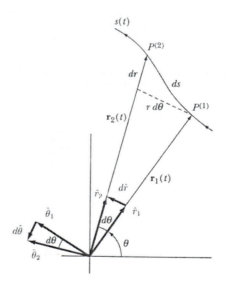

Table $C.2$: Differential relations plus a diagram of the unit vectors for 2-dimensional polar coordinates.

C.2.2   Cylindrical Coordinates (ρ, ϕ, z)

The three-dimensional cylindrical coordinates (ρ, ϕ, z) are obtained by adding the motion along the symmetry axis $\hat{\mathbf{z}}$ to the case for polar coordinates. The unit basis vectors are shown in Table C.3 where the angular unit vector $\hat{\boldsymbol{\phi}}$ is taken to be tangential corresponding to the direction a point on the circumference would move. The distance and volume elements, the cartesian coordinate components of the cylindrical unit basis vectors, and the unit vector time derivatives are shown in Table C.3. The time dependence of the unit vectors is used to derive the acceleration. As for the two-dimensional polar coordinates, the $\hat{\boldsymbol{\rho}}$ and $\hat{\boldsymbol{\theta}}$ direction components of the acceleration for cylindrical coordinates are coupled functions of $\rho, \dot{\rho}, \ddot{\rho}, \phi,$ and $\ddot{\phi}$.

Coordinates	ρ, ϕ, z
Distance element	$d\mathbf{s} = d\rho\hat{\boldsymbol{\rho}} + \rho d\phi\hat{\boldsymbol{\phi}} + dz\hat{\mathbf{z}}$
Volume element	$dv = \rho d\rho d\phi dz$
Unit vectors	$\hat{\boldsymbol{\rho}} = \hat{\imath}\cos\phi + \hat{\jmath}\sin\phi$ $\hat{\boldsymbol{\phi}} = -\hat{\imath}\sin\phi + \hat{\jmath}\cos\phi$ $\hat{\mathbf{z}} = \hat{\mathbf{k}}$
Time derivatives of unit vectors	$\frac{d\hat{\boldsymbol{\rho}}}{dt} = \dot{\phi}\hat{\boldsymbol{\phi}}$ $\frac{d\hat{\boldsymbol{\phi}}}{dt} = -\dot{\phi}\hat{\boldsymbol{\rho}}$ $\frac{d\hat{\mathbf{z}}}{dt} = 0$
Velocity	$\mathbf{v} = \dot{\rho}\hat{\boldsymbol{\rho}} + \rho\dot{\phi}\hat{\boldsymbol{\phi}} + \dot{z}\hat{\mathbf{z}}$
Kinetic energy	$\frac{m}{2}\left(\dot{\rho}^2 + \rho^2\dot{\phi}^2 + \dot{z}^2\right)$
Acceleration	$\mathbf{a} = \left(\ddot{\rho} - \rho\dot{\phi}^2\right)\hat{\boldsymbol{\rho}}$ $+ \left(\rho\ddot{\phi} + 2\dot{\rho}\dot{\phi}\right)\hat{\boldsymbol{\phi}} + \ddot{z}\hat{\mathbf{z}}$

Table C.3: Differential relations plus a diagram of the unit vectors for cylindrical coordinates.

C.2.3   Spherical Coordinates (r, θ, ϕ)

The three dimensional spherical coordinates, can be treated the same way as for cylindrical coordinates. The unit basis vectors are shown in Table C.4 where the angular unit vectors $\hat{\boldsymbol{\theta}}$ and $\hat{\boldsymbol{\phi}}$ are taken to be tangential corresponding to the direction a point on the circumference moves for a positive rotation angle.

Coordinates	r, θ, ϕ
Distance element	$ds = dr\hat{\mathbf{r}} + rd\theta\hat{\boldsymbol{\theta}} + r\sin\theta d\phi\hat{\boldsymbol{\phi}}$
Volume element	$dv = r^2\sin\theta dr d\theta d\phi$
Unit vectors	$\hat{\mathbf{r}} = \hat{\imath}\sin\theta\cos\phi + \hat{\jmath}\sin\theta\sin\phi + \hat{\mathbf{k}}\cos\theta$ $\hat{\boldsymbol{\theta}} = \hat{\imath}\cos\theta\cos\phi + \hat{\jmath}\cos\theta\sin\phi - \hat{\mathbf{k}}\sin\theta$ $\hat{\boldsymbol{\phi}} = -\hat{\imath}\sin\phi + \hat{\jmath}\cos\phi$
Time derivatives of unit vectors	$\frac{d\hat{\mathbf{r}}}{dt} = \hat{\boldsymbol{\theta}}\dot{\theta} + \hat{\boldsymbol{\phi}}\dot{\phi}\sin\theta$ $\frac{d\hat{\boldsymbol{\theta}}}{dt} = -\hat{\mathbf{r}}\dot{\theta} + \hat{\boldsymbol{\phi}}\dot{\phi}\cos\theta$ $\frac{d\hat{\boldsymbol{\phi}}}{dt} = -\hat{\mathbf{r}}\dot{\phi}\sin\theta - \hat{\boldsymbol{\theta}}\dot{\phi}\cos\theta$
Velocity	$\mathbf{v} = \dot{r}\hat{\mathbf{r}} + r\dot{\theta}\hat{\boldsymbol{\theta}} + r\dot{\phi}\sin\theta\hat{\boldsymbol{\phi}}$
Kinetic energy	$\frac{m}{2}\left(\dot{r}^2 + r^2\dot{\theta}^2 + r^2\sin^2\theta\dot{\phi}^2\right)$
Acceleration	$\mathbf{a} = \left(\ddot{r} - r\dot{\theta}^2 - r\dot{\phi}^2\sin^2\theta\right)\hat{\mathbf{r}}$ $+ \left(r\ddot{\theta} + 2\dot{r}\dot{\theta} - r\dot{\phi}^2\sin\theta\cos\theta\right)\hat{\boldsymbol{\theta}}$ $+ \left(r\ddot{\phi}\sin\theta + 2\dot{r}\dot{\phi}\sin\theta + 2r\dot{\theta}\dot{\phi}\cos\theta\right)\hat{\boldsymbol{\phi}}$

Table C.4 Differential relations plus a diagram of the unit vectors for spherical coordinates.

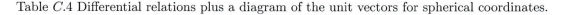

The distance and volume elements, the cartesian coordinate components of the spherical unit basis vectors, and the unit vector time derivatives are shown in the table given in figure $C.4$. The time dependence of the unit vectors is used to derive the acceleration. As for the case of cylindrical coordinates, the $\hat{\mathbf{r}}, \hat{\boldsymbol{\theta}}$, and $\hat{\boldsymbol{\phi}}$ components of the acceleration involve coupling of the coordinates and their time derivatives.

It is important to note that the angular unit vectors $\hat{\boldsymbol{\theta}}$ and $\hat{\boldsymbol{\phi}}$ are taken to be tangential to the circles of rotation. However, for discussion of angular velocity of angular momentum it is more convenient to use the axes of rotation defined by $\hat{\mathbf{r}} \times \hat{\boldsymbol{\theta}}$ and $\hat{\mathbf{r}} \times \hat{\boldsymbol{\phi}}$ for specifying the vector properties which is perpendicular to the unit vectors $\hat{\boldsymbol{\theta}}$ and $\hat{\boldsymbol{\phi}}$. Be careful not to confuse the unit vectors $\hat{\boldsymbol{\theta}}$ and $\hat{\boldsymbol{\phi}}$ with those used for the angular velocities $\dot{\theta}$ and $\dot{\phi}$.

C.3 Frenet-Serret coordinates

The cartesian, polar, cylindrical, or spherical curvilinear coordinate systems, all are orthogonal coordinate systems that are fixed in space. There are situations where it is more convenient to use the Frenet-Serret coordinates which comprise an orthogonal coordinate system that is fixed to the particle that is moving along a continuous, differentiable, trajectory in three-dimensional Euclidean space. Let $s(t)$ represent a monotonically increasing arc-length along the trajectory of the particle motion as a function of time t. The Frenet-Serret coordinates, shown in figure $C.5$, are the three instantaneous orthogonal unit vectors $\hat{\mathbf{t}}, \hat{\mathbf{n}}$, and $\hat{\mathbf{b}}$ where the tangent unit vector $\hat{\mathbf{t}}$ is the instantaneous tangent to the curve, the normal unit vector $\hat{\mathbf{n}}$ is in the plane of curvature of the trajectory pointing towards the center of the instantaneous radius of curvature and is perpendicular to the tangent unit vector $\hat{\mathbf{t}}$, while the binormal unit vector is $\hat{\mathbf{b}} = \hat{\mathbf{t}} \times \hat{\mathbf{n}}$ which is the perpendicular to the plane of curvature and is mutually perpendicular to the other two Frenet-Serrat unit vectors. The Frenet-Serret unit vectors are defined by the relations

$$\frac{d\hat{\mathbf{t}}}{ds} = \kappa \hat{\mathbf{n}} \tag{C.16}$$

$$\frac{d\hat{\mathbf{b}}}{ds} = -\tau \hat{\mathbf{n}} \tag{C.17}$$

$$\frac{d\hat{\mathbf{n}}}{ds} = -\kappa \hat{\mathbf{t}} + \tau \hat{\mathbf{b}} \tag{C.18}$$

The curvature $\kappa = \frac{1}{\rho}$ where ρ is the radius of curvature and τ is the torsion that can be either positive or negative. For increasing s, a non-zero curvature κ implies that the triad of unit vectors rotate in a right-handed sense about $\hat{\mathbf{b}}$. If the torsion τ is positive (negative) the triad of unit vectors rotates in right (left) handed sense about $\hat{\mathbf{t}}$.

Distance element	$d\mathbf{s}(t) = \hat{\mathbf{t}} \left\lvert \frac{d\mathbf{r}(t)}{dt} \right\rvert dt = \hat{\mathbf{t}} v(t) dt$	
Unit vectors	$\hat{\mathbf{t}}(t) = \frac{\mathbf{v}(t)}{\lvert v(t) \rvert}$ $\hat{\mathbf{n}}(t) = \frac{d\hat{\mathbf{t}}/dt}{\lvert d\hat{\mathbf{t}}/dt \rvert}$ $\hat{\mathbf{b}}(t) = \hat{\mathbf{t}} \times \hat{\mathbf{n}}$	
Time derivatives of unit vectors	$\frac{d}{dt} \begin{pmatrix} \hat{\mathbf{t}} \\ \hat{\mathbf{n}} \\ \hat{\mathbf{b}} \end{pmatrix} = \lvert v \rvert \begin{pmatrix} 0 & \kappa & 0 \\ -\kappa & 0 & \tau \\ 0 & -\tau & 0 \end{pmatrix} \begin{pmatrix} \hat{\mathbf{t}} \\ \hat{\mathbf{n}} \\ \hat{\mathbf{b}} \end{pmatrix}$	
Velocity	$\mathbf{v}(t) = \frac{d\mathbf{r}(t)}{dt}$	
Acceleration	$\mathbf{a}(t) = \frac{dv}{dt} \hat{\mathbf{t}} + \kappa v^2 \hat{\mathbf{n}}$	

Table $C.5$. The differential relations plus a diagram of the corresponding unit vectors for the Frenet-Serret coordinate system.

The above equations also can be rewritten in the form using a new unit rotation vector $\boldsymbol{\omega}$ where

$$\boldsymbol{\omega} = \tau\hat{\mathbf{t}} + \kappa\hat{\mathbf{b}} \tag{C.19}$$

Then equations $C.16 - C.18$ are transformed to

$$\frac{d\hat{\mathbf{t}}}{ds} = \boldsymbol{\omega} \times \hat{\mathbf{t}} \tag{C.20}$$

$$\frac{d\hat{\mathbf{n}}}{ds} = \boldsymbol{\omega} \times \hat{\mathbf{n}} \tag{C.21}$$

$$\frac{d\hat{\mathbf{b}}}{ds} = \boldsymbol{\omega} \times \hat{\mathbf{b}} \tag{C.22}$$

In general the Frenet-Serret unit vectors are time dependent. If the curvature $\kappa = 0$ then the curve is a straight line and $\hat{\mathbf{n}}$ and $\hat{\mathbf{b}}$ are not well defined. If the torsion is zero then the trajectory lies in a plane. Note that a helix has constant curvature and constant torsion.

The rate of change of a general vector field $\mathbf{E}$ along the trajectory can be written as

$$\frac{d\mathbf{E}}{ds} = \left(\frac{dE_t}{ds}\hat{\mathbf{t}} + \frac{dE_n}{ds}\hat{\mathbf{n}} + \frac{dE_b}{ds}\hat{\mathbf{b}} \right) + \boldsymbol{\omega} \times \mathbf{E} \tag{C.23}$$

The Frenet-Serret coordinates are used in the life sciences to describe the motion of a moving organism in a viscous medium. The Frenet-Serret coordinates also have applications to General Relativity.

Workshop exercises

1. The goal of this problem is to help you understand the origin of the equations that relate two different coordinate systems. Refer to diagrams for cylindrical and spherical coordinates as your teaching assistant explains how to arrive at expressions for $x_1, x_2,$ and x_3 in terms of $\rho, \phi,$ and z and how to derive expressions for the velocity and acceleration vectors in cylindrical coordinates. Now try to relate spherical and rectangular coordinate systems. Your group should derive expressions relating the coordinates of the two systems, expressions relating the unit vectors and their time derivatives of the two systems, and finally, expressions for the velocity and acceleration in spherical coordinates.

Appendix D

Coordinate transformations

Coordinate systems can be translated, or rotated with respect to each other as well as being subject to spatial inversion or time reversal. Scalars, vectors, and tensors are defined by their transformation properties under rotation, spatial inversion and time reversal, and thus such transformations play a pivotal role in physics.

D.1 Translational transformations

Translational transformations are involved frequently for transforming between the center of mass and laboratory frames for reaction kinematics as well as when performing vector addition of central forces for the cases where the centers are displaced. Both the classical Galilean transformation or the relativistic Lorentz transformation are handled the same way. Consider two parallel orthonormal coordinate frames where the origin of $F'(x', y', z')$ is displaced by a time dependent vector $\mathbf{a}(t)$ from the origin of frame $F(x, y, z)$. Then the Galilean transformation for a vector $\mathbf{r}$ in frame F to $\mathbf{r}'$ in frame F' is given by

$$\mathbf{r}(x', y', z') = \mathbf{r}(x, y, z) + \mathbf{a}(t) \tag{D.1}$$

The velocities for a moving frame are given by the vector difference of the velocity in a stationary frame, and the velocity of the origin of the moving frame. Linear accelerations can be handled similarly.

D.2 Rotational transformations

D.2.1 Rotation matrix

Rotational transformations of the coordinate system are used extensively in physics. The transformation properties of fields under rotation define the scalar and vector properties of fields, as well as rotational symmetry and conservation of angular momentum.

Rotation of the coordinate frame does not change the value of any scalar observable such as mass, temperature etc. That is, transformation of a scalar quantity is invariant under coordinate rotation from $x, y, z \rightarrow x', y', z'$.

$$\phi(x'y'z') = \phi(xyz) \tag{D.2}$$

By contrast, the components of a vector along the coordinate axes change under rotation of the coordinate axes. This difference in transformation properties under rotation between a scalar and a vector is important and defines both scalars and a vectors.

Matrix mechanics, described in appendix A, provides the most convenient way to handle coordinate rotations. The transformation matrix, between coordinate systems having differing orientations is called the **rotation matrix**. This transforms the components of any vector with respect to one coordinate frame to the components with respect to a second coordinate frame rotated with respect to the first frame.

Assume a point P has coordinates (x_1, x_2, x_3) with respect to a certain coordinate system. Consider rotation to another coordinate frame for which the point P has coordinates (x'_1, x'_2, x'_3) and assume that the origins of both frames coincide. Rotation of a frame does not change the vector, only the vector components of the unit basis states. Therefore

$$\mathbf{x} = \hat{\mathbf{e}}_1' x_1' + \hat{\mathbf{e}}_2' x_2' + \hat{\mathbf{e}}_3' x_3' = \hat{\mathbf{e}}_1 x_1 + \hat{\mathbf{e}}_2 x_2 + \hat{\mathbf{e}}_3 x_3 \tag{D.3}$$

Note that if one designates that the unit vectors for the unprimed coordinate frame are $(\hat{\mathbf{e}}_1, \hat{\mathbf{e}}_2, \hat{\mathbf{e}}_3)$ and for the primed coordinate frame $(\hat{\mathbf{e}}_1', \hat{\mathbf{e}}_2', \hat{\mathbf{e}}_3')$, then taking the scalar product of equation $D.3$ sequentially with each of the unit base vectors $(\hat{\mathbf{e}}_1', \hat{\mathbf{e}}_2', \hat{\mathbf{e}}_3')$ leads to the following three relations

$$\begin{aligned}
x_1' &= (\hat{\mathbf{e}}_1' \cdot \hat{\mathbf{e}}_1) x_1 + (\hat{\mathbf{e}}_1' \cdot \hat{\mathbf{e}}_2) x_2 + (\hat{\mathbf{e}}_1' \cdot \hat{\mathbf{e}}_3) x_3 \\
x_2' &= (\hat{\mathbf{e}}_2' \cdot \hat{\mathbf{e}}_1) x_1 + (\hat{\mathbf{e}}_2' \cdot \hat{\mathbf{e}}_2) x_2 + (\hat{\mathbf{e}}_2' \cdot \hat{\mathbf{e}}_3) x_3 \\
x_3' &= (\hat{\mathbf{e}}_3' \cdot \hat{\mathbf{e}}_1) x_1 + (\hat{\mathbf{e}}_3' \cdot \hat{\mathbf{e}}_2) x_2 + (\hat{\mathbf{e}}_3' \cdot \hat{\mathbf{e}}_3) x_3
\end{aligned} \tag{D.4}$$

Note that the $(\hat{\mathbf{e}}_i' \cdot \hat{\mathbf{e}}_j)$ are the direction cosines as defined by the scalar product of two unit vectors for axes i, j, that is, they are the cosine of the angle between the two unit vectors.

Equation $D.4$ can be written in matrix form as

$$\mathbf{x}' = \boldsymbol{\lambda} \cdot \mathbf{x} \tag{D.5}$$

where the "$\cdot$" means the *inner matrix product* of the rotation matrix $\boldsymbol{\lambda}$ and the vector $\mathbf{x}$ where

$$\mathbf{x}' \equiv \begin{pmatrix} x_1' \\ x_2' \\ x_3' \end{pmatrix} \qquad \mathbf{x} \equiv \begin{pmatrix} x_1 \\ x_2 \\ x_3 \end{pmatrix} \qquad \boldsymbol{\lambda} \equiv \begin{pmatrix} \hat{\mathbf{e}}_1' \cdot \hat{\mathbf{e}}_1 & \hat{\mathbf{e}}_1' \cdot \hat{\mathbf{e}}_2 & \hat{\mathbf{e}}_1' \cdot \hat{\mathbf{e}}_3 \\ \hat{\mathbf{e}}_2' \cdot \hat{\mathbf{e}}_1 & \hat{\mathbf{e}}_2' \cdot \hat{\mathbf{e}}_2 & \hat{\mathbf{e}}_2' \cdot \hat{\mathbf{e}}_3 \\ \hat{\mathbf{e}}_3' \cdot \hat{\mathbf{e}}_1 & \hat{\mathbf{e}}_3' \cdot \hat{\mathbf{e}}_2 & \hat{\mathbf{e}}_3' \cdot \hat{\mathbf{e}}_3 \end{pmatrix} \tag{D.6}$$

The inverse procedure is obtained by multiplying equation $D.3$ successively by one of the unit basis vectors $(\hat{\mathbf{e}}_1, \hat{\mathbf{e}}_2, \hat{\mathbf{e}}_3)$ leading to three equations

$$\begin{aligned}
x_1 &= (\hat{\mathbf{e}}_1 \cdot \hat{\mathbf{e}}_1') x_1' + (\hat{\mathbf{e}}_1 \cdot \hat{\mathbf{e}}_2') x_2' + (\hat{\mathbf{e}}_1 \cdot \hat{\mathbf{e}}_3') x_3' \\
x_2 &= (\hat{\mathbf{e}}_2 \cdot \hat{\mathbf{e}}_1') x_1' + (\hat{\mathbf{e}}_2 \cdot \hat{\mathbf{e}}_2') x_2' + (\hat{\mathbf{e}}_2 \cdot \hat{\mathbf{e}}_3') x_3' \\
x_3 &= (\hat{\mathbf{e}}_3 \cdot \hat{\mathbf{e}}_1') x_1' + (\hat{\mathbf{e}}_3 \cdot \hat{\mathbf{e}}_2') x_2' + (\hat{\mathbf{e}}_3 \cdot \hat{\mathbf{e}}_3') x_3'
\end{aligned} \tag{D.7}$$

Equation $D.7$ can be written in matrix form as

$$\mathbf{x} = \boldsymbol{\lambda}^T \cdot \mathbf{x}' \tag{D.8}$$

where $\boldsymbol{\lambda}^T$ is the transpose of $\boldsymbol{\lambda}$.

Note that substituting equation $D.5$ into equation $D.8$ gives

$$\mathbf{x} = \boldsymbol{\lambda}^T \cdot (\boldsymbol{\lambda} \cdot \mathbf{x}) = \left(\boldsymbol{\lambda}^T \cdot \boldsymbol{\lambda} \right) \cdot \mathbf{x} \tag{D.9}$$

Thus

$$\left(\boldsymbol{\lambda}^T \cdot \boldsymbol{\lambda} \right) = \mathbb{I}$$

where $\mathbb{I}$ is the identity matrix. This implies that the rotation matrix $\boldsymbol{\lambda}$ is orthogonal with $\boldsymbol{\lambda}^T = \boldsymbol{\lambda}^{-1}$.

It is convenient to rename the elements of the rotation matrix to be

$$\lambda_{ij} \equiv (\hat{\mathbf{e}}_i' \cdot \hat{\mathbf{e}}_j) \tag{D.10}$$

so that the rotation matrix is written more compactly as

$$\boldsymbol{\lambda} \equiv \begin{pmatrix} \lambda_{11} & \lambda_{12} & \lambda_{13} \\ \lambda_{21} & \lambda_{22} & \lambda_{23} \\ \lambda_{31} & \lambda_{32} & \lambda_{33} \end{pmatrix}$$

and equation $D.4$ becomes

$$\begin{aligned}
x_1' &= \lambda_{11} x_1 + \lambda_{12} x_2 + \lambda_{13} x_3 \\
x_2' &= \lambda_{21} x_1 + \lambda_{22} x_2 + \lambda_{23} x_3 \\
x_3' &= \lambda_{31} x_1 + \lambda_{32} x_2 + \lambda_{33} x_3
\end{aligned} \tag{D.11}$$

Consider an arbitrary rotation through an angle θ. Equations $(B.10)$ and $(B.11)$ can be used to relate six of the nine quantities λ_{ij} in the rotation matrix, so only three of the quantities are independent. That is, because of equation $(B.11)$ we have three equations which ensure that the transformation is unitary.

$$\lambda_{i1}^2 + \lambda_{i2}^2 + \lambda_{i3}^2 = 1 \tag{D.12}$$

Also requiring that the axes be orthogonal gives three equations

$$\sum_j \lambda_{ij}\lambda_{kj} = 0, \qquad i \neq k \tag{D.13}$$

These six relations can be expressed as

$$\sum_j \lambda_{ij}\lambda_{kj} = \delta_{ik} \tag{D.14}$$

The fact that the rotation matrix should have three independent quantities is due to the fact that all rotations can be expressed in terms of rotations about three orthogonal axes.

D.1 *Example: Rotation matrix:*

Consider a point $P(x_1, x_2, x_3) = P(3, 4, 5)$ in the unprimed coordinate system. Consider the same point $P(x_1', x_2', x_3')$ in the primed coordinate system which has been rotated by an angle $60°$ about the x_1 axis as shown. The direction cosines $\lambda_{i'j} = \cos(\theta_{i'j})$ can be determined from the figure to be the following

i'	j	$\theta_{i'j}$	$\lambda_{i'j}=\cos(\theta_{i'j})$
1	1	0	1
1	2	90	0
1	3	90	0
2	1	90	0
2	2	60	0.500
2	3	$90-60$	0.866
3	1	90	0
3	2	$90+60$	-0.866
3	3	60	0.500

Thus the rotation matrix is

$$\lambda = \begin{pmatrix} 1. & 0 & 0 \\ 0 & 0.500 & 0.866 \\ 0 & -0.866 & 0.500 \end{pmatrix}$$

The transform point $P'(x_1', x_2', x_3')$ therefore is given by

$$\begin{pmatrix} x_1' \\ x_2' \\ x_3' \end{pmatrix} = \begin{pmatrix} 1. & 0 & 0 \\ 0 & 0.500 & 0.866 \\ 0 & -0.866 & 0.500 \end{pmatrix} \cdot \begin{pmatrix} 3 \\ 4 \\ 5 \end{pmatrix} = \begin{pmatrix} 3 \\ 6.330 \\ -0.964 \end{pmatrix}$$

Note that the radial coordinate $r_P = r_P' = \sqrt{50}$. That is, the rotational transformation is unitary and thus the magnitude of the vector is unchanged.

D.2 *Example: Proof that a rotation matrix is orthogonal*

Consider the rotation matrix

$$\lambda = \frac{1}{9}\begin{pmatrix} 4 & 7 & -4 \\ 1 & 4 & 8 \\ 8 & -4 & 1 \end{pmatrix}$$

The product

$$\lambda^T \cdot \lambda = \frac{1}{81}\begin{pmatrix} 4 & 1 & 8 \\ 7 & 4 & -4 \\ -4 & 8 & 1 \end{pmatrix} \cdot \begin{pmatrix} 4 & 7 & -4 \\ 1 & 4 & 8 \\ 8 & -4 & 1 \end{pmatrix} = \frac{1}{81}\begin{pmatrix} 81 & 0 & 0 \\ 0 & 81 & 0 \\ 0 & 0 & 81 \end{pmatrix} = 1$$

which implies that λ is orthogonal.

D.2.2   Finite rotations

Consider two finite 90° rotations λ_A and λ_B illustrated in figure $D.1$. The λ_A rotation is 90° around the x_3 axis in a right-handed direction as shown. In such a rotation the axes transform to $x_1' = x_2$, $x_2' = -x_1$, $x_3' = x_3$ and the rotation matrix is

$$\lambda_A = \begin{pmatrix} 0 & 1 & 0 \\ -1 & 0 & 0 \\ 0 & 0 & 1 \end{pmatrix} \qquad (D.15)$$

The second rotation λ_B is a right-handed rotation about the x_1' axis which formerly was the x_2 axis. Then $x_1'' = x_2'$, $x_2'' = -x_1'$, $x_3'' = x_3'$ and the rotation matrix is

$$\lambda_B = \begin{pmatrix} 1 & 0 & 0 \\ 0 & 0 & 1 \\ 0 & -1 & 0 \end{pmatrix} \qquad (D.16)$$

Consider the product of these two finite rotations which corresponds to a single rotation matrix λ_{AB}

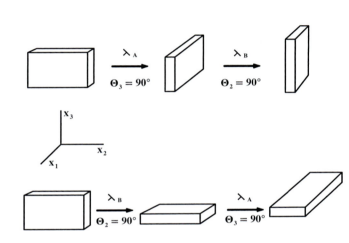

Figure D.1: Order of two finite rotations for a parallelepiped.

$$\lambda_{AB} = \lambda_B \lambda_A \qquad (D.17)$$

That is:

$$\lambda_{AB} = \begin{pmatrix} 1 & 0 & 0 \\ 0 & 0 & 1 \\ 0 & -1 & 0 \end{pmatrix}\begin{pmatrix} 0 & 1 & 0 \\ -1 & 0 & 0 \\ 0 & 0 & 1 \end{pmatrix} = \begin{pmatrix} 0 & 1 & 0 \\ 0 & 0 & 1 \\ 1 & 0 & 0 \end{pmatrix} \qquad (D.18)$$

Now consider that the order of these two rotations is reversed.

$$\lambda_{BA} = \lambda_A \lambda_B \qquad (D.19)$$

That is:

$$\lambda_{BA} = \begin{pmatrix} 0 & 1 & 0 \\ -1 & 0 & 0 \\ 0 & 0 & 1 \end{pmatrix}\begin{pmatrix} 1 & 0 & 0 \\ 0 & 0 & 1 \\ 0 & -1 & 0 \end{pmatrix} = \begin{pmatrix} 0 & 0 & 1 \\ -1 & 0 & 0 \\ 0 & -1 & 0 \end{pmatrix} \neq \lambda_{AB} \qquad (D.20)$$

An entirely different orientation results as illustrated in figure $D.1$.

This behavior of finite rotations is a consequence of the fact that *finite rotations do not commute*, that is, reversing the order does not give the same answer. Thus, if we associate the vectors **A** and **B** with these rotations, then it implies that the vector product **AB** $\neq$ **BA**. That is, for finite rotation matrices, the product does not behave like for true vectors since they do not commute.

D.2.3 Infinitessimal rotations

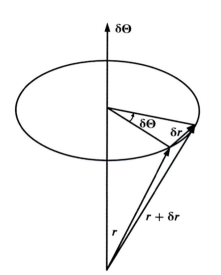

Infinitessimal rotations do not suffer from the noncommutation defect of finite rotations. If the position vector of a point changes from $\mathbf{r}$ to $\mathbf{r} + \delta\mathbf{r}$ then the geometrical situation is represented correctly by

$$\delta\mathbf{r} = \delta\boldsymbol{\theta} \times \mathbf{r} \tag{D.21}$$

where $\delta\boldsymbol{\theta}$ is a quantity whose magnitude is equal to the infinitessimal rotation angle and which has a direction along the instantaneous axis of rotation as illustrated in figure $D.2$.

The infinitessimal angle $\delta\boldsymbol{\theta}$ is a vector which is shown by proving that two infinitessimal rotations $\delta\boldsymbol{\theta}_1$ and $\delta\boldsymbol{\theta}_2$ commute. The change in position vectors of the point are

$$\delta\mathbf{r}_1 = \delta\boldsymbol{\theta}_1 \times \mathbf{r} \tag{D.22}$$

and

$$\delta\mathbf{r}_2 = \delta\boldsymbol{\theta}_2 \times (\mathbf{r} + \delta\mathbf{r}_1) \tag{D.23}$$

Thus the final position vector for $\delta\boldsymbol{\theta}_1$ followed by $\delta\boldsymbol{\theta}_2$ is

$$\mathbf{r} + \delta\mathbf{r}_1 + \delta\mathbf{r}_2 = \mathbf{r} + \delta\boldsymbol{\theta}_1 \times \mathbf{r} + \delta\boldsymbol{\theta}_2 \times (\mathbf{r} + \delta\mathbf{r}_1) \tag{D.24}$$

Assuming that the second-order infinitesimals can be ignored gives

$$\mathbf{r} + \delta\mathbf{r}_1 + \delta\mathbf{r}_2 = \mathbf{r} + \delta\boldsymbol{\theta}_1 \times \mathbf{r} + \delta\boldsymbol{\theta}_2 \times \mathbf{r} \tag{D.25}$$

Figure D.2: Infinitessimal rotation

Consider now the inverse order of rotations.

$$\mathbf{r} + \delta\mathbf{r}_2 + \delta\mathbf{r}_1 = \mathbf{r} + \delta\boldsymbol{\theta}_2 \times \mathbf{r} + \delta\boldsymbol{\theta}_1 \times (\mathbf{r} + \delta\mathbf{r}_2) \tag{D.26}$$

Again, neglecting the second-order infinitesimals gives

$$\mathbf{r} + \delta\mathbf{r}_2 + \delta\mathbf{r}_1 = \mathbf{r} + \delta\boldsymbol{\theta}_2 \times \mathbf{r} + \delta\boldsymbol{\theta}_1 \times \mathbf{r} \tag{D.27}$$

Note that the products of these two infinitessimal rotations, $D25$ and $D27$ are identical. That is, assuming that second-order infinitesimals can be neglected, then the infinitessimal rotations commute, and thus $\delta\boldsymbol{\theta}_1$ and $\delta\boldsymbol{\theta}_2$ are correctly represented by vectors.

The fact that $\delta\boldsymbol{\theta}$ is a vector allows angular velocity to be represented by a vector. That is, angular velocity is the ratio of an infinitessimal rotation to an infinitessimal time.

$$\boldsymbol{\omega} = \frac{\delta\boldsymbol{\theta}}{\delta t} \tag{D.28}$$

Note that this implies that the velocity of the point can be expressed as

$$\mathbf{v} = \frac{\delta\mathbf{r}}{\delta t} = \frac{\delta\boldsymbol{\theta}}{\delta t} \times \mathbf{r} = \boldsymbol{\omega} \times \mathbf{r} \tag{D.29}$$

D.2.4 Proper and improper rotations

The requirement that the coordinate axes be orthogonal, and that the transformation be unitary, leads to the relation between the components of the rotation matrix.

$$\sum_j \lambda_{ij}\lambda_{kj} = \delta_{ik} \tag{D.30}$$

It was shown in equation $A.12$ that, for such an orthogonal matrix, the inverse matrix λ^{-1} equals the transposed matrix λ^T

$$\lambda^{-1} = \lambda^T$$

Inserting the orthogonality relation for the rotation matrix leads to the fact that the square of the determinant of the rotation matrix equals one,

$$|\lambda|^2 = 1 \tag{D.31}$$

that is

$$|\lambda| = \pm 1 \tag{D.32}$$

A **proper rotation** is the rotation of a normal vector and has

$$|\lambda| = +1 \tag{D.33}$$

An **improper rotation** corresponds to

$$|\lambda| = -1 \tag{D.34}$$

An improper rotation implies a rotation plus a spatial reflection which cannot be achieved by any combination of only rotations.

Consider the cross product of two vectors $\mathbf{c} = \mathbf{a} \times \mathbf{b}$. It can be shown that the cross product behaves under rotation as:

$$c_i' = |\lambda| \sum_j \lambda_{ij} c_j \tag{D.35}$$

For all proper rotations the determinant of $\lambda = +1$ and thus the cross product also acts like a proper vector under rotation. This is not true for improper rotations where $|\lambda| = -1$.

D.3   Spatial inversion transformation

Spatial inversion, that is, mirror reflection, corresponds to reflection of all coordinate vectors, $\widehat{\mathbf{i}} = - \widehat{\mathbf{i}}, \widehat{\mathbf{j}} = - \widehat{\mathbf{j}}$, and $\widehat{\mathbf{k}} = - \widehat{\mathbf{k}}$. Such a transformation corresponds to the transformation matrix

$$\boldsymbol{\lambda} = \begin{pmatrix} -1 & 0 & 0 \\ 0 & -1 & 0 \\ 0 & 0 & -1 \end{pmatrix} = - \begin{pmatrix} 1 & 0 & 0 \\ 0 & 1 & 0 \\ 0 & 0 & 1 \end{pmatrix} \tag{D.36}$$

Thus $|\lambda| = -1$, that is, it corresponds to an improper rotation. A spatial inversion for two vectors $\mathbf{A}(r)$ and $\mathbf{B}(r)$ correspond to

$$\begin{aligned} \mathbf{A}(r) &= -\mathbf{A}(-r) \\ \mathbf{B}(r) &= -\mathbf{B}(-r) \end{aligned} \tag{D.37}$$

That is, normal polar vectors change sign under spatial reflection. However, the cross product $\mathbf{C} = \mathbf{A} \times \mathbf{B}$ does not change sign under spatial inversion since the product of the two minus signs is positive. That is,

$$\mathbf{C}(r) = +\mathbf{C}(-r) \tag{D.38}$$

Thus the cross product behaves differently from a polar vector. This improper behavior is characteristic of an **axial vector**, which also is called a **pseudovector**.

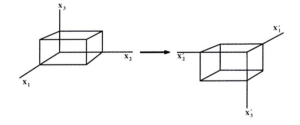

Figure D.3: Inversion of an object corresponds to reflection about the origin of all axes.

Examples of pseudovectors are angular momentum, spin, magnetic field etc. These pseudovectors are defined using the right-hand rule and thus have handedness. For a right-handed system

$$\mathbf{C}_R = \mathbf{A} \times \mathbf{B} \tag{D.39}$$

Changing to a left-handed system leads to

$$\mathbf{C}_L = \mathbf{B} \times \mathbf{A} = -\mathbf{A} \times \mathbf{B} \tag{D.40}$$

That is, handedness corresponds to a definite ordering of the cross product. Proper orthogonal transformations are said to preserve chirality (Greek for handedness) of a coordinate system.

An example of the use of the right-handed system is the usual definition of cartesian unit vectors,

$$\hat{\mathbf{i}} \times \hat{\mathbf{j}} = \hat{\mathbf{k}} \tag{D.41}$$

An obvious question to be asked, is the handedness of a coordinate system merely a mathematical curiosity or does it have some deep underlying significance? Consider the Lorentz force

$$\mathbf{F} = q\left(\mathbf{E} + \mathbf{v} \times \mathbf{B}\right) \tag{D.42}$$

Since force and velocity are proper vectors then the magnetic $\mathbf{B}$ field must be a pseudo vector. Note that calculation of the $\mathbf{B}$ field occurs only in cross products such as,

$$\nabla \times \mathbf{B} = \mu\mathbf{j} \tag{D.43}$$

where the current density $\mathbf{j}$ is a proper vector. Another example is the Biot-Savart Law which expresses $\mathbf{B}$ as

$$d\mathbf{B} = \frac{\mu_o I}{4\pi} \frac{d\mathbf{l} \times \mathbf{r}}{r^2} \tag{D.44}$$

Thus even though $\mathbf{B}$ is a pseudo vector, the force $\mathbf{F}$ remains a proper vector. Thus if a left-handed coordinate definition of $\mathbf{B}_L = \frac{\mu_o I}{4\pi} \frac{\mathbf{r} \times d\mathbf{l}}{r^2}$ is used in $D.44$, and $\mathbf{F} = q\left(\mathbf{E} + \mathbf{B}_L \times \mathbf{v}\right)$ in $D.42$, then the same final physical result would be obtained.

It was long thought that the laws of physics were symmetric with respect to spatial inversion (i.e. mirror reflection), meaning that the choice between a left-handed and right-handed representations (chirality) was arbitrary. This is true for gravitational, electromagnetic and the strong force, and is called the conservation of parity. The fourth fundamental force in nature, the weak force, violates parity and favours handedness. It turns out that right-handed ordinary matter is symmetrical with left-handed antimatter.

In addition to the two flavours of vectors, one has scalars and pseudoscalars defined by:

$$\phi\left(r\right) = +\phi\left(-r\right) \tag{D.45}$$
$$\phi\left(r\right) = -\phi\left(-r\right) \tag{D.46}$$

An example of a pseudoscalar is the scalar product $\mathbf{A} \cdot \left(\mathbf{B} \times \mathbf{C}\right)$

D.4   Time reversal transformation

The basic laws of classical mechanics are invariant to the sense of the direction of time. Under time reversal the vector $\mathbf{r}$ is unchanged while both momentum $\mathbf{p}$ and time t change sign under time reversal, thus the time derivative $\mathbf{F} = \frac{d\mathbf{p}}{dt}$ is invariant to time reversal; that is, the force is unchanged and Newton's Laws $\mathbf{F} = \frac{d\mathbf{p}}{dt}$ are invariant under time reversal. Since the force can be expressed as the gradient of a scalar potential for a conservative field, then the potential also remains unchanged. That is

$$\frac{d\mathbf{p}}{dt} = -\nabla U(r) = \mathbf{F} \tag{D.47}$$

It is necessary to introduce tensor algebra, given in appendix E, prior to discussion of the transformation properties of observables which is the topic of appendix $E5$.

Workshop exercises

1. Suppose the x_2-axis of a rectangular coordinate system is rotated by $30°$ away from the x_3-axis around the x_1-axis.

   (a) Find the corresponding transformation matrix. Try to do this by drawing a diagram instead of going to the book or the notes for a formula.

(b) Is this an orthogonal matrix? If so, show that it satisfies the main properties of an orthogonal matrix. If not, explain why it fails to be orthogonal.

(c) Does this matrix represent a proper or an improper rotation? How do you know?

2. When you were first introduced to vectors, you most likely were told that a scalar is a quantity that is defined by a magnitude, while a vector has both a magnitude and a direction. While this is certainly true, there is another, more sophisticated way to define a scalar quantity and a vector quantity: through their transformation properties. A scalar quantity transforms as $\phi' = \phi$ while a vector quantity transforms as $A'_i = \sum_j \lambda_{ij} A_j$. To show that the scalar product does indeed transform as a scalar, note that:

$$\mathbf{A'}\cdot\mathbf{B'} \;=\; \sum_i A'_i B'_i = \sum_i \left(\sum_j \lambda_{ij} A_j\right)\left(\sum_k \lambda_{ik} B_k\right) = \sum_{j,k}\left(\sum_i \lambda_{ij}\lambda_{ik}\right) A_j B_k$$

$$=\; \sum_j \left(\sum_k \delta_{jk} A_j B_k\right) = \sum_j A_j B_j = \mathbf{A}\cdot\mathbf{B}$$

Now you will show that the vector product transforms as a vector. Begin by writing out what you are trying to show explicitly and show it to the teaching assistant. Once the teaching assistant has confirmed that you have the correct expression, try to prove it. The vector product is a bit more difficult to work with than the scalar product, so your teaching assistant is prepared to give you a hint if you get stuck.

3. Suppose you have two rectangular coordinate systems that share a common origin, but one system is rotated by an angle θ with respect to the other. To describe this rotation, you have made use of the rotation matrix $\lambda(\theta)$. (I'm changing the notation slightly to put the emphasis on the angle of rotation.)

(a) Verify that the product of two rotation matrices $\lambda(\theta_1)\lambda(\theta_2)$ is in itself a rotation matrix.

(b) In abstract algebra, a group G is defined as a set of elements g together with a binary operation $*$ acting on that set such that four properties are satisfied:

    i. (Closure) For any two elements g_i and g_j in the group G, the product of the elements, $g_i * g_j$ is also in the group G.

    ii. (Associativity) For any three elements g_i, g_j, g_k of the group G, $(g_i * g_j) * g_k = g_i * (g_j * g_k)$.

    iii. (Existence of Identity) The group G contains an identity element e such that $g * e = e * g = g$ for all $g \in G$.

    iv. (Existence of Inverses) For each element $g \in G$, there exists an inverse element $g^{-1} \in G$ such that $g * g^{-1} = g^{-1} * g = e$.

Show that if the product $*$ denotes the product of two matrices, then the set of rotation matrices together with $*$ forms a group. This group is known as the special orthogonal group in two dimensions, also known as $SO(2)$.

(c) Is this group commutative? In abstract algebra, a commutative group is called an abelian group.

4. When you look in a mirror the image of you appears left-to-right reversed, that is, the image of your left ear appears to be the right ear of the image and vise versa. Explain why the image is left-right reversed rather than up-down reversed or reversed about some other axis; i.e. explain what breaks the symmetry that leads to these properties of the mirror image.

Problems

[1] Find the transformation matrix that rotates the axis x_3 of a rectangular coordinate system 45° toward x_1 around the x_2 axis.

[2] For simplicity, take λ to be a two-dimensional transformation matrix. Show by direct expansion that $|\lambda|^2 = 1$.

Appendix E

Tensor algebra

E.1   Tensors

Mathematically scalars and vectors are the first two members of a hierarchy of entities, called **tensors**, that behave under coordinate transformations as described in appendix D. The use of the tensor notation provides a compact and elegant way to handle transformations in physics.

A scalar is a rank 0 tensor with one component, that is invariant under change of the coordinate system.

$$\phi(x'y'z') = \phi(xyz) \tag{E.1}$$

A vector is a rank 1 tensor which has three components, that transform under rotation according to matrix relation

$$\mathbf{x}' = \boldsymbol{\lambda} \cdot \mathbf{x} \tag{E.2}$$

where $\boldsymbol{\lambda}$ is the rotation matrix. Equation $E2$ can be written in the suffix form as

$$x_i' = \sum_{j=1}^{3} \lambda_{ij} x_j \tag{E.3}$$

The above definitions of scalars and vectors can be subsumed into a class of entities called tensors of rank n that have 3^n components. A scalar is a tensor of rank $r = 0$, with only $3^0 = 1$ component, whereas a vector has rank $r = 1$, that is, the vector $\mathbf{x}$ has one suffix i and $3^1 = 3$ components.

A second-order tensor T_{ij} has rank $r = 2$ with two suffixes, that is, it has $3^2 = 9$ components that transform under rotation as

$$T_{ij}' = \sum_{k=1}^{3} \sum_{l=1}^{3} \lambda_{ik} \lambda_{jl} T_{kl} \tag{E.4}$$

For second-order tensors, the transformation formula given by equation $E.4$ can be written more compactly using matrices. Thus the second-order tensor can be written as a 3×3 matrix

$$\mathbf{T} \equiv \begin{pmatrix} T_{11} & T_{12} & T_{13} \\ T_{21} & T_{22} & T_{23} \\ T_{31} & T_{32} & T_{33} \end{pmatrix} \tag{E.5}$$

The rotational transformation given in equation $E.4$ can be written in the form

$$T_{ij}' = \sum_{l=1}^{3} \left(\sum_{k=1}^{3} \lambda_{ik} T_{kl} \right) \lambda_{jl} = \sum_{l=1}^{3} \left(\sum_{k=1}^{3} \lambda_{ik} T_{kl} \right) \lambda_{lj}^{T} \tag{E.6}$$

where λ_{lj}^{T} are the matrix elements of the transposed matrix $\boldsymbol{\lambda}^{T}$. The summations in $E.6$ can be expressed in both the tensor and conventional matrix form as the matrix product

$$\mathbf{T}' = \boldsymbol{\lambda} \cdot \mathbf{T} \cdot \boldsymbol{\lambda}^{T} \tag{E.7}$$

Equation $E7$ defines the rotational properties of a spherical tensor.

E.2   Tensor products

E.2.1   Tensor outer product

Tensor products feature prominently when using tensors to represent transformations. A second-order tensor $\mathbf{T}$ can be formed by using the **tensor product,** also called **outer product,** of two vectors $\mathbf{a}$ and $\mathbf{b}$ which, written in suffix form, is

$$\mathbf{T} \equiv \mathbf{a} \otimes \mathbf{b} = \begin{pmatrix} a_1 b_1 & a_1 b_2 & a_1 b_3 \\ a_2 b_1 & a_2 b_2 & a_2 b_3 \\ a_3 b_1 & a_3 b_2 & a_3 b_3 \end{pmatrix} \tag{E.8}$$

In component form the matrix elements of this matrix are given by

$$T_{ij} = a_i b_j \tag{E.9}$$

This second-order **tensor product** has a rank $r = 2$, that is, it equals the sum of the ranks of the two vectors. Equation $E8$ is called a *dyad* since it was derived by taking the dyadic product of two vectors. In general, multiplication, or division, of two vectors leads to second-order tensors. Note that this second-order tensor product completes the triad of tensors possible taking the product of two vectors. That is, the scalar product $\mathbf{a} \cdot \mathbf{b}$, has rank $r = 0$, the vector product $\mathbf{a} \times \mathbf{b}$, rank $r = 1$ and the tensor product $\mathbf{a} \otimes \mathbf{b}$ has rank[1] $r = 2$.

Higher-order tensors can be created by taking more complicated tensor products. For example, a rank-3 tensor can be created by taking the tensor outer product of the rank-2 tensor T_{ij} and a vector c_k which, for a dyadic tensor, can be written as the tensor product of three vectors. That is,

$$T_{ijk} = T_{ij} c_k = a_i b_j c_k \tag{E.10}$$

In summary, the rank of the tensor product equals the sum of the ranks of the tensors included in the tensor product.

E.2.2   Tensor inner product

The lowest rank tensor product, which is called the **inner product,** is obtained by taking the tensor product of two tensors for the special case where one index is repeated, and taking the sum over this repeated index. Summing over this repeated index, which is called **contraction**, removes the two indices for which the index is repeated, resulting in a tensor that has rank r equal to the sum of the ranks minus 2 for one contraction. That is, the product tensor has rank $r = r_1 + r_2 - 2$.

The simplest example is the inner product of two vectors which has rank $r = 1 + 1 - 2 = 0$, that is, it is the scalar product that equals the trace of the inner product matrix, and this inner product is commutative.

An especially important case is the inner product of a rank-2 dyad $\mathbf{a} \otimes \mathbf{b}$, given by equation $E8$, with a vector $\mathbf{c}$, that is, the inner product $\mathbf{T} = \mathbf{a} \otimes \mathbf{b} \cdot \mathbf{c}$. Written in component form, the inner product is

$$\sum_i^3 a_i b_i c_j = \left(\sum_i^3 a_i b_i \right) c_j = (\mathbf{a} \cdot \mathbf{b}) \, c_j \tag{E.11}$$

The scalar product $\mathbf{a} \cdot \mathbf{b}$ is a scalar number, and thus the inner-product tensor is the vector $\mathbf{c}$ renormalized by the magnitude of the scalar product $\mathbf{a} \cdot \mathbf{b}$. That is, it has a rank $r = 2 + 1 - 2 = 1$. Thus the inner product of this rank-2 tensor with a vector gives a vector. The inner product of a rank-2 tensor with a rank-1 tensor is used in this book for handling the rotation matrix, the inertia tensor for rigid-body rotation, and for the stress and the strain tensors used to describe elasticity in solids.

E.1 *Example: Displacement gradient tensor*

The displacement gradient tensor provides an example of the use of the matrix representation to manipulate tensors. Let $\phi(x_1, x_2, x_3)$ be a vector field expressed in a cartesian basis. The definition of the gradient $\mathbf{G} = \nabla \phi$ gives that

$$d\phi = \mathbf{G} \cdot d\mathbf{x}$$

---

[1] The common convention is to denote the scalar product as $\mathbf{a} \cdot \mathbf{b}$, the vector product as $\mathbf{a} \times \mathbf{b}$, and tensor product as $\mathbf{a} \otimes \mathbf{b}$.

Calculating the components of $d\phi$ in terms of $\mathbf{x}$ gives

$$d\phi_1 = \frac{\partial\phi_1}{\partial x_1}dx_1 + \frac{\partial\phi_1}{\partial x_2}dx_2 + \frac{\partial\phi_1}{\partial x_3}dx_3$$

$$d\phi2 = \frac{\partial\phi_2}{\partial x_1}dx_1 + \frac{\partial\phi_2}{\partial x_2}dx_2 + \frac{\partial\phi_2}{\partial x_3}dx_3$$

$$d\phi_3 = \frac{\partial\phi_3}{\partial x_1}dx_1 + \frac{\partial\phi_3}{\partial x_2}dx_2 + \frac{\partial\phi_3}{\partial x_3}dx_3$$

Using index notation this can be written as

$$d\phi_i = \frac{\partial\phi_i}{\partial x_j}dx_j$$

The second-rank gradient tensor $\mathbf{G}$ can be represented in the matrix form as

$$\mathbf{G} = \begin{vmatrix} \frac{\partial\phi_1}{\partial x_1} & \frac{\partial\phi_1}{\partial x_2} & \frac{\partial\phi_1}{\partial x_3} \\ \frac{\partial\phi_2}{\partial x_1} & \frac{\partial\phi_2}{\partial x_2} & \frac{\partial\phi_2}{\partial x_3} \\ \frac{\partial\phi_3}{\partial x_1} & \frac{\partial\phi_3}{\partial x_2} & \frac{\partial\phi_3}{\partial x_3} \end{vmatrix}$$

Then the vector ϕ can be expressed compactly as the inner product of $\mathbf{G}$ and $\mathbf{x}$, that is

$$d\phi = \mathbf{G}\cdot d\mathbf{x}$$

E.3   Tensor properties

In principle one must distinguish between a 3×3 square matrix, and the tensor component representations of a rank-2 tensor. However, as illustrated by the previous discussion, for orthogonal transformations, the tensor components of the second rank tensor transform identically with the matrix components. Thus functionally, the matrix formulation and tensor representations are identical. As a consequence, all the terminology and operations used in matrix mechanics are equally applicable to the tensor representation.

The tensor representation of the rotation matrix provides the simplest example of the equivalence of the matrix and tensor representations of transformations. Appendix $D.2$ showed that the unitary rotation matrix $\boldsymbol{\lambda}$, acting on a vector $\mathbf{x}$ transforms it to the vector $\mathbf{x}'$ that is rotated with respect to $\mathbf{x}$. That is, the transformation is

$$\mathbf{x}' = \boldsymbol{\lambda}\cdot\mathbf{x} \tag{D5}$$

where

$$\mathbf{x}' \equiv \begin{pmatrix} x_1' \\ x_2' \\ x_3' \end{pmatrix} \qquad \mathbf{x} \equiv \begin{pmatrix} x_1 \\ x_2 \\ x_3 \end{pmatrix} \qquad \boldsymbol{\lambda} \equiv \begin{pmatrix} \hat{\mathbf{e}}_1'\cdot\hat{\mathbf{e}}_1 & \hat{\mathbf{e}}_1'\cdot\hat{\mathbf{e}}_2 & \hat{\mathbf{e}}_1'\cdot\hat{\mathbf{e}}_3 \\ \hat{\mathbf{e}}_2'\cdot\hat{\mathbf{e}}_1 & \hat{\mathbf{e}}_2'\cdot\hat{\mathbf{e}}_2 & \hat{\mathbf{e}}_2'\cdot\hat{\mathbf{e}}_3 \\ \hat{\mathbf{e}}_3'\cdot\hat{\mathbf{e}}_1 & \hat{\mathbf{e}}_3'\cdot\hat{\mathbf{e}}_2 & \hat{\mathbf{e}}_3'\cdot\hat{\mathbf{e}}_3 \end{pmatrix} \tag{D6}$$

Appendix $D.2$ showed that the rotation matrix $\boldsymbol{\lambda}$ requires 9 components to fully specify the transformation from the initial 3-component vector $\mathbf{x}$ to the rotated vector $\mathbf{x}'$. The rotation tensor is a dyad as well as being unitary and dimensionless. Note that equation $D5$ is an example of the inner product of a rank-2 rotation tensor acting on a vector leading to a another vector that is rotated with respect to the first vector.

In general, rank-2 tensors have dimensions and are not unitary. For example, the angular velocity vector $\boldsymbol{\omega}$ and the angular momentum vector $\mathbf{L}$ are related by the inner product of the inertia tensor $\{\mathbf{I}\}$ and $\boldsymbol{\omega}$. That is

$$\mathbf{L} = \{\mathbf{I}\}\cdot\boldsymbol{\omega} \tag{11.6}$$

The inertia tensor has dimensions of $mass \times length^2$ and relates two very different vector observables. The stress tensor and the strain tensor, discussed in chapter 15, provide another example of second-order tensors that are used to transform one vector observable to another vector observable analogous to the case of the rotation matrix or the inertia tensor.

Note that pseudo-tensors can be used to make a rotational transformation plus a change in the sign. That is, they lead to a parity inversion.

The tensor notation is used extensively in physics since it provides a powerful, elegant, and compact representation for describing transformations.

E.4 Contravariant and covariant tensors

In general the configuration space used to specify a dynamical system is not a Euclidean space in that there may not be a system of coordinates for which the distance between any two neighboring points can be represented by the sum of the squares of the coordinate differentials. For example, a set of cartesian coordinate does not exist for the two-dimension motion of a single particle constrained to the curved surface of a fixed sphere. Such curved spaces need to be represented in terms of Riemannian geometry rather than Euclidean geometry. Curved configuration spaces occur in some branches of physics such as Einstein's General Theory of Relativity.

Tensors have transformation properties that can be either contravariant or covariant. Consider a set of generalized coordinates q' that are a function of the coordinates q. Then infinitessimal changes dq^m will lead to infinitessimal changes dq'^n where

$$dq'^n = \sum_m \frac{\partial q'^n}{\partial q^m} dq^m \tag{E.12}$$

Contravariant components of a tensor transform according to the relation

$$\lambda'^n = \sum_m \frac{\partial q'^n}{\partial q^m} \lambda^m \tag{E.13}$$

Equation $E13$ relates the contravariant components in the unprimed and primed frames.

Derivatives of a scalar function ϕ, such as

$$\lambda'_n = \frac{\partial \phi}{\partial q^n} = \sum_m \frac{\partial \phi}{\partial q^m} \frac{\partial q^m}{\partial q^n} = \sum_m \frac{\partial q^m}{\partial q^n} \lambda^m \tag{E.14}$$

That is, **covariant** components of the tensor transform according to the relation

$$\lambda'_n = \sum_m \frac{\partial q^m}{\partial q^n} \lambda^m \tag{E.15}$$

It is important to differentiate between contravariant and covariant vectors. The Einstein superscript/subscript convention for distinguishing between these two flavours of tensors is given in table $E1$

Table E.1. Einstein notation for tensors.

x^μ	denotes a contravariant vector
x_ν	denotes a covariant vector

In linear algebra one can map from one coordinate system to another as illustrated in appendix D. That is, the tensor $\mathbf{x}$ can be expressed as components with respect to either the unprimed or primed coordinate frames

$$\mathbf{x} = \hat{\mathbf{e}}'_1 x'_1 + \hat{\mathbf{e}}'_2 x'_2 + \hat{\mathbf{e}}'_3 x'_3 = \hat{\mathbf{e}}_1 x_1 + \hat{\mathbf{e}}_2 x_2 + \hat{\mathbf{e}}_3 x_3 \tag{E.16}$$

For a $n-$dimensional manifold the unit basis column vectors $\hat{\mathbf{e}}$ transform according to the transformation matrix $\boldsymbol{\lambda}$

$$\hat{\mathbf{e}}' = \boldsymbol{\lambda} \cdot \hat{\mathbf{e}} \tag{E.17}$$

Since the tensor $\mathbf{x}$ is independent of the coordinate basis, the components of $\mathbf{x}$ must have the opposite transform

$$\mathbf{x}' = \left(\boldsymbol{\lambda}^{-1}\right)^T \cdot \mathbf{x} \tag{E.18}$$

This normal vector $\mathbf{x}$ is called a "contravariant vector" because it transforms contrary to the basis column vector transformation.

The inverse of equation $E.18$ gives that the column vector element

$$x_\mu = \sum_\nu \lambda_{\mu\nu} x'_\nu \tag{E.19}$$

Consider the case of a gradient with respect to the coordinate **x** in both the unprimed and primed bases. Using the chain rule for the partial derivative then the component of the gradient in the primed frame can be expanded as

$$(\nabla f)'_\mu = \frac{\partial f}{\partial x'_\mu} = \sum_\nu \frac{\partial f}{\partial x_\nu}\frac{\partial x_\nu}{\partial x'_\mu} = \sum_\nu \frac{\partial f}{\partial x_\nu}\lambda_{\nu\mu}\delta_{\mu\nu} = \lambda_{\mu\mu}\frac{\partial f}{\partial x_\mu} \qquad (E.20)$$

That is, the gradient transforms as

$$\nabla' f = \boldsymbol{\lambda}\cdot\nabla f \qquad (E.21)$$

That is, *a gradient transforms as a covariant vector, like the unit vectors, whereas a vector x is contravariant under transformation.*

Normally the basis is orthonormal, $\left(\boldsymbol{\lambda}^{-1}\right)^T = \boldsymbol{\lambda}$, and thus there is no difference between contravariant and covariant vectors. However, for curved coordinate systems, such as non-Euclidean geometry in the General Theory of Relativity, the covariant and contravariant vectors behave differently.

The Einstein convention is extended to apply to matrices by writing the elements of the matrix **A** as A^μ_ν while the elements of the transposed matrix $\mathbf{A}^{-1}$ are written as $A_\nu{}^\mu$. The matrix product for **A** with a contravariant vector **X** is written as

$$X'^\mu = \sum_\nu A^\mu_\nu X^\nu \qquad (E.22)$$

where the summation over ν effectively cancels the identical superscript and subscript ν.

Similarly a covariant vector, such as a gradient, is written as,

$$\left(\nabla' f\right)_\mu = \sum_\nu \left(A^{-1}\right)^{T}{}_{\mu}{}^{\nu}\left(\nabla f\right)_\nu = \sum_\nu \left(A^{-1}\right)^{\nu}{}_{\mu}\left(\nabla f\right)_\nu \qquad (E.23)$$

Again the summation cancels the ν superscript and subscript. The Kronecker delta symbol is written as

$$\sum_\nu \delta^\mu{}_\nu X^\nu = X^\mu \qquad (E.24)$$

E.5   Generalized inner product

The generalized definition of an *inner product* is

$$S = \sum_{\mu\nu} g_{\mu\nu}X^\mu Y^\nu \qquad (E.25)$$

where $g_{\mu\nu}$ is a unitary matrix called a covariant metric. The covariant metric transforms a contravariant to a covariant tensor. For example the matrix element of a covariant tensor X_ν can be written as

$$X_\nu = \sum_\mu g_{\mu\nu}X^\mu \qquad (E.26)$$

By association of the *covariant metric* with either of the vectors in the inner product gives

$$S = \sum_{\mu\nu} g_{\mu\nu}X^\mu Y^\nu = \sum_\nu X_\nu Y^\nu = \sum_\mu X^\mu Y_\mu \qquad (E.27)$$

Similarly it can be defined in terms of an *orthogonal contravariant metric* $g^{\mu\nu}$ where

$$S = \sum_{\mu\nu} g^{\mu\nu}X_\mu Y_\nu \qquad (E.28)$$

Then

$$X^\nu = \sum_\mu g^{\mu\nu}X_\mu \qquad (E.29)$$

Association of the contravariant metric with one of the vectors in the inner product gives the inner product

$$S = \sum_{\mu\nu} g^{\mu\nu}X_\mu Y_\nu = \sum_\nu X^\nu Y_\nu = \sum_\mu X_\mu Y^\mu \qquad (E.30)$$

For most situations in this book the metric $g_{\mu\nu}$ is diagonal and unitary.

E.6 Transformation properties of observables

In physics, observables can be represented by spherical tensors which specify the angular momentum and parity characteristics of the observable, and the tensor rank is independent of the time dependence. The transformation properties of these tensors, coupled with their time-reversal invariance, specify the fundamental characteristics of the observables.

Table $E.2$ summarizes the transformation properties under rotation, spatial inversion and time reversal for observables encountered in classical mechanics and electrodynamics. Note that observables can be scalar, vector, pseudovector, or second-order tensors, under rotation, and even or odd under either space inversion or time inversion. For example, in classical mechanics the inertia tensor $\mathbf{I}$ relates the angular velocity vector $\boldsymbol{\omega}$ to the angular momentum vector $\mathbf{L}$ by taking the inner product $\mathbf{L} = \mathbf{I} \cdot \boldsymbol{\omega}$. In general $\mathbf{I}$ is not diagonal and thus the angular momentum is not parallel to the angular velocity $\boldsymbol{\omega}$. A similar example in electrodynamics is the dielectric tensor $\mathbf{K}$ which relates the displacement field $\mathbf{D}$ to the electric field $\mathbf{E}$ by $\mathbf{D} = \mathbf{K} \cdot \mathbf{E}$. For anisotropic crystal media $\mathbf{K}$ is not diagonal leading to the electric field vectors $\mathbf{E}$ and $\mathbf{D}$ not being parallel.

As discussed in chapter 7, Noether's Theorem states that symmetries of the transformation properties lead to important conservation laws. The behavior of classical systems under rotation relates to the conservation of angular momentum, the behavior under spatial inversion relates to parity conservation, and time-reversal invariance relates to conservation of energy. That is, conservative forces conserve energy and are time-reversal invariant.

Table $E.2$: Transformation properties of scalar, vector, pseudovector, and tensor observables under rotation, spatial inversion, and time reversal[2]

Physical Observable		Rotation (Tensor rank)	Space inversion	Time reversal	Name
1) Classical Mechanics					
Mass density	ρ	0	Even	Even	Scalar
Kinetic energy	$p^2/2m$	0	Even	Even	Scalar
Potential energy	$U(r)$	0	Even	Even	Scalar
Lagrangian	L	0	Even	Even	Scalar
Hamiltonian	H	0	Even	Even	Scalar
Gravitational potential	ϕ	0	Even	Even	Scalar
Coordinate	$\mathbf{r}$	1	Odd	Even	Vector
Velocity	$\mathbf{v}$	1	Odd	Odd	Vector
Momentum	$\mathbf{p}$	1	Odd	Odd	Vector
Angular momentum	$\mathbf{L} = \mathbf{r} \times \mathbf{p}$	1	Even	Odd	Pseudovector
Force	$\mathbf{F}$	1	Odd	Even	Vector
Torque	$\mathbf{N} = \mathbf{r} \times \mathbf{F}$	1	Even	Even	Pseudovector
Gravitational field	$\mathbf{g}$	1	Odd	Even	Vector
Inertia tensor	$\mathbf{I}$	2	Even	Even	Tensor
Elasticity stress tensor	$\mathbf{T}_{ik}$	2	Even	Even	Tensor
2) Electromagnetism					
Charge density	ρ	0	Even	Even	Scalar
Current density	$\mathbf{j}$	1	Odd	Odd	Vector
Electric field	$\mathbf{E}$	1	Odd	Even	Vector
Polarization	$\mathbf{P}$	1	Odd	Even	Vector
Displacement	$\mathbf{D}$	1	Odd	Even	Vector
Magnetic B field	$\mathbf{B}$	1	Even	Odd	Pseudovector
Magnetization	$\mathbf{M}$	1	Even	Odd	Pseudovector
Magnetic H field	$\mathbf{H}$	1	Even	Odd	Pseudovector
Poynting vector	$\mathbf{S} = \mathbf{E} \times \mathbf{H}$	1	Odd	Odd	Vector
Dielectric tensor	$\mathbf{K}$	2	Even	Even	Tensor
Maxwell stress tensor	$\mathbf{T}_{ik}$	2	Even	Even	Tensor

---

[2]Based on table 6.1 in *"Classical Electrodynamics"* 2nd edition, by J.D. Jackson [?]

Appendix F

Aspects of multivariate calculus

Multivariate calculus provides the framework for handling systems having many variables associated with each of several bodies. It is assumed that the reader has studied linear differential equations plus multivariate calculus and thus has been exposed to the calculus used in classical mechanics. Chapter 5 of this book introduced variational calculus which covers several important aspects of multivariate calculus such as Euler's variational calculus and Lagrange multipliers. This appendix provides a brief review of a selection of other aspects of multivariate calculus that feature prominently in classical mechanics.

F.1 Partial differentiation

The extension of the derivative to multivariate calculus involves use of partial derivatives. The partial derivative with respect to the variable x_i of a multivariate function $f(x_1, x_2, ..., x_N)$ involves taking the normal one-variable derivative with respect to x_i assuming that the other $N-1$ variables are held constant. That is,

$$\frac{\partial f(x_1, x_2, ...x_N)}{\partial x_i} = \lim_{h_i \to 0} \left[\frac{f(x_1, x_2, ..x_{i-1}, (x_i + h_i), ..x_N) - f(x_1, x_2, .., x_N)}{h_i} \right] \quad (F.1)$$

where it will be assumed that the function $f(x)$ is a continuously-differentiable function to n^{th} order, then all partial derivatives of that order or less are independent of the order in which they are performed. That is,

$$\frac{\partial^2 f(x)}{\partial x_i \partial x_j} = \frac{\partial^2 f(x)}{\partial x_j \partial x_i} \quad (F.2)$$

The chain rule for partial differentiation gives that

$$\frac{\partial f(y_1, y_2, ..., y_N)}{\partial y_j} = \sum_{k=1}^{N} \frac{\partial f(x)}{\partial x_k} \frac{\partial x_k(y)}{\partial y_j} \quad (F.3)$$

The total differential of a multivariate function $f(x)$ is

$$df = \sum_{k=1}^{N} \frac{\partial f(x)}{\partial x_k} dx_k \quad (F.4)$$

This can be extended to higher-order derivatives using the operator formalism

$$d^n f(x) = \left(dx_1 \frac{\partial}{\partial x_1} + ... + dx_N \frac{\partial}{\partial x_N} \right)^n f(x) = \sum dx_{j_1}...dx_{j_n} \frac{\partial^n f(x)}{\partial x_{j_1}...\partial x_{j_n}} \quad (F.5)$$

F.2 Linear operators

The linear operator notation provides a powerful, elegant, and compact way to express, and apply, the equations of multivariate calculus; it is used extensively in mathematics and physics. The linear operators typically comprise partial derivatives that act on scalar, vector, or tensor fields. Table $F1$ lists a few elementary examples of the use of linear operators in this textbook. The first four linear operators involve the widely used del operator $\boldsymbol{\nabla}$ to generate the gradient, divergence and curl as described in appendices G and H. The fifth and sixth linear operators act on the Lagrangian in Lagrangian mechanics applications. The final two linear operators act on the wavefunction for wave mechanics.

Name	Partial derivative	Field	Action
Gradient	$\boldsymbol{\nabla} \equiv \hat{\imath}\frac{\partial}{\partial x} + \hat{\jmath}\frac{\partial}{\partial y} + \hat{\mathbf{k}}\frac{\partial}{\partial z}$	Scalar potential V	$\mathbf{E} = \boldsymbol{\nabla} V$
Divergence	$\boldsymbol{\nabla}\cdot \equiv \left(\hat{\imath}\frac{\partial}{\partial x} + \hat{\jmath}\frac{\partial}{\partial y} + \hat{\mathbf{k}}\frac{\partial}{\partial z}\right)\cdot$	Vector field $\mathbf{E}$	$\boldsymbol{\nabla}\cdot\mathbf{E}$
Curl	$\boldsymbol{\nabla}\times \equiv \left(\hat{\imath}\frac{\partial}{\partial x} + \hat{\jmath}\frac{\partial}{\partial y} + \hat{\mathbf{k}}\frac{\partial}{\partial z}\right)\times$	Vector field $\mathbf{E}$	$\boldsymbol{\nabla}\times\mathbf{E}$
Laplacian	$\nabla^2 = \boldsymbol{\nabla}\cdot\boldsymbol{\nabla} \equiv \frac{\partial^2}{\partial x^2} + \frac{\partial^2}{\partial y^2} + \frac{\partial^2}{\partial z^2}$	Scalar potential V	$\nabla^2 V$
Euler-Lagrange	$\Lambda_j \equiv \frac{d}{dt}\frac{\partial}{\partial \dot{q}_j} - \frac{\partial}{\partial q_j}$	Scalar Lagrangian L	$\Lambda L = 0$
Canonical momentum	$p_j \equiv \frac{\partial}{\partial \dot{q}_j}$	Scalar Lagrangian L	$p_j \equiv \frac{\partial L}{\partial \dot{q}_j}$
Canonical momentum	$p_j \equiv \frac{\hbar}{i}\frac{\partial}{\partial \dot{q}_j}$	Wavefunction Ψ	$p_j\Psi \equiv \frac{\hbar}{i}\frac{\partial \Psi}{\partial \dot{q}_j}$
Hamiltonian	$H = i\hbar\frac{\partial}{\partial t}$	Wavefunction Ψ	$H\Psi = i\hbar\frac{\partial \Psi}{\partial t} = E\Psi$

Table $F.1$, examples of linear operators used in this textbook.

There are three ways of expressing operations such as addition, multiplication, transposition or inversion of operations that are completely equivalent because they all are based on the same principles of linear algebra. For example, a transformation $\mathbf{O}$ acting on a vector $\mathbf{A}$ can produced the vector $\mathbf{B}$. The simplest way to express this transformation is in terms of components

$$B_i = \sum_{j=1}^{3} O_{ij} A_j \tag{F.6}$$

Another way is to use matrix mechanics where the 3×3 matrix $(\mathbf{O})$ transforms the column vector $(\mathbf{A})$ to the column vector $(\mathbf{B})$, that is,

$$(\mathbf{B}) = (\mathbf{O})\,(\mathbf{A}) \tag{F.7}$$

The third approach is to assume an operator $\mathbf{O}$ acts on the vector $\mathbf{A}$

$$\mathbf{B} = \mathbf{OA} \tag{F.8}$$

In classical mechanics, and quantum mechanics, these three equivalent approaches are used and exploited extensively and interchangeably. In particular the rules of matrix manipulation, that are given in appendix A, are synonymous, and equivalent to, those that apply for operator manipulation. If the operator is complex then the operator properties are summarized as follows.

The generalization of the transpose for complex operators is the *Hermitian conjugate* $O^\dagger$

$$O_{ij}^\dagger = O_{ji}^* \tag{F.9}$$

Note also that

$$\mathbf{O}^\dagger = (O^*)^T = (O^T)^* \tag{F.10}$$

The generalization of a symmetric matrix is *Hermitian*, that is, O is equal to its Hermitian conjugate

$$O_{ij}^\dagger = O_{ji}^* = O_{ij} \tag{F.11}$$

For a real matrix the complex conjugation has no effect so the matrix is real and symmetric.

The generalization of orthogonal is *unitary* for which the operator is unitary if it is non-singular and

$$O^{-1} = O^\dagger \tag{F.12}$$

which implies

$$OO^\dagger = U = O^\dagger O \tag{F.13}$$

F.3 Transformation Jacobian

The Jacobian determinant, which is usually called the Jacobian, is used extensively in mechanics for both rotational and translational coordinate transformations. The Jacobian determinant is defined as being the ratio of the n-dimensional volume element $dx_1 dx_2 ... dx_n$ in one coordinate system, to the volume element $dy_1 dy_2 ... dy_n$ in the second coordinate system. That is

$$J(y_1 y_2 ... y_n) \equiv \frac{\partial x_1 \partial x_2 ... \partial x_n}{\partial y_1 \partial y_2 ... \partial y_n} = \begin{vmatrix} \frac{\partial x_1}{\partial y_1} & \frac{\partial x_1}{\partial y_2} & \cdots & \frac{\partial x_1}{\partial y_n} \\ \frac{\partial x_2}{\partial y_1} & \frac{\partial x_2}{\partial y_2} & \cdots & \frac{\partial x_2}{\partial y_n} \\ \vdots & \vdots & \vdots & \vdots \\ \frac{\partial x_n}{\partial y_1} & \frac{\partial x_n}{\partial y_2} & \cdots & \frac{\partial x_n}{\partial y_n} \end{vmatrix} \qquad (F.14)$$

F.3.1 Transformation of integrals:

Consider a coordinate transformation for the integral of the function $f(x_1, x_2, ..x_n)$ to the integral of a function $g(y_1, y_2, ... y_n)$ where $y_i = h(x_1, x_2, ... x_n)$. The coordinate transformation of the integral equation can be expressed in terms of the Jacobian $J(y_1 y_2 ... y_n)$

$$\int f(x_1, x_2, ..x_n) dx_1 dx_2 ... dx_n = \int g(y_1, y_2, ... y_n) dy_1 dy_2 ... dy_n = \qquad (F.15)$$

$$\int f(x_1, x_2, ..x_n) \frac{\partial x_1 \partial x_2 ... \partial x_n}{\partial y_1 \partial y_2 ... \partial y_n} dy_1 dy_2 ... dy_n = \int f(y_1, y_2, ..y_n) J(y_1, y_2, ... y_n) dy_1 dy_2 ... dy_n$$

F.3.2 Transformation of differential equations:

The differential cross sections for scattering can be defined either by the number of a definite kind of particle/per event, going into the volume element in momentum space $dp_1 dp_2 dp_3$, or by the number going into the solid angle element having momentum between p and $p + dp$. That is, the first definition can be written as a differential equation

$$\frac{\partial^3 S(p_1, p_2, p_3)}{\partial p_1 \partial p_2 \partial p_3} dp_1 dp_2 dp_3 = \frac{\partial^3 S(p_1(p\theta\phi), p_2(p\theta\phi), p_3(p\theta\phi))}{\partial p_1 \partial p_2 \partial p_3} \frac{\partial(p_1, p_2, p_3)}{\partial(p, \theta, \phi)} dp d\theta d\phi \qquad (F.16)$$

As shown in table $C.4$, $dp_1 dp_2 dp_3 = p^2 \sin\theta dp d\theta d\phi$, that is, the Jacobian equals $p^2 \sin\theta$. Thus equation $F.16$ can be written as

$$\frac{\partial^3 S(p_1, p_2, p_3)}{\partial p_1 \partial p_2 \partial p_3} dp_1 dp_2 dp_3 = \left[\frac{\partial^3 S}{\partial p_1 \partial p_2 \partial p_3} p^2 \right] (\sin\theta dp d\theta d\phi) = \frac{\partial^2 \sigma(p, \theta, \phi)}{\partial p \partial \Omega} dp d\Omega \qquad (F.17)$$

The differential cross section is defined by

$$\frac{\partial^2 \sigma(p, \theta, \phi)}{\partial p \partial \Omega} \equiv \frac{\partial^3 S}{\partial p_1 \partial p_2 \partial p_3} p^2 \qquad (F.18)$$

where the p^2 factor is absorbed into the cross section and the solid angle term is factored out

F.3.3 Properties of the Jacobian:

In classical mechanics the Jacobian often is extended from 3 dimensions to n-dimensional transformations. The Jacobian is unity for unitary transformations such as rotations and linear translations which implies that the volume element is preserved. It will be shown that this also is true for a certain class of transformations in classical mechanics that are called canonical transformations. The Jacobian transforms the local density to be correct for any scale transformations such as transforming linear dimensions from centimeters to inches.

F.1 *Example: Jacobian for transform from cartesian to spherical coordinates*

Consider the transform in the three-dimensional integral $\int f(x_1, x_2, x_3) dx_1 dx_2 dx_3$ under transformation from cartesian coordinates (x_1, x_2, x_3) to spherical coordinates (r, θ, ϕ). The transformation is governed by

the geometric relations $x_1 = r \sin\theta \cos\phi$, $x_2 = r \sin\theta \sin\phi$, $x_3 = r \cos\theta$. *For this transformation the Jacobian determinant equals*

$$J(r,\theta,\phi) = \begin{vmatrix} \sin\theta\cos\phi & r\cos\theta\cos\phi & -r\sin\theta\sin\phi \\ \sin\theta\sin\phi & r\cos\theta\sin\phi & r\sin\theta\cos\phi \\ \cos\theta & -r\sin\theta & 0 \end{vmatrix} = r^2 \sin\theta$$

Thus the three-dimensional volume integral transforms to

$$\int f(x_1, x_2, x_3)dx_1 dx_2 dx_3 = \int f(r,\theta,\phi)J(r,\theta,\phi)drd\theta d\phi = \int f(r,\theta,\phi)r^2 \sin\theta drd\theta d\phi$$

which is the well-known volume integral in spherical coordinates.

F.4   Legendre transformation

Hamiltonian mechanics can be derived directly from Lagrange mechanics by considering the Legendre transformation between the conjugate variables $(\mathbf{q}, \dot{\mathbf{q}}, t)$ and $(\mathbf{q}, \mathbf{p}, t)$. Such a derivation is of considerable importance in that it shows that Hamiltonian mechanics is based on the same variational principles as those used to derive Lagrangian mechanics; that is d'Alembert's Principle or Hamilton's Principle. The general problem of converting Lagrange's equations into the Hamiltonian form hinges on the inversion of equation (8.3) that defines the generalized momentum $\mathbf{p}$. This inversion is simplified by the fact that (8.3) is the first partial derivative of the Lagrangian $L(\mathbf{q}, \dot{\mathbf{q}}, \mathbf{t})$ which is a scalar function.

Consider transformations between two functions $F(\mathbf{u}, \mathbf{w})$ and $G(\mathbf{v}, \mathbf{w})$ where $\mathbf{u}$ and $\mathbf{v}$ are the active variables related by the functional form

$$\mathbf{v} = \boldsymbol{\nabla}_{\mathbf{u}}F(\mathbf{u}, \mathbf{w}) \tag{F.19}$$

and where $\mathbf{w}$ designates passive variables and $\boldsymbol{\nabla}_{\mathbf{u}}F(\mathbf{u}, \mathbf{w})$ is the first-order derivative of $F(\mathbf{u}, \mathbf{w})$, i.e. the gradient, with respect to the components of the vector $\mathbf{u}$. The Legendre transform states that the inverse formula can always be written in the form

$$\mathbf{u} = \boldsymbol{\nabla}_{\mathbf{v}}G(\mathbf{v}, \mathbf{w}) \tag{F.20}$$

where the function $G(\mathbf{v}, \mathbf{w})$ is related to $F(\mathbf{u}, \mathbf{w})$ by the symmetric relation

$$G(\mathbf{v}, \mathbf{w}) + \mathbf{F}(\mathbf{u}, \mathbf{w}) = \mathbf{u} \cdot \mathbf{v} \tag{F.21}$$

and where the scalar product $\mathbf{u} \cdot \mathbf{v} = \sum_{i=1}^{N} u_i v_i$.

Furthermore the derivatives with respect to all the passive variables $\{w_i\}$ are related by

$$\boldsymbol{\nabla}_{\mathbf{w}}F(\mathbf{u}, \mathbf{w}) = -\boldsymbol{\nabla}_{\mathbf{w}}G(\mathbf{v}, \mathbf{w}) \tag{F.22}$$

The relationship between the functions $F(\mathbf{u}, \mathbf{w})$ and $G(\mathbf{v}, \mathbf{w})$ is symmetrical and each is said to be the Legendre transform of the other.

Workshop exercises

1. Below you will find a set of integrals. Your teaching assistant will divide you into groups and each group will be assigned one integral to work on. Once your group has solved the integral, write the solution on the board in the space provided by the teaching assistant.

   (a) $\int_0^{2\pi} \int_0^{\pi/4} \int_0^{\cos\theta} r^2 \sin\theta drd\theta d\phi$

   (b) $\int \left(\frac{\dot{r}}{r} - \frac{r\dot{r}}{r^2} \right) dt$

   (c) $\int_S \mathbf{A} \cdot d\mathbf{a}$ where $\mathbf{A} = x\hat{\imath} + y\hat{\jmath} + z\hat{\mathbf{k}}$ and S is the sphere $x^2 + y^2 + z^2 = 9$.

   (d) $\int_S (\boldsymbol{\nabla} \times \mathbf{A}) \cdot d\mathbf{a}$ where $\mathbf{A} = y\hat{\imath} + z\hat{\jmath} + x\hat{\mathbf{k}}$ and S is the surface defined by the paraboloid $z = 1 - x^2 - y^2$, where $z \geq 0$.

Appendix G

Vector differential calculus

This appendix reviews vector differential calculus which is used extensively in both classical mechanics and electromagnetism.

G.1 Scalar differential operators

G.1.1 Scalar field

Differential operators like time $\left(\frac{d}{dt}\right)$ do not change the rotational properties of scalars or proper vectors. A scalar operator $\frac{d}{ds}$ acting on a scalar field $\phi(xyz)$, in a rotated coordinated frame $\phi'(x'y'z')$ is unchanged.

$$\frac{d\phi'}{ds} = \frac{d\phi}{ds} \tag{G.1}$$

G.1.2 Vector field

Similarly for a proper vector field

$$\frac{dA'_i}{ds} = \sum_j \lambda_{ij} \frac{dA_j}{ds} \tag{G.2}$$

That is, differentiation of scalar or vector fields with respect to a scalar operator does not change the rotational behavior. In particular, the scalar differentials of vectors continue to obey the rules of ordinary proper vectors. The scalar operator $\frac{\partial}{\partial t}$ is used for calculation of velocity or acceleration.

G.2 Vector differential operators in cartesian coordinates

Vector differential operators, such as the gradient operator, are important in physics. The action of vector operators differ along different orthogonal axes.

G.2.1 Scalar field

Consider a continuous, single-valued scalar function $\phi(x_i, x_j, x_k)$. Since

$$\phi' = \phi \tag{G.3}$$

then the partial differential with respect to one component x_i of the vector $\mathbf{x}'$ gives

$$\frac{\partial \phi'}{\partial x'_i} = \sum_j \frac{\partial \phi}{\partial x_j} \frac{\partial x_j}{\partial x'_i} \tag{G.4}$$

The inverse rotation gives that

$$x_j = \sum_k \lambda_{kj} x'_k \tag{G.5}$$

543

Therefore

$$\frac{\partial x_j}{\partial x_i'} = \sum_k \lambda_{kj} \frac{\partial x_k'}{\partial x_i'} = \sum_k \lambda_{kj} \delta_{ik} = \lambda_{ij} \tag{G.6}$$

Thus

$$\frac{\partial \phi'}{\partial x_i'} = \sum_j \lambda_{ij} \frac{\partial \phi}{\partial x_j} \tag{G.7}$$

That is the vector derivative acting of a scalar field transforms like a proper vector.

Define the gradient, or $\boldsymbol{\nabla}$ operator, as

$$\boldsymbol{\nabla} \equiv \sum_i \widehat{\boldsymbol{e}}_i \frac{\partial}{\partial x_i} \tag{G.8}$$

where $\widehat{\boldsymbol{e}}_i$ is the unit vector along the x_i axis. In cartesian coordinates, the del vector operator is,

$$\boldsymbol{\nabla} \equiv \widehat{\mathbf{i}} \frac{\partial}{\partial x} + \widehat{\mathbf{j}} \frac{\partial}{\partial y} + \widehat{\mathbf{k}} \frac{\partial}{\partial z} \tag{G.9}$$

The gradient was applied to the gravitational and electrostatic potential to derive the corresponding field. For example, for electrostatics it was shown that the gradient of the scalar electrostatic potential field V can be written in cartesian coordinates as

$$\mathbf{E} = -\boldsymbol{\nabla} V \tag{G.10}$$

Note that the gradient of a scalar field produces a vector field. You are familiar with this if you are a skier in that the gravitational force pulls you down the line of steepest descent for the ski slope.

G.2.2   Vector field

Another possible operation for the del operator is the scalar product with a vector. Using the definition of a scalar product in cartesian coordinates gives

$$\boldsymbol{\nabla} \cdot \mathbf{A} = \widehat{\mathbf{i}} \cdot \widehat{\mathbf{i}} \frac{\partial A_x}{\partial x} + \widehat{\mathbf{j}} \cdot \widehat{\mathbf{j}} \frac{\partial A_y}{\partial y} + \widehat{\mathbf{k}} \cdot \widehat{\mathbf{k}} \frac{\partial A_z}{\partial z} = \frac{\partial A_x}{\partial x} + \frac{\partial A_y}{\partial y} + \frac{\partial A_z}{\partial z} \tag{G.11}$$

This scalar derivative of a vector field is called the divergence. Note that the scalar product produces a scalar field which is invariant to rotation of the coordinate axes.

The vector product of the del operator with another vector, is called the curl which is used extensively in physics. It can be written in the determinant form

$$\boldsymbol{\nabla} \times \mathbf{A} = \begin{vmatrix} \widehat{\mathbf{i}} & \widehat{\mathbf{j}} & \widehat{\mathbf{k}} \\ \frac{\partial}{\partial x} & \frac{\partial}{\partial y} & \frac{\partial}{\partial z} \\ A_x & A_y & A_z \end{vmatrix} \tag{G.12}$$

By contrast to the scalar product, both the gradient of a scalar field, and the vector product, are vector fields for which the components along the coordinate axes transform in a specific manner, such as to keep the length of the vector constant, as the coordinate frame is rotated. The gradient, scalar and vector products with the $\boldsymbol{\nabla}$ operator are the first order derivatives of fields that occur most frequently in physics.

Second derivatives of fields also are used. Let us consider some possible combinations of the product of two del operators.

1) $\boldsymbol{\nabla} \cdot (\boldsymbol{\nabla} V) = \nabla^2 V$

The scalar product of two del operators is a scalar under rotation. Evaluating the scalar product in cartesian coordinates gives

$$\left(\widehat{\mathbf{i}} \frac{\partial}{\partial x} + \widehat{\mathbf{j}} \frac{\partial}{\partial y} + \widehat{\mathbf{k}} \frac{\partial}{\partial z} \right) \cdot \left(\widehat{\mathbf{i}} \frac{\partial V}{\partial x} + \widehat{\mathbf{j}} \frac{\partial V}{\partial y} + \widehat{\mathbf{k}} \frac{\partial V}{\partial z} \right) = \frac{\partial^2 V}{\partial x^2} + \frac{\partial^2 V}{\partial y^2} + \frac{\partial^2 V}{\partial z^2} \tag{G.13}$$

This also can be obtained without confusion by writing this product as;

$$\boldsymbol{\nabla} \cdot (\boldsymbol{\nabla} V) = \boldsymbol{\nabla} \cdot \boldsymbol{\nabla} V = (\boldsymbol{\nabla} \cdot \boldsymbol{\nabla}) V \tag{G.14}$$

where the scalar product of the del operator is a scalar, called the Laplacian ∇^2, given by

$$\boldsymbol{\nabla} \cdot \boldsymbol{\nabla} = \nabla^2 \equiv \frac{\partial^2}{\partial x^2} + \frac{\partial^2}{\partial y^2} + \frac{\partial^2}{\partial z^2} \tag{G.15}$$

The Laplacian operator is encountered frequently in physics.

2) $\boldsymbol{\nabla} \times (\boldsymbol{\nabla} V) = 0$
Note that the vector product of two identical vectors

$$\mathbf{A} \times \mathbf{A} = 0 \tag{G.16}$$

Therefore

$$\boldsymbol{\nabla} \times (\boldsymbol{\nabla} V) = 0 \tag{G.17}$$

This can be confirmed by evaluating the separate components along each axis.

3) $\boldsymbol{\nabla} \cdot (\boldsymbol{\nabla} \times \mathbf{A}) = 0$
This is zero because the cross-product is perpendicular to $\boldsymbol{\nabla} \times \mathbf{A}$ and thus the dot product is zero.

4) $\boldsymbol{\nabla} \times (\boldsymbol{\nabla} \times \mathbf{A}) = \boldsymbol{\nabla} \cdot (\boldsymbol{\nabla} \cdot \mathbf{A}) - \nabla^2 \mathbf{A}$
The identity

$$\mathbf{A} \times (\mathbf{B} \times \mathbf{C}) = \mathbf{B} (\mathbf{A} \cdot \mathbf{C}) - (\mathbf{A} \cdot \mathbf{B}) \mathbf{C} \tag{G.18}$$

can be used to give

$$\boldsymbol{\nabla} \times (\boldsymbol{\nabla} \times \mathbf{A}) = \boldsymbol{\nabla} \cdot (\boldsymbol{\nabla} \cdot \mathbf{A}) - \nabla^2 \mathbf{A} \tag{G.19}$$

since $\boldsymbol{\nabla} \cdot \boldsymbol{\nabla} = \nabla^2$.

There are pitfalls in the discussion of second derivatives in that it is assumed that both del operators operate on the same variable, otherwise the results are different.

G.3 Vector differential operators in curvilinear coordinates

As discussed in Appendix C there are many situations where the symmetries make it more convenient to use orthogonal curvilinear coordinate systems rather than cartesian coordinates. Thus it is necessary to extend vector derivatives from cartesian to curvilinear coordinates. Table C.1 can be used for expressing vector derivatives in curvilinear coordinate systems.

G.3.1 Gradient:

The gradient in curvilinear coordinates is

$$\boldsymbol{\nabla} f = \frac{1}{h_1} \frac{\partial f}{\partial q_1} \hat{\mathbf{q}}_1 + \frac{1}{h_2} \frac{\partial f}{\partial q_2} \hat{\mathbf{q}}_2 + \frac{1}{h_3} \frac{\partial f}{\partial q_3} \hat{\mathbf{q}}_3 \tag{G.20}$$

where the coefficients h_i are listed in table C.1.

For cylindrical coordinates this becomes

$$\boldsymbol{\nabla} f = \frac{\partial f}{\partial \rho} \hat{\boldsymbol{\rho}} + \frac{1}{\rho} \frac{\partial f}{\partial \varphi} \hat{\boldsymbol{\varphi}} + \frac{\partial f}{\partial z} \hat{\mathbf{z}} \tag{G.21}$$

In spherical coordinates

$$\boldsymbol{\nabla} f = \frac{\partial f}{\partial r} \hat{\mathbf{r}} + \frac{1}{r} \frac{\partial f}{\partial \theta} \hat{\boldsymbol{\theta}} + \frac{1}{r \sin \theta} \frac{\partial f}{\partial \varphi} \hat{\boldsymbol{\varphi}} \tag{G.22}$$

G.3.2   Divergence:

The divergence can be expressed as

$$\boldsymbol{\nabla} \cdot \mathbf{A} = \frac{1}{h_1 h_2 h_3} \left[\frac{\partial}{\partial q_1} \left(A_1 h_2 h_3 \right) + \frac{\partial}{\partial q_2} \left(A_2 h_3 h_1 \right) + \frac{\partial}{\partial q_3} \left(A_3 h_1 h_2 \right) \right] \tag{G.23}$$

In cylindrical coordinates the divergence is

$$\boldsymbol{\nabla} \cdot \mathbf{A} = \frac{1}{\rho} \frac{\partial}{\partial \rho} \left(\rho A_\rho \right) + \frac{1}{\rho} \frac{\partial A_\varphi}{\partial \varphi} + \frac{\partial A_z}{\partial z} = \frac{A_\rho}{\rho} + \frac{\partial A_\rho}{\partial \rho} + \frac{1}{\rho} \frac{\partial A_\varphi}{\partial \varphi} + \frac{\partial A_z}{\partial z} \tag{G.24}$$

In spherical coordinates the divergence is

$$\boldsymbol{\nabla} \cdot \mathbf{A} = \frac{1}{r^2 \sin \theta} \left[\frac{\partial}{\partial r} \left(A_r r^2 \sin \theta \right) + \frac{\partial}{\partial \theta} \left(A_\theta r \sin \theta \right) + \frac{\partial}{\partial \varphi} \left(A_\varphi r \right) \right] \tag{G.25}$$

G.3.3   Curl:

$$\boldsymbol{\nabla} \times \mathbf{A} = \frac{1}{h_1 h_2 h_3} \begin{vmatrix} h_1 \hat{\mathbf{q}}_1 & h_2 \hat{\mathbf{q}}_2 & h_3 \hat{\mathbf{q}}_3 \\ \frac{\partial}{\partial q_1} & \frac{\partial}{\partial q_2} & \frac{\partial}{\partial q_3} \\ h_1 A_1 & h_2 A_2 & h_3 A_3 \end{vmatrix} \tag{G.26}$$

In cylindrical coordinates the curl is

$$\boldsymbol{\nabla} \times \mathbf{A} = \frac{1}{\rho} \begin{vmatrix} \hat{\boldsymbol{\rho}} & \rho \hat{\boldsymbol{\varphi}} & \hat{\mathbf{z}} \\ \frac{\partial}{\partial \rho} & \frac{\partial}{\partial \varphi} & \frac{\partial}{\partial z} \\ A_\rho & \rho A_\varphi & A_z \end{vmatrix} \tag{G.27}$$

In spherical coordinates the curl is

$$\boldsymbol{\nabla} \times \mathbf{A} = \frac{1}{r^2 \sin \theta} \begin{vmatrix} \hat{\mathbf{r}} & r \hat{\boldsymbol{\theta}} & r \sin \theta \hat{\boldsymbol{\varphi}} \\ \frac{\partial}{\partial r} & \frac{\partial}{\partial \theta} & \frac{\partial}{\partial \varphi} \\ A_r & r \rho A_\theta & r \sin \theta A_\varphi \end{vmatrix} \tag{G.28}$$

G.3.4   Laplacian:

Taking the divergence of the gradient of a scalar gives

$$\nabla^2 f = \boldsymbol{\nabla} \cdot \boldsymbol{\nabla} f = \frac{1}{h_1 h_2 h_3} \left[\frac{\partial}{\partial q_1} \left(\frac{h_2 h_3}{h_1} \frac{\partial f}{\partial q_1} \right) + \frac{\partial}{\partial q_2} \left(\frac{h_3 h_1}{h_2} \frac{\partial f}{\partial q_2} \right) + \frac{\partial}{\partial q_3} \left(\frac{h_1 h_2}{h_3} \frac{\partial f}{\partial q_3} \right) \right] \tag{G.29}$$

The Laplacian of a scalar function f in cylindrical coordinates is

$$\nabla^2 f = \frac{1}{\rho} \frac{\partial}{\partial \rho} \left(\rho \frac{\partial f}{\partial \rho} \right) + \frac{1}{\rho^2} \frac{\partial^2 f}{\partial \varphi^2} + \frac{\partial^2 f}{\partial z^2} \tag{G.30}$$

The Laplacian of a scalar function f in spherical coordinates is

$$\nabla^2 f = \frac{1}{r^2} \frac{\partial}{\partial r} \left(r^2 \frac{\partial f}{\partial r} \right) + \frac{1}{r^2 \sin \theta} \frac{\partial}{\partial \theta} \left(\sin \theta \frac{\partial f}{\partial \theta} \right) + \frac{1}{r^2 \sin \theta} \frac{\partial^2 f}{\partial \varphi^2} \tag{G.31}$$

The gradient, divergence, curl and Laplacian are used extensively in curvilinear coordinate systems when dealing with vector fields in Newtonian mechanics, electromagnetism, and fluid flow.

Appendix H

Vector integral calculus

Field equations, such as for electromagnetic and gravitational fields, require both line integrals, and surface integrals, of vector fields to evaluate potential, flux and circulation. These require use of the gradient, the Divergence Theorem and Stokes Theorem which are discussed in the following sections.

H.1   Line integral of the gradient of a scalar field

The change ΔV in a scalar field for an infinitessimal step $d\mathbf{l}$ along a path can be written as

$$\Delta V = (\boldsymbol{\nabla} V) \cdot d\mathbf{l} \tag{H.1}$$

since the gradient of V, that is, $\boldsymbol{\nabla} V$, is the rate of change of V with $d\mathbf{l}$. Discussions of gravitational and electrostatic potential show that the line integral between points a and b is given in terms of the del operator by

$$V_b - V_a = \int_a^b (\boldsymbol{\nabla} V) \cdot d\mathbf{l} \tag{H.2}$$

This relates the difference in values of a scalar field at two points to the line integral of the dot product of the gradient with the element of the line integral.

H.2   Divergence theorem

H.2.1   Flux of a vector field for Gaussian surface

Consider the flux Φ of a vector field $\mathbf{F}$ for a closed surface, usually called a Gaussian surface, S shown in figure $H.1$.

$$\Phi = \oint_S \mathbf{F} \cdot d\mathbf{S} \tag{H.3}$$

If the enclosed volume is cut in to two pieces enclosed by surfaces $S_1 = S_a + S_{ab}$ and $S_2 = S_b + S_{ab}$. The flux through the surface S_{ab} common to both S_1 and S_2 are equal and in the same direction. Then the net flux through the sum of S_1 and S_2 is given by

$$\oint_{S_1} \mathbf{F} \cdot d\mathbf{S} + \oint_{S_2} \mathbf{F} \cdot d\mathbf{S} = \oint_S \mathbf{F} \cdot d\mathbf{S} \tag{H.4}$$

since the contributions of the common surface S_{ab} cancel in that the flux out of S_1 is equal and opposite to the flux into S_2 over the surface S_{ab}. That is, independent of how many times the volume enclosed by S is subdivided, the net flux for the sum of all the Gaussian surfaces enclosing these subdivisions of the volume, still equals $\oint_S \mathbf{F} \cdot d\mathbf{S}$.

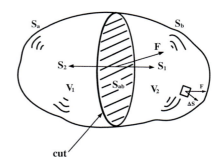

Figure H.1: A volume V enclosed by a closed surface S is cut into two pieces at the surface S_{ab}. This gives V_1 enclosed by S_1 and V_1 enclosed by S_2.

Consider that the volume enclosed by S is subdivided into N subdivisions where $N \to \infty$, then even though $\oint_{S_i} \mathbf{F} \cdot d\mathbf{S} \to 0$ as $N \to \infty$, the sum over surfaces of all the infinitessimal volumes remains unchanged

$$\Phi = \oint_S \mathbf{F} \cdot d\mathbf{S} = \sum_i^{N \to \infty} \oint_{S_i} \mathbf{F} \cdot d\mathbf{S} \tag{H.5}$$

Thus we can take the limit of a sum of an infinite number of infinitessimal volumes as is needed to obtain a differential form. The surface integral for each infinitessimal volume will equal zero which is not useful, that is $\oint_{S_i} \mathbf{F} \cdot d\mathbf{S} \to 0$ as $N \to \infty$. However, the flux per unit volume has a finite value as $N \to \infty$. This ratio is called the *divergence* of the vector field;

$$div\mathbf{F} = Lim_{\Delta \tau_i \to 0} \frac{\oint_{S_i} \mathbf{F} \cdot d\mathbf{S}}{\Delta \tau_i} \tag{H.6}$$

where $\Delta \tau_i$ is the infinitessimal volume enclosed by surface S_i. The divergence of the vector field is a scalar quantity.

Thus the sum of flux over all infinitessimal subdivisions of the volume enclosed by a closed surface S equals

$$\Phi = \oint_S \mathbf{F} \cdot d\mathbf{S} = \sum_i^{N \to \infty} \frac{\oint_{S_i} \mathbf{F} \cdot d\mathbf{S}}{\Delta \tau_i} \Delta \tau_i = \sum_i^{N \to \infty} div\mathbf{F} \Delta \tau_i \tag{H.7}$$

In the limit $N \to \infty$, $\Delta \tau_i \to 0$, this becomes the integral;

$$\Phi = \oint_S \mathbf{F} \cdot d\mathbf{S} = \int_{\substack{Enclosed \\ volume}} div\mathbf{F} d\tau \tag{H.8}$$

This is called the *Divergence Theorem* or Gauss's Theorem. To avoid confusion with Gauss's law in electrostatics, it will be referred to as the Divergence theorem.

H.2.2 Divergence in cartesian coordinates.

Consider the special case of an infinitessimal rectangular box, size $\Delta x, \Delta y, \Delta z$ shown in figure $H.2$. Consider the net flux for the z component F_z *entering* the surface $\Delta x \Delta y$ at location (x, y, z).

$$\Delta \Phi_z^{in} = \left(F_z + \frac{\Delta x}{2} \frac{\partial F_z}{\partial x} + \frac{\Delta y}{2} \frac{\partial F_z}{\partial y} \right) \Delta x \Delta y \tag{H.9}$$

The net flux of the z component *out* of the surface at $z + \Delta z$ is

$$\Delta \Phi_z^{out} = \left(F_z + \Delta z \frac{\partial F_z}{\partial z} + \frac{\Delta x}{2} \frac{\partial F_z}{\partial x} + \frac{\Delta y}{2} \frac{\partial F_z}{\partial y} \right) \Delta x \Delta y \tag{H.10}$$

Thus the net flux out of the box due to the z component of F is

$$\Delta \Phi_z = \Delta \Phi_z^{out} - \Delta \Phi_z^{in} = \frac{\partial F_z}{\partial z} \Delta x \Delta y \Delta z \tag{H.11}$$

Adding the similar x and y components for $\Delta \Phi$ gives

$$\Delta \Phi = \left(\frac{\partial F_x}{\partial x} + \frac{\partial F_y}{\partial y} + \frac{\partial F_z}{\partial z} \right) \Delta x \Delta y \Delta z \tag{H.12}$$

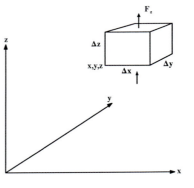

Figure H.2: Computation of flux out of an infinitessimal rectangular box, Δx, Δy, Δz.

This gives that the divergence of the vector field $\mathbf{F}$ is

$$div\mathbf{F} = Lim_{\Delta \tau_i \to 0} \frac{\oint_{S_i} \mathbf{F} \cdot d\mathbf{S}}{\Delta \tau_i} = \left(\frac{\partial F_x}{\partial x} + \frac{\partial F_y}{\partial y} + \frac{\partial F_z}{\partial z} \right) \tag{H.13}$$

since $\Delta\tau = \Delta x \Delta y \Delta z$. But the right hand side of the equation equals the scalar product $\boldsymbol{\nabla}\cdot\mathbf{F}$, that is,

$$div\mathbf{F} = \boldsymbol{\nabla}\cdot\mathbf{F} \tag{H.14}$$

The divergence is a scalar quantity. The physical meaning of the divergence is that it gives the net flux per unit volume flowing out of an infinitessimal volume. A positive divergence corresponds to a net outflow of flux from the infinitessimal volume at any location while a negative divergence implies a net inflow of flux to this infinitessimal volume.

It was shown that for an infinitessimal rectangular box

$$\Delta\Phi = \left(\frac{\partial F_x}{\partial x} + \frac{\partial F_y}{\partial y} + \frac{\partial F_z}{\partial z}\right)\Delta x \Delta y \Delta z = \boldsymbol{\nabla}\cdot\mathbf{F}\Delta\tau \tag{H.15}$$

Integrating over the finite volume enclosed by the surface S gives

$$\Phi = \oint_S \mathbf{F}\cdot d\mathbf{S} = \int_{\substack{Enclosed \\ volume}} \boldsymbol{\nabla}\cdot\mathbf{F}d\tau \tag{H.16}$$

This is another way of expressing the Divergence theorem

$$\Phi = \oint_S \mathbf{F}\cdot d\mathbf{S} = \int_{\substack{Enclosed \\ volume}} div\mathbf{F}d\tau \tag{H.17}$$

The divergence theorem, developed by Gauss, is of considerable importance, it relates the surface integral of a vector field, that is, the outgoing flux, to a volume integral of $\boldsymbol{\nabla}\cdot\mathbf{F}$ over the enclosed volume.

H.1 *Example: Maxwell's Flux Equations*

As an example of the usefulness of this relation, consider the Gauss's law for the flux in Maxwell's equations.

Gauss' Law for the electric field

$$\Phi_E = \oint_{\substack{Closed \\ surface}} E\cdot dS = \frac{1}{\varepsilon_0}\int_{\substack{enclosed \\ volume}} \rho d\tau$$

But the divergence relation gives that

$$\Phi_E = \oint_S \mathbf{E}\cdot d\mathbf{S} = \int_{\substack{Enclosed \\ volume}} \boldsymbol{\nabla}\cdot\mathbf{E}d\tau$$

Combining these gives

$$\oint_{\substack{Closed \\ surface}} \mathbf{E}\cdot d\mathbf{S} = \int_{\substack{Enclosed \\ volume}} \boldsymbol{\nabla}\cdot\mathbf{E}d\tau = \frac{1}{\varepsilon_0}\int_{\substack{enclosed \\ volume}} \rho d\tau$$

This is true independent of the shape of the surface or enclosed volume, leading to the differential form of Maxwell's first law, that is Gauss's law for the electric field.

$$\nabla\cdot E = \frac{\rho}{\varepsilon_0}$$

The differential form of Gauss's law relates $\boldsymbol{\nabla}\cdot\mathbf{E}$ to the charge density ρ at that same location. This is much easier to evaluate than a surface and volume integral required using the integral form of Gauss's law.

Gauss's law for magnetism

$$\Phi_B = \oint_{\substack{Closed \\ surface}} \mathbf{B}\cdot d\mathbf{S} = 0$$

Using the divergence theorem gives that

$$\Phi_B = \oint_{\substack{Closed \\ surface}} \mathbf{B}\cdot d\mathbf{S} = \int_{\substack{Enclosed \\ volume}} \boldsymbol{\nabla}\cdot\mathbf{B}d\tau = 0$$

This is true independent of the shape of the Gaussian surface leading to the differential form of Gauss's law for **B**

$$\mathbf{\nabla} \cdot \mathbf{B} = 0$$

That is, the local value of the divergence of **B** *is zero everywhere.*

H.2 *Example: Buoyancy forces in fluids*

Buoyancy in fluids provides an example of the use of flux in physics. Consider a fluid of density $\rho(z)$ in a gravitational field $\bar{g}(z) = -g(z)\hat{z}$ where the z axis points in the opposite direction to the gravitational force. Pressure equals force per unit area and is a scalar quantity. For a conservative fluid system, in static equilibrium, the net work done per unit area for an infinitessimal displacement dr is zero. The net pressure force per unit area is the difference $P(r+dr) - P(r) = \nabla P \cdot dr$ while the net change in gravitational potential energy is $\rho(z)\bar{g}(z) \cdot dr$. Thus energy conservation gives

$$[\mathbf{\nabla} P + \rho(z)\bar{\mathbf{g}}(\mathbf{z})] \cdot d\mathbf{r} = 0$$

which can be expanded as

$$\frac{dP}{dz} = -\rho(z)g(z) \qquad (A)$$
$$\frac{dP}{dx} = \frac{dP}{dy} = 0$$

Integrating the net forces normal to the surface over any closed surface enclosing an empty volume, inside the fluid, gives a net buoyancy force on this volume that simplifies using the Divergence theorem

$$\oint \mathbf{F} \cdot d\mathbf{S} = \oint P d\hat{\mathbf{S}} \cdot d\mathbf{S} = \oint P dS = \int_{\substack{Enclosed \\ vol}} \left(\frac{dP}{dx} + \frac{dP}{dy} + \frac{dP}{dz} \right) d\tau$$

Using equations A leads to the net buoyancy force

$$\oint \mathbf{F} \cdot d\mathbf{S} = \int_{\substack{Enclosed \\ vol}} \frac{dP}{dz} d\tau = - \int_{\substack{Enclosed \\ vol}} \rho(z)g(z) d\tau$$

The right hand side of this equation equals minus the weight of the displaced fluid. That is, the buoyancy force equals the weight of the fluid displaced by the empty volume. Note that this proof applies both to compressible fluids, where the density depends on pressure, as well as to incompressible fluids where the density is constant. It also applies to situations where local gravity g is position dependent. If an object of mass M is completely submerged then the net force on the object is $Mg - \int_{\substack{Enclosed \\ vol}} \rho(z)g(z)d\tau$. If the object floats on the surface of a fluid then the buoyancy force must be calculated separately for the volume under the fluid surface and the upper volume above the fluid surface. The buoyancy due to displaced air usually is negligible since the density of air is about 10^{-3} times that of fluids such as water.

H.3   Stokes Theorem

H.3.1   The curl

Maxwell's laws relate the circulation of the field around a closed loop to the rate of change of flux through the surface bounded by the closed loop. It is possible to write these integral equations in a differential form as follows.

Consider the line integral around a closed loop C shown in figure $H.3$.

If this area is subdivided into two areas enclosed by loops C_1 and C_2, then the sum of the line integrals is the same

$$\oint_C \mathbf{F} \cdot d\mathbf{l} = \oint_{C_1} \mathbf{F} \cdot d\mathbf{l} + \oint_{C_2} \mathbf{F} \cdot d\mathbf{l} \qquad (H.18)$$

because the contributions along the common boundary cancel since they are taken in opposite directions if C_1 and C_2 both are taken in the same direction. Note that the line integral, and corresponding enclosed area, are vector quantities related by the right-hand rule and this must be taken into account when subdividing the area. Thus the area can be subdivided into an infinite number of pieces for which

$$\oint_C \mathbf{F} \cdot dl = \sum_i^{N \to \infty} \oint_{C_i} \mathbf{F} \cdot dl = \sum_i^{N \to \infty} \frac{\oint_{C_i} \mathbf{F} \cdot dl}{\Delta \mathbf{S}_i \cdot \widehat{\mathbf{n}}} \Delta \mathbf{S}_i \cdot \widehat{\mathbf{n}} \tag{H.19}$$

where $\Delta \mathbf{S}_i$ is the infinitesimal area bounded by the closed sub-loop C_i and $\Delta \mathbf{S}_i \cdot \widehat{\mathbf{n}}$ is the normal component of this area pointing along the $\widehat{\mathbf{n}}$ direction which is the direction along which the line integral points.

The component of the curl of the vector function along the direction $\widehat{\mathbf{n}}$ is defined to be

$$(curl\mathbf{F}) \cdot \widehat{\mathbf{n}} \equiv Lim_{\Delta S \to 0} \sum_i^{N \to \infty} \frac{\oint_{C_i} \mathbf{F} \cdot dl}{\Delta \mathbf{S}_i \cdot \widehat{\mathbf{n}}} \tag{H.20}$$

Thus the line integral can be written as

$$\oint_C \mathbf{F} \cdot dl = \sum_i^{N \to \infty} \frac{\oint_{C_i} \mathbf{F} \cdot dl}{\Delta \mathbf{S}_i \cdot \widehat{\mathbf{n}}} \Delta \mathbf{S}_i \cdot \widehat{\mathbf{n}} \tag{H.21}$$

$$= \int [(curl\mathbf{F}) \cdot \widehat{\mathbf{n}}] \, d\mathbf{S}_i \cdot \widehat{\mathbf{n}}$$

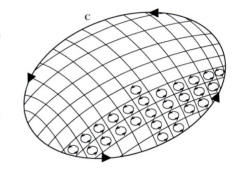

The product $\widehat{\mathbf{n}} \cdot \widehat{\mathbf{n}} = 1$, that is, this is true independent of the direction of the infinitessimal loop. Thus the above relation leads to *Stokes Theorem*

$$\oint_C \mathbf{F} \cdot dl = \int_{\substack{Area \\ bounded \\ by \\ C}} (curl\mathbf{F}) \cdot d\mathbf{S} \tag{H.22}$$

Figure H.3: The circulation around a path is equal to the sum of the circulations around subareas made by subdividing the area.

This relates the line integral to a surface integral over a surface bounded by the loop.

H.3.2   Curl in cartesian coordinates

Consider the infinitessimal rectangle $\Delta x \Delta y$ pointing in the $\widehat{\mathbf{k}}$ direction shown in figure $H.4$.

The line integral, taken in a right-handed way around $\widehat{\mathbf{k}}$ gives

$$\oint_C \mathbf{F} \cdot dl = F_x \Delta x + \left(F_y + \frac{\partial F_y}{\partial x} \Delta x \right) - \left(F_x + \frac{\partial F_x}{\partial y} \Delta y \right) - F_y \Delta y = \left(\frac{\partial F_y}{\partial x} - \frac{\partial F_x}{\partial y} \right) \Delta x \Delta y \tag{H.23}$$

Thus since $\Delta x \Delta y = \Delta \mathbf{S}_z$ the z component of the curl is given by

$$(curl\mathbf{F}) \cdot \widehat{\mathbf{k}} = \frac{\oint_{C_i} \mathbf{F} \cdot dl}{\Delta \mathbf{S}_i \cdot \widehat{\mathbf{n}}} = \left(\frac{\partial F_y}{\partial x} - \frac{\partial F_x}{\partial y} \right) \tag{H.24}$$

The same argument for the component of the curl in the y direction is given by

$$(curl\mathbf{F}) \cdot \widehat{\mathbf{j}} = \left(\frac{\partial F_x}{\partial z} - \frac{\partial F_z}{\partial x} \right) \tag{H.25}$$

Similarly the same argument for the component of the curl in the x direction is given by

$$(curl\mathbf{F}) \cdot \widehat{\mathbf{i}} = \left(\frac{\partial F_z}{\partial y} - \frac{\partial F_y}{\partial z} \right) \tag{H.26}$$

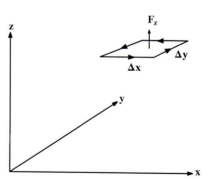

Figure H.4: Circulation around an infinitessimal rectangle $\Delta x \Delta y$ in the z direction.

Thus combining the three components of the curl gives

$$curl\mathbf{F} = \left(\frac{\partial F_z}{\partial y} - \frac{\partial F_y}{\partial z}\right)\widehat{\mathbf{i}} + \left(\frac{\partial F_x}{\partial z} - \frac{\partial F_z}{\partial x}\right)\widehat{\mathbf{j}} + \left(\frac{\partial F_y}{\partial x} - \frac{\partial F_x}{\partial y}\right)\widehat{\mathbf{k}} \qquad (H.27)$$

Note that cross-product of the del operator with the vector $\mathbf{F}$ is

$$\boldsymbol{\nabla} \times \mathbf{F} = \begin{vmatrix} \widehat{\mathbf{i}} & \widehat{\mathbf{j}} & \widehat{\mathbf{k}} \\ \frac{\partial}{\partial x} & \frac{\partial}{\partial y} & \frac{\partial}{\partial z} \\ F_x & F_y & F_z \end{vmatrix} \qquad (H.28)$$

which is identical to the right hand side of the relation for the curl in cartesian coordinates. That is;

$$\boldsymbol{\nabla} \times \mathbf{F} = curl\overrightarrow{\mathbf{F}} \qquad (H.29)$$

Therefore *Stokes Theorem* can be rewritten as

$$\oint_C \mathbf{F} \cdot d\mathbf{l} = \int_{\substack{Area \\ bounded \\ by \\ C}} (curl\mathbf{F}) \cdot d\mathbf{S} = \int_{\substack{Area \\ bounded \\ by \\ C}} (\boldsymbol{\nabla} \times \mathbf{F}) \cdot d\mathbf{S} \qquad (H.30)$$

The physics meaning of the curl is that it is the circulation, or rotation, for an infinitessimal loop at any location. The word curl is German for rotation.

H.3 *Example: Maxwell's circulation equations*

As an example of the use of the curl, consider Faraday's Law

$$\oint_{\substack{Closed \\ loop \\ C}} \mathbf{E} \cdot d\mathbf{l} = -\int_{\substack{surface \\ bounded \\ by \\ C}} \frac{\partial \mathbf{B}}{\partial t} \cdot \partial \mathbf{S}$$

Using Stokes Theorem gives

$$\oint_C \mathbf{E} \cdot d\mathbf{l} = \int_{\substack{Surface \\ bounded \\ by \\ C}} (\boldsymbol{\nabla} \times \mathbf{E}) \cdot d\mathbf{S}$$

These two relations are independent of the shape of the closed loop, thus we obtain Faraday's Law in the differential form

$$(\boldsymbol{\nabla} \times \mathbf{E}) = -\frac{\partial \mathbf{B}}{\partial t}$$

A differential form of the Ampère-Maxwell law also can be obtained from

$$\oint_{\substack{Closed \\ loop \\ C}} \mathbf{B} \cdot d\mathbf{l} = \mu_0 \int_{\substack{Bounded \\ by \\ C}} (\mathbf{j} + \varepsilon_0 \frac{\partial \mathbf{E}}{\partial t}) \cdot d\mathbf{S}$$

Using Stokes Theorem

$$\oint_C \mathbf{B} \cdot d\mathbf{l} = \int_{\substack{Surface \\ bounded \\ by \\ C}} (\boldsymbol{\nabla} \times \mathbf{B}) \cdot d\mathbf{S}$$

Again this is independent of the shape of the loop and thus we obtain
  Ampère-Maxwell law in differential form

$$\boldsymbol{\nabla} \times \mathbf{B} = \mu_0 \mathbf{j} + \mu_0 \varepsilon_0 \frac{\partial \mathbf{E}}{\partial t}$$

The differential forms of Maxwell's circulation relations are easier to apply than the integral equations because the differential form relates the curl to the time derivatives at the same specific location.

H.4 Potential formulations of curl-free and divergence-free fields

Interesting consequences result from the Divergence theorem and Stokes Theorem for vector fields that are either curl-free or divergence-free. In particular two theorems result from the second derivatives of a vector field.

Theorem 1; Curl-free (irrotational) fields:

For curl-free fields

$$\boldsymbol{\nabla} \times \mathbf{F} = 0 \tag{H.31}$$

everywhere. This is automatically obeyed if the vector field is expressed as the gradient of a scalar field

$$\mathbf{F} = \boldsymbol{\nabla}\phi \tag{H.32}$$

since

$$\boldsymbol{\nabla} \times (\boldsymbol{\nabla}\phi) = 0 \tag{H.33}$$

That is, any curl-free vector field can be expressed in terms of the gradient of a scalar field.

The scalar field ϕ is not unique, that is, any constant α can be added to ϕ since $\boldsymbol{\nabla}\alpha = 0$, that is, the addition of the constant α does not change the gradient. This independence to addition of a number to the scalar potential is called a gauge invariance discussed in chapter 13.2, for which

$$\mathbf{F} = \boldsymbol{\nabla}\phi' = \boldsymbol{\nabla}\left(\phi + \alpha\right) = \boldsymbol{\nabla}\phi \tag{H.34}$$

That is, this gauge-invariant transformation does not change the observable $\mathbf{F}$. The electrostatic field $\mathbf{E}$ and the gravitation field $\mathbf{g}$ are examples of irrotational fields that can be expressed as the gradient of scalar potentials.

Theorem 2; Divergence-free (solenoidal) fields:

For divergence-free fields

$$\boldsymbol{\nabla} \cdot \mathbf{F} = 0 \tag{H.35}$$

everywhere. This is automatically obeyed if the field $\mathbf{F}$ is expressed in terms of the curl of a vector field $\mathbf{G}$ such that

$$\mathbf{F} = \boldsymbol{\nabla} \times \mathbf{G} \tag{H.36}$$

since $\boldsymbol{\nabla} \cdot \boldsymbol{\nabla} \times \mathbf{G} = \mathbf{0}$. That is, any divergence-free vector field can be written as the curl of a related vector field.

As discussed in chapter 13.2, the vector potential $\mathbf{G}$ is not unique in that a gauge transformation can be made by adding the gradient of any scalar field, that is, the gauge transformation $\mathbf{G}' = \mathbf{G} + \boldsymbol{\nabla}\varphi$ gives

$$\mathbf{F} = \boldsymbol{\nabla} \times \mathbf{G}' = \boldsymbol{\nabla} \times (\mathbf{G} + \boldsymbol{\nabla}\varphi) = \boldsymbol{\nabla} \times \mathbf{G}. \tag{H.37}$$

This gauge invariance for transformation to the vector potential $\mathbf{G}'$ does not change the observable vector field $\mathbf{F}$. The magnetic field $\mathbf{B}$ is an example of a solenoidal field that can be expressed in terms of the curl of a vector potential $\mathbf{A}$.

H.4 *Example: Electromagnetic fields:*

Electromagnetic interactions are encountered frequently in classical mechanics so it is useful to discuss the use of potential formulations of electrodynamics.

For electrostatics, Maxwell's equations give that

$$\boldsymbol{\nabla} \times \mathbf{E} = 0$$

Therefore theorem 1 states that it is possible to express this static electric field as the gradient of the scalar electric potential V, where

$$\mathbf{E} = -\boldsymbol{\nabla}V$$

For *electrodynamics, Maxwell's equations give that*

$$(\nabla \times \mathbf{E}) + \frac{\partial \mathbf{B}}{\partial t} = 0$$

Assume that the magnetic field can be expressed in the terms of the vector potential $\mathbf{B} = \nabla \times \mathbf{A}$, *then the above equation becomes*

$$\nabla \times (\mathbf{E} + \frac{\partial \mathbf{A}}{\partial t}) = 0$$

Theorem 1 gives that this curl-less field can be expressed as the gradient of a scalar field, here taken to be the electric potential V.

$$(\mathbf{E} + \frac{\partial \mathbf{A}}{\partial t}) == -\nabla V$$

that is

$$\mathbf{E} = -(\nabla V + \frac{\partial \mathbf{A}}{\partial t})$$

Gauss' law states that

$$\nabla \cdot \mathbf{E} = \frac{\rho}{\varepsilon_0}$$

which can be rewritten as

$$\nabla \cdot \mathbf{E} = -\nabla^2 V - \frac{\partial (\nabla \cdot \mathbf{A})}{\partial t} = \frac{\rho}{\varepsilon_0} \qquad (X)$$

Similarly insertion of the vector potential $\mathbf{A}$ *in Ampère's Law gives*

$$\nabla \times \mathbf{B} = \nabla \times (\nabla \times \mathbf{A}) = \mu_0 \mathbf{j} + \mu_0 \varepsilon_0 \frac{\partial \mathbf{E}}{\partial t} = \mu_0 \mathbf{j} - \mu_0 \varepsilon_0 \nabla \left(\frac{\partial V}{\partial t} \right) - \mu_0 \varepsilon_0 \left(\frac{\partial^2 \mathbf{A}}{\partial t^2} \right)$$

Using the vector identity $\nabla \times (\nabla \times \mathbf{A}) = \nabla (\nabla \cdot \mathbf{A}) - \nabla^2 A$ *allows the above equation to be rewritten as*

$$\left(\nabla^2 \mathbf{A} - \mu_0 \varepsilon_0 \left(\frac{\partial^2 \mathbf{A}}{\partial t^2} \right) \right) - \nabla \left(\nabla \cdot \mathbf{A} + \mu_0 \varepsilon_0 \left(\frac{\partial V}{\partial t} \right) \right) = -\mu_0 \mathbf{j} \qquad (Y)$$

The use of the scalar potential V *and vector potential* $\mathbf{A}$ *leads to two coupled equations X and Y. These coupled equations can be transformed into two uncoupled equations by exploiting the freedom to make a gauge transformation for the vector potential such that the middle brackets in both equations X and Y are zero. That is, choosing the Lorentz gauge*

$$\nabla \cdot \mathbf{A} = -\mu_0 \varepsilon_0 \left(\frac{\partial V}{\partial t} \right)$$

simplifies equations X and Y to be

$$\nabla^2 V - \mu_0 \varepsilon_0 \frac{\partial^2 V}{\partial t^2} = -\frac{\rho}{\varepsilon_0}$$

$$\nabla^2 \mathbf{A} - \mu_0 \varepsilon_0 \left(\frac{\partial^2 \mathbf{A}}{\partial t^2} \right) = -\mu_0 \mathbf{j}$$

The virtue of using the Lorentz gauge, rather than the Coulomb gauge $\nabla \cdot \mathbf{A} = 0$, *is that it separates the equations for the scalar and vector potentials. Moreover, these two equations are the wave equations for these two potential fields corresponding to a velocity* $c = \frac{1}{\sqrt{\mu_0 \varepsilon_0}}$. *This example illustrates the power of using the concept of potentials in describing vector fields.*

Appendix I

Waveform analysis

I.1 Harmonic waveform decomposition

Any linear system that is subject to a time-dependent forcing function $F(t)$, can be expressed as a linear superposition of frequency-dependent solutions of the individual harmonic decomposition $a(\omega)$ of the forcing function. Similarly, any linear system subject to a spatially-dependent forcing function $F(x)$ can be expressed as a linear superposition of the wavenumber-dependent solutions of the individual harmonic decomposition $a(k_x)$ of the forcing function. Fourier analysis provides the mathematical procedure for the transformation between the periodic waveforms and the harmonic content, that is, $F(t) \Leftrightarrow a(\omega)$, or $F(x) \Leftrightarrow a(k_x)$. Fourier's theorem states that any arbitrary forcing function $F(t)$ can be decomposed into a sum of harmonic terms. For example for a time-dependent periodic forcing function the decomposition can be a cosine series of the form

$$F(t) = \sum_{n=1}^{\infty} \alpha_n \cos(n\omega_0 t + \phi_n) \tag{I.1}$$

where ω_0 is the lowest (fundamental) frequency solution. For an aperiodic function a cosine decomposition can be of the form

$$F(t) = \int_0^{\infty} \alpha(\omega) \cos(\omega t + \phi(\omega)) d\omega \tag{I.2}$$

Either of the complementary functions $F(t) \Leftrightarrow a(\omega)$, or $F(x) \Leftrightarrow a(k_x)$ are equivalent representations of the harmonic content that can be used to describe signals and waves. The following two sections give an introduction to Fourier analysis.

I.1.1 Periodic systems and the Fourier series

Discrete solutions occur for systems when periodic boundary conditions exist. The response of periodic systems can be described in either the time versus angular frequency domains, or equivalently, the spatial coordinate x versus the corresponding wave number k_x. For periodic systems this decomposition leads to the Fourier series where a generalized phase coordinate ϕ can be used to represent either the time or spatial coordinates, that is, with $\phi = \omega_0 t$ or $\phi = k_x x$ respectively. The Fourier series relates the two representations of the discrete wave solutions for such periodic systems.

Fourier's theorem states that for a general periodic system any arbitrary forcing function $F(\phi)$ can be decomposed into a sum of sinusoidal or cosinusoidal terms. The summation can be represented by three equivalent series expansions given below, where $\phi = \omega_0 t$ or $\phi = \mathbf{k}_0 \cdot \mathbf{r}$, and where $\omega_0, \mathbf{k}_0$ are the fundamental angular frequency and fundamental wave number respectively.

$$f(\phi) = \frac{a_0}{2} + \sum_{n=1}^{\infty} [a_n \cos(n\phi) + b_n \sin(n\phi)] \tag{I.3}$$

$$f(\phi) = \frac{a_0}{2} + \sum_{n=0}^{\infty} c_n \cos(n\phi + \varphi_n) \tag{I.4}$$

555

$$f\left(\phi\right) = \frac{a_0}{2} + \sum_{n=0}^{\infty} d_n \sin\left(n\phi + \theta_n\right) \tag{I.5}$$

where n is an integer, and φ_n, θ_n are phase shifts fit to the initial conditions.

The normal modes of a discrete system form a complete set of solutions that satisfy the following orthogonality relation

$$\int_0^{2\pi} f_n\left(\phi\right) f_m\left(\phi\right) d\phi = c_n \delta_{mn} \tag{I.6}$$

where δ_{mn} is the Kronecker delta symbol defined in equation $(A.10)$. Orthogonality can be used to determine the coefficients for equations $(I.3)$ to be

$$a_0 = \frac{1}{\pi} \int_{-\pi}^{+\pi} f\left(\phi\right) d\phi \tag{I.7}$$

$$a_n = \frac{1}{\pi} \int_{-\pi}^{+\pi} f\left(\phi\right) \cos\left(n\phi\right) d\phi \tag{I.8}$$

$$b_n = \frac{1}{\pi} \int_{-\pi}^{+\pi} f\left(\phi\right) \sin\left(n\phi\right) d\phi \tag{I.9}$$

Similarly the coefficients for $(I.4)$ and $(I.5)$ are related to the above coefficients by

$$c_n^2 = d_n^2 = a_n^2 + b_n^2$$

Instead of the simple trigonometric form used in equations $(I.3 - I.5)$ the cosine and sine functions can be expanded into the exponential form where

$$\cos\phi = \frac{1}{2}\left(e^{i\phi} + e^{-i\phi}\right) \tag{I.10}$$

$$\sin\phi = \frac{-i}{2}\left(e^{i\phi} - e^{-i\phi}\right)$$

then equation $(I.3)$ becomes

$$f\left(\phi\right) = \sum_{n=-\infty}^{\infty} g_n e^{in\phi} \tag{I.11}$$

where n is any integer and, from the orthogonality, the Fourier coefficients are given by

$$g_n = \frac{1}{2\pi} \int_{-\pi}^{+\pi} f\left(\phi\right) e^{n\phi} d\phi \tag{I.12}$$

These coefficients are related to the cosine plus sine series amplitudes by

$$g_n = \frac{1}{2}\left(a_n - ib_n\right) \qquad \text{(when } n \text{ is positive)}$$

$$g_n = \frac{1}{2}\left(a_n + ib_n\right) \qquad \text{(when } n \text{ is negative)}$$

These results show that the coefficients of the exponential series are in general *complex*, and that they occur in conjugate pairs (that is, the imaginary part of a coefficient a_n is equal but opposite in sign to that for the coefficient a_{-n}). Although the introduction of complex coefficients may appear unusual, it should be remembered that the real part of a pair of coefficients denotes the magnitude of the cosine wave of the relevant frequency, and that the imaginary part denotes the magnitude of the sine wave. If a particular pair of coefficients a_n and a_{-n} are real, then the component at the frequency $n\omega_0$ is simply a cosine; if a_n and a_{-n} are purely imaginary, the component is just a sine; and if, as is the general case, a_n and a_{-n} are complex, both cosine and a sine terms are present.

The use of the exponential form of the Fourier series gives rise to the notion of 'negative frequency'. Of course, $f\left(t\right) = a_n \cos\omega_n t$ is a wave of a single frequency $\omega_n = n\omega_0$ radians/second, and may be represented by a single line of height a_n in a normal spectral diagram. However, using the exponential form of the Fourier series results in both positive and negative ω components.

The coexistence of both negative and positive angular frequencies $\pm\omega$ can be understood by consideration of the Argand diagram where the real component is plotted along the x-axis and the imaginary component along the y-axis. The function $g_n e^{+i\omega t}$ represents a vector of length g_n that rotates with an angular velocity ω in a positive direction, that is counterclockwise, whereas, $g_n e^{-i\omega t}$ represents the vector rotating in a negative direction, that is clockwise. Thus the sum of the two rotating vectors, according to equations $(I.3)$, leads to cancellation of the opposite components on the imaginary y axis and addition of the two $g_n \cos \omega t$ real components on the x axis. Subtraction leads to cancellation of the real x components and addition of the imaginary y axis components.

I.1.2 Aperiodic systems and the Fourier Transform

The Fourier transform (also called the Fourier integral) does for the non-repetitive signal waveform what the Fourier series does for the repetitive signal. It was shown that the line spectrum of a recurrent periodic pulse waveform is modified as the pulse duration decreases, assuming the period of the waveform (and hence its fundamental component) remains unchanged. Suppose now that the duration of the pulses remain fixed but the separation between them increases, giving rise to an increasing period. In the limit, only a single rectangular pulse remains, its neighbors having moved away on either side towards $\pm\infty$. In this case, the fundamental frequency ω_0 tends towards zero and the harmonics become extremely closely spaced and of vanishingly small amplitudes, that is, the system approximates a continuous spectrum.

Mathematically, this situation may be expressed by modifications to the exponential form of the Fourier series already derived. Let the phase factor $\phi = \omega_0 t$ in equation $(I.11)$ then

$$g_n = \frac{\omega_0}{2\pi} \int_{-\pi}^{+\pi} f(t) e^{n\omega_0 t} dt = \frac{1}{\tau} \int_{-\frac{\tau}{2}}^{+\frac{\tau}{2}} f(t) e^{n\omega_0 t} dt \tag{I.13}$$

where τ is the period of the periodic force. Let $G(\omega) = \tau g_n$, $\omega = n\omega_0$, and take the limit for $\tau \to \infty$, then equation $(I.12)$ can be written as

$$G(\omega) = \int_{-\infty}^{+\infty} f(t) e^{\omega t} dt \tag{I.14}$$

Similarly making the same limit for $\tau \to \infty$ then $\omega_0 = \frac{2\pi}{\tau} \to d\omega$ and equation $(I.11)$ becomes

$$f(t) = \sum_{n=-\infty}^{\infty} \frac{G(\omega)}{\tau} e^{in\omega_0 t} = \sum_{n=-\infty}^{\infty} G(\omega) \frac{\omega_0}{2\pi} e^{i\omega t} = \frac{1}{2\pi} \int_{-\infty}^{+\infty} G(\omega) e^{i\omega t} d\omega \tag{I.15}$$

Equation $(I.15)$ shows how a non-repetitive time-domain wave form is related to its continuous spectrum. These are known as Fourier integrals or Fourier transforms. They are of central importance for signal processing. For convenience the transforms often are written in the operator formalism using the $\mathcal{F}$ symbol in the form

$$f(t) = \frac{1}{2\pi} \int_{-\infty}^{+\infty} G(\omega) e^{i\omega t} d\omega \equiv \mathcal{F}^{-1}\left[\frac{1}{2\pi} G(\omega)\right] \tag{I.16}$$

$$G(\omega) = \int_{-\infty}^{+\infty} f(t) e^{-i\omega t} dt \equiv \mathcal{F} f(t) \tag{I.17}$$

It is very important to grasp the significance of these two equations. The first tells us that the Fourier transform of the waveform $f(t)$ is continuously distributed in the frequency range between $\omega = \pm\infty$, whereas the second shows how, in effect, the waveform may be synthesized from an infinite set of exponential functions of the form $e^{\pm i\omega t}$, each weighted by the relevant value of $G(\omega)$. It is crucial to realize that this transformation can go either way equally, that is, from $G(\omega)$ to $f(t)$ or vice versa.[1]

---

[1] The only asymmetry in the Fourier transform relations comes from the 2π factor originating from the fact that by convention physicists use the angular frequency $\omega = 2\pi\nu$ rather than the frequency ν. In order to restore symmetry many papers use the factor $\frac{1}{\sqrt{2\pi}}$ in both relations rather than using the $\frac{1}{2\pi}$ factor in equation $I.16$ and unity in equation $I.17$.

I.1 *Example: Fourier transform of a single isolated square pulse:*

Consider a single isolated square pulse of width τ that is described by the rectangular function Π defined as

$$\Pi(t) = \left\{ \begin{array}{ll} 1 & |t| < \frac{\tau}{2} \\ 0 & |t| > \frac{\tau}{2} \end{array} \right.$$

That is, assume that the amplitude of the pulse is unity between $-\frac{\tau}{2} \leq t \leq \frac{\tau}{2}$. Then the Fourier transform

$$G(\omega) = \int_{-\tau}^{+\tau} 1.e^{-i\omega t} dt = \tau \left(\frac{\sin \frac{\omega\tau}{2}}{\frac{\omega\tau}{2}} \right)$$

which is an unnormalized $sinc(\omega\tau)$ function. Note that the width of the pulse $\Delta t = \pm\frac{\tau}{2}$ leads to a frequency envelope that has the first zeros at $\Delta\omega = \pm\frac{\pi}{\tau}$. Thus the product of these widths $\Delta t \cdot \Delta\omega = \pm\pi$ which is independent of the width of the pulse, that is $\Delta\omega = \frac{\pi}{\Delta t}$ which is an example of the uncertainty principle which is applicable to all forms of wave motion.

I.2 *Example: Fourier transform of the Dirac delta function:*

The Dirac delta function, $\delta(t - t')$, is a pulse of extremely short duration and unit area at $t = t'$ and is zero at all other times. That is,

$$1 = \int_{-\infty}^{+\infty} \delta(t - t') \, dt$$

The Dirac function, which is sometimes referred to as the impulse function, has many important applications to physics and signal processing. For example, a shell shot from a gun is given a mechanical impulse imparting a certain momentum to the shell in a very short time. Other things being equal, one is interested only in the impulse imparted to the shell, that is, the time integral of the force accelerating the shell in the gun, rather than the details of the time dependence of the force. Since the force acts for a very short time the Dirac delta function can be employed in such problems.

As described in section 3.11 and appendix J, the Dirac delta function is employed in signal processing when signals are sampled for short time intervals. The Fourier transform of the delta function is needed for discussion of sampling of signals

$$G(\omega) = \int_{-\infty}^{+\infty} \delta(t - t') e^{-i\omega t} dt = e^{-i\omega t'}$$

Since $e^{-i\omega t}$ essentially is constant over the infinitesimal time duration of the $\delta(t - t')$ function, and the time integral of the δ function is unity, thus the term $e^{-i\omega t}$ has unit magnitude for any value of ω and has a phase shift of $-\omega(t - t')$ radians. For $t' = 0$ the phase shift is zero and thus the Fourier transform of a Dirac $\delta(t)$ function is $G(\omega) = 1$. That is, this is a uniform white spectrum for all values of ω.

I.2 Time-sampled waveform analysis

An alternative approach for unloosing periodic signals, that is complementary to the Fourier analysis harmonic decomposition, is time-sampled (discrete-sample) waveform analysis where the signal amplitude is measured repetitively at regular time intervals in a time-ordered sequence, that is, a sequence of samples of the instantaneous delta-function amplitudes is recorded. Typically an amplitude-to-digital converter is used to digitize the amplitude for each measured sample and the digital numbers are recorded; this process is called **digital signal processing**.

The general principles are best explained by first considering the response of a linear system to a step function impulse, followed by a square impulse, and leading to the response of a δ-function impulsive driving force.

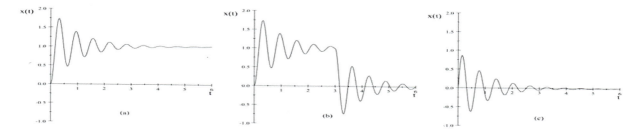

Figure I.1: Response of a underdamped linear oscillator with $\omega = 10$, and $\Gamma = 2$ to the following impulsive force. (a) Step function force $F = 0$ for $t < 0$ and $F = m$ for $t > 0$. (b) Square-wave force where $F = m$ for $0 < t < \tau$ for $\tau = 3$, and $F = 0$ at other times. (c) Delta-function impulse $P = 1$.

I.2.1 Delta-function impulse response

Consider the damped oscillator equation

$$\ddot{x} + \Gamma\dot{x} + \omega_0^2 x = \frac{F(t)}{m} \tag{I.18}$$

and assume that a step function is applied at time $t = 0$. That is;

$$\frac{F(t)}{m} = 0 \qquad t < 0 \qquad\qquad\qquad \frac{F(t)}{m} = a \qquad t > 0 \tag{I.19}$$

where a is a constant. The initial conditions are that $x(0) = \dot{x}(0) = 0$.

The transient or complementary solution is the solution of the linearly-damped harmonic oscillator

$$\ddot{x} + \Gamma\dot{x} + \omega_0^2 x = 0 \tag{I.20}$$

This is independent of the driving force and the solution is given in the chapter 3.5 discussion of the linearly-damped harmonic oscillator.

The particular, steady-state, solution is easy to obtain just by inspection since the force is a constant, that is, the particular solution is

$$x_S = \frac{a}{\omega_0^2} \qquad t > 0 \qquad\qquad\qquad x_S = 0 \qquad t < 0$$

Taking the sum of the transient and particular solutions, using the initial conditions, gives the final solution to be

$$x(t) = \frac{a}{\omega_0^2}\left[1 - e^{-\frac{\Gamma}{2}t}\cos\omega_1 t - \frac{\Gamma e^{-\frac{\Gamma}{2}t}}{2\omega_1}\sin\omega_1 t\right] \tag{I.21}$$

where $\omega_1 \equiv \sqrt{\omega_0^2 - \left(\frac{\Gamma}{2}\right)^2}$. This functional form is shown in figure $I.1a$. Note that the amplitude of the transient response equals $-a$ at $t = 0$ to cancel the particular solution when it jumps to $+a$. The oscillatory behavior then is just that of the transient response.

A square impulse can be generated by the superposition of two opposite-sign stepfunctions separated by a time τ as shown in figure $I.1b$.

The square impulse can be taken to the limit where the width τ is negligibly small relative to the response times of the system. It can be shown that letting $\tau \to 0$, but keeping the magnitude of the total impulse $P = a\tau$ finite for the impulse at time t_0, leads to the solution for the δ-function impulse occurring at t_0

$$x(t) = \frac{P}{\omega_1}e^{-\frac{\Gamma}{2}(t-t_0)}\sin\omega_1(t - t_0) \qquad t > t_0 \tag{I.22}$$

This response to a delta function impulse is shown in figure $I.1c$ for the case where $t_0 = 0$. An example is the response when the hammer strikes a piano string at $t = 0$.

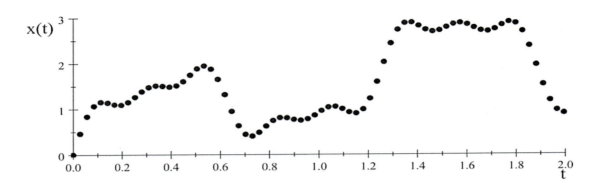

Figure I.2: Decomposition of the function $x(t) = 2\sin(t) + \sin(5t) + \frac{1}{3}\sin(15t) + \frac{1}{5}\sin(25t)$ into a time-ordered sequence of δ-function samples.

I.2.2   Green's function waveform decomposition

The response of the linearly-damped linear oscillator to an delta function impulse, that has been expressed above, can be used to exploit the powerful Green's technique for decomposition of any general forcing function. That is, if the driven system is linear, then the principle of superposition is applicable and allowing expression of the inhomogeneous part of the differential equation as the sum of individual delta functions. That is;

$$\ddot{x} + \Gamma\dot{x} + \omega_0^2 x = \sum_{n=-\infty}^{\infty} \frac{F_n(t)}{m} = \sum_{n=-\infty}^{\infty} I_n(t) \tag{I.23}$$

As illustrated in figure *I.2* discrete-time waveform analysis involves repeatedly sampling the instantaneous amplitude in a regular and repetitive sequence of δ-function impulses. Since the superposition principle applies for this linear system then the waveform can be described by a sum of an ordered series of delta-function impulses where t' is the time of an impulse. Integrating over all the δ-function responses that have occurred at time t', that is prior to the time of interest t, leads to

$$x(t) = \int_{-\infty}^{t} \frac{F(t')}{m\omega_1} e^{-\frac{\Gamma}{2}(t-t')} \sin\omega_1(t-t')\, dt' \qquad t \geq t' \tag{I.24}$$

The Green's function $G(t-t')$ is defined by

$$\begin{aligned} G(t-t') &= \frac{1}{m\omega_1} e^{-\frac{\Gamma}{2}(t-t')} \sin\omega_1(t-t') \qquad t \geq t' \\ &= 0 \qquad t < t' \end{aligned} \tag{I.25}$$

Superposition allows the summed response of the system to be written in an integral form

$$x(t) = \int_{-\infty}^{t} F(t')G(t-t')dt' \tag{I.26}$$

which gives the final time dependence of the forced system. This repetitive time-sampling approach avoids the need of using Fourier analysis. Note that the Green's function $G(t-t')$ includes implicitly the frequency of the free undamped linear oscillator ω_0, the free damped linear oscillator $\omega_1 \equiv \sqrt{\omega_0^2 - \left(\frac{\Gamma}{2}\right)^2}$, as well as the damping coefficient Γ. Access to the combination of fast microcomputers coupled to fast digital sampling techniques has made digital signal sampling the pre-eminent technique for signal recording of audio, video, and detector signal processing.

Bibliography

[1] **SELECTION OF TEXTBOOKS ON CLASSICAL MECHANICS**

[Ar78]   V. I. Arnold, *"Mathematical methods of Classical Mechanics"*, 2nd edition, Springer-Verlag (1978)

This textbook provides an elegant and advanced exposition of classical mechanics expressed in the language of differential topology.

[Co50]   H.C. Corben and P. Stehle, *"Classical Mechanics"*, John Wiley (1950)

This classic textbook covers the material at the same level and comparable scope as the present textbook.

[Fo05]   G. R. Fowles, G. L. Cassiday, *"Analytical Mechanics"*. Thomson Brookes/Cole, Belmont, (2005)

An elementary undergraduate text that emphasizes computer simulations.

[Go50]   H. Goldstein, *"Classical Mechanics"*, Addison-Wesley, Reading (1950)

This has remained the gold standard graduate textbook in classical mechanics since 1950. Goldstein's book is the best graduate-level reference to supplement the present textbook. The lack of worked examples is an impediment to using Goldstein for undergraduate courses. The 3rd edition, published by Goldstein, Poole, and Safko (2002), uses the symplectic notation that makes the book less friendly to undergraduates. The Cline book adopts the nomenclature used by Goldstein to provide a consistent presentation of the material.

[Gr06]   R. D. Gregory, *"Classical Mechanics"*, Cambridge University Press

This outstanding, and original, introduction to analytical mechanics was written by a mathematician. It is ideal for the undergraduate, but the breadth of the material covered is limited.

[Gr10]   W. Greiner, *"Classical Mechanics, Systems of particles and Hamiltonian Dynamics"*, 2nd edition, Springer (2010). This excellent modern graduate textbook is similar in scope and approach to the present text. Greiner includes many interesting worked examples, as well as a reproduction of the Struckmeier[Str08] presentation of the extended Lagrangian and Hamiltonian mechanics formalism of Lanczos[La49].

[Jo98]   J. V. José and E. J. Saletan, *"Classical Dynamics, A Contemporary Approach"*, Cambridge University Press (1998)

This modern advanced graduate-level textbook emphasizes configuration manifolds and tangent bundles which makes it unsuitable for use by most undergraduate students.

[Jo05]   O. D. Johns, *"Analytical Mechanics for Relativity and Quantum Mechanics"*, 2nd edition, Oxford University Press (2005). Excellent modern graduate text that emphasizes the Lanczos[La49] parametric approach to Special Relativity. The Johns and Cline textbooks were developed independently but are similar in scope and approach. For consistency, the name "generalized energy", which was introduced by Johns, has been adopted in the Cline textbook.

[Ki85]   T.W.B. Kibble, F.H. Berkshire. *"Classical Mechanics, (5th edition)"*, Imperial College Press, London, 2004. Based on the textbook written by Kibble that was published in 1966 by McGraw-Hill. The 4th and 5th editions were published jointly by Kibble and Berkshire. This excellent and well-established textbook addresses the same undergraduate student audience as the present textbook. This book covers the variational principles and applications with minimal discussion of the philosophical implications of the variational approach.

[La49]   C. Lanczos, *"The Variational Principles of Mechanics"*, University of Toronto Press, Toronto, (1949)

          An outstanding graduate textbook that has been one of the founding pillars of the field since 1949. It gives an excellent introduction to the philosophical aspects of the variational approach to classical mechanics, and introduces the extended formulations of Lagrangian and Hamiltonian mechanics that are applicable to relativistic mechanics.

[La60]   L. D. Landau, E. M. Lifshitz, *"Mechanics"*, Volume 1 of a *Course in Theoretical Physics*, Pergamon Press (1960)

          An outstanding, succinct, description of analytical mechanics that is devoid of any superfluous text. This Course in Theoretical Physics is a masterpiece of scientific writing and is an essential component of any physics library. The compactness and lack of examples makes this textbook less suitable for most undergraduate students.

[Li94]   Yung-Kuo Lim, *"Problems and Solutions on Mechanics"* (1994)

          This compendium of 408 solved problems, which are taken from graduate qualifying examinations in physics at several U.S. universities, provides an invaluable resource that complements this textbook for study of Lagrangian and Hamiltonian mechanics.

[Ma65]   J. B. Marion, *"Classical Dynamics of Particles and Systems"*, Academic Press, New York, (1965)

          This excellent undergraduate text played a major role in introducing analytical mechanics to the undergraduate curriculum. It has an outstanding collection of challenging problems. The 5th edition has been published by S. T. Thornton and J. B. Marion, Thomson, Belmont, (2004).

[Me70]   L. Meirovitch, *"Methods of Analytical Dynamics"*, McGraw-Hill New York, (1970)

          An advanced engineering textbook that emphasizes solving practical problems, rather than the underlying theory.

[Mu08]   H. J. W. Müller-Kirsten, *"Classical Mechanics and Relativity"*, World Scientific, Singapore, (2008)

          This modern graduate-level textbook emphasizes relativistic mechanics making it an excellent complement to the present textbook.

[Pe82]   I. Percival and D. Richards, *"Introduction to Dynamics"* Cambridge University Press, London, (1982)

          Provides a clear presentation of Lagrangian and Hamiltonian mechanics, including canonical transformations, Hamilton-Jacobi theory, and action-angle variables.

[Sy60]   J.L. Synge, *"Principles of Classical Mechanics and Field Theory"* , Volume III/I of *"Handbuck der Physik"* Springer-Verlag, Berlin (1960).

          A classic graduate-level presentation of analytical mechanics.

[Ta05]   J. R. Taylor, *"Classical Mechanics"*, University Science Books, Sausalito, (2006)

          This undergraduate book gives a well-written descriptive introduction to analytical mechanics. The scope of the book is limited and the problems are easy.

[2] GENERAL REFERENCES

[Bak96]   L. Baker, J.P. Gollub, *Chaotic Dynamics*, 2nd edition, 1996 (Cambridge University Press)

[Bat31]   H. Bateman, *Phys. Rev.* **38** (1931) 815

[Bau31]   P.S. Bauer, *Proc. Natl. Acad. Sci.* **17** (1931) 311

[Bor25a] M. Born and P. Jordan, *Zur Quantenmechanik*, Zeitschrift für Physik, 34, (1925) 858-888.

[Bor25b] M. Born, W. Heisenberg, and P. Jordan, *Zur Quantenmechanik II*, Zeitschrift für Physik, 35, (1925), 557-615,

[Boy08]   R. W. Boyd, *Nonlinear Optics*, 3rd edition, 2008 (Academic Press, NY)

[Bri14]   L. Brillouin, Ann. Physik **44**(1914)

[Bri60]   L. Brillouin, Wave Propagation and Group Velocity, 1960 (Academic Press, New York)

[Cay1857] A. Cayley, Proc. Roy. Soc. London **8** (1857) 506

[Cei10]   J.L. Cieśliński, T. Nikiciuk, J. Phys. A:Math. Theor. **43** (2010) 175205

[Cio07]   Ciocci and Langerock, *Regular and Chaotic Dynamics*, **12** (2007) 602

[Cli71]   D. Cline, Proc. Orsay Coll. on Intermediate Nuclei, Ed. Foucher, Perrin, Veneroni, 4 (1971).

[Cli72]   D. Cline and C. Flaum, Proc. of the Int. Conf. on Nuclear Structure Studies Using Electron Scattering, Sendai, Ed. Shoa, Ui, 61 (1972).

[Cli86]   D. Cline, Ann. Rev. Nucl. Part. Sci. 36, (1986) 683.

[Coh77]   R.J. Cohen, Amer. J. of Phys. **45** (1977) 12

[Cra65]   F.S. Crawford, *Berkeley Physics Course 3; Waves,* 1970 (Mc Graw Hill, New York)

[Cum07]   D. Cumin, C.P. Unsworth, Physica D 226 (2007) 181

[Dav58]   A. S. Davydov and G. F. Filippov. Nuclear Physics, 8 (1958) 237

[Dek75]   H. Dekker, Z. Physik, **B21** (1975) 295

[Dep67]   A. Deprit, American J. of Phys **35,** no.5 424 (1967)

[Dir30]   P.A.M. Dirac, *Quantum Mechanics*, Oxford University Press, (1930).

[Dou41]   D. Douglas, Trans. Am. Math. Soc. **50** (1941) 71

[Fey84]   R.P. Feynman, R.B. Leighton, M. Sands, The Feynman Lectures, (Addison-Wesley, Reading, MA,1984) Vol. 2, p17.5

[Fro80]   C. Frohlich, Scientific American, **242** (1980) 154

[Gal13]   C. R. Galley, Physical Review Letters, **11** (2013) 174301

[Gal14]   C. R. Galley, D. Tsang, L.C. Stein, arXiv:1412.3082v1 [math-phys] 9 Dec 2014

[Har03]   James B. Hartle, *Gravity: An Introduction to Einstein's General Relativity* (Addison Wesley, 2003)

[Kur75]   International Symposium on Math. Problems in Theoretical Physics, Lecture Notes in Physics, Vol39 Springer, NY (1975)

[Mus08a] Z.E. Musielak, J. Phys. A. Math. Theor. **41** (2008) 055205

[Mus08b] Z.E. Musielak, D. Rouy, L.D. Swift, Chaos, Solitons, Fractals **38** (2008) 894

[Ray1881] J.W. Strutt, 3rd Baron Rayleigh, Proc. London Math. Soc., s1-4 (1), (1881) 357

[Ray1887] J.W. Strutt, 3rd Baron Rayleigh, *The Theory of Sound,* 1887 (Macmillan, London)

[Rou1860] E.J. Routh, *Treatise on the dynamics of a system of rigid bodies*, MacMillan (1860)

[Sim98]   M. Simon, D. Cline, K. Vetter, et al, Unpublished

[Sta05]   T. Stachowiak and T. Okada, Chaos, Solitons, and Fractals, **29** (2006) 417.

[Str00]   S.H. Strogatz, Physica **D43** (2000) 1

[Str05]   J. Struckmeier, J. Phys. A: Math; Gen. **38** (2005) 1257

[Str08]   J. Struckmeier, Int. J. of Mod. Phys. **E18** (2008) 79

[Vir15]   E.G. Virga, Phys, Rev. **E91** (2015) 013203

[Win67]   A.T. Winfree, J. Theoretical Biology **16** (1967) 15

Index

Two dramatically different philosophical approaches to classical mechanics were proposed during the 17th – 18th centuries. Newton developed his vectorial formulation that uses time-dependent differential equations of motion to relate vector observables like force and rate of change of momentum. Euler, Lagrange, Hamilton, and Jacobi, developed powerful alternative variational formulations based on the assumption that nature follows the principle of least action. These variational formulations now play a pivotal role in science and engineering.

This book introduces variational principles and their application to classical mechanics. The relative merits of the intuitive Newtonian vectorial formulation, and the more powerful variational formulations are compared. Applications to a wide variety of topics illustrate the intellectual beauty, remarkable power, and broad scope provided by use of variational principles in physics.

This second edition adds discussion of the use of variational principles applied to the following topics:

(1) Systems subject to initial boundary conditions
(2) The hierarchy of related formulations based on action, Lagrangian, Hamiltonian, and equations of motion, to systems that involve symmetries
(3) Variational principles to non-conservative systems
(4) Variable-mass systems
(5) The General Theory of Relativity

Douglas Cline is a Professor of Physics in the Department of Physics and Astronomy, University of Rochester, Rochester, New York.

UNIVERSITY *of* ROCHESTER

RIVER CAMPUS
LIBRARIES

Bibliography

"Variational Principles in Classical Mechanics." Cline, Douglas. In **Variational Principles in Classical Mechanics,** by Cline, Douglas. pp. 1–600. **River Campus Libraries,** 2017. (600 pages).

These course materials were produced by XanEdu
and are intended for your individual use. If you have
any questions regarding these materials, please contact:

XanEdu Customer Support
support@xanedu.com
1-888-212-3121

XanEdu works to inspire the education community
with innovative solutions to build affordable, accessible
learning experiences that drive student success.

XanEdu, Inc.

www.xanedu.com

CPID: 903385

9 781711 483252